I0055020

Parkinson's Disease

SECOND EDITION

A subject collection from *Cold Spring Harbor Perspectives in Medicine*

OTHER SUBJECT COLLECTIONS FROM *COLD SPRING HARBOR PERSPECTIVES IN MEDICINE*

Developmental Oncology: Principles and Therapy of Cancers of Children and Young Adults

Type 1 Diabetes: Advances in Understanding and Treatment 100 Years after the Discovery of Insulin, Second Edition

Cancer Metabolism: Historical Landmarks, New Concepts, and Opportunities

Modeling Cancer in Mice

Angiogenesis: Biology and Pathology, Second Edition

Retinal Disorders: Approaches to Diagnosis and Treatment, Second Edition

Breast Cancer: From Fundamental Biology to Therapeutic Strategies

Aging: Geroscience as the New Public Health Frontier, Second Edition

Combining Human Genetics and Causal Inference to Understand Human Disease and Development

Lung Cancer: Disease Biology and Its Potential for Clinical Translation

Influenza: The Cutting Edge

Leukemia and Lymphoma: Molecular and Therapeutic Revolution

Addiction, Second Edition

Hepatitis C Virus: The Story of a Scientific and Therapeutic Revolution

The PTEN Family

Metastasis: Mechanism to Therapy

Genetic Counseling: Clinical Practice and Ethical Considerations

Bioelectronic Medicine

SUBJECT COLLECTIONS FROM *COLD SPRING HARBOR PERSPECTIVES IN BIOLOGY*

Evolution and Development of Neural Circuits

Machine Learning for Protein Science and Engineering

Glia, Second Edition

Speciation

The Biology of Lipids: Trafficking, Regulation, and Function, Second Edition

Synthetic Biology and Greenhouse Gases

Wound Healing: From Bench to Bedside

The Endoplasmic Reticulum, Second Edition

Sex Differences in Brain and Behavior

Regeneration

The Nucleus, Second Edition

Auxin Signaling: From Synthesis to Systems Biology, Second Edition

Stem Cells: From Biological Principles to Regenerative Medicine

Heart Development and Disease

Cell Survival and Cell Death, Second Edition

Calcium Signaling, Second Edition

Engineering Plants for Agriculture

Protein Homeostasis, Second Edition

Translation Mechanisms and Control

Parkinson's Disease

SECOND EDITION

A subject collection from *Cold Spring Harbor Perspectives in Medicine*

EDITED BY

Serge Przedborski

Vagelos College of Physicians and Surgeons
Columbia University

Stanley Fahn

Vagelos College of Physicians and Surgeons
Columbia University

COLD SPRING HARBOR LABORATORY PRESS
Cold Spring Harbor, New York • www.cshlpress.org

Parkinson's Disease, Second Edition

A subject collection from *Cold Spring Harbor Perspectives in Medicine*
Articles online at www.cshperspectives.org

All rights reserved
© 2025 by Cold Spring Harbor Laboratory Press, Cold Spring Harbor, New York

Executive Editor	Richard Sever
Project Supervisor	Barbara Acosta
Editorial Assistant	Danett Gil
Permissions Administrator	Carol Brown
Production Editor	Diane Schubach
Production Manager/Cover Designer	Denise Weiss
Publisher	John Inglis

Front cover artwork: Three-dimensional illustration of a human brain with highlighted substantia nigra (orange) inside a transparent purple midbrain, showcasing its role in movement control. (Image by Kateryna Kon © Shutterstock; Image ID 2578252497.)

Library of Congress Cataloging-in-Publication Data

Names: Przedborski, Serge editor | Fahn, Stanley, 1933- editor.
Title: Parkinson's disease/edited by Serge Przedborski, Vagelos College of Physicians and Surgeons, Columbia University and Stanley Fahn, Vagelos College of Physicians and Surgeons, Columbia University.
Other titles: Parkinson's disease (Cold Spring Harbor Laboratory) | Cold Spring Harbor perspectives in medicine.
Description: Second edition. | Cold Spring Harbor : Cold Spring Harbor Laboratory Press, [2025] | "A subject collection from Cold Spring Harbor perspectives in medicine". | Includes bibliographical references and index. | Summary: "Parkinson's Disease is a progressive neurodegenerative disorder whose underlying causes have proven difficult to establish. This second edition of the Cold Spring Harbor volume reviews recent progress towards our understanding of the underlying biology, genetic and environmental contributions, and progress in development of new therapies for the condition"--Provided by publisher.
Identifiers: LCCN 2025012517 (print) | LCCN 2025012518 (ebook) | ISBN 9781621825036 hardcover | ISBN 9781621825043 epub
Subjects: LCSH: Parkinson's disease.
Classification: LCC RC382 .P373 2025 (print) | LCC RC382 (ebook) | DDC 616.8/33-dc23/eng/20250606
LC record available at https://lccn.loc.gov/2025012517
LC ebook record available at https://lccn.loc.gov/2025012518

All World Wide Web addresses are accurate to the best of our knowledge at the time of printing.

Authorization to photocopy items for internal or personal use, or the internal or personal use of specific clients, is granted by Cold Spring Harbor Laboratory Press, provided that the appropriate fee is paid directly to the Copyright Clearance Center (CCC). Write or call CCC at 222 Rosewood Drive, Danvers, MA 01923 (978-750-8400) for information about fees and regulations. Prior to photocopying items for educational classroom use, contact CCC at the above address. Additional information on CCC can be obtained at CCC Online at www.copyright.com.

For a complete catalog of all Cold Spring Harbor Laboratory Press publications, visit our website at www.cshlpress.org.

Contents

Contents

Preface

ARKINSON'S DISEASE (PD) WAS ONCE A TABOO SUBJECT, with affected individuals making every effort to conceal the physical manifestations of their movement disorder. This is no longer the case. Today, rather than hiding their condition, politicians, artists, and world leaders with PD openly acknowledge their diagnosis and share how they cope with its medical, social, and emotional challenges.

However, while public awareness of PD is now widespread, our understanding of why and how the disease develops and progresses still lags behind. For every question answered, clinical and basic researchers in the field uncover many more unanswered ones. If we are to develop effective therapies for this debilitating disorder, we must first unravel the pathobiology of PD. Achieving this will require bringing together talented individuals with diverse skill sets and perspectives to work collaboratively across disciplines. Clinicians should be encouraged to engage with the basic physiology, molecular mechanisms, and cellular biology of PD, while basic researchers must gain exposure to the finer clinical aspects of the disease and develop an understanding of the day-to-day realities faced by affected individuals.

We often hear from colleagues—both clinicians and basic scientists—who want to contribute to PD research but struggle to find accessible resources. Clinicians frequently find basic science texts filled with technical jargon and abstract concepts, while basic scientists find clinical textbooks too cryptic or detailed for nonspecialists. This gap in resources led us to wonder: Is there a single book we could recommend to both groups? A book that bridges the clinical and basic science aspects of PD under one cover—one that provides a broad yet insightful overview rather than an exhaustive compendium of every detail? Such a book would serve as a didactic tool, offering fundamental knowledge alongside key takeaways and practical insights.

With this vision in mind, Cold Spring Harbor Laboratory Press partnered with us in 2012 to publish the first edition of this monograph, aimed at providing a comprehensive bench-to-bedside understanding of PD. Encouraged by its reception, we were recently invited to produce this second edition, which includes both updated chapters reflecting advancements since 2012 and new chapters on critical topics such as disease progression, cell-based replacement therapies, biomarkers, lipid dysregulation, and the gut–brain axis. These subjects were either absent from the first edition or have since emerged as essential areas of study. Each expert contributor was asked to write their chapter as if guiding a new student or faculty member—whether clinician or basic scientist—who is eager to enter the field.

Readers will begin their journey with Goetz's chapter on the history of PD, setting the stage for understanding how this neurological disorder was initially identified and defined. From there, Patel et al., Irmady and Przedborski, along with Dickson, provide the clinical foundations, disease progression insights, and neuropathological underpinnings of PD. Among other key points, they highlight that PD's clinical features are not exclusive to the disease but are shared by a number of other conditions. They also emphasize that while PD is primarily known for its motor symptoms and the loss of dopaminergic neurons, it also involves nonmotor symptoms and degeneration of nondopaminergic neurons, all of which play a crucial role in disease expression and disability.

As with other prominent adult-onset neurodegenerative diseases such as Alzheimer's and amyotrophic lateral sclerosis, PD is primarily sporadic. However, in a small subset of cases, PD follows a familial pattern, inherited as a dominant or recessive trait linked to various gene mutations. These rare genetic forms of PD are under intense study, as understanding the normal functions of these genes and how mutations disrupt them may provide critical insights into the pathobiology of sporadic PD.

Several chapters explore the genetics of PD, beginning with an introduction by Westenberger et al., followed by discussions on specific genes and their products, including α-synuclein (Vekrellis et al.) and LRRK2 (Pfeffer and Alessi). The section concludes with a key discussion by Menon on genomics, offering a broader perspective that considers all genes and their interactions to provide insights into complex diseases such as PD.

Beyond genetics, PD research also seeks to understand how neurodegeneration alters the functional neuroanatomy of the basal ganglia, leading to the motor abnormalities observed in patients. To explore this, Lanciego and Obeso, Wichmann, and Nakano et al. examine the disease through the lenses of functional neuroanatomy and brain imaging, highlighting how therapeutic strategies that modulate specific neural pathways provide symptomatic relief for PD patients. The section concludes with a discussion and an addendum by Kayhanian and Barker on dopamine cell-based replacement therapies, which, like neuromodulation, aim to provide symptomatic benefit. Lastly, Lodge and Agin-Liebes contribute a chapter on biomarkers, exploring their potential role in early diagnosis, disease monitoring, and the development of targeted therapies.

Because experimental models are essential to studying the neurobiology of PD, a series of chapters by Bezard et al., Tieu et al., and Dawson and Dawson cover the topic, ranging from primate to rodent models and from toxin-based to genetic-based models.

The final section of the book explores emerging and significant pathogenic mechanisms in Parkinson's disease. Chapters by Schon et al., Area-Gómez et al., Martinez-Vicente and Vila, Gao et al., Elyaman, and Oludipe et al. examine key topics, including the roles of mitochondria, lipids, autophagy and protein quality control, the gut–brain axis, and the innate and adaptive immune systems in the disease process.

This is the roadmap of the book. Before we begin, we invite you to embark on this journey from bench to bedside with curiosity and enthusiasm. We hope you find as much value in reading this volume as our distinguished colleagues and we did in preparing it.

Finally, we extend our gratitude to the authors who contributed their time and expertise, and to Danett Gil, Barbara Acosta, and Richard Sever at Cold Spring Harbor Laboratory Press, whose invaluable assistance and guidance made this book possible.

SERGE PRZEDBORSKI
STANLEY FAHN

Historical Perspectives of Parkinson's Disease: Early Clinical Descriptions and Neurological Therapies

Christopher G. Goetz

Department of Neurological Studies, Rush University Medical Center, Chicago, Illinois 60612, USA

Correspondence: cgoetz@rush.edu

Although components of possible Parkinson's disease can be found in earlier documents, the first clear medical description was written in 1817 by James Parkinson. In the mid-1800s, Jean-Martin Charcot was particularly influential in refining and expanding this early description and in disseminating information internationally about Parkinson's disease. He separated the clinical spectrum of Parkinson's disease from multiple sclerosis and other disorders characterized by tremor, and he recognized cases that later would likely be classified among the parkinsonism-plus syndromes. Early treatments of Parkinson's disease were based on empirical observation, and anticholinergic drugs were used as early as the nineteenth century. The discovery of dopaminergic deficits in Parkinson's disease and the synthetic pathway of dopamine led to the first human trials of levodopa. Further historically important anatomical, biochemical, and physiological studies identified additional pharmacological and neurosurgical targets for Parkinson's disease and allow modern clinicians to offer an array of therapies aimed at improving function in this still incurable disease.

Important historical anchors for the study of Parkinson's disease concern the early descriptions of the disorder, its separation from other neurological conditions, and the evolution of therapy from empirical observations to rational treatment designs based on the growing knowledge of anatomy, biochemistry, and physiology of the basal ganglia.[1] This work provides the background history of Parkinson's disease, highlighting persons and discoveries primarily from the nineteenth and early twentieth centuries.

EARLY CLINICAL DESCRIPTIONS

Defining Parkinson's Disease

Parkinson's disease was first medically described as a neurological syndrome by James Parkinson in 1817, although fragments of parkinsonism can be found in earlier descriptions (Parkinson 1817). As examples, Sylvius de la Boë wrote of rest tremor, and Sauvages described festination (Sylvius de la Boë 1680; de Sauvages 1768; Tyler 1992). Much earlier, traditional Indian texts from ~1000 BC and ancient Chinese sources

---

[1]This is an update to a previous article published in *Cold Spring Harbor Perspectives in Medicine* [Goetz (2011). *Cold Spring Harb Perspect Med* **1**: a008862. doi:10.1101/cshperspect.a008862].

Copyright © 2025 Cold Spring Harbor Laboratory Press; all rights reserved
Cite this article as *Cold Spring Harb Perspect Med* doi: 10.1101/cshperspect.a041642

also provide descriptions that suggest Parkinson's disease (Manyam 1990; Zhang et al. 2006). In succinct and pithy English, Parkinson (1817) captured the clinical picture:

> Involuntary tremulous motion, with lessened muscular power, in parts not in action and even when supported; with a propensity to bend the trunk forward, and to pass from a walking to a running pace: the senses and intellects being uninjured.

Parkinson reported on six case sketches, three of the patients observed in the streets of London and one only seen from a distance (Fig. 1).

Jean-Martin Charcot, in his teaching at the Salpêtrière more than 50 years later, was more thorough in his descriptions and distinguished bradykinesia as a separate cardinal feature of the illness (Charcot 1872):

> Long before rigidity actually develops, patients have significant difficulty performing ordinary activities: this problem relates to another cause. In some of the various patients I showed you, you can easily recognize how difficult it is for them to do things even though rigidity or tremor is not the limiting feature. Instead, even a cursory exam demonstrates that their problem relates more to slowness in execution of movement rather than to real weakness. In spite of tremor, a patient is still able to do most things, but he performs them with remarkable slowness. Between the thought and the action there is a considerable time lapse. One would think neural activity can only be effected after remarkable effort.

Charcot and his students described the clinical spectrum of this disease, noting two prototypes: the tremorous and the rigid/akinetic form. They described in full detail the arthritic changes, dysautonomia, and pain that can accompany Parkinson's disease. Charcot (1872) was also the first to suggest the use of the term "Parkinson's disease" rejecting the earlier designation of paralysis agitans or shaking palsy, because he recognized that Parkinson's disease patients are not markedly weak and do not necessarily have tremor.

William Gowers, working in London, contributed an important study of Parkinson's disease demographics in his *A Manual of Diseases of the Nervous System*, describing his personal experience with 80 patients in the 1880s. He correctly identified the slight male predominance of the dis-

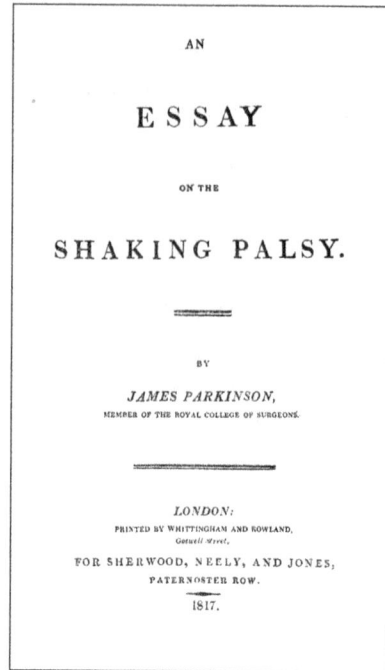

AN

ESSAY

ON THE

SHAKING PALSY.

BY

JAMES PARKINSON,
MEMBER OF THE ROYAL COLLEGE OF SURGEONS.

LONDON:
PRINTED BY WHITTINGHAM AND ROWLAND,
Goswell Street,
FOR SHERWOOD, NEELY, AND JONES,
PATERNOSTER ROW.
1817.

Figure 1. *Essay on the Shaking Palsy.* James Parkinson's short monograph is the first clear medical document dealing with Parkinson's disease (Parkinson 1817). (Reprinted, with permission, from Goetz 2011, © Cold Spring Harbor Laboratory Press.)

order and studied the joint deformities typical of the disease. Known for his descriptive prose, Gowers (1888) offered one of the most memorable similes regarding parkinsonian tremor:

> The movement of the fingers at the metacarpal-phalangeal joints is similar to that by which Orientals beat their small drums.

Further clinical descriptions and studies of the pathologic changes related to Parkinson's disease were predominantly reported by the French neurologic school. Richer and Meige (1895) provided clinical and morphologic details of the progressive stages of parkinsonian disability, and the former provided drawings and statues that remain among the most important pictorial documents related to Parkinson's disease. Babinski (1921) commented on the strange motor fluctuations intrinsic to the disease itself. Brissaud (1925) first proposed damage to the substantia nigra as the anatomical seat of Parkinson's dis-

Cite this article as *Cold Spring Harb Perspect Med* doi: 10.1101/cshperspect.a041642

ease, and Trétiakoff (1921) and Foix and Nico-lesco (1925) pursued further pathologic studies of the midbrain in relationship to the disease during the 1920s.

The most complete pathologic analysis of Parkinson's disease and the clear delineation of the brainstem lesions was performed by Green-field and Bosanquet (1953). The morbidity and clinical progression of Parkinson's disease was studied in the important article by Hoehn and Yahr (1967) in which their internationally recog-nized staging system was first introduced. This time-honored staging system is anchored in the distinction between unilateral (stage I) disease and bilateral disease (stages II–V) and the devel-opment of postural reflex impairment (stage III) as a key turning point in the disease's clinical significance.

Although nonmotor elements of Parkinson's disease were recognized by Parkinson (sleep dis-ruption, constipation), Charcot (pain, fatigue), and Gowers (cognitive decline, depression), Lewy (1923) focused special attention on these symptoms and emphasized that they often pre-sent years ahead of the defining motor symptoms and signs. Hallucinations, typically considered as a late disorder associated with progressive disease and medication, were recognized to be within the spectrum of Parkinson's disease long before the levodopa era (Fénelon et al. 2006).

Separating Parkinson's Disease from Other Disorders

Before Charcot, the classification system, or no-sology, of neurological disease was primitive, and disorders were largely grouped by primary symp-toms—for instance, tremors or weakness. Char-cot's first important contribution to the study of Parkinson's disease was his differentiation of this disorder from other tremorous disorders, specif-ically multiple sclerosis (Charcot 1872). Exam-ining large numbers of patients within the vast Salpêtrière Hospital in Paris, he developed a pro-tocol to observe tremor at rest and then during action. He noted that the patients with action tremor had accompanying features of weakness, spasticity, and visual disturbance. In contrast, those with rest tremor differed in having rigidity,

slowed movements, a typical hunched posture, and very soft speech. His early tremor studies were highly publicized and helped to establish Parkinson's disease as a distinct neurological en-tity that could be confidently diagnosed (Fig. 2).

Once the archetype of Parkinson's disease was established, Charcot and his students iden-tified variants with features that were atypical of classical Parkinson's disease. These were termed Parkinson's disease without tremor, Parkinson's disease with extended posture, and Parkinson's disease with hemiplegia. These cases are of his-torical interest, because they are likely examples of disorders that would later be grouped under the term parkinsonism-plus syndromes, includ-ing progressive supranuclear palsy, corticobasal degeneration, and multiple system atrophy. As one example, Charcot presented a patient named Bachère on several occasions. On June 12, 1888, Charcot emphasized that Bachère did not have marked tremor, and in contrast to the usual arm flexion of typical Parkinson's disease, he had a stiff and extended posture (Fig. 3; Charcot 1888a; Goetz 1987):

> Look how he stands. I present him in profile so you can see the inclination of the head and trunk, well described by Parkinson. All this is typical. What is atypical, however, is that Bachère's forearms and legs are extended, making the extremities like rigid bars, whereas in the ordinary case, the same body parts are partly flexed. One can say then that in the typical case of Parkinson's disease, flexion is the predominant feature, whereas here, extension predominates and accounts for this unusual pres-entation. The difference is even more evident when the patients walk.

In addition to extended posture, this patient had particular facial bradykinesia and contracted forehead muscles. Charcot commented that the patient had the perpetual look of surprise, be-cause the eyes remained widely opened and the forehead continually wrinkled (Fig. 3; Charcot 1888a; Goetz 1987). In a modern setting, Jan-kovic (1984) has detailed similar facial morphol-ogy in parkinsonism-plus patients, specifically those with progressive supranuclear palsy. No specific supranuclear eye movement abnormali-ties were described. Another Salpêtrière patient with "Parkinson's disease in extension" was de-

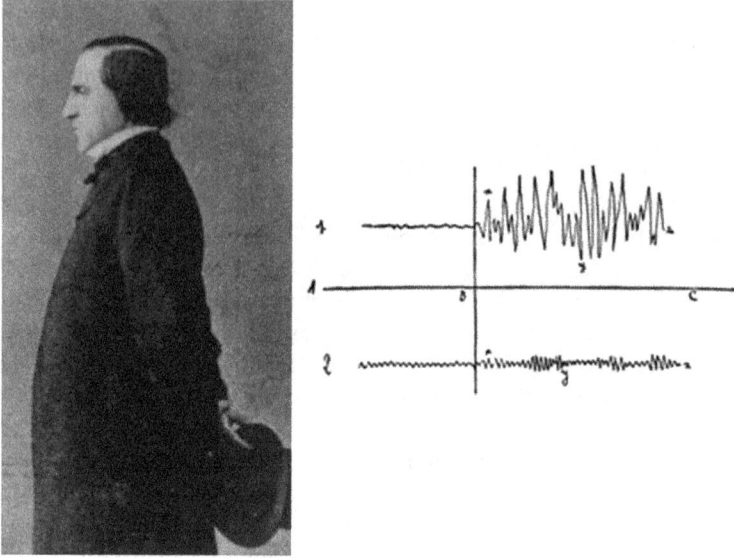

Figure 2. Charcot and "Myographic Curves." (*Left*) French neurologist Jean-Martin Charcot (1825–1893). (*Right*) Semidiagrammatic "myographic curves" published by Charcot (1887). The *top* tracing represents an intention tremor in multiple sclerosis. Segment AB indicates "at rest," and BC indicates increasing oscillations during voluntary movement. The lower tracing represents a Parkinsonian tremor, with segment AB indicating a tremor at rest, which persists in segment BC during voluntary movement. Charcot's graphical recording method upon which these drawings were based is not described, but in other circumstances, he relied on various pneumatic tambour-like mechanisms (Charcot 1872; Goetz 1987). (Reprinted, with permission, from Goetz 2011, © Cold Spring Harbor Laboratory Press.)

scribed by Dutil (1889) and eye movement abnormalities are mentioned, although a supranuclear lesion is not documented clinically (Goetz 1996). This case also had highly asymmetric rigidity of the extremities, a feature more reminiscent of corticobasal degeneration than progressive supranuclear palsy. In this case, the extended neck posture was graphically emphasized (Dutil 1889; Goetz 1996):

> The face is masked, the forehead wrinkled, the eyebrows raised, the eyes immobile.... This facies, associated with the extended posture of the head and trunk, gives the patient a singularly majestic air.

With clinical features reminiscent of both progressive supranuclear palsy and corticobasal degeneration, this patient was mentioned in several articles from the Salpêtrière school, although no autopsy was apparently performed. Collectively, these cases demonstrate that even the earliest diagnosticians recognized classic Parkinson's dis-

ease and cases that needed to be distinguished from it. Today, these parkinsonism-plus diagnoses are known to have additional distinctive features, including poor response to dopaminergic therapies and different pathological lesions than seen in Parkinson's disease.

Another important entity to be distinguished from Parkinson's disease was postencephalitic parkinsonism, today a rare cause of parkinsonism, but a very frequent disorder in the period after 1916. Following the influenza epidemic of 1916–1917, a neurologic syndrome that included parkinsonism, along with multiple other signs, occurred in alarming numbers (von Economo 1919). The additional behavioral, ocular, and motor problems of these patients attested to more diffuse neurologic disease than seen typically in Parkinson's disease. This condition has largely disappeared in the twenty-first century, because the survivors have died and no recurrence of an epidemic of this magnitude has recurred. Other important forms of atypical par-

A

B

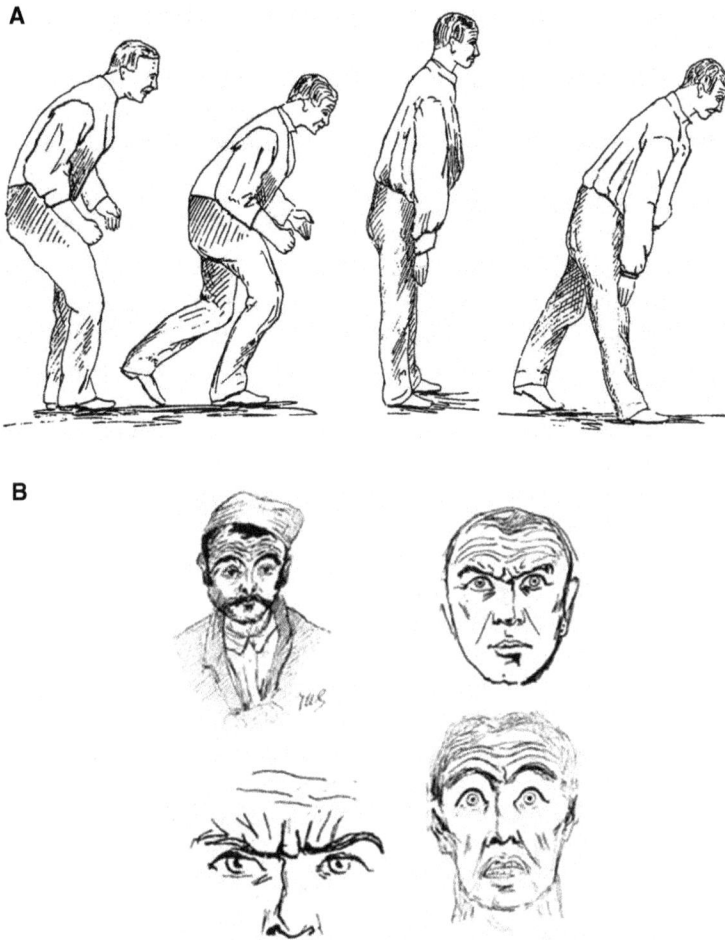

Figure 3. Atypical parkinsonism. (*A*) Drawing from Charcot's original lesson, given on June 12, 1888, in which he contrasted a typical Parkinson's disease showing a flexed posture (*left*) with a parkinsonian variant that included the absence of tremor and extended posture (*right*). Charcot regularly taught his students by comparing and contrasting cases of patients from the Salpêtrière inpatient and outpatient services. (*B*) Four drawings by Charcot from his lesson on atypical Parkinson's disease, dated June 12, 1888, showing the distinctive facial features of his patient, Bachère, showing forehead muscles and superior orbicularis in simultaneous contraction, activation of the palpebral portion of the orbicularis, and combined activation of the frontalis superior portion of the orbicularis and platysma, giving a frightened expression in contrast to the placid, blank stare of typical Parkinson's disease patients. This case is a compelling case of likely progressive supranuclear palsy (Charcot 1888a; Goetz 1987). (Reprinted, with permission, from Goetz 2011, © Cold Spring Harbor Laboratory Press.)

kinsonism to be distinguished from Parkinson's disease include a juvenile form of Parkinson's disease, originally described by Willige (1911), with a more full description and its association with atrophy of the globus pallidus provided by Hunt (1917) and van Bogaert (1930).

In the years after these pioneering papers, the concepts of neural circuits evolved with key nu-

clei of importance to the clinical presentation of parkinsonism being the substantia nigra, the globus pallidus, and the caudate nucleus and putamen (striatum). Involvement of the striatum resulting in parkinsonism was documented in a variety of neurological disorders. Striatal–nigral degeneration was described by Adams et al. (1964), and, although originally classified as a single disease, it has since been merged into the larger diagnosis of multiple system atrophy. Parkinsonian states related to striatal pathology were later identified in the form of Huntington's disease, in which a parkinsonian presentation is referred to as the Westphal variant (Westphal 1883) and in cases of striatal calcification, either on a hereditary basis (Bruyn et al. 1964) or as an acquired metabolic disorder often related to hypoparathyroidism (Muenter and Whisnant 1968).

The historical discussion of parkinsonian disorders that are frequently confused with Parkinson's disease includes drug-induced and toxin-induced cases as well. The introduction of the antipsychotic agents, originally termed neuroleptics, led to dramatic improvements in schizophrenic and other psychotic behaviors, but induced parkinsonism largely indistinguishable from Parkinson's disease itself (Steck 1954). Later understanding that these drugs block dopamine receptors in the striatum explained this clinical presentation and led to the development of antipsychotic drugs with lower proclivity to block striatal receptors and less propensity to induce parkinsonism. The landmark observation on a cluster of young patients who presented with severe parkinsonism that appeared to be typical Parkinson's disease except for the young onset and severity of signs led to the discovery that the causative agent was a self-administered narcotic derivative, 5 1-methyl-4-phenyl-1,2,3, 6-tetrahydropyridine (MPTP), that selectively damages the substantia nigra (Langston et al. 1983). This product has provided a means to induce parkinsonism in experimental animals and remains the "gold standard" model to study Parkinson's disease in preclinical studies of new treatments for Parkinson's disease.

The recent pandemic involving COVID-19 has left many patients with neurological residua, and researchers recall the postencephalitic cases of parkinsonism that occurred in the wake of the 1918 flu epidemic, termed von Economo's disease. Although post-COVID-19 parkinsonism has not developed in a systematic worldwide pattern, the post-1918 syndrome was marked by the classic features of parkinsonism with marked dystonia and other movement disorders occurring in the aftermath of the acute sleeping sickness. The observations provoke a vigilance of careful follow-up of all post-COVID syndromes with an awareness that postencephalitic parkinsonism occurred usually between 1 and 5 years after the acute infections (Pearce 1996).

THE EVOLUTION OF TREATMENTS

The history of Parkinson's disease is tightly linked to therapeutic interventions, ranging from serendipitous observations to controlled clinical trials of specifically designed agents.

Parkinson devoted a chapter of his monograph to "considerations respecting the means of cure" (Parkinson 1817). In humility and perhaps with a vision toward current concepts of neuroprotection, he hoped for the identification of a treatment by which "the progress of the disease may be stopped" (Parkinson 1817). To this end, he advocated very early therapeutic intervention when signs were largely confined to the arms without balance and gait impairments. Reflecting therapeutic approaches of the early nineteenth century, Parkinson recommended venesection, specifically advocating bloodletting from the neck, followed by vesicatories to induce blistering and inflammation of the skin. Small pieces of cork were purposefully inserted into the blisters to cause a "sufficient quantity" of purulent discharge (Parkinson 1817). All these efforts were designed to divert blood and inflammatory pressure away from the brain and spinal cord and, in this way, decompress the medulla that Parkinson considered the seat of neurological dysfunction.

Pharmacological Advances: Charcot and Gowers

Being the two most celebrated clinical neurologists of the nineteenth century, Jean-Martin Charcot and William Gowers serve as important icons for the study of standard and emerging treatments for Parkinson's disease. Charcot's intern, Leopold Ordenstein, wrote his medical thesis (Ordenstein 1868) on the treatment of parkinsonian tremor with belladonna alkaloids, the first well-established treatment of Parkinson's disease. The credit of this observation surely be-

Cite this article as *Cold Spring Harb Perspect Med* doi: 10.1101/cshperspect.a041642

longs to Charcot himself who managed his Salpêtrière School with strict centralized supervision and oversaw every aspect of the neurological program. As with other young and aspiring students like Gilles de la Tourette and Pierre Marie, Ordenstein profited from publishing the observation with his name as the sole author, but contemporaries would not have been deluded into thinking of it as coming from anyone besides Professor Charcot. Of the many centrally active anticholinergic agents of the era, Charcot's preferred product for Parkinson's disease was hyoscyamine. This plant-based agent was prepared as pills, usually powder rolled into bits of white bread or as a syrup. As demonstrated in a prescription located in the Philadelphia College of Physicians, Charcot's anticholinergic treatment was sometimes combined with rye-based ergot products that in fact are the pharmacological basis of early modern dopamine agonists (Fig. 4). Although Tyler (1992) has aptly documented that Charcot was not the first interventionist to advocate hyoscyamine, Charcot's name became linked to the drug because of the widespread international publication of his lectures and classroom demonstrations. Another term widely adopted for hyoscyamine in the early twentieth century was Bulgarian therapy (Cura Bulgara, curabulgarin), because of the widely publicized and commercialized tonics of the Bulgarian herbal therapist, Ivan Raeff, supported strongly by Queen Elena of Italy (W Poewe, pers. comm.).

A unique historical opportunity to examine the early treatment of Parkinson's disease is provided by a series of 18 unpublished letters in the Charcot collection at the Bibliothèque Charcot in Paris (Portfolio MA VIII: Parkinson's disease). These letters cover a period of at least 15 months from January 1863 to March 1864. Although the collection only contains the patient's letters and not Charcot's replies, one can follow the doctor–patient interaction because of Charcot's technique of closing his letters traditionally with: "I would be most obliged Monsieur, if you would remind me of this prescription the next time you write." The patient's letters, therefore, systematically begin with a summary of the prescribed therapy and follow with the patient's own observations. In addition to hyoscyamine and ergot-

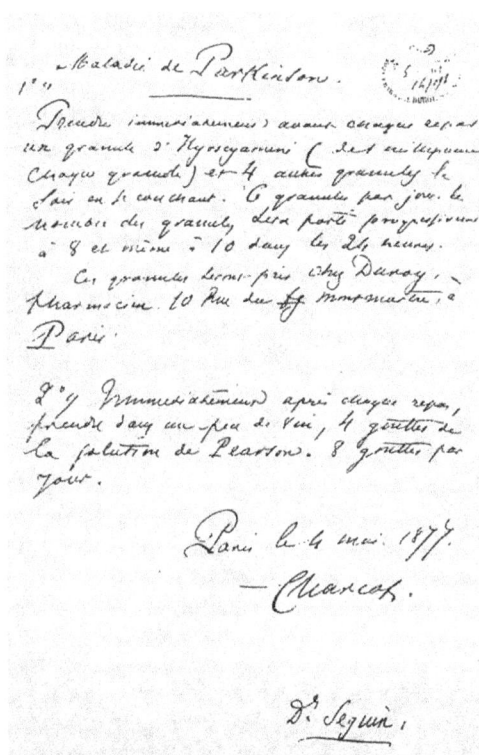

Figure 4. Early treatment of Parkinson's disease. Prescription dated 1877 from the College of Physicians of Philadelphia Library. In treating Parkinson's disease, Charcot used belladonna alkaloids (agents with potent anticholinergic properties) as well as rye-based products that had ergot activity, a feature of some currently available dopamine agonists. Charcot's advice was empiric and preceded the recognition of the well-known dopaminergic/cholinergic balance that is implicit to normal striatal neurochemical activity (Charcot 1872). (Reprinted, with permission, from Goetz 2011, © Cold Spring Harbor Laboratory Press.)

based products, Charcot advocated an overall program of rest and reduced stress. This type of therapy was generally advocated for many primary neurological disorders (Mitchell 1908). For this patient, he added camphor, silver nitrate, iron compounds, henbane pills, and zinc oxide. The use of iron may have been based on Romberg's earlier observation that carbonate of iron in association with warm baths and cold affusions to the head and back induced "a marked diminution of symptoms" (Romberg 1846). Whether based on his own experience or Romberg's warning against trying strychnine,

Charcot steered away from this therapy for Parkinson's disease patients. Charcot was highly specific in his instructions, insisting that quinquina must be diluted with syrup made from orange rind and each dose of silver nitrate must be impregnated in 9 g of soft bread to form an ingestible pill. The letters communicate encouragement to the patient, reinforce the need for patience in facing chronic illness, and a willingness to consider new treatment strategies if traditional ones were unsuccessful. However, his enthusiasm to try new interventions never clouded his objective vision of efficacy. In reviewing pharmacologic treatments for Parkinson's disease in 1872, Charcot stated (Charcot 1872):

> Everything, or almost everything, has been tried against this disease. Among the medicinal substances that have been extolled and which I have myself administered to no avail, I need only enumerate a few.

In rejecting most medicines, Charcot advocated vibratory therapy for the management of Parkinson's disease. Charcot (1892a) had observed that after long carriage, train, or horseback rides, patients with Parkinson's disease experienced marked symptom amelioration. He, therefore, developed a replication device to provide rhythmic movement by an electrically powered "shaking chair (*fauteuil trépidant*)" (Fig. 5). His student, Gilles de la Tourette, fashioned a helmet that was more easily transported and vibrated the brain rather than the body (Goetz et al. 1995). Other used therapies included hydrotherapy, spa treatments, and light exercise. Electrical stimulation by faradic, galvanic, or direct spark (franklinization) therapy was used to stimulate weakened muscles. Charcot was, however, adamant that patients with Parkinson's disease were not particularly weak, having tested them with dynamometers and finding their strength to be normal for most of the duration of the illness. It was partly for this reason that he dismissed the terms, paralysis agitans and shaking palsy, and advocated instead the designation, Parkinson's disease.

A more unusual and hazardous early treatment of Parkinson's disease involved the use of a suspension apparatus to stretch the spinal cord (Goetz et al. 1995). Developed in 1883 in Russia, the apparatus gained celebrity when Charcot examined its safety and efficacy in a variety of disorders, including Parkinson's disease. Using gravity and the patient's weight to put excessive vertical traction on the spinal cord and nerves, the therapist hoisted the subject in midair with a pulley and a harness that slipped under the chin and occiput. In Parkinson's disease patients, rigidity and some sensory symptoms improved, but tremor was not

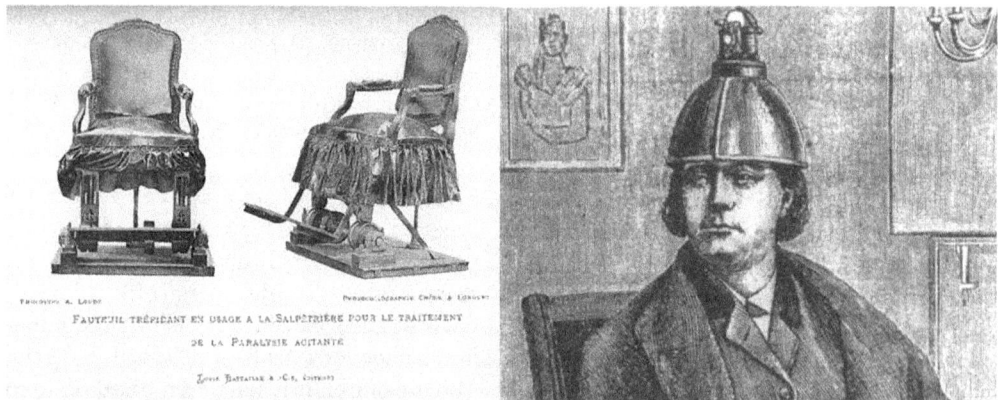

Figure 5. Vibratory therapy. Charcot observed that patients with Parkinson's disease experienced a reduction in their rest tremor after taking a carriage ride or after horseback riding. He developed a therapeutic vibratory chair that simulated the rhythmic shaking of a carriage (Goetz 1996). A vibratory helmet to shake the head and brain was later developed. Such therapies were not used widely but the availability of modern medical vibratory chairs offers an opportunity to confirm or refute Charcot's observation. (Reprinted, with permission, from Goetz 2011, © Cold Spring Harbor Laboratory Press.)

Cite this article as *Cold Spring Harb Perspect Med* doi: 10.1101/cshperspect.a041642

ameliorated. de Goncourt and de Goncourt (1887–1889) described the therapy with allusions to the macabre artwork of Goya, and the serious side effects and stress on patients led Charcot to abandon this strategy shortly after its introduction in France (1887–1889).

Charcot's British contemporary, W.R. Gowers, followed similar treatment strategies. He stressed the negative effects of mental strain and physical exhaustion, advocating that "life should be quiet and regular, freed, as far as may be, from care and work" (Gowers 1899). For tremor, he used hyoscyamine and also noted arsenic, morphia, conium (hemlock), and "Indian hemp" (*Cannabis*) as effective agents for temporary tremor abatement. Writing specifically of the power of *Cannabis* and opium in combination, he stated: "I have several times seen a very distinct improvement for a considerable time under their use" (Gowers 1899).

Levodopa and Dopamine-Based Therapies

Through the mid-twentieth century, the treatment of Parkinson's disease remained largely that of the nineteenth century, and although a wide variety of centrally active anticholinergic drugs were developed and used, they all were similar in their efficacy and side effect profiles. In the *Handbook of Clinical Neurology*, the chapter, "Drug treatment of parkinsonism and its assessment" (published in 1968) discusses 10 synthetic anticholinergic compounds and a potpourri of agents under the designation "Other drugs which have been recommended, some of them without any justification" (Onuaguluchi 1968). The emphasis of this period remained on supportive physical therapy and the management of hypersalivation, seborrhea, decubiti, and infections. In the context of this relative stagnation, the impact of levodopa was magnified.

As summarized by Oleh Hornykiewicz (2002), dopamine was first synthesized in 1910 by George Barger and James Ewens. In the same year, Henry Dale discovered its weak sympathomimetic qualities. These observations were later remembered when P. Holtz discovered the enzyme, dopa decarboxylase, and documented that levodopa was synthesized to dopamine through its action. At this time, dopamine was relegated to a simple intermediate compound for the synthesis of noradrenaline and adrenaline. The consistent identification of substantial amounts of dopamine in various tissues, however, prompted the search for a more primary role. Working in Hermann Blaschko's Cambridge University laboratory, Hornykiewicz studied blood pressure control in experimental animals and clearly confirmed that dopamine had distinct effects independent of other catecholamines. Shortly thereafter, in the late 1950s, two seminal discoveries occurred: dopamine localization within the brain, specifically in the striatum; and the development of the reserpine model, later to be used as the first model of parkinsonism that was reversed by levodopa treatment. In concert, these discoveries rapidly advanced hypotheses on the role of dopamine loss in the pathogenesis of Parkinson's disease itself (Carlsson et al. 1958; Sano et al. 1959), and led Bertler and Rosengren (1959) to conclude that "dopamine is concerned with the function of the striatum and thus with the control of movement." Ehringer and Hornykiewicz (1960) turned to human brain and after examining a series of control specimens, discovered the striatal dopamine depletion in Parkinson's disease and postencephalitic parkinsonism brains. With the knowledge that levodopa was the natural precursor to dopamine, Hornykiewicz was now prepared to suggest human trials in Parkinson's disease patients.

Walther Birkmayer received Hornykiewicz's supply of laboratory levodopa and injected it intravenously for the first time to parkinsonian patients in 1961. The antiakinetic effects were quickly published (Birkmayer and Hornykiewicz 1961):

> Bed-ridden patients who were unable to sit up, patients who could not stand up when seated, and patients who when standing could not start walking performed all these activities with ease after L-dopa. They walked around with normal associated movements and they could even run and jump. The voiceless, aphonic speech, blurred by palilalia and unclear articulation, became forceful and clear as in a normal person.

Subsequent open-label levodopa trials with oral preparations confirmed both short and long-term benefits, and a double-blind placebo-controlled trial followed (Barbeau 1969; Cotzias et al.

1969; Yahr et al. 1969). These reports launched levodopa's establishment as the premier agent to treat Parkinson's disease symptoms and signs. Although new formulations and peripherally acting dopa-decarboxylase inhibitors have added new dimensions to the therapy, none of these events rival the first discoveries.

Given that levodopa is a naturally occurring amino acid, researchers have reexamined older therapies to search for possible discoveries of levodopa-containing compounds in early medicine. Of note, cowage or cowitch plant (*Mucuna pruriens*) is known under the name of Atmagupta in Sanskrit and contains levodopa (Manyam 1990). One of the remedies used to treat the condition thought to be possible Parkinson's disease in traditional Indian medicine is called Masabaldi Pacana, which contains beans of *M. pruriens*. These observations offer interesting, albeit indirect, evidence that patients with parkinsonism may have experienced the benefit of levodopa early in the history of medicine.

The more modern discoveries of dopamine agonists, monoamine oxidase inhibitors, and catechol-*O*-methyl transferase inhibitors date to the contemporary period and are of less importance to this historical review that emphasizes early discoveries. These developments have been based on the logical understanding of the dopamine system, metabolic pathways, and receptor populations. Further discoveries of modulating influences by serotonin, adenosine, GABA, and glutamate systems have opened horizons for further pharmacological developments. The history of amantadine is of interest because of its serendipitous discovery as an antiparkinsonian agent. Developed as an antiviral agent, it was used widely in nursing home populations, and Schwab noted its unexpected benefit on tremor, balance, and akinesia in both Parkinson's disease and postencephalitic parkinsonian patients (Schwab et al. 1969).

SURGERY

In the early 1900s, surgery for movement disorders was pioneered by Victor Horsley and his engineering colleague, Robert Henry Clarke (Fig. 6). They developed early stereotaxic equipment to target brain nuclei, although their early surgeries dealt with hyperkinetic disorders rather than Parkinson's disease (Horsley and Clarke 1908). Bucy and Case (1939) and Klemme (1940) excised the cerebral cortex to treat parkinsonian tremor, but this type of ablative surgery induced hemiparesis and was abandoned. Meyers (1940) first focused on the basal ganglia as a lesion target for abating parkinsonian tremor in the 1940s and noted that rigidity improved as well as tremor. Importantly, spasticity and paresis did not compromise the improvement (Meyers 1940). In 1953, by accident, Irving S. Cooper cut the ante-

Figure 6. Early surgical interventions. (*Left*) Victor Horsley (1857–1916) was a celebrated British surgeon who attempted a surgical intervention on a movement disorder patient with athetosis in 1909. He excised the motor cortex with substantial improvement in involuntary movements. (*Middle*) Working in London with his physiologist colleague, Robert Henry Clarke (1850–1926), he developed early stereotaxic equipment, first for animal experiments and then for humans. (*Right*) This daunting surgical apparatus taken from their reports in *Brain* in 1908 guided them to deep brain centers including the basal ganglia and the cerebellum (Horsley and Clarke 1908). (Reprinted, with permission, from Goetz 2011, © Cold Spring Harbor Laboratory Press.)

Cite this article as *Cold Spring Harb Perspect Med* doi: 10.1101/cshperspect.a041642

rior choroidal artery during surgery on a parkinsonian patient and was forced to ligate it to prevent a hematoma. The unexpected and remarkable relief of tremor and rigidity on the contralateral side led to more widespread use of this procedure, although mortality was ~10% (Cooper 1953). Electrical coagulation procedures involving the globus pallidus, thalamus, and ansa lenticularis (ansotomy) were performed with early stereotaxic procedures (Spiegel and Wycis 1954). Hassler (1955) and Reichert (1962) focused more directly on the ventrolateral nucleus of the thalamus, also referred to as the ventral oralis anterior (Voa) nucleus. All these reports were hampered by the lack of involvement by medical neurologists with resultant concerns of incomplete reporting, lack of long-term follow-up, and potential minimalization of morbidity. Further, the role of surgery was eclipsed by the advent of levodopa in the 1960s, so that a long hiatus occurred when surgery was not extensively used in Parkinson's disease. During this time, however, more advanced surgical techniques were developed, and these innovations would be later applied to Parkinson's disease patients near the end of the twentieth century. Such treatments date to the contemporary period and include pallidotomy, subthalamic nucleus ablation, deep brain stimulation to the thalamus, pallidum, and subthalamic, and various transplantation procedures. Most recent are the developments of gene-based therapies that have entered clinical trials.

Placebo Therapy

The relationships between dopamine release and positive motivation, novelty-seeking behaviors, and attention have allowed researchers to understand the long-acknowledged placebo impact on Parkinson's disease. The Charcot letters cited above suggest that Charcot too understood clearly the importance of his presence and command over the patient's well-being. As anchored as he remained in neuroanatomical concepts through the end of his career, Charcot's last monograph was titled *Faith Cure* (1892b) and dealt with the profound improvements that some patients with neurological disease experienced through non-

traditional therapies. Placebo-controlled trials have become standard in Parkinson's disease, even in the surgical arena, mainly because a large percentage of patients on placebo treatment experience objective improvement in parkinsonism (Goetz et al. 2008). The facilitation of striatal dopaminergic activity in these settings has been demonstrated by neuroimaging techniques (de la Fuente-Fernández et al. 2001). The funding of federal grants for the specific study of placebo effects in Parkinson's disease is, in itself, of historical significance (Goetz et al. 2008).

CONCLUDING REMARKS

Historical documents on Parkinson's disease and descriptions that evoke parkinsonism from eras before the first full medical delineation of the disease provide a continuing source of potential neurological insights. As only one example, summarized in a review of traditional and complementary therapies for Parkinson's disease (Manyam and Sánchez-Ramos 1999), in 1928, Lewin isolated an alkaloid from the *Banisteriopsis caapi* vine used in ceremonial medicine among Amazonian tribes. He provided purified banisterine to his colleague, Beringer, who tested it on patients with Parkinson's disease with reported marked benefit. The data were presented to the Berlin Medical Association along with a film documenting the changes in rigidity, bradykinesia, and gait impairment. Although this agent was not pursued further, the example underscores the potential lessons from careful reading of traditional medicine sources and the prospects for new discoveries based on older observations. Charcot's advocacy for vibratory therapy has been tested in a modern setting (Kapur et al. 2011), but Gowers' encouraged use of *Cannabis* has yet to be systematically evaluated with strong clinical trial methodology. Numerous other therapies have suggestive roles in Parkinson's disease but have not been rigorously tested, including specific forms of physical exercise, massage therapy, and relaxation techniques. The active participation of the parkinsonian subject in these treatments complicates a controlled study design, but these interventions scientifically deserve to be tested with the same rigor as new pharmacological or surgical treatments. In the continuing search for therapies to ameliorate current disability and to slow the natural deterioration that is implicit to Parkinson's disease today, the guiding words of Charcot remain modern and applicable: "If you do not have a proven treatment for certain illnesses, bide your time, do what you can, but do not harm your patients" (Charcot 1888b).

ACKNOWLEDGMENTS

I acknowledge the Parkinson's Foundation, which supports the Rush University Medical Center Parkinson's Disease and Movement Disorder Program as a Clinical Center of Excellence. I also thank the expert archival staff of the Bibliothèque Charcot at the Salpêtrière Hospital in Paris, France.

REFERENCES

Adams RD, van Bogaert L, Vander Eecken H. 1964. Striatonigral degeneration. *J Neuropathol Exp Neurol* 23: 584–608.

Babinski J. 1921. Kinésie parodoxale. *Rev Neurol* 37: 1266–1270.

Barbeau A. 1969. L-dopa therapy in Parkinson's disease. *Can Med Assoc J* 101: 59–68.

Bertler Å, Rosengren E. 1959. Occurrence and distribution of dopamine in brain and other tissues. *Experientia* 15: 10–11. doi:10.1007/BF02157069

Birkmayer W, Hornykiewicz O. 1961. Der L-Dioxyphenylalanin-Effekt bei der Parkinson-Akinese. *Wien Klin Wschr* 73: 787–788.

Brissaud E. 1925. *Leçons sur les maladies nerveuses.* Masson, Paris.

Bruyn GW, Bots GT, Staal A. 1964. Familial bilateral vascular calcification in the central nervous system. *Psychiatr Neurol Neurochir* 67: 342–376.

Bucy PC, Case JT. 1939. Tremor: physiologic mechanism and abolition by surgical means. *Arch Neurol Psychiatr* 41: 721–746. doi:10.1001/archneurpsyc.1939.02270160077007

Carlsson A, Lindqvist M, Magnusson T, Waldeck B. 1958. On the presence of 3-hydroxytyramine in brain. *Science* 127: 471. doi:10.1126/science.127.3296.471.a

Charcot JM. 1872. De la paralysie agitante. In *Oeuvres Complètes (t 1) Leçons sur les maladies du système nerveux,* pp. 155–188. A Delahaye, Paris. [In English: Charcot JM. 1877. On Parkinson's disease. In *Lectures on diseases of the nervous system delivered at the Salpêtrière* (transl. Sigerson G), pp. 129–156. New Sydenham Society, London.]

Charcot JM. 1888a. *Leçons du Mardi: Policlinique de la Salpêtrière, 1887–1888.* Bureaux du Progrès Médical, Paris. Lesson of June 12, 1888.

Charcot JM. 1888b. *Leçons du Mardi: Policlinique de la Salpêtrière, 1887–1888.* Bureaux du Progrès Médical, Paris. Lesson of November 15, 1887.

Charcot JM. 1892a. La médicine vibratoire: application des vibrations rapides et continues a traitement de quelques maladies du système nerveux. *Prog Méd* 16: 149–151. [In English: Charcot JM. 1892. Vibratory therapeutics: the application of rapid and continuous vibrations to the treatment of certain diseases of the nervous system. *J Nerv Ment Dis* 19: 880–886.]

Charcot JM. 1892b. Faith-cure. *New Rev* 11: 244–262. [In French: Charcot JM. 1892. La foi qui guérit. *Rev Hebdomadaire* 5: 112–132.]

Cooper IS. 1953. Ligation of the anterior choroidal artery for involuntary movements-parkinsonism. *Psychiatr Q* 27: 317–319. doi:10.1007/BF01562492

Cotzias GC, Papavasiliou PS, Gellene R. 1969. Modification of parkinsonism—chronic treatment with L-dopa. *N Engl J Med* 280: 337–345. doi:10.1056/NEJM196902132800701

de Goncourt E, de Goncourt D. 1887–1889. *Journal: Mémoires de la vie littéraire,* Vol. 3. Flammarion, Paris.

de la Fuente-Fernández R, Ruth TJ, Sossi V, Schulzer M, Calne DB, Stoessl AJ. 2001. Expectation and dopamine release: mechanism of the placebo effect in Parkinson's disease. *Science* 293: 1164–1166. doi:10.1126/science.1060937

de Sauvages de la Croix FB. 1763. *Nosologia methodica.* Sumptibus Fratrum de Tournes, Amsterdam.

Dutil A. 1889. Sur un cas de paralysie agitante à forme hémiplégique avec attitude anormale de la tête et du tronc (extension). *Nouvelle Iconographie de la Salpêtrière* 2: 165–169.

Ehringer H, Hornykiewicz O. 1960. Verteilung von noradrenalin und dopamin (3-hydroxytyramin) im gehirn des menschen und ihr verhalten bei erkrankungen des extrapyramidalen systems. *Klin Wschr* 38: 1236–1239. doi:10.1007/BF01485901

Fénelon G, Goetz CG, Karenberg A. 2006. Hallucinations in Parkinson's disease in the pre-levodopa era. *Neurology* 66: 93–98. doi:10.1212/01.wnl.0000191325.31068.c4

Foix MC, Nicolesco J. 1925. *Les noyaux gris centraux et la région mesencéphalo-sous-optique.* Masson, Paris.

Goetz CG. 1987. *Charcot the clinician: the Tuesday lessons.* Raven, New York.

Goetz CG. 1996. An early photographic case of probable progressive supranuclear palsy. *Mov Disord* 11: 617–618. doi:10.1002/mds.870110604

Goetz CG. 2011. The history of Parkinson's disease: early clinical descriptions and neurological therapies. *Cold Spring Harb Perspect Med* 1: a008862. doi:10.1101/cshperspect.a008862

Goetz CG, Bonduelle M, Gelfand T. 1995. *Charcot: constructing neurology.* Oxford University Press, New York.

Goetz CG, Wuu J, McDermott MP, Adler CH, Fahn S, Freed CR, Hauser RA, Olanow WC, Shoulson I, Tandon PK, et al. 2008. Placebo response in Parkinson's disease: comparisons among 11 trials covering medical and surgical interventions. *Mov Disord* 23: 690–699. doi:10.1002/mds.21894

Gowers WR. 1888. *A manual of diseases of the nervous system.* J & A Churchill, London.

 Cite this article as *Cold Spring Harb Perspect Med* doi: 10.1101/cshperspect.a041642

Gowers WR. 1899. Paralysis agitans. In *A system of medicine* (ed. Allbutt A, Rolleston T), pp. 156–178. Macmillan, London.

Greenfield JG, Bosanquet FD. 1953. The brain-stem lesions in Parkinsonism. *J Neurol Neurosurg Psychiatry* **16:** 213–226. doi:10.1136/jnnp.16.4.213

Hassler R. 1955. The influence of stimulations and coagulations in the human thalamus on the tremor at rest and its physiopathologic mechanism. In *Proceedings of the Second International Congress of Neuropathology*, Vol. 1, pp. 637–642, London.

Hoehn MM, Yahr MD. 1967. Parkinsonism: onset, progression and mortality. *Neurology* **17:** 427–427. doi:10.1212/WNL.17.5.427

Hornykiewicz O. 2002. Dopamine miracle: from brain homogenate to dopamine replacement. *Mov Disord* **17:** 501–508. doi:10.1002/mds.10115

Horsley V, Clarke RH. 1908. The structure and functions of the cerebellum examined by a new method. *Brain* **31:** 45–124. doi:10.1093/brain/31.1.45

Hunt JR. 1917. Progressive atrophy of the globus pallidus. *Brain* **40:** 58–148. doi:10.1093/brain/40.1.58

Jankovic J. 1984. Progressive supranuclear palsy. *Neurol Clin* **2:** 473–486. doi:10.1016/S0733-8619(18)31085-5

Kapur SS, Stebbins GT, Goetz CG. 2011. Vibration therapy and Parkinson's disease. *Mov Disord* **26:** S132.

Klemme RM. 1940. Surgical treatment of dystonia, paralysis agitans and athetosis. *Arch Neurol Psychiatry* **44:** 926.

Langston JW, Ballard P, Tetrud JW, Irwin I. 1983. Chronic parkinsonism in humans due to a product of meperidine-analog synthesis. *Science* **219:** 979–980. doi:10.1126/science.6823561

Lewy FH. 1923. *Die Lehr von Tonus und der Bewegung.* Springer, Berlin.

Manyam BV. 1990. Paralysis agitans and levodopa in "Ayurveda": ancient Indian medical treatise. *Mov Disord* **5:** 47–48. doi:10.1002/mds.870050112

Manyam BV, Sánchez-Ramos JR. 1999. Traditional and complementary therapies in Parkinson's disease. *Adv Neurol* **80:** 565–574.

Meyers R. 1940. The modification of alternating tremors, rigidity and festination by surgery of the basal ganglia. *Proc Assoc Nerv Ment Dis* **21:** 602–665.

Mitchell SW. 1908. Rest and psychotherapy. *J Am Med Assoc* **50:** 2034.

Muenter MD, Whisnant JP. 1968. Basal ganglia calcification, hypoparathyroidism, and extrapyramidal motor manifestations. *Neurology* **18:** 1075–1083. doi:10.1212/wnl.18.11.1075

Onuaguluchi G. 1968. Drug treatment of parkinsonism and its assessment. In *Handbook of clinical neurology* (ed. Vinken PJ, Bruyn GW), Vol. 6, pp. 218–226. North-Holland, Amsterdam.

Ordenstein L. 1868. *Sur la paralysie agitante et la sclérose en plaque généralisée.* E Martinet, Paris.

Parkinson J. 1817. *An essay on the shaking palsy.* Whittingham and Rowland for Sherwood, Needly and Jones, London.

Pearce JM. 1996. Baron Constantin von Economo and encephalitis lethargica. *J Neurol Neurosurg Psychiatry* **60:** 167. doi:10.1136/jnnp.60.2.167

Reichert T. 1962. Long term follow-up of results of stereotaxic treatment in extrapyramidal disorders. *Confin Neurol* **22:** 336–363. doi:10.1159/000104386

Richer P, Meige H. 1895. Etude morphologique sur la maladie de Parkinson. *Nouvelle Iconographie de la Salpêtrière* **8:** 361–371.

Romberg M. 1846. *Lehrbuch der nervenkrankheiten des menschen.* A Duncker, Berlin. [In English: Romberg M. 1853. *A manual of the nervous diseases of man* (trans. Sieveking EH). New Sydenham Society, London]

Sano I, Gamo T, Kakimoto Y. 1959. Distribution of catechol compounds in human brain. *Biochim Biophys Acta* **32:** 586–587. doi:10.1016/0006-3002(59)90652-3

Schwab RS, England AC, Poskanzer DC. 1969. Amantadine in the treatment of Parkinson's disease. *J Am Med Assoc* **208:** 1168–1170. doi:10.1001/jama.1969.03160070046011

Spiegel EA, Wycis HT. 1954. Ansotomy in paralysis agitans. *Arch Neurol Psychiatry* **71:** 598–614. doi:10.1001/archneurpsyc.1954.02320410060005

Steck H. 1954. Le syndrôme extra-pyramidale et diencéphalique au cours des traitements au Largactil et au Serpasil. *Ann Méd-Psychchiatr* **112:** 737–743.

Sylvius de la Boë F. 1680. *Opera medica.* Danielem Elsevirium et Abrahamum Wolfgang, Amsterdam.

Trétiakoff C. 1921. Contribution à l'étude de l'anatomie du locus Niger. *Rev Neurol (Paris)* **37:** 592–608.

Tyler K. 1992. A history of Parkinson's disease. In *Handbook of Parkinson's disease* (ed. Koller WC), pp. 1–34. Marcel Dekker, New York.

van Bogaert L. 1930. Contribution clinique et anatomique à l'etude de la paralysie agitante juvenile primitive. *Rev Neurol* **2:** 315–326.

von Economo C. 1919. Grippe-encephalitis und Encephalitis lethargic. *Wien Klin Wschr* **32:** 393–396.

Westphal ACO. 1883. Über eine dem Bilde der cerebrospinalen grauen Degeneration ähnliche Erkrankung des centralen Nervensystems ohne anatomischen Befund, nebst einigen Bemerkungen über paradoxe Contraction. *Arch Psychiatr Nervenkr* **14:** 87–95, 767–773. doi:10.1007/BF02004266

Willige V. 1911. Ueber Paralysis Agiotans in Jugendlichen alter. *Ztschr ges Neurol u Psychiatt* **7:** 263–265.

Yahr MD, Duvoisin RC, Schear MJ, Barrett RE, Hoehn MM. 1969. Treatment of parkinsonism with levodopa. *Arch Neurol* **21:** 343–354. doi:10.1001/archneur.1969.00480160015001

Zhang ZX, Dong ZH, Román GC. 2006. Early descriptions of Parkinson's disease in ancient China. *Arch Neurol* **63:** 782–784. doi:10.1001/archneur.63.5.782

Clinical Management of Parkinson's Disease: Features, Diagnosis, and Principles of Treatment

Bhavana Patel,[1,2] Ashley Rawls,[1,2] Tracy Tholanikunnel,[1,2] and Michael S. Okun[1,2]

[1]Department of Neurology, University of Florida College of Medicine, Gainesville, Florida 32611, USA
[2]Norman Fixel Institute for Neurological Diseases, Gainesville, Florida 32608, USA

Correspondence: bhavana.patel@neurology.ufl.edu

Parkinson's disease (PD) is a progressive, neurological syndrome that is associated with a plethora of motor and nonmotor symptoms. Recognizing prodromal symptoms and diagnosing PD early and accurately as well as employing timely management strategies targeting motor and nonmotor symptoms across all disease stages will have the potential to improve clinical outcomes. The application of critical advances in the field including the development of biomarkers, pharmacological treatments, exercise, and surgical therapies will be important for clinical practitioners. In this review, we will address differential diagnoses and disease mimics, as well as provide critical updates on clinical diagnosis and management strategies.

The English physician, James Parkinson, first described Parkinson's disease (PD) more than 200 years ago in *An Essay on the Shaking Palsy* (Parkinson 1817). Parkinson described several individuals with resting tremor, shuffling gait, stooped posture, sleep problems, and constipation (Parkinson 1817). It was Jean Martin Charcot, the famous neurologist, who later renamed the condition Parkinson's disease. Charcot (1872) added bradykinesia and rigidity to the constellation of possible symptoms.

PD is the second most common neurodegenerative disorder after Alzheimer's disease (de Lau and Breteler 2006), and it was estimated that there will be 6.1 million individuals affected by PD by 2016 (Dorsey et al. 2018). Recently, this number has been updated to reflect a new diagnosis approximately every 6 minutes (Willis et al. 2022). The number of people living with PD is projected to double by 2040 to nearly 13 million (Dorsey and Bloem 2018). The global burden of PD has more than doubled since 1990 and is expected to continue to increase as the population ages (Dorsey et al. 2018).

CLINICAL FEATURES

Motor Symptoms

There are four cardinal motor symptoms of PD: bradykinesia, rest tremor, rigidity, postural instability, and gait impairment. These features characterize a clinical syndrome known as parkinsonism. Additionally, secondary motor symptoms that may be present include diminished arm swing, decreased blink rate, masked facies (hypomimia), and decreased voice volume (hypophonia) (Jankovic 2008).

Copyright © 2025 Cold Spring Harbor Laboratory Press; all rights reserved
Cite this article as *Cold Spring Harb Perspect Med* doi: 10.1101/cshperspect.a041638

1. Bradykinesia refers to slowness of movement with a decrement of amplitude or speed during attempted rapid alternating movements of body segments (Jankovic 2008). While simple slowness may be frequently observed in patients with decreased muscle power, spasticity, or depression, it is important to differentiate this sign from true bradykinesia. The difference can be distinguished clinically through the use of repetitive movements, which commonly demonstrate the emergence of progressive slowness and/or loss of amplitude (Berardelli et al. 2001). More globally, bradykinesia can be observed with hypomimia (decreased facial expression and eye blinking), hypophonia (softer voice), and micrographia (progressively smaller handwriting).

2. Rest tremor refers to an involuntary, rhythmic oscillation around a fixed point in the nonpostural or supported position, thus removing the action of gravitational forces. Tremor is frequently the first motor symptom of PD and affects ~90% of patients at some point in their condition (Obeso et al. 2017). This resting tremor commonly vanishes with active movements, and can recur when the arms are stretched outward; referred to as a reemergent tremor. In PD, the frequency of the rest tremor is usually in the low range (3–6 Hz) and amplitudes can range from slight (<1 cm) to severe (>10 cm) (Goetz et al. 2008). Tremors commonly present asymmetrically and are characterized by supination and pronation, leading to a "pill-rolling" phenomenon. Other types of tremors that can be observed in PD include finger flexion–extension, abduction-adduction, lower limb, and jaw tremors. Head tremor is not typical in PD and thus should lead to careful diagnostic reassessment. Tremor may be less responsive or even unresponsive to pharmacologic treatment, including levodopa (Fishman 2008).

3. Rigidity refers to stiffness or resistance of a limb when it is flexed passively and activating both agonist and antagonist muscles (Samii et al. 2004). This resistance is appreciated throughout the full range of movement, and

thus does not increase at higher mobilization speeds; distinguishing it from spasticity related to upper motor neuron structural lesions. The classic "cogwheel rigidity" can be appreciated with passive limb movement. The examiner should assess for rigidity in all extremities and at the neck for axial rigidity. To detect mild rigidity, voluntary movement of other limbs can be helpful (Jankovic 2008).

4. Postural and gait impairment are typically experienced later in the disease course, often ~5–10 years following initial diagnosis; however, they may occur earlier in some individuals. Parkinson's patients commonly develop a stooped posture due to a loss of postural reflexes, which can be a major contributor to falls (Jankovic 2008). Parkinsonian gait is characterized by short, shuffling steps with decreased arm swing and stooped posture. To an observer, these features give an impression that the patient is "chasing their own center of gravity." Festination is also an important phenomenon manifested by a fast succession of steps and difficulty in voluntarily stopping (Edwards et al. 2008). Postural instability does not respond to levodopa in contrast to the other cardinal symptoms (Sethi 2008). It is a major cause of falls, commonly contributing to fractures, loss of independence, and nursing-home placement in PD.

There are other motor phenomena that can be observed in PD including dyskinesia, dystonia, and motor fluctuations. Dyskinesia is commonly induced with levodopa use, and refers to abnormal, involuntary, choreiform movements that may affect the limbs, head, and/or torso. Levodopa-induced dyskinesias can be classified as peak-dose dyskinesia, wearing-off dyskinesia, or diphasic dyskinesia. Diphasic dyskinesia begins shortly after levodopa ingestion and is followed by improvement in parkinsonism and dissipation of dyskinesia, with a subsequent return of dyskinesia as dopamine levels decline (Obeso et al. 2017).

Dystonia refers to involuntary, prolonged muscle contractions with abnormal postures, usually in the limbs. Common manifestations of dystonia in PD include toe curling or foot inversion, and may result in pain and gait impairment. Dystonia frequently occurs in association with low dopamine levels such as the early morning hours or periods of motor fluctuations (Tolosa and Compta 2006).

Motor fluctuations refer to "off times," when poor response to levodopa alternates with "on times," or periods of improved symptoms. Motor fluctuations are more commonly observed in advancing PD. Off periods may be either predictable or unpredictable. Dose failure refers to episodes where levodopa has a delayed or absent clinical effect.

Freezing is a phenomenon where there is an episodic reduced ability to continue steppage. This is most commonly experienced during turning and step initiation but can also be observed when patients encounter spatial constraints, stress, or distraction (Giladi and Nieuwboer 2008).

Nonmotor Symptoms

Traditionally, PD has been regarded as a motor disorder, perhaps because motor symptoms often are the first noticed, even for untrained observers. However, in recent years, there has been increasing recognition of the importance of nonmotor symptoms in diagnosing and treating PD. Nonmotor symptoms, further detailed in Table 1, describe symptoms other than those involved in movement, such as tremor, rigidity, and bradykinesia. The impact of nonmotor symptoms is often greater than that of motor symptoms and they can be a major source of deterioration of quality of life (Poewe 2008).

Prodromal Parkinsonism

There are specific nonmotor symptoms that are commonly present before any of the classical motor signs develop, sometimes years or decades prior. These premotor symptoms confer some potential diagnostic utility in the early stages of the disease. Premotor symptoms include constipation, anosmia, rapid eye movement (REM) sleep behavior disorder, and depression.

Olfactory loss has been described as a reduction in smell identification, discrimination, and

Table 1. Nonmotor symptoms in Parkinson's disease (PD)

Neuropsychiatric features	Dysautonomia
Depression	Orthostatic hypotension
Anxiety	Constipation
Apathy	Urinary dysfunction
Impulse control disorder	Sexual dysfunction
Psychosis	Sialorrhea
Sleep disorders	**Sensory abnormalities**
Insomnia	Hyposmia or anosmia
REM sleep behavior disorder	Paresthesia
Restless leg syndrome	Numbness
Excessive daytime sleepiness	
Periodic limb movements of sleep	

Table based on data from Seppi et al. (2019).
(REM) Rapid eye movement.

increased threshold for smelling with a prevalence of up to 90% (Doty et al. 1988). These changes may precede the diagnosis by years or even decades (Ponsen et al. 2004; Ross et al. 2008; Fullard et al. 2017).

Symptoms of autonomic dysfunction can be early manifestations of PD. Constipation is defined as a spontaneous bowel movement frequency of less than once every 2 days, or at least weekly, and a need for a laxative. Constipation is commonly a prodromal feature observed in PD (Berg et al. 2015).

Rapid eye movement sleep behavior disorder (RBD) is a parasomnia marked by the occurrence of dream enactment behavior and vivid dreams. Physiologically, RBD is marked by a loss of the normal atonia that characterizes REM sleep. When RBD is isolated and idiopathic, it is highly specific for portending the future risk of α-synucleinopathies such as PD (Berg et al. 2015). There are other sleep disorders observed in the prodrome of PD; however, these disorders are common in older adults and are significantly less specific than RBD for later risk of PD. These disorders include excessive daytime sleepiness, restless leg syndrome, and insomnia.

The duration of the prodromal period of parkinsonism is largely unknown. While some studies have demonstrated prodromal durations of >20 years, many patients report significantly shorter prodromal intervals. Interestingly, different individual prodromal markers have different lead times. In general, diagnosing prodromal parkinsonism becomes easier as patients approach clinical PD onset (Berg et al. 2015).

DIAGNOSIS OF PD

The gold standard diagnosis for PD is neuropathological confirmation (Adler et al. 2021). Numerous studies have revealed that tissue confirmation of a clinical diagnosis of PD ranges from 65% to 93%, based largely on which criteria are used and the stage of disease (Adler et al. 2021). In 2015, the International Parkinson and Movement Disorder Society (MDS) proposed a set of criteria for clinicians to be more objective and accurate in diagnosis (Postuma et al. 2018). Briefly, the MDS PD clinical criteria were based on expert neurological examination, and were characterized by the presence of bradykinesia with at least one additional cardinal motor feature of PD: resting tremor (4–6 Hz) or rigidity (Postuma et al. 2018).

The diagnosis of PD currently relies on three key elements: (1) a thorough history, (2) a clinical examination, and (3) the use of appropriate ancillary testing. The history should include a detailed interview of the patient, and also the family to collect vital information about symptom onset, disease progression, and effects on activities of daily living. Motor symptoms are usually the most obvious to recognize, including resting tremor, rigidity (e.g., asking about limb stiffness, difficulty turning in bed, challenges arising from seated position), bradykinesia (e.g., moving more slowly independent of other musculoskeletal factors including arthritis), and gait disturbance (e.g., shuffling gait, decreased arm swing, difficulty with turning). There are 20% of cases of PD that present without resting tremor, and the clinician should keep this in mind when collecting information (Gupta et al. 2020). Nonmotor symptoms should also be carefully evaluated, as patients commonly report hyposmia or anosmia, sleep disorders (e.g., RBD), neuropsychiatric features (e.g., depression, anxiety, or apathy), autonomic

changes (e.g., constipation, bladder dysfunction), or cognitive changes (Tolosa et al. 2021). These features have been referred to as prodromal symptoms. A thorough drug history should be collected inclusive of potential exposure to dopamine-blocking agents. Also critical is determining exposure to environmental toxins (e.g., manganese in welders, paraquat in farmers, and trichloroethylene in dry cleaning and other industries). Collecting a family history and ethnic ancestry may possibly inform the presence of monogenic forms of PD in specific populations (e.g., Ashkenazi Jewish and North African Arab populations have a higher frequency of leucine-rich repeat kinase 2 [LRRK2] genetic PD) (Tolosa et al. 2021).

The PD clinical examination commonly begins with evaluation of a person in the seated position. Observation of the feet, both off and on the floor, may reveal lower extremity tremor, which when present is most commonly associated with a diagnosis of PD (Rajalingam et al. 2019). We recommend that the tremor be evaluated at rest, posture, and action and in all extremities. The clinician should observe for the presence of a masked face and possible facial tremor. There may be associated hypophonia, and this may be revealed by asking the patient to read aloud a passage (i.e., the rainbow passage). Eye movements are a critical aspect of the examination and should include a full range of motion and examination of latency, speed, and accuracy. Rapid alternating movements (e.g., finger tapping, foot tapping) must be performed to assess for bradykinesia (i.e., slowness of movement with associated decrement). Motor symptoms are typically asymmetrical. To test rigidity, the examiner should slowly move the patient's head, elbows, wrists, knees, and ankles in flexion and in extension while the patient is fully relaxed. Rigidity is commonly referred to as cogwheel in PD and has a "click-click-click" presentation that can be augmented by distraction maneuvers (i.e., use of tapping of a contralateral extremity while the examiner performs a slow movement). Cogwheel rigidity is thought to be associated with underlying tremor that is "mixed in" with the stiffness. The patient's ability to rise from a chair without the use of his or her arms is important to assess disability and to guide treatment. The patient should be asked to walk down an open hallway, and the examiner should observe posture, arm swing, stride length, and, especially, the ability to turn. Walking may enhance tremor or reveal dystonia. Stooped posture, shuffling gait, decreased arm swing, decreased stride length, freezing, and en bloc turning (turning of the head, trunk, and pelvis as a unit rather than turning in a top-down sequence) are all clues to a possible diagnosis of PD. Eye movement abnormalities may include hypometric saccades both horizontal and vertical, impaired smooth pursuit, and convergence insufficiency (Jung and Kim 2019). Potential exclusionary criteria for PD (e.g., cerebellar signs, oculomotor abnormalities, significant cognitive impairment) and other movement disorders (e.g., chorea, myoclonus, tics) should be considered (McFarland and Hess 2017). Dystonia, particularly of the lower extremity with foot turning or toe curling is common in PD (Shetty et al. 2019). If the examination is performed when the patient is on dopamine replacement therapy, dyskinesias may manifest. In undiagnosed cases, the first examination is usually performed without dopaminergic medications. Subsequently, the examination can be repeated "ON" and "OFF" dopaminergic therapy, and the percent change as well as improvement in individual features may guide both diagnostic certainty and treatment changes. Clinicians with many PD and parkinsonism cases in their practice should be encouraged to learn to administer the MDS-Unified Parkinson's Disease Rating Scale, as most of the above features are captured and can be compared to track progression and levodopa responsiveness over time.

In select cases, ancillary testing may be required to evaluate for PD "mimickers." Laboratory testing for hypothyroidism, anemia, and Wilson disease (particularly for young-onset PD) may be considered. Structural pathology contributing to clinical parkinsonism can be evaluated by brain imaging, usually a magnetic resonance imaging (MRI) scan. However, in many cases, MRI is more useful for parkinsonism rather than idiopathic PD (Brücke and

Brücke 2022). Dopamine transporter imaging using single-photon emission tomography (DAT-SPECT) has become more widely used in clinical practice to assess for a dopaminergic deficit, although these images do not differentiate between PD and other neurodegenerative parkinsonisms (Brücke and Brücke 2022). Alternatively, positron emission tomography (PET), using a fluorodopa approach, can be used to assess the scan pattern indicative of a degenerative parkinsonism. However, expertise, insurance coverage, and access to a cyclotron facility have restricted its use and adoption into broad clinical practice (Brücke and Brücke 2022). While both the DAT-SPECT and PET scans are very sensitive and can distinguish a dopaminergic-related disease state from a non-dopaminergic disease state (e.g., essential tremor or drug-induced parkinsonism), these scans struggle to distinguish PD from other causes of degenerative parkinsonism (e.g., corticobasal syndrome, dementia with Lewy bodies [DLB], multiple system atrophy, progressive supranuclear palsy [PSP]) (Brücke and Brücke 2022). A skin biopsy test is now available to aid in the diagnosis of "synucleinopathy" (Kim et al. 2019). In most kits, the practitioner samples three body regions, and the results will reveal the presence or absence of phosphorylated α-synuclein and nerve fiber loss. Investigations are ongoing to determine whether distribution patterns of phosphorylated α-synuclein can help differentiate among the synucleinopathies (Gibbons et al. 2023). Most practitioners can deliver a diagnosis of PD without an MRI, DAT-SPECT, or PET scan. Adherence to the criteria for diagnosis, coupled with levodopa responsiveness and long-term follow-up, will usually lead to accurate diagnosis.

Since PD diagnosis relies on the presence or absence of clinical symptoms, this can lead to a delay in detection in early disease. It will become increasingly important to identify PD cases earlier, as more disease-modifying agents become available for testing (Parnetti et al. 2019). Clinicians should inquire for REM sleep behavioral disorder, loss of smell and the presence of constipation as potential prodromal or early PD symptoms. In the past decade, clinical biomarker research has been working toward enhancing diagnostic accuracy, including sensitivity for the detection of early and prodromal disease states. Hyposmia or anosmia has been observed in 90% of patients with PD, and olfactory testing with the University of Pennsylvania Smell Identification Test (UPSIT) or Sniffin Stick Test, can be used in the initial clinical testing for suspected early cases of PD (Tolosa et al. 2021). Many patients with PD will not be aware they have a deficit in their ability to identify specific odors. Cerebrospinal fluid (CSF) is a useful, but not a practical test for obtaining PD biomarkers in the research setting. The need for repeated lumbar punctures limits its wide application. CSF has not to date been shown to be capable of monitoring disease progression (Parnetti et al. 2019). CSF α-synuclein seeding activity has been shown to distinguish different α-synuclein strains in PD versus multiple system atrophy (MSA), while CSF neurofilament light chain has proven to be a nonspecific marker for neurodegeneration—a marker that may possibly differentiate PD from atypical parkinsonism at least in some case series (Tolosa et al. 2021). Serum neurofilament light chain is currently considered too nonspecific to separate different neurodegenerative diseases (Tolosa et al. 2021). Skin biopsy analyses for α-synuclein using real-time quaking induced conversion (RT-QuIC) or protein misfolding cyclic amplification (PMCA) have recently revealed high specificity and sensitivity for the presence of "synucleinopathies," although this test is not specific to PD versus other parkinsonisms, and may miss some monogenic causes of PD such as the LRRK2 gene mutations (Tolosa et al. 2021). Many studies have continued to investigate tissue biopsies including the olfactory mucosa, the gastrointestinal (GI) tract, and the salivary tissues, and data continues to emerge on their potential usefulness in clinical practice and in clinical trials (Tolosa et al. 2021).

Suspected PD patients should in most cases be administered an oral levodopa challenge. Initially, a clinician may titrate levodopa to 600 mg total daily dose (200 mg per dose) and if no response may go higher per dose. It is important to rate each clinical symptom by performing an MDS-UPDRS clinical scale before and after levodopa is administered (200–250 mg as a single dose) to determine "levodopa responsiveness" (Schade et al. 2017). This is important as sometimes PD tremor does not respond to levodopa, and features such as bradykinesia and rigidity will only be "picked up" if formally assessed. Improvement in motor symptoms by at least 30% with levodopa compared to baseline is clinically significant and has high sensitivity and specificity for a clinical diagnosis of PD (Schade et al. 2017). In cases where there is a lack of response to levodopa, the clinician should consider pursuing a gastric emptying study to ensure optimal medication absorption (Marrinan et al. 2014).

DIFFERENTIAL DIAGNOSIS

The diagnostic accuracy for PD will improve with clinical experience and by carefully reexamining the patient over time. The differential diagnosis should start broadly for any patient presenting with symptoms of parkinsonism. Radiographic findings may support differential diagnoses (key imaging findings are depicted in Table 2).

1. *Vascular parkinsonism:* Classically vascular parkinsonism is considered when patients present with predominantly lower body parkinsonian symptoms, such as a slightly more broad-based gait disturbance along with postural instability (Raccagni et al. 2020). Patients with vascular parkinsonism more commonly have a postural rather than an action tremor, can have pyramidal signs, and are less likely to have anosmia preceding motor symptoms. Many individuals are not responsive to levodopa (Narasimhan et al. 2022). Structural brain imaging is imperative to assess the burden of white matter disease and to correlate the clinical findings, as well as weigh the vascular risk factors (e.g., hypertension, diabetes, hyperlipidemia, prior strokes, etc.). MRI brain should be included when considering this diagnosis (Raccagni et al. 2020). Recently, experts have questioned how common vascular parkinsonism really is, and have stressed the im-

Table 2. Key radiographic findings in differential diagnosis of parkinsonisms

Parkinsonisms	Image modality	Description	Image example	References
Parkinson's disease (PD) or dementia with Lewy bodies (DLB)	Magnetic resonance imaging (MRI)—T2*/ susceptibility-weighted imaging (SWI)	(A) Absence of swallow tail sign, which is the normal appearance of a hyperintensity within the substantia nigra, can be seen in PD or DLB (B) Non-PD patient, normal hyperintensity present within the substantia nigra		Gupta et al. (2022) (image reprinted from Schwarz et al. 2014, under the terms of the Creative Commons Attribution License)

Continued

Table 2. *Continued*

Parkinsonisms	Image modality	Description	Image example	References
Vascular parkinsonism	MRI–fluid-attenuated inversion recovery (FLAIR)	Periventricular white matter lesions, lacunar infarcts in the basal ganglia, dilatation of lateral and third ventricles, presence of microbleeds, subcortical, or cortical atrophy		Ma et al. (2019) (image reprinted from George et al. 2024, under the terms of the Creative Commons Attribution 4.0 International License)
DLB	F-fluorodeoxyglucose positron emission tomography (FDG PET)	Cingulate island sign: occipital hypometabolism with relatively spared posterior cingulate cortex (A) Cingulate island sign: individual with DLB with normal ^{18}F-FDG uptake in the posterior cingulate region (yellow arrowhead) surrounded by reduced ^{18}F-FDG uptake in the occipital cortex (B) Healthy control with normal ^{18}F-FDG uptake in the posterior cingulate, occipital, and other cortical regions		Lim et al. (2009); Asahara et al. (2023) (image reprinted from McKeith et al. 2017, under the terms of the Creative Commons Attribution License 4.0 (CC BY))

Continued

Cite this article as *Cold Spring Harb Perspect Med* doi: 10.1101/cshperspect.a041638

Table 2. *Continued*

Parkinsonisms	Image modality	Description	Image example	References
Multiple systems atrophy (MSA)	MRI	Hot cross bun sign: T2 hyperintensities in the pontocerebellar tracts; can also be seen in spinocerebellar ataxias		Watanabe et al. (2018) (image reprinted from Gaillard et al. 2011, found at Radiopaedia, doi:10.53347/rID-13008)
Progressive supranuclear palsy (PSP)	MRI	Hummingbird sign: midbrain atrophy with relatively preserved pons in sagittal planes; measure with midbrain to pons ratio		Whitwell et al. (2017) (image reprinted from under the terms of the CC BY 4.0 Attribution 4.0 International License)

Continued

Table 2. *Continued*

Parkinsonisms	Image modality	Description	Image example	References
PSP	MRI	Morning Glory sign: concavity of lateral margin of midbrain tegmentum in axial planes Mickey mouse sign: midbrain atrophy resulting in rounded midbrain peduncles (red arrow)		Whitwell et al. (2017; Mueller et al. 2018) (image reprinted under the terms of the CC BY 4.0 Attribution 4.0 International License)
Corticobasal degeneration (CBD)	MRI	Asymmetric frontoparietal cortex and basal ganglia atrophy		Shir et al. (2023) (image reprinted from Saeed et al. 2017, under the terms of the Creative Commons Attribution 4.0 International License)
Fragile X tremor/ ataxia syndrome (FXTAS)	MRI	Middle cerebellar peduncle (MCP) sign: T2/FLAIR hyperintensity and atrophy in both middle cerebellar peduncles; can also be seen in MSA and in carriers without neurologic manifestations		Lee et al. (2019) (image reprinted from Famula et al. 2018, under the terms of the Creative Commons Attribution 4.0 International License, © 2018 Famula, McKenzie, McLennan, Grigsby, Tassone, Hessl, Rivera, Martinez-Cerdeno, and Hagerman)

(RBD) Rapid eye movement behavior disorder, (MDS) International Parkinson and Movement Disorder Society, (CT) computed tomography, (DAT) dopamine transporter, (DAT-SPECT) dopamine transporter imaging with single-photon emission, (CSF) cerebrospinal fluid, (CNS) central nervous system, (RT-QuIC) real-time quaking induced conversion, (PMCA) protein misfolding cyclic amplification, (PDD) Parkinson disease dementia, (MSA-C) multiple systems atrophy-cerebellar, (SCA) spinocerebellar ataxia, (PSP-RS) progressive supranuclear palsy Richardson syndrome, (CBS) corticobasal syndrome, (FMR1) Fragile X mental retardation 1, (MCP) middle cerebellar peduncles, (EEG) electroencephalography, (NREM) nonrapid eye movement, (CASPR2) contactin-associated protein-like 2, (DPPX) dipeptidyl-peptidase-like protein-6, (LRRK2) leucine-rich repeat kinase 2, (PINK1) PTEN-induced putative kinase.

Cite this article as *Cold Spring Harb Perspect Med* doi: 10.1101/cshperspect.a041638

portance of levodopa trials since many cases of PD will also have white matter disease on MRI.

2. Drug-induced parkinsonism: The clinician should carefully collect a history of exposure to dopamine-blocking agents, in particular antipsychotic medications. Symmetric parkinsonian symptoms may be a clue, with the possible absence of rest tremor and none to little response to levodopa (Brigo et al. 2014). Other medication classes have been recognized as contributing to drug-induced parkinsonism, such as antidepressants, GI prokinetics, and antiepileptic drugs (Shiraiwa et al. 2018). Specifically, antiemetics such as promethazine, prochlorperazine, and metoclopramide frequently administered to address GI symptoms and as part of migraine regimens may result in parkinsonism. Furthermore, a review of recent hospitalizations may yield inadvertent exposures to dopamine-blocking drugs such as haloperidol. Discontinuing the offending medication early in the disease process can, in select cases, lead to improvement of parkinsonian symptoms, although the condition could overall persist (Shiraiwa et al. 2018). In cases where the offending medication cannot be stopped, the prescribing clinician should consider decreasing the dose or finding an alternative treatment (Shiraiwa et al. 2018). The clinician should be aware of the possibility that the dopamine-blocking medication may have unmasked a previously undiagnosed case of PD. Additionally, other drug-induced movement disorders may also be present in these cases, such as orolingual dyskinesias, tardive dystonia, or akathisia (Brigo et al. 2014; Shiraiwa et al. 2018).

3. Essential tremor (ET): Essential tremor cases may be misinterpreted as tremor-predominant PD, especially since up to 20% of individuals with ET may have rest tremor (Cohen et al. 2003). In contrast to rest tremor, which primarily involves more flexion-extension movements at the metacarpal-phalangeal or phalangeal joints, essential tremor typically causes more shaking in the wrist region

(Shanker 2019). Family history of tremor (which accounts for ~50% of patients) and tremor responsiveness to alcohol may provide more circumstantial evidence, although alcohol responsiveness is not unique to essential tremor (Shanker 2019). Additional distinctions that favor an ET diagnosis include an asymmetric postural or kinetic hand tremor reaching frequencies up to 12 Hz, less frequent rest tremor, and absence of other signs of parkinsonism (Shanker 2019). Due to the possible presence of a rest tremor among individuals with ET along with instances of individuals with ET later developing PD, the use of DAT-SPECT or skin biopsy can be helpful when management strategies include selecting a target for ablative surgery or deep brain stimulation.

4. Dementia with Lewy bodies (DLB): The cardinal feature of DLB is dementia, presenting with deficits in attention, executive function, and visual processing (McKeith et al. 2017). The core clinical features include delirium-like cognitive fluctuations with variations in attention and alertness, well-formed visual hallucinations (i.e., people or animals), RBD, and parkinsonism with motor features co-occurring or beginning after 1 year of cognitive dysfunction (McKeith et al. 2017). Supportive clinical features include severe sensitivity to antipsychotic agents that can pose a challenging problem, especially in those who require treatment for psychosis (McKeith et al. 2017). Additional supportive findings are postural instability, repeated falls, syncope, severe autonomic dysfunction (e.g., constipation, orthostatic hypotension, etc.), hypersomnia, hyposmia, and mood disturbances (McKeith et al. 2017). When the distinction must be determined between DLB and PD dementia (PDD, dementia occurring later in the disease duration of a well-established case of levodopa-responsive PD), the clinician will need to carefully consider the historical features of the illness. Many experts use the "1-year" rule of onset between dementia and parkinsonism. Although this is not an absolute criterion when dementia oc-

curs at the same time or 1 year before the motor symptoms the diagnosis is more likely DLB (McKeith et al. 2017). Imaging biomarkers include reduced dopamine transporter uptake on DAT-SPECT, polysomnography confirming REM sleep without atonia, and an abnormal [123]iodine-metaiodobenzylguanidine (MIBG) myocardial scintigraphy (McKeith et al. 2017).

5. Multiple system atrophy (MSA): MSA is characterized by autonomic dysfunction, parkinsonism, and a cerebellar syndrome all occurring in varying degrees. It can be further divided into phenotypes: MSA-parkinsonian variant (MSA-P, striatonigral degeneration) and MSA-cerebellar variant (MSA-C, olivopontocerebellar atrophy) based on the predominant motor syndrome at the time of evaluation (Fanciulli and Wenning 2015). Symptoms of autonomic failure occur early in the disease, and include orthostatic hypotension, urinary incontinence, constipation, erectile dysfunction, and sweat gland dysfunction (Wenning et al. 2022). Parkinsonism is often poorly responsive to dopaminergic therapy; however, up to nearly 57% of individuals with MSA-P and up to 25% of individuals with MSA-C may have a beneficial response for a mean of 3.5 ± 2.7 years and 3.2 ± 2.3 years, respectively (Wenning et al. 2013; Low et al. 2015). Dopaminergic therapy in MSA may lead to craniocervical dystonia, providing another clue to diagnosis (Boesch et al. 2002). Other motor features may include rapid progression within 3 years of onset of motor symptoms, early postural instability, pyramidal dysfunction (Babinski sign, hyperreflexia), early severe dysarthria and dysphagia, myoclonic postural or kinetic tremor, anterocollis, and camptocormia (Wenning et al. 2022). The presence of stridor, if identified, can serve as another diagnostic clue. Stridor may occur in MSA due to laryngeal dysfunction most commonly occurring during inspiration and is frequently noted during sleep (Cortelli et al. 2019). MRI may reveal atrophy of the pons, cerebellum, putamen, and middle cerebellar peduncle (Wenning et al. 2022). The "hot cross bun sign" is a cruciform hyperintensity in the pons on T2 MRI images, although present in a minority of cases, may reveal selective loss of pontocerebellar fibers in the raphe while sparing the corticospinal tracts and tegmentum (Massey et al. 2012; Wenning et al. 2022). This radiographic finding can be observed in patients with MSA, but it is not specific to this disease and has been reported in secondary parkinsonism along with spinocerebellar ataxia type 2 and 3 (Bürk et al. 2001; Muqit et al. 2001).

6. Progressive supranuclear palsy (PSP): PSP encompasses a large clinical spectrum and has been separated into numerous subtypes, with the most classic presentation being PSP-Richardson syndrome (PSP-RS). PSP-RS presents with a symmetric akinetic-rigid syndrome primarily involving the trunk and neck, postural instability within 3 years of symptom onset, and ocular motor dysfunction (Höglinger et al. 2017). Ocular motor dysfunction includes square wave jerks, eyelid opening apraxia, slowing of vertical saccadic eye movements, and limitation in voluntary vertical gaze. Cognitive dysfunction in PSP often presents as a dysexecutive/frontal syndrome or language disorder in the form of a primary progressive aphasia or progressive apraxia of speech (Höglinger et al. 2017). Additional supportive features include poor levodopa response, hypokinetic spastic dysarthria, dysphagia, and photophobia (Höglinger et al. 2017). The "procerus sign," or abnormal vertical wrinkling of the forehead, can be a helpful clinical sign for early PSP (Bhattacharjee 2018). Imaging may be helpful in distinguishing PSP from other parkinsonisms. Brain MRI may reveal midbrain atrophy in relation to the pons, which on sagittal sections is referred to as the "hummingbird sign" and on axial images as the "Mickey Mouse sign" (Virhammar et al. 2022). Additionally, quantitative measurements of the midbrain and pons may assist in differentiating PSP and MSA (Massey et al. 2013).

7. Corticobasal degeneration (CBD): Individuals with CBD have widespread deposition of hyperphosphorylated four-repeat tau in neurons and glia that can present as four different phenotypes: (1) corticobasal syndrome (CBS), (2) frontal behavioral–spatial syndrome, (3) nonfluent, agrammatic variant of primary progressive aphasia, or (4) progressive supranuclear palsy (PSP) syndrome (Armstrong et al. 2013). CBS is defined as an asymmetric presentation of limb rigidity or akinesia, dystonia, and/or myoclonus, along with ideomotor apraxia affecting the limb or face, cortical sensory deficits, and alien limb phenomenon (Armstrong et al. 2013). The alien limb phenomenon is described as an inability to recognize one's own hand along with involuntary movements, for example, limb elevation or abnormal posturing (Graff-Radford et al. 2013). MRI findings in CBD include asymmetric cortical atrophy, which may evolve as the disease progresses (Constantinides et al. 2019). PSP and CBD may manifest similarly making it challenging to distinguish clinically.

8. Fragile X tremor/ataxia syndrome (FXTAS): FXTAS is defined by the expansion of the CGG trinucleotide repeat (>200) in the promotor region of the fragile X mental retardation 1 (FMR1) gene (Cabal-Herrera et al. 2020). It is the most common cause of inherited intellectual disability and autism spectrum disorder (Mila et al. 2018). Cerebellar ataxia and intention tremor are the core presenting features of FXTAS, along with parkinsonism, neuropathy, and executive dysfunction (Cabal-Herrera et al. 2020). Initial presentation typically includes action or intention tremor followed by an ataxic gait several years later (Cabal-Herrera et al. 2020). The peak onset of the action or intention tremor is usually early in the sixth decade, with a resting tremor component occurring in ~13%–26% of patients (Cabal-Herrera et al. 2020). Mild parkinsonism is present in ~29%–60% of FXTAS carriers (Cabal-Herrera et al. 2020). Impaired optokinetic nystagmus vertically, slowing of vertical saccades, saccadic pursuits, and square wave jerks can also be observed in these patients, and many of the ocular manifestations may be similar to the eye movement abnormalities seen in PSP (Cabal-Herrera et al. 2020). Individuals with FXTAS can have executive dysfunction, which has been correlated with the number of CGG repeats and typically worsens over time (Cabal-Herrera et al. 2020). MRI may reveal bilateral hyperintensities in the middle cerebellar peduncles, coined the "MCP sign"; however, this imaging sign can also be observed in MSA and in carriers without neurologic manifestations (Cabal-Herrera et al. 2020). While not specific to FXTAS, hyperintensities can also be observed in the periventricular white matter, the corpus callosum, and the brainstem (Cabal-Herrera et al. 2020). FXTAS should be included in the differential diagnosis in patients with parkinsonism with ataxia and tremor (Cabal-Herrera et al. 2020). Diagnosis can be confirmed by genetic testing.

9. Autoimmune parkinsonism: Autoimmune parkinsonism refers to disorders resulting from autoantibodies targeting neuronal antigens. Clinical presentation may vary based on the autoantibodies causing disease (Gövert et al. 2020). Anti-leucine-rich glioma inactivated 1 (LGI1) encephalitis typically presents with subacute limbic encephalitis, cognitive dysfunction, fasciobrachial dystonic seizures, and/or parkinsonism, chorea, or myoclonus (Gövert et al. 2020; Sturchio et al. 2022). MRI may reveal T2 hyperintensities in the medial temporal lobe and hippocampus and electroencephalography (EEG) with coincident epileptiform activity or slowing. The CSF has been reported to be associated with increased protein and pleocytosis (Giannoccaro et al. 2023). Anti-IgLON5 disease can present with various neurologic symptoms, including parkinsonism, chorea and orofacial dyskinesias, dystonia, and ataxia (Sturchio et al. 2022). This entity should be considered in patients who manifest prominent sleep disorders such as nonrapid eye movement (NREM) sleep parasomnias, excessive daytime sleepiness, or insomnia. Individuals

with anti-IgLON5 disease may also experience bulbar symptoms, respiratory problems, cognitive, or psychiatric dysfunction (Gövert et al. 2020; Sturchio et al. 2022). Although rare, individuals with anti-IgLON5 disease can have a PSP-like presentation with postural instability and eye movement abnormalities (Gövert et al. 2020). MRI may be nonspecific or normal, and ~60% of CSF samples have pleocytosis (Giannoccaro et al. 2023). In anti-contactin-associated protein-like 2 (CASPR2) disease, patients can present with parkinsonism with ataxia, limbic encephalitis, neuropathic pain, anterograde, and episodic memory dysfunction (Gövert et al. 2020; Giannoccaro et al. 2023). MRI and CSF are frequently normal, while EEG may reveal epileptic changes or slowing (Giannoccaro et al. 2023). Another autoimmune syndrome called, anti-dipeptidyl-peptidase-like protein-6 (DPPX) disease presents in middle-aged males with symptoms of cognitive and behavioral changes, prolonged diarrhea, weight loss, and movement symptoms of parkinsonism, tremors, and/or myoclonus (Sturchio et al. 2022). The EEG is usually abnormal, while the MRI and CSF are nonspecific (Giannoccaro et al. 2023). Other autoantibodies that may present with parkinsonism include ANNA-1/Hu, ANNA-2/Ri, NMDA receptor, and CRMP5/CV2 (Sturchio et al. 2022).

GENETIC PARKINSONISM

Monogenic forms of PD account for ~10%–20% of cases (Shadrina and Slominsky 2023). At this time, genetic testing is not routinely performed as part of a PD evaluation, due to lack of immediate impact and cost (Jia et al. 2022). However, as more clinical trials target specific gene mutations, testing may become more common (Tolosa et al. 2021). Table 3 is an overview of the most common monogenic forms of PD. This section expands further on the most common genetic risk factor for PD, *GBA1*, followed by the most common monogenic form of PD, *PARK-LRRK2*.

GBA1 encodes GCase, which is a widely expressed lysosomal enzyme that metabolizes glucocerebrosidase into ceramide and glucose (Chen et al. 2023). PD and *GBA1* mutations were first linked in patients with Gaucher's disease, which is an autosomal-recessive lysosomal disorder, potentially due to α-synuclein aggregation and lysosome dysregulation (Sidransky and Lopez 2012; Koros et al. 2023). There are hundreds of pathogenic mutations in GBA1 described in previous literature that lead to the production of misfolded GCase that contributes to PD pathogenesis (Chen et al. 2023). About 5%–10% of PD patients carry *GBA1* mutations, which can confer a 10- to 30-fold increased penetrance and lifetime risk of PD development (Chen et al. 2023). *GBA1* mutation is more common in patients with early-onset disease and has a more intense motor deterioration and more rapid development of dementia and psychiatric symptoms such as psychosis (Koros et al. 2023).

PARK-LRRK2, also referred to as PARK 8, involves a missense mutation in LRRK2 gene, and is inherited in an autosomal-dominant manner (Koros et al. 2023). LRRK2 is used in the regulation of vesicular trafficking (Koros et al. 2023). The clinical phenotype resembles late-onset PD with fewer nonmotor symptoms, such as RBD and cognitive issues, in comparison to idiopathic PD (Koros et al. 2023). The mutation may be present in 13% of sporadic Ashkenazi Jewish and 30% of sporadic Arab-Berber patients (Trinh et al. 2018). Neuropathology has been reported as heterogenous, including α-synuclein Lewy body pathology, nigral degeneration, and Alzheimer's disease tau pathology (Kalia et al. 2015; Henderson et al. 2019).

Many experts consider the use of genetic testing in those presenting with early-onset disease (onset before 50 years of age), positive family history, and persons from high-risk populations with a high prevalence of monogenic forms of PD (e.g., Ashkenazi Jewish, North African Berber Arabic populations) (Cook et al. 2021; Pal et al. 2023). Although there are many commercial genetic testing services most experts recommend genetic counseling for both the patient and possibly the family. Genetic counseling for family members is critical as many will choose not to be tested, once they weigh the risks and benefits of knowing their genetic status (Cook et al. 2023).

MANAGEMENT OF PARKINSON'S DISEASE

Management of PD ideally includes a multi- or interdisciplinary approach to care including neurorehabilitation with physical, occupational, and speech as well as swallowing therapy (Radder et al. 2020; Goldman et al. 2024). Additional disciplines that may enhance care include dieticians, social workers, genetic counselors, psychiatrists, neuropsychologists, and counseling psychologists (Radder et al. 2020). Additionally, most experts have validated the expanding evidence base by prescribing exercise as a therapy. High-intensity exercise for ~150 min per week is the current recommendation for people with PD for motor symptoms and quality of life (Alberts and Rosenfeldt 2020). Exercise has been shown to improve many motor and nonmotor features of the disease (Schenkman et al. 2018; van der Kolk et al. 2019; Johansson et al. 2022). Finally, prescribing specific gait, balance training, aerobic exercise, strength, resistance training, Tai Chi, music, and dance therapies have all been employed to benefit the person with PD (Ernst et al. 2023).

Management of Motor Symptoms

Dopamine replacement (levodopa) and various other dopaminergic therapies have for decades been the mainstay of treatment for the motor symptoms of PD (Fahn 2006). These therapies have been shown to be associated with significant benefits, especially in tremor, bradykinesia, and rigidity (Fahn 1999). Over time, symptomatic benefits may be lessened due to disease progression. Commonly, gait and balance symptoms usually become progressively less responsive to medication therapy (Smulders et al. 2016).

In general, pharmacologic therapies for PD commonly include levodopa combined with a peripheral dopa decarboxylase inhibitor, dopamine agonists, monoamine oxidase inhibitors, catechol-O-methyltransferase (COMT) inhibitors, or a combination of these (Fox et al. 2018). Recently, another class of therapy achieved U.S. Food and Drug Administration

(FDA) approval, the adenosine A2A antagonists (Hauser et al. 2021). Although collectively these therapies are not disease modifying, evidence has revealed that clinicians should not delay initiation when any symptom is impacting function, quality of life, or ability to work (Fox and Lang 2014). Patients should be counseled that "levodopa phobia" can lead to undertreatment and that the evidence strongly suggests there is no benefit to this approach and that there could be a risk of withholding medications (Titova et al. 2018). Finally, a recent large study revealed that an early start of levodopa was associated with an improvement in quality of life in the Levodopa in EArly Parkinson's disease (LEAP) study, further emphasizing the benefits of "not delaying" dopaminergic therapy (Verschuur et al. 2019).

Levodopa therapy is available in various formulations, and these can vary by country and include carbidopa/levodopa (CD/LD) or benserazide/levodopa preparations available in immediate and extended-release tablets. Oral disintegrating CD/LD tablets and inhaled levodopa powder are also now available (LeWitt et al. 2019). In the United States, scored CD/LD tablets were FDA approved in 2021 for more precise dosing, with each segment containing carbidopa 6.25 mg and levodopa 25 mg (Riverside Pharmaceuticals 2021). Most recently, continuous, subcutaneous infusions of levodopa have been approved in the United States, several European countries, and Japan (AbbVie News Center 2024a,b). Available dopamine agonists (nonergot derivatives) include oral preparations of ropinirole and pramipexole in both immediate and extended-release formulations, the rotigotine patch, and apomorphine injections (Fox et al. 2018). It is critical, especially when using dopamine agonists, for clinicians to monitor for adverse effects, particularly monitoring for impulse control disorders, excessive daytime sleepiness, sleep attacks, and hallucinations (Hobson et al. 2002; Weintraub et al. 2006; Staubo et al. 2024). These side effects have a larger association with dopamine agonists than with levodopa therapy (Pringsheim et al. 2021).

The monoamine oxidase B inhibitors selegiline, safinamide, and rasagiline can be considered as initial, monotherapy treatment; however, most

Table 3. Overview of the most common monogenic forms of Parkinson's disease (PD) and deep brain stimulation (DBS) responsiveness

Gene	Alternate name	Pattern of inheritance	Frequency in PD	Ethnic population distribution	Age of onset	DBS responsivity for motor symptoms
GBA1	n/a	Genetic risk factor	5%–10% of total PD patients	N370S polymorphism is the most common GBA1 mutation in the Ashkenazi Jewish population	Early onset	Effective but must discuss with the patient about the potential of accelerated cognitive decline potentially associated with the procedure (Artusi and Lopiano 2023)
PARK-LRRK2	PARK 8	Autosomal dominant	1% of total PD patients	European; present in 13% of sporadic Ashkenazi Jewish and 30% of sporadic Arab–Berber patients	Late onset	Effective
PARK-SNCA	PARK 1, 4	Autosomal dominant	0.045%	European > Asian and Hispanic	Early onset (triplication), late onset (duplication)	Few examples of good response for duplications
PARK-VPS35	PARK 17	Autosomal dominant	0.115%	European, Asian, and Ashkenazi Jewish	Late onset	Small numbers reported
PARK-CHCHD2	PARK 22	Autosomal dominant	Rare	Japanese and Chinese descent	Late onset	One study using subthalamic nucleus (STN)-DBS reported clinical improvement (Kamo et al. 2022)
PARK-PRKN	PARK 2	Autosomal recessive	12.5% of recessive PD	Asian > Caucasian and Hispanic	Early onset	Effective
PARK-PINK1	PARK 6	Autosomal recessive	1.9% of recessive PD	European, Asian, Arab–Berber, and Polynesian populations	Early onset	Effective

Continued

Cite this article as *Cold Spring Harb Perspect Med* doi: 10.1101/cshperspect.a041638

Table 3. *Continued*

Gene	Alternate name	Pattern of inheritance	Frequency in PD	Ethnic population distribution	Age of onset	DBS responsivity for motor symptoms
PARK-DJ1	PARK 7	Autosomal recessive	0.16% of recessive PD	Patients from Italy, Iran, and Turkey	Early onset	None reported
TAF1	XDP	X-linked	Rare	Patients of Filipino descent, particularly with maternal ancestry from the Panay Island	Median onset of 40 years	Effective for dystonia > parkinsonism

Table compiled from data in de Oliveira et al. (2019), Rizzone et al. (2019), Tolosa et al. (2021), Jia et al. (2022), Saunders-Pullman et al. (2022), and Chin et al. (2023).

(RBD) Rapid eye movement behavior disorder, (MDS) International Parkinson and Movement Disorder Society, (CT) computed tomography, (MRI) magnetic resonance imaging, (PET) positron emission tomography, (DAT) dopamine transporter, (DAT-SPECT) dopamine transporter imaging with single-photon emission tomography, (CSF) cerebral spinal fluid, (CNS) central nervous system, (RT-qIC) real-time quaking induced conversion, (PMCA) protein misfolding cyclic amplification, (DLB) dementia with Lewy bodies, (PDD) Parkinson disease dementia, (MSA) multiple systems atrophy, (MSA-P) multiple systems atrophy-parkinsonism, (MSA-C) multiple systems atrophy-cerebellar, (SCA) spinocerebellar ataxia, (PSP) progressive supranuclear palsy, (PSP-RS) progressive supranuclear palsy Richardson syndrome, (CBS) corticobasal syndrome, (FXTAS) Fragile X tremor/ataxia syndrome, (FMR1) Fragile X mental retardation 1, (MCP) middle cerebellar peduncles, (EEG) electroencephalography, (CSF) cerebrospinal fluid, (NREM) nonrapid eye movement, (CASPR2) contactin-associated protein-like 2, (DPPX) dipeptidyl-peptidase-like protein-6, (LRRK2) leucine-rich repeat kinase 2, (PINK1) PTEN-induced putative kinase.

individuals will require additional dopaminergic therapy (Pringsheim et al. 2021). Safinamide is a newer MAO-B inhibitor approved as an adjunctive therapy for use with CD/LD (Schapira et al. 2017). It is critical to monitor for drug–drug interactions when prescribing MAO-B inhibitors; however, occurrence is rare with low doses.

Adjunctive therapy may also include catecholamine-*O*-methyltransferase (COMT) inhibitors, which prevents the peripheral degradation of levodopa potentially extending its effects but also inducing more dyskinesia and possibly necessitating a decrease in levodopa dose (Fox et al. 2018; Fabbri et al. 2022). COMTs include entacapone, tolcapone, and opicapone (Song et al. 2021; Fabbri et al. 2022). Tolcapone is no longer widely used due to its small risk of hepatotoxicity (Olanow 2000; Olanow and Watkins 2007). Entacapone is available individually and in combination with CD/LD. Opicapone is the newest COMT inhibitor and can be dosed once daily (Fabbri et al. 2022).

Amantadine, an antiviral drug that is an *N*-methyl-D-aspartate receptor (NMDAR) antagonist has been approved for the treatment of levodopa-induced dyskinesia and is available in immediate and extended-release formulations (Sharma et al. 2018). Amantadine may also treat some early motor symptoms beyond dyskinesia, although many experts reserve its use for later in the disease (Fox et al. 2018).

Furthermore, a novel adenosine A2A receptor antagonist, known as istradefylline, was introduced as an additional adjunctive treatment option to improve dopaminergic "on" time in 2013 in Japan and in 2019 in the United States (Isaacson et al. 2022).

Management of Nonmotor Symptoms

Nonmotor symptoms are common manifestations in PD and have been associated with worse health-related quality of life (Duncan 2013). Some studies have suggested that nonmotor symptoms can be more disabling than the PD motor manifestations (Hermanowicz et al. 2019). Therefore, screening and management of these symptoms should be prioritized when discussing treatment options. Nonmotor symptoms may include neuropsychiatric, autonomic dysfunction, sleep disturbance, and sensory dysfunction including visual, olfactory, and pain symptoms (Barone et al. 2009; Seppi et al. 2019).

Currently, for mild cognitive impairment, there is insufficient evidence to routinely initiate an acetylcholinesterase inhibitor; however, this class of drugs has been shown to have benefits in PD-related dementia (Seppi et al. 2019). Management of psychosis in PD should always include a review of medications and a potential reduction or removal of potential offending agents, such as anticholinergics, amantadine, pain medications, and dopamine agonists. There are three second-generation antipsychotics that are used in the management of hallucinations and these include clozapine, pimavanserin, and quetiapine (Seppi et al. 2019). Experts use these medications as they have the potential to control hallucinations without worsening motor symptoms in PD. Furthermore, autonomic symptoms that can occur early in the disease can range from orthostatic hypotension, constipation, delayed gastric emptying, gastroparesis, erectile dysfunction, excessive sweating, and drooling. A summary of available strategies for motor and nonmotor symptom management is provided in Figure 1.

Advanced Therapies

Continuous subcutaneous infusion of apomorphine for management of motor fluctuations has been available in Europe and recently approved in the United States (Katzenschlager et al. 2018; Supernus Pharmaceuticals 2025). Doses for subcutaneous apomorphine are gradually titrated every few days to weeks to achieve a stable ON response and to improve the number of hours of on time (~2 h improvement) (Katzenschlager et al. 2018).

Levodopa/carbidopa intestinal gel (LCIG) is available in many countries for the management of severe motor fluctuations and dyskinesias (Fernandez et al. 2015; Freire-Alvarez et al. 2021). LCIG involves the placement of a percutaneous endoscopic gastrostomy tube with a jejunal extension to directly deliver levodopa into the proximal jejunum. More recently, a newer

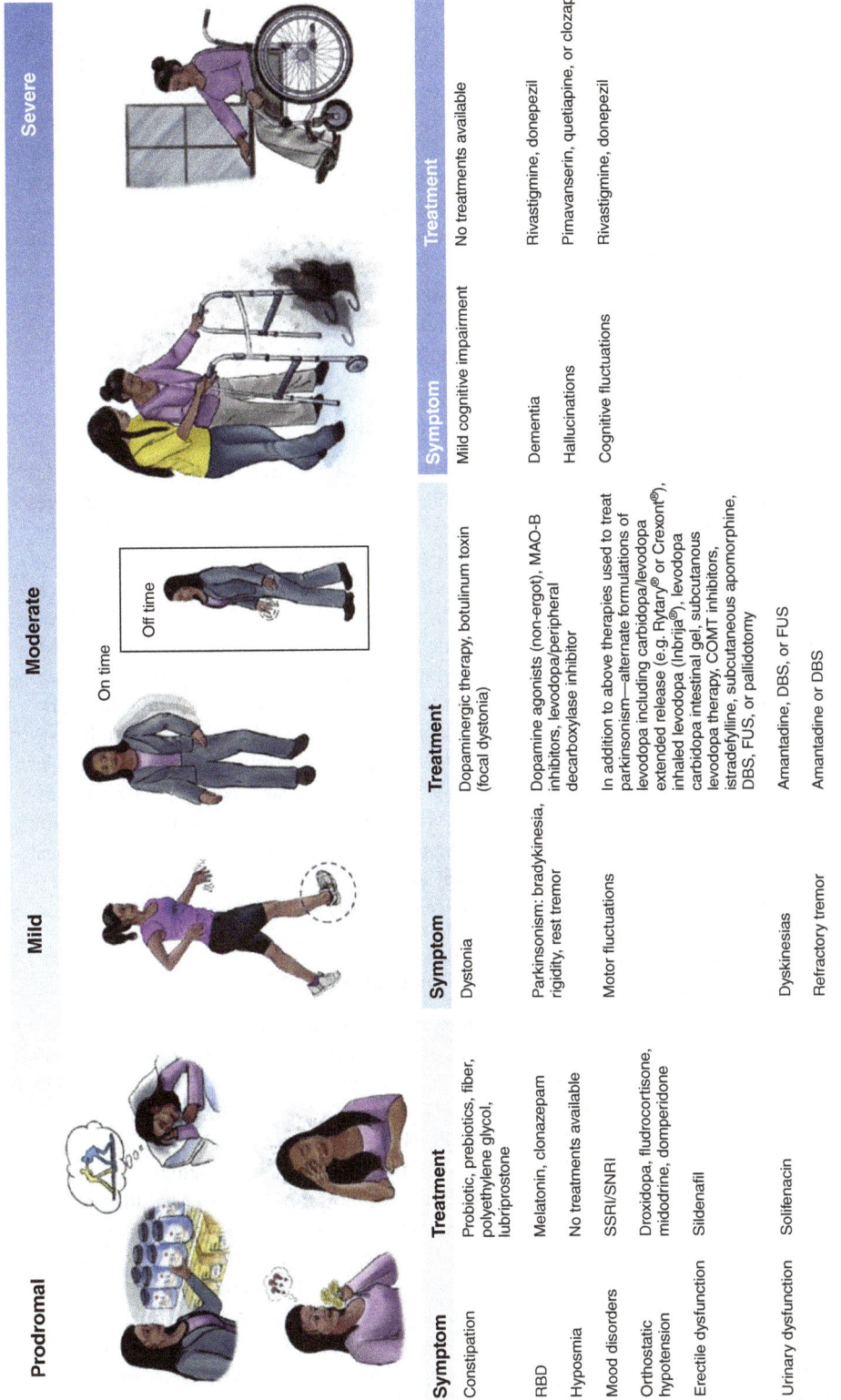

Prodromal

Symptom	Treatment
Constipation	Probiotic, prebiotics, fiber, polyethylene glycol, lubriprostone
RBD	Melatonin, clonazepam
Hyposmia	No treatments available
Mood disorders	SSRI/SNRI
Orthostatic hypotension	Droxidopa, fludrocortisone, midodrine, domperidone
Erectile dysfunction	Sildenafil
Urinary dysfunction	Solifenacin

Mild / Moderate

Symptom	Treatment
Dystonia	Dopaminergic therapy, botulinum toxin (focal dystonia)
Parkinsonism: bradykinesia, rigidity, rest tremor	Dopamine agonists (non-ergot), MAO-B inhibitors, levodopa/peripheral decarboxylase inhibitor
Motor fluctuations	In addition to above therapies used to treat parkinsonism—alternate formulations of levodopa including carbidopa/levodopa extended release (e.g. Rytary® or Crexont®), inhaled levodopa (Inbrija®), levodopa carbidopa intestinal gel, subcutaneous levodopa therapy, COMT inhibitors, istradefylline, subcutaneous apomorphine, DBS, FUS, or pallidotomy
Dyskinesias	Amantadine, DBS, or FUS
Refractory tremor	Amantadine or DBS

Severe

Symptom	Treatment
Mild cognitive impairment	No treatments available
Dementia	Rivastigmine, donepezil
Hallucinations	Pimavanserin, quetiapine, or clozapine
Cognitive fluctuations	Rivastigmine, donepezil

Figure 1. Identification and management of common motor and nonmotor symptoms in prodromal and clinically diagnosed Parkinson's disease (PD). These illustrations are designed to be general representations of various symptoms and stages of PD; however, these should not be interpreted as matching to individual person profiles and symptoms may occur at different time points. (RBD) Rapid eye movement behavior disorder, (ER) extended release, (COMT) catecholamine-O-methyltransferase, (DBS) deep brain stimulation. (Figure based on data in Fox et al. 2018 and Seppi et al. 2019.)

formulation of LCIG with entacapone has been approved in some European countries (Senek et al. 2017). LCIG may offer benefits in ∼3–4 hours of enhanced on time; however, clinicians should be aware of the higher rate of tube-related adverse events (Viljaharju et al. 2024).

Deep brain stimulation of the subthalamic nucleus or globus pallidus internus is an effective therapy to manage select cases of PD-related motor fluctuations, medically refractory tremor, dystonia, and also for levodopa-induced dyskinesias (Ramirez-Zamora and Ostrem 2018). Technological advances include the development of directional leads consisting of two middle rings that are segmented into thirds. These electrodes may facilitate the shaping of the electrical field and can be used by the clinician to push stimulation away from side effects in an effort to optimize benefit(s) (Patel et al. 2021). There is also a platform available for multiple independent source programming that may allow the clinician to steer or titrate current. Furthermore, sensing of brain signals that correlate with symptoms and treatment response is currently being tested for use in PD and may offer more personalized stimulation. Additionally, advancements include image-based programming, remote programming, and novel programming strategies including closed-loop and adaptive programming (Sandoval-Pistorius et al. 2023). DBS generally improves on time by ∼4+ hours and is the most effective therapy against refractory dyskinesia (Weaver et al. 2009). Finally, VIM thalamic DBS is occasionally chosen as a target in PD when resting tremor is severe and manifests a pronounced intentional component (Cury et al. 2017).

Unilateral magnetic resonance-guided high-intensity focused ultrasound (FUS) targeting the (1) thalamus has been approved for the treatment of tremor-dominant PD, and (2) globus pallidus internus for motor symptoms of bradykinesia, rigidity, and dyskinesia (Sinai et al. 2022; Krishna et al. 2023). The procedure is usually performed in a single session, during which high-intensity ultrasound waves are delivered across the skull to create a thermal lesion in the targeted structure. At this time, this procedure is approved for unilateral targeting as

studies to evaluate whether bilateral lesions will result in speech, swallowing, and cognitive deficits are ongoing (Stieglitz et al. 2022; Krishna et al. 2023). In this procedure, an incision is not made in the skull, and there are no wires or devices, thus it is associated with less infections. Limitations of this procedure include its approval only to treat unilateral symptoms and no adjustments can be made over time (Krishna et al. 2023). Recent data have revealed it can be a powerful addition to the treatment armamentarium, particularly in elderly patients at high risk for open surgery.

CONCLUSION

While clinical features of PD have been described for more than 200 years, there continues to be advances in our understanding of the disease process. Improvements in understanding of prodromal PD will facilitate earlier diagnosis, as well as identification of at-risk populations, which has important implications for possible therapeutic clinical trials. Diagnostic advancements include the use of imaging biomarkers, as well as DAT-SPECT or skin biopsy. Likewise, treatment options have continued to advance in recent years. While remarkable progress has been made over the last two decades, the diagnosis of PD ultimately still relies on clinical skill and investigation, highlighting the importance of strong clinical knowledge of the disease.

COMPETING INTEREST STATEMENT

B.P. has received research support from the American Brain Foundation, Mary E. Groff Charitable Trust, National Institutes of Health (NIHK23 AG073575), and University of Florida (UF) Harry T. Mangurian Jr. Foundation for Pilot Studies in Lewy Body Dementia. A.R. has received an honorarium from Mediflix and the Michael J. Fox Foundation, and serves as the editor for the *Journal of American Medical Association* (*JAMA*) Neurology Diversity, Equity, and Inclusion. T.T. has no disclosures. M.S.O. serves as Medical Advisor to the Parkinson's Foundation, and has received research grants from NIH, the Parkinson's Foundation, the Michael J. Fox Foundation, the Parkinson Alliance, the Smallwood Foundation, the Bachmann-Strauss Foundation, the Tourette Syndrome Association, and the UF Foundation. M.S.O. research is supported by NIH R01NR014852, R01NS096008, UH3NS119844, and U01NS119 562. M.S.O. is PI of the NIH R25NS108939 Training Grant. M.S.O. has received royalties for publications with Demos, Manson, Amazon, Smashwords, Books4Patients, Perseus, Robert Rose, Oxford, and Cambridge (movement disorders books). M.S.O. is an associate editor for the *New England Journal of Medicine, Journal Watch Neurology,* and *JAMA Neurology.* M.S.O. has participated in CME and educational activities (past 12–24 mo) on movement disorders sponsored by WebMD/Medscape, RMEI Medical Education, American Academy of Neurology, Movement Disorders Society, Mediflix, and Vanderbilt University. The institution and not M.S.O. receives grants from industry. M.S.O. has participated as a site PI and/or co-I for several NIH, foundation, and industry-sponsored trials over the years but has not received honoraria. Research projects at the University of Florida receive device and drug donations.

ACKNOWLEDGMENTS

We acknowledge the Parkinson's Foundation-funded University of Florida (UF) Center of Excellence and the UF Foundation, and illustrator Erica Rodriguez.

REFERENCES

AbbVie News Center. 2024a. Abbvie launches PRODUO-DOPA (foslevodopa/foscarbidopa) for people living with advanced Parkinson disease in the European union. https://news.abbvie.com/2024-01-09-AbbVie-Launches-PRODUODOPA-R-foslevodopa-foscarbidopa-for-Peopl e-Living-with-Advanced-Parkinsons-Disease-in-the-Eur opean-Union

AbbVie News Center. 2024b. U.S. FDA approves VYALEV (foscarbidopa and foslevodopa) for adults living with advanced Parkinson's disease. https://news.abbvie.com/2024-10-17-U-S-FDA-Approves-VYALEV-TM-foscarbidopa-and-foslevodopa-for-Adults-Living-with-Advanced-Parkinsons-Disease

Adler CH, Beach TG, Zhang N, Shill HA, Driver-Dunckley E, Mehta SH, Atri A, Caviness JN, Serrano G, Shprecher DR, et al. 2021. Clinical diagnostic accuracy of early/advanced Parkinson disease: an updated clinicopathologic study. *Neurol Clin Pract* 11: e414–e421. doi:10.1212/CPJ .0000000000001016

Alberts JL, Rosenfeldt AB. 2020. The universal prescription for Parkinson's disease: exercise. *J Parkinsons Dis* 10: S21–S27. doi:10.3233/JPD-202100

Armstrong MJ, Litvan I, Lang AE, Bak TH, Bhatia KP, Borroni B, Boxer AL, Dickson DW, Grossman M, Hallett M, et al. 2013. Criteria for the diagnosis of corticobasal degeneration. *Neurology* 80: 496–503. doi:10.1212/WNL .0b013e31827f0fd1

Artusi CA, Lopiano L. 2023. Should we offer deep brain stimulation to Parkinson's disease patients with GBA mutations? *Front Neurol* 14: 1158977. doi:10.3389/fneur .2023.1158977

Asahara Y, Kameyama M, Ishii K, Ishibashi K. 2023. Diagnostic performance of the cingulate island sign ratio for differentiating dementia with Lewy bodies from Alzheimer's disease changes depending on the mini-mental state examination score. *J Neurol Sci* 455: 122782. doi:10 .1016/j.jns.2023.122782

Barone P, Antonini A, Colosimo C, Marconi R, Morgante L, Avarello TP, Bottacchi E, Cannas A, Ceravolo G, Ceravolo R, et al. 2009. The PRIAMO study: a multicenter assessment of nonmotor symptoms and their impact on quality of life in Parkinson's disease. *Mov Disord* 24: 1641–1649. doi:10.1002/mds.22643

Berardelli A, Rothwell JC, Thompson PD, Hallett M. 2001. Pathophysiology of bradykinesia in Parkinson's disease. *Brain* 124: 2131–2146. doi:10.1093/brain/124 .11.2131

Berg D, Postuma RB, Adler CH, Bloem BR, Chan P, Dubois B, Gasser T, Goetz CG, Halliday G, Joseph L, et al. 2015. MDS research criteria for prodromal Parkinson's disease. *Mov Disord* 30: 1600–1611. doi:10.1002/mds.26431

Bhattacharjee S. 2018. Procerus sign: mechanism, clinical usefulness, and controversies. *Ann Indian Acad Neurol* 21: 164–165. doi:10.4103/aian.AIAN_408_17

Boesch SM, Wenning GK, Ransmayr G, Poewe W. 2002. Dystonia in multiple system atrophy. *J Neurol Neurosurg Psychiatry* 72: 300–303. doi:10.1136/jnnp.72.3.300

Brigo F, Erro R, Marangi A, Bhatia K, Tinazzi M. 2014. Differentiating drug-induced parkinsonism from Parkinson's disease: an update on non-motor symptoms and investigations. *Parkinsonism Relat Disord* 20: 808–814. doi:10.1016/j.parkreldis.2014.05.011

Brücke T, Brücke C. 2022. Dopamine transporter (DAT) imaging in Parkinson's disease and related disorders. *J Neural Transm (Vienna)* 129: 581–594. doi:10.1007/ s00702-021-02452-7

Bürk K, Skalej M, Dichgans J. 2001. Pontine MRI hyperintensities ("the cross sign") are not pathognomonic for multiple system atrophy (MSA). *Mov Disord* 16: 535. doi:10.1002/mds.1107

Cabal-Herrera AM, Tassanakijpanich N, Salcedo-Arellano MJ, Hagerman RJ. 2020. Fragile X-associated tremor/ ataxia syndrome (FXTAS): pathophysiology and clinical implications. *Int J Mol Sci* 21: 4391. doi:10.3390/ijms 21124391

Charcot J-M. 1872. De la paralysie agitante. In *Oeuvres Complètes (t 1) Leçons sur les maladies du système nerveux,* pp. 155–188. A Delahaye, Paris [In English: Charcot J-M.

1877. On Parkinson's disease. In *Lectures on diseases of the nervous system delivered at the Salpêtrière* (transl. Sigerson G), pp. 129–156. New Sydenham Society, London.

Chen D, Zheng Y, Zhang G, Huang Y, Zheng B, Zhang J, Xiong F, Su Q. 2023. The loss of function GBA1 c.231C>G mutation associated with Parkinson disease. *J Neural Transm (Vienna)* 130: 905–913. doi:10.1007/s00702-023-02651-4

Chin HL, Lin CY, Chou OH. 2023. X-linked dystonia parkinsonism: epidemiology, genetics, clinical features, diagnosis, and treatment. *Acta Neurol Belg* 123: 45–55. doi:10.1007/s13760-022-02144-3

Claeys T, Boogers A, Vanneste D. 2020. MRI findings in fragile X-associated tremor/ataxia syndrome. *Acta Neurol Belg* 120: 181–183. doi:10.1007/s13760-019-01237-w

Cohen O, Pullman S, Jurewicz E, Watner D, Louis ED. 2003. Rest tremor in patients with essential tremor: prevalence, clinical correlates, and electrophysiologic characteristics. *Arch Neurol* 60: 405–410. doi:10.1001/archneur.60.3.405

Constantinides VC, Paraskevas GP, Paraskevas PG, Stefanis L, Kapaki E. 2019. Corticobasal degeneration and corticobasal syndrome: a review. *Clin Park Relat Disord* 1: 66–71. doi:10.1016/j.prdoa.2019.08.005

Cook L, Schulze J, Kopil C, Hastings T, Naito A, Wojcieszek J, Payne K, Alcalay RN, Klein C, Saunders-Pullman R, et al. 2021. Genetic testing for Parkinson disease: are we ready? *Neurol Clin Pract* 11: 69–77. doi:10.1212/CPJ.0000000000000831

Cook L, Verbrugge J, Schwantes-An TH, Schulze J, Beck JC, Naito A, Hall A, Chan AK, Casaceli CJ, Marder K, et al. 2023. Providing genetic testing and genetic counseling for Parkinson's disease to the community. *Genet Med* 25: 100907. doi:10.1016/j.gim.2023.100907

Cortelli P, Calandra-Buonaura G, Benarroch EE, Giannini G, Iranzo A, Low PA, Martinelli P, Provini F, Quinn N, Tolosa E, et al. 2019. Stridor in multiple system atrophy: consensus statement on diagnosis, prognosis, and treatment. *Neurology* 93: 630–639. doi:10.1212/WNL.0000000000008208

Cury RG, Fraix V, Castrioto A, Pérez Fernández MA, Krack P, Chabardes S, Seigneuret E, Alho EJL, Benabid AL, Moro E. 2017. Thalamic deep brain stimulation for tremor in Parkinson disease, essential tremor, and dystonia. *Neurology* 89: 1416–1423. doi:10.1212/WNL.0000000000004295

de Lau LM, Breteler MM. 2006. Epidemiology of Parkinson's disease. *Lancet Neurol* 5: 525–535. doi:10.1016/S1474-4422(06)70471-9

de Oliveira LM, Barbosa ER, Aquino CC, Munhoz RP, Fasano A, Cury RG. 2019. Deep brain stimulation in patients with mutations in Parkinson's disease-related genes: a systematic review. *Mov Disord Clin Pract* 6: 359–368. doi:10.1002/mdc3.12795

Dorsey ER, Bloem BR. 2018. The Parkinson pandemic—a call to action. *JAMA Neurol* 75: 9–10. doi:10.1001/jamaneurol.2017.3299

Dorsey ER, Sherer T, Okun MS, Bloem BR. 2018. The emerging evidence of the Parkinson pandemic. *J Parkinsons Dis* 8: S3–S8. doi:10.3233/JPD-181474

Doty RL, Deems DA, Stellar S. 1988. Olfactory dysfunction in parkinsonism: a general deficit unrelated to neurologic signs, disease stage, or disease duration. *Neurology* 38: 1237–1244. doi:10.1212/wnl.38.8.1237

Edwards M, Quinn N, Bhatia K. 2008. *Parkinson's disease and other movement disorders.* Oxford University Press, Oxford.

Ernst M, Folkerts AK, Gollan R, Lieker E, Caro-Valenzuela J, Adams A, Cryns N, Monsef I, Dresen A, Roheger M, et al. 2023. Physical exercise for people with Parkinson's disease: a systematic review and network meta-analysis. *Cochrane Database Syst Rev* 1: CD013856. doi:10.1002/14651858.CD013856.pub2

Fabbri M, Ferreira JJ, Rascol O. 2022. COMT inhibitors in the management of Parkinson's disease. *CNS Drugs* 36: 261–282. doi:10.1007/s40263-021-00888-9

Fahn S. 1999. Parkinson disease, the effect of levodopa, and the ELLDOPA trial. Earlier vs later L-DOPA. *Arch Neurol* 56: 529–535. doi:10.1001/archneur.56.5.529

Fahn S. 2006. Levodopa in the treatment of Parkinson's disease. *J Neural Transm Suppl* 71: 1–15. doi:10.1007/978-3-211-33328-0_1

Famula JL, McKenzie F, McLennan YA, Grigsby J, Tassone F, Hessl D, Rivera SM, Martinez-Cerdeno V, Hagerman RJ. 2018. Presence of middle cerebellar peduncle sign in FMR1 premutation carriers without tremor and ataxia. *Front Neurol* 9: 695. doi:10.3389/fneur.2018.00695

Fanciulli A, Wenning GK. 2015. Multiple-system atrophy. *N Engl J Med* 372: 249–263. doi:10.1056/NEJMra1311488

Fernandez HH, Standaert DG, Hauser RA, Lang AE, Fung VS, Klostermann F, Lew MF, Odin P, Steiger M, Yakupov EZ, et al. 2015. Levodopa-carbidopa intestinal gel in advanced Parkinson's disease: final 12-month, open-label results. *Mov Disord* 30: 500–509. doi:10.1002/mds.26123

Fishman PS. 2008. Paradoxical aspects of parkinsonian tremor. *Mov Disord* 23: 168–173. doi:10.1002/mds.21736

Fox SH, Lang AE. 2014. "Don't delay, start today": delaying levodopa does not delay motor complications. *Brain* 137: 2628–2630. doi:10.1093/brain/awu212

Fox SH, Katzenschlager R, Lim SY, Barton B, de Bie RMA, Seppi K, Coelho M, Sampaio C; Committee MDSE-BM. 2018. International Parkinson and movement disorder society evidence-based medicine review: update on treatments for the motor symptoms of Parkinson's disease. *Mov Disord* 33: 1248–1266. doi:10.1002/mds.27372

Freire-Alvarez E, Kurča E, Lopez Manzanares L, Pekkonen E, Spanaki C, Vanni P, Liu Y, Sánchez-Soliño O, Barbato LM. 2021. Levodopa-carbidopa intestinal gel reduces dyskinesia in Parkinson's disease in a randomized trial. *Mov Disord* 36: 2615–2623. doi:10.1002/mds.28703

Fullard ME, Morley JF, Duda JE. 2017. Olfactory dysfunction as an early biomarker in Parkinson's disease. *Neurosci Bull* 33: 515–525. doi:10.1007/s12264-017-0170-x

Gaillard F, Yu Y, Sharma R, et al. 2011. Hot cross bun sign (pons). *Radiopaedia* doi:10.53347/rID-13008

Gasser T, Schwarz J, Arnold G, Trenkwalder C, Oertel WH. 1992. Apomorphine test for dopaminergic responsiveness in patients with previously untreated Parkinson's disease. *Arch Neurol* 49: 1131–1134. doi:10.1001/archneur.1992.00530350045017

George P, Roushdy T, Fathy M, Hamid E, Ibrahim YA, El-Belkimy M, Abdulghani MO, Shalash A. 2024. The clinical and neuroimaging differences between vascular par-

Cite this article as *Cold Spring Harb Perspect Med* doi: 10.1101/cshperspect.a041638

kinsonism and Parkinson's disease: a case-control study. *BMC Neurol* **24**: 56. doi:10.1186/s12883-024-03556-9

Giannoccaro MP, Verde F, Morelli L, Rizzo G, Ricciardiello F, Liguori R. 2023. Neural surface antibodies and neurodegeneration: clinical commonalities and pathophysiological relationships. *Biomedicines* **11**: 666. doi:10.3390/biomedicines11030666

Gibbons C, Wang N, Rajan S, Kern D, Palma JA, Kaufmann H, Freeman R. 2023. Cutaneous α-synuclein signatures in patients with multiple system atrophy and Parkinson disease. *Neurology* **100**: e1529–e1539. doi:10.1212/WNL.0000000000206772

Giladi N, Nieuwboer A. 2008. Understanding and treating freezing of gait in parkinsonism, proposed working definition, and setting the stage. *Mov Disord* **23** (Suppl 2): S423–S425. doi:10.1002/mds.21927

Gjerum L, Frederiksen KS, Henriksen OM, Law I, Anderberg L, Andersen BB, Bjerregaard E, Hejl AM, Høgh P, Hasselbalch SG. 2020. A visual rating scale for cingulate island sign on 18F-FDG-PET to differentiate dementia with Lewy bodies and Alzheimer's disease. *J Neurol Sci* **410**: 116645. doi:10.1016/j.jns.2019.116645

Goetz CG, Tilley BC, Shaftman SR, Stebbins GT, Fahn S, Martinez-Martin P, Poewe W, Sampaio C, Stern MB, Dodel R, et al. 2008. Movement Disorder Society-sponsored revision of the Unified Parkinson's Disease Rating Scale (MDS-UPDRS): scale presentation and clinimetric testing results. *Mov Disord* **23**: 2129–2170.

Goldman JG, Volpe D, Ellis TD, Hirsch MA, Johnson J, Wood J, Aragon A, Biundo R, Di Rocco A, Kasman GS, et al. 2024. Delivering multidisciplinary rehabilitation care in Parkinson's disease: an international consensus statement. *J Parkinsons Dis* **14**: 135–166. doi:10.3233/JPD-230117

Gövert F, Leypoldt F, Junker R, Wandinger KP, Deuschl G, Bhatia KP, Balint B. 2020. Antibody-related movement disorders—a comprehensive review of phenotype-autoantibody correlations and a guide to testing. *Neurol Res Pract* **2**: 6. doi:10.1186/s42466-020-0053-x

Graff-Radford J, Rubin MN, Jones DT, Aksamit AJ, Ahlskog JE, Knopman DS, Petersen RC, Boeve BF, Josephs KA. 2013. The alien limb phenomenon. *J Neurol* **260**: 1880–1888. doi:10.1007/s00415-013-6898-y

Gupta DK, Marano M, Zweber C, Boyd JT, Kuo SH. 2020. Prevalence and relationship of rest tremor and action tremor in Parkinson's disease. *Tremor Other Hyperkinet Mov* **10**: 58. doi:10.5334/tohm.552

Gupta R, Kumar G, Kumar S, Thakur B, Tiwari R, Verma AK. 2022. The swallow tail sign of substantia nigra: a case-control study to establish its role in diagnosis of Parkinson disease on 3T MRI. *J Neurosci Rural Pract* **13**: 181–185. doi:10.1055/s-0041-1740578

Hauser RA, Hattori N, Fernandez H, Isaacson SH, Mochizuki H, Rascol O, Stocchi F, Li J, Mori A, Nakajima Y, et al. 2021. Efficacy of istradefylline, an adenosine A2A receptor antagonist, as adjunctive therapy to levodopa in Parkinson's disease: a pooled analysis of 8 phase 2b/3 trials. *J Parkinsons Dis* **11**: 1663–1675. doi:10.3233/JPD-212672

Henderson MX, Sengupta M, Trojanowski JQ, Lee VMY. 2019. Alzheimer's disease tau is a prominent pathology in LRRK2 Parkinson's disease. *Acta Neuropathol Commun* **7**: 183. doi:10.1186/s40478-019-0836-x

Hermanowicz N, Jones SA, Hauser RA. 2019. Impact of nonmotor symptoms in Parkinson's disease: a PMDAlliance survey. *Neuropsychiatr Dis Treat* **15**: 2205–2212. doi:10.2147/NDT.S213917

Hobson DE, Lang AE, Martin WR, Razmy A, Rivest J, Fleming J. 2002. Excessive daytime sleepiness and sudden-onset sleep in Parkinson disease: a survey by the Canadian movement disorders group. *JAMA* **287**: 455–463. doi:10.1001/jama.287.4.455

Höglinger GU, Respondek G, Stamelou M, Kurz C, Josephs KA, Lang AE, Mollenhauer B, Müller U, Nilsson C, Whitwell JL, et al. 2017. Clinical diagnosis of progressive supranuclear palsy: the movement disorder society criteria. *Mov Disord* **32**: 853–864. doi:10.1002/mds.26987

Isaacson SH, Betté S, Pahwa R. 2022. Istradefylline for OFF episodes in Parkinson's disease: a US perspective of common clinical scenarios. *Degener Neurol Neuromuscul Dis* **12**: 97–109.

Jankovic J. 2008. Parkinson's disease: clinical features and diagnosis. *J Neurol Neurosurg Psychiatry* **79**: 368–376. doi:10.1136/jnnp.2007.131045

Jia F, Fellner A, Kumar KR. 2022. Monogenic Parkinson's disease: genotype, phenotype, pathophysiology, and genetic testing. *Genes (Basel)* **13**: 471. doi:10.3390/genes13030471

Johansson ME, Cameron IGM, Van der Kolk NM, de Vries NM, Klimars E, Toni I, Bloem BR, Helmich RC. 2022. Aerobic exercise alters brain function and structure in Parkinson's disease: a randomized controlled trial. *Ann Neurol* **91**: 203–216. doi:10.1002/ana.26291

Jung I, Kim JS. 2019. Abnormal eye movements in parkinsonism and movement disorders. *J Mov Disord* **12**: 1–13. doi:10.14802/jmd.18034

Kalia LV, Lang AE, Hazrati LN, Fujioka S, Wszolek ZK, Dickson DW, Ross OA, Van Deerlin VM, Trojanowski JQ, Hurtig HI, et al. 2015. Clinical correlations with Lewy body pathology in LRRK2-related Parkinson disease. *JAMA Neurol* **72**: 100–105. doi:10.1001/jamaneurol.2014.2704

Kamo H, Oyama G, Nishioka K, Funayama M, Hattori N. 2022. Deep brain stimulation for a patient with familial Parkinson's disease harboring CHCHD2 p.T61I. *Mov Disord Clin Pract* **9**: 407–409. doi:10.1002/mdc3.13428

Katzenschlager R, Poewe W, Rascol O, Trenkwalder C, Deuschl G, Chaudhuri KR, Henriksen T, van Laar T, Spivey K, Vel S, et al. 2018. Apomorphine subcutaneous infusion in patients with Parkinson's disease with persistent motor fluctuations (TOLEDO): a multicentre, double-blind, randomised, placebo-controlled trial. *Lancet Neurol* **17**: 749–759. doi:10.1016/S1474-4422(18)30239-4

Kim JY, Illigens BM, McCormick MP, Wang N, Gibbons CH. 2019. Alpha-synuclein in skin nerve fibers as a biomarker for α-synucleinopathies. *J Clin Neurol* **15**: 135–142. doi:10.3988/jcn.2019.15.2.135

Korczyn AD. 2015. Vascular parkinsonism—characteristics, pathogenesis and treatment. *Nat Rev Neurol* **11**: 319–326. doi:10.1038/nrneurol.2015.61

Koros C, Bougea A, Simitsi AM, Papagiannakis N, Angelopoulou E, Pachi I, Antonelou R, Bozi M, Stamelou M, Stefanis L. 2023. The landscape of monogenic Parkinson's disease in populations of non-European ancestry: a narrative review. *Genes (Basel)* **14**: 2097. doi:10.3390/genes14112097

Krishna V, Fishman PS, Eisenberg HM, Kaplitt M, Baltuch G, Chang JW, Chang WC, Martinez Fernandez R, Del Alamo M, Halpern CH, et al. 2023. Trial of globus pallidus focused ultrasound ablation in Parkinson's disease. *N Engl J Med* **388**: 683–693. doi:10.1056/NEJMoa2202721

Lee C, Park KW, Choi N, Ryu HS, Chung SJ. 2019. Fragile X-associated tremor/ataxia syndrome: an illustrative case. *J Mov Disord* **12**: 184–186. doi:10.14802/jmd.18060

LeWitt PA, Hauser RA, Pahwa R, Isaacson SH, Fernandez HH, Lew M, Saint-Hilaire M, Pourcher E, Lopez-Manzanares L, Waters C, et al. 2019. Safety and efficacy of CVT-301 (levodopa inhalation powder) on motor function during off periods in patients with Parkinson's disease: a randomised, double-blind, placebo-controlled phase 3 trial. *Lancet Neurol* **18**: 145–154. doi:10.1016/S1474-4422(18)30405-8

Lim SM, Katsifis A, Villemagne VL, Best R, Jones G, Saling M, Bradshaw J, Merory J, Woodward M, Hopwood M, et al. 2009. The ^{18}F-FDG PET cingulate island sign and comparison to ^{123}I-β-CIT SPECT for diagnosis of dementia with Lewy bodies. *J Nucl Med* **50**: 1638–1645. doi:10.2967/jnumed.109.065870

Low PA, Reich SG, Jankovic J, Shults CW, Stern MB, Novak P, Tanner CM, Gilman S, Marshall FJ, Wooten F, et al. 2015. Natural history of multiple system atrophy in the USA: a prospective cohort study. *Lancet Neurol* **14**: 710–719. doi:10.1016/S1474-4422(15)00058-7

Ma KKY, Lin S, Mok VCT. 2019. Neuroimaging in vascular parkinsonism. *Curr Neurol Neurosci Rep* **19**: 102. doi:10.1007/s11910-019-1019-7

Marrinan S, Emmanuel AV, Burn DJ. 2014. Delayed gastric emptying in Parkinson's disease. *Mov Disord* **29**: 23–32. doi:10.1002/mds.25708

Massey LA, Micallef C, Paviour DC, O'Sullivan SS, Ling H, Williams DR, Kallis C, Holton JL, Revesz T, Burn DJ, et al. 2012. Conventional magnetic resonance imaging in confirmed progressive supranuclear palsy and multiple system atrophy. *Mov Disord* **27**: 1754–1762. doi:10.1002/mds.24968

Massey LA, Jäger HR, Paviour DC, O'Sullivan SS, Ling H, Williams DR, Kallis C, Holton J, Revesz T, Burn DJ, et al. 2013. The midbrain to pons ratio: a simple and specific MRI sign of progressive supranuclear palsy. *Neurology* **80**: 1856–1861. doi:10.1212/WNL.0b013e318292a2d2

McFarland NR, Hess CW. 2017. Recognizing atypical parkinsonisms: "Red Flags" and therapeutic approaches. *Semin Neurol* **37**: 215–227. doi:10.1055/s-0037-1602422

McKeith IG, Boeve BF, Dickson DW, Halliday G, Taylor JP, Weintraub D, Aarsland D, Galvin J, Attems J, Ballard CG, et al. 2017. Diagnosis and management of dementia with Lewy bodies: fourth consensus report of the DLB consortium. *Neurology* **89**: 88–100. doi:10.1212/WNL.0000000000004058

Mila M, Alvarez-Mora MI, Madrigal I, Rodriguez-Revenga L. 2018. Fragile X syndrome: an overview and update of the *FMR1* gene. *Clin Genet* **93**: 197–205. doi:10.1111/cge.13075

Mueller C, Hussl A, Krismer F, Heim B, Mahlknecht P, Nocker M, Scherfler C, Mair K, Esterhammer R, Schocke

M, et al. 2018. The diagnostic accuracy of the hummingbird and morning glory sign in patients with neurodegenerative parkinsonism. *Parkinsonism Relat Disord* **54**: 90–94. doi:10.1016/j.parkreldis.2018.04.005

Muqit MM, Mort D, Miskiel KA, Shakir RA. 2001. "Hot cross bun" sign in a patient with parkinsonism secondary to presumed vasculitis. *J Neurol Neurosurg Psychiatry* **71**: 565–566. doi:10.1136/jnnp.71.4.565

Narasimhan M, Schwartz R, Halliday G. 2022. Parkinsonism and cerebrovascular disease. *J Neurol Sci* **433**: 120011. doi:10.1016/j.jns.2021.120011

Obeso JA, Stamelou M, Goetz CG, Poewe W, Lang AE, Weintraub D, Burn D, Halliday GM, Bezard E, Przedborski S, et al. 2017. Past, present, and future of Parkinson's disease: a special essay on the 200th anniversary of the shaking palsy. *Mov Disord* **32**: 1264–1310. doi:10.1002/mds.27115

Olanow CW. 2000. Tolcapone and hepatotoxic effects. Tasmar advisory panel. *Arch Neurol* **57**: 263–267. doi:10.1001/archneur.57.2.263

Olanow CW, Watkins PB. 2007. Tolcapone: an efficacy and safety review (2007). *Clin Neuropharmacol* **30**: 287–294. doi:10.1097/wnf.0b013e318038d2b6

Pal G, Cook L, Schulze J, Verbrugge J, Alcalay RN, Merello M, Sue CM, Bardien S, Bonifati V, Chung SJ, et al. 2023. Genetic testing in Parkinson's disease. *Mov Disord* **38**: 1384–1396. doi:10.1002/mds.29500

Parkinson J. 1817. *An essay on the shaking palsy.* Sherwood, Neely, and Jones, Paternoster Row, London.

Parnetti L, Gaetani L, Eusebi P, Paciotti S, Hansson O, El-Agnaf O, Mollenhauer B, Blennow K, Calabresi P. 2019. CSF and blood biomarkers for Parkinson's disease. *Lancet Neurol* **18**: 573–586. doi:10.1016/S1474-4422(19)30024-9

Patel B, Chiu S, Wong JK, Patterson A, Deeb W, Burns M, Zeilman P, Wagle-Shukla A, Almeida L, Okun MS, et al. 2021. Deep brain stimulation programming strategies: segmented leads, independent current sources, and future technology. *Expert Rev Med Devices* **18**: 875–891. doi:10.1080/17434440.2021.1962286

Poewe W. 2008. Non-motor symptoms in Parkinson's disease. *Eur J Neurol* **15** (Suppl 1): 14–20. doi:10.1111/j.1468-1331.2008.02056.x

Ponsen MM, Stoffers D, Booij J, van Eck-Smit BL, Wolters E, Berendse HW. 2004. Idiopathic hyposmia as a preclinical sign of Parkinson's disease. *Ann Neurol* **56**: 173–181.

Portet M, Filyridou M, Howlett DC. 2019. Hot cross bun sign. *J Neurol* **266**: 2573–2574. doi:10.1007/s00415-019-09439-1

Postuma RB, Poewe W, Litvan I, Lewis S, Lang AE, Halliday G, Goetz CG, Chan P, Slow E, Seppi K, et al. 2018. Validation of the MDS clinical diagnostic criteria for Parkinson's disease. *Mov Disord* **33**: 1601–1608. doi:10.1002/mds.27362

Pringsheim T, Day GS, Smith DB, Rae-Grant A, Licking N, Armstrong MJ, de Bie RMA, Roze E, Miyasaki JM, Hauser RA, et al. 2021. Dopaminergic therapy for motor symptoms in early Parkinson disease practice guideline summary: a report of the AAN guideline subcommittee. *Neurology* **97**: 942–957. doi:10.1212/WNL.0000000000012868

Cite this article as *Cold Spring Harb Perspect Med* doi: 10.1101/cshperspect.a041638

Raccagni C, Nonnekes J, Bloem BR, Peball M, Boehme C, Seppi K, Wenning GK. 2020. Gait and postural disorders in parkinsonism: a clinical approach. *J Neurol* **267**: 3169–3176. doi:10.1007/s00415-019-09382-1

Radder DLM, Nonnekes J, van Nimwegen M, Eggers C, Abbruzzese G, Alves G, Browner N, Chaudhuri KR, Ebersbach G, Ferreira JJ, et al. 2020. Recommendations for the organization of multidisciplinary clinical care teams in Parkinson's disease. *J Parkinsons Dis* **10**: 1087–1098. doi:10.3233/JPD-202078

Rajalingam R, Breen DP, Chen R, Fox S, Kalia LV, Munhoz RP, Slow E, Strafella AP, Lang AE, Fasano A. 2019. The clinical significance of lower limb tremors. *Parkinsonism Relat Disord* **65**: 165–171. doi:10.1016/j.parkreldis.2019.06.007

Ramirez-Zamora A, Ostrem JL. 2018. Globus pallidus interna or subthalamic nucleus deep brain stimulation for Parkinson disease: a review. *JAMA Neurol* **75**: 367–372. doi:10.1001/jamaneurol.2017.4321

Riverside Pharmaceuticals. 2021. Dhivy (scored carbidopa/levodopa) [package insert]. U.S. Food and Drug Administration, Riverside Pharmaceuticals Corporation, Washington, DC.

Rizzone MG, Martone T, Balestrino R, Lopiano L. 2019. Genetic background and outcome of deep brain stimulation in Parkinson's disease. *Parkinsonism Relat Disord* **64**: 8–19. doi:10.1016/j.parkreldis.2018.08.006

Ross GW, Petrovitch H, Abbott RD, Tanner CM, Popper J, Masaki K, Launer, White LR. 2008. Association of olfactory dysfunction with risk for future Parkinson's disease. *Ann Neurol* **63**: 167–173.

Saeed U, Compagnone J, Aviv RI, Strafella AP, Black SE, Lang AE, Masellis M. 2017. Imaging biomarkers in Parkinson's disease and parkinsonian syndromes: current and emerging concepts. *Transl Neurodegener* **6**: 8. doi:10.1186/s40035-017-0076-6

Samii A, Nutt JG, Ransom BR. 2004. Parkinson's disease. *Lancet* **363**: 1783–1793. doi:10.1016/S0140-6736(04)16305-8

Sandoval-Pistorius SS, Hacker ML, Waters AC, Wang J, Provenza NR, de Hemptinne C, Johnson KA, Morrison MA, Cernera S. 2023. Advances in deep brain stimulation: from mechanisms to applications. *J Neurosci* **43**: 7575–7586. doi:10.1523/JNEUROSCI.1427-23.2023

Saunders-Pullman R, Ortega RA, Wang C, Raymond D, Elango S, Leaver K, Urval N, Katsnelson V, Gerber R, Swan M, et al. 2022. Association of olfactory performance with motor decline and age at onset in people with Parkinson disease and the *LRRK2* G2019S variant. *Neurology* **99**: e814–e823. doi:10.1212/WNL.0000000000200737

Schade S, Sixel-Döring F, Ebentheuer J, Schulz X, Trenkwalder C, Mollenhauer B. 2017. Acute levodopa challenge test in patients with de novo Parkinson's disease: data from the DeNoPa cohort. *Mov Disord Clin Pract* **4**: 755–762. doi:10.1002/mdc3.12511

Schapira AH, Fox SH, Hauser RA, Jankovic J, Jost WH, Kenney C, Kulisevsky J, Pahwa R, Poewe W, Anand R. 2017. Assessment of safety and efficacy of safinamide as a levodopa adjunct in patients with Parkinson disease and motor fluctuations: a randomized clinical trial. *JAMA Neurol* **74**: 216–224. doi:10.1001/jamaneurol.2016.4467

Schenkman M, Moore CG, Kohrt WM, Hall DA, Delitto A, Comella CL, Josbeno DA, Christiansen CL, Berman BD, Kluger BM, et al. 2018. Effect of high-intensity treadmill exercise on motor symptoms in patients with de novo Parkinson disease: a phase 2 randomized clinical trial. *JAMA Neurol* **75**: 219–226. doi:10.1001/jamaneurol.2017.3517

Schwarz ST, Afzal M, Morgan PS, Bajaj N, Gowland PA, Auer DP. 2014. The "swallow tail" appearance of the healthy nigrosome—a new accurate test of Parkinson's disease: a case-control and retrospective cross-sectional MRI study at 3T. *PLoS ONE* **9**: e93814. doi:10.1371/journal.pone.0093814

Senek M, Nielsen EI, Nyholm D. 2017. Levodopa-entacapone-carbidopa intestinal gel in Parkinson's disease: a randomized crossover study. *Mov Disord* **32**: 283–286. doi:10.1002/mds.26855

Seppi K, Ray Chaudhuri K, Coelho M, Fox SH, Katzenschlager R, Perez Lloret S, Weintraub D, Sampaio C; the collaborators of the Parkinson's Disease Update on Non-Motor Symptoms Study Group on behalf of the Movement Disorders Society Evidence-Based Medicine Committee. 2019. Update on treatments for nonmotor symptoms of Parkinson's disease—an evidence-based medicine review. *Mov Disord* **34**: 180–198. doi:10.1002/mds.27602

Sethi K. 2008. Levodopa unresponsive symptoms in Parkinson disease. *Mov Disord* **23** (Suppl 3): S521–S533. doi:10.1002/mds.22049

Shadrina MI, Slominsky PA. 2023. Genetic architecture of Parkinson's disease. *Biochemistry (Mosc)* **88**: 417–433. doi:10.1134/S0006297923030100

Shams S, Fällmar D, Schwarz S, Wahlund LO, van Westen D, Hansson O, Larsson EM, Haller S. 2017. MRI of the swallow tail sign: a useful marker in the diagnosis of Lewy body dementia? *AJNR Am J Neuroradiol* **38**: 1737–1741. doi:10.3174/ajnr.A5274

Shanker V. 2019. Essential tremor: diagnosis and management. *BMJ* **366**: l4485. doi:10.1136/bmj.l4485

Sharma VD, Lyons KE, Pahwa R. 2018. Amantadine extended-release capsules for levodopa-induced dyskinesia in patients with Parkinson's disease. *Ther Clin Risk Manag* **14**: 665–673. doi:10.2147/TCRM.S144481

Shetty AS, Bhatia KP, Lang AE. 2019. Dystonia and Parkinson's disease: what is the relationship? *Neurobiol Dis* **132**: 104462. doi:10.1016/j.nbd.2019.05.001

Shir D, Pham NTT, Botha H, Koga S, Kouri N, Ali F, Knopman DS, Petersen RC, Boeve BF, Kremers WK, et al. 2023. Clinicoradiologic and neuropathologic evaluation of corticobasal syndrome. *Neurology* **101**: e289–e299. doi:10.1212/WNL.0000000000207397

Shiraiwa N, Tamaoka A, Ohkoshi N. 2018. Clinical features of drug-induced parkinsonism. *Neurol Int* **10**: 7877. doi:10.4081/ni.2018.7877

Sidransky E, Lopez G. 2012. The link between the GBA gene and parkinsonism. *Lancet Neurol* **11**: 986–998. doi:10.1016/S1474-4422(12)70190-4

Sinai A, Nassar M, Sprecher E, Constantinescu M, Zaaroor M, Schlesinger I. 2022. Focused ultrasound thalamotomy in tremor dominant Parkinson's disease: long-term results. *J Parkinsons Dis* **12**: 199–206. doi:10.3233/JPD-212810

Smulders K, Dale ML, Carlson-Kuhta P, Nutt JG, Horak FB. 2016. Pharmacological treatment in Parkinson's disease: effects on gait. *Parkinsonism Relat Disord* **31**: 3–13. doi:10 .1016/j.parkreldis.2016.07.006

Song Z, Zhang J, Xue T, Yang Y, Wu D, Chen Z, You W, Wang Z. 2021. Different catechol-*O*-methyl transferase inhibitors in Parkinson's disease: a Bayesian network meta-analysis. *Front Neurol* **12**: 707723. doi:10.3389/fneur.2021.707723

Staubo SC, Fuskevåg OM, Toft M, Lie IH, Alvik KMJ, Jostad P, Tingvoll SH, Lilleng H, Rosqvist K, Størset E, et al. 2024. Dopamine agonist serum concentrations and impulse control disorders in Parkinson's disease. *Eur J Neurol* **31**: e16144. doi:10.1111/ene.16144

Stieglitz LH, Mahendran S, Oertel MF, Baumann CR. 2022. Bilateral focused ultrasound pallidotomy for Parkinson-related facial dyskinesia—a case report. *Mov Disord Clin Pract* **9**: 647–651. doi:10.1002/mdc3.13462

Sturchio A, Dwivedi AK, Gastaldi M, Grimberg MB, Businaro P, Duque KR, Vizcarra JA, Abdelghany E, Balint B, Marsili L, et al. 2022. Movement disorders associated with neuronal antibodies: a data-driven approach. *J Neurol* **269**: 3511–3521. doi:10.1007/s00415-021-10934-7

Titova N, Levin O, Katunina E, Ray Chaudhuri K. 2018. "Levodopa phobia": a review of a not uncommon and consequential phenomenon. *NPJ Parkinsons Dis* **4**: 31. doi:10.1038/s41531-018-0067-z

Tolosa E, Compta Y. 2006. Dystonia in Parkinson's disease. *J Neurol* **253** (Suppl 7): VII7–VII13. doi:10.1007/s00415-006-7003-6

Tolosa E, Garrido A, Scholz SW, Poewe W. 2021. Challenges in the diagnosis of Parkinson's disease. *Lancet Neurol* **20**: 385–397. doi:10.1016/S1474-4422(21)00030-2

Trinh J, Zeldenrust FMJ, Huang J, Kasten M, Schaake S, Petkovic S, Madoev H, Grünewald A, Almuammar S, König IR, et al. 2018. Genotype–phenotype relations for the Parkinson's disease genes SNCA, LRRK2, VPS35: MDSGene systematic review. *Mov Disord* **33**: 1857–1870. doi:10.1002/mds.27527

van der Kolk NM, de Vries NM, Kessels RPC, Joosten H, Zwinderman AH, Post B, Bloem BR. 2019. Effectiveness of home-based and remotely supervised aerobic exercise in Parkinson's disease: a double-blind, randomised controlled trial. *Lancet Neurol* **18**: 998–1008. doi:10.1016/S1474-4422(19)30285-6

Verschuur CVM, Suwijn SR, Boel JA, Post B, Bloem BR, van Hilten JJ, van Laar T, Tissingh G, Munts AG, Deuschl G, et al. 2019. Randomized delayed-start trial of levodopa in Parkinson's disease. *N Engl J Med* **380**: 315–324. doi:10 .1056/NEJMoa1809983

Viljaharju V, Mertsalmi T, Pauls KAM, Koivu M, Eerola-Rautio J, Udd M, Pekkonen E. 2024. Levodopa-entacapone-carbidopa intestinal gel treatment in advanced Parkinson's disease: a single-center study of 30 patients. *Mov Disord Clin Pract* **11**: 159–165. doi:10.1002/mdc3.13926

Virhammar J, Blohmé H, Nyholm D, Georgiopoulos C, Fällmar D. 2022. Midbrain area and the hummingbird sign from brain MRI in progressive supranuclear palsy and idiopathic normal pressure hydrocephalus. *J Neuroimaging* **32**: 90–96. doi:10.1111/jon.12932

Watanabe H, Riku Y, Hara K, Kawabata K, Nakamura T, Ito M, Hirayama M, Yoshida M, Katsuno M, Sobue G. 2018. Clinical and imaging features of multiple system atrophy: challenges for an early and clinically definitive diagnosis. *J Mov Disord* **11**: 107–120. doi:10.14802/jmd.18020

Weaver FM, Follett K, Stern M, Hur K, Harris C, Marks WJ, Rothlind J, Sagher O, Reda D, Moy CS, et al. 2009. Bilateral deep brain stimulation vs best medical therapy for patients with advanced Parkinson disease: a randomized controlled trial. *JAMA* **301**: 63–73. doi:10.1001/jama.2008.929

Weintraub D, Siderowf AD, Potenza MN, Goveas J, Morales KH, Duda JE, Moberg PJ, Stern MB. 2006. Association of dopamine agonist use with impulse control disorders in Parkinson disease. *Arch Neurol* **63**: 969–973. doi:10.1001/archneur.63.7.969

Wenning GK, Geser F, Krismer F, Seppi K, Duerr S, Boesch S, Köllensperger M, Goebel G, Pfeiffer KP, Barone P, et al. 2013. The natural history of multiple system atrophy: a prospective European cohort study. *Lancet Neurol* **12**: 264–274. doi:10.1016/S1474-4422(12)70327-7

Wenning GK, Stankovic I, Vignatelli L, Fanciulli A, Calandra-Buonaura G, Seppi K, Palma JA, Meissner WG, Krismer F, Berg D, et al. 2022. The movement disorder society criteria for the diagnosis of multiple system atrophy. *Mov Disord* **37**: 1131–1148. doi:10.1002/mds.29005

Whitwell JL, Höglinger GU, Antonini A, Bordelon Y, Boxer AL, Colosimo C, van Eimeren T, Golbe LI, Kassubek J, Kurz C, et al. 2017. Radiological biomarkers for diagnosis in PSP: where are we and where do we need to be? *Mov Disord* **32**: 955–971. doi:10.1002/mds.27038

Willis AW, Roberts E, Beck JC, Fiske B, Ross W, Savica R, Van Den Eeden SK, Tanner CM, Marras C, Group PsFP. 2022. Incidence of Parkinson disease in North America. *NPJ Parkinsons Dis* **8**: 170. doi:10.1038/s41531-022-00410-y

Progression in Parkinson's Disease

Krithi Irmady[1] and Serge Przedborski[2]

[1]Laboratory of Molecular Neuro-oncology, The Rockefeller University, New York, New York 10065, USA

[2]Departments of Neurology, Pathology and Cell Biology and Neuroscience, Columbia University, New York, New York 10032, USA

Correspondence: kirmady@rockefeller.edu; sp30@cumc.columbia.edu

Parkinson's disease (PD) is a common neurodegenerative disorder characterized by relentlessly progressive motor and nonmotor clinical features. In this paper, we offer a comprehensive overview of progression in PD, covering the heterogeneous symptomatology crucial for monitoring progression from clinical, pathological, and biomarker perspectives. We also discuss prevailing theories concerning the underlying pathobiology driving progression in PD and summarize the literature on emerging biomarkers that are expected to facilitate early prognosis and effective monitoring of disease progression.

Parkinson's disease (PD) is an adult-onset neurodegenerative disorder characterized by motor symptoms, such as tremor, slowness of movement, stiffness, and poor balance, primarily attributed to the degeneration of dopaminergic pathways in the ascending ventral midbrain. Additionally, patients with PD often experience nonmotor symptoms, including fatigue, anxiety, depression, cognitive decline, and sleep disturbances. Due to the redundancy within these dopaminergic pathways and the compensatory capacity of dopaminergic neurons, patients may exhibit prodromal PD symptoms—sometimes decades before the appearance of the cardinal motor signs typically used for clinical diagnosis.

Among these prodromal symptoms, idiopathic rapid eye movement (REM) sleep behavior disorder (RBD), as detected by polysomnography, is considered the strongest indicator of future PD. The risk of developing PD or a related synucleinopathy after an RBD diagnosis ranges from 19% to 38% at 5 years, 40% to 65% at 10 years, and up to 90% by 15 years (Postuma et al. 2009; Iranzo et al. 2017; Yao et al. 2018).

While the specifics of prodromal PD symptoms are beyond the scope of this paper, their presence supports the view that PD may begin years before it meets clinical diagnostic criteria. The prevailing hypothesis suggests that PD progresses slowly, starting with an "undercover stage," advancing to a "manifesting stage" when the cardinal motor features emerge, and continuing along a progressive trajectory. Although PD is diagnosed based on signs and symptoms that initially affect specific body parts, these symptoms often increase in severity and spread to other areas of the body. This progression is a hallmark of PD, and the absence of substantial clinical worsening over 5–10 years may suggest a parkinsonian syndrome other than PD (Postuma et al. 2015).

This paper will focus on the critical question of how and why PD progresses. We will discuss the clinical tools commonly used to assess pro-

Copyright © 2025 Cold Spring Harbor Laboratory Press; all rights reserved
Cite this article as *Cold Spring Harb Perspect Med* doi: 10.1101/cshperspect.a041641

gression, what progression entails in terms of PD signs and symptoms, treatment response, and the emergence of late-stage effects, as well as the cellular and molecular underpinnings of disease progression.

CLINICAL ASSESSMENT OF PD PROGRESSION

Although today, different imaging methodology may be used to assess PD severity and predict progression (see Niethammer et al. 2025) clinicians have traditionally used various scales that evaluate both motor and nonmotor dysfunction. In 1967, Hoehn and Yahr introduced a five-stage classification system based on the degree of clinical disability, which was later modified to include 0.5-point increments to better capture intermediate stages of PD progression (Hoehn and Yahr 1967). The Unified Parkinson's Disease Rating scale (UPDRS), developed to complement the Hoehn and Yahr scale (Fahn et al. 1987), was later revised by the Movement Disorder Society (MDS). The MDS-UPDRS, now the standard for evaluating PD severity in clinical and research settings, is a comprehensive four-part scale that assesses PD signs and symptoms (Goetz et al. 2008). Part I evaluates nonmotor experiences of daily living, part II addresses motor experiences of daily living, part III examines motor function, and part IV focuses on motor complications.

However, using these scales for PD assessment can be complicated by interrater variability and nonlinear scoring effects, while questionnaires may be prone to recall bias, especially in patients experiencing cognitive dysfunction. Newer wearable devices designed to monitor motor activity continuously could enhance the sensitivity and consistency of tracking disease progression, although further validation is necessary to confirm these devices' effectiveness (Moreau et al. 2023).

CLINICAL PROGRESSION IN PD: BEFORE AND AFTER THE ADVENT OF LEVODOPA

At the beginning of the twentieth century, *Vicia faba*, a legume crop, became a focus of scientific study by Markus Guggenheim, who isolated levodopa (L-DOPA, l-3,4-dihydroxyphenylalanine) but found no immediate applications. It was not until 1957, when Arvid Carlsson demonstrated that 3,4-hydroxyphenylalanine could reverse the effects of reserpine in rabbits, that interest in levodopa intensified. Subsequently, Hornykiewicz and Birkmayer administered intravenous levodopa to patients with severe PD, observing dramatic improvements. A few years later, neurologist George Cotzias began early trials of oral levodopa formulations in PD patients.

Levodopa was approved by the Food and Drug Administration (FDA) in 1970, and, five decades later, it remains the gold standard in PD pharmacological treatment. Unlike dopamine, which cannot cross the blood–brain barrier, levodopa can, and it is converted into dopamine by the enzyme dopa-decarboxylase in the central nervous system (CNS). Peripheral dopa-decarboxylase inhibitors, such as carbidopa or benserazide, are typically combined with levodopa to increase its therapeutic effect by preventing the conversion of levodopa to dopamine prior gaining access to the CNS.

Levodopa is usually initiated when PD patients experience motor symptoms that impair activities of daily living. In the early 2000s, concerns about levodopa-induced dyskinesias led clinicians to recommend alternatives such as dopamine agonists (Jankovic and Aguilar 2008). However, current guidelines now recommend levodopa as first-line therapy when indicated (Pringsheim et al. 2021). For younger patients expected to need long-term treatment, dopamine agonists may be considered, but, generally, the lowest effective dose of levodopa is recommended. Ultimately, most PD patients will require levodopa therapy, and a lack of response to it may suggest a diagnosis other than PD (Postuma et al. 2015). Given the near inevitability of levodopa use, understanding whether it impacts disease progression is critical.

Before levodopa's development, most PD patients progressed to Hoehn and Yahr stage 3, marked by worsening motor symptoms without loss of ambulation, after a median of 5–7 years (Hoehn and Yahr 1967). Stage 5, the most advanced stage, was typically reached after 10–14 years. Compared to the UPDRS, the

Hoehn and Yahr scale is relatively insensitive to levodopa's effects, and studies reported similar progression rates between patients with and without levodopa treatment (Hely et al. 1999). On average, patients progress to the next Hoehn and Yahr stage every 2 years, except for the transition from stage 2 to 2.5, which averages ~5 years (Kashihara and Kitayama 2023).

Studies prior to levodopa's development offered limited insights into PD's natural progression beyond Hoehn and Yahr staging, making it challenging to evaluate untreated disease progression over extended periods. As previously summarized (Poewe and Mahlknecht 2009), placebo-controlled clinical trials provide some of the best data on short-term progression in untreated PD. In the deprenyl and tocopherol anti-oxidative therapy of parkinsonism (DATATOP) trial, a significant 1-year decline in UPDRS scores was observed (average of 14 points, including a 9-point motor score decline) (Parkinson Study Group 1993). Other studies showed smaller early-stage changes, typically 8–10 points in overall UPDRS score and 5–6 points in motor score (Parkinson Study Group 1993,2004; Kieburtz 1996; Shults 2002; Fahn et al. 2004; Olanow et al. 2006). Notably, a study involving community-based surveys found that patients on levodopa experienced an average 1-year motor score decline of only 3.3 points (Schrag et al. 2007). For newly diagnosed PD patients, the focus is often on when progression will start to affect quality of life. Within the first 5 years, about half of newly diagnosed patients develop clinically significant disability, such as functional dependence, cognitive impairment, dysautonomia, or postural instability (Brumm et al. 2023).

The question of whether dopamine replacement therapy with levodopa or dopamine agonists has positive impacts on motor scores by merely improving symptoms or by delaying disease progression remains debated. If levodopa does delay progression, it would support early treatment initiation and reassure hesitant patients. Studies like the Earlier versus Later Levodopa (ELLDOPA) trial aimed to separate levodopa's symptom control from its potential effects on progression. In ELLDOPA, patients received either carbidopa/levodopa or placebo for 40

weeks, followed by a 2-week washout (Fahn et al. 2004). At the end of 42 weeks, those on levodopa had better UPDRS scores than those on placebo, suggesting levodopa either slowed progression or had lasting symptom benefits. However, in the Levodopa in Early Parkinson's (LEAP) study, which compared early versus delayed levodopa initiation, there was no significant difference between groups after 80 weeks (Verschuur et al. 2019). A 5-year follow-up of these patients confirmed no impact on progression from early levodopa treatment (Frequin et al. 2024). These findings indicate that while levodopa is safe for early-stage treatment, it may not modify disease progression. Beyond levodopa, novel disease-modifying treatments are under investigation but are beyond the scope of this paper.

Changes in motor performance alone fail to capture the true decline observed in patients diagnosed with PD (Fig. 1; Poewe 2006). Patients with PD who receive levodopa therapy may experience daily fluctuations in motor performance (so-called "on–off" periods) and treatment-induced involuntary movements (levodopa-induced dyskinesia) that can progressively worsen over time. Approximately 20%–50% of patients with PD who receive levodopa experience treatment-related complications during the first 5 years (Poewe et al. 1986; Schrag and Quinn 2000). In addition, some motor features often seen in advanced PD such as falls and freezing of gait do not respond well to levodopa treatment, and tend to worsen over time (Kalia and Lang 2015). Nearly 80% of patients develop freezing of gait or falls within 20 years of PD diagnosis (Hely et al. 2005; Kumar 2009). Similarly, within 15–20 years of diagnosis, half of patients diagnosed with PD develop difficulty with swallowing and choking, serious risk factors for death secondary to aspiration pneumonia (Hely et al. 2005; Kumar 2009).

Moreover, nonmotor aspects of PD, including dementia, sleep disorders, pain, fatigue, psychosis, and dysautonomia, also progressively impact quality of life, even for patients receiving dopamine replacement therapies. Whether these nonmotor symptoms are caused by extraventral midbrain dopaminergic pathway pathol-

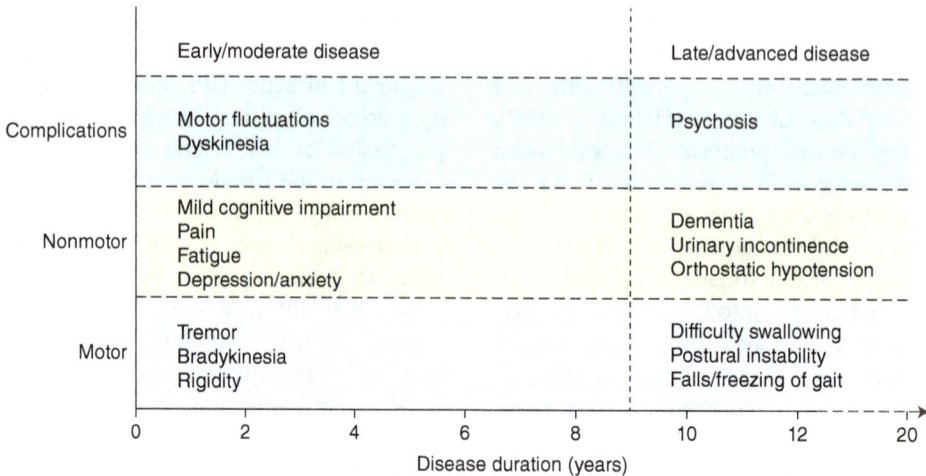

Figure 1. Clinical symptoms associated with Parkinson's disease progression. (Figure based on data in Kalia and Lang 2015. Created in BioRender. Irmady, K. (2025), https://BioRender.com/e66r229.)

ogy with poor levodopa response remains to be determined. Studies suggest that among patients diagnosed with PD who present with normal cognition at the time of diagnosis, approximately half experience cognitive decline within 6 years (Pigott et al. 2015), and dementia was reported in 83% of individuals who survive for 20 years (Kumar 2009). Within 20 years of diagnosis, approximately half of patients diagnosed with PD develop symptomatic orthostatic hypotension, and more than 70% experience urinary incontinence (Kumar 2009). These nonmotor symptoms significantly influence the need for nursing home care and impact mortality rates (Goetz and Stebbins 1993, 1995).

Before the advent of levodopa therapy, Hoehn and Yahr (1967) reported a mean survival time of 9.4 years after diagnosis, with an average age at death of 67 years. Mortality estimates suggested that PD significantly reduced life expectancy, with patients three times more likely to die than age-matched individuals without PD (Hoehn and Yahr 1967). Following the introduction of levodopa, studies reported reduced mortality rates and longer life expectancies in patients with PD. However, mortality remains higher among those with PD compared to the general population (Marttila et al. 1977; Maier Hoehn 1983; Ebmeier et al. 1990; Shoulson and Parkinson Study Group 1998; Hely et al.

1999, 2005; Montastruc et al. 2001; Marras et al. 2005), with mortality ratios (PD mortality relative to controls) ranging from 0.9 to 3.8 (Macleod et al. 2014). On average, survival for patients with PD decreases by ~5% per year of follow-up (Macleod et al. 2014). Certain PD complications, including falls and swallowing difficulties, respond poorly to levodopa and significantly increase mortality risk. Some studies suggest that deep brain stimulation (DBS) may help reduce PD-related mortality (Ngoga et al. 2014). However, these findings may be influenced by confounding factors, such as closer follow-up after DBS and selection bias favoring healthier patients for DBS. The gap in life expectancy between individuals with and without PD narrows with age. At older ages, people with PD are more likely to die from causes unrelated to PD than from PD-related complications (Dommershuijsen et al. 2023).

CLINICAL VARIABLES AFFECTING PD PROGRESSION

Clinical Factors

Despite currently available treatments, progression in PD remains inevitable for all patients. Research indicates that the rate of progression may be most rapid during the early stages of the disease, with a 5.1% decline in UPDRS scores as patients advance from Hoehn and Yahr stage 1 to stage 2.5, compared to a slower decline of 0.35% from stage 3 to stage 5. This pattern suggests the possibility of a nonlinear functional and/or structural neurodegenerative process in PD (Nurmi et al. 2003; Pirker et al. 2003; Schrag et al. 2007). While these metrics indicate that clinical progression may plateau after a certain disease duration, clinical motor scales often lack sensitivity and do not accurately capture progression at advanced stages. Moreover, even within the MDS-UPDRS, a 1-point change affecting gait can lead to a significantly different functional outcome compared to a 1-point change in tremor intensity. The development and implementation of new disease-modifying treatments, especially those that can be administered during prodromal or early stages of the disease, may offer the greatest potential for slowing progression.

Young-onset patients, defined as those diagnosed with PD before the age of 40 (with some studies using a cutoff of 50 or 55), do not typically exhibit rapid progression in the early stages of the disease (Quinn et al. 1987; Schrag and Schott 2006). Generally, these patients present with milder symptoms at onset and experience a slower rate of disease progression than those diagnosed at older ages, although they may encounter treatment-related motor complications (Kostić et al. 2002; Schrag and Schott 2006; Bega et al. 2015; Raket et al. 2022). This observation suggests that age plays a cumulative role in the advancement of PD pathology.

The age at onset significantly impacts disease progression during the early and middle stages. Despite the initial slower progression observed in young-onset patients, they may still experience malignant progression in later stages, which is often accompanied by severe symptoms such as hallucinations, frequent falls, and cognitive decline. These ominous signs typically precede death by about 5 years, regardless of the age at onset or the timing of their presentation (Kempster et al. 2010).

Studies have also shown that the female sex is associated with a lower risk of developing PD (Baba et al. 2005; Cerri et al. 2019), and female patients exhibit slower motor and nonmotor progression compared to their male counterparts (Cholerton et al. 2018; Lewis et al. 2024).

Another area of interest is the predominant symptom at the onset of the disease. The two most common PD subtypes are tremor-dominant (TD) and postural instability and gait disorder (PIGD). These subtypes demonstrate different rates of progression, with the TD subtype generally progressing more slowly in the early stages and requiring a later initiation of levodopa, as well as exhibiting fewer cognitive complications (Jankovic et al. 1990; Hershey et al. 1991; Rajput et al. 1993). Classifying PD by subtype can be valuable for patients and their families seeking to understand the likely trajectory of the disease. While individual progressions cannot be predicted with certainty, patients with TD symptoms may find reassurance in knowing they statistically have a better chance of slower progression. Such classifications could also aid future studies on disease-modifying therapies and biomarker identification. However, a recent study indicates that early classification (within 1 year of symptom onset) may not reliably predict these subtypes, as some patients display poor classification fidelity over time (Simuni et al. 2016). In this longitudinal study from the Parkinson's Disease Progression Markers Initiative (PPMI), 20% of individuals initially classified as TD and 40% classified as PIGD transitioned to the other subtype within 12 months. Thus, a more reliable classification may emerge after a few clinic visits, allowing symptoms to stabilize. Additionally, studies have shown that patients with a history of RBD tend to experience faster motor progression rates compared to those without a history of RBD (Postuma et al. 2012; Kawada et al. 2015; Pagano et al. 2018).

Rather than classifying patients solely based on the presence of tremor or postural issues in the early stages of the disease, researchers have suggested that prognostication could be enhanced by incorporating both motor and nonmotor features and employing hierarchical cluster analysis to define subtypes (Fereshtehnejad et al. 2015, 2017). One study identified three distinct subgroups based on the severity of motor and nonmotor symptoms at diagnosis: mild motor, diffuse malignant, and intermediate (De Pablo-Fernández et al. 2019). The average rates of progression and survival were closely aligned with the subtype-specific characteristics observed in early-stage disease. Specifically, the mild motor-predominant group demonstrated the longest survival times and the slowest progression rates, while the diffuse malignant group exhibited the shortest survival times and the fastest progression rates. The intermediate group fell between the other two subgroups, showing moderate rates of both progression and survival. Therefore, a more comprehensive approach to clinical subtyping at diagnosis may facilitate more accurate estimations of disease progression and survival.

Genetic Factors

Since the discovery of a mutation encoding α-synuclein (SNCA) in a Greek-Italian family in 1996, several additional genes have been associated with familial PD, including LRRK2, PRKN, and PINK1. Additionally, PD-associated risk factors, such as GBA1 heterozygote mutations, have been identified (more details in Westenberger et al. 2024).

Progression of PD varies depending on the associated genetic mutation (Aasly 2020). Patients with SNCA triplication or SNCA mutations typically experience accelerated progression of motor and cognitive symptoms (Polymeropoulos et al. 1997; Singleton et al. 2003). In contrast, individuals with PRKN mutations tend to have a slower disease progression with a lower likelihood of developing cognitive impairment (Lücking et al. 2000). Those with GBA1 mutations are at a higher risk of cognitive dysfunction (Brockmann et al. 2015; Malek et al. 2018; Stoker et al. 2020). Despite the identification of several PD-onset risk genes, familial PD represents a small proportion of the overall PD population. Therefore, identifying genetic factors that may influence PD progression in sporadic cases remains an important goal.

Polygenic risk scores may help predicting disease progression in patients with sporadic PD (Paul et al. 2018; Liu et al. 2021; Park and Lee 2024), and recent studies have sought to identify genetic variants that may impact symptom progression. Interestingly, the genetics associated

with PD risk do not substantially overlap with those influencing PD progression (Tan et al. 2021). For example, mutations in the mitochondrial lipid synthesis gene lysophosphatidic acid phosphatase ACP6 and the solute carrier SLC44A1 have been linked to motor symptom progression (Iwaki et al. 2019; Martínez Carrasco et al. 2023). The APOE ε4 genetic variant is also associated with cognitive dysfunction in PD (Tan et al. 2021), potentially due to increased amyloid burden, as Alzheimer's-like pathology is commonly observed in patients with PD dementia. Various loci, including RIMS2, TMEM108, and WWOX, have been associated with an increased risk of progression to PD dementia (Liu et al. 2021). However, cohort heterogeneity poses challenges in this research, but similar analyses across different clinical aspects of PD would likely provide valuable insights for clinicians. For instance, patients identified as being at high risk for dementia may be advised to avoid certain medications, such as anticholinergics commonly prescribed for motor symptoms or bladder dysfunction (Hong et al. 2019; Harnod et al. 2021). Invasive procedures like subthalamic nucleus DBS have also shown an increased association with dementia (Razmkon et al. 2024), necessitating caution for those genetically at risk. Additionally, researchers have identified clinicogenetic predictors of impulse control disorder (ICD) in PD, linking polymorphisms in genes, such as DAB1, PRKAG2, MEFV, and PRKCE to ICD (Weintraub et al. 2022). ICD behaviors—such as gambling, compulsive buying, and hypersexuality—are common side effects of dopaminergic agonist medications (Weintraub et al. 2006). Early identification of risk factors for ICD in PD patients can guide clinicians in avoiding dopamine agonists in these cases.

Given the multifaceted nature of PD, relying on a single clinical variable for accurate progression predictions is unrealistic. Advanced biostatistical methods that incorporate multiple clinical variables—such as age, sex, motor and nonmotor clinical features, genetics, and disease markers—offer promising approaches for developing data-driven prediction algorithms that can be applied clinically at an individual level (Fereshtehnejad and Postuma 2017).

PATHOLOGICAL UNDERPINNINGS OF PD PROGRESSION

Multiple lines of evidence indicate that symptom progression in PD arises from degenerative changes affecting both presynaptic and postsynaptic regions in dopaminergic and nondopaminergic pathways. In PD, the loss of dopaminergic neurons in the substantia nigra pars compacta (SNpc)—a region of the ventral midbrain that bears the brunt of pathology—reduces dopamine input to medium spiny neurons in the striatum, a critical area for motor control. To date, several mechanisms have been proposed to drive neuronal death in the SNpc, including impaired clearance of abnormal proteins, α-synuclein accumulation, dysfunctions in autophagy and lysosomal and mitochondrial activities, and neuroinflammation.

Positron emission tomography studies with [^{11}C]RTI-32 of the posterior striatum reveal that by the time PD becomes clinically evident, up to 50% of dopamine innervation is already lost (Guttman et al. 1997). Postmortem studies further indicate that within the first 5 years after symptom onset, up to 80% of striatal dopamine innervation is lost (Kordower et al. 2013).

Presynaptic Mechanisms, from Cell to Pathway to Function

We have previously mentioned (Przedborski et al. 2003) that most patients with neurodegenerative disorders, including those with PD, can recall when their symptoms first appeared. However, due to redundancies in neural networks, the link between clinical symptom progression—which some researchers suggest may accelerate exponentially during the early years (Schrag et al. 2007)—and the remaining neuron count is likely nonlinear. Symptoms typically emerge only when neuron numbers in pathways such as the ventral midbrain dopaminergic pathways drop below the threshold necessary to sustain normal function. Consequently, a patient may remain clinically stable for an extended period, only to experience rapid clinical deterioration once neuron loss surpasses this functional threshold.

All neurodegenerative disorders, including PD, progress slowly, often taking years to reach advanced stages. In our earlier publication (Przedborski et al. 2003), we questioned whether this gradual progression implies that diseased neurons undergo a prolonged decline before dying. However, neurodegeneration involves asynchronous cell death, with neurons deteriorating at different rates (Przedborski et al. 2003). Consequently, at any given time, affected neurons are in various stages of degeneration, with only a subset actually dying. Conventional clinical, radiological, and biochemical measures provide a snapshot of the overall condition of a cell population but offer limited insight into the death rates of individual neurons in the brain. Since in cell cultures, neurons often die rapidly once compromised, we posit the slow clinical progression reflect a slow decay of neural pathways due to the quick deterioration of small subsets of neurons over time.

Historically, PD research has concentrated on identifying cell-autonomous mechanisms underlying neuronal susceptibility. However, recent evidence supports an immune hypothesis, suggesting that neuroinflammation may influence disease onset and progression. It remains uncertain, however, whether neuroinflammation is a primary or secondary driver of neurodegeneration, or whether it might exert either beneficial or harmful effects. Both resident and peripheral immune cells appear to be involved, as evidenced by activated microglia, T lymphocytes, and elevated cytokine levels in PD brains (McGeer et al. 1988; Gerhard et al. 2006; Sommer et al. 2019). Furthermore, aggregated α-synuclein and markers of peripheral inflammation have been observed in the gut during early stages of the disease, indicating that PD pathology may propagate along the gut–brain axis (Challis et al. 2020; Schaeffer et al. 2020).

Postsynaptic Mechanisms

During the early stages of PD, dopamine loss can be managed effectively with dopamine replacement therapies. However, as the disease progresses, patients require increasingly higher doses, which eventually lose efficacy in controlling symptoms. This escalation in dosage leads to shorter response durations (known as "wearing-off") and can result in motor complications, including levodopa-induced dyskinesia (Zhuang et al. 2013). The diminishing effectiveness of dopamine therapy despite dose increases points to the involvement of postsynaptic homeostatic mechanisms in the brain. Supporting this, studies of postmortem PD brains show dendritic remodeling and a loss of dendritic spines from striatal medium spiny neurons (McNeill et al. 1988; Zaja-Milatovic et al. 2005).

Molecular changes in nondopaminergic neurons in the striatum and other regions also contribute to PD progression. For example, loss of cholinergic neurons in the nucleus basalis of Meynert and the pedunculopontine nucleus in the pons is associated with cognitive dysfunction in PD (Pasquini et al. 2021). Many of the molecular pathways implicated in the degeneration of the ventral midbrain dopaminergic neurons are not confined to the SNpc (Irmady et al. 2023), and this broad involvement suggests these pathways contribute to disease progression beyond dopamine depletion and reflect the wider vulnerability of affected brain regions. Although animal models have shed light on postsynaptic changes in the striatum and other brain regions, the precise processes that differentiate mild from advanced stages of PD are yet to be identified.

α-Synuclein in PD Progression

Misfolded α-synuclein is a hallmark of PD. In PD, α-synuclein adopts a β-sheet, amyloid-like structure, forming aggregates and inclusion bodies. When located intracellularly, these inclusions are called Lewy bodies; in axons and dendrites, they are known as Lewy neurites. The Braak staging system, which classifies postmortem PD brains, initially suggested that α-synuclein aggregation is integral to PD progression (Braak et al. 2003). Synuclein aggregates initially form in nondopaminergic structures in the lower brainstem or, according to this model, α-synuclein aggregates first appear in nondopaminergic structures in the lower brainstem or olfactory bulb and gradually advance to neocortical regions, eventually corresponding to clinical PD.

In Braak stages I and II, Lewy pathology is found in the olfactory regions and lower brainstem. In stages III and IV, α-synuclein aggregates are detectable in the midbrain—particularly the SNpc—as well as in the basal forebrain, transentorhinal cortex, and hippocampus. By stages V and VI, α-synuclein inclusions reach cortical association areas, eventually affecting the entire neocortex. Despite this progression, the role of Lewy bodies in PD neuron health remains debated (Tompkins and Hill 1997). Braak's original study showed that staging did not correlate directly with Hoehn and Yahr scores for disease severity (Braak et al. 2003); some individuals with advanced Braak stages IV and beyond showed no overt parkinsonian symptoms, while others with severe PD (Hoehn and Yahr stage 5) were categorized as Braak stages III and IV, typically associated with only early parkinsonism. Clarifying the exact role of Lewy body pathology in PD onset and progression requires further research.

The relationship between Lewy pathology and dementia in PD is similarly debated, with some studies finding positive correlations between Lewy pathology and cognitive decline, while others show no correlation (Braak et al. 2005; Parkkinen et al. 2008; Irwin et al. 2012).

Additionally, the underlying pathophysiological mechanisms may differ across PD stages, suggesting that a single intervention may not effectively slow progression throughout all stages. Identifying the molecular drivers of continued clinical progression is therefore essential to improving long-term disease management for PD patients.

BIOMARKERS AS PROGNOSTIC INDICATORS OF PD PROGRESSION

As reported by the Biomarkers Definitions Working Group, a biomarker is defined as "a characteristic that is objectively measured and evaluated as an indicator of normal biological processes, pathogenic processes, or pharmacologic responses to a therapeutic intervention" (Biomarkers Definitions Working Group 2001). In PD, there is an urgent need to identify biomarkers that can objectively track disease progression over time. As for other neurological disorders, an ideal biomarker would be easily quantifiable, highly sensitive and specific, and would have prognostic value for predicting the disease course or monitoring therapeutic response. Additionally, biomarkers linked to pathways targeted by specific therapeutic candidates could validate the efficacy of treatments being tested. Moreover, the most desirable biomarker would not only fulfill these criteria but also be detectable in the earliest stages of PD, allowing for preventive interventions. While several biomarkers show promise for diagnosing PD or differentiating it from other parkinsonian disorders (Vijiaratnam and Foltynie 2023), this section will specifically highlight those biomarkers with the potential to serve as indicators of disease severity and progression (Table 1).

In recent decades, advances in imaging and omics technologies—such as transcriptomics, genomics, proteomics, lipidomics, and metabolomics—have significantly boosted the search for biomarkers in PD. Researchers now have access to extensive longitudinal data from biospecimens and clinical records, which are housed in large-scale, well-annotated international databases. Notable resources include the Parkinson's Progression Markers Initiative or PPMI (Marek et al. 2011), which has collected longitudinal data for more than a decade, the Harvard Biomarkers Study (Ding et al. 2011), BioFIND (Kang et al. 2016), and the Parkinson's Disease Biomarkers Program (PDBP) within the Accelerating Medicines Partnership: Parkinson's Disease (AMP-PD, www.amp-pd.org).

Imaging-Based Biomarkers in PD

Imaging studies have used a range of techniques to track PD progression. These include single-photon emission computed tomography (SPECT) scans to measure striatal dopaminergic inputs, functional imaging, and magnetic resonance imaging (MRI) to assess brain structure and biochemical composition. Other modalities have also been explored, as summarized in previous review articles (Saeed et al. 2017; Mitchell et al. 2021).

In early PD, the most widely used imaging modality for diagnosis is dopamine transporter (DAT) imaging via SPECT. Besides DAT imaging, other markers such as $[^{18}F]$fluorodopa ($[^{18}F]$DOPA) and vesicular monoamine transporter imaging offer valuable insights into presynaptic dopaminergic denervation. However, measuring presynaptic input alone to track disease progression poses challenges: As the disease advances, the signal-to-noise ratio diminishes, making it increasingly difficult to quantify remaining striatal innervation in later PD stages. This limitation reduces the utility of these markers for monitoring disease progression. Despite this, some studies involving small patient cohorts have found correlations between striatal dopamine input levels and PD severity and progression (Benamer et al. 2000; Takahashi et al. 2019). Additionally, the activity of aromatic amino acid decarboxylase, which converts $[^{18}F]$DOPA to $[^{18}F]$dopamine, is often used as an indicator of presynaptic nigrostriatal axon density. Decreased putaminal $[^{18}F]$DOPA uptake has been linked to more severe motor symptoms but does not correlate with cognitive or mood dysfunction, suggesting these nonmotor symptoms may involve extradopaminergic mechanisms (Broussolle et al. 1999). Conversely, radionuclide imaging studies of cholinergic and noradrenergic activity have shown associations with cognitive function in PD (Shimada et al. 2009; Sommerauer et al. 2018), although the predictive value of radionuclide imaging for progression to PD dementia remains uncertain.

The motor symptoms of PD are associated with an aberrant metabolic pattern, known as the PD-related motor pattern (PDRP), characterized by increased activity in the pallidothalamic and pontine regions and decreased activity in cortical motor areas (Eidelberg et al. 1994, 1995). A longitudinal imaging study on PDRP and regional glucose metabolism used serial $[^{18}F]$fluorodeoxyglucose PET (FDG-PET) scans to monitor individuals in early PD over 4 years postdiagnosis (Huang et al. 2007b). This study revealed that elevated PDRP correlated with worsening motor ratings and decreased putaminal DAT binding. Additional research identified a PD-related cognitive pattern (PDCP), marked

Table 1. Biomarkers associated with disease severity or predictive of motor and cognitive progression in Parkinson's disease (PD)

Biomarker modality	Disease severity	Disease progression	References (PMIDs)
Imaging			
Dopamine transporter (SPECT)	+		10928580;30995093
Fluorodopa	+		10475108
Positron emission tomography (noradrenaline/acetylcholine)	+		19474411;29272343;16344512
FDG-PET	+		8063874;7884498;17470495
MRI	+	+	26888982;31514113;26810522
DTI	+	+	37965163;28899020
CSF			
Glucocerebrosidase		+	36253102
Neurogranin	+		28649607
Pentraxins	+	+	36504237
Dopamine metabolites	+		33942926;28108196
Neurofilament light chain (NfL)	+	+	34480363
GFAP	+	+	37475029
α-Synuclein			
Oligomeric α-synuclein	+		36096686;22344688;2754884926782965;33978256
Phosphorylated α-synuclein	+		25637461;26782965
Amyloid β (1–42)		+	20720189;24744728;26330275
Blood/serum/plasma/exosome			
DJ-1	+		17720313
GFAP	+	+	36759508;37932720;38429295
Inflammatory profile			
Lymphocyte profile	+		23054369;32688028l;2623942934717679;31930038
Neutrophil/lymphocyte ratio	+		29713303;36912400
Cytokine profile	+	+	26999434;17986095;25587378
Neurofilament light chain	+	+	33639572;33964785;3480661935577512;32503574; 34480363

Table based on data in Vijiaratnam and Foltynie (2023).

(CSF) Cerebrospinal fluid, (DTI) diffusion tensor imaging, (GFAP) glial fibrillary acidic protein, (FDG-PET) fluorodeoxyglucose positron emission tomography, (MRI) magnetic resonance imaging, (SPECT) single-photon emission computed tomography.

by decreased metabolism in medial frontal and parietal association areas and relative increases in the cerebellar cortex and dentate nuclei (Huang et al. 2007a). The presence of PDCP correlated with cognitive dysfunction severity, suggesting that PDRP and PDCP networks may serve as progression markers and aid in evaluating treatment efficacy.

PET imaging has also been used to assess microglial activity and neuroinflammation in various brain regions, including the basal ganglia, frontal cortex, and cingulate cortex, however with mixed results regarding correlations with disease severity or striatal dopaminergic loss (Ouchi et al. 2005; Gerhard et al. 2006).

MRI studies have demonstrated reduced cortical and striatal volumes in PD patients compared to controls and identified differences between PD and other parkinsonian syndromes (Schulz et al. 1999; Tinaz et al. 2011; Takahashi et al. 2019). However, these volumetric changes have yet to be systematically studied for progression tracking, leaving their utility as progression biomarkers unclear. Frontal and parietal atrophy in PD has been associated with longer disease duration (Sterling et al. 2016), and a PD-specific network atrophy pattern was shown to predict motor, cognitive, and global progression more effectively than other tested biomarkers, including cerebrospinal fluid (CSF) markers (Zeighami et al. 2019). MRI studies examining SNpc volume in PD have yielded inconsistent results, with some studies reporting no significant changes, while others have shown either reductions or increases in SNpc volume. One 3year longitudinal study linked early changes in the SNpc to the onset of gait freezing (Wieler et al. 2016), and a quantitative MRI study found a moderate correlation between motor severity and neuromelanin-iron levels in the SNpc (Hartono et al. 2023).

Diffusion tensor imaging (DTI), a noninvasive method to evaluate brain structure by tracking water molecule movement, has shown potential in PD progression studies. In a 2-year DTI study, patients with the more severe PIGD subtype exhibited greater increases in free water content in the brain than those with the milder TD subtype (Bower et al. 2023). Using a logistic regression model to analyze motor scores and free water content, researchers accurately differentiated between TD and PIGD subtypes, achieving higher accuracy than other clinical or imaging assessments. Another multisite study found that increased free water in the posterior substantia nigra during the first year predicted motor symptom progression over 4 years (Burciu et al. 2017).

CSF Biomarkers in PD

The α-synuclein seed amplification assay (SAA) is one of the most promising recent breakthroughs in PD diagnostics (Siderowf et al. 2023). However, its potential as an independent tool to assess disease severity or predict progression has not been fully explored. When paired with an enzyme-linked immunosorbent assay measuring oligomeric α-synuclein levels in CSF, SAA may help stage disease severity (Majbour et al. 2022). Associations between total levels of free α-synuclein or α-synuclein derived from extracellular vesicles and disease progression remain inconsistent (Hall et al. 2015, 2016; Stuendl et al. 2016; Wang et al. 2018; Niu et al. 2020). Phosphorylated α-synuclein has been shown to correlate with disease severity (Wang et al. 2012; Majbour et al. 2016a), although one study found this correlation may vary by disease stage, with lower phosphorylated α-synuclein levels indicating worse clinical severity in early stages and better clinical severity in later stages (Stewart et al. 2015). If these findings hold, a U-shaped correlation may complicate its use as a biomarker in therapeutic trials requiring extended patient monitoring. Oligomeric α-synuclein levels in CSF also correlate with disease severity (Majbour et al. 2016a, b, 2021), and Tau and amyloid β (1–42) copathology in CSF has been linked to cognitive decline in PD, with low amyloid β (1–42) levels correlating with cognitive impairment in longitudinal studies (Siderowf et al. 2010; Alves et al. 2014; Parnetti et al. 2014; Terrelonge et al. 2016). Low glucocerebrosidase activity in CSF at PD diagnosis has been associated with an increased risk of PD dementia (Oftedal et al. 2023). In addition, low levels of pentraxins, extracellular scaffolding proteins that mark synaptic degeneration, are predictive of cognitive and motor performance in PD (Nilsson et al. 2023). A CSF panel of dopaminergic metabolite levels also correlates with disease progression (Stefani et al. 2017; Kremer et al. 2021). Thus, while many promising avenues exist, more formal longitudinal studies with large sample sizes are warranted to better understand the value of these CSF biomarkers beyond helping in diagnosis.

Blood-Based Biomarkers

Blood-based biomarkers provide practical advantages due to their accessibility, affordability, and quantifiability. A recent study demonstrated that RNA changes in blood align with RNA changes observed in PD patients' brains (Irmady et al. 2023) with correlations to clinical features such as severity and dementia. Although this study did not address longitudinal blood marker changes or associations with disease progression, the alignment between brain and blood RNA signals suggests a promising avenue for blood-based prognostic indicators. Whether this alignment reflects PD as a multisystem disease or brain RNAs being secreted into peripheral blood remains unclear.

Studies have also identified other blood-based biomarker changes associated with specific PD endophenotypes. For instance, RNA differences related to nucleic acid metabolism, mitochondrial function, and immune responses have been found between the milder TD and more severe PIGD subtypes (Pinho et al. 2016). Blood proinflammatory markers, such as apolipoprotein A2, C-reactive protein, uric acid, and vitamin D, have been linked to more severe motor phenotypes (Lawton et al. 2020). Independent validation is necessary to confirm these findings and assess their prognostic value.

Plasma levels of glial fibrillary acidic protein (GFAP), an astrocytic protein, may help differentiate patients with PD dementia from those without dementia and predict cognitive decline (Lin et al. 2023; Tang et al. 2023; Che et al. 2024). Immune profiling in serum has revealed associations between proinflammatory lymphocytic and cytokine profiles and more rapid motor and cognitive decline (Rentzos et al. 2007; Saunders et al. 2012; Tang et al. 2014; Chen et al. 2015; Williams-Gray et al. 2016; Sanjari Moghaddam et al. 2018; Sun et al. 2019; Magistrelli et al. 2020; Bhatia et al. 2021). Nonetheless, neither GFAP nor cytokine profiling has demonstrated reliable prognostic capability on an individual basis, and definitive cutoff values remain elusive.

One promising prognostic blood-based biomarker is neurofilament light chain (NfL), a cytoskeletal protein subunit released upon axonal injury—a feature of PD. Although elevated NfL levels are nonspecific and can occur in other neurological conditions, serum NfL levels correlate with both CSF NfL (Hansson et al. 2017) and presynaptic dopamine loss (Diekämper et al. 2021; Ye et al. 2021). NfL cutoff values can distinguish between healthy controls and

PD patients with moderate sensitivity and high specificity. NfL levels also correlate with motor and cognitive scores and hold prognostic potential for predicting the more severe PIGD phenotype and dementia (Ng et al. 2020; Aamodt et al. 2021; Ma et al. 2021; Pilotto et al. 2021; Ye et al. 2021; Vijiaratnam et al. 2022; Ygland Rödström et al. 2022). One study found that patients with high serum NfL levels (>21.84 pg/mL) exhibited faster progression (Lin et al. 2019). The utility of NfL as a progression biomarker may be enhanced by combining it with other biomarkers, genetic data, or clinical variables (Vijiaratnam et al. 2022).

To date, no single biomarker effectively serves as a standalone measure for tracking clinical progression or forecasting the future trajectory of PD. A combinatorial approach—integrating biofluid analyses with imaging techniques—holds the most promise for biomarker development. Biomarkers derived from large, well-annotated clinical consortia and validated in independent cohorts are likely to yield the most robust indicators, especially given the current accessibility of these databases.

CONCLUDING REMARKS

Progression in PD remains a challenging aspect of the disease, highlighting the urgent need to deepen our understanding of its underlying mechanisms. Investigating factors that contribute to both the onset and progression of PD is essential to uncover critical insights. There is a pressing need to develop improved clinical tools capable of accurately tracking disease trajectory. Finally, identifying imaging and biofluid biomarkers with the potential to predict clinical progression and signal therapeutic benefits will be crucial as newer therapies become available for managing PD progression.

REFERENCES

*Reference is also in this subject collection.

Aamodt WW, Waligorska T, Shen J, Tropea TF, Siderowf A, Weintraub D, Grossman M, Irwin D, Wolk DA, Xie SX, et al. 2021. Neurofilament light chain as a biomarker for

Cite this article as *Cold Spring Harb Perspect Med* doi: 10.1101/cshperspect.a041641

cognitive decline in Parkinson disease. *Mov Disord* **36:** 2945–2950. doi:10.1002/mds.28779

Aasly JO. 2020. Long-term outcomes of genetic Parkinson's disease. *J Mov Disord* **13:** 81–96. doi:10.14802/jmd.19080

Alves G, Lange J, Blennow K, Zetterberg H, Andreasson U, Førland MG, Tysnes OB, Larsen JP, Pedersen KF. 2014. CSF Aβ42 predicts early-onset dementia in Parkinson disease. *Neurology* **82:** 1784–1790. doi:10.1212/WNL.0000000000000425

Baba Y, Putzke JD, Whaley NR, Wszolek ZK, Uitti RJ. 2005. Gender and the Parkinson's disease phenotype. *J Neurol* **252:** 1201–1205. doi:10.1007/s00415-005-0835-7

Bega D, Kim S, Zhang Y, Elm J, Schneider J, Hauser R, Fraser A, Simuni T. 2015. Predictors of functional decline in early Parkinson's disease: NET-PD LS1 cohort. *J Parkinsons Dis* **5:** 773–782. doi:10.3233/JPD-150668

Benamer HT, Patterson J, Wyper DJ, Hadley DM, Macphee GJ, Grosset DG. 2000. Correlation of Parkinson's disease severity and duration with 123I-FP-CIT SPECT striatal uptake. *Mov Disord* **15:** 692–698. doi:10.1002/1531-8257(200007)15:4<692::AID-MDS1014>3.0.CO;2-V

Bhatia D, Grozdanov V, Ruf WP, Kassubek J, Ludolph AC, Weishaupt JH, Danzer KM. 2021. T-cell dysregulation is associated with disease severity in Parkinson's disease. *J Neuroinflammation* **18:** 250. doi:10.1186/s12974-021-02296-8

Biomarkers Definitions Working Group. 2001. Biomarkers and surrogate endpoints: preferred definitions and conceptual framework. *Clin Pharmacol Ther* **69:** 89–95. doi:10.1067/mcp.2001.113989

Bower AE, Crisomia SJ, Chung JW, Martello JP, Burciu RG. 2023. Free water imaging unravels unique patterns of longitudinal structural brain changes in Parkinson's disease subtypes. *Front Neurol* **14:** 1278065. doi:10.3389/fneur.2023.1278065

Braak H, Del Tredici K, Rüb U, de Vos RAI, Jansen Steur ENH, Braak E. 2003. Staging of brain pathology related to sporadic Parkinson's disease. *Neurobiol Aging* **24:** 197–211. doi:10.1016/S0197-4580(02)00065-9

Braak H, Rüb U, Jansen Steur ENH, Del Tredici K, de Vos RAI. 2005. Cognitive status correlates with neuropathologic stage in Parkinson disease. *Neurology* **64:** 1404–1410. doi:10.1212/01.WNL.0000158422.41380.82

Brockmann K, Srulijes K, Pflederer S, Hauser A-K, Schulte C, Maetzler W, Gasser T, Berg D. 2015. GBA-associated Parkinson's disease: reduced survival and more rapid progression in a prospective longitudinal study. *Mov Disord* **30:** 407–411. doi:10.1002/mds.26071

Broussolle E, Dentresangle C, Landais P, Garcia-Larrea L, Pollak P, Croisile B, Hibert O, Bonnefoi F, Galy G, Froment JC, et al. 1999. The relation of putamen and caudate nucleus 18F-Dopa uptake to motor and cognitive performances in Parkinson's disease. *J Neurol Sci* **166:** 141–151. doi:10.1016/S0022-510X(99)00127-9

Brumm MC, Siderowf A, Simuni T, Burghardt E, Choi SH, Caspell-Garcia C, Chahine LM, Mollenhauer B, Foroud T, Galasko D, et al. 2023. Parkinson's progression markers initiative: a milestone-based strategy to monitor Parkinson's disease progression. *J Parkinsons Dis* **13:** 899–916. doi:10.3233/JPD-223433

Burciu RG, Ofori E, Archer DB, Wu SS, Pasternak O, McFarland NR, Okun MS, Vaillancourt DE. 2017. Progression

marker of Parkinson's disease: a 4-year multi-site imaging study. *Brain* **140:** 2183–2192. doi:10.1093/brain/awx146

Cerri S, Mus L, Blandini F. 2019. Parkinson's disease in women and men: what's the difference? *J Parkinsons Dis* **9:** 501–515. doi:10.3233/JPD-191683

Challis C, Hori A, Sampson TR, Yoo BB, Challis RC, Hamilton AM, Mazmanian SK, Volpicelli-Daley LA, Gradinaru V. 2020. Gut-seeded α-synuclein fibrils promote gut dysfunction and brain pathology specifically in aged mice. *Nat Neurosci* **23:** 327–336. doi:10.1038/s41593-020-0589-7

Che N, Ou R, Li C, Zhang L, Wei Q, Wang S, Jiang Q, Yang T, Xiao Y, Lin J, et al. 2024. Plasma GFAP as a prognostic biomarker of motor subtype in early Parkinson's disease. *NPJ Parkinsons Dis* **10:** 48. doi:10.1038/s41531-024-00664-8

Chen Y, Qi B, Xu W, Ma B, Li L, Chen Q, Qian W, Liu X, Qu H. 2015. Clinical correlation of peripheral CD4+ cell subsets, their imbalance and Parkinson's disease. *Mol Med Rep* **12:** 6105–6111. doi:10.3892/mmr.2015.4136

Cholerton B, Johnson CO, Fish B, Quinn JF, Chung KA, Peterson-Hiller AL, Rosenthal LS, Dawson TM, Albert MS, Hu SC, et al. 2018. Sex differences in progression to mild cognitive impairment and dementia in Parkinson's disease. *Parkinsonism Relat Disord* **50:** 29–36. doi:10.1016/j.parkreldis.2018.02.007

De Pablo-Fernández E, Lees AJ, Holton JL, Warner TT. 2019. Prognosis and neuropathologic correlation of clinical subtypes of Parkinson disease. *JAMA Neurol* **76:** 470–479. doi:10.1001/jamaneurol.2018.4377

Diekämper E, Brix B, Stöcker W, Vielhaber S, Galazky I, Kreissl MC, Genseke P, Düzel E, Körtvelyessy P. 2021. Neurofilament levels are reflecting the loss of presynaptic dopamine receptors in movement disorders. *Front Neurosci* **15:** 690013. doi:10.3389/fnins.2021.690013

Ding H, Sarokhan AK, Roderick SS, Bakshi R, Maher NE, Ashourian P, Kan CG, Chang S, Santarlasci A, Swords KE, et al. 2011. Association of SNCA with Parkinson: replication in the Harvard NeuroDiscovery Center Biomarker Study. *Mov Disord* **26:** 2283–2286. doi:10.1002/mds.23934

Dommershuijsen LJ, Darweesh SKL, Ben-Shlomo Y, Kluger BM, Bloem BR. 2023. The elephant in the room: critical reflections on mortality rates among individuals with Parkinson's disease. *NPJ Parkinsons Dis* **9:** 145. doi:10.1038/s41531-023-00588-9

Ebmeier KP, Calder SA, Crawford JR, Stewart L, Besson JA, Mutch WJ. 1990. Mortality and causes of death in idiopathic Parkinson's disease: results from the Aberdeen whole population study. *Scott Med J* **35:** 173–175. doi:10.1177/003693309003500605

Eidelberg D, Moeller JR, Dhawan V, Spetsieris P, Takikawa S, Ishikawa T, Chaly T, Robeson W, Margouleff D, Przedborski S. 1994. The metabolic topography of parkinsonism. *J Cereb Blood Flow Metab* **14:** 783–801. doi:10.1038/jcbfm.1994.99

Eidelberg D, Moeller JR, Ishikawa T, Dhawan V, Spetsieris P, Chaly T, Robeson W, Dahl JR, Margouleff D. 1995. Assessment of disease severity in parkinsonism with fluorine-18-fluorodeoxyglucose and PET. *J Nucl Med* **36:** 378–383.

Fahn S, Elton RL, Fahn S, Marsden CD, Calne DB, Goldstein M. 1987. *Unified Parkinsons's disease rating scale: Recent developments in Parkinson's disease*, pp. 153–163. Macmillan Healthcare Information, Florham Park, NJ.

Fahn S, Oakes D, Shoulson I, Kieburtz K, Rudolph A, Lang A, Olanow CW, Tanner C, Marek K; Parkinson Study Group. 2004. Levodopa and the progression of Parkinson's disease. *N Engl J Med* **351:** 2498–2508. doi:10.1056/NEJMoa033447

Fereshtehnejad SM, Postuma RB. 2017. Subtypes of Parkinson's disease: what do they tell us about disease progression? *Curr Neurol Neurosci Rep* **17:** 34. doi:10.1007/s11910-017-0738-x

Fereshtehnejad SM, Romenets SR, Anang JBM, Latreille V, Gagnon JF, Postuma RB. 2015. New clinical subtypes of Parkinson disease and their longitudinal progression: a prospective cohort comparison with other phenotypes. *JAMA Neurol* **72:** 863–873. doi:10.1001/jamaneurol.2015.0703

Fereshtehnejad SM, Zeighami Y, Dagher A, Postuma RB. 2017. Clinical criteria for subtyping Parkinson's disease: biomarkers and longitudinal progression. *Brain* **140:** 1959–1976. doi:10.1093/brain/awx118

Frequin HL, Verschuur CVM, Suwijn SR, Boel JA, Post B, Bloem BR, van Hilten JJ, van Laar T, Tissingh G, Munts AG, et al. 2024. Long-term follow-up of the LEAP study: early versus delayed levodopa in early Parkinson's disease. *Mov Disord* **39:** 975–982. doi:10.1002/mds.29796

Gerhard A, Pavese N, Hotton G, Turkheimer F, Es M, Hammers A, Eggert K, Oertel W, Banati RB, Brooks DJ. 2006. In vivo imaging of microglial activation with [11C](R)-PK11195 PET in idiopathic Parkinson's disease. *Neurobiol Dis* **21:** 404–412. doi:10.1016/j.nbd.2005.08.002

Goetz CG, Stebbins GT. 1993. Risk factors for nursing home placement in advanced Parkinson's disease. *Neurology* **43:** 2227–2229. doi:10.1212/WNL.43.11.2227

Goetz CG, Stebbins GT. 1995. Mortality and hallucinations in nursing home patients with advanced Parkinson's disease. *Neurology* **45:** 669–671. doi:10.1212/WNL.45.4.669

Goetz CG, Tilley BC, Shaftman SR, Stebbins GT, Fahn S, Martinez-Martin P, Poewe W, Sampaio C, Stern MB, Dodel R, et al. 2008. Movement disorder society-sponsored revision of the unified Parkinson's disease rating scale (MDS-UPDRS): scale presentation and clinimetric testing results. *Mov Disord* **23:** 2129–2170. doi:10.1002/mds.22340

Guttman M, Burkholder J, Kish SJ, Hussey D, Wilson A, DaSilva J, Houle S. 1997. [11c]RTI-32 PET studies of the dopamine transporter in early dopa-naive Parkinson's disease: implications for the symptomatic threshold. *Neurology* **48:** 1578–1583. doi:10.1212/WNL.48.6.1578

Hall S, Surova Y, Öhrfelt A, Zetterberg H, Lindqvist D, Hansson O. 2015. CSF biomarkers and clinical progression of Parkinson disease. *Neurology* **84:** 57–63. doi:10.1212/WNL.0000000000001098

Hall S, Surova Y, Öhrfelt A; Swedish BioFINDER Study; Blennow K, Zetterberg H, Hansson O. 2016. Longitudinal measurements of cerebrospinal fluid biomarkers in Parkinson's disease. *Mov Disord* **31:** 898–905. doi:10.1002/mds.26578

Hansson O, Janelidze S, Hall S, Magdalinou N, Lees AJ, Andreasson U, Norgren N, Linder J, Forsgren L, Constantinescu R, et al. 2017. Blood-based NfL: a biomarker for differential diagnosis of parkinsonian disorder. *Neurology* **88:** 930–937. doi:10.1212/WNL.0000000000003680

Harnod T, Yang YC, Chiu LT, Wang JH, Lin SZ, Ding DC. 2021. Use of bladder antimuscarinics is associated with an increased risk of dementia: a retrospective population-based case-control study. *Sci Rep* **11:** 4827. doi:10.1038/s41598-021-84229-2

Hartono S, Chen RC, Welton T, Tan AS, Lee W, Teh PY, Chen C, Hou W, Tham WP, Lim EW, et al. 2023. Quantitative iron-neuromelanin MRI associates with motor severity in Parkinson's disease and matches radiological disease classification. *Front Aging Neurosci* **15:** 1287917. doi:10.3389/fnagi.2023.1287917

Hely MA, Morris JG, Traficante R, Reid WG, O'Sullivan DJ, Williamson PM. 1999. The Sydney multicentre study of Parkinson's disease: progression and mortality at 10 years. *J Neurol Neurosurg Psychiatry* **67:** 300–307. doi:10.1136/jnnp.67.3.300

Hely MA, Morris JGL, Reid WGJ, Trafficante R. 2005. Sydney multicenter study of Parkinson's disease: non-L-dopa-responsive problems dominate at 15 years. *Mov Disord* **20:** 190–199. doi:10.1002/mds.20324

Hershey LA, Feldman BJ, Kim KY, Commichau C, Lichter DG. 1991. Tremor at onset. Predictor of cognitive and motor outcome in Parkinson's disease? *Arch Neurol* **48:** 1049–1051. doi:10.1001/archneur.1991.00530220069021

Hoehn MM, Yahr MD. 1967. Parkinsonism: onset, progression and mortality. *Neurology* **17:** 427–442. doi:10.1212/WNL.17.5.427

Hong CT, Chan L, Wu D, Chen WT, Chien LN. 2019. Antiparkinsonism anticholinergics increase dementia risk in patients with Parkinson's disease. *Parkinsonism Relat Disord* **65:** 224–229. doi:10.1016/j.parkreldis.2019.06.022

Huang C, Mattis P, Tang C, Perrine K, Carbon M, Eidelberg D. 2007a. Metabolic brain networks associated with cognitive function in Parkinson's disease. *Neuroimage* **34:** 714–723. doi:10.1016/j.neuroimage.2006.09.003

Huang C, Tang C, Feigin A, Lesser M, Ma Y, Pourfar M, Dhawan V, Eidelberg D. 2007b. Changes in network activity with the progression of Parkinson's disease. *Brain* **130:** 1834–1846. doi:10.1093/brain/awm086

Iranzo A, Santamaría J, Valldeoriola F, Serradell M, Salamero M, Gaig C, Niñerola-Baizán A, Sánchez-Valle R, Lladó A, De Marzi R, et al. 2017. Dopamine transporter imaging deficit predicts early transition to synucleinopathy in idiopathic rapid eye movement sleep behavior disorder. *Ann Neurol* **82:** 419–428. doi:10.1002/ana.25026

Irmady K, Hale CR, Qadri R, Fak J, Simelane S, Carroll T, Przedborski S, Darnell RB. 2023. Blood transcriptomic signatures associated with molecular changes in the brain and clinical outcomes in Parkinson's disease. *Nat Commun* **14:** 3956. doi:10.1038/s41467-023-39652-6

Irwin D, White M, Toledo J, Xie S, Robinson J, Van Deerlin V, Leverenz J, Montine T, Lee V, Duda J, et al. 2012. Neuropathologic substrates of Parkinson's disease dementia (S52.002). *Neurology* **78:** S52.002.

Iwaki H, Blauwendraat C, Leonard HL, Kim JJ, Liu G, Maple-Grødem J, Corvol JC, Pihlstrøm L, van Nimwegen M, Hutten SJ, et al. 2019. Genomewide association study of Parkinson's disease clinical biomarkers in 12 longitudinal

Cite this article as *Cold Spring Harb Perspect Med* doi: 10.1101/cshperspect.a041641

patients' cohorts. *Mov Disord* **34:** 1839–1850. doi:10 .1002/mds.27845

Jankovic J, Aguilar LG. 2008. Current approaches to the treatment of Parkinson's disease. *Neuropsychiatr Dis Treat* **4:** 743–757. doi:10.2147/NDT.S2006

Jankovic J, McDermott M, Carter J, Gauthier S, Goetz C, Golbe L, Huber S, Koller W, Olanow C, Shoulson I. 1990. Variable expression of Parkinson's disease: a baseline analysis of the DATATOP cohort. The Parkinson study group. *Neurology* **40:** 1529–1534. doi:10.1212/ WNL.40.10.1529

Kalia LV, Lang AE. 2015. Parkinson's disease. *Lancet* **386:** 896–912. doi:10.1016/S0140-6736(14)61393-3

Kang UJ, Goldman JG, Alcalay RN, Xie T, Tuite P, Henchcliffe C, Hogarth P, Amara AW, Frank S, Rudolph A, et al. 2016. The BioFIND study: characteristics of a clinically typical Parkinson's disease biomarker cohort. *Mov Disord* **31:** 924–932. doi:10.1002/mds.26613

Kashihara K, Kitayama M. 2023. Time taken for and causes of a decline to Hoehn and Yahr stage 5 in patients with Parkinson's disease. *Intern Med* **62:** 711–716. doi:10 .2169/internalmedicine.8922-21

Kawada T, Anang JBM, Postuma R. 2015. Predictors of dementia in Parkinson disease: a prospective cohort study. *Neurology* **84:** 1285. doi:10.1212/WNL.00000000 00001408

Kempster PA, O'Sullivan SS, Holton JL, Revesz T, Lees AJ. 2010. Relationships between age and late progression of Parkinson's disease: a clinico-pathological study. *Brain* **133:** 1755–1762. doi:10.1093/brain/awq059

Kieburtz. 1996. Effect of lazabemide on the progression of disability in early Parkinson's disease. *Ann Neurol* **40:** 99–107. doi:10.1002/ana.410400116

Kordower JH, Olanow CW, Dodiya HB, Chu Y, Beach TG, Adler CH, Halliday GM, Bartus RT. 2013. Disease duration and the integrity of the nigrostriatal system in Parkinson's disease. *Brain* **136:** 2419–2431. doi:10.1093/ brain/awt192

Kostić VS, Marinković J, Svetel M, Stefanova E, Przedborski S. 2002. The effect of stage of Parkinson's disease at the onset of levodopa therapy on development of motor complications. *Eur J Neurol* **9:** 9–14. doi:10.1046/j.1468-1331 .2002.00346.x

Kremer T, Taylor KI, Siebourg-Polster J, Gerken T, Staempfli A, Czech C, Dukart J, Galasko D, Foroud T, Chahine LM, et al. 2021. Longitudinal analysis of multiple neurotransmitter metabolites in cerebrospinal fluid in early Parkinson's disease. *Mov Disord* **36:** 1972–1978. doi:10.1002/ mds.28608

Kumar N. 2009. The Sydney multicenter study of Parkinson's disease: the inevitability of dementia at 20 years. *Year B Neurol Neurosurg* **2009:** 94–95.

Lawton M, Baig F, Toulson G, Morovat A, Evetts SG, Ben-Shlomo Y, Hu MT. 2020. Blood biomarkers with Parkinson's disease clusters and prognosis: the Oxford discovery cohort. *Mov Disord* **35:** 279–287. doi:10.1002/mds.27888

Lewis MM, Cheng XV, Du G, Zhang L, Li C, De Jesus S, Tabbal SD, Mailman R, Li R, Huang X. 2024. Clinical progression of Parkinson's disease in the early 21st century: insights from AMP-PD dataset. bioRxiv doi:10 .1101/2024.01.29.24301950

Lin CH, Li CH, Yang KC, Lin FJ, Wu CC, Chieh JJ, Chiu MJ. 2019. Blood NfL: a biomarker for disease severity and progression in Parkinson disease. *Neurology* **93:** e1104– e1111. doi:10.1212/WNL.0000000000008088

Lin J, Ou R, Li C, Hou Y, Zhang L, Wei Q, Pang D, Liu K, Jiang Q, Yang T, et al. 2023. Plasma glial fibrillary acidic protein as a biomarker of disease progression in Parkinson's disease: a prospective cohort study. *BMC Med* **21:** 420. doi:10.1186/s12916-023-03120-1

Liu G, Peng J, Liao Z, Locascio JJ, Corvol JC, Zhu F, Dong X, Maple-Grødem J, Campbell MC, Elbaz A, et al. 2021. Genome-wide survival study identifies a novel synaptic locus and polygenic score for cognitive progression in Parkinson's disease. *Nat Genet* **53:** 787–793. doi:10 .1038/s41588-021-00847-6

Lücking CB, Dürr A, Bonifati V, Vaughan J, De Michele G, Gasser T, Harhangi BS, Meco G, Denèfle P, Wood NW, et al. 2000. Association between early-onset Parkinson's disease and mutations in the parkin gene. *N Engl J Med* **342:** 1560–1567. doi:10.1056/NEJM200005253422103

Ma LZ, Zhang C, Wang H, Ma YH, Shen XN, Wang J, Tan L, Dong Q, Yu JT. 2021. Serum neurofilament dynamics predicts cognitive progression in de novo Parkinson's disease. *J Parkinsons Dis* **11:** 1117–1127. doi:10.3233/JPD-212535

Macleod AD, Taylor KSM, Counsell CE. 2014. Mortality in Parkinson's disease: a systematic review and meta-analysis. *Mov Disord* **29:** 1615–1622. doi:10.1002/mds.25898

Magistrelli L, Storelli E, Rasini E, Contaldi E, Comi C, Cosentino M, Marino F. 2020. Relationship between circulating $CD4^+$ T lymphocytes and cognitive impairment in patients with Parkinson's disease. *Brain Behav Immun* **89:** 668–674. doi:10.1016/j.bbi.2020.07.005

Maier Hoehn MM. 1983. Parkinsonism treated with levodopa: progression and mortality. *J Neural Transm Suppl* **19:** 253–264.

Majbour NK, Vaikath NN, Eusebi P, Chiasserini D, Ardah M, Varghese S, Haque ME, Tokuda T, Auinger P, Calabresi P, et al. 2016a. Longitudinal changes in CSF α-synuclein species reflect Parkinson's disease progression. *Mov Disord* **31:** 1535–1542. doi:10.1002/mds.26754

Majbour NK, Vaikath NN, van Dijk KD, Ardah MT, Varghese S, Vesterager LB, Montezinho LP, Poole S, Safieh-Garabedian B, Tokuda T, et al. 2016b. Oligomeric and phosphorylated α-synuclein as potential CSF biomarkers for Parkinson's disease. *Mol Neurodegener* **11:** 7. doi:10 .1186/s13024-016-0072-9

Majbour NK, Abdi IY, Dakna M, Wicke T, Lang E, Ali Moussa HY, Thomas MA, Trenkwalder C, Safieh-Garabedian B, Tokuda T, et al. 2021. Cerebrospinal α-synuclein oligomers reflect disease motor severity in DeNoPa longitudinal cohort. *Mov Disord* **36:** 2048–2056. doi:10 .1002/mds.28611

Majbour N, Aasly J, Abdi I, Ghanem S, Erskine D, van de Berg W, El-Agnaf O. 2022. Disease-associated α-synuclein aggregates as biomarkers of Parkinson disease clinical stage. *Neurology* **99:** e2417–e2427. doi:10.1212/WNL .0000000000201199

Malek N, Weil RS, Bresner C, Lawton MA, Grosset KA, Tan M, Bajaj N, Barker RA, Burn DJ, Foltynie T, et al. 2018. Features of GBA-associated Parkinson's disease at presentation in the UK tracking Parkinson's study. *J Neurol*

Neurosurg Psychiatry **89**: 702–709. doi:10.1136/jnnp-2017-317348

Marek K, Jennings D, Lasch S, Siderowf A, Tanner C, Simuni T, Coffey C, Kieburtz K, Flagg E, Chowdhury S, et al. 2011. The Parkinson progression marker initiative (PPMI). *Prog Neurobiol* **95**: 629–635. doi:10.1016/j.pneurobio.2011.09.005

Marras C, McDermott MP, Rochon PA, Tanner CM, Naglie G, Rudolph A, Lang AE; the Parkinson Study Group. 2005. Survival in Parkinson disease: thirteen-year follow-up of the DATATOP cohort. *Neurology* **64**: 87–93. doi:10.1212/01.WNL.0000148603.44618.19

Martínez Carrasco A, Real R, Lawton M, Hertfelder Reynolds R, Tan M, Wu L, Williams N, Carroll C, Corvol J-C, Hu M, et al. 2023. Genome-wide analysis of motor progression in Parkinson disease. *Neurol Genet* **9**: e200092. doi:10.1212/NXG.0000000000200092

Marttila RJ, Rinne UK, Siirtola T, Sonninen V. 1977. Mortality of patients with Parkinson's disease treated with levodopa. *J Neurol* **216**: 147–153. doi:10.1007/BF00313615

McGeer PL, Itagaki S, Boyes BE, McGeer EG. 1988. Reactive microglia are positive for HLA-DR in the substantia nigra of Parkinson's and Alzheimer's disease brains. *Neurology* **38**: 1285–1291. doi:10.1212/WNL.38.8.1285

McNeill TH, Brown SA, Rafols JA, Shoulson I. 1988. Atrophy of medium spiny I striatal dendrites in advanced Parkinson's disease. *Brain Res* **455**: 148–152. doi:10.1016/0006-8993(88)90124-2

Mitchell T, Lehéricy S, Chiu SY, Strafella AP, Stoessl AJ, Vaillancourt DE. 2021. Emerging neuroimaging biomarkers across disease stage in Parkinson disease: a review. *JAMA Neurol* **78**: 1262–1272. doi:10.1001/jamaneurol.2021.1312

Montastruc JL, Desboeuf K, Lapeyre-Mestre M, Senard JM, Rascol O, Brefel-Courbon C. 2001. Long-term mortality results of the randomized controlled study comparing bromocriptine to which levodopa was later added with levodopa alone in previously untreated patients with Parkinson's disease. *Mov Disord* **16**: 511–514. doi:10.1002/mds.1093

Moreau C, Rouaud T, Grabli D, Benatru I, Remy P, Marques AR, Drapier S, Mariani LL, Roze E, Devos D, et al. 2023. Overview on wearable sensors for the management of Parkinson's disease. *NPJ Parkinsons Dis* **9**: 153. doi:10.1038/s41531-023-00585-y

Ng ASL, Tan YJ, Yong ACW, Saffari SE, Lu Z, Ng EY, Ng SYE, Chia NSY, Choi X, Heng D, et al. 2020. Utility of plasma neurofilament light as a diagnostic and prognostic biomarker of the postural instability gait disorder motor subtype in early Parkinson's disease. *Mol Neurodegener* **15**: 33. doi:10.1186/s13024-020-00385-5

Ngoga D, Mitchell R, Kausar J, Hodson J, Harries A, Pall H. 2014. Deep brain stimulation improves survival in severe Parkinson's disease. *J Neurol Neurosurg Psychiatry* **85**: 17–22. doi:10.1136/jnnp-2012-304715

* Niethammer M, Nakano Y, Eidelberg D. 2025. 20. Imaging of disease-related networks in Parkinson's disease. *Cold Spring Harb Perspect Med* doi: 10.1101/cshperspect.a041841

Nilsson J, Constantinescu J, Nellgård B, Jakobsson P, Brum WS, Gobom J, Forsgren L, Dalla K, Constantinescu R, Zetterberg H, et al. 2023. Cerebrospinal fluid biomarkers of synaptic dysfunction are altered in Parkinson's disease and related disorders. *Mov Disord* **38**: 267–277. doi:10.1002/mds.29287

Niu M, Li Y, Li G, Zhou L, Luo N, Yao M, Kang W, Liu J. 2020. A longitudinal study on α-synuclein in plasma neuronal exosomes as a biomarker for Parkinson's disease development and progression. *Eur J Neurol* **27**: 967–974. doi:10.1111/ene.14208

Nurmi E, Bergman J, Eskola O, Solin O, Vahlberg T, Sonninen P, Rinne JO. 2003. Progression of dopaminergic hypofunction in striatal subregions in Parkinson's disease using [^{18}F]CFT PET. *Synapse* **48**: 109–115. doi:10.1002/syn.10192

Oftedal L, Maple-Grødem J, Dalen I, Tysnes OB, Pedersen KF, Alves G, Lange J. 2023. Association of CSF glucocerebrosidase activity with the risk of incident dementia in patients with Parkinson disease. *Neurology* **100**: e388–e395. doi:10.1212/WNL.0000000000201418

Olanow CW, Schapira AHV, LeWitt PA, Kieburtz K, Sauer D, Olivieri G, Pohlmann H, Hubble J. 2006. TCH346 as a neuroprotective drug in Parkinson's disease: a double-blind, randomised, controlled trial. *Lancet Neurol* **5**: 1013–1020. doi:10.1016/S1474-4422(06)70602-0

Ouchi Y, Yoshikawa E, Sekine Y, Futatsubashi M, Kanno T, Ogusu T, Torizuka T. 2005. Microglial activation and dopamine terminal loss in early Parkinson's disease. *Ann Neurol* **57**: 168–175. doi:10.1002/ana.20338

Pagano G, De Micco R, Yousaf T, Wilson H, Chandra A, Politis M. 2018. REM behavior disorder predicts motor progression and cognitive decline in Parkinson disease. *Neurology* **91**: e894–e905. doi:10.1212/WNL.0000000000006134

Park M, Lee YG. 2024. Association of family history and polygenic risk score with longitudinal prognosis in Parkinson disease. *Neurol Genet* **10**: e200115. doi:10.1212/NXG.0000000000200115

Parkinson Study Group. 1993. Effects of tocopherol and deprenyl on the progression of disability in early Parkinson's disease. *N Engl J Med* **328**: 176–183. doi:10.1056/NEJM199301213280305

Parkinson Study Group. 2004. A controlled, randomized, delayed-start study of rasagiline in early Parkinson disease. *Arch Neurol* **61**: 561–566. doi:10.1001/archneur.61.4.561

Parkkinen L, Pirttilä T, Alafuzoff I. 2008. Applicability of current staging/categorization of α-synuclein pathology and their clinical relevance. *Acta Neuropathol* **115**: 399–407. doi:10.1007/s00401-008-0346-6

Parnetti L, Farotti L, Eusebi P, Chiasserini D, De Carlo C, Giannandrea D, Salvadori N, Lisetti V, Tambasco N, Rossi A, et al. 2014. Differential role of CSF α-synuclein species, tau, and Aβ42 in Parkinson's disease. *Front Aging Neurosci* **6**: 53. doi:10.3389/fnagi.2014.00053

Pasquini J, Brooks DJ, Pavese N. 2021. The cholinergic brain in Parkinson's disease. *Mov Disord Clin Pract* **8**: 1012–1026. doi:10.1002/mdc3.13319

Paul KC, Schulz J, Bronstein JM, Lill CM, Ritz BR. 2018. Association of polygenic risk score with cognitive decline and motor progression in Parkinson disease. *JAMA Neurol* **75**: 360–366. doi:10.1001/jamaneurol.2017.4206

Cite this article as *Cold Spring Harb Perspect Med* doi: 10.1101/cshperspect.a041641

Pigott K, Rick J, Xie SX, Hurtig H, Chen-Plotkin A, Duda JE, Morley JF, Chahine LM, Dahodwala N, Akhtar RS, et al. 2015. Longitudinal study of normal cognition in Parkinson disease. *Neurology* 85: 1276–1282. doi:10.1212/WNL.0000000000002001

Pilotto A, Imarisio A, Conforti F, Scalvini A, Masciocchi S, Nocivelli S, Turrone R, Gipponi S, Cottini E, Borroni B, et al. 2021. Plasma NfL, clinical subtypes and motor progression in Parkinson's disease. *Parkinsonism Relat Disord* 87: 41–47. doi:10.1016/j.parkreldis.2021.04.016

Pinho R, Guedes LC, Soreq L, Lobo PP, Mestre T, Coelho M, Rosa MM, Gonçalves N, Wales P, Mendes T, et al. 2016. Gene expression differences in peripheral blood of Parkinson's disease patients with distinct progression profiles. *PLoS ONE* 11: e0157852. doi:10.1371/journal.pone.0157852

Pirker W, Holler I, Gerschlager W, Asenbaum S, Zettinig G, Brücke T. 2003. Measuring the rate of progression of Parkinson's disease over a 5-year period with β-CIT SPECT. *Mov Disord* 18: 1266–1272. doi:10.1002/mds.10531

Poewe W. 2006. The natural history of Parkinson's disease. *J Neurol* 253(Suppl 7): VII2–VII6.

Poewe W, Mahlknecht P. 2009. The clinical progression of Parkinson's disease. *Parkinsonism Relat Disord* 15(Suppl 4): S28–S32. doi:10.1016/S1353-8020(09)70831-4

Poewe WH, Lees AJ, Stern GM. 1986. Low-dose L-dopa therapy in Parkinson's disease: a 6-year follow-up study. *Neurology* 36: 1528–1530. doi:10.1212/WNL.36.11.1528

Polymeropoulos MH, Lavedan C, Leroy E, Ide SE, Dehejia A, Dutra A, Pike B, Root H, Rubenstein J, Boyer R, et al. 1997. Mutation in the α-synuclein gene identified in families with Parkinson's disease. *Science* 276: 2045–2047. doi:10.1126/science.276.5321.2045

Postuma RB, Gagnon JF, Vendette M, Fantini ML, Massicotte-Marquez J, Montplaisir J. 2009. Quantifying the risk of neurodegenerative disease in idiopathic REM sleep behavior disorder. *Neurology* 72: 1296–1300. doi:10.1212/01.wnl.0000340980.19702.6e

Postuma RB, Bertrand JA, Montplaisir J, Desjardins C, Vendette M, Rios Romenets S, Panisset M, Gagnon JF. 2012. Rapid eye movement sleep behavior disorder and risk of dementia in Parkinson's disease: a prospective study. *Mov Disord* 27: 720–726. doi:10.1002/mds.24939

Postuma RB, Berg D, Stern M, Poewe W, Olanow CW, Oertel W, Obeso J, Marek K, Litvan I, Lang AE, et al. 2015. MDS clinical diagnostic criteria for Parkinson's disease. *Mov Disord* 30: 1591–1601. doi:10.1002/mds.26424

Pringsheim T, Day GS, Smith DB, Rae-Grant A, Licking N, Armstrong MJ, de Bie RMA, Roze E, Miyasaki JM, Hauser RA, et al. 2021. Dopaminergic therapy for motor symptoms in early Parkinson disease practice guideline summary: a report of the AAN guideline subcommittee. *Neurology* 97: 942–957. doi:10.1212/WNL.0000000000012868

Przedborski S, Vila M, Jackson-Lewis V. 2003. Neurodegeneration: what is it and where are we? *J Clin Invest* 111: 3–10. doi:10.1172/JCI200317522

Quinn N, Critchley P, Marsden CD. 1987. Young onset Parkinson's disease. *Mov Disord* 2: 73–91. doi:10.1002/mds.870020201

Rajput AH, Pahwa R, Pahwa P, Rajput A. 1993. Prognostic significance of the onset mode in parkinsonism. *Neurology* 43: 829–830. doi:10.1212/WNL.43.4.829

Raket LL, Oudin Åström D, Norlin JM, Kellerborg K, Martinez-Martin P, Odin P. 2022. Impact of age at onset on symptom profiles, treatment characteristics and health-related quality of life in Parkinson's disease. *Sci Rep* 12: 526. doi:10.1038/s41598-021-04356-8

Razmkon A, Abdollahifard S, Rezaei H, Bahadori AR, Eskandarzadeh P, Rastegar Kazerooni A. 2024. Effect of deep brain stimulation on Parkinson disease dementia: a systematic review and meta-analysis. *Basic Clin Neurosci* 15: 157–164. doi:10.32598/bcn.2021.3420.1

Rentzos M, Nikolaou C, Andreadou E, Paraskevas GP, Rombos A, Zoga M, Tsoutsou A, Boufidou F, Kapaki E, Vassilopoulos D. 2007. Circulating interleukin-15 and RANTES chemokine in Parkinson's disease. *Acta Neurol Scand* 116: 374–379. doi:10.1111/j.1600-0404.2007.00894.x

Saeed U, Compagnone J, Aviv RI, Strafella AP, Black SE, Lang AE, Masellis M. 2017. Imaging biomarkers in Parkinson's disease and parkinsonian syndromes: current and emerging concepts. *Transl Neurodegener* 6: 8. doi:10.1186/s40035-017-0076-6

Sanjari Moghaddam H, Ghazi Sherbaf F, Mojtahed Zadeh M, Ashraf-Ganjouei A, Aarabi MH. 2018. Association between peripheral inflammation and DATSCAN data of the striatal nuclei in different motor subtypes of Parkinson disease. *Front Neurol* 9: 234. doi:10.3389/fneur.2018.00234

Saunders JAH, Estes KA, Kosloski LM, Allen HE, Dempsey KM, Torres-Russotto DR, Meza JL, Santamaria PM, Bertoni JM, Murman DL, et al. 2012. CD4+ regulatory and effector/memory T cell subsets profile motor dysfunction in Parkinson's disease. *J Neuroimmune Pharmacol* 7: 927–938. doi:10.1007/s11481-012-9402-z

Schaeffer E, Kluge A, Böttner M, Zunke F, Cossais F, Berg D, Arnold P. 2020. α-Synuclein connects the gut-brain axis in Parkinson's disease patients—a view on clinical aspects, cellular pathology and analytical methodology. *Front Cell Dev Biol* 8: 573696. doi:10.3389/fcell.2020.573696

Schrag A, Quinn N. 2000. Dyskinesias and motor fluctuations in Parkinson's disease. A community-based study. *Brain* 123: 2297–2305. doi:10.1093/brain/123.11.2297

Schrag A, Schott JM. 2006. Epidemiological, clinical, and genetic characteristics of early-onset parkinsonism. *Lancet Neurol* 5: 355–363. doi:10.1016/S1474-4422(06)70411-2

Schrag A, Dodel R, Spottke A, Bornschein B, Siebert U, Quinn NP. 2007. Rate of clinical progression in Parkinson's disease. A prospective study. *Mov Disord* 22: 938–945. doi:10.1002/mds.21429

Schulz JB, Skalej M, Wedekind D, Luft AR, Abele M, Voigt K, Dichgans J, Klockgether T. 1999. Magnetic resonance imagingbased volumetry differentiates idiopathic Parkinson's syndrome from multiple system atrophy and progressive supranuclear palsy. *Ann Neurol* 45: 65–74. doi:10.1002/1531-8249(199901)45:1<65::AID-ART12>3.0.CO;2-1

Shimada H, Hirano S, Shinotoh H, Aotsuka A, Sato K, Tanaka N, Ota T, Asahina M, Fukushi K, Kuwabara S, et al.

2009. Mapping of brain acetylcholinesterase alterations in Lewy body disease by PET. *Neurology* **73:** 273–278. doi:10.1212/WNL.0b013e3181ab2b58

Shoulson I; Parkinson Study Group. 1998. Mortality in DATATOP: a multicenter trial in early Parkinson's disease. *Ann Neurol* **43:** 318–325. doi:10.1002/ana.410430309

Shults CW. 2002. Effects of coenzyme Q10 in early Parkinson disease. *Arch Neurol* **59:** 1541. doi:10.1001/archneur.59.10.1541

Siderowf A, Xie SX, Hurtig H, Weintraub D, Duda J, Chen-Plotkin A, Shaw LM, Van Deerlin V, Trojanowski JQ, Clark C. 2010. CSF amyloid β 1-42 predicts cognitive decline in Parkinson disease. *Neurology* **75:** 1055–1061. doi:10.1212/WNL.0b013e3181f39a78

Siderowf A, Concha-Marambio L, Lafontant DE, Farris CM, Ma Y, Urenia PA, Nguyen H, Alcalay RN, Chahine LM, Foroud T, et al. 2023. Assessment of heterogeneity among participants in the Parkinson's progression markers initiative cohort using α-synuclein seed amplification: a cross-sectional study. *Lancet Neurol* **22:** 407–417. doi:10.1016/S1474-4422(23)00109-6

Simuni T, Caspell-Garcia C, Coffey C, Lasch S, Tanner C, Marek K. 2016. How stable are Parkinson's disease subtypes in de novo patients: analysis of the PPMI cohort? *Parkinsonism Relat Disord* **28:** 62–67. doi:10.1016/j.parkreldis.2016.04.027

Singleton AB, Farrer M, Johnson J, Singleton A, Hague S, Kachergus J, Hulihan M, Peuralinna T, Dutra A, Nussbaum R, et al. 2003. α-synuclein locus triplication causes Parkinson's disease. *Science* **302:** 841. doi:10.1126/science.1090278

Sommer A, Marxreiter F, Krach F, Fadler T, Grosch J, Maroni M, Graef D, Eberhardt E, Riemenschneider MJ, Yeo GW, et al. 2019. Th17 lymphocytes induce neuronal cell death in a human iPSC-based model of Parkinson's disease. *Cell Stem Cell* **24:** 1006. doi:10.1016/j.stem.2019.04.019

Sommerauer M, Fedorova TD, Hansen AK, Knudsen K, Otto M, Jeppesen J, Frederiksen Y, Blicher JU, Geday J, Nahimi A, et al. 2018. Evaluation of the noradrenergic system in Parkinson's disease: an 11C-MeNER PET and neuromelanin MRI study. *Brain* **141:** 496–504. doi:10.1093/brain/awx348

Stefani A, Pierantozzi M, Olivola E, Galati S, Cerroni R, D'Angelo V, Hainsworth AH, Saviozzi V, Fedele E, Liguori C. 2017. Homovanillic acid in CSF of mild stage Parkinson's disease patients correlates with motor impairment. *Neurochem Int* **105:** 58–63. doi:10.1016/j.neuint.2017.01.007

Sterling NW, Wang M, Zhang L, Lee EY, Du G, Lewis MM, Styner M, Huang X. 2016. Stage-dependent loss of cortical gyrification as Parkinson disease "unfolds." *Neurology* **86:** 1143–1151. doi:10.1212/WNL.0000000000002492

Stewart T, Sossi V, Aasly JO, Wszolek ZK, Uitti RJ, Hasegawa K, Yokoyama T, Zabetian CP, Leverenz JB, Stoessl AJ, et al. 2015. Phosphorylated α-synuclein in Parkinson's disease: correlation depends on disease severity. *Acta Neuropathol Commun* **3:** 7. doi:10.1186/s40478-015-0185-3

Stoker TB, Camacho M, Winder-Rhodes S, Liu G, Scherzer CR, Foltynie T, Evans J, Breen DP, Barker RA, Williams-Gray CH. 2020. Impact of GBA1 variants on long-term clinical progression and mortality in incident Parkinson's disease. *J Neurol Neurosurg Psychiatry* **91:** 695–702. doi:10.1136/jnnp-2020-322857

Stuendl A, Kunadt M, Kruse N, Bartels C, Moebius W, Danzer KM, Mollenhauer B, Schneider A. 2016. Induction of α-synuclein aggregate formation by CSF exosomes from patients with Parkinson's disease and dementia with Lewy bodies. *Brain* **139:** 481–494. doi:10.1093/brain/awv346

Sun C, Zhao Z, Yu W, Mo M, Song C, Si Y, Liu Y. 2019. Abnormal subpopulations of peripheral blood lymphocytes are involved in Parkinson's disease. *Ann Transl Med* **7:** 637. doi:10.21037/atm.2019.10.105

Takahashi H, Watanabe Y, Tanaka H, Mochizuki H, Kato H, Hatazawa J, Tomiyama N. 2019. Quantifying the severity of Parkinson disease by use of dopaminergic neuroimaging. *AJR Am J Roentgenol* **213:** 163–168. doi:10.2214/AJR.18.20655

Tan MMX, Lawton MA, Jabbari E, Reynolds RH, Iwaki H, Blauwendraat C, Kanavou S, Pollard MI, Hubbard L, Malek N, et al. 2021. Genome-wide association studies of cognitive and motor progression in Parkinson's disease. *Mov Disord* **36:** 424–433. doi:10.1002/mds.28342

Tang P, Chong L, Li X, Liu Y, Liu P, Hou C, Li R. 2014. Correlation between serum RANTES levels and the severity of Parkinson's disease. *Oxid Med Cell Longev* **2014:** 208408.

Tang Y, Han L, Li S, Hu T, Xu Z, Fan Y, Liang X, Yu H, Wu J, Wang J. 2023. Plasma GFAP in Parkinson's disease with cognitive impairment and its potential to predict conversion to dementia. *NPJ Parkinsons Dis* **9:** 23. doi:10.1038/s41531-023-00447-7

Terrelonge M Jr, Marder KS, Weintraub D, Alcalay RN. 2016. CSF β-amyloid 1-42 predicts progression to cognitive impairment in newly diagnosed Parkinson disease. *J Mol Neurosci* **58:** 88–92. doi:10.1007/s12031-015-0647-x

Tinaz S, Courtney MG, Stern CE. 2011. Focal cortical and subcortical atrophy in early Parkinson's disease. *Mov Disord* **26:** 436–441. doi:10.1002/mds.23453

Tompkins MM, Hill WD. 1997. Contribution of somal Lewy bodies to neuronal death. *Brain Res* **775:** 24–29. doi:10.1016/S0006-8993(97)00874-3

Verschuur CVM, Suwijn SR, Boel JA, Post B, Bloem BR, van Hilten JJ, van Laar T, Tissingh G, Munts AG, Deuschl G, et al. 2019. Randomized delayed-start trial of levodopa in Parkinson's disease. *N Engl J Med* **380:** 315–324. doi:10.1056/NEJMoa1809983

Vijiaratnam N, Foltynie T. 2023. How should we be using biomarkers in trials of disease modification in Parkinson's disease? *Brain* **146:** 4845–4869. doi:10.1093/brain/awad265

Vijiaratnam N, Lawton M, Heslegrave AJ, Guo T, Tan M, Jabbari E, Real R, Woodside J, Grosset K, Chelban V, et al. 2022. Combining biomarkers for prognostic modelling of Parkinson's disease. *J Neurol Neurosurg Psychiatry* **93:** 707–715. doi:10.1136/jnnp-2021-328365

Wang Y, Shi M, Chung KA, Zabetian CP, Leverenz JB, Berg D, Srulijes K, Trojanowski JQ, Lee VMY, Siderowf AD, et al. 2012. Phosphorylated α-synuclein in Parkinson's disease. *Sci Transl Med* **4:** 121ra20.

Wang H, Atik A, Stewart T, Ginghina C, Aro P, Kerr KF, Seibyl J, Jennings D, Investigators PARS, Jensen PH, et al. 2018. Plasma α-synuclein and cognitive impairment in the Parkinson's associated risk syndrome: a pilot study. *Neurobiol Dis* **116**: 53–59. doi:10.1016/j.nbd.2018.04.015

Weintraub D, Siderowf AD, Potenza MN, Goveas J, Morales KH, Duda JE, Moberg PJ, Stern MB. 2006. Association of dopamine agonist use with impulse control disorders in Parkinson disease. *Arch Neurol* **63**: 969–973. doi:10.1001/archneur.63.7.969

Weintraub D, Posavi M, Fontanillas P, Tropea TF, Mamikonyan E, Suh E, Trojanowski JQ, Cannon P, Van Deerlin VM, 23andMe Research Team, et al. 2022. Genetic prediction of impulse control disorders in Parkinson's disease. *Ann Clin Transl Neurol* **9**: 936–949. doi:10.1002/acn3.51569

* Westenberger A, Brüggemann N, Klein C. 2024. Genetics of Parkinson's disease: from causes to treatment. *Cold Spring Harb Perspect Med* **12**: a041774. doi:10.1101/cshperspect.a041774

Wieler M, Gee M, Camicioli R, Martin WRW. 2016. Freezing of gait in early Parkinson's disease: nigral iron content estimated from magnetic resonance imaging. *J Neurol Sci* **361**: 87–91. doi:10.1016/j.jns.2015.12.008

Williams-Gray CH, Wijeyekoon R, Yarnall AJ, Lawson RA, Breen DP, Evans JR, Cummins GA, Duncan GW, Khoo TK, Burn DJ, et al. 2016. Serum immune markers and disease progression in an incident Parkinson's disease cohort (ICICLE-PD). *Mov Disord* **31**: 995–1003. doi:10.1002/mds.26563

Yao C, Fereshtehnejad SM, Dawson BK, Pelletier A, Gan-Or Z, Gagnon JF, Montplaisir JY, Postuma RB. 2018. Long-standing disease-free survival in idiopathic REM sleep behavior disorder: is neurodegeneration inevitable? *Parkinsonism Relat Disord* **54**: 99–102. doi:10.1016/j.parkreldis.2018.04.010

Ye R, Locascio JJ, Goodheart AE, Quan M, Zhang B, Gomperts SN. 2021. Serum NFL levels predict progression of motor impairment and reduction in putamen dopamine transporter binding ratios in de novo Parkinson's disease: an 8-year longitudinal study. *Parkinsonism Relat Disord* **85**: 11–16. doi:10.1016/j.parkreldis.2021.02.008

Ygland Rödström E, Mattsson-Carlgren N, Janelidze S, Hansson O, Puschmann A. 2022. Serum neurofilament light chain as a marker of progression in Parkinson's disease: long-term observation and implications of clinical subtypes. *J Parkinsons Dis* **12**: 571–584. doi:10.3233/JPD-212866

Zaja-Milatovic S, Milatovic D, Schantz AM, Zhang J, Montine KS, Samii A, Deutch AY, Montine TJ. 2005. Dendritic degeneration in neostriatal medium spiny neurons in Parkinson disease. *Neurology* **64**: 545–547. doi:10.1212/01.WNL.0000150591.33787.A4

Zeighami Y, Fereshtehnejad S-M, Dadar M, Collins DL, Postuma RB, Dagher A. 2019. Assessment of a prognostic MRI biomarker in early de novo Parkinson's disease. *NeuroImage Clin* **24**: 101986. doi:10.1016/j.nicl.2019.101986

Zhuang X, Mazzoni P, Kang UJ. 2013. The role of neuroplasticity in dopaminergic therapy for Parkinson disease. *Nat Rev Neurol* **9**: 248–256. doi:10.1038/nrneurol.2013.57

Neuropathology of Parkinson's Disease and Parkinsonism

Dennis W. Dickson

Department of Neuroscience, Mayo Clinic, Jacksonville, Florida 32224, USA

Correspondence: dickson.dennis@mayo.edu

Parkinsonism, the clinical term for a disorder with prominent bradykinesia and variably associated extrapyramidal signs and symptoms, is virtually always accompanied by degeneration of the nigrostriatal dopaminergic system, with neuronal loss and gliosis in the substantia nigra at autopsy. Neuronal loss is particularly marked in the ventrolateral cell groups of the substantia nigra, which project to the putamen via the nigrostriatal pathway. Parkinsonism is pathologically heterogeneous, with the most common pathologic substrates related to abnormalities in the presynaptic protein α-synuclein or the microtubule-binding protein tau. In idiopathic Parkinson's disease (PD), α-synuclein accumulates in neuronal perikarya (Lewy bodies) and neuronal processes (Lewy neurites). The disease process is multifocal and involves select central nervous system neurons, as well as neurons in the peripheral autonomic nervous system. The particular set of neurons affected determines nonmotor clinical presentations. Multiple system atrophy (MSA) is the other major α-synucleinopathy. It is also associated with autonomic dysfunction and in some cases with cerebellar signs. The hallmark histopathologic feature of MSA is an accumulation of α-synuclein within glial cytoplasmic inclusions (GCIs). The most common of the Parkinsonian tauopathies is progressive supranuclear palsy (PSP), which is clinically associated with severe postural instability leading to early falls. The tau pathology of PSP also affects both neurons and glia.

Given the population frequency of Parkinson's disease (PD), α-synuclein pathology similar to that in PD, but not accompanied by neuronal loss, is common (10% of people older than 65 years of age) in neurologically normal individuals. This is often referred to as incidental Lewy body disease. Braak and coworkers proposed a staging scheme for PD progression based upon analysis of brains of individuals with a range of clinical presentations, including neurologically normal people and those with end-stage Parkinsonism. Although MSA-like and progressive supranuclear palsy (PSP)-like pathology can be detected in neurologically normal individuals, such cases are too infrequent to permit assessment of patterns of disease progression, comparable to those for PD.[1]

PD is a progressive neurological disorder defined by a characteristic clinical syndrome with bradykinesia, tremor, rigidity, and postural

---

[1]This is an update to a previous article published in *Cold Spring Harbor Perspectives in Medicine* [Dickson (2012). *Cold Spring Harb Perspect Med* **2**: a009258. doi:10.1101/cshperspect.a009258].

Copyright © 2025 Cold Spring Harbor Laboratory Press; all rights reserved
Cite this article as *Cold Spring Harb Perspect Med* doi: 10.1101/cshperspect.a041610

instability. There are a large number of different disorders that can have some or all of these clinical features, and the clinical syndrome is referred to as "parkinsonism." Disorders with parkinsonism as a prominent feature are referred to as "parkinsonian" disorders (Table 1). Some parkinsonian disorders are chronic and progressive and caused by an unknown degenerative disease process, whereas others may have a clear genetic cause, such as those driven by autosomal-dominant mutations in the gene for α-synuclein (*SNCA*) or leucine-rich repeat kinase 2 (*LRRK2*). There are a number of other less common genetic causes of parkinsonism, some of which are autosomal recessive (for review, see Funayama et al. 2023). Others can be transient and caused by effects of toxins, metabolic disturbances, or drugs. The latter may have no telltale signs with standard pathological methods and can be considered "functional" rather than "structural" disorders. Some toxins that cause parkinsonism (e.g., MPTP-induced parkinsonism) produce lasting brain damage that leaves structural changes. It is important to emphasize that it is not possible to diagnose parkinsonism with neuropathologic methods; it is only possible to describe pathologic findings—histologic, neurochemical, and molecular—that are frequently associated with parkinsonism.

Table 1. Parkinsonian disorders

Classification	Examples
Degenerative parkinsonism	
α-Synuclein	Parkinson's disease, multiple system atrophy
Tau	Progressive supranuclear palsy, corticobasal degeneration
TDP-43	Frontotemporal degeneration, Perry syndrome
Nondegenerative parkinsonism	
Vascular	Vascular parkinsonism
Toxic	MPTP, manganese poisoning
Drug-induced	Antipsychotic medications
Infectious	Influenza virus (postencephalitic parkinsonism)

Table reprinted and modified from Dickson (2012), © Cold Spring Harbor Laboratory Press.

Degenerative parkinsonian disorders can be inherited or sporadic, but are all characterized by neuronal loss in selective populations of vulnerable neurons. The common denominator of all degenerative parkinsonian disorders is the loss of dopaminergic neurons of the substantia nigra that project to the putamen (i.e., dopaminergic nigrostriatal pathway) leading loss of pigment due to neuromelanin in the substantia nigra (Fig. 1). Degenerative diseases can be classified a number of ways, but increasingly they are classified based on molecular pathomechanisms. Many of the most common degenerative diseases have onset in mid-to-late life and have pathologic accumulations of abnormal conformers of normal cellular proteins in select vulnerable neuronal populations. Most degenerative parkinsonian disorder fall into one of two molecular classes—tauopathies and α-synucleinopathies—based on pathologic accumulation of the microtubule-associated protein tau or the presynaptic protein α-synuclein within vulnerable neurons and often within glial cells.

PD is a degenerative parkinsonian disorder characterized by neuronal inclusions composed of α-synuclein. The inclusions are located in neuronal perikarya and referred to as Lewy bodies (Fig. 2) or in neuronal cell processes and referred to as Lewy neurites. The combination of Lewy bodies and Lewy neurites is sometimes referred to as "Lewy-related pathology" (Dickson et al. 2009), because it is increasingly clear that abnormal α-synuclein accumulation in neuronal perikarya may be the tip of the iceberg, with evidence of accumulation of α-synuclein not only in neuronal cell processes (Irizarry et al. 1998) but also with the synaptic compartment (Kramer and Schulz-Schaeffer 2007; Muntané et al. 2008; Schulz-Schaeffer 2010).

The other major degenerative parkinsonian disorder characterized by inclusions composed of α-synuclein is multiple system atrophy (MSA), a parkinsonian disorder that affects not only the nigrostriatal dopaminergic pathway, but also the cerebellar afferent pathology (pontocerebellar and olivocerebellar fibers). Neuronal inclusions in MSA (Fig. 3), however, are a minor component of the pathology. In contrast, α-synuclein inclusions within the

Cite this article as *Cold Spring Harb Perspect Med* doi: 10.1101/cshperspect.a041610

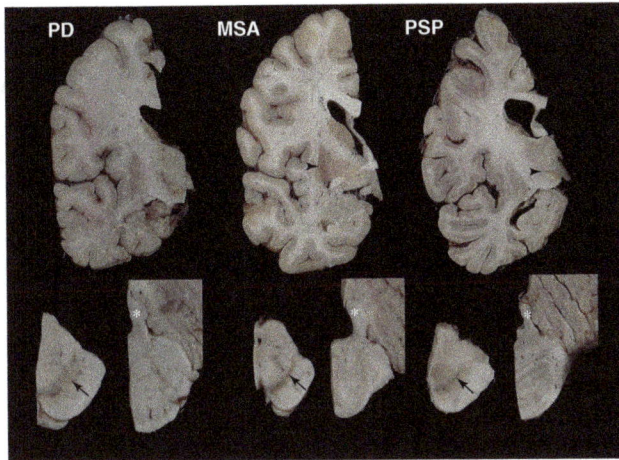

Figure 1. Coronal sections of cerebrum in Parkinson's disease (PD), multiple system atrophy (MSA), and progressive supranuclear palsy (PSP), as well as transverse sections of the brainstem at the level of the midbrain and the pons, the latter at the level of the superior cerebellar peduncle. The lateral ventricle is slightly enlarged in all three disorders, most marked in PSP. There is striking atrophy of the subthalamic nucleus in PSP (arrowhead) and atrophy and red-brown discoloration of the putamen in MSA. Both subthalamic and putamen are normal in PD. All three disorders have a loss of dark neuromelanin pigment in the substantia nigra (black arrows), which is more marked in PD and MSA than in PSP. There is also variable atrophy of the superior cerebellar peduncle (asterisks), which is marked in PSP, but minimal in PD and MSA.

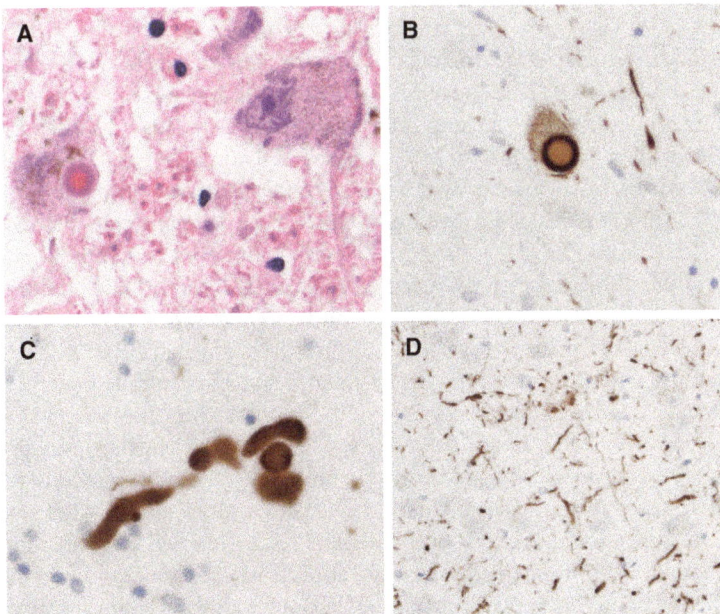

Figure 2. Parkinson's disease (PD) pathology. (*A*) Hematoxylin and eosin stain of the substantia nigra in PD shows a hyaline-type Lewy body in a neuromelanin-containing neuron. (*B*) α-Synuclein immunohistochemistry shows a Lewy body and Lewy neurites. (*C*) α-Synuclein immunohistochemistry shows intra-axonal Lewy bodies in the brainstem. (*D*) α-Synuclein immunohistochemistry shows many neurites in the CA2/3 sector of the hippocampus in PD.

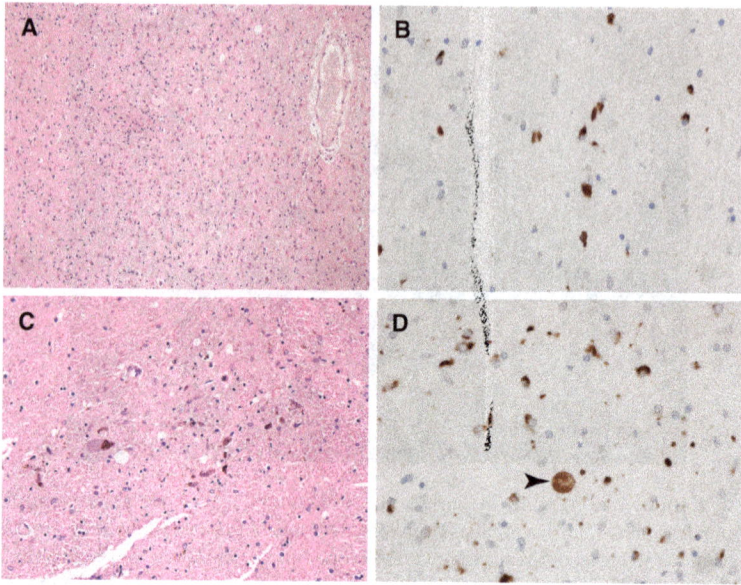

Figure 3. Multiple system atrophy (MSA) pathology. (*A*) Low power image of hematoxylin and eosin stained putamen shows severe neuronal loss and astrocytic gliosis. (*B*) α-Synuclein immunohistochemistry of the white matter in MSA shows many glial cytoplasmic inclusions (GCIs). (*C*) High power image of the ventrolateral part of the substantia nigra shows neuronal loss and gliosis with extraneuronal neuromelanin pigment, as well as brown granular iron-type pigment. (*D*) α-Synuclein immunohistochemistry of the pontine base shows many GCIs and a few neuronal cytoplasmic inclusions (arrowhead).

cytoplasm of oligodendroglial cells, so-called glial cytoplasmic inclusions (GCIs) (Lantos 1998), are the major pathologic finning in MSA (Fig. 3). Interestingly, neurons in MSA may also have α-synuclein inclusions within their nuclei (Fig. 3; Lin et al. 2004), a feature not seen in affected neurons in PD.

The most common of the degenerative parkinsonian disorders associated with neuronal inclusions composed of tau protein is PSP, in which there are also tau inclusions within glial cells (both astrocytes and oligodendrocytes) (Fig. 4; Dickson et al. 2007). PSP is sometimes referred to as a "parkinsonism-plus" disorder in that the clinical features consistently include other neurologic features not clearly related to parkinsonism, such as eye movement disorder and dementia (Steele et al. 1964). This corresponds to the involvement of brain regions beyond the dopaminergic neurons of the nigrostriatal pathway. MSA is also a "parkinsonism-plus" disorder because patients invariably have evidence of autonomic dysfunction and

often signs of cerebellar dysfunction, such as nystagmus and ataxia. The concept of MSA as a "parkinsonism-plus" disorder has become increasingly muddled as it clear that most patients with PD also have non-parkinsonian clinical features, such as autonomic dysfunction, sleep disorders, and even dementia in late stages of the disease (Langston 2006). Indeed, Lewy bodies and Lewy neurites are not confined to the nigrostriatal system in PD but can be widespread in peripheral and central autonomic neurons, even the cerebral cortex (Braak and Del Tredici 2009).

The focus of this article will be on the pathology of the most common of the degenerative parkinsonian disorders—PD, MSA, and PSP. The discussion will compare and contrast their clinical, macroscopic, and microscopic features. The rationale for limiting the discussion to these disorders is that in an autopsy series of parkinsonism, these are the most common diagnoses and clinically the most important differential diagnosis. Another common cause of parkinsonism found in autopsy series is cerebrovascu-

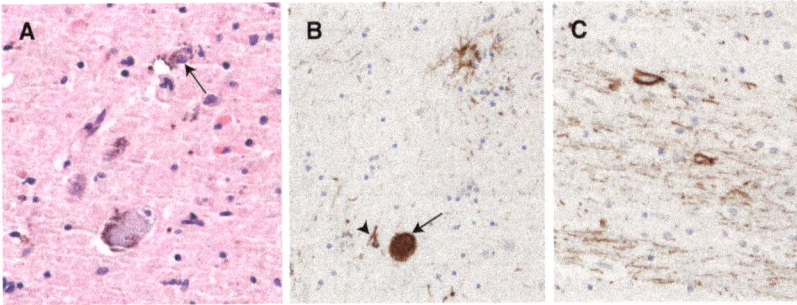

Figure 4. Progressive supranuclear palsy (PSP) pathology. (*A*) Globus neurofibrillary tangle in the substantia nigra. Note extraneuronal neuromelanin (arrow). (*B*) Tau immunohistochemistry of putamen shows a globus neurofibrillary tangle (arrow), an oligodendroglial coiled body (arrowhead), and a tufted astrocyte.

lar disease, which produces a disorder referred to as "vascular parkinsonism" (Zijlmans et al. 2004; Kalaria 2018). Given the heterogeneity of cerebrovascular pathology (e.g., infarcts, hemorrhages, and white matter pathology) and the neuroanatomical distribution of these lesions (e.g., basal ganglia, thalamus, and brainstem), the clinical and pathological criteria of vascular parkinsonism are not well established. The last section provides a brief overview of pathologic features of other degenerative parkinsonian disorders.

CLINICAL COMPARISON OF PD, MSA, AND PSP

Characteristic Clinical Features of PD

PD can be diagnosed with considerable accuracy, particularly by neurologists specializing in diagnosis and management of movement disorders (Hughes et al. 2002), when robust clinical criteria are used such as those of the Queen Square Parkinson Disease Brain Bank, which has inclusion criteria (bradykinesia and at least one of rigidity, tremor, or postural instability) and exclusion criteria (the absence of strokes, head injury encephalitis, neuroleptic treatment, supranuclear gaze palsy, cerebellar signs, early severe autonomic dysfunction, early dementia pyramidal tract, exposure to toxins signs) as well as the presence of supportive features (chronic progressive disease course, unilateral onset and asymmetry of signs during disease course, excellent and prolonged response to levodopa, late levodopa-induced dyskinesia). Asymmetry is an important supportive feature in that the other major degenerative parkinsonian disorders MSA and PSP are usually symmetrical. Response to dopamine replacement therapy (e.g., levodopa or dopamine agonists) is typical of PD, whereas MSA and PSP have a limited response to such therapy. The exclusion criteria also include the absence of family history of movement disorder, but this criterion is often ignored today given increasing evidence of genetic determinants of PD (Farrer 2006; Blauwendraat et al. 2020). Some of the exclusion criteria are meant to rule out PSP (e.g., supranuclear gaze palsy) and MSA (e.g., cerebellar signs and early severe autonomic dysfunction) (Table 2), but it is increasingly recognized that autonomic dysfunction is common in PD and that it may also be an early feature of the disease, given recent evidence that peripheral nerves and ganglia of the autonomic nervous system are affected early in PD and may actually be affected prior to significant brain involvement (Langston 2006; Lang 2007). It is of interest that epidemiologic studies indicate that autonomic symptoms may precede clinical PD by more than a decade (Abbott et al. 2001). Another clinical syndrome that may be a harbinger of PD is rapid eye movement behavior disorder (RBD), a condition that appears a number of years before PD (Schenck et al. 1996). The RBD syndrome appears to have its anatomic origins within lower brainstem nuclei (Kayama and Koyama 2003) that are consistently affected in PD. Olfactory

D.W. Dickson

Table 2. Clinical differential diagnoses

	Parkinson's disease (PD)	Multiple system atrophy (MSA)	Progressive supranuclear palsy (PSP)
Asymmetrical motor signs	Almost always	Sometimes	Sometimes
Levodopa response	Almost always	Sometimes	Sometimes
Autonomic dysfunction	Sometimes	Always	Almost never
Dementia	Often late in disease	Not common	Subcortical type

Table reprinted and modified from Dickson (2012), © Cold Spring Harbor Laboratory Press.

dysfunction is common in PD (Hawkes et al. 1997), and it may also precede motor symptoms (Berendse et al. 2001). The late stages of PD have involvement of the cerebral cortex and at this state of the disease, PD is characterized by cognitive dysfunction or frank dementia, referred to as PD dementia. PD dementia is distinguished from dementia with Lewy bodies (McKeith et al. 2004), in which dementia is an early and prominent clinical feature (Table 3).

Characteristic Clinical Features of MSA

MSA is a nonheritable neurodegenerative disease characterized by parkinsonism, cerebellar ataxia, and idiopathic orthostatic hypotension (also known as Shy–Drager syndrome), a syndrome complex first recognized by Oppenheimer, who noted overlap in the pathology of sporadic olivopontocerebellar atrophy (OPCA) and striatonigral degeneration (SND) (Oppenheimer 1976). Depending on the predominant signs and symptoms, MSA is subdivided into MSA-C, for those with predominant degeneration in cerebellar circuitry and ataxia, and MSA-P, for those with predominant degeneration in the basal ganglia with parkinsonism (Gilman et al. 1999). Autonomic dysfunction is required for the clinical diagnosis of MSA, but as noted above autonomic dysfunction can also be seen in PD, while it is rare in PSP.

Characteristic Clinical Features of PSP

One of the earliest clinical features of PSP is unexplained falls. Eventually, most patients with PSP develop postural instability, vertical gaze paresis, nuchal and axial rigidity, and dys-

arthria. Despite many differences in clinical presentation, it is not uncommon for an individual to carry a diagnosis of PD for years before a correct diagnosis of PSP is made (Rajput et al. 1991; Josephs and Dickson 2003). Recently, it has been suggested that a subset of cases of pathologically confirmed PSP have parkinsonism with many similarities to PD, including asymmetry, tremor, and partial response to levodopa, PSP-P (Williams et al. 2005). Many patients with PSP have cognitive problems or dementia, but this does not help to differentiate PSP from PD, because late in the disease process PD patients also frequently develop dementia (Hely et al. 2008), and even early in the disease, PD patients may have mild cognitive deficits compared to healthy individuals. On the other hand, most MSA patients have better preservation of cognition.

NEUROPATHOLOGY OF PARKINSONISM

Macroscopic Pathology PD, MSA, PSP

PD is often unremarkable, with mild frontal atrophy in some cases. There is no significant atrophy of brainstem, and this can be useful in the differential diagnosis PSP and MSA, in which there is midbrain atrophy in PSP and pontine atrophy in MSA. Sections of the brainstem usually reveal loss of the normally dark black pigment in the substantia nigra (Fig. 1) and locus ceruleus, but pigment loss in the substantia nigra is also characteristic of PSP and MSA. The loss of pigmentation correlates with neuronal loss of dopaminergic neurons in the substantia nigra and noradrenergic neurons in the locus

Table 3. Pathologic comparison of Parkinson's disease (PD), multiple system atrophy (MSA), and progressive supranuclear palsy (PSP)

Region	PD	MSA	PSP
Amygdala	Consistent/severe	Spared	Spared
Hippocampus	Variable/moderate	Spared	Spared
Temporal cortex	Variable/moderate	Spared	Spared
Cingulate cortex	Variable/moderate	Uncommon/mild	Spared
Superior frontal gyrus	Uncommon/mild	Spared	Variable/moderate
Motor cortex	Spared	Variable/moderate	Consistent/severe
Caudate/putamen	Uncommon/mild	Consistent/severe	Consistent/severe
Globus pallidus	Spared	Variable/moderate	Consistent/severe
Basal nucleus of Meynert	Consistent/severe	Uncommon/mild	Consistent/severe
Hypothalamus	Consistent/severe	Uncommon/mild	Consistent/severe
Thalamus	Spared	Uncommon/mild	Consistent/severe
Subthalamic nucleus	Spared	Spared	Consistent/severe
Red nucleus	Spared	Spared	Variable/moderate
Substantia nigra	Consistent/severe	Consistent/severe	Consistent/severe
Oculomotor complex	Variable/moderate	Spared	Consistent/severe
Midbrain tectum	Spared	Spared	Consistent/severe
Locus ceruleus	Consistent/severe	Uncommon/mild	Consistent/severe
Pontine tegmentum	Variable/moderate	Uncommon/mild	Consistent/severe
Pontine nuclei	Spared	Consistent/severe	Variable/moderate
Medullary tegmentum	Consistent/severe	Consistent/severe	Consistent/severe
Inferior olivary nucleus	Spared	Consistent/severe	Variable/moderate
Dentate nucleus	Spared	Spared	Consistent/severe
Cerebellar white matter	Spared	Consistent/severe	Uncommon/mild

Table reprinted and modified from Dickson (2012), © Cold Spring Harbor Laboratory Press.

ceruleus. Pigment loss in the locus ceruleus is consistent in PD, but less predictable in PSP and MSA. MSA-P has atrophy and brownish discoloration of the posterolateral putamen (Fig. 1), the brown color correlating with increased iron pigment. In cases with significant cerebellar signs, there is also atrophy of the pontine base and atrophy and gray discoloration of the cerebellar white matter. More subtle atrophy is noted in the medulla (e.g., inferior olive) and the cerebellar cortex. PSP has mild frontal cortical atrophy and often-marked atrophy of the midbrain. The latter is uncommon in PD and MSA. The cerebellar dentate nucleus usually has atrophy and discoloration of the white matter in the dentate hilus, with similar atrophy and discoloration in the cerebellar outflow pathway. This produces marked atrophy of the superior

cerebellar peduncle (Fig. 1). The basal ganglia and thalamus are usually macroscopically unremarkable, but the subthalamic nucleus is almost always smaller than normal and often discolored (Fig. 1). The subthalamic nucleus and the superior cerebellar peduncle are not affected in PD or MSA (Fig. 1).

MICROSCOPIC PATHOLOGY

Lewy Bodies and Lewy-Related Pathology in PD

Classical Lewy bodies have a hyaline appearance on hematoxylin and eosin (H&E), whereas α-synuclein immunoreactive inclusions in less vulnerable neuronal populations, such as the amygdala and cortex, are pale staining and poorly circumscribed (Fig. 2). These lesions are referred to as "cortical Lewy bodies" (Ikeda et al. 1978). A related pale staining neuronal cytoplasmic inclusion found in pigmented brainstem neurons of the substantia nigra and locus ceruleus is the "pale body" (Pappolla et al. 1988; Dale et al. 1992). Evidence suggests that cortical Lewy bodies and pale bodies may be early cytologic alterations that precede the classical Lewy bodies, so-called pre–Lewy bodies. In some cases with severe pathology, hyaline-type inclusions consistent with classical Lewy bodies can be detected in the amygdala and cortex, particularly the limbic cortex. Although most of the α-synuclein immunoreactive cytopathology in PD is within neurons, α-synuclein immunoreactive glia, particularly oligodendroglia, can be detected in small numbers in the midbrain and basal ganglia (Wakabayashi and Takahashi 1996; Wakabayashi et al. 2000). At the ultrastructural level, Lewy bodies are composed of dense granular material and straight filaments that are ∼10–15 nm in diameter (Forno 1969; Tiller-Borcich and Forno 1988; Galloway et al. 1992). Similar filaments can be created in the test tube with recombinant α-synuclein (Conway et al. 2000; Crowther et al. 2000). The presence of α-synuclein in cytoplasmic inclusions represents aberrant cytologic localization, because α-synuclein is normally a component of presynaptic terminals. The factors that give rise to the abnormal conformation remain to be determined, but several posttranslational modifications, including phosphorylation, truncation, and oxidative damage are implicated (for review, see Dickson 2001). The composition of the dense granular material in Lewy bodies is unknown, but perhaps related to other components that have been shown to be present in Lewy bodies. Antibodies to neurofilament (Galvin et al. 1997), ubiquitin (Kuzuhara et al. 1988), and the ubiquitin-binding protein p62 (Kuusisto et al. 2003) are among the most consistently detected proteins in Lewy bodies. A subset of Lewy bodies shows immunoreactivity with antibodies to tau protein (Ishizawa et al. 2003), but this is a small subset of Lewy bodies and almost always in neuronal populations that are inherently vulnerable to tau pathology. It is rare to find tau immunoreactiv-

ity in cortical-type Lewy bodies in PD. Many other antibodies that inconsistently label Lewy bodies have been reported (Pollanen et al. 1993).

Glial Cytoplasmic Inclusions in MSA

Lantos and coworkers first described glial (oligodendroglial) cytoplasmic inclusions in MSA. GCIs can be detected with silver stains, in particular, the Gallyas silver stain, but are best seen with antibodies to α-synuclein (Fig. 3) and ubiquitin, in which they appear as flame- or sickle-shaped inclusions in oligodendrocytes. At the ultrastructural level, GCIs are non-membrane-bound cytoplasmic inclusions composed of 10–20 nm diameter-coated filament similar to the filaments in Lewy bodies (Lin et al. 2004). Although most α-synuclein inclusions in MSA are in oligodendroglial cells, certain neuronal populations are vulnerable to neuronal cytoplasmic and intranuclear inclusions, particularly those in the pontine base, inferior olivary nucleus, and putamen (Fig. 3). A few of the neuronal inclusions in MSA resemble Lewy bodies, but their anatomical distribution is distinct from neuronal populations vulnerable to Lewy bodies. Intranuclear α-synuclein-immunoreactive inclusions (Lin et al. 2004) are not found in PD.

Neuronal and Glial Tau Pathology in PSP

PSP is a degenerative tauopathy characterized by the accumulation of filamentous tau inclusions within neurons. Tau is a microtubule-associated protein that is biochemically composed of six major isoforms related to alternative mRNA splicing, including three isoforms with four 32-amino acid conserved repeats (4R-tau) in the microtubule-binding domain and three isoforms with three repeats (3R tau). In PSP, 4R tau preferentially accumulates, whereas in Alzheimer's disease tau inclusions are composed on a nearly equal mixture of 3R and 4R tau. Monoclonal antibodies specific to 3R and 4R tau now permit assessment of the type of tau that accumulates within neuronal lesions with routine immunohistochemistry (de Silva et al. 2003). In addition to neurofibrillary tangles

(Fig. 4), tau pathology of PSP is characterized by inclusions in astrocytes (so-called "tufted astrocytes") and in oligodendroglia (so-called "coiled bodies"). The latter glial lesions are distinct from the GCIs of MSA, not only based on their immunoreactivity with tau, but also on their characteristic distribution and morphology. Tau also accumulates in cell processes (both neuronal and glial), referred to as tau-positive "threads."

Distribution of Pathology

The hallmark of any neurodegenerative disease is selective neuronal loss. Accompanying neuronal loss in all neurodegenerative disorders are reactive changes in astrocytes and microglia. Microglia express markers of activation, such as the class II major histocompatibility antigen HLA-DR (McGeer et al. 1988), and astrocytes become hypertrophic and accumulate the intermediate filament protein, glial fibrillary acidic protein. Dying neurons undergo phagocytosis by microglia, a term referred to as neuronophagia. In the substantia nigra and locus ceruleus, evidence of neuronophagia is neuromelanin pigment in the cytoplasm of microglia. In cases with very long disease duration, microglia migrate to blood vessels and exit the brain along with the neuromelanin pigment. Neuronal loss in the substantia nigra is most marked in the ventrolateral tier of neurons of the pars compacta (A9) in all parkinsonian disorders. In contrast, the dorsal and medial neuronal cell groups are less vulnerable. Loss of medial neuronal cell groups (e.g., ventral tegmental region or A10) may be increased in parkinsonian disorders with dementia (Rinne et al. 1989).

Braak PD Staging Scheme

It has been known for many years that Lewy bodies in PD extend well beyond the substantia nigra (Jellinger 1991). Based on the distribution of α-synuclein pathology, Braak and coworkers have proposed a staging scheme for PD (Braak et al. 2004). In this scheme, neuronal pathology occurs early in the dorsal motor nucleus of the vagus in the medulla and the anterior olfactory nucleus in the olfactory bulb. As the disease progresses, locus ceruleus neurons in the pons and then dopaminergic neurons in the substantia nigra are affected. In later stages, pathology extends to the basal forebrain, amygdala and the medial temporal lobe structures, with convexity cortical areas affected in the last stages. Although the staging scheme is attractive, it should be remembered that this scheme is not based on the distribution of neuronal loss, but of distribution of abnormal synuclein deposits and how it relates to the progression of neuronal loss has not been rigorously studied.

The scheme was originally based on evaluation of brains of individuals that were not necessarily well characterized in life, and cases were chosen for further study if they had pathology in the medulla (Del Tredici et al. 2002), thus biasing the results in favor of "early" pathology in the medulla. In more recent studies of prospectively studied individuals who have come to autopsy, the scheme proposed by Braak and coworkers has not always held true. Some elderly individuals have Lewy bodies confined to the olfactory bulb (Fujishiro et al. 2008a; Beach et al. 2009) or the amygdala, the latter particularly true if associated with concurrent Alzheimer-type pathology (Uchikado et al. 2006). Moreover, some neurologically normal individuals have sparse but widespread Lewy body pathology, even involving the cortex (Parkkinen et al. 2005; Frigerio et al. 2011), which would seem to violate the theory of progression from brainstem and perhaps fit better with a multicentric disease process from the onset. Clearly, the observed distribution of Lewy bodies is dependent on case selection (Parkkinen et al. 2001). Although the staging scheme of Braak and coworkers has exceptions, it nevertheless has prompted considerable debate in the field and reawakened recognition of early nonmotor clinical features in PD (Jain 2011). Subsequent iterations of the Braak scheme proposed that autonomic neurons in the peripheral nervous system may be affected before involvement of the central nervous system (Braak and Del Tredici 2009), and this prompted recognition that PD should be considered a multiorgan disease process, not merely a disorder of the central nervous system (Beach et al. 2010). Moreover, it has fed the debate on cell-to-cell transmission of unknown putative disease factors (prion-like α-synuclein species) (Hawkes et al. 2009), especially given the fact that fetal mesencephalic intrastriatal transplants to treat PD have been shown to develop Lewy body pathology (Kordower et al. 2008), possibly by cell-to-cell transmission (Kordower et al. 2011).

Jellinger Staging Scheme for MSA

It has been more challenging to stage pathology in MSA and PSP because of the rarity of these disorders and because of their inherent variability. Nevertheless, Jellinger has proposed a staging scheme for MSA that scores the severity of SND and OPCA, each on a 3-point scale. The final classification is indicated by an OPCA/SND score (e.g., OPCA 1/SND 3 for a typical MSA-P case and OPCA 3/SND 1 for a typical MSA-C case). Halliday and coworkers (2011) proposed a similar scheme and graphically illustrated the two major MSA types, as well as the overlap in OPCA and SND system degenerations. Ozawa and colleagues used a semiquantitative scoring scheme for lesion density and found differences in the proportion of MSA types in Japanese compared to European autopsy cohorts, with far more OPCA in Japanese (Ozawa et al. 2004, 2010). Detection of MSA in neurologically normal individuals (incidental MSA) is extremely uncommon (Fujishiro et al. 2008b), and large numbers of such cases would be needed to determine the earliest sites of involvement to develop a staging scheme for MSA analogous to the Braak staging scheme for PD.

Distribution of Pathology in PSP

The distribution of neuronal loss and neurofibrillary degeneration in PSP was beautifully documented in the original report by Steele, Richardson, and Olszewski based on classic silver staining methods (Dickson 2004). Improved neuropathologic methods with tau immunohistochemistry have extended these observations, by recognizing glial involvement, as well greater cortical pathology than noted in the original report, particularly affecting motor and premo-

tor cortices of the frontal lobe (Hauw et al. 1990). Nevertheless, the cardinal nuclei affected in PSP remain those originally described by Steele et al. and include the globus pallidus, subthalamic nucleus, substantia nigra, midbrain tectum, periaqueductal gray, locus ceruleus, and the cerebellar dentate nucleus. Other regions that are consistently affected include corpus striatum, ventrolateral thalamus, red nucleus, pontine and medullary tegmentum, pontine base, and inferior olivary nucleus. Spinal cord involvement is also common, where neuronal inclusions can be found in intermediolateral cell columns. Heterogeneity in the distribution of tau pathology in PSP is increasingly recognized (Williams et al. 2005; Dickson et al. 2010). While neuroimaging staging schemes have been proposed (Planche et al. 2024), a neuropathologic staging scheme remains to be defined. The presence of PSP-like pathology in neurologically normal individuals (incidental PSP) is uncommon (Evidente et al. 2011). As in MSA, the paucity of such cases precludes the development of a staging scheme for PSP.

OTHER DEGENERATIVE PARKINSONIAN DISORDERS

Corticobasal Degeneration

Corticobasal degeneration (CBD), which is also known as cortical basal ganglionic degeneration, is a parkinsonism-plus disorder with characteristic focal cortical signs in addition to atypical levodopa-nonresponsive parkinsonism (Litvan et al. 1997; Boeve et al. 1999). Patients with CBD may present with progressive asymmetrical rigidity and apraxia (i.e., the corticobasal syndrome), but other clinical syndromes are also reported, such as progressive aphasia and progressive frontal lobe dementia (Litvan 1999). Parkinsonism is characterized by bradykinesia, rigidity, and dystonia, but most patients do not have tremor. The pathologic correlate of focal cortical findings on clinical evaluations is focal cortical atrophy, which is uncommon in PD, MSA, and PSP. Cortical atrophy in CBD is often most marked in the superior frontal gyrus, and the motor cortex may be severely affected. The midbrain does not have atrophy as in PSP, but pigment loss is common in the substantia nigra. In contrast to PSP, the superior cerebellar peduncle and the subthalamic nucleus are grossly normal (Dickson 1999). Microscopically, the affected cortical areas have neuronal loss, spongiosis, and gliosis with swollen achromatic or ballooned neurons. Cortical neurons in affected areas have pleomorphic tau-immunoreactive inclusions, and there are invariably numerous tau-positive threads in both gray and white matter of affected cortices, as well as in the basal ganglia (Dickson et al. 2002). The most characteristic lesion in CBD is an annular cluster of short, stubby processes with fuzzy outlines that represent tau accumulation in distal processes of astrocytes. A lesion referred to as an "astrocytic plaque" (Fig. 5; Feany and Dickson 1995). Astrocytic plaques differ from the tufted astrocytes seen in PSP, and the two lesions do not coexist in the same brain (Komori 1999). The globus pal-

lidus and putamen show mild neuronal loss with gliosis. Thalamic nuclei may also be affected. The substantia nigra usually shows moderate to severe neuronal loss with extraneuronal neuromelanin, gliosis, and tau immunoreactive neuronal lesions. The lower brainstem is less affected than in PSP (Dickson 1999).

Chronic Traumatic Encephalopathy

Individuals that suffer repeated closed head trauma may develop parkinsonism as well as dementia, a disorder currently referred to as chronic traumatic encephalopathy (CTE). In the past, this syndrome was referred to as dementia pugilistica or "punch drunk" syndrome because it was often associated with dementia and parkinsonism in professional boxers. There may be evidence of increasing frequency of the syndrome in other contact sports. In addition to other signs of chronic head trauma, such as

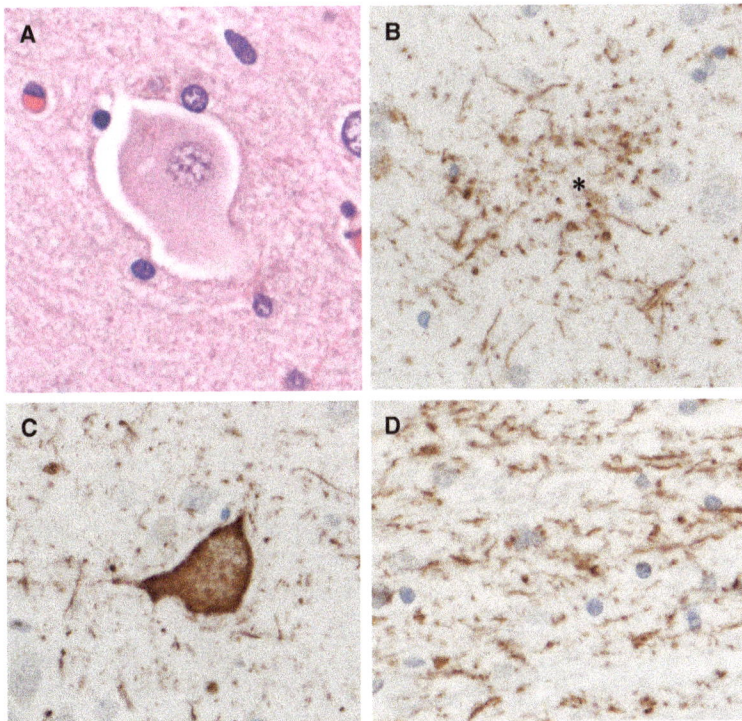

Figure 5. Corticobasal degeneration (CBD) pathology. (A) Ballooned neurons in the cerebral cortex (hematoxylin and eosin [H&E]), (B) astrocytic plaque (asterisk) (C), ballooned neurons, and (D) neuropil threads with tau immunohistochemistry.

small contusions or chronic subdural membrane, patients with CTE also had tau pathology that is patchy and predominant in gray matter at the depths of cortical sulci and in superficial cerebral white matter (Fig. 6; McKee et al. 2009). In these areas, tau accumulates in both neurons and astrocytes. The tau protein that accumulates is biochemically similar to that found in Alzheimer's disease.

Guam Parkinson–Dementia Complex

A characteristic Parkinsonism with dementia (Parkinson dementia complex [PDC]) with a number of features that overlap with PSP (Steele et al. 2002; Steele 2005) is common in the native Chamorro population of Guam and in the Kii peninsula of Japan (Kuzuhara and Kokubo 2005). The gross findings in PDC are notable for cortical atrophy affecting especially the medial temporal lobe, as well as atrophy of the hippocampus and the tegmentum of the rostral brainstem, which overlaps with atrophy seen in Alzheimer's disease. These areas typically have neuronal loss and gliosis with many neurofibrillary tangles in residual neurons and extracellular neurofibrillary tangle are numerous (Hirano et al. 1961). The substantia nigra and locus ceruleus have neuronal loss and neurofibrillary tangles. The basal nucleus and large

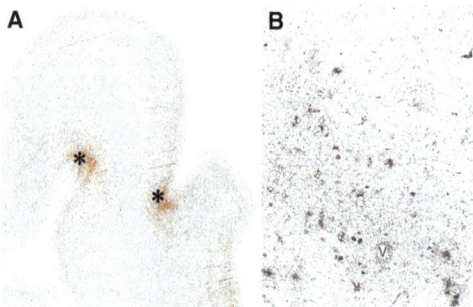

Figure 6. Chronic traumatic encephalopathy (CTE) pathology. (*A*) Low power view of cerebral cortex stained with tau immunohistochemistry shows foci tau pathology in the superficial cortices at the depths of sulci (asterisks), the characteristic lesions of CTE. (*B*) Higher magnification shows tau-positive astrocytes and neuropil threads clustered around blood vessels in the cortex at the depths of the sulci.

neurons in the striatum are also vulnerable to neurofibrillary tangle. Biochemically and morphologically, neurofibrillary tangles in Guam PDC like those in postencephalitic Parkinsonism (see below) are indistinguishable from those in Alzheimer's disease (Buee-Scherrer et al. 1995; Morris et al. 1999).

TDP-43-Related Parkinsonism

In addition to α-synuclein and tau, a third major protein has been found to accumulate within neurons and glial cells causing a number of neurodegenerative disorders, including amyotrophic lateral sclerosis, frontotemporal lobar degeneration (FTLD), Alzheimer's disease, and even some parkinsonian disorders. This protein is termed TDP-43 after TAR DNA-binding protein of 43 kDa molecular weight, a protein originally found to bind to an HIV transactive response DNA-binding protein. It is now known to be an RNA/DNA-binding protein that has a number of functions, not all of which are currently known (Buratti and Baralle 2010).

It is normally a nuclear protein and its accumulation in cytoplasmic inclusions is decidedly abnormal. Frontotemporal lobar degenerations are clinically and pathologically heterogeneous and importantly fall into two major classes—TDP-43 proteinopathies and tauopathies (Mackenzie et al. 2010). In some classification schemes, CBD and PSP are included among FTLD-tau, although, as noted above, frontal lobe clinical features in both CBD and PSP may be overshadowed by atypical parkinsonism in individual cases. On the other hand, parkinsonism is often a minor component of FTLD-TDP. Nevertheless, in an autopsy series of atypical parkinsonism, some cases will inevitably have pathology of FTLD-TDP. These patients often present with mixed clinical syndromes: dementia with parkinsonism, parkinsonism-plus syndrome, or corticobasal syndrome (Josephs et al. 2007). Pathologic findings will be those of FTLD-TDP—focal cortical atrophy with neuronal loss, gliosis, spongiosis, and neuronal inclusions of TDP-43—with the additional findings of significant neuronal loss in the substantia nigra associated with TDP-43 neuro-

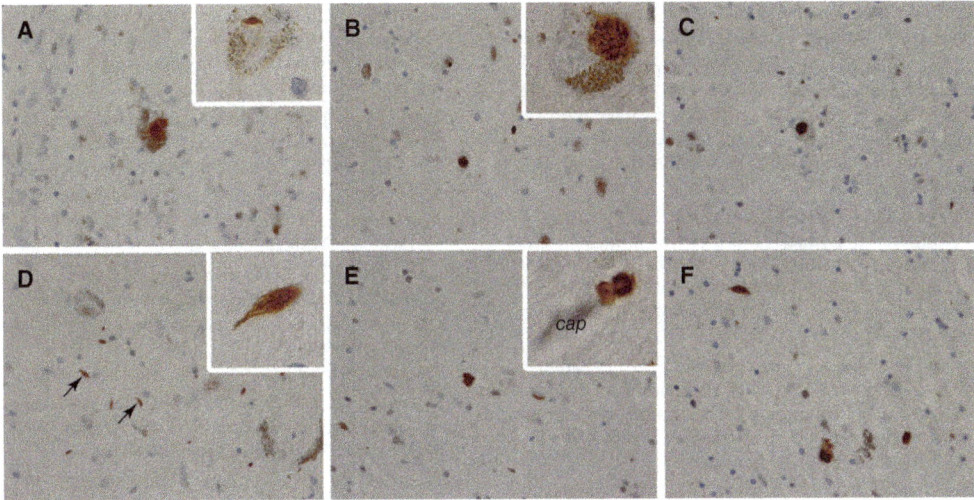

Figure 7. Perry syndrome. A range of TDP-43 pathologies are detected, including neuronal inclusions (*A,B*), some with intranuclear inclusions (*inset* in *B*), globose intracytoplasmic inclusions in the substantia nigra (*B*), and globus pallidus (*C*), stubby neurites in the substantia nigra (*D*), and perivascular glial inclusions (*E*). Some of the TDP-43-positive dystrophic neurites are torpedo-shaped or coarsely granular (*F*).

nal inclusions. In many cases, there is also TDP-43 pathology in the basal ganglia, which may contribute to the movement disorder.

A rare autosomal-dominant Parkinsonian disorder associated with TDP-43 pathology is Perry syndrome (Fig. 7; Wider et al. 2009). Perry syndrome is an early-onset, autosomal-dominant parkinsonism, associated with hypoventilation, depression, and severe weight loss. It is caused by mutations in *DCTN1*, and is typically a rapidly progressive disease in part due to central hypoventilation (Konno et al. 2017). Recently, the preferred term for this disorder has been changed to *DCTN1*-related neurodegeneration (Dulski et al. 2010).

CONCLUSIONS

Parkinsonian disorders are increasingly classified according to underlying molecular pathology (Dickson et al. 2009), with α-synucleinopathies (PD, MSA) and tauopathies (PSP, CBD, Guam PDC, CTE) being the most common. Recently, another category as been recognized —Parkinsonism associated with TDP-43 proteinopathy. As genetic and molecular studies are increasingly used to further refine underly-

ing disease processes, it is likely that other molecular forms of Parkinsonism will be identified. Currently, the one feature that unifies Parkinsonian disorders is nigrostriatal dopaminergic degeneration, but intriguing evidence from genetic studies (Höglinger et al. 2011; Nalls et al. 2011) suggest that there may also be shared genetic risk factors (e.g., MAPT) between tauopathies and α-synucleinopathies, but these remain to be defined with future studies.

ACKNOWLEDGMENTS

Supported by National Institutes of Health (NIH) grants. Dr. John Steele is acknowledged for his efforts to provide samples from his patients on Guam for neuropathologic study. The clinicians and geneticists involved in familial PD studies, especially Dr. Zbigniew Wszolek, is acknowledged. These studies would not be possible without the generous donation of patients and their families toward research on Parkinsonism.

REFERENCES

Abbott RD, Petrovitch H, White LR, Masaki KH, Tanner CM, Curb JD, Grandinetti A, Blanchette PL, Popper JS,

Ross GW. 2001. Frequency of bowel movements and the future risk of Parkinson's disease. *Neurology* **57**: 456–462.

Beach TG, Adler CH, Lue L, Sue LI, Bachalakuri J, Henry-Watson J, Sasse J, Boyer S, Shirohi S, Brooks R, et al. 2009. Unified staging system for Lewy body disorders: correlation with nigrostriatal degeneration, cognitive impairment and motor dysfunction. *Acta Neuropathol* **117**: 613–634.

Beach TG, Adler CH, Sue LI, Vedders L, Lue L, White CL, III, Akiyama H, Caviness JN, Shill HA, Sabbagh MN, et al. 2010. Multi-organ distribution of phosphorylated α-synuclein histopathology in subjects with Lewy body disorders. *Acta Neuropathol* **119**: 689–702. doi:10.1007/s00401-010-0664-3

Berendse HW, Booij J, Francot CM, Bergmans PL, Hijman R, Stoof JC, Wolters EC. 2001. Subclinical dopaminergic dysfunction in asymptomatic Parkinson's disease patients' relatives with a decreased sense of smell. *Ann Neurol* **50**: 34–41. doi:10.1002/ana.1049

Blauwendraat C, Nalls MA, Singleton AB. 2020. The genetic architecture of Parkinson's disease. *Lancet Neurol* **19**: 170–178. doi:10.1016/S1474-4422(19)30287-X

Boeve BF, Maraganore DM, Parisi JE, Ahlskog JE, Graff-Radford N, Caselli RJ, Dickson DW, Kokmen E, Petersen RC. 1999. Pathologic heterogeneity in clinically diagnosed corticobasal degeneration. *Neurology* **53**: 795–795. doi:10.1212/WNL.53.4.795

Braak H, Del Tredici K. 2009. Neuroanatomy and pathology of sporadic Parkinson's disease. *Adv Anat Embryol Cell Biol* **201**: 1–119.

Braak H, Ghebremedhin E, Rüb U, Bratzke H, Del Tredici K. 2004. Stages in the development of Parkinson's disease-related pathology. *Cell Tissue Res* **318**: 121–134. doi:10.1007/s00441-004-0956-9

Buee-Scherrer V, Buee L, Hof PR, Leveugle B, Gilles C, Loerzel AJ, Perl DP, Delacourte A. 1995. Neurofibrillary degeneration in amyotrophic lateral sclerosis/parkinsonism-dementia complex of Guam. Immunochemical characterization of tau proteins. *Am J Pathol* **146**: 924–932.

Buratti E, Baralle FE. 2010. The multiple roles of TDP-43 in pre-mRNA processing and gene expression regulation. *RNA Biol* **7**: 420–429. doi:10.4161/rna.7.4.12205

Conway KA, Harper JD, Lansbury PT Jr. 2000. Fibrils formed in vitro from α-synuclein and two mutant forms linked to Parkinson's disease are typical amyloid. *Biochemistry* **39**: 2552–2563. doi:10.1021/bi991447r

Crowther RA, Daniel SE, Goedert M. 2000. Characterisation of isolated α-synuclein filaments from substantia nigra of Parkinson's disease brain. *Neurosci Lett* **292**: 128–130. doi:10.1016/S0304-3940(00)01440-3

Dale GE, Probst A, Luthert P, Martin J, Anderton BH, Leigh PN. 1992. Relationships between Lewy bodies and pale bodies in Parkinson's disease. *Acta Neuropathol (Berl)* **83**: 525–529. doi:10.1007/BF00310030

Del Tredici K, Rüb U, De Vos RA, Bohl JR, Braak H. 2002. Where does Parkinson disease pathology begin in the brain? *J Neuropathol Exp Neurol* **61**: 413–426. doi:10.1093/jnen/61.5.413

de Silva R, Lashley T, Gibb G, Hanger D, Hope A, Reid A, Bandopadhyay R, Utton M, Strand C, Jowett T, et al. 2003. Pathological inclusion bodies in tauopathies contain dis-tinct complements of tau with three or four microtubule-binding repeat domains as demonstrated by new specific monoclonal antibodies. *Neuropathol Appl Neurobiol* **29**: 288–302. doi:10.1046/j.1365-2990.2003.00463.x

Dickson DW. 1999. Neuropathologic differentiation of progressive supranuclear palsy and corticobasal degeneration. *J Neurol* **246**: II6–II15. doi:10.1007/BF03161076

Dickson DW. 2001. α-Synuclein and the Lewy body disorders. *Curr Opin Neurol* **14**: 423–432. doi:10.1097/00019052-200108000-00001

Dickson DW. 2004. Sporadic tauopathies: Pick's disease, corticobasal degeneration, progressive supranuclear palsy and argyrophilic grain disease. In *The neuropathology of dementia*, 2nd ed. (ed. Esiri MM, Lee VMY, Trojanowski JQ), pp. 227–256. Cambridge University Press, New York.

Dickson DW. 2012. Parkinson's disease and parkinsonism: neuropathology. *Cold Spring Harb Perspect Med* **2**: a009258. doi:10.1101/cshperspect.a009258

Dickson DW, Bergeron C, Chin SS, Duyckaerts C, Horoupian D, Ikeda K, Jellinger K, Lantos PL, Lippa CF, Mirra SS, et al. 2002. Office of rare diseases neuropathologic criteria for corticobasal degeneration. *J Neuropathol Exp Neurol* **61**: 935–946. doi:10.1093/jnen/61.11.935

Dickson DW, Rademakers R, Hutton ML. 2007. Progressive supranuclear palsy: pathology and genetics. *Brain Pathol* **17**: 74–82. doi:10.1111/j.1750-3639.2007.00054.x

Dickson DW, Braak H, Duda JE, Duyckaerts C, Gasser T, Halliday GM, Hardy J, Leverenz JB, Del Tredici K, Wszolek ZK, et al. 2009. Neuropathological assessment of Parkinson's disease: refining the diagnostic criteria. *Lancet Neurol* **8**: 1150–1157. doi:10.1016/S1474-4422(09)70238-8

Dickson DW, Ahmed Z, Algom AA, Tsuboi Y, Josephs KA. 2010. Neuropathology of variants of progressive supranuclear palsy. *Curr Opin Neurol* **23**: 394–400. doi:10.1097/WCO.0b013e32833be924

Dulski J, Konno T, Wszolek Z. 2010. DCTN1-Related neurodegeneration. In *GeneReviews* (ed. Adam MP, Feldman J, Mirzaa GM, et al.). University of Washington, Seattle.

Evidente VG, Adler CH, Sabbagh MN, Connor DJ, Hentz JG, Caviness JN, Sue LI, Beach TG. 2011. Neuropathological findings of PSP in the elderly without clinical PSP: possible incidental PSP? *Parkinsonism Relat Disord* **17**: 365–371. doi:10.1016/j.parkreldis.2011.02.017

Farrer MJ. 2006. Genetics of Parkinson disease: paradigm shifts and future prospects. *Nat Rev Genet* **7**: 306–318. doi:10.1038/nrg1831

Farrer MJ, Hulihan MM, Kachergus JM, Dächsel JC, Stoessl AJ, Grantier LL, Calne S, Calne DB, Lechevalier B, Chapon F, et al. 2009. DCTN1 mutations in Perry syndrome. *Nat Genet* **41**: 163–165. doi:10.1038/ng.293

Feany MB, Dickson DW. 1995. Widespread cytoskeletal pathology characterizes corticobasal degeneration. *Am J Pathol* **146**: 1388–1396.

Forno LS. 1969. Concentric hyalin intraneuronal inclusions of Lewy type in the brains of elderly persons (50 incidental cases): relationship to parkinsonism. *J Am Geriatr Soc* **17**: 557–575. doi:10.1111/j.1532-5415.1969.tb01316.x

Frigerio R, Fujishiro H, Ahn TB, Josephs KA, Maraganore DM, DelleDonne A, Parisi JE, Klos KJ, Boeve BF, Dickson DW, et al. 2011. Incidental Lewy body disease: do some cases represent a preclinical stage of dementia with Lewy bodies? *Neurobiol Aging* **32:** 857–863. doi:10.1016/j.neurobiolaging.2009.05.019

Fujishiro H, Tsuboi Y, Lin WL, Uchikado H, Dickson DW. 2008a. Co-localization of tau and α-synuclein in the olfactory bulb in Alzheimer's disease with amygdala Lewy bodies. *Acta Neuropathol* **116:** 17–24. doi:10.1007/s00401-008-0383-1

Fujishiro H, Ahn TB, Frigerio R, DelleDonne A, Josephs KA, Parisi JE, Eric Ahlskog J, Dickson DW. 2008b. Glial cytoplasmic inclusions in neurologically normal elderly: prodromal multiple system atrophy? *Acta Neuropathol* **116:** 269–275. doi:10.1007/s00401-008-0398-7

Funayama M, Nishioka K, Li Y, Hattori N. 2023. Molecular genetics of Parkinson's disease: contributions and global trends. *J Hum Genet* **68:** 125–130. doi:10.1038/s10038-022-01058-5

Galloway PG, Mulvihill P, Perry G. 1992. Filaments of Lewy bodies contain insoluble cytoskeletal elements. *Am J Pathol* **140:** 809–822.

Galvin JE, Lee VM, Baba M, Mann DM, Dickson DW, Yamaguchi H, Schmidt ML, Iwatsubo T, Trojanowski JQ. 1997. Monoclonal antibodies to purified cortical Lewy bodies recognize the mid-size neurofilament subunit. *Ann Neurol* **42:** 595–603. doi:10.1002/ana.410420410

Gilman S, Low PA, Quinn N, Albanese A, Ben-Shlomo Y, Fowler CJ, Kaufmann H, Klockgether T, Lang AE, Lantos PL, et al. 1999. Consensus statement on the diagnosis of multiple system atrophy. *J Neurol Sci* **163:** 94–98. doi:10.1016/S0022-510X(98)00304-9

Halliday GM, Holton JL, Revesz T, Dickson DW. 2011. Neuropathology underlying clinical variability in patients with synucleinopathies. *Acta Neuropathol* **122:** 187–204. doi:10.1007/s00401-011-0852-9

Hauw JJ, Verny M, Delaère P, Cervera P, He Y, Duyckaerts C. 1990. Constant neurofibrillary changes in the neocortex in progressive supranuclear palsy. Basic differences with Alzheimer's disease and aging. *Neurosci Lett* **119:** 182–186. doi:10.1016/0304-3940(90)90829-X

Hawkes CH, Shephard BC, Daniel SE. 1997. Olfactory dysfunction in Parkinson's disease. *J Neurol Neurosurg Psychiatry* **62:** 436–446. doi:10.1136/jnnp.62.5.436

Hawkes CH, Del Tredici K, Braak H. 2009. Parkinson's disease: the dual hit theory revisited. *Ann NY Acad Sci* **1170:** 615–622. doi:10.1111/j.1749-6632.2009.04365.x

Hely MA, Reid WG, Adena MA, Halliday GM, Morris JG. 2008. The Sydney multicenter study of Parkinson's disease: the inevitability of dementia at 20 years. *Mov Disord* **23:** 837–844. doi:10.1002/mds.21956

Hirano A, Kurland LT, Krooth RS, Lessell S. 1961. Parkinsonism-dementia complex, an endemic disease on the island of Guam. I: Clinical features. *Brain* **84:** 642–661. doi:10.1093/brain/84.4.662

Höglinger GU, Melhem NM, Dickson DW, Sleiman PM, Wang LS, Klei L, Rademakers R, de Silva R, Litvan I, Riley DE, et al. 2011. Identification of common variants influencing risk of the tauopathy progressive supranuclear palsy. *Nat Genet* **43:** 699–705. doi:10.1038/ng.859

Hughes AJ, Daniel SE, Ben-Shlomo Y, Lees AJ. 2002. The accuracy of diagnosis of parkinsonian syndromes in a specialist movement disorder service. *Brain* **125:** 861–870. doi:10.1093/brain/awf080

Ikeda K, Ikeda S, Yoshimura T, Kato H, Namba M. 1978. Idiopathic parkinsonism with Lewy-type inclusions in cerebral cortex. A case report. *Acta Neuropathol (Berl)* **41:** 165–168. doi:10.1007/BF00689769

Irizarry MC, Growdon W, Gomez-Isla T, Newell K, George JM, Clayton DF, Hyman BT. 1998. Nigral and cortical Lewy bodies and dystrophic nigral neurites in Parkinson's disease and cortical Lewy body disease contain α-synuclein immunoreactivity. *J Neuropathol Exp Neurol* **57:** 334–337. doi:10.1097/00005072-199804000-00005

Ishizawa T, Mattila P, Davies P, Wang D, Dickson DW. 2003. Colocalization of tau and α-synuclein epitopes in Lewy bodies. *J Neuropathol Exp Neurol* **62:** 389–397. doi:10.1093/jnen/62.4.389

Jain S. 2011. Multi-organ autonomic dysfunction in Parkinson disease. *Parkinsonism Relat Disord* **17:** 77–83. doi:10.1016/j.parkreldis.2010.08.022

Jellinger KA. 1991. Pathology of Parkinson's disease. Changes other than the nigrostriatal pathway. *Mol Chem Neuropathol* **14:** 153–197. doi:10.1007/BF03159935

Josephs KA, Dickson DW. 2003. Diagnostic accuracy of progressive supranuclear palsy in the society for progressive supranuclear palsy brain bank. *Mov Disord* **18:** 1018–1026. doi:10.1002/mds.10488

Josephs KA, Ahmed Z, Katsuse O, Parisi JF, Boeve BF, Knopman DS, Petersen RC, Davies P, Duara R, Graff-Radford NR, et al. 2007. Neuropathologic features of frontotemporal lobar degeneration with ubiquitin-positive inclusions with progranulin gene (PGRN) mutations. *J Neuropathol Exp Neurol* **66:** 142–151. doi:10.1097/nen.0b013e31803020cf

Kalaria RN. 2018. The pathology and pathophysiology of vascular dementia. *Neuropharmacology* **134:** 226–239. doi:10.1016/j.neuropharm.2017.12.030

Kayama Y, Koyama Y. 2003. Control of sleep and wakefulness by brainstem monoaminergic and cholinergic neurons. *Acta Neurochir* **87:** 3–6. doi:10.1007/978-3-7091-6081-7_1

Komori T. 1999. Tau-positive glial inclusions in progressive supranuclear palsy, corticobasal degeneration and Pick's disease. *Brain Pathol* **9:** 663–679. doi:10.1111/j.1750-3639.1999.tb00549.x

Konno T, Ross OA, Teive HAG, Sławek J, Dickson DW, Wszolek ZK. 2017. DCTN1-related neurodegeneration: Perry syndrome and beyond. *Parkinsonism Relat Disord* **41:** 14–24. doi:10.1016/j.parkreldis.2017.06.004

Kordower JH, Chu Y, Hauser RA, Freeman TB, Olanow CW. 2008. Lewy body-like pathology in long-term embryonic nigral transplants in Parkinson's disease. *Nat Med* **14:** 504–506. doi:10.1038/nm1747

Kordower JH, Dodiya HB, Kordower AM, Terpstra B, Paumier K, Madhavan L, Sortwell C, Steece-Collier K, Collier TJ. 2011. Transfer of host-derived α synuclein to grafted dopaminergic neurons in rat. *Neurobiol Dis* **43:** 552–557. doi:10.1016/j.nbd.2011.05.001

Kramer ML, Schulz-Schaeffer WJ. 2007. Presynaptic α-synuclein aggregates, not Lewy bodies, cause neurodegeneration in dementia with Lewy bodies. *J Neurosci* **27:** 1405–1410. doi:10.1523/JNEUROSCI.4564-06.2007

Kuusisto E, Parkkinen L, Alafuzoff I. 2003. Morphogenesis of Lewy bodies: dissimilar incorporation of α-synuclein, ubiquitin, and 62. *J Neuropathol Exp Neurol* **62**: 1241–1253. doi:10.1093/jnen/62.12.1241

Kuzuhara S, Kokubo Y. 2005. Atypical parkinsonism of Japan: amyotrophic lateral sclerosis-parkinsonism-dementia complex of the Kii peninsula of Japan (Muro disease): an update. *Mov Disord* **20**: S108–S113. doi:10.1002/mds.20548

Kuzuhara S, Mori H, Izumiyama N, Yoshimura M, Ihara Y. 1988. Lewy bodies are ubiquitinated. A light and electron microscopic immunocytochemical study. *Acta Neuropathol (Berl)* **75**: 345–353. doi:10.1007/BF00687787

Lang AE. 2007. The progression of Parkinson disease: a hypothesis. *Neurology* **68**: 948–952. doi:10.1212/01.wnl.0000257110.91041.5d

Langston JW. 2006. The Parkinson's complex: parkinsonism is just the tip of the iceberg. *Ann Neurol* **59**: 591–596. doi:10.1002/ana.20834

Lantos PL. 1998. The definition of multiple system atrophy: a review of recent developments. *J Neuropathol Exp Neurol* **57**: 1099–1111. doi:10.1097/00005072-199812000-00001

Lin WL, DeLucia MW, Dickson DW. 2004. α-Synuclein immunoreactivity in neuronal nuclear inclusions and neurites in multiple system atrophy. *Neurosci Lett* **354**: 99–102. doi:10.1016/j.neulet.2003.09.075

Litvan I. 1999. Recent advances in atypical parkinsonian disorders. *Curr Opin Neurol* **12**: 441–446. doi:10.1097/00019052-199908000-00011

Litvan I, Agid Y, Goetz C, Jankovic J, Wenning GK, Brandel JP, Lai EC, Verny M, Ray-Chaudhuri K, McKee A, et al. 1997. Accuracy of the clinical diagnosis of corticobasal degeneration: a clinicopathologic study. *Neurology* **48**: 119–125. doi:10.1212/WNL.48.1.119

Mackenzie IR, Neumann M, Bigio EH, Cairns NJ, Alafuzoff I, Kril J, Kovacs GG, Ghetti B, Halliday G, Holm IE, et al. 2010. Nomenclature and nosology for neuropathologic subtypes of frontotemporal lobar degeneration: an update. *Acta Neuropathol* **119**: 1–4. doi:10.1007/s00401-009-0612-2

McGeer PL, Itagaki S, Boyes BE, McGeer EG. 1988. Reactive microglia are positive for HLA-DR in the substantia nigra of Parkinson's and Alzheimer's disease brains. *Neurology* **38**: 1285–1291. doi:10.1212/WNL.38.8.1285

McKee AC, Cantu RC, Nowinski CJ, Hedley-Whyte ET, Gavett BE, Budson AE, Santini VE, Lee HS, Kubilus CA, Stern RA. 2009. Chronic traumatic encephalopathy in athletes: progressive tauopathy after repetitive head injury. *J Neuropathol Exp Neurol* **68**: 709–735. doi:10.1097/NEN.0b013e3181a9d503

McKeith I, Mintzer J, Aarsland D, Burn D, Chiu H, Cohen-Mansfield J, Dickson D, Dubois B, Duda JE, Feldman H, et al. 2004. Dementia with Lewy bodies. *Lancet Neurol* **3**: 19–28. doi:10.1016/S1474-4422(03)00619-7

Morris HR, Lees AJ, Wood NW. 1999. Neurofibrillary tangle parkinsonian disorders—tau pathology and tau genetics. *Mov Disord* **14**: 731–736. doi:10.1002/1531-8257(199909)14:5<731::AID-MDS1004>3.0.CO;2-J

Muntané G, Dalfó E, Martinez A, Ferrer I. 2008. Phosphorylation of tau and α-synuclein in synaptic-enriched fractions of the frontal cortex in Alzheimer's disease, and in Parkinson's disease and related α-synucleinopathies. *Neuroscience* **152**: 913–923. doi:10.1016/j.neuroscience.2008.01.030

Nalls MA, Plagnol V, Hernandez DG, Sharma M, Sheerin UM, Saad M, Simon-Sanchez J, Schulte C, Lesage S, et al. 2011. Imputation of sequence variants for identification of genetic risks for Parkinson's disease: a meta-analysis of genome-wide association studies. *Lancet* **377**: 641–649. doi:10.1016/S0140-6736(10)62345-8

Oppenheimer DR. 1976. Diseases of the basal ganglia, cerebellum and motor neurons. In *Greenfield's neuropathology*, 3rd ed. (ed. Blackwood W, Corsellis JAN), pp. 608–651. Edward Arnold, London.

Ozawa T, Paviour D, Quinn NP, Josephs KA, Sangha H, Kilford L, Healy DG, Wood NW, Lees AJ, Holton JL, et al. 2004. The spectrum of pathological involvement of the striatonigral and olivopontocerebellar systems in multiple system atrophy: clinicopathological correlations. *Brain* **127**: 2657–2671.

Ozawa T, Tada M, Kakita A, Onodera O, Ishihara T, Morita T, Shimohata T, Wakabayashi K, Takahashi H, Nishizawa M. 2010. The phenotype spectrum of Japanese multiple system atrophy. *J Neurol Neurosurg Psychiatry* **81**: 1253–1255.

Pappolla MA, Shank DL, Alzofon J, Dudley AW. 1988. Colloid (hyaline) inclusion bodies in the central nervous system: their presence in the substantia nigra is diagnostic of Parkinson's disease. *Hum Pathol* **19**: 27–31.

Parkkinen L, Soininen H, Laakso M, Alafuzoff I. 2001. α-Synuclein pathology is highly dependent on the case selection. *Neuropathol Appl Neurobiol* **27**: 314–325. doi:10.1046/j.0305-1846.2001.00342.x

Parkkinen L, Pirttilä T, Tervahauta M, Alafuzoff I. 2005. Widespread and abundant α-synuclein pathology in a neurologically unimpaired subject. *Neuropathology* **25**: 304–314. doi:10.1111/j.1440-1789.2005.00644.x

Planche V, Mansencal B, Manjon JV, Meissner WG, Tourdias T, Coupé P. 2024. Staging of progressive supranuclear palsy-Richardson syndrome using MRI brain charts for the human lifespan. *Brain Commun* **6**: fcae055. doi:10.1093/braincomms/fcae055

Pollanen MS, Dickson DW, Bergeron C. 1993. Pathology and biology of the Lewy body. *J Neuropathol Exp Neurol* **52**: 183–191. doi:10.1097/00005072-199305000-00001

Rajput AH, Rozdilsky B, Rajput A. 1991. Accuracy of clinical diagnosis in parkinsonism—a prospective study. *Can J Neurol Sci* **18**: 275–278. doi:10.1017/S0317167100031814

Rinne JO, Rummukainen J, Paljarvi L, Rinne UK. 1989. Dementia in Parkinson's disease is related to neuronal loss in the medial substantia nigra. *Ann Neurol* **26**: 47–50. doi:10.1002/ana.410260107

Schenck CH, Bundlie SR, Mahowald MW. 1996. Delayed emergence of a parkinsonian disorder in 38% of 29 older men initially diagnosed with idiopathic rapid eye movement sleep behaviour disorder. *Neurology* **46**: 388–393. doi:10.1212/WNL.46.2.388

Schulz-Schaeffer WJ. 2010. The synaptic pathology of α-synuclein aggregation in dementia with Lewy bodies, Parkinson's disease and Parkinson's disease dementia. *Acta Neuropathol* **120**: 131–143. doi:10.1007/s00401-010-0711-0

Steele JC. 2005. Parkinsonism-dementia complex of Guam. *Mov Disord* **20**: S99–S107. doi:10.1002/mds.20547

Cite this article as *Cold Spring Harb Perspect Med* doi: 10.1101/cshperspect.a041610

Steele JC, Richardson JC, Olszewski J. 1964. Progressive supranuclear palsy. A heterogeneous degeneration involving the brain stem, basal ganglia and cerebellum with vertical gaze and pseudobulbar palsy, nuchal dystonia and dementia. *Arch Neurol* **10:** 333–359.

Steele JC, Caparros-Lefebvre D, Lees AJ, Sacks OW. 2002. Progressive supranuclear palsy and its relation to pacific foci of the parkinsonism-dementia complex and Guadeloupean parkinsonism. *Parkinsonism Relat Disord* **9:** 39–54.

Tiller-Borcich JK, Forno LS. 1988. Parkinson's disease and dementia with neuronal inclusions in the cerebral cortex: Lewy bodies or pick bodies. *J Neuropathol Exp Neurol* **47:** 526–535.

Uchikado H, Lin WL, DeLucia MW, Dickson DW. 2006. Alzheimer disease with amygdala Lewy bodies: a distinct form of α-synucleinopathy. *J Neuropathol Exp Neurol* **65:** 685–697.

Wakabayashi K, Takahashi H. 1996. Gallyas-positive, tau-negative glial inclusions in Parkinson's disease midbrain. *Neurosci Lett* **217:** 133–136.

Wakabayashi K, Hayashi S, Yoshimoto M, Kudo H, Takahashi H. 2000. NACP/α-synuclein-positive filamentous inclusions in astrocytes and oligodendrocytes of Parkinson's disease brains. *Acta Neuropathol (Berl)* **99:** 14–20. doi:10.1007/PL00007400

Wider C, Dickson DW, Stoessl AJ, Tsuboi Y, Chapon F, Gutmann L, Lechevalier B, Calne DB, Personett DA, Hulihan M, et al. 2009. Pallidonigral TDP-43 pathology in Perry syndrome. *Parkinsonism Relat Disord* **15:** 281–286. doi:10.1016/j.parkreldis.2008.07.005

Williams DR, de Silva R, Paviour DC, Pittman A, Watt HC, Kilford L, Holton JL, Revesz T, Lees AJ. 2005. Characteristics of two distinct clinical phenotypes in pathologically proven progressive supranuclear palsy: Richardson's syndrome and PSP-parkinsonism. *Brain* **128:** 1247–1258. doi:10.1093/brain/awh488

Zijlmans JC, Daniel SE, Hughes AJ, Révész T, Lees AJ. 2004. Clinicopathological investigation of vascular parkinsonism, including clinical criteria for diagnosis. *Mov Disord* **19:** 630–640. doi:10.1002/mds .20083

Genetics of Parkinson's Disease: From Causes to Treatment

Ana Westenberger,[1] Norbert Brüggemann,[1,2] and Christine Klein[1]

[1]Institute of Neurogenetics; [2]Department of Neurology, University of Lübeck and University Hospital Schleswig-Holstein, Campus Lübeck, 23538 Lübeck, Germany

Correspondence: ana.westenberger@uni-luebeck.de; norbert.brueggemann@uni-luebeck.de; christine.klein@uni-luebeck.de

The genetic architecture of Parkinson's disease (PD) comprises five autosomal dominantly inherited forms with a clinical picture overall resembling idiopathic disease (PARK-*SNCA*, PARK-*LRRK2*, PARK-*VPS35*, PARK-*CHCHD2*, and PARK-*RAB32*) and three recessive types (PARK-*PRKN*, PARK-*PINK1*, and PARK-*PARK7*), several monogenic forms causing atypical parkinsonism, as well as a plethora of known genetic risk factors, most notably *SNCA* and *GBA1* including a recently discovered risk variant unique to individuals of African descent, as well as polygenic scores. The Movement Disorder Society Genetic mutation database (MDSGene) (www.mdsgene.org) provides PD genotype–phenotype relationships, whereas global PD genetics networks, such as the Global Parkinson's Genetics Program (www.gp2.org) elucidate PD genetic factors at an unprecedented scale. Two large studies in relatively unselected, multicenter PD samples estimate the frequency of genetic forms, including PARK-*GBA1*, at ~15%. PD genetics are becoming increasingly actionable, with the first gene-targeted clinical trials underway. Furthermore, PD genetics has recently been incorporated into a new biological classification of PD.

More than a quarter of a century after the identification of a pathogenic missense variant in the *alpha-synuclein* (*SNCA*) gene as the first recognized monogenic cause of Parkinson's disease (PD) (Polymeropoulos et al. 1997), the genetic contribution to PD is becoming increasingly evident and actionable. Perhaps the most obvious manifestation of practical consequences of an established genetic cause or strong risk factor is the advent of gene-targeted clinical trials, several of which are currently underway or in the advanced planning phase (Prasuhn and Brüggemann 2021; Cavallieri et al. 2023) and possible restorative and gene therapies are on the horizon, for patients with both genetic and idiopathic PD (Barker and Björklund 2023; Ng et al. 2023).

A considerable fraction of ~15% of all PD patients unselected for early age at onset or positive family history carry a pathogenic variant in one of the seven established genes for monogenic PD (i.e., *SNCA*, *LRRK2*, *VPS35*, *CHCHD2*, *PRKN*, *PINK1*, and *DJ1*) (Lange et al. 2022), the newly linked *RAB32* gene (Gustavsson et al. 2024), the strong risk factor gene *GBA1*, or, rarely, other genes linked to parkinsonian phenotypes (Cook et al. 2024; Westenberger et al.

Copyright © 2025 Cold Spring Harbor Laboratory Press; all rights reserved
Cite this article as *Cold Spring Harb Perspect Med* doi: 10.1101/cshperspect.a041774

2024). It has even been argued that PD is not a single entity but that there are several PDs, with well-described monogenic forms serving as frontrunners of these individual types of PD (Bloem et al. 2021).

While PARK-*LRRK2* and PARK-*VPS35* tend to be clinically indistinguishable from idiopathic PD, more distinct phenotype–genotype relationships have been established for other monogenic forms, such as early age of onset, milder disease course, and preserved cognition in PARK-*PRKN* or a more severe clinical course in many patients with PARK-*SNCA*, especially in carriers of triplications and missense variants (Kasten et al. 2018; Trinh et al. 2018; Vollstedt et al. 2023b). Despite a broad clinical spectrum, sometimes even within pedigrees harboring the same familial pathogenic variant, these insights in phenotype–genotype relationships may cautiously inform genetic counseling of PD patients and their relatives (Pal et al. 2023a) and even treatment choices (Pal et al. 2022; Avenali et al. 2023). Importantly, the role of PD genetics extends well beyond that of monogenic forms of PD, where the importance of polygenic risk for our understanding of the genetic architecture of PD and possible resulting patient stratification for research studies and clinical trials is widely recognized (Nalls et al. 2019; Prasuhn et al. 2019, 2021; Koch et al. 2021; Tunold et al. 2024).

Very recently, two consortia attempted to establish a biological definition, staging, and classification of PD. The first system introduces an integrated staging system allowing the definition of a prodromal phase based on α-synuclein pathology and pathologic dopaminergic imaging (Simuni et al. 2024). A considerable proportion of patients with certain genetic subtypes, in particular PARK-*PRKN* and PARK-*LRRK2*, however, do not fall into the neuronal synuclein disease (NSD) category (Simuni et al. 2024), as no (or rarely) Lewy body pathology is evident in postmortem studies (Kalia et al. 2015; Henderson et al. 2019; Madsen et al. 2021). This is of great importance as these subgroups may not benefit from α-synuclein-targeted approaches to the same extent as those with α-synuclein pathology. The second biological classification, termed "SynNeurGe," is based on the presence or ab-

sence of α-synuclein, neurodegeneration, and genetics to define PD research diagnostic criteria (Höglinger et al. 2024). Of further note, the genetic classification of "SynNeurGe" takes variable penetrance into account (i.e., the likelihood of manifesting PD in the presence of a pathogenic variant in a specific PD gene) (Höglinger et al. 2024).

Despite these exciting advances within or resulting from the field of PD genetics, many open questions remain that will be highlighted in the respective parts of this article below. They include, but are not limited to (1) discovering novel genetic PD causes and contributions, for example through long-read genome sequencing or the extension of genetic studies to underrepresented populations; (2) placing a stronger focus on protective and compensatory genetic factors to overcome the current preoccupation with genetic cause and risk; (3) conducting more comprehensive studies to establish genotype–phenotype relationships both cross-sectionally and longitudinally; (4) gaining further insights in gene–gene and gene–environment interactions, as well as functional consequences and pathways; (5) generating clinical trial–ready cohorts; and (6) stratifying the risk of yet unaffected mutation carriers to convert to PD pathogenesis and diagnosis.

GENOTYPE–PHENOTYPE CORRELATION EFFORTS AND RESEARCH INITIATIVES TO ESTABLISH GENETIC TESTING PRACTICES

In the following, we will list efforts to establish genotype–phenotype relationships for monogenic PD and genetic testing practices.

The Movement Disorder Society Genetic Mutation Database (MDSGene)

The increased availability and affordability of genetic testing technology have been closely followed by the expansion of scientific publications reporting the patients' genetic and clinical data. However, for movement disorders specialists, this multitude of multifaceted data and literature is becoming increasingly difficult to follow, interpret, and use. The Movement Disorder Society

Genetic mutation database (MDSGene) arose from this critical need for an all-inclusive genotype–phenotype data correlation and assessment. The goal of this online resource (available at www.mdsgene.org) is to provide a comprehensive, systematic overview of published data on patients with movement disorders and causative genetic variants, including typical and atypical forms of monogenic PD (Kasten et al. 2018; Klein et al. 2018; Trinh et al. 2018; Wittke et al. 2021).

The Michael J. Fox Foundation Global Monogenic PD Project

The Michael J. Fox Foundation Global Monogenic PD (MJFF GMPD) project aimed to embrace monogenic PD worldwide and leverage MDSGene. Corresponding authors of PD-related articles were asked to provide additional information on previously published or newly identified mutation carriers, resulting in a resource of almost 4000 affected and unaffected carriers of pathogenic variants in genes implicated in monogenic PD. However, most of the provided information was on individuals of European/White ancestry (Vollstedt et al. 2019, 2023a,b).

The Global Parkinson's Genetics Program

The Global Parkinson's Genetics Program (GP2; gp2.org) (Tan et al. 2021) is the largest PD genetics project worldwide. This 10-year program (2020–2029) aims to elucidate the genetic architecture of PD, prodromal, and atypical PD using as inclusive an approach as possible. It specifically targets sites and centers that did not have access to genetic testing or that have not participated in PD research before to expand the global network and include individuals with diverse ancestry (Schumacher-Schuh et al. 2022). These efforts have already resulted in the first exciting discoveries, such as the recently identified risk variant in individuals of African ancestry in the GBA1 gene (Rizig et al. 2023). The Monogenic Network of GP2 strives for a better understanding of known monogenic forms of PD and the discovery of known genetic causes of PD (Lange et al. 2023; Junker et al. 2024). In the latter capacity, it was recently involved in the validation of the newly detected RAB32 pathogenic variant causing PD (Gustavsson et al. 2024). Another important focus of GP2 is to unravel the causes of sporadic PD (Towns et al. 2023).

The Rostock Parkinson's Disease (ROPAD) Study

The Rostock Parkinson's Disease (ROPAD) study is an ongoing observational clinical study (NCT03866603) initiated in April 2019 (www.centogene.com/pharma/clinical-trial-support/rostock-international-parkinsons-disease-study-ropad). Within the first 2 years, 12,580 reportedly unrelated individuals (index patients) with a clinical diagnosis of PD and an age of ≥18 years were enrolled at movement disorders centers in 16 different countries in Europe, the Middle East, and North and South America (Westenberger et al. 2024). Among 1864 (~15%) ROPAD participants, genetic testing was positive based on GBA1 risk variants (>10%), or pathogenic/likely pathogenic variants in LRRK2 (~3%), PRKN (~1%), and SNCA, PINK1, or a combination of two genetic findings in two genes (~0.5%, combined).

The PD GENEration Study

PD GENEration is a multicenter, observational, and registry study (NCT04057794), offering genetic testing and counseling to individuals with PD (www.parkinson.org/advancing-research/our-research/pdgeneration). Among 8300 participants for whom genetic testing has been completed, the study has found that 1111 participants (>13%) have a genetic form of PD, based on variants in GBA1 (10%), LRRK2 (>2%), PRKN (<1%), and a very few variants in SNCA/PINK1/PARK7/VPS35/combination of two genes (<1%) (Cook et al. 2024). In collaboration with GP2, PD GENEration is currently extending its genetic testing and counseling efforts to all of South America.

MONOGENIC FORMS OF PD

By definition, monogenic diseases are caused by rare pathogenic variants of large effect size in a single gene. Currently, there are eight genes related to monogenic classical PD forms. Monoallelic (heterozygous) variants in SNCA, LRRK2, VPS35,

CHCHD2, and *RAB32* are considered responsible for autosomal-dominant monogenic PD forms that mostly have a late onset, while biallelic (homozygous/compound-heterozygous) variants in *PRKN*, *PINK1*, and *PARK7* represent causes of early-onset autosomal-recessive monogenic forms of PD. Of note, penetrance (i.e., the likelihood that a carrier of the well-established gene mutation(s) will indeed develop the disease—at all or within a certain age interval), may be significantly reduced in autosomal-dominant forms of PD apart from PARK-*SNCA*. This is due to an intricate interplay between monogenic causative variants, likely many additional genetic variants with a smaller-effect size, mitochondrial, and/or environmental factors. In this section, we will discuss the clinical and genetic features of the aforementioned PD forms, designated according to The Movement Disorders Society Task Force for the Nomenclature of Genetic Movement Disorders recommendations that use the abbreviated phenotype prefix (PARK) followed by the respective gene name (Marras et al. 2016).

Monogenic Autosomal-Dominant Forms of PD

PARK-SNCA

PARK-*SNCA*, caused by variants in the *SNCA* gene, was the very first genetic PD form identified (Polymeropoulos et al. 1997). In large multiethnic patient groups, it accounts for PD in 0.1%–0.2% of individuals (Cook et al. 2024; Westenberger et al. 2024). To date, only a small number of missense changes and two structural variants, found in approximately 200 patients, have been reported in *SNCA* (Trinh et al. 2018; Guo et al. 2021; Diaw et al. 2023). The most common missense variant is p.Ala53Thr, whereas whole-gene duplications and triplications represent the known structural variants. Interestingly, *SNCA* dosage correlates with disease onset as patients with triplications have the earliest onset (median of below 35 years), those with duplications have the latest onset (median of >45 years), while individuals with the p. Ala53Thr variant have an intermediate age at onset (AAO) (close to 45 years) (Book et al. 2018; Trinh et al. 2018). When considering the PARK-

SNCA patients as a group, regardless of the type of *SNCA* variants they carry, cognitive decline, dyskinesias, and motor fluctuations are present in about three-quarters of the reported patients with a good response of motor symptoms to dopaminergic drugs (Trinh et al. 2018). A study considering only patients with structural variants reported cognitive impairment in 60% of the patients (Book et al. 2018).

The *SNCA* gene encodes α-synuclein, a small protein expressed predominantly in the brain, rich in posttranslational modifications, and concentrated in presynaptic terminals where it interacts with membranes of synaptic vesicles. In PD, α-synuclein monomers adopt a pathological β-sheet conformation, leading to the formation of toxic α-synuclein oligomers and amyloid fibrils, and, finally, Lewy body pathology, which has been neuropathologically confirmed as the core neuropathology in PARK-*SNCA* but also nongenetic PD. The role of α-synuclein in PD and other synucleinopathies has recently been summarized in Spillantini et al. (1997) and Calabresi et al. (2023a,b). Briefly, pathological α-synuclein forms slow vesicle exocytosis at the presynaptic terminals and alters postsynaptic responses to transmitters, and point mutations in α-synuclein may have greater potential to activate microglia compared to wild-type α-synuclein (Fig. 1; Table 1; Hoenen et al. 2016; Sánchez and Maguire-Zeiss 2020; Badanjak et al. 2021). As mentioned above, it was suggested to redefine PD and dementia with Lewy bodies as neuronal α-synuclein disease rather than as clinical syndromes (Simuni et al. 2024). This new biological definition leverages the availability of tools to assess neuronal α-synuclein in patients and at-risk individuals during life. However, it does not provide a framework for patients presenting with clinical PD without evidence of α-synuclein pathology, which has been documented, for example, for PARK-*PRKN* (Madsen et al. 2021) and PARK-*LRRK2* (Kalia et al. 2015; Siderowf et al. 2023).

PARK-LRRK2

PARK-*LRRK2*, due to heterozygous and rarely homozygous mutations in the *LRRK2* gene, is

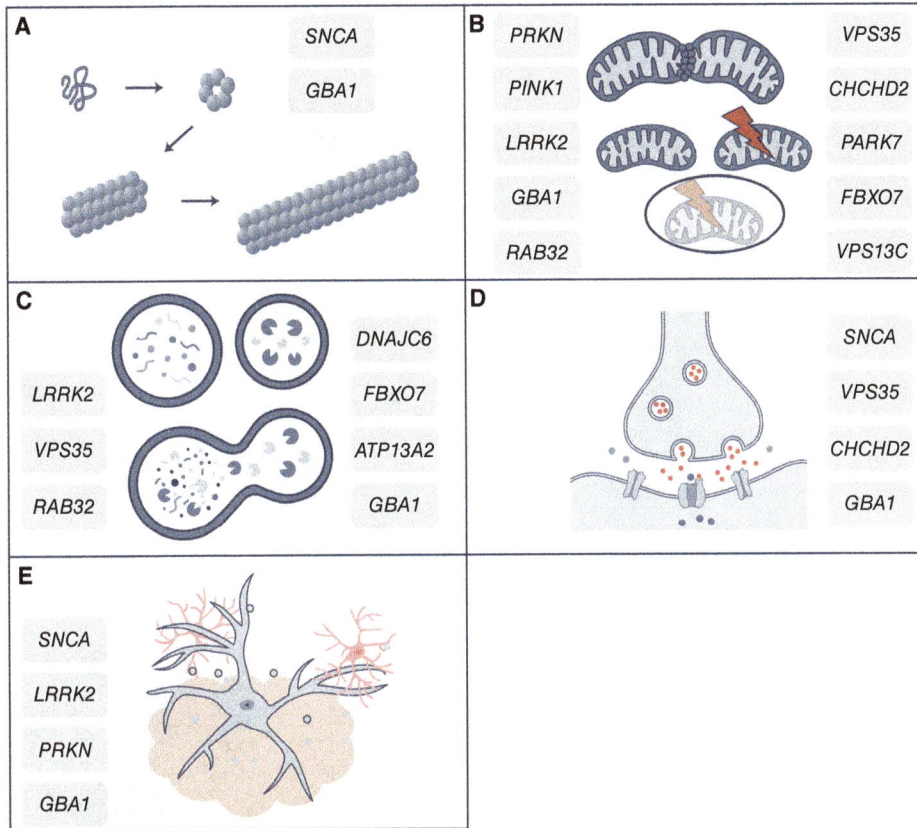

Figure 1. Overview of the main pathological parkinsonism pathways and the contributing genes. (*A*) Oligomerization and fibrillation of α-synuclein, (*B*) mitochondrial dysfunction and clearance, (*C*) endosomal/lysosomal dysfunction, (*D*) impaired neurotransmission, and (*E*) neuroinflammation. Of note, α-synuclein pathology is not limited to PARK-*SNCA* and PARK-*GBA1*. However, these two forms are more strongly connected to the development of α-synuclein oligomers and fibrils.

the most frequent monogenic PD form affecting 2.4%–2.9% of PD patients in general, 2.3% of sporadic PD patients, and 4.9% of PD patients with a positive family history (Cook et al. 2024; Westenberger et al. 2024). Furthermore, these pathogenic or likely pathogenic *LRRK2* variants are found in 17.6%–19.7% of patients with a monogenic etiology of PD (Cook et al. 2024; Westenberger et al. 2024).

More than 100 missense variants in *LRRK2* have been reported in PD (Kalogeropulou et al. 2022); however, only a few are repeatedly found in patients. In specific populations, such as Ashkenazi Jewish people or North African Berbers, the c.6055G>A (p.Gly2019Ser) *LRRK2* variant is present with a very high frequency of 29%

(Ozelius et al. 2006) and 37% (Lesage et al. 2006; Bouhouche et al. 2017) of patients with familial PD, respectively. Similarly, another *LRRK2* variant, c.4321C>G (p.Arg1441Gly) is very common in the Basque population (46% of all familial PD), due to a founder effect (Simón-Sánchez et al. 2006; Gorostidi et al. 2009). Furthermore, the c.4321C>T (p.Arg1441Cys) *LRRK2* change was detected in 2%–5% of PD patients from Belgium or Southern Italy, but is infrequent in other regions of the world (Nuytemans et al. 2008; De Rosa et al. 2014).

PARK-*LRRK2* typically manifests as late-onset PD with slower disease progression, better cognitive performance, a lower risk for dementia, and a lower rate of rapid eye movement (REM)

Table 1. List of the main genetic Parkinson's disease (PD)/atypical parkinsonism forms discussed in this review and their clinical, genetic, and molecular determinants

Designation	Clinical features	Type of pathogenic variants	Potential pathogenic mechanism	Functional/physical interaction or colocalization with other PD/atypical parkinsonism-related proteins
Monogenic late-onset autosomal-dominant PD forms				
PARK-*SNCA*	Missense mutations and triplications cause early-onset parkinsonism with prominent cognitive decline. Duplications cause a clinical picture similar to idiopathic PD, apart from an ~10-yr earlier AAO, but ~10-yr later than in those carrying missense mutations or triplications)	Missense, whole-gene multiplication	α-synuclein aggregation, synaptic dysfunction, neuroinflammation	GCase, LRRK2, VPS35
PARK-*LRRK2*	Typical levodopa-responsive parkinsonism	Missense	Lysosomal and mitochondrial dysfunction, neuroinflammation	Parkin, RAB32, α-synuclein, VPS35
PARK-*VPS35*	Typical levodopa-responsive parkinsonism	Missense	Lysosomal, mitochondrial, and synaptic dysfunction	α-synuclein, Parkin, LRRK2
PARK-*CHCHD2*	Typical levodopa-responsive parkinsonism	Missense	Mitochondrial and synaptic dysfunction	None
PARK-*RAB32*	Typical levodopa-responsive parkinsonism	Missense	Lysosomal and mitochondrial dysfunction	LRRK2, PINK1
Monogenic early-onset autosomal-recessive PD forms				
PARK-*PRKN*	Early-onset parkinsonism	Missense, nonsense, splice-site, small indels, structural	Mitochondrial dysfunction, neuroinflammation	PINK1, FBXO7, LRRK2, VPS35
PARK-*PINK1*	Early-onset parkinsonism	Missense, nonsense, splice-site, small indels, structural	Mitochondrial dysfunction	Parkin, RAB32, FBXO7
PARK-*PARK7*	Early-onset parkinsonism	Missense, nonsense, splice-site, small indels, structural	Mitochondrial dysfunction	None
Monogenic autosomal-dominant atypical parkinsonism				
PARK-*DCTN1*	Adult-onset (atypical) parkinsonism with depression or apathy, followed by weight loss and respiratory hypoventilation/failure (referred to as *Perry syndrome*)	Missense		None

Continued

Cite this article as *Cold Spring Harb Perspect Med* doi: 10.1101/cshperspect.a041774

Table 1. *Continued*

Designation	Clinical features	Type of pathogenic variants	Potential pathogenic mechanism	Functional/physical interaction or colocalization with other PD/atypical parkinsonism-related proteins
Monogenic autosomal-recessive atypical parkinsonism				
PARK-DNAJC6	Atypical parkinsonism	Missense, nonsense, splice-site	Lysosomal dysfunction	None
PARK-FBXO7	Atypical parkinsonism	Missense, nonsense, splice-site	Mitochondrial and lysosomal dysfunction	Parkin, PINK1
PARK-SYNJ1	Atypical parkinsonism	Missense, nonsense, small indels	Lysosomal dysfunction	None
PARK-VPS13C	Atypical parkinsonism	Missense, nonsense, splice-site, small indels, structural	Mitochondrial dysfunction	None
MxMD-ATP13A2	Atypical parkinsonism with vertical gaze palsy and mini-myoclonus	Missense, nonsense, splice-site, small indels	Lysosomal dysfunction	None
Genetic risk factor				
PARK-GBA1	Classical parkinsonism with lower AAO, faster rate of motor progression, and the presence of nonmotor symptoms	Missense, nonsense, splice-site, small indels, structural	Lysosomal dysfunction, neuroinflammation	α-synuclein

This prefix denotes that variants in *ATP13A2* may contribute to multiple equally prominent movement disorder phenotypes (i.e., adult-onset spasticity, progressive ataxia, or action myoclonus in addition to the here-relevant juvenile-onset atypical dystonia–parkinsonism) (Lange et al. 2022).

(AAO) Age at onset, (GCase) glucocerebrosidase A, (MxMD) mixed movement disorders.

sleep behavior disorder and hyposmia than in idiopathic PD (Wszolek et al. 2004; Alcalay et al. 2015b; Kalia et al. 2015; Saunders-Pullman et al. 2018, 2022; Simuni et al. 2020). The milder phenotypic presentation is also reflected by a smaller degree of dopaminergic neuronal loss than idiopathic PD (Simuni et al. 2020). About two-thirds of the reported PARK-*LRRK2* patients show dyskinesias and motor fluctuations and nearly 40% suffer from dystonia (Trinh et al. 2018). The median AAO is 57–58 years (Trinh et al. 2018; Vollstedt et al. 2023b; Westenberger et al. 2024). Due to their high frequency, concurrent *GBA1* variants are not uncommon, and a combination of a disease-relevant *LRRK2* and *GBA1* variant can be found in 0.2% of PD patients (Cook et al. 2024; Westenberger et al. 2024). The *LRRK2* gene encodes leucine-rich repeat kinase 2 (LRRK2), a large enzyme whose dimers are located in the cytoplasm or at the outer membrane of the mitochondria. LRRK2 kinase activity includes autophosphorylation and phosphorylation of its numerous interactors. Very recently, a pathogenic variant in a direct LRRK2 interactor, Rab32 (Waschbüsch et al. 2014), was found to cause a new form of autosomal-dominant PD (Gustavsson et al. 2024). In addition, among LRRK2's interactors are some of the well-established PD-related proteins (i.e., Parkin, α-synuclein, and VPS35) (Table 1; Mir et al. 2018; Bonello et al. 2019; O'Hara et al. 2020). Considered pathogenic mechanisms of *LRRK2* variants include alterations in vesicular trafficking, cytoskeletal dynamics, autophagy, lysosomal and mitochondrial function, immune response, and neurotransmission (Fig. 1). Of the 26 *LRRK2* variants scored as definitely/probably/possibly pathogenic in the MDSGene, by far, the most frequent *LRRK2* variant is c.6055G>A (p. Gly2019Ser) in the kinase domain (Trinh et al. 2018). This and other pathogenic *LRRK2* alleles result in biochemically active proteins, and numerous studies showed an increased kinase activity (Kalogeropulou et al. 2022; Borsche et al. 2023). Thus, the currently proposed mechanism of action of disease-causing *LRRK2* changes is a toxic gain of function. Accordingly, the first LRRK2 kinase-inhibiting compounds were able to demonstrate target and pathway engagement in patients with PD and healthy volunteers (Jennings et al. 2022, 2023). Along the same lines, LRRK2 loss-of-function/loss-of-protein (i.e., nonsense/frame-shift/copy-number) variants seem to have no phenotypic impact in the heterozygous or even homozygous state, as they have been found in control individuals (Whiffin et al. 2020; Beetz et al. 2021). The fact that even the complete absence of LRRK2 may have no phenotypic consequence represents conceptually novel support for the safety of currently pursued therapeutic strategies for patients with kinase activity–boosting LRRK2 mutations. Importantly, recent experimental evidence indicates that up to 23 variants act by robustly stimulating LRRK2 kinase activity, and PD patients with these changes may particularly benefit from LRRK2 inhibitor treatments (Kalogeropulou et al. 2022).

The age-related penetrance of *LRRK2* mutations is low, and it differs among different populations (Tolosa et al. 2020). For example, in Tunisia, *LRRK2* p.G2019S carriers have a median AAO that is 10 years earlier than in Norway (e.g., 86% vs. 43%, respectively, at the age of 70 years) (Hentati et al. 2014), while the penetrance of the same variant was 25%–43% at age 80 years in Ashkenazi Jewish carriers of the same variant versus their and non-Ashkenazi Jewish relatives

(Lee et al. 2017). As mentioned above (age-related), penetrance is influenced by genetic, mitochondrial, and environmental factors. The associations of single variants or polygenic risk score with a higher penetrance of PD among p.G2019S *LRRK2* carriers reported to date, illustrates the complex interplay between genetic risk variants and causative *LRRK2* mutations (Trinh et al. 2014; Iwaki et al. 2020; Lai et al. 2021). In addition, several mitochondria-related features (the presence of large mitochondrial DNA [mtDNA] deletions, reduced NADH dehydrogenase activity, elevated mitochondrial mass, and mtDNA copy numbers) correlated with PD status in manifesting versus nonmanifesting carriers of the *LRRK2* mutations, suggesting the role in mtDNA dyshomeostasis in the penetrance of PARK-*LRRK2* (Ouzren et al. 2019; Delcambre et al. 2020). Finally, tobacco use was associated with a later median AAO (8 years), while black tea consumption had an independent but additive effect when combined with tobacco use, indicating that both of these lifestyle factors have a delaying effect on AAO in PARK-*LRRK2* (Lüth et al. 2020).

PARK-VPS35

PARK-*VPS35* is a very rare autosomal dominant PD type, the clinical presentation of which is very similar to idiopathic PD with a median AAO of 52 years (Wider et al. 2008; Trinh et al. 2018). To date, <20 different potentially disease-causing missense variants in the *VPS35* gene have been reported with the c.858G>A transition (p.Asp620Asn) being by far the most frequent and the only variant proven to be definitely pathogenic (Trinh et al. 2018).

The vacuolar protein sorting 35 (VPS35) protein is nearly ubiquitously expressed in human tissues and is a component of the retromer complex that transports endosomal cargo proteins between endosomal compartments and the Golgi apparatus or cell surface. The pathogenic mechanism of *VPS35* variants in PD is not clear and the current knowledge and hypotheses are summarized in Sassone et al. (2021) and Williams et al. (2022). Briefly, *VPS35* mutations might lead to impaired lysosomal and autophagy function, neurotransmission, and mitochondrial homeostasis (Fig. 1; Zavodszky et al. 2014; Tang et al. 2015a,b; Wang et al. 2016). Furthermore, VPS35 seems to interact with the α-synuclein, Parkin, and LRRK2 proteins (Table 1; Zavodszky et al. 2014; Williams et al. 2018; Ma et al. 2021; Pal et al. 2023b).

PARK-CHCHD2

PARK-CHCHD2 is caused by heterozygous missense and loss-of-protein variants in the CHCHD2 gene. After the initial description of a large Japanese family harboring the c.182C>T (p.Thr61Ile) pathogenic variant, only a handful of mostly Asian PARK-CHCHD2 patients have been described worldwide (Funayama et al. 2015; Koschmidder et al. 2016; Lee et al. 2018). Thus, variants in CHCHD2 are considered a confirmed cause of PD, although overall very rare (Lange et al. 2022) and not found, for example, in the ~12,500 patients included in the ROPAD study (Westenberger et al. 2024). PARK-CHCHD2 manifests as a typical levodopa-responsive PD form similar to idiopathic PD, with a mean AAO of more than 55 years.

CHCHD2 encodes coiled-coil-helix–coiled-coil-helix domain containing 2 (CHCHD2), a small protein localized in the mitochondrial intermembrane space, where it regulates mitochondrial electron transport and maintains mitochondrial crista integrity (Meng et al. 2017).

PARK-RAB32

Through a hypothesis-driven candidate approach that considered the role of RAB GTPases as both regulators and substrates of LRRK2, RAB32 was identified as the most recent autosomal dominant late-onset PD gene (Gustavsson et al. 2024). Namely, the c.213C>G RAB32 variant encoding the p.Ser71Arg missense mutation cosegregated with PD in three families and was detected in 13 additional patients (Gustavsson et al. 2024). Remarkably, this variant seems to be a founder variant as all carriers, although originating from at least seven different countries, shared a common haplotype. The average AAO was 55 years, and the clinical presentation of most patients included tremor as the initial symptom and, in general, resembled idiopathic PD (Gustavsson et al. 2024).

The RAB32 protein interacts with LRRK2, and the initial functional evidence suggests that the p.Ser71Arg variant enhances RAB32-mediated LRRK2 kinase activation (Table 1; Gustavsson et al. 2024). In addition, it seems that this missense variant decreases the ability of RAB32 to colocalize with the PINK1 protein (Gustavsson et al. 2024).

As currently, only one RAB32 variant from a single founder was described, it is unclear whether this variant arose independently in (other) independent patients like in carriers of the most frequent LRRK2 and VPS35 variants (Zabetian et al. 2006; Ando et al. 2012) and whether there are other pathogenic variants in this small gene. Thus, the extent of genotypic variability of PARK-RAB32 remains to be investigated in future studies.

Monogenic Autosomal-Recessive Forms of PD

PARK-PRKN

PARK-PRKN was the first early-onset PD form described, and with over 1000 patients reported in the literature, it remains the most prominent recessively inherited PD type (Kasten et al. 2018). Furthermore, after variants in GBA1 and LRRK2, changes in PRKN are the third most common cause of genetic PD (Cook et al. 2024; Westenberger et al. 2024). Pathogenic or likely pathogenic PRKN variants are found in 0.7%–1.0% of PD patients in general, in 0.7% of sporadic PD patients, and in 1.7% of PD patients with a positive family history (Cook et al. 2024; Westenberger et al. 2024).

The median AAO of PARK-PRKN is ~30 years (Kasten et al. 2018; Zhao et al. 2020; Vollstedt et al. 2023b; Menon et al. 2024) with a wide range between the ages of 3 and 81 years, which in part could be explained by the type and position of mutation, as well as the number of exons involved (Menon et al. 2024). Juvenile PD as defined by an AAO below the age of 20 years is present in 16%–20% of PARK-PRKN patients

(Kasten et al. 2018; Vollstedt et al. 2023b; Menon et al. 2024); dystonia, dyskinesias, and motor fluctuations were reported in 15%–20%. The disease course is slower, and motor fluctuations occur later than in idiopathic PD (Menon et al. 2024). Cognitive decline is uncommon, whereas other nonmotor symptoms are not infrequent (Kasten et al. 2018; Menon et al. 2024). A good treatment response is usually observed (>95%). Approximately 150 various, mostly loss-of-protein and missense *PRKN* sequence variants have been identified in PARK-*PRKN* (Kasten et al. 2018; Menon et al. 2024). Copy-number variants, including deletions of single or multiple *PRKN* exons, may be found more frequently than pathogenic small nucleotide changes in PARK-*PRKN* (Kasten et al. 2018; Zhao et al. 2020; Menon et al. 2024; Westenberger et al. 2024), highlighting the importance of comprehensive variant testing. Very recently, a large 7 Mb inversion involving breakpoints outside of the *PRKN* gene has been identified by long-read sequencing in monozygotic twins in whom only a single heterozygous *PRKN* mutation was identified by initial standard genetic testing, despite their typical *PRKN* clinical presentation (Daida et al. 2023). This indicates that contemporary technological approaches that allow for advanced analysis of complex structural variants that were undetectable by conventional methods are instrumental in molecular diagnosis of previously unresolved early-onset PD cases.

The *PRKN* gene encodes an E3 ubiquitin ligase Parkin that is involved in the ubiquitination of numerous proteins. Together with another early-onset PD protein, PINK1, Parkin functions in a common pathway for sensing and selectively eliminating damaged mitochondria from the mitochondrial network through the autophagy–lysosome pathway as recently reviewed in Pickrell and Youle (2015) and Ge et al. (2020). Parkin depends on PINK1 for the recruitment to dysfunctional mitochondria (Fig. 1; Table 1). Once recruited, Parkin becomes enzymatically active and exhibits its E3 ligase role needed for ubiquitinating the damaged mitochondria and tagging them in this way for autophagic clearance by lysosomes. Furthermore, Parkin is selectively activated at human dopaminergic synapses and neurons from PD patients with *PRKN* mutations display defective recycling of synaptic vesicles, leading to the accumulation of toxic oxidized dopamine. However, Parkin recruitment to synaptic vesicles is independent of PINK1 (Song et al. 2023).

PARK-PINK1

With >150 patients described in the literature, PARK-*PINK1* represents the second most frequent recessively inherited PD type. The median AAO of PARK-*PINK1* is comparable to PARK-*PRKN* (Lange and Klein 2010; Kasten et al. 2018). However, *PINK1* variants appear to be more frequently associated with dystonia, dyskinesias, and motor fluctuations (in about one-third of the patients) and with cognitive decline and psychotic symptoms (14% and 8%). Almost all individuals had a good response to levodopa treatment (Kasten et al. 2018).

To date, >200 different disease-causing sequence variants have been reported in *PINK1* (www.mdsgene.org). Over 60% of the variants reported in the unrelated patients were small sequence changes that were mostly missense while 20% were structural variants (Kasten et al. 2018).

The *PINK1* gene encodes protein phosphatase and tensin homolog (PTEN)-induced kinase 1 that consists of an amino-terminal mitochondrial-targeting sequence and a carboxy-terminal kinase domain. As mentioned above, PINK1 acts together with Parkin to clear away damaged mitochondria from cells. It senses the misfolded proteins in the mitochondrial matrix and becomes stabilized and accumulated on dysfunctional mitochondria with lower membrane potential (Fig. 1). Subsequently, PINK1 phosphorylates Parkin inducing Parkin's translocation to mitochondria, which in further cascade of events leads to mitophagy (Pickrell and Youle 2015; Ge et al. 2020).

PARK-PARK7

PARK-*PARK7*, with <50 patients reported to date, represents an extremely rare genetic type of PD. Most patients reported to date were from Italy, Iran, or Turkey. The median AAO is <30 years (Kasten et al. 2018), with over 95% of patients having an early (>80%) or even juvenile (>10%) PD. Apart from PD cardinal signs, about half manifested with dystonia, approximately one-quarter had dyskinesias, and one-fifth cognitive decline or psychotic symptoms (Kasten et al. 2018). To date, only about 20, mostly missense (50%) or loss-of-protein (25%) *PARK7* variants have been reported (Kasten et al. 2018). The wild-type PARK7 protein, also known as protein deglycase DJ-1, functions as an oxidative stress sensor and an antioxidant. It is also involved in multiple mechanisms that are protective against neurodegeneration, such as its role as a molecular chaperone, protease, glyoxalase, and transcriptional regulator (Dolgacheva et al. 2019; Skou et al. 2024).

ATYPICAL MONOGENIC PARKINSONISM

From a genetic standpoint, atypical PD or atypical parkinsonism can be divided into two distinct entities: (1) monogenic forms clinically overlapping with PD, but typically characterized by an early age at onset, a rapidly progressive clinical course, and numerous additional clinical features not compatible with a diagnosis of idiopathic PD, and (2) late-onset forms, such as multiple system atrophy (MSA), progressive supranuclear palsy (PSP), or corticobasal degeneration (CBD), the etiology of which is considered to include a genetic contribution, but no monogenic causes have been established to date.

The spectrum of genes that, when mutated, can cause atypical parkinsonism or parkinsonian features as part of other neurological syndromes is very broad, especially when including parkinsonism and dystonia–parkinsonism in childhood-onset conditions (Lange et al. 2022; Pérez-Dueñas et al. 2022; Weissbach et al. 2022). Therefore, we here focus on six well-established forms of atypical monogenic parkinsonism with an average AAO in the second decade of life or later. A detailed description of these six monogenic forms can be found on the MDSGene website (www.mdsgene.org) and in Wittke et al. (2021), including pathogenic variants in *ATP13A2*, *DNAJC6*, *FBXO7*, *SYNJ1*, *VPS13C*, and *DCTN1*. These forms, even when taken together, are very rare,

and for PARK-*SYNJ1* and PARK-*VPS13C*, fewer than 10 families each were included in the systematic literature review (Wittke et al. 2021). The median AAO across all forms is 24 years, which is considerably lower for the five recessive forms than in the only dominant form caused by *DCTN1* pathogenic variants, which presents later (49 years). Indeed, when estimating the clinical relevance using 67 variables to distinguish between these six forms, AAO has the highest relative importance. The first 10 of these 67 clinical features contribute 87% of the classification accuracy and, besides AAO, include upper motor neuron involvement, hypoventilation, decreasing body weight, mini-myoclonus, vertical gaze palsy, autonomic symptoms, other nonmotor signs, such as cognitive decline, as well as a limited response to L-dopa treatment (Wittke et al. 2021). Some of these features are shared across several forms, such as cognitive decline in most of the monogenic atypical forms, as well as in PSP and CBD, or autonomic symptoms that are found in both monogenic atypical parkinsonism and in MSA. Some specific clinical features, such as vertical gaze palsy, are typical of patients with pathogenic variants in *ATP13A2*. Yet other clinical signs can be considered typical of certain conditions, such as mini-myoclonus in PARK-*ATP13A2* or rapid weight loss combined with respiratory complications in PARK-*DCTN1* (Table 1).

MSA-like clinical features may result from pathogenic variants in various genes linked to either PD (*SNCA*, *LRRK2*, *GBA1*) or repeat expansion disorders (SCA-related genes, *C9orf72*, *RFC1*, *NOTCH2NLC*) or genes related to inflammation (Tseng et al. 2023). Genes implicated in MSA risk have been identified through genome-wide association study (GWAS) approaches and include *FBXO47*, *ELOVL7*, *EDN1*, and *MAPT* (Sailer et al. 2016). A more recent autopsy-confirmed GWAS revealed common variants near *ZIC1* and *ZIC4* (Hopfner et al. 2022). Unlike the synucleinopathy MSA, PSP belongs to the group of tauopathies. Pathogenic variants in *MAPT*—first identified as causative of frontotemporal dementia and parkinsonism in 1998 (Hutton et al. 1998)—were later linked to rare, familial forms of PSP or PSP-like conditions

(Wen et al. 2021). GWAS also identified *MAPT*, along with *STX6*, *EIF2AK3*, and *MOBP*, to play a role in PSP genetic risk (Höglinger et al. 2011). A more recent GWAS implicated genetic variation at the *LRRK2* locus in the survival of PSP (Jabbari et al. 2021). Interestingly, the neuropathology of PARK-*LRRK2* can vary and may include tau deposition (Wider et al. 2010). Furthermore, shared genetic risk was discovered across CBD and PSP (Kouri et al. 2015; Yokoyama et al. 2017). Although monogenic and nonmonogenic atypical parkinsonism share several clinical features, the vastly different age of onset serves as a clear discriminator (Wittke et al. 2021).

PD GENETIC RISK FACTORS

PARK-GBA1

Biallelic mutations in the *GBA1* gene encoding the glucocerebrosidase A (GCase) protein cause Gaucher's disease (GD), a recessively inherited lysosomal storage disorder (Tsuji et al. 1987). However, it has been observed that patients with GD and asymptomatic carriers of heterozygous variants have an increased risk for developing PD (Sidransky et al. 2009). Heterozygous pathogenic *GBA1* variants are now considered the most common genetic and strongest risk factor for PD, found in ~10% of all PD patients and >12% of those with a positive family history (Cook et al. 2024; Westenberger et al. 2024). Although *GBA1* variants are considered a risk factor with a relatively "high penetrance" with increasing age, they might also be viewed as acting in a dominant manner with overall highly reduced penetrance. Of note, the penetrance of some *GBA1* variants may be close to that of *LRRK2* mutations (Gan-Or et al. 2015; Lee et al. 2017; Shukla et al. 2019). Compared to idiopathic PD, the AAO is lower in PARK-*GBA1* and the rate of motor progression is faster, and they usually more frequently manifest nonmotor symptoms (Cilia et al. 2016; Gan-Or et al. 2018; Westenberger et al. 2024) although this depends on the specific variant (Usnich et al. 2023).

GCase is a lysosomal hydrolase the role of which is to degrade glucocerebroside into ceramide and glucose. In general, it is considered that *GBA1* variants exhibit their neurotoxic effect by reducing the activity or levels of GCase, and the various known and hypothesized mechanisms have been recently reviewed (Menozzi et al. 2023). Interestingly, reduced GCase levels were found in various tissues of PD patients with or without *GBA1* variants when compared to controls (for review, see Gan-Or et al. 2018). Of note, glucocerebroside and α-synuclein form a bidirectional pathogenic loop, as glucocerebroside leads to accumulation of aggregated α-synuclein, and α-synuclein accumulation results in reduced GCase activity (Mazzulli et al. 2011; Zunke et al. 2018).

To date, over 80 studies reported 371 unique *GBA1* variants in PD (Parlar et al. 2023). These variants are listed in the PARK-*GBA1* browser (pdgenetics.shinyapps.io/GBA1Browser) and classified as (1) mild variants (causing the mildest form of GD; type I), (2) severe variants (causing the more severe GD forms: type II and type III), (3) risk variants for PD (*GBA1* changes, associated only with PD and not with GD), or (4) variants of unknown severity (Parlar et al. 2023). The above-mentioned variants include missense, nonsense, splice-site, and structural DNA changes (Gabbert et al. 2023; Parlar et al. 2023). Importantly, these types of variants have different penetrance (i.e., severe GD-related variants have higher penetrance than PD risk factors) (Balestrino et al. 2020; Westenberger et al. 2024) and seem to have different effects on PD AAO and progression (motor symptoms and cognition decline faster in carriers of severe *GBA1* variants in comparison to mild) (Cilia et al. 2016; Liu et al. 2016; Thaler et al. 2018; Ren et al. 2022; Zhou et al. 2023). Interestingly, a novel common risk factor for risk of PD and AAO was recently found at the *GBA1* locus in individuals of African or African admixed ancestry, which was rare in non-African populations and associated with decreased glucocerebrosidase activity (Rizig et al. 2023).

Genetic Risk Factors Contributing to Complex PD

Over 80% of patients are affected by nonmonogenic, complex PD arising from the interplay of a polygenic contribution of multiple smaller-effect genetic factors and environmental influences. The results of numerous GWAS, meta-GWAS, and candidate-gene association studies are summarized in several recent reviews (Billingsley et al. 2018; Blauwendraat et al. 2020; Grenn et al. 2020; Khani et al. 2024) and available using a PD GWAS browser tool (pdgenetics.shinyapps.io/GWASBrowser). Interestingly, the most statistically significant signals resulting from these studies are common variants located within or in the proximity of some of the major genes involved in monogenic neurodegenerative disorders: *SNCA*, *LRRK2*, and *MAPT* (Simón-Sánchez et al. 2009; Nalls et al. 2019; Foo et al. 2020). Similarly, *LRRK2* and *GBA1* came up in the first large-scale PD rare variant (burden) analysis (Makarious et al. 2023). The most recent PD GWAS meta-analysis investigated ~50,000 multi-ancestry PD patients, ~20,000 proxy cases, and 2.5 million controls. The 78 identified independent genome-wide significant loci included 12 potentially novel ones (*MTF2*, *PIK3CA*, *ADD1*, *SYBU*, *IRS2*, *USP8*, *PIGL*, *FASN*, *MYLK2*, *USP25*, *EP300*, and *PPP6R2*) and fine-mapped six putative causal variants at known PD loci (Kim et al. 2024).

FUNCTIONAL VALIDATION, INTERPRETATION, AND CLASSIFICATION OF PD-RELATED GENETIC VARIANTS

Benefits that are currently and potentially available to a PD patient upon receiving a result of genetic testing hinge on correctly interpreting the pathogenicity of uncovered variants. This interpretation and subsequent classification of genetic variants is a complex and challenging process standardized by a set of guidelines issued by the American College of Medical Genetics and Genomics and the Association for Molecular Pathology (ACMG/AMP). According to these recommendations, a five-tier terminology system to categorize variants as "pathogenic," "likely pathogenic," "of uncertain significance," "likely benign," and "benign" was introduced (Richards et al. 2015). And while the designations containing words "pathogenic" or

"benign" offer a self-explanatory meaning to the genetic testing result and potentially the counseling and treatment direction, the variants of uncertain significance (VUSs) are a somewhat of a confusing gray zone to patients and their physicians as they may or may not have a role in the disease pathogenesis. The correct understanding of their genotype–phenotype correlations and functional roles (through experimental validation) represents a groundwork for the potential reclassification of VUSs into a disease-relevant or irrelevant DNA change. Indeed, the lack of PD-related phenotype in carriers of *LRRK2* loss-of-protein variants (that are otherwise frequently pathogenic when found in other genes), indicates that these are more likely VUSs or benign as opposed to disease-causing (Whiffin et al. 2020; Beetz et al. 2021). Furthermore, based on a recent large-scale functional testing effort, the effect of ~20 *LRRK2* variants was shown, aiding their classification to "likely pathogenic" and even more importantly suggesting that carriers of these variants may benefit from LRRK2 inhibitory therapies (Kalogeropulou et al. 2022). Thus, the novel evidence allows the pathogenicity level of a variant to be reevaluated and potentially reclassified, which might have far-reaching impacts on medical decision-making and patient well-being.

GENE-TARGETED TRIALS AND THERAPIES

The characterization of gene-specific pathways has facilitated the development of targeted therapies, the first of which have been tested in early phase clinical trials in patients. Due to their relative frequency, most therapies so far have been developed for patients with *GBA1* and *LRRK2* variants, which are presented in the next two sections.

Clinical Trials in PARK-*GBA1*

Pathogenic variants in *GBA1* cause reduced levels of GCase activity predisposing patients to lysosomal dysfunction, formation of α-synuclein fibrils, and impairment of multiple distinct pathways involved in PD pathogenesis, for example, mitochondrial dysfunction (Kim et al. 2021; Baden et al. 2023) and activation of neuro-inflammatory responses (Bo et al. 2022). GCase activity reduction is strongest in carriers of severe *GBA1* variants, less pronounced in carriers of mild variants, and only marginally altered in most patients with idiopathic PD (Alcalay et al. 2015a).

The therapeutic strategies in PARK-*GBA1* currently under investigation are substrate reduction for GCase to lessen the burden of its compromised function, the enhancement of GCase activity, GCase enzyme replacement, and GCase overexpression. Substrate reduction therapy is an established therapeutic principle in GD to suppress glycosphingolipid accumulation, which otherwise predisposes patients to neuronal α-synuclein aggregation (Fredriksen et al. 2021). Although the administration of the brain-penetrating glucosylceramide synthase inhibitor venglustat was safe, well tolerated, and led to target engagement, no clinical beneficial effects have yet been demonstrated (Table 2). The well-established over-the-counter cough expectorant and chaperone molecule ambroxol was shown to increase neuronal GCase activity, reduce α-synuclein levels, and rescue the loss of dopaminergic neurons in previous preclinical studies (Gegg et al. 2022). In the first small phase 2a trial in PD patients with and without *GBA1* variants, very high dosages of oral ambroxol led to penetration into the CSF, target engagement, and potential clinical benefit although this study was not controlled (Table 2). The efficacy of ambroxol and another small molecule enhancing GCase function is currently investigated in clinical trials enrolling patients with PARK-*GBA1* and Lewy body dementia (LBD). The major challenge for enzyme replacement therapy (ERT) in lysosomal diseases involving the brain is the inability of ERTs to cross the blood–brain barrier. One attempt to potentially allow penetration into the central nervous system is the transient opening of the blood–brain barrier by MR-guided focused ultrasound (MRgFUS) (Table 2). Another ongoing gene therapeutic trial investigates the intracisternal administration of a viral vector to facilitate GCase overexpression (Table 2).

Aside from *GBA1*-targeted therapies, patients with PARK-*GBA1* may also particularly benefit from α-synuclein and LRRK2-targeted strategies due to the high burden of α-synuclein aggregation in many *GBA1* variant carriers and the potential interaction between LRRK2 kinase activity and GCase function (Ysselstein et al. 2019). The exact mechanisms of the LRRK2–GBA1 interaction are still to be explored and may be cell-type-specific (Kedariti et al. 2022).

Clinical Trials in PARK-*LRRK2*

The main therapeutic strategy in PARK-*LRRK2* currently under investigation is the down-regulation of LRRK2 kinase activity by administering small molecules orally or antisense oligonucleotides intrathecally (Table 3). The first studies indicate safety and tolerability in healthy volunteers and PD patients, successful enrichment of BIIB122/DNL151 and DNL201 in the cerebrospinal fluid, inhibition of peripheral LRRK2 kinase activity, target inhibition, and downstream effects on lysosomal biomarkers. Aberrant LRRK2 signaling also seems to play a role in the pathogenesis of idiopathic PD contributing to neuroinflammation, lysosomal, and mitochondrial dysfunction (Di Maio et al. 2018). This is reflected by the inclusion of PD patients without pathogenic *LRRK2* variants in LRRK2-targeted trials. Recently, a phase 3 trial investigating BIIB122/DNL151 in patients with PARK-*LRRK2* was stopped due to the anticipated long recruitment phase, in favor of a parallel phase 2b trial that does not take the genotype into account. A specific emphasis in LRRK2 kinase-down-regulating trials is laid on pulmonary function due to the observation of morphological changes in the lungs of nonhuman primates (Fuji et al. 2015). These changes have been reversible after drug withdrawal and were not associated with compromised pulmonary function preclinically (Baptista et al. 2020). Furthermore, no safety concerns were raised in the first early studies in humans (Table 3), and rare loss-of-function/loss-of-protein variants have not yet been associated with a pulmonary phenotype (Whiffin et al. 2020).

Table 2. Completed and ongoing clinical trials in PARK-GBA1

Compound	Registration number	Strategy	Phase	Outcomes	Target population (number)	References
Completed						
Ambroxol	NCT02941822	GCase activation	Phase 2a	Safety and tolerability+, biomarkers+ (efficacy+)	PARK-GBA1 ($n = 8$), PD ($n = 10$)	Mullin et al. 2020
Cerezyme	NCT04370665	Enzyme replacement therapy, disruption of putaminal BBB by MRgFUS	1	Safety and feasibility+	PARK-GBA1 ($n = 4$)	Meng et al. 2022
LTI-291	NL6574	Allosteric modulator and activator of GCase	1b	Safety and tolerability+, biomarkers+	PARK-GBA1 ($n = 40$)	den Heijer et al. 2023
Venglustat	NCT02906020	Substrate reduction, glucosylceramide synthase inhibition	2, part 1	Safety and tolerability+, target engagement+	PARK-GBA1 ($n = 29$)	Peterschmitt et al. 2022
Venglustat	NCT02906020	Substrate reduction, glucosylceramide synthase inhibition	2, part 2	Safety and tolerability+, no clinical benefit	PARK-GBA1 ($n = 221$)	Giladi et al. 2023
Ongoing						
Ambroxol	NCT02914366	GCase activation	2	Efficacy, biomarkers	LBD ($n = 75$)	Silveira et al. 2019[a]
Ambroxol	NCT05830396	GCase activation	2/3	Efficacy, safety, tolerability, biomarkers	PARK-GBA1 ($n = 80$)	None
BIA 28-6156	NCT05819359	GCase activation	2	Efficacy, safety, tolerability	PARK-GBA1 ($n = 237$)	None
LY3884961	NCT04127578	Gene therapy, GCase overexpression by intracisternal injection of viral AAV9 vector	1/2a	Safety, tolerability, immunogenicity, biomarkers	PARK-GBA1 ($n = 20$)	None

(BBB) Blood–brain barrier, (LBD) Lewy body dementia, (MRgFUS) MR-guided focused ultrasound, (PD) Parkinson's disease.
[a]Published study protocol, active, not recruiting.

Table 3. Completed and ongoing clinical trials in PARK-*LRRK2*

Compound	Registration number	Strategy	Phase	Outcomes	Target population (number)	References
Completed						
BIIB094	NCT03976349	Intrathecal injection of LRRK2-inhibiting ASO	1	Safety and tolerability	PARK-*LRRK2* and idiopathic PD (*n* = 82)	[a]
BIIB122 (DNL151)	NCT04056689	LRRK2 kinase inhibition	1b	Safety and tolerability+, biomarkers+	PD (*n* = 36)	Jennings et al. 2023
DNL201	NCT03710707	LRRK2 kinase inhibition	1b	Safety and tolerability+, pathway engagement+	PD (*n* = 28)	Jennings et al. 2022
Ongoing						
BIIB122 (DNL151)	NCT05348785	LRRK2 kinase inhibition	2b	Efficacy, safety	PD (*n* = 640)	None

(ASO) Antisense oligonucleotide, (PD) Parkinson's disease.
[a]Study completed, results not yet published.

CONCLUDING REMARKS

The field of PD genetics has seen unprecedented growth in the past 5 years. This includes, but is not limited to, the discovery of novel genes and risk factors, the establishment of global networks to elucidate the genetic architecture of PD at a multiethnic level, the completion of large-scale studies to determine the genotypic and phenotypic spectrum of PD, and the first gene-targeted treatment approaches. A recent, new classification of PD is based on biology and takes both genetic factors and penetrance into account with the aim of identifying at-risk individuals in the earliest possible stages for potential inclusion in future therapeutic trials of disease-modifying trials. For this, the GLP-1 agonist Lixenatide has recently emerged as the first potential candidate (Meissner et al. 2024).

ACKNOWLEDGMENTS

A.W., N.B., and C.K. are supported by the DFG (FOR2488) and C.K. receives funding for Parkinson's disease (PD) genetics work through ASAP (Global Parkinson's Genetics Program).

REFERENCES

Alcalay RN, Levy OA, Waters CC, Fahn S, Ford B, Kuo SH, Mazzoni P, Pauciulo MW, Nichols WC, Gan-Or Z, et al. 2015a. Glucocerebrosidase activity in Parkinson's disease with and without *GBA* mutations. *Brain* 138: 2648–2658. doi:10.1093/brain/awv179

Alcalay RN, Mejia-Santana H, Mirelman A, Saunders-Pullman R, Raymond D, Palmese C, Caccappolo E, Ozelius L, Orr-Urtreger A, Clark L, et al. 2015b. Neuropsychological performance in LRRK2 G2019S carriers with Parkinson's disease. *Parkinsonism Relat Disord* 21: 106–110. doi:10.1016/j.parkreldis.2014.09.033

Ando M, Funayama M, Li Y, Kashihara K, Murakami Y, Ishizu N, Toyoda C, Noguchi K, Hashimoto T, Nakano N, et al. 2012. *VPS35* mutation in Japanese patients with typical Parkinson's disease. *Mov Disord* 27: 1413–1417. doi:10.1002/mds.25145

Avenali M, Zangaglia R, Cuconato G, Palmieri I, Albanese A, Artusi CA, Bozzali M, Calandra-Buonaura G, Cavallieri F, Cilia R, et al. 2023. Are patients with GBA-Parkinson disease good candidates for deep brain stimulation? A longitudinal multicentric study on a large Italian cohort. *J Neurol Neurosurg Psychiatry* 95: 309–315.

Badanjak K, Fixemer S, Smajić S, Skupin A, Grünewald A. 2021. The contribution of microglia to neuroinflammation in Parkinson's disease. *Int J Mol Sci* 22: 4676. doi:10.3390/ijms22094676

Baden P, Perez MJ, Raji H, Bertoli F, Kalb S, Illescas M, Spanos F, Giuliano C, Calogero AM, Oldrati M, et al. 2023. Glucocerebrosidase is imported into mitochondria and preserves complex I integrity and energy metabolism. *Nat Commun* 14: 1930. doi:10.1038/s41467-023-37454-4

Balestrino R, Tunesi S, Tesei S, Lopiano L, Zecchinelli AL, Goldwurm S. 2020. Penetrance of glucocerebrosidase (*GBA*) mutations in Parkinson's disease: a kin cohort study. *Mov Disord* 35: 2111–2114. doi:10.1002/mds.28 200

Baptista MAS, Merchant K, Barrett T, Bhargava S, Bryce DK, Michael Ellis J, Estrada AA, Fell MJ, Fiske BK, Fuji RN, et al. 2020. LRRK2 inhibitors induce reversible changes in nonhuman primate lungs without measurable pulmonary deficits. *Sci Transl Med* 12: eaav0820. doi:10.1126/scitranslmed.aav0820

Barker RA, Björklund A. 2023. Restorative cell and gene therapies for Parkinson's disease. *Handb Clin Neurol* 193: 211–226. doi:10.1016/B978-0-323-85555-6.00012-6

Beetz C, Westenberger A, Al-Ali R, Ameziane N, Alhashmi N, Boustany RM, Al Mutairi F, Alfadhel M, Al-Hassnan Z, AlSayed M, et al. 2021. *LRRK2* loss-of-function variants in patients with rare diseases: no evidence for a phenotypic impact. *Mov Disord* 36: 1029–1031. doi:10.1002/mds.28452

Billingsley KJ, Bandres-Ciga S, Saez-Atienzar S, Singleton AB. 2018. Genetic risk factors in Parkinson's disease. *Cell Tissue Res* 373: 9–20. doi:10.1007/s00441-018-2817-y

Blauwendraat C, Nalls MA, Singleton AB. 2020. The genetic architecture of Parkinson's disease. *Lancet Neurol* 19: 170–178. doi:10.1016/S1474-4422(19)30287-X

Bloem BR, Okun MS, Klein C. 2021. Parkinson's disease. *Lancet* 397: 2284–2303. doi:10.1016/S0140-6736(21)00218-X

Bo RX, Li YY, Zhou TT, Chen NH, Yuan YH. 2022. The neuroinflammatory role of glucocerebrosidase in Parkinson's disease. *Neuropharmacology* 207: 108964. doi:10.1016/j.neuropharm.2022.108964

Bonello F, Hassoun SM, Mouton-Liger F, Shin YS, Muscat A, Tesson C, Lesage S, Beart PM, Brice A, Krupp J, et al. 2019. LRRK2 impairs PINK1/Parkin-dependent mitophagy via its kinase activity: pathologic insights into Parkinson's disease. *Hum Mol Genet* 28: 1645–1660. doi:10.1093/hmg/ddz004

Book A, Guella I, Candido T, Brice A, Hattori N, Jeon B, Farrer MJ, SNCA Multiplication Investigators of the GEoPD Consortium. 2018. A meta-analysis of α-synuclein multiplication in familial parkinsonism. *Front Neurol* 9: 1021. doi:10.3389/fneur.2018.01021

Borsche M, Pratuseviciute N, Schaake S, Hinrichs F, Morel G, Uter J, Lohmann K, Klein C, Alessi DR, Hagenah J, et al. 2023. The new p.F1700L LRRK2 variant causes Parkinson's disease by extensively increasing kinase activity. *Mov Disord* 38: 1105–1107. doi:10.1002/mds.29385

Bouhouche A, Tibar H, El Haj RB, El Bayad K, Razine R, Tazrout S, Skalli A, Bouslam N, Elouardi L, Benomar A, et al. 2017. *LRRK2* g2019s mutation: prevalence and clinical features in Moroccans with Parkinson's disease. *Parkinsons Dis* 2017: 2412486. doi:10.1155/2017/2412486

Calabresi P, Di Lazzaro G, Marino G, Campanelli F, Ghiglieri V. 2023a. Advances in understanding the function of α-synuclein: implications for Parkinson's disease. *Brain* 146: 3587–3597. doi:10.1093/brain/awad150

Calabresi P, Mechelli A, Natale G, Volpicelli-Daley L, Di Lazzaro G, Ghiglieri V. 2023b. α-Synuclein in Parkinson's disease and other synucleinopathies: from overt neuro-degeneration back to early synaptic dysfunction. *Cell Death Dis* 14: 176. doi:10.1038/s41419-023-05672-9

Cavallieri F, Cury RG, Guimarães T, Fioravanti V, Grisanti S, Rossi J, Monfrini E, Zedde M, Di Fonzo A, Valzania F, et al. 2023. Recent advances in the treatment of genetic forms of Parkinson's disease: hype or hope? *Cells* 12: 764. doi:10.3390/cells12050764

Cilia R, Tunesi S, Marotta G, Cereda E, Siri C, Tesei S, Zecchinelli AL, Canesi M, Mariani CB, Meucci N, et al. 2016. Survival and dementia in *GBA*-associated Parkinson's disease: the mutation matters. *Ann Neurol* 80: 662–673. doi:10.1002/ana.24777

Cook L, Verbrugge J, Schwantes-An TH, Foroud T, Marder KS, Mata I, Mencacci NE, Alcalay RN. 2024. Parkinson's disease variant detection and disclosure: PD GENEration, a North American study. *Brain*. In press.

Daida K, Funayama M, Billingsley KJ, Malik L, Miano-Burkhardt A, Leonard HL, Makarious MB, Iwaki H, Ding J, Gibbs JR, et al. 2023. Long-read sequencing resolves a complex structural variant in *PRKN* Parkinson's disease. *Mov Disord* 38: 2249–2257. doi:10.1002/mds.29610

Delcambre S, Ghelfi J, Ouzren N, Grandmougin L, Delbrouck C, Seibler P, Wasner K, Aasly JO, Klein C, Trinh J, et al. 2020. Mitochondrial mechanisms of LRRK2 G2019S penetrance. *Front Neurol* 11: 881. doi:10.3389/fneur.2020.00881

den Heijer JM, Kruithof AC, Moerland M, Walker M, Dudgeon L, Justman C, Solomini I, Splitalny L, Leymarie N, Khatri K, et al. 2023. A phase 1B trial in *GBA1*-associated Parkinson's disease of BIA-28-6156, a glucocerebrosidase activator. *Mov Disord* 38: 1197–1208. doi:10.1002/mds.29346

De Rosa A, De Michele G, Guacci A, Carbone R, Lieto M, Peluso S, Picillo M, Barone P, Salemi F, Laiso A, et al. 2014. Genetic screening for the LRRK2 R1441C and G2019S mutations in Parkinsonian patients from Campania. *J Parkinsons Dis* 4: 123–128. doi:10.3233/JPD-130312

Diaw SH, Borsche M, Streubel-Gallasch L, Dulovic-Mahlow M, Hermes J, Lenz I, Seibler P, Klein C, Brüggemann N, Vos M, et al. 2023. Characterization of the pathogenic α-synuclein variant V15A in Parkinson's disease. *NPJ Park Dis* 9: 148. doi:10.1038/s41531-023-00584-z

Di Maio R, Hoffman EK, Rocha EM, Keeney MT, Sanders LH, De Miranda BR, Zharikov A, Van Laar A, Stepan AF, Lanz TA, et al. 2018. LRRK2 activation in idiopathic Parkinson's disease. *Sci Transl Med* 10: eaar5429. doi:10.1126/scitranslmed.aar5429

Dolgacheva LP, Berezhnov AV, Fedotova EI, Zinchenko VP, Abramov AY. 2019. Role of DJ-1 in the mechanism of pathogenesis of Parkinson's disease. *J Bioenerg Biomembr* 51: 175–188. doi:10.1007/s10863-019-09798-4

Foo JN, Chew EGY, Chung SJ, Peng R, Blauwendraat C, Nalls MA, Mok KY, Satake W, Toda T, Chao Y, et al. 2020. Identification of risk loci for Parkinson disease in Asians and comparison of risk between Asians and Europeans: a genome-wide association study. *JAMA Neurol* 77: 746–754. doi:10.1001/jamaneurol.2020.0428

Fredriksen K, Aivazidis S, Sharma K, Burbidge KJ, Pitcairn C, Zunke F, Gelyana E, Mazzulli JR. 2021. Pathological α-syn aggregation is mediated by glycosphingolipid chain length and the physiological state of α-syn in vivo. *Proc Natl Acad Sci* **118:** e2108489118. doi:10.1073/pnas.2108489118

Fuji RN, Flagella M, Baca M, Baptista MAS, Brodbeck J, Chan BK, Fiske BK, Honigberg L, Jubb AM, Katavolos P, et al. 2015. Effect of selective LRRK2 kinase inhibition on nonhuman primate lung. *Sci Transl Med* **7:** 273ra15.

Funayama M, Ohe K, Amo T, Furuya N, Yamaguchi J, Saiki S, Li Y, Ogaki K, Ando M, Yoshino H, et al. 2015. CHCHD2 mutations in autosomal dominant late-onset Parkinson's disease: a genome-wide linkage and sequencing study. *Lancet Neurol* **14:** 274–282. doi:10.1016/S1474-4422(14)70266-2

Gabbert C, Schaake S, Lüth T, Much C, Klein C, Aasly JO, Farrer MJ, Trinh J. 2023. GBA1 in Parkinson's disease: variant detection and pathogenicity scoring matters. *BMC Genomics* **24:** 322. doi:10.1186/s12864-023-09417-y

Gan-Or Z, Amshalom I, Kilarski LL, Bar-Shira A, Gana-Weisz M, Mirelman A, Marder K, Bressman S, Giladi N, Orr-Urtreger A. 2015. Differential effects of severe vs mild GBA mutations on Parkinson disease. *Neurology* **84:** 880–887. doi:10.1212/WNL.0000000000001315

Gan-Or Z, Liong C, Alcalay RN. 2018. GBA-associated Parkinson's disease and other synucleinopathies. *Curr Neurol Neurosci Rep* **18:** 44. doi:10.1007/s11910-018-0860-4

Ge P, Dawson VL, Dawson TM. 2020. PINK1 and Parkin mitochondrial quality control: a source of regional vulnerability in Parkinson's disease. *Mol Neurodegener* **15:** 1–18. doi:10.1186/s13024-020-00367-7

Gegg ME, Menozzi E, Schapira AHV. 2022. Glucocerebrosidase-associated Parkinson disease: pathogenic mechanisms and potential drug treatments. *Neurobiol Dis* **166:** 105663. doi:10.1016/j.nbd.2022.105663

Giladi N, Alcalay RN, Cutter G, Gasser T, Gurevich T, Höglinger GU, Marek K, Pacchetti C, Schapira AHV, Scherzer CR, et al. 2023. Safety and efficacy of venglustat in GBA1-associated Parkinson's disease: an international, multicentre, double-blind, randomised, placebo-controlled, phase 2 trial. *Lancet Neurol* **22:** 661–671. doi:10.1016/S1474-4422(23)00205-3

Gorostidi A, Ruiz-Martínez J, Lopez De Munain A, Alzualde A, Martí Massó JF. 2009. LRRK2 g2019s and R1441G mutations associated with Parkinson's disease are common in the Basque Country, but relative prevalence is determined by ethnicity. *Neurogenetics* **10:** 157–159. doi:10.1007/s10048-008-0162-0

Grenn FP, Kim JJ, Makarious MB, Iwaki H, Illarionova A, Brolin K, Kluss JH, Schumacher-Schuh AF, Leonard H, Faghri F, et al. 2020. The Parkinson's disease genome-wide association study locus browser. *Mov Disord* **35:** 2056–2067. doi:10.1002/mds.28197

Guo Y, Sun Y, Song Z, Zheng W, Xiong W, Yang Y, Yuan L, Deng H. 2021. Genetic analysis and literature review of SNCA variants in Parkinson's disease. *Front Aging Neurosci* **13:** 648151. doi:10.3389/fnagi.2021.648151

Gustavsson EK, Follett J, Trinh J, Barodia SK, Real R, Liu Z, Grant-Peters M, Fox JD, Appel-Cresswell S, Stoessl AJ, et al. 2024. RAB32 ser71arg in autosomal dominant Parkinson's disease: linkage, association, and functional analyses. *Lancet Neurol* **23:** 603–614. doi:10.1016/S1474-4422(24)00121-2

Henderson MX, Sengupta M, Trojanowski JQ, Lee VMY. 2019. Alzheimer's disease tau is a prominent pathology in LRRK2 Parkinson's disease. *Acta Neuropathol Commun* **7:** 183. doi:10.1186/s40478-019-0836-x

Hentati F, Trinh J, Thompson C, Nosova E, Farrer MJ, Aasly JO. 2014. LRRK2 parkinsonism in Tunisia and Norway: a comparative analysis of disease penetrance. *Neurology* **83:** 568–569. doi:10.1212/WNL.0000000000000675

Hoenen C, Gustin A, Birck C, Kirchmeyer M, Beaume N, Felten P, Grandbarbe L, Heuschling P, Heurtaux T. 2016. α-Synuclein proteins promote pro-inflammatory cascades in microglia: stronger effects of the a53t mutant. *PLoS ONE* **11:** e0162717. doi:10.1371/journal.pone.0162717

Höglinger GU, Melhem NM, Dickson DW, Sleiman PMA, Wang LS, Klei L, Rademakers R, De Silva R, Litvan I, Riley DE, et al. 2011. Identification of common variants influencing risk of the tauopathy progressive supranuclear palsy. *Nat Genet* **43:** 699–705. doi:10.1038/ng.859

Höglinger GU, Adler CH, Berg D, Klein C, Outeiro TF, Poewe W, Postuma R, Stoessl AJ, Lang AE. 2024. A biological classification of Parkinson's disease: the SynNeurGe research diagnostic criteria. *Lancet Neurol* **23:** 191–204. doi:10.1016/S1474-4422(23)00404-0

Hopfner F, Tietz AK, Ruf VC, Ross OA, Koga S, Dickson D, Aguzzi A, Attems J, Beach T, Beller A, et al. 2022. Common variants near ZIC1 and ZIC4 in autopsy-confirmed multiple system atrophy. *Mov Disord* **37:** 2110–2121. doi:10.1002/mds.29164

Hutton M, Lendon CL, Rizzu P, Baker M, Froelich S, Houlden HH, Pickering-Brown S, Chakraverty S, Isaacs A, Grover A, et al. 1998. Association of missense and 5′-splice-site mutations in tau with the inherited dementia FTDP-17. *Nature* **393:** 702–705. doi:10.1038/31508

Iwaki H, Blauwendraat C, Makarious MB, Bandrés-Ciga S, Leonard HL, Gibbs JR, Hernandez DG, Scholz SW, Faghri F, Nalls MA, et al. 2020. Penetrance of Parkinson's disease in LRRK2 p.G2019S carriers is modified by a polygenic risk score. *Mov Disord* **35:** 774–780. doi:10.1002/mds.27974

Jabbari E, Koga S, Valentino RR, Reynolds RH, Ferrari R, Tan MMX, Rowe JB, Dalgard CL, Scholz SW, Dickson DW, et al. 2021. Genetic determinants of survival in progressive supranuclear palsy: a genome-wide association study. *Lancet Neurol* **20:** 107–116. doi:10.1016/S1474-4422(20)30394-X

Jennings D, Huntwork-Rodriguez S, Henry AG, Sasaki JC, Meisner R, Diaz D, Solanoy H, Wang X, Negrou E, Bondar VV, et al. 2022. Preclinical and clinical evaluation of the LRRK2 inhibitor DNL201 for Parkinson's disease. *Sci Transl Med* **14:** eabj2658. doi:10.1126/scitranslmed.abj2658

Jennings D, Huntwork-Rodriguez S, Vissers MFJM, Daryani VM, Diaz D, Goo MS, Chen JJ, Maciuca R, Fraser K, Mabrouk OS, et al. 2023. LRRK2 inhibition by BIIB122 in healthy participants and patients with Parkinson's disease. *Mov Disord* **38:** 386–398. doi:10.1002/mds.29297

Junker J, Lange LM, Vollstedt EJ, Roopnarain K, Doquenia MLM, Annuar AA, Avenali M, Bardien S, Bahr N, Ellis M, et al. 2024. Understanding monogenic Parkinson's disease at a global scale. *medRxiv* doi:10.1101/2024.03.12.24304154

    Cite this article as *Cold Spring Harb Perspect Med* doi: 10.1101/cshperspect.a041774

Kalia LV, Lang AE, Hazrati LN, Fujioka S, Wszolek ZK, Dickson DW, Ross OA, Van Deerlin VM, Trojanowski JQ, Hurtig HI, et al. 2015. Clinical correlations with Lewy body pathology in *LRRK2*-related Parkinson disease. *JAMA Neurol* **72**: 100–105. doi:10.1001/jamaneurol.2014.2704

Kalogeropoulou AF, Purlyte E, Tonelli F, Lange SM, Wightman M, Prescott AR, Padmanabhan S, Sammler E, Alessi DR. 2022. Impact of 100 LRRK2 variants linked to Parkinson's disease on kinase activity and microtubule binding. *Biochem J* **479**: 1759–1783. doi:10.1042/BCJ20220161

Kasten M, Hartmann C, Hampf J, Schaake S, Westenberger A, Vollstedt EJ, Balck A, Domingo A, Vulinovic F, Dulovic M, et al. 2018. Genotype-phenotype relations for the Parkinson's disease genes *parkin, PINK1, DJ1*: MDSGene systematic review. *Mov Disord* **33**: 730–741. doi:10.1002/mds.27352

Kedariti M, Frattini E, Baden P, Cogo S, Civiero L, Ziviani E, Zilio G, Bertoli F, Aureli M, Kaganovich A, et al. 2022. LRRK2 kinase activity regulates GCase level and enzymatic activity differently depending on cell type in Parkinson's disease. *NPJ Park Dis* **8**: 92. doi:10.1038/s41531-022-00354-3

Khani M, Cerquera-Cleves C, Kekenadze M, Wild Crea P, Singleton AB, Bandres-Ciga S. 2024. Towards a global view of Parkinson's disease genetics. *Ann Neurol* **95**: 831–842. doi:10.1002/ana.26905

Kim S, Wong YC, Gao F, Krainc D. 2021. Dysregulation of mitochondria-lysosome contacts by GBA1 dysfunction in dopaminergic neuronal models of Parkinson's disease. *Nat Commun* **12**: 1807. doi:10.1038/s41467-021-22113-3

Kim JJ, Vitale D, Otani DV, Lian MM, Heilbron K, Aslibekyan S, Auton A, Babalola E, Bell RK, Bielenberg J, et al. 2024. Multi-ancestry genome-wide association meta-analysis of Parkinson's disease. *Nat Genet* **56**: 27–36. doi:10.1038/s41588-023-01584-8

Klein C, Hattori N, Marras C. 2018. MDSGene: closing data gaps in genotype-phenotype correlations of monogenic Parkinson's disease. *J Parkinsons Dis* **8**: S25–S30. doi:10.3233/JPD-181505

Koch S, Laabs BH, Kasten M, Vollstedt EJ, Becktepe J, Brüggemann N, Franke A, Krämer UM, Kuhlenbäumer G, Lieb W, et al. 2021. Validity and prognostic value of a polygenic risk score for Parkinson's disease. *Genes (Basel)* **12**: 1859. doi:10.3390/genes12121859

Koschmidder E, Weissbach A, Brüggemann N, Kasten M, Klein C, Lohmann K. 2016. A nonsense mutation in *CHCHD2* in a patient with Parkinson disease. *Neurology* **86**: 577–579. doi:10.1212/WNL.0000000000002361

Kouri N, Ross OA, Dombroski B, Younkin CS, Serie DJ, Soto-Ortolaza A, Baker M, Finch NCA, Yoon H, Kim J, et al. 2015. Genome-wide association study of corticobasal degeneration identifies risk variants shared with progressive supranuclear palsy. *Nat Commun* **6**: 7247. doi:10.1038/ncomms8247

Lai D, Alipanahi B, Fontanillas P, Schwantes-An TH, Aasly J, Alcalay RN, Beecham GW, Berg D, Bressman S, Brice A, et al. 2021. Genomewide association studies of *LRRK2* modifiers of Parkinson's disease. *Ann Neurol* **90**: 76–88. doi:10.1002/ana.26094

Lange LM, Klein C. 2010. PINK1 type of young-onset Parkinson disease. In *GeneReviews* (ed. Adam MP, Feldman J, Mirzaa GM, et al.). University of Washington, Seattle.

Lange LM, Gonzalez-Latapi P, Rajalingam R, Tijssen MAJ, Ebrahimi-Fakhari D, Gabbert C, Ganos C, Ghosh R, Kumar KR, Lang AE, et al. 2022. Nomenclature of genetic movement disorders: recommendations of the international Parkinson and movement disorder society task force—an update. *Mov Disord* **37**: 905–935. doi:10.1002/mds.28982

Lange LM, Avenali M, Ellis M, Illarionova A, Keller Sarmiento IJ, Tan AH, Madoev H, Galandra C, Junker J, Roopnarain K, et al. 2023. Elucidating causative gene variants in hereditary Parkinson's disease in the Global Parkinson's Genetics Program (GP2). *NPJ Park Dis* **9**: 100. doi:10.1038/s41531-023-00526-9

Lee AJ, Wang Y, Alcalay RN, Mejia-Santana H, Saunders-Pullman R, Bressman S, Corvol JC, Brice A, Lesage S, Mangone G, et al. 2017. Penetrance estimate of *LRRK2* p.G2019S mutation in individuals of non-Ashkenazi Jewish ancestry. *Mov Disord* **32**: 1432–1438. doi:10.1002/mds.27059

Lee RG, Sedghi M, Salari M, Shearwood AMJ, Stentenbach M, Kariminejad A, Goullee H, Rackham O, Laing NG, Tajsharghi H, et al. 2018. Early-onset Parkinson disease caused by a mutation in CHCHD2 and mitochondrial dysfunction. *Neurol Genet* **4**: e276. doi:10.1212/NXG.0000000000000276

Lesage S, Dürr A, Tazir M, Lohmann E, Leutenegger AL, Janin S, Pollak P, Brice A. 2006. *LRRK2* g2019s as a cause of Parkinson's disease in North African Arabs. *N Engl J Med* **354**: 422–423. doi:10.1056/NEJMc055540

Liu G, Boot B, Locascio JJ, Jansen IE, Winder-Rhodes S, Eberly S, Elbaz A, Brice A, Ravina B, van Hilten JJ, et al. 2016. Specifically neuropathic Gaucher's mutations accelerate cognitive decline in Parkinson's. *Ann Neurol* **80**: 674–685. doi:10.1002/ana.24781

Lüth T, König IR, Grünewald A, Kasten M, Klein C, Hentati F, Farrer M, Trinh J. 2020. Age at onset of LRRK2 p. Gly2019Ser is related to environmental and lifestyle factors. *Mov Disord* **35**: 1854–1858. doi:10.1002/mds.28238

Ma KY, Fokkens MR, Reggiori F, Mari M, Verbeek DS. 2021. Parkinson's disease-associated VPS35 mutant reduces mitochondrial membrane potential and impairs PINK1/Parkin-mediated mitophagy. *Transl Neurodegener* **10**: 19. doi:10.1186/s40035-021-00243-4

Madsen DA, Schmidt SI, Blaabjerg M, Meyer M. 2021. Interaction between parkin and α-synuclein in park2-mediated Parkinson's disease. *Cells* **10**: 283. doi:10.3390/cells10020283

Makarious MB, Lake J, Pitz V, Fu AY, Guidubaldi JL, Solsberg CW, Bandres-Ciga S, Leonard HL, Kim JJ, Billingsley KJ, et al. 2023. Large-scale rare variant burden testing in Parkinson's disease. *Brain* **146**: 4622–4632. doi:10.1093/brain/awad214

Marras C, Lang A, van de Warrenburg BP, Sue CM, Tabrizi SJ, Bertram L, Mercimek-Mahmutoglu S, Ebrahimi-Fakhari D, Warner TT, Durr A, et al. 2016. Nomenclature of genetic movement disorders: recommendations of the international Parkinson and movement disorder society task force. *Mov Disord* **31**: 436–457. doi:10.1002/mds.26527

Mazzulli JR, Xu YH, Sun Y, Knight AL, McLean PJ, Caldwell GA, Sidransky E, Grabowski GA, Krainc D. 2011. Gaucher disease glucocerebrosidase and α-synuclein form a bidirectional pathogenic loop in synucleinopathies. *Cell* **146:** 37–52. doi:10.1016/j.cell.2011.06.001

Meissner WG, Remy P, Giordana C, Maltête D, Derkinderen P, Houéto JL, Anheim M, Benatru I, Boraud T, Brefel-Courbon C, et al. 2024. Trial of lixisenatide in early Parkinson's disease. *N Engl J Med* **390:** 1176–1185. doi:10.1056/NEJMoa2312323

Meng H, Yamashita C, Shiba-Fukushima K, Inoshita T, Funayama M, Sato S, Hatta T, Natsume T, Umitsu M, Takagi J, et al. 2017. Loss of Parkinson's disease-associated protein CHCHD2 affects mitochondrial crista structure and destabilizes cytochrome c. *Nat Commun* **8:** 15500. doi:10.1038/ncomms15500

Meng Y, Pople CB, Huang Y, Jones RM, Ottoy J, Goubran M, Oliveira LM, Davidson B, Lawrence LSP, Lau AZ, et al. 2022. Putaminal recombinant glucocerebrosidase delivery with magnetic resonance-guided focused ultrasound in Parkinson's disease: a phase I study. *Mov Disord* **37:** 2134–2139. doi:10.1002/mds.29190

Menon PJ, Sambin S, Criniere-Boizet B, Courtin T, Tesson C, Casse F, Ferrien M, Mariani LL, Carvalho S, Lejeune FX, et al. 2024. Genotype-phenotype correlation in PRKN-associated Parkinson's disease. *NPJ Park Dis* **10:** 72. doi:10.1038/s41531-024-00677-3

Menozzi E, Toffoli M, Schapira AHV. 2023. Targeting the GBA1 pathway to slow Parkinson disease: insights into clinical aspects, pathogenic mechanisms and new therapeutic avenues. *Pharmacol Ther* **246:** 108419. doi:10.1016/j.pharmthera.2023.108419

Mir R, Tonelli F, Lis P, Macartney T, Polinski NK, Martinez TN, Chou MY, Howden AJM, König T, Hotzy C, et al. 2018. The Parkinson's disease VPS35[D620N] mutation enhances LRRK2-mediated Rab protein phosphorylation in mouse and human. *Biochem J* **475:** 1861–1883. doi:10.1042/BCJ20180248

Mullin S, Smith L, Lee K, D'Souza G, Woodgate P, Elflein J, Hällqvist J, Toffoli M, Streeter A, Hosking J, et al. 2020. Ambroxol for the treatment of patients with Parkinson disease with and without glucocerebrosidase gene mutations: a nonrandomized, noncontrolled trial. *JAMA Neurol* **77:** 427–434. doi:10.1001/jamaneurol.2019.4611

Nalls MA, Blauwendraat C, Vallerga CL, Heilbron K, Bandres-Ciga S, Chang D, Tan M, Kia DA, Noyce AJ, Xue A, et al. 2019. Identification of novel risk loci, causal insights, and heritable risk for Parkinson's disease: a meta-analysis of genome-wide association studies. *Lancet Neurol* **18:** 1091–1102. doi:10.1016/S1474-4422(19)30320-5

Ng J, Barral S, Waddington SN, Kurian MA. 2023. Gene therapy for dopamine dyshomeostasis: from Parkinson's to primary neurotransmitter diseases. *Mov Disord* **38:** 924–936. doi:10.1002/mds.29416

Nuytemans K, Rademakers R, Theuns J, Pals P, Engelborghs S, Pickut B, de Pooter T, Peeters K, Mattheijssens M, Van den Broeck M, et al. 2008. Founder mutation p.R1441C in the leucine-rich repeat kinase 2 gene in Belgian Parkinson's disease patients. *Eur J Hum Genet* **16:** 471–479. doi:10.1038/sj.ejhg.5201986

O'Hara DM, Pawar G, Kalia SK, Kalia LV. 2020. LRRK2 and α-synuclein: distinct or synergistic players in Parkinson's disease? *Front Neurosci* **14:** 577. doi:10.3389/fnins.2020.00577

Ouzren N, Delcambre S, Ghelfi J, Seibler P, Farrer MJ, König IR, Aasly JO, Trinh J, Klein C, Grünewald A. 2019. Mitochondrial DNA deletions discriminate affected from unaffected LRRK2 mutation carriers. *Ann Neurol* **86:** 324–326. doi:10.1002/ana.25510

Ozelius LJ, Senthil G, Saunders-Pullman R, Ohmann E, Deligtisch A, Tagliati M, Hunt AL, Klein C, Henick B, Hailpern SM, et al. 2006. *LRRK2* g2019s as a cause of Parkinson's disease in Ashkenazi Jews. *N Engl J Med* **354:** 424–425. doi:10.1056/NEJMc055509

Pal G, Mangone G, Hill EJ, Ouyang B, Liu Y, Lythe V, Ehrlich D, Saunders-Pullman R, Shanker V, Bressman S, et al. 2022. Parkinson disease and subthalamic nucleus deep brain stimulation: cognitive effects in GBA mutation carriers. *Ann Neurol* **91:** 424–435. doi:10.1002/ana.26302

Pal G, Cook L, Schulze J, Verbrugge J, Alcalay RN, Merello M, Sue CM, Bardien S, Bonifati V, Chung SJ, et al. 2023a. Genetic testing in Parkinson's disease. *Mov Disord* **38:** 1384–1396. doi:10.1002/mds.29500

Pal P, Taylor M, Lam PY, Tonelli F, Hecht CA, Lis P, Nirujogi RS, Phung TK, Yeshaw WM, Jaimon E, et al. 2023b. Parkinson's VPS35[D620N] mutation induces LRRK2-mediated lysosomal association of RILPL1 and TMEM55B. *Sci Adv* **9:** eadj1205. doi:10.1126/sciadv.adj1205

Parlar SC, Grenn FP, Kim JJ, Baluwendraat C, Gan-Or Z. 2023. Classification of *GBA1* variants in Parkinson's disease: the *GBA1*-PD browser. *Mov Disord* **38:** 489–495. doi:10.1002/mds.29314

Pérez-Dueñas B, Gorman K, Marcé-Grau A, Ortigoza-Escobar JD, Macaya A, Danti FR, Barwick K, Papandreou A, Ng J, Meyer E, et al. 2022. The genetic landscape of complex childhood-onset hyperkinetic movement disorders. *Mov Disord* **37:** 2197–2209. doi:10.1002/mds.29182

Peterschmitt MJ, Saiki H, Hatano T, Gasser T, Isaacson SH, Gaemers SJM, Minini P, Saubadu S, Sharma J, Walbillic S, et al. 2022. Safety, pharmacokinetics, and pharmacodynamics of oral venglustat in patients with Parkinson's disease and a GBA mutation: results from part 1 of the randomized, double-blinded, placebo-controlled MOVES-PD trial. *J Parkinsons Dis* **12:** 557–570. doi:10.3233/JPD-212714

Pickrell AM, Youle RJ. 2015. The roles of PINK1, Parkin, and mitochondrial fidelity in Parkinson's disease. *Neuron* **85:** 257–273. doi:10.1016/j.neuron.2014.12.007

Polymeropoulos MH, Lavedan C, Leroy E, Ide SE, Dehejia A, Dutra A, Pike B, Root H, Rubenstein J, Boyer R, et al. 1997. Mutation in the α-synuclein gene identified in families with Parkinson's disease. *Science* **276:** 2045–2047. doi:10.1126/science.276.5321.2045

Prasuhn J, Brüggemann N. 2021. Genotype-driven therapeutic developments in Parkinson's disease. *Mol Med* **27:** 42. doi:10.1186/s10020-021-00281-8

Prasuhn J, Brüggemann N, Hessler N, Berg D, Gasser T, Brockmann K, Olbrich D, Ziegler A, König IR, Klein C, et al. 2019. An omics-based strategy using coenzyme Q10 in patients with Parkinson's disease: concept evaluation in a double-blind randomized placebo-controlled parallel group trial. *Neurol Res Pract* **1:** 31. doi:10.1186/s42466-019-0033-1

Prasuhn J, Kasten M, Vos M, König IR, Schmid SM, Wilms B, Klein C, Brüggemann N. 2021. The use of vitamin K2 in patients with Parkinson's disease and mitochondrial dysfunction (PD-K2): a theranostic pilot study in a placebo-controlled parallel group design. *Front Neurol* **11:** 592104. doi:10.3389/fneur.2020.592104

Ren J, Zhang R, Pan C, Xu J, Sun H, Hua P, Zhang L, Zhang W, Xu P, Ma C, et al. 2022. Prevalence and genotype-phenotype correlations of *GBA*-related Parkinson disease in a large Chinese cohort. *Eur J Neurol* **29:** 1017–1024. doi:10.1111/ene.15230

Richards S, Aziz N, Bale S, Bick D, Das S, Gastier-Foster J, Grody WW, Hegde M, Lyon E, Spector E, et al. 2015. Standards and guidelines for the interpretation of sequence variants: a joint consensus recommendation of the American College of Medical Genetics and Genomics and the Association for Molecular Pathology. *Genet Med* **17:** 405–424. doi:10.1038/gim.2015.30

Rizig M, Bandres-Ciga S, Makarious MB, Ojo OO, Crea PW, Abiodun OV, Levine KS, Abubakar SA, Achoru CO, Vitale D, et al. 2023. Identification of genetic risk loci and causal insights associated with Parkinson's disease in African and African admixed populations: a genome-wide association study. *Lancet Neurol* **22:** 1015–1025. doi:10.1016/S1474-4422(23)00283-1

Sailer A, Scholz SW, Nalls MA, Schulte C, Federoff M, Price TR, Lees A, Ross OA, Dickson DW, Mok K, et al. 2016. A genome-wide association study in multiple system atrophy. *Neurology* **87:** 1591–1598. doi:10.1212/WNL.0000000000003221

Sánchez K, Maguire-Zeiss K. 2020. MMP13 expression is increased following mutant α-synuclein exposure and promotes inflammatory responses in microglia. *Front Neurosci* **14:** 585544. doi:10.3389/fnins.2020.585544

Sassone J, Reale C, Dati G, Regoni M, Pellecchia MT, Garavaglia B. 2021. The role of VPS35 in the pathobiology of Parkinson's disease. *Cell Mol Neurobiol* **41:** 199–227. doi:10.1007/s10571-020-00849-8

Saunders-Pullman R, Mirelman A, Alcalay RN, Wang C, Ortega RA, Raymond D, Mejia-Santana H, Orbe-Reilly M, Johannes BA, Thaler A, et al. 2018. Progression in the *LRRK2*-associated Parkinson disease population. *JAMA Neurol* **75:** 312–319. doi:10.1001/jamaneurol.2017.4019

Saunders-Pullman R, Ortega RA, Wang C, Raymond D, Elango S, Leaver K, Urval N, Katsnelson V, Gerber R, Swan M, et al. 2022. Association of olfactory performance with motor decline and age at onset in people with Parkinson disease and the *LRRK2* G2019S variant. *Neurology* **99:** E814–E823. doi:10.1212/WNL.0000000000200737

Schumacher-Schuh AF, Bieger A, Okunoye O, Mok KY, Lim SY, Bardien S, Ahmad-Annuar A, Santos-Lobato BL, Strelow MZ, Salama M, et al. 2022. Underrepresented populations in Parkinson's genetics research: current landscape and future directions. *Mov Disord* **37:** 1593–1604. doi:10.1002/mds.29126

Shukla LC, Schulze J, Farlow J, Pankratz ND, Wojcieszek J, Foroud T. 2019. Parkinson disease overview. In *GeneReviews*. University of Washington, Seattle.

Siderowf A, Concha-Marambio L, Lafontant DE, Farris CM, Ma Y, Urenia PA, Nguyen H, Alcalay RN, Chahine LM, Foroud T, et al. 2023. Assessment of heterogeneity among participants in the Parkinson's Progression Markers Initiative cohort using α-synuclein seed amplification: a cross-sectional study. *Lancet Neurol* **22:** 407–417. doi:10.1016/S1474-4422(23)00109-6

Sidransky E, Nalls MA, Aasly JO, Aharon-Peretz J, Annesi G, Barbosa ER, Bar-Shira A, Berg D, Bras J, Brice A, et al. 2009. Multicenter analysis of glucocerebrosidase mutations in Parkinson's disease. *N Engl J Med* **361:** 1651–1661. doi:10.1056/NEJMoa0901281

Silveira CRA, MacKinley J, Coleman K, Li Z, Finger E, Bartha R, Morrow SA, Wells J, Borrie M, Tirona RG, et al. 2019. Ambroxol as a novel disease-modifying treatment for Parkinson's disease dementia: protocol for a single-centre, randomized, double-blind, placebo-controlled trial. *BMC Neurol* **19:** 20. doi:10.1186/s12883-019-1252-3

Simón-Sánchez J, Martí-Massó JF, Sánchez-Mut JV, Paisán-Ruiz C, Martínez-Gil A, Ruiz-Martínez J, Sáenz A, Singleton AB, Lopéz de Munain A, Pérez-Tur J. 2006. Parkinson's disease due to the R1441G mutation in Dardarin: a founder effect in the Basques. *Mov Disord* **21:** 1954–1959. doi:10.1002/mds.21114

Simón-Sánchez J, Schulte C, Bras JM, Sharma M, Gibbs JR, Berg D, Paisan-Ruiz C, Lichtner P, Scholz SW, Hernandez DG, et al. 2009. Genome-wide association study reveals genetic risk underlying Parkinson's disease. *Nat Genet* **41:** 1308–1312. doi:10.1038/ng.487

Simuni T, Brumm MC, Uribe L, Caspell-Garcia C, Coffey CS, Siderowf A, Alcalay RN, Trojanowski JQ, Shaw LM, Seibyl J, et al. 2020. Clinical and dopamine transporter imaging characteristics of leucine rich repeat kinase 2 (LRRK2) and glucosylceramidase β (GBA) Parkinson's disease participants in the Parkinson's Progression Markers Initiative: a cross-sectional study. *Mov Disord* **35:** 833–844. doi:10.1002/mds.27989

Simuni T, Chahine LM, Poston K, Brumm M, Buracchio T, Campbell M, Chowdhury S, Coffey C, Concha-Marambio L, Dam T, et al. 2024. A biological definition of neuronal α-synuclein disease: towards an integrated staging system for research. *Lancet Neurol* **23:** 178–190. doi:10.1016/S1474-4422(23)00405-2

Skou LD, Johansen SK, Okarmus J, Meyer M. 2024. Pathogenesis of DJ-1/PARK7-mediated Parkinson's disease. *Cells* **13:** 296. doi:10.3390/cells13040296

Skrahina V, Gaber H, Vollstedt EJ, Förster TM, Usnich T, Curado F, Brüggemann N, Paul J, Bogdanovic X, Zülbahar S, et al. 2021. The Rostock international Parkinson's disease (ROPAD) study: protocol and initial findings. *Mov Disord* **36:** 1005–1010. doi:10.1002/mds.28416

Song P, Peng W, Sauve V, Fakih R, Xie Z, Ysselstein D, Krainc T, Wong YC, Mencacci NE, Savas JN, et al. 2023. Parkinson's disease-linked parkin mutation disrupts recycling of synaptic vesicles in human dopaminergic neurons. *Neuron* **111:** 3775–3788.e7. doi:10.1016/j.neuron.2023.08.018

Spillantini MG, Schmidt ML, Lee VMY, Trojanowski JQ, Jakes R, Goedert M. 1997. α-Synuclein in Lewy bodies. *Nature* **388:** 839–840. doi:10.1038/42166

Tan AH, Noyce A, Carrasco AM, Brice A, Reimer A, Illarionova A, Singleton A, Schumacher-Schuh A, Stecher B, Siddiqi B, et al. 2021. GP2: the Global Parkinson's Genetics Program. *Mov Disord* **36:** 842–851. doi:10.1002/mds.28494

Tang FL, Erion JR, Tian Y, Liu W, Yin DM, Ye J, Tang B, Mei L, Xiong WC. 2015a. VPS35 in dopamine neurons is required for endosome-to-golgi retrieval of lamp2a, a receptor of chaperone-mediated autophagy that is critical for α-synuclein degradation and prevention of pathogenesis of Parkinson's disease. *J Neurosci* 35: 10613–10628. doi:10.1523/JNEUROSCI.0042-15.2015

Tang FL, Liu W, Hu JX, Erion JR, Ye J, Mei L, Xiong WC. 2015b. VPS35 deficiency or mutation causes dopaminergic neuronal loss by impairing mitochondrial fusion and function. *Cell Rep* 12: 1631–1643. doi:10.1016/j.celrep.2015.08.001

Thaler A, Bregman N, Gurevich T, Shiner T, Dror Y, Zmira O, Gan-Or Z, Bar-Shira A, Gana-Weisz M, Orr-Urtreger A, et al. 2018. Parkinson's disease phenotype is influenced by the severity of the mutations in the GBA gene. *Parkinsonism Relat Disord* 55: 45–49. doi:10.1016/j.parkreldis.2018.05.009

Tolosa E, Vila M, Klein C, Rascol O. 2020. LRRK2 in Parkinson disease: challenges of clinical trials. *Nat Rev Neurol* 16: 97–107. doi:10.1038/s41582-019-0301-2

Towns C, Richer M, Jasaityte S, Stafford EJ, Joubert J, Antar T, Martinez-Carrasco A, Makarious MB, Casey B, Vitale D, et al. 2023. Defining the causes of sporadic Parkinson's disease in the global Parkinson's genetics program (GP2). *NPJ Park Dis* 9: 131. doi:10.1038/s41531-023-00533-w

Trinh J, Guella I, Farrer MJ. 2014. Disease penetrance of late-onset parkinsonism: a meta-analysis. *JAMA Neurol* 71: 1535–1539. doi:10.1001/jamaneurol.2014.1909

Trinh J, Zeldenrust FMJ, Huang J, Kasten M, Schaake S, Petkovic S, Madoev H, Grünewald A, Almuammar S, König IR, et al. 2018. Genotype-phenotype relations for the Parkinson's disease genes SNCA, LRRK2, VPS35: MDSGene systematic review. *Mov Disord* 33: 1857–1870. doi:10.1002/mds.27527

Tseng FS, Foo JQX, Mai AS, Tan EK. 2023. The genetic basis of multiple system atrophy. *J Transl Med* 21: 104. doi:10.1186/s12967-023-03905-1

Tsuji S, Choudary PV, Martin BM, Stubblefield BK, Mayor JA, Barranger JA, Ginns EI. 1987. A mutation in the human glucocerebrosidase gene in neuronopathic Gaucher's disease. *N Engl J Med* 316: 570–575. doi:10.1056/NEJM198703053161002

Tunold JA, Tan MMX, Toft M, Ross O, van de Berg WDJ, Pihlstrøm L. 2024. Lysosomal polygenic burden drives cognitive decline in Parkinson's disease with low Alzheimer risk. *Mov Disord* 39: 596–601. doi:10.1002/mds.29698

Usnich T, Olmedillas M, Schell N, Paul JJ, Curado F, Skobalj S, Csoti I, Ertan S, Gruber D, Zittel S, et al. 2023. Frequency of non-motor symptoms in Parkinson's disease patients carrying the E326K and T369M GBA risk variants. *Park Relat Disord* 107: 105248. doi:10.1016/j.parkreldis.2022.105248

Vollstedt EJ, Kasten M, Klein C, Aasly J, Adler C, Ahmad-Annuar A, Albanese A, Alcalay RN, Al-Mubarak B, Alvarez V, et al. 2019. Using global team science to identify genetic Parkinson's disease worldwide. *Ann Neurol* 86: 153–157. doi:10.1002/ana.25514

Vollstedt EJ, Madoev H, Aasly A, Ahmad-Annuar A, Al-Mubarak B, Alcalay RN, Alvarez V, Amorin I, Annesi G, Arkadir D, et al. 2023a. Establishing an online resource to facilitate global collaboration and inclusion of underrepresented populations: experience from the MJFF Global Genetic Parkinson's Disease Project. *PLoS ONE* 18: e0292180. doi:10.1371/journal.pone.0292180

Vollstedt EJ, Schaake S, Lohmann K, Padmanabhan S, Brice A, Lesage S, Tesson C, Vidailhet M, Wurster I, Hentati F, et al. 2023b. Embracing monogenic Parkinson's disease: the MJFF Global Genetic PD Cohort. *Mov Disord* 38: 286–303. doi:10.1002/mds.29288

Wang W, Wang X, Fujioka H, Hoppel C, Whone AL, Caldwell MA, Cullen PJ, Liu J, Zhu X. 2016. Parkinson's disease-associated mutant VPS35 causes mitochondrial dysfunction by recycling DLP1 complexes. *Nat Med* 22: 54–63. doi:10.1038/nm.3983

Waschbüsch D, Michels H, Strassheim S, Ossendorf E, Kessler D, Gloeckner CJ, Barnekow A. 2014. LRRK2 transport is regulated by its novel interacting partner Rab32. *PLoS ONE* 9: e111632. doi:10.1371/journal.pone.0111632

Weissbach A, Pauly MG, Herzog R, Hahn L, Halmans S, Hamami F, Bolte C, Camargos S, Jeon B, Kurian MA, et al. 2022. Relationship of genotype, phenotype, and treatment in dopa-responsive dystonia: MDSGene review. *Mov Disord* 37: 237–252. doi:10.1002/mds.28874

Wen Y, Zhou Y, Jiao B, Shen L. 2021. Genetics of progressive supranuclear palsy: a review. *J Parkinsons Dis* 11: 93–105. doi:10.3233/JPD-202302

Westenberger A, Skrahina V, Usnich T, Beetz C, Vollstedt E, Laabs B, Paul J, Curado F. 2024. Relevance of genetic testing in the gene-targeted trial era: the Rostock Parkinson's Disease Study. *Brain*. In press.

Whiffin N, Armean IM, Kleinman A, Marshall JL, Minikel EV, Goodrich JK, Quaife NM, Cole JB, Wang Q, Karczewski KJ, et al. 2020. The effect of LRRK2 loss-of-function variants in humans. *Nat Med* 26: 869–877. doi:10.1038/s41591-020-0893-5

Wider C, Skipper L, Solida A, Brown L, Farrer M, Dickson D, Wszolek ZK, Vingerhoets FJG. 2008. Autosomal dominant dopa-responsive parkinsonism in a multigenerational Swiss family. *Park Relat Disord* 14: 465–470. doi:10.1016/j.parkreldis.2007.11.013

Wider C, Dickson DW, Wszolek ZK. 2010. Leucine-rich repeat kinase 2 gene-associated disease: redefining genotype-phenotype correlation. *Neurodegener Dis* 7: 175–179. doi:10.1159/000289232

Williams ET, Glauser L, Tsika E, Jiang H, Islam S, Moore DJ. 2018. Parkin mediates the ubiquitination of VPS35 and modulates retromer-dependent endosomal sorting. *Hum Mol Genet* 27: 3189–3205. doi:10.1093/hmg/ddy224

Williams ET, Chen X, Otero PA, Moore DJ. 2022. Understanding the contributions of VPS35 and the retromer in neurodegenerative disease. *Neurobiol Dis* 170: 105768. doi:10.1016/j.nbd.2022.105768

Wittke C, Petkovic S, Dobricic V, Schaake S, Arzberger T, Compta Y, Englund E, Ferguson LW, Gelpi E, Roeber S, et al. 2021. Genotype-phenotype relations for the atypical Parkinsonism genes: MDSGene systematic review. *Mov Disord* 36: 1499–1510. doi:10.1002/mds.28517

Wszolek ZK, Pfeiffer RF, Tsuboi Y, Uitti RJ, McComb RD, Stoessl AJ, Strongosky AJ, Zimprich A, Müller-Myhsok B, Farrer MJ, et al. 2004. Autosomal dominant parkinsonism associated with variable synuclein and tau pathology.

Cite this article as *Cold Spring Harb Perspect Med* doi: 10.1101/cshperspect.a041774

Neurology **62:** 1619–1622. doi:10.1212/01.WNL.0000125 015.06989.DB

Yokoyama JS, Karch CM, Fan CC, Bonham LW, Kouri N, Ross OA, Rademakers R, Kim J, Wang Y, Höglinger GU, et al. 2017. Shared genetic risk between corticobasal degeneration, progressive supranuclear palsy, and frontotemporal dementia. *Acta Neuropathol* **133:** 825–837. doi:10.1007/s00401-017-1693-y

Ysselstein D, Nguyen M, Young TJ, Severino A, Schwake M, Merchant K, Krainc D. 2019. LRRK2 kinase activity regulates lysosomal glucocerebrosidase in neurons derived from Parkinson's disease patients. *Nat Commun* **10:** 5570. doi:10.1038/s41467-019-13413-w

Zabetian CP, Hutter CM, Yearout D, Lopez AN, Factor SA, Griffith A, Leis BC, Bird TD, Nutt JG, Higgins DS, et al. 2006. LRRK2 g2019s in families with Parkinson disease who originated from Europe and the Middle East: evidence of two distinct founding events beginning two millennia ago. *Am J Hum Genet* **79:** 752–758. doi:10.1086/508025

Zavodszky E, Seaman MNJ, Moreau K, Jimenez-Sanchez M, Breusegem SY, Harbour ME, Rubinsztein DC. 2014. Mutation in VPS35 associated with Parkinson's disease impairs WASH complex association and inhibits autophagy. *Nat Commun* **5:** 5570. doi:10.1038/ncomms4828

Zhao Y, Qin L, Pan H, Liu Z, Jiang L, He Y, Zeng Q, Zhou X, Zhou X, Zhou Y, et al. 2020. The role of genetics in Parkinson's disease: a large cohort study in Chinese mainland population. *Brain* **143:** 2220–2234. doi:10.1093/brain/awaa167

Zhou Y, Wang Y, Wan J, Zhao Y, Pan H, Zeng Q, Zhou X, He R, Zhou X, Xiang Y, et al. 2023. Mutational spectrum and clinical features of GBA1 variants in a Chinese cohort with Parkinson's disease. *NPJ Park Dis* **9:** 129. doi:10.1038/s41531-023-00571-4

Zunke F, Moise AC, Belur NR, Gelyana E, Stojkovska I, Dzaferbegovic H, Toker NJ, Jeon S, Fredriksen K, Mazzulli JR. 2018. Reversible conformational conversion of α-synuclein into toxic assemblies by glucosylceramide. *Neuron* **97:** 92–107.e10. doi:10.1016/j.neuron.2017.12.012

α-Synuclein in Parkinson's Disease: 12 Years Later

Kostas Vekrellis,[1] Evangelia Emmanouilidou,[2] Maria Xilouri,[3] and Leonidas Stefanis[3,4]

[1]Center for Basic Research, Biomedical Research Foundation of the Academy of Athens, Athens 11527, Greece

[2]Laboratory of Biochemistry, Department of Chemistry, National and Kapodistrian University of Athens, Athens 15784, Greece

[3]Center for Clinical, Experimental Surgery and Translational Research, Biomedical Research Foundation of the Academy of Athens, Athens 11527, Greece

[4]First Department of Neurology, National and Kapodistrian University of Athens Medical School, Athens 11528, Greece; and Center for Clinical, Experimental Surgery and Translational Research, Biomedical Research Foundation of the Academy of Athens, Athens 11527, Greece

Correspondence: lstefanis@bioacaddemy.gr

α-Synuclein (AS) is a small presynaptic protein that is genetically, biochemically, and neuropathologically linked to Parkinson's disease (PD) and related synucleinopathies. We present here a review of the topic of this relationship, focusing on more recent knowledge. In particular, we review the genetic evidence linking AS to familial and sporadic PD, including a number of recently identified point mutations in the SNCA gene. We briefly go over the relevant neuropathological findings, stressing the evidence indicating a correlation between aberrant AS deposition and nervous system dysfunction. We analyze the structural characteristics of the protein, in relation to both its physiologic and pathological conformations, with particular emphasis on posttranslational modifications, aggregation properties, and secreted forms. We review the interrelationship of AS with various cellular compartments and functions, with particular focus on the synapse and protein degradation systems. We finally go over the recent exciting data indicating that AS can provide the basis for novel robust biomarkers in the field of synucleinopathies, while at the same time results from the first clinical trials specifically targeting AS are being reported.

Since the initial discovery in 1997 of the genetic link between Parkinson's Disease (PD) and SNCA, the gene encoding for the presynaptic neuronal protein α-synuclein (AS), the evidence supporting the importance of AS in PD pathogenesis and evolution has continued to mount. The combination of further genetic discoveries and the understanding of AS biology and its impact on cellular processes, through the development of relevant cellular and animal models, has been instrumental in this regard. This has culminated recently in the emergence of the first wet biomarker for PD, based on the aggregation properties of AS, and in the execution of the first clinical trials targeting AS. In this paper, which follows a similar article published by Stefanis (2012), we attempt to summarize the main aspects of AS biology, focusing on its normal function, its aggregation properties and cellular pathogenic effects, its genetic and neuropathological link to PD, its potential as a biomarker, and, finally its targeting in clinical trials.

Copyright © 2025 Cold Spring Harbor Laboratory Press; all rights reserved
Cite this article as Cold Spring Harb Perspect Med doi: 10.1101/cshperspect.a041645

THE ORIGINS OF THE LINK: *SNCA* AS AN IMPORTANT GENETIC CONTRIBUTOR TO PARKINSON'S DISEASE

In PD, unlike Alzheimer's Disease (AD) where the neuropathological discovery of β-amyloid deposition preceded the genetic discovery of mutations in the amyloid precursor protein (APP) leading to autosomal-dominant AD, genetics came first. In 1997, Polymeropoulos et al. (1997) reported for the first time a specific genetic defect leading to familial PD. This involved a large family of Italian origin, the Contursi kindred, with an autosomal-dominant pattern of inheritance. The genetic defect identified was a missense p.A53T mutation in the *SNCA* gene. This was conceptually very important, as it ran against the perceived notion of PD as a sporadic disease initiated by environmental factors. Importantly, in the same publication (Polymeropoulos et al. 1997), Greek PD patients with an autosomal-dominant inheritance pattern from seemingly unrelated families were identified with the same mutation, solidifying the etiological link and demonstrating a founder effect, likely thousands of years old. Since then, further research on carriers of this particular mutation has refined the clinical and other manifestations of the disease. Despite noticeable heterogeneity, ranging from incomplete penetrance to extremely aggressive manifestations, the general pattern is that of a disease that presents early, with a mean age of onset of 45, and is more severe compared to idiopathic PD (iPD)

(Papadimitriou et al. 2016). Over this period of 27 years, a number of other point mutations in the *SNCA* gene have been identified, all leading to autosomal-dominant PD (Fig. 1; Krüger et al. 1998; Zarranz et al. 2004; Appel-Cresswell et al. 2013; Kiely et al. 2013; Lesage et al. 2013; Proukakis et al. 2013; Pasanen et al. 2014; Kapasi et al. 2020; Fevga et al. 2021; Liu et al. 2021; Daida et al. 2022; Diaw et al. 2023). Such cases are rarer, compared to those harboring the p.A53T mutation, so their clinical picture is not as well defined, but seems to vary by the specific mutation. An insertion of seven amino acids leading to an elongated peptide conferring novel aggregation properties has also been reported in a case of juvenile onset (Yang et al. 2023). Although a toxic gain of function is a common denominator, not all mutations lead to enhanced aggregation propensity; for example, the clinically aggressive G51D mutant form decreases the rate of fibrillization (Rutherford et al. 2014), but G51D fibrils, once formed, have altered properties that may lead to enhanced seeding and neurotoxicity (Hayakawa et al. 2020; Sun et al. 2021). It is fair to say that no single mechanism has been identified through which *SNCA* point mutations lead to PD; this could either indicate that such a common mechanism has yet to be identified, or that such point mutations lead to the disease through different mechanisms.

Yet another conceptual leap was the discovery in 2003 that excess copies of the *SNCA* gene could lead to autosomal-dominant disease. In particular, the Iowa kindred was the first to be

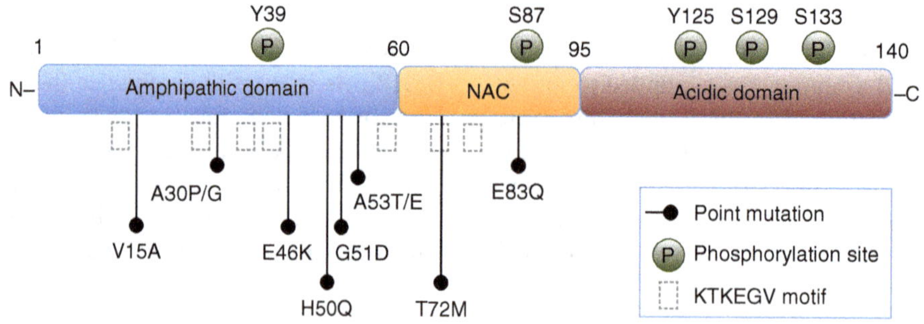

Figure 1. Simplified schematic illustration of α-synuclein primary structure. Within the three basic protein domains (amphipathic, hydrophobic nonamyloid component [NAC], and acidic), the KTKEGV motifs, the known point mutations and the phosphorylation sites are depicted.

discovered to harbor a triplication of the *SNCA* locus (Singleton et al. 2003). Subsequently, duplications were also identified (Chartier-Harlin et al. 2004). Intriguingly, there is an obvious gene dosage effect, in that cases with the *SNCA* triplication versus duplication demonstrate a much higher, basically complete, penetrance, earlier age of onset, and enhanced disease severity, including prominent cognitive decline. This brings home the important point that excess levels of the normal AS protein are sufficient to lead, in a dose-dependent fashion, to the neuropathological and clinical manifestations of PD, establishing the importance of AS levels in PD pathogenesis.

The genome-wide association studies (GWAS) in populations of predominant Caucasian origin have collectively and cumulatively shown that the highest hit most strongly associated with sporadic PD is in the *SNCA* locus (Nalls et al. 2019). This has been recently confirmed in two large non-Caucasian ethnic groups in China and India (Pan et al. 2023; Andrews et al. 2024). Thus, irrespective of ethnic origin, genetic alterations within the *SNCA* locus significantly influence the risk of development of sporadic PD, proving that sporadic PD is linked genetically to *SNCA* and AS. The GWAS approach thus provided an unbiased platform to confirm prior targeted association studies (Mueller et al. 2005). The polymorphisms associated with the disease appear to be associated with higher *SNCA* mRNA and AS protein levels (Fuchs et al. 2008), but more work is needed to substantiate this. This genetic association, not surprisingly, extends to other synucleinopathies, such as dementia with Lewy bodies (DLBs) (Guerreiro et al. 2018; Chia et al. 2021) or REM sleep behavior disorder (RBD) (Krohn et al. 2022), although the exact sites of association may not be identical. These studies overall clearly establish *SNCA* as a pleomorphic gene locus involved both in rare genetic and sporadic forms of PD and other synucleinopathies.

NEUROPATHOLOGICAL FINDINGS LINKING α-SYNUCLEIN TO PARKINSON'S DISEASE

Very soon after the genetic discovery of the p.A53T *SNCA* mutation, studies were performed to assess whether deposition of AS could be discerned within neuronal Lewy bodies (LBs) and Lewy neurites (LNs), the characteristic neuropathological features of PD. This proved to be the case, not only in the rare genetic synucleinopathies, but also in the vast majority of iPD brains examined, and even in a range of related conditions, termed collectively LB diseases, such as DLB (Spillantini et al. 1997, 1998; Baba et al. 1998). AS antibodies (Abs) label the filamentous portion of LBs, consisting of a single protofilament, as identified by cryo-electron microscopy, thus having different properties from AS filaments identified in the oligodendrocytic synucleinopathy multiple system atrophy (MSA) (Yang et al. 2022). AS filaments may not be homogeneous across PD cases, suggesting different strains that may confer variable pathogenic effects, accounting partially for disease heterogeneity and distinct subtypes (Strohäker et al. 2019). Of note, nonfilamentous AS, that may be quite abundant, also exists within LB (Shahmoradian et al. 2019). There is likely also considerable astrocytic AS pathology in the spectrum of LB diseases, which is just beginning to be appreciated (Altay et al. 2022).

Immunostaining was used for the groundbreaking neuropathological study of Braak et al. (2003), which provided a basis for the staging of the disease. LNs, mainly, and also LBs, were present in various brain regions even in asymptomatic individuals. According to this staging scheme, aberrant AS deposition follows a stereotypical pattern from initial sites of involvement in the olfactory bulb and the dorsal motor nucleus of the vagus to more rostral areas of the brainstem, involving the substantia nigra pars compact at a third stage, and eventually to higher order association cortical areas in stages 5 and 6 (Braak et al. 2003). This AS immunohistochemical staging scheme has been controversial but has been borne out by most subsequent studies (e.g., Coughlin et al. 2019). Alternative staging schemes have been proposed to account for more rostral AS deposition in the absence of obligatory brainstem involvement in initial disease stages (Beach et al. 2009; Borghammer et al. 2021). Overall, there is consensus that aberrant AS deposition, as assessed by traditional AS immunostaining, is associated with regional brain dysfunction, which however is not always obvious clinically, as seemingly healthy individuals may harbor advanced stages of AS pathology (Parkkinen et al. 2008). Whether AS aberrant deposition is actually the cause of neuronal dysfunction is a question of debate, as some consider that such deposition could be secondary and incidental, and this notion cannot be excluded at this time (Espay et al. 2019). Of great interest is the newly developed technique of proximity ligation assay (PLA), which labels preferentially intermediate, oligomeric, and not fully fibrillar forms of AS (Roberts et al. 2015). In a recent application of this technique in PD brains, it was found that PLA-identified AS pathology in the hippocampus correlated much better with cognitive dysfunction than staining for classical Lewy pathology (Sekiya et al. 2022).

A related issue is that some genetic PD cases do not manifest Lewy pathology. In particular, biallelic *PRKN* mutation carriers rarely manifest such pathology (for review, see Madsen et al. 2021), while LRRK2 mutation carriers manifest quite significant variability in this regard. Among LRRK2 mutation carriers, those with the G2019S mutation most often show evidence of synucleinopathy, in 60%–70% of cases. Interestingly, there is an association between the existence of Lewy pathology and more widespread nonmotor disease manifestations, suggesting again that AS pathology is linked to brain dysfunction (Kalia et al. 2015).

Overall, the combination of genetic and neuropathological evidence, but also the evidence provided by the cell and animal models below, provides a strong argument for the pathogenicity of AS abnormal conformations in the context of PD and related synucleinopathies.

STRUCTURE, PHYSIOLOGICAL FUNCTION, AND SECRETION

Full-length AS is a small 140 aa protein primarily expressed in the presynaptic nerve endings of the adult brain in a region-specific manner. Except from the full-length protein, truncated AS fragments of 126, 112, and 98 aa are produced by alternative splicing (Beyer et al. 2006; McLean et al. 2012). The primary structure of AS consists of three distinct well-characterized domains, each conferring different physicochemical properties to the protein (Fig. 1). The basic amino terminus (1–60 aa) carries seven 11 aa repeats containing the consensus KTKEGV sequence, which is well-conserved among species and among all the members of the synuclein family, α, β, and γ-synucleins (Bussell and Eliezer 2003). Due to this repeated motif, AS can adopt a helical secondary structure that can take the form of either two interconnected antiparallel α-helices in solution or one contiguous α-helix upon binding to acidic lipid membranes (Fusco et al. 2014). The highly hydrophobic nonamyloid component (NAC) core domain (61–95 aa) provides, at least to a great extent, the inherent property of AS to self-aggregate, generating high-order fibrillar or low molecular weight (LMW) oligomeric assemblies (Giasson et al. 2001). Last, the acidic carboxy-terminal tail (96–140 aa) carries most of the posttranslational modifications (PTMs) and is capable of Ca^{2+} binding (Oueslati et al. 2010). This domain underlies the flexible nature of the protein since it hosts the majority of the molecular interactions of AS with other proteins, metals, or small molecules (Uversky et al. 2001). Importantly, the carboxyl-terminus region can interact transiently with the amino-terminus domain forming compact structures that are resistant to aggregation (Hong et al. 2011; Burré et al. 2012). In addition, carboxy-terminally truncated AS (CT-AS) products tend to aggregate faster than the full-length protein, supporting a role of the C-end in preserving the normal structure of the protein (Hoyer et al. 2004; Li et al. 2005).

Despite extensive investigation, the exact native secondary structure of AS remains largely unresolved. AS has been characterized as an intrinsically unfolded protein retaining minimal ordered structure in simple solutions (Uversky et al. 2001; Fauvet et al. 2012). In a cellular environment, it can physiologically adopt multiple conformations ranging from α-helical LMW multimers to β-sheet rich high molecular weight oligomers and aggregates (Uversky 2003). Even though it lacks a *trans*-membrane or a lipid-anchor domain, AS can peripherally associate with cellular membranes showing a preference to acidic or high curvature membranes. Membrane

 Cite this article as *Cold Spring Harb Perspect Med* doi: 10.1101/cshperspect.a041645

binding drives a conformational change toward the α-helical structure and can either promote or prevent multimerization. In this context, previous studies have shown that the molecular crowding conditions in the cytosol encourage the endogenous formation of tetrameric assemblies, which are characterized by a helical conformation and appear resistant to further aggregation (Bartels et al. 2011). However, subsequent studies failed to confirm that the tetramer is a predominant conformation of AS in the cytosol, concluding that cytosolic AS remains mostly as a mobile monomeric protein constantly binding to various interactors (Binolfi et al. 2012; Theillet et al. 2016).

The primary physiological function of AS has not been fully elucidated, but it appears to have a key role in the regulation of synaptic transmission and dopamine synthesis. Even though it is not required for neuronal development, synapse formation, or neurotransmission per se, AS can potently modulate synaptic activity through different modes of action. A plethora of studies have elaborated on the role of AS in synaptic vesicle (SV) trafficking, particularly in SV clustering (Diao et al. 2013; Wang et al. 2014) and distribution (Scott and Roy 2012; Sun et al. 2019). Further elaboration on this role suggested that AS could modulate exocytosis in a dose-dependent manner through dilation of the exocytic fusion pore during the "kiss-and-run" process, a mechanism that applies to both regulated protein secretion and neurotransmission (Logan et al. 2017; Nellikka et al. 2021). AS can directly associate with the SV membrane via its interaction with the chaperone cysteine-string protein a (CSPa) and the vesicle SNARE protein synaptobrevin-2 (VAMP2) to either facilitate SNARE complex assembly or prevent the disassembly of the SNARE complex until neurotransmitter release is completed (Burré et al. 2010; Garcia-Reitboeck et al. 2010). Further supporting a chaperone-like activity, 14-3-3 chaperone protein and its binding partners can also bind to AS (Ostrerova et al. 1999; Williams et al. 2021). Finally, AS can act as a negative modulator of dopamine synthesis and recycling as suggested by its interaction with the dopamine synthesis enzymes, tyrosine hydroxylase (TH), and aromatic amino acid decarboxylase as well as with dopamine transporter (DAT) (Tehranian et al. 2006; Swant et al. 2011; Butler et al. 2015, 2017; Sivakumar et al. 2023).

Apart from these well-defined physiological functions, AS has also been implicated in suppression of apoptosis by inhibiting PKC activity (Jin et al. 2011; Guo et al. 2021), regulation of glucose levels (Rodriguez-Araujo et al. 2013; Wijesekara et al. 2021), regulation of calmodulin activity (Martinez et al. 2003; Ueda et al. 2023), maintenance of polyunsaturated fatty acid levels, and neuronal differentiation (Surguchov 2024).

Despite the lack of a signal sequence, AS is physiologically secreted in the extracellular milieu suggesting a yet unidentified paracrine role for this protein (Lee et al. 2005; Emmanouilidou et al. 2010a,b; Wu et al. 2023). In support for such a modulatory role, the mechanism of AS secretion is Ca^{2+}-dependent and seems to be precisely regulated by neuronal activity in the brain (Emmanouilidou et al. 2016; Yamada and Iwatsubo 2018). Still, our understanding about the mechanisms that regulate AS release is largely incomplete. Although initially proposed, passive diffusion cannot account for such release, as cell-produced AS cannot freely diffuse out from the cell interior (Lee et al. 2008). Instead, insights from cell culture systems indicate that AS follows an unconventional pathway of release that involves, at least in part, the externalization of exosomes, nano-sized extracellular vesicles (EVs) of endosomal origin that participate in targeted intercellular communication (Emmanouilidou et al. 2010a,b; Alvarez-Erviti et al. 2011; Fussi et al. 2018). Since exosome-associated AS is only a minor part of externalized AS, conventional ER-Golgi exocytosis could mediate AS export, as indicated by the association of the protein with secretory vesicles, although direct evidence that these vesicles are responsible for AS secretion is missing (Lee et al. 2005; Logan et al. 2017). In the context of the living brain, the secretion of AS from glutamatergic terminals in the striatum is tightly controlled by the levels of the neurotransmitter GABA through an intercellular mechanism that involves presynaptic Ca^{2+} channels, further suggesting that the maintenance of extracellular AS levels in the brain parenchyma is critical for neuronal homeostasis (Emmanouilidou et al. 2016).

It is unclear whether the different conformations (normal, misfolded, or fibrillar) are released using common secretory pathways. Part of the misfolded cytoplasmic AS can escape cells using an unconventional pathway of release called misfolding-associated protein secretion (MAPS) that is mediated by the selective sorting of cargos to late endosomes and fusion of these endosomes with the plasma membrane (Lee et al. 2016). Acting as a co-chaperone of Hsc70, CSPa forms a high-order oligomer that captures AS upon palmitoylation and mediates its translocation to the late endosome lumen. The multivesicular body (MVB) that is subsequently generated carries soluble AS cargo, which is released upon fusion of the MVB with the plasma membrane (Wu et al. 2023). Alternatively, AS multimers that can accumulate within lysosomes can be released from neurons via SNARE-dependent lysosomal exocytosis (Xie et al. 2022).

Increased levels of AS oligomeric species have been observed to be associated with exosomes (Delenclos et al. 2017; Guo et al. 2020). Furthermore, AS has been detected in exosomes from patients with synucleinopathies (Stuendl et al. 2016; Harischandra et al. 2019), and AS mutations have been reported to aid the packaging of aggregated protein in exosomes (Gustafsson et al. 2018). In this regard, exosomes derived from PD patient tissue or from inflamed cells induce AS aggregation and pathology in vitro and in vivo (Grey et al. 2015; Lee et al. 2016; Huang et al. 2022; Jin et al. 2023). Endogenous AS appears to be essential for the ability of exosomes to propagate pathology in vivo (Melachroinou et al. 2024). In general, research to date is confirming a role for exosomes in the transmission of AS pathology in synucleinopathies; however, the exact mechanisms related to their packaging, release, and uptake have not been elucidated yet.

α-SYNUCLEIN AGGREGATION STATES: FOCUS ON STRAINS

It is widely considered that the toxic potential of AS is linked to its propensity to assume under certain circumstances abnormal conformations, such as intermediate soluble oligomers, also termed protofibrils, and eventually mature fibrils. The exact nature of the toxic species remains elusive. Importantly, fibrillar forms can transform soluble monomeric AS into an aggregated conformation. This forms the basis for the presumed disease propagation across brain regions (Lee et al. 2011) and the first wet biomarker for PD (see below).

So far, the findings from in vitro and in vivo experiments, and the observation in human tissue samples suggest that oligomers play a critical role in the initiation and progression of α-synucleinopathies (Kalia et al. 2013; Cremades et al. 2017). The oligomers that lead to fibril formation are known as "on-pathway" species. However, there are also "off-pathway" species that do not progress into fibrils (Miraglia et al. 2018). De Giorgi et al. (2020) demonstrated that, during fibril formation from monomeric AS, newly generated fibrils could be ThT-negative despite exhibiting a clear β-sheet structure in ssNMR. Interestingly though, injection of such fibrils into the SN of mice caused pS129 AS accumulation and spreading to other interconnected brain structures (De Giorgi et al. 2020). The different disease phenotypes observed in synucleinopathies suggest that each disorder may be caused by a different "strain" of AS conformations. Importantly, these different strains appear to affect specific cellular populations in the brain and maintain an ability to be serially transmitted, reminiscent of prions (Lau et al. 2020). This observation possibly suggests that strain-specific information is carried by the structure of the aggregated AS and can be transmitted, akin to prion molecules. It has also been proposed that the cellular environment drives one conformation over another (Woerman 2021). In this respect, Peng and colleagues demonstrated in a formative study that oligodendrocytes present a specific type of AS conformations, which differs from that in neurons (Peng et al. 2018). It is possible that different assemblies also exist between different neuronal types, which could in part explain their differential vulnerability in PD. In agreement with different conformers of AS having distinct biological and structural properties, distinct AS strains named ribbons and fibrils could propagate in human iPSC (induced pluripotent stem cell)-derived neurons; ribbons were more potent in recruiting and seeding endogenous AS, and resulted in more pS129-positive AS inclusions (Gribaudo et al. 2019).

Importantly, neurons from AS knockout mice exposed to AS fibrils do not develop intracellular inclusions and have intact neuronal function. Thus, endogenous AS templating to form insoluble fibrillar aggregates is crucial for pathology initiation and progression (Rey et al. 2018). Despite these major advances, a crucial unanswered question is whether these recombinant oligomers have different properties to the actual aggregates found in the brain in terms of heterogeneity and toxicity. Innovative approaches are urgently needed to detect and separate specific strains in the brains of individuals with synucleinopathies, as existing methods fall short in this regard.

CELLULAR AND ANIMAL PARKINSON'S DISEASE MODELS

Various cellular and animal PD models have been developed to investigate the pathological roles of the protein, including transgenic approaches, use of viral vectors, and, more recently, inoculation with recombinant preformed AS fibrils or LB extracts. However, none of these models recapitulates faithfully all aspects of PD pathophysiology and the choice of the appropriate model depends on the question being addressed.

Cellular models have been instrumental in the identification of the pathological roles of AS on various intracellular processes, such as mitochondrial function, oxidative stress, and proteasomal/lysosomal degradation pathways (for review, see Delenclos et al. 2019). Their main advantage is that they enable modeling of the mechanisms controlling the folding, oligomerization, aggregation, and cell-to-cell propagation of the protein, as well as the high-throughput screening of potential modifiers of these processes. From the first and simplest yeast models (Outeiro and Lindquist 2003) to more complex cellular systems (for review, see Delenclos et al. 2019) using mammalian neuronal and non-

neuronal cell lines and, more recently, humanized iPSC-derived cultures generated from patient fibroblasts (Mohamed et al. 2019), these cell-based systems provide a unique opportunity to model the disease in a dish and test novel pharmacological interventions. The generation of fluorescent reporter lines enabled the dynamic monitoring of AS–AS interactions and subsequent aggregation, even though it is questioned whether the modified AS behaves similarly to its nonmodified counterpart (Delenclos et al. 2019). The newest models that use patient-derived iPSCs allow investigation of the contribution of protein aggregation to early axonal dysfunction and provide novel mechanistic insights related to patient-specific risk factors or disease-specific mutations, thus paving the way for personalized treatments. Finally, several studies are now using iPSC-derived organoids or assembloids from PD patients to model disease pathophysiology in a more integrative manner that recapitulates better the brain's microenvironment (Bose et al. 2022; Calabresi et al. 2023a,b).

On the other hand, PD animal models offer the potential to model early alterations associated with AS overexpression and aggregation, that precede dopaminergic cell loss, such as synaptic dysfunction (synaptopathy) and nigrostriatal plasticity (Cenci and Björklund 2020). These models gradually develop LB-like inclusions of aggregated AS, which usually leads to neuronal loss, thus recapitulating major aspects of PD pathology. The simplest invertebrate PD models are particularly useful for high-throughput screening applications, whereas mammalian models are required to explore complex motor/nonmotor features and behavioral alterations. Transgenic animal models involve the expression of wild-type (WT) or PD-linked mutant forms of AS through different promoters, thus enabling regional and temporal control of expression. Viral vector–mediated models, on the other hand, offer many advantages, including the targeted injection into selective brain areas, the capacity to transduce both neurons and glia depending on the serotype used, and the ability of injection at any age of the animal.

A major breakthrough in understanding the mechanisms underlying the cell-to-cell propagation of AS-related pathology originated from studies where human (or mouse) recombinant AS preformed fibrils (PFFs) or PD brain extracts containing LBs or extracts from AS transgenic mice are injected into the brain (striatum, substantia nigra, and olfactory bulb), muscles, peritoneal cavity, or in the periphery of AS-over-expressing or WT rodents or nonhuman primates (for review, see Recasens et al. 2018). These studies are highly reproducible and are characterized by the presence of widespread Ser129-phosphorylated AS inclusions, mirroring aspects of the spread and staging of the human disease. Similar approaches have also been very fruitful in cellular and, in particular, neuronal models, where it has been possible to model the maturation of seeded AS fibrils into LB-like structures through the engagement of various compensatory but also detrimental neuronal processes (Mahul-Mellier et al. 2020).

α-SYNUCLEIN POSTTRANSLATIONAL MODIFICATIONS

AS exhibits a number of PTMs, including phosphorylation, ubiquitination, nitration, acetylation, truncation, SUMOylation, and O-GlcNA-cylation. Of these, phosphorylated AS is thought to be the major pathological form (Fig. 1; Anderson et al. 2006).

Phosphorylation

Several studies have shown that AS phosphorylated at serine 129 (pSer129) is a marker of mature AS aggregates. Examination of postmortem tissue from PD and MSA patients at different disease stages showed that pSer129 AS is the major and earliest PTM along PD progression (Wakabayashi 2020; Sonustun et al. 2022). Nevertheless, pSer129 seems to occur after the initial aggregation process (Pantazopoulou et al. 2021; Ghanem et al. 2022) and may have a physiological role at the synapse (see below). An additional caveat is that in models of AS overexpression, such as for example viral models, phosphorylated AS is not necessarily aggregated and should not be used as a sole readout of aggregation in these circumstances.

Whether pSer129 is a driving force for AS aggregation and neurotoxicity remains a subject of debate. Mice inoculated with pSer129 AS PFFs exhibited enhanced AS pathological deposition and dopaminergic cell loss associated with motor deficits (Karampetsou et al. 2017). The Lashuel and Li groups, using semisynthetic approaches to synthesize pSer129 AS, showed that the phosphorylated fibers were toxic and less resistant to proteolysis by proteinase K compared to WT fibers, suggesting that S129 phosphorylation induces a distinct strain of AS species (Fauvet and Lashuel 2016; Ma et al. 2016). However, other studies have reported that pSer129 phosphorylation does not influence AS aggregation and can reduce its toxicity (Weston et al. 2021; Ghanem et al. 2022).

The role of pSer87 is controversial as this PTM falls within the NAC region of AS, which is crucial for its aggregation and fibrillogenesis. However, pSer87 AS viral overexpression in the nigrostriatal system of rats caused reduced accumulation and no dopaminergic neuron loss or motor impairment, in contrast to WT AS overexpression (Oueslati et al. 2012). Collectively, data on the phosphorylation sites at S129, S87, and Y39 support the notion that phosphorylation decreases AS binding to membranes (Dikiy et al. 2016; Reimer et al. 2022).

Nitration, Acetylation

Nitrated AS has been reported in various in vivo and in vitro experimental models of PD and also in association with LB pathology (Przedborski et al. 2001; He et al. 2019; Manzanza et al. 2021; Magalhães and Lashuel 2022). Increased levels of nitrated AS have been detected in LBs and SN neurons, as well as in peripheral blood monocytes of PD patients (Prigione et al. 2010). Nitration of AS in mice was shown to elicit macrophage activation and T-cell responses that lead to exacerbated nigrostriatal degeneration (Benner et al. 2008). Interestingly, recent in vivo studies in an AAV–AS mouse model showed that AS nitration induces loss of neurons and increased cell–cell transfer of AS pathology (Barrett and Timothy Greenamyre 2015; Musgrove et al. 2019).

Cite this article as *Cold Spring Harb Perspect Med* doi: 10.1101/cshperspect.a041645

Studies using label-free single molecule detection methods, as well as recombinant acetylated AS, have shown that an amino-terminal acetylation has a protective effect, as it can significantly decrease oligomerization (Iyer et al. 2016; Bu et al. 2017). Bell et al. (2023) studied five amino-terminal acetylated familial variants (A30P, E46K, H50Q, G51D, and A53T) of AS and found that each variant responds to amino-terminal acetylation in unique ways, highlighting the great complexity of the behavior of AS and its high susceptibility to chemical modifications (Bell et al. 2022, 2023). In general, acetylation reduces the oligomerization capacity of AS, as well as the rate of fibril formation.

Truncation

Studies indicate a strong link between CT-AS and its aggregation. CT-AS is present in the brain and colon of PD patients and has increased ability to form fibrils and increased toxicity. Blocking carboxy-terminal truncation using antibodies to the carboxyl-terminus of the protein in an AS transgenic animal model reduced PD symptoms and reversed AS accumulation (Games et al. 2014). However, in vitro and in vivo studies with CT-AS fibrils have produced mixed results, with some reporting increased capacity to induce prion-like seeding of full-length AS by CT fibrils and others observing a decreased ability compared to WT AS fibrils (Sorrentino and Giasson 2020; Ohgita et al. 2022). It is possible that these controversies may stem from the different length of truncated forms used in the different studies.

Sumoylation, Ubiquitination

Impairment of AS SUMOylation in vitro by mutations of SUMO residues increased its aggregation propensity and neuronal toxicity. Interestingly, increased SUMOylated AS has been detected in PD brains (Rott et al. 2017; Rousseaux et al. 2018). Verma et al. (2020) showed in cellular models that SUMOylation is neuroprotective against MPP$^+$ or AS PFFs. Similar results were obtained in vivo. Other reports show that SUMOylation competes with ubiquitination of AS, thus potentially blocking ubiquitin-dependent degradation pathways. Increased SUMOylation also increased extracellular AS levels and its association with exosomes. It is, therefore, possible that SUMOylation may in this way affect the spreading of AS between cells (Kunadt et al. 2015; Stuendl et al. 2016). A number of studies have shown that ubiquitination by ligases such as Nedd4 enhanced the protein's clearance through an endosomal–lysosomal pathway (Liani et al. 2004; Tofaris et al. 2011); however, recent in vitro and in vivo ubiquitination studies have suggested that AS ubiquitination may promote the production of aggregated forms (Rott et al. 2017; Zhang et al. 2017; Wang et al. 2019). In contrast, the in vitro ubiquitination of WT AS at different sites was found to produce structurally different aggregates but with reduced aggregation ability (Moon et al. 2020).

O-GlcNAcylation

The O-GlcNAcylation of AS decreases its aggregation propensity and toxicity in cultured primary neurons without affecting its membrane binding affinity (Marotta et al. 2015; Levine et al. 2017). Moreover, O-GlcNAcylation hampers the cleavage of AS by calpain in vitro, a process involved in the formation of aggregates. It is possible that O-GlcNAc could similarly inhibit the cleavage of AS by as yet unidentified proteases that generate aggregation-prone protein fragments. In addition, pharmacological inhibition of glycoside hydrolase O-GlcNAcase (OGA) (which thus increases O-GlcNAcylation) blunts AS PFF cellular uptake (Tavassoly et al. 2021) and alleviates the degeneration and pathology in dopaminergic neurons caused by AS overexpression in an AAV mouse model (Lee et al. 2020). Similarly, small inhibitors to OGA after daily dosing improved motor impairment, reduced astrogliosis, and facilitated dopamine neurotransmission in mouse modes of PD (Permanne et al. 2022).

PATHOGENIC EFFECTS OF α-SYNUCLEIN IN VARIOUS CELLULAR COMPARTMENTS AND FUNCTIONS

AS has a pathogenic potential within neurons and possibly astrocytes in the context of LB diseases, and this may occur by aberrant effects at various cellular sites. Due to its predominant localization in presynaptic terminals and its physiological role at the synapse, major efforts have been undertaken to characterize its pathogenic role at synaptic terminals (Fig. 2); however, other aberrant cellular effects are also considered (Fig. 3).

α-Synuclein at the Synapse: a Love and Hate Affair

Among its multiple functions, the physiological role of AS in SV homeostasis and neurotransmitter release, as well as its aberrant effects on synaptic transmission and SNARE complex assembly, are the most well-documented (Scott

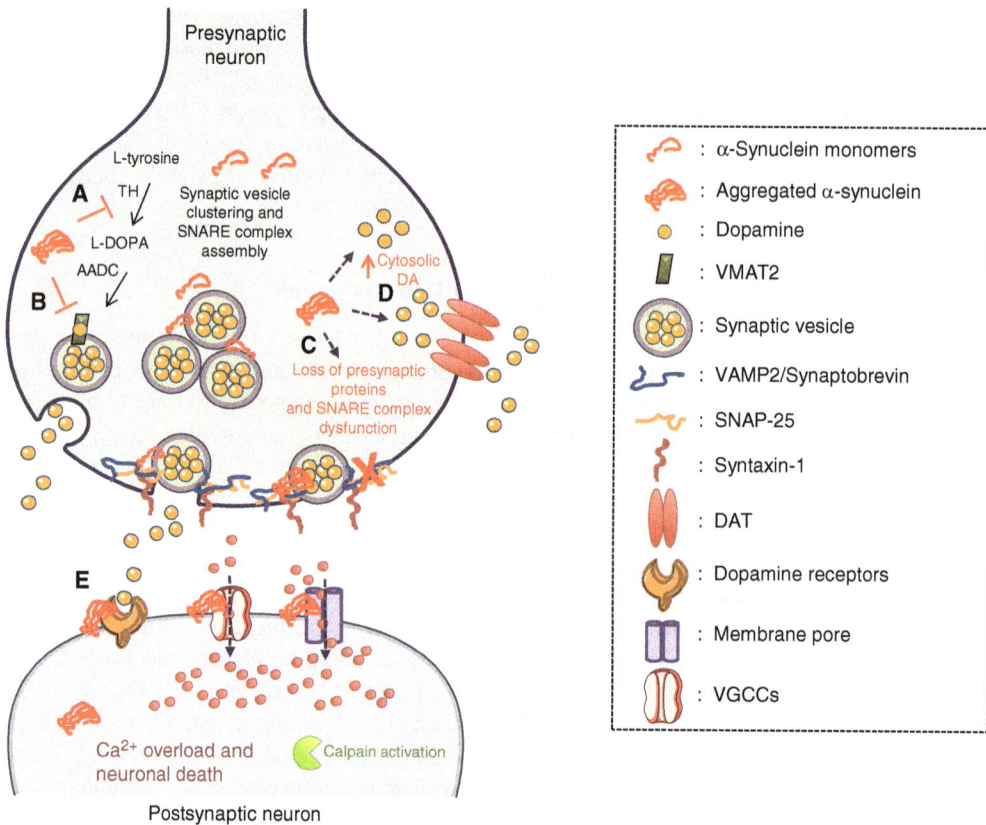

Figure 2. The pathological effects of α-synuclein (AS) at the synapse. (*A*) Aggregated AS reduces the activity of tyrosine hydroxylase (TH), the enzyme responsible for catalyzing the conversion of L-tyrosine to L-DOPA, thus impairing dopamine biosynthesis. (*B*) Increased levels of AS inhibit VMAT2, which is responsible for the uptake of monoamines transmitters (such as dopamine) into SVs; therefore, it modulates the neurotransmitter storage. (*C*) Disease-related AS conformations alter the levels of presynaptic proteins and evoke SNARE complex dysfunction, interfering with the SV fusion and dopamine release. (*D*) AS aggregates trigger dopamine transporter (DAT) recruitment to the plasma membrane, leading to increased entry of dopamine and increased cytosolic dopamine levels, which may be neurotoxic by facilitating further protein aggregation through the generation of dopamine-modified AS adducts. (*E*) Aberrant AS conformations may affect the activity of dopamine receptors and voltage-gated calcium channels (VGCCs), as well as promote the formation of ion-permeable pores in the plasma membrane; this may lead to intracellular Ca^{2+} overload and calpain activation, facilitating further protein aggregation and triggering a Ca^{2+}-dependent signaling cascade leading to neuronal demise.

   Cite this article as *Cold Spring Harb Perspect Med* doi: 10.1101/cshperspect.a041645

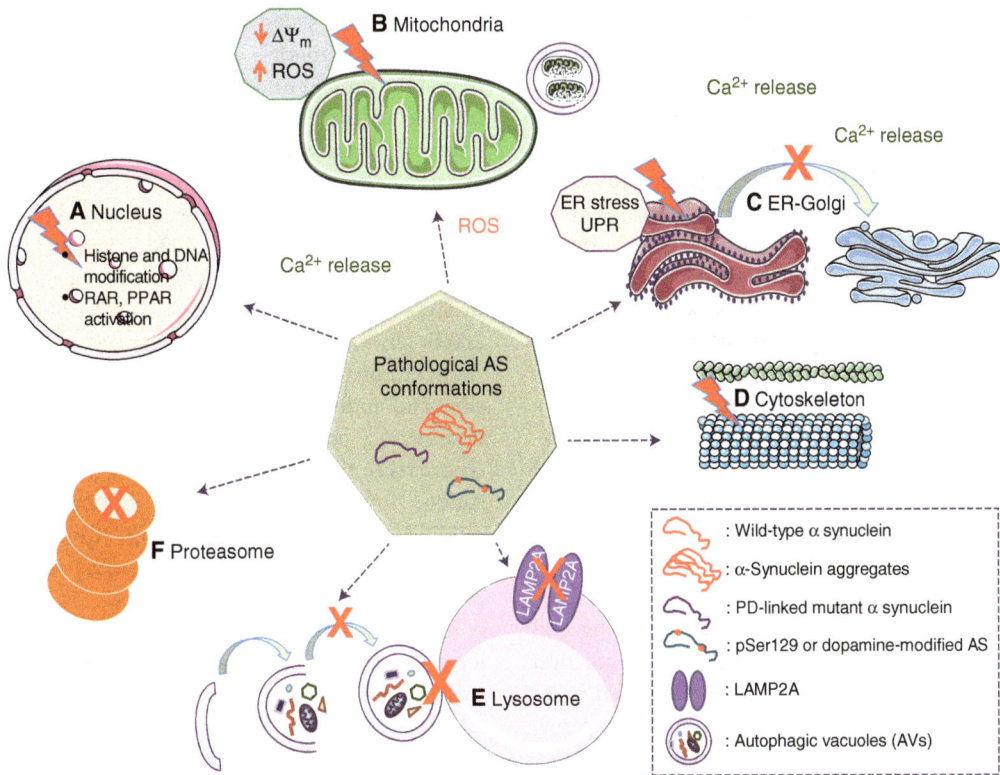

Figure 3. Aberrant effects of disease-related α-synuclein (AS) conformations (aggregated, dopamine modified, and phosphorylated) on the various cellular functions/compartments. (*A*) Within the nucleus, AS inhibits histone acetylation via its direct binding to histones or by inhibiting the action of histone acetyltransferase enzymes, thus interfering with the process of gene transcription. Through interaction with RA and peroxisome proliferator–activated receptors (PPARs), AS can alter Nurr1 transcription, thus affecting dopaminergic neuron survival. (*B*) AS mutations of aggregated species can transiently and dynamically engage with mitochondria, causing their depolarization, reduced energy production, fragmentation, and destruction through mitophagy. (*C*) Within the endoplasmic reticulum (ER), pathological AS assemblies can induce ER stress and unfolded protein response (UPR), impairment of Ca^{2+} homeostasis, and alterations in the vesicle-dependent protein trafficking, the latter affecting the ER–Golgi protein transport. (*D*) Aggregated AS may affect the structure and function of the neuronal microtubule cytoskeleton (through interaction with actin and tubulin), leading to axonal transport defects. (*E*) Disease-related conformations of AS may impair chaperone-mediated autophagy (CMA) activity through aberrant interaction with the LAMP2A receptor, as well as macroautophagic activity, affecting the formation of autophagosomes or their maturation and fusion with the lysosome. (*F*) Increased levels or pathological forms of AS may inhibit proteasomal function, thus leading to AS accumulation and formation of insoluble protein aggregates.

et al. 2010; Gao et al. 2023). PD-linked mutations, as well as aberrant AS conformations may exert pathological effects at the presynaptic terminal, including loss of presynaptic proteins (Chung et al. 2009), redistribution of SNARE proteins and impairment of neurotransmitter release (Garcia-Reitboeck et al. 2010), and inhibition of SV recycling pool size and mobility (Nemani et al. 2010). At the presynaptic terminal, AS can exist both in a cytosolic and a membrane-bound form, and this localization may alter the propensity for aggregation. Membrane binding seems to exert a protective effect against aggregation, based on observations that some PD-linked AS missense mutations may inhibit this (Jo et al. 2002; Fares et al. 2014; Ghosh et al.

2014; Liu et al. 2021). Other reports, however, suggest the opposite (Lee et al. 2002; Perni et al. 2017; Limbocker et al. 2021). This controversy may reflect the coexistence of monomeric and oligomeric membrane-bound species that accelerate further the protein aggregation thus exerting a neurotoxic function. It has been also found that AS oligomers are preferably bound to synapsin 1 and VAMP2 (Betzer et al. 2015) and attenuate SNARE complex assembly. Intriguingly, recent findings support a rather physiologic role for the pathology-related pSer129 phosphorylated AS at synapses, advocating a model where activity-induced pSer129-AS triggers the interaction of AS with a network of synaptic proteins that eventually leads to physiologic attenuation of neurotransmitter release (Parra-Rivas et al. 2023).

Not only the conformation but also the protein dosage may alter the vesicle recycling and docking, as both gain- and loss-of-AS function can impair SV recycling. AS is expressed in all presynaptic terminals; however, in PD, dopaminergic neurons are the most vulnerable, possibly because AS regulates dopamine synthesis and turnover by altering the activity of critical components of the pathway, such as TH, DAT, and VMAT2 (Calabresi et al. 2023a,b). Differences between neuronal subtypes and neurotransmitter systems affected in early (norepinephrinergic, serotonergic, cholinergic) or late (dopaminergic) stages of the disease may also account for the differential vulnerability to AS-related synaptopathy. Common features between these neurons are prominent calcium currents, low intrinsic calcium buffering capacity, sustained spontaneous spiking, and broad spikes (Surmeier and Schumacker 2013).

Furthermore, a great wealth of data spanning from primary neuronal cultures, iPSC-derived neurons, and animal PD models, pinpoints a pathogenic effect of aggregated AS (established by overexpression of the WT or PD-linked AS mutations or PFF inoculation) on the levels of presynaptic proteins, such as SNAPs, VAMP2, Synapsins, Syntaxins, Synaptotagmins, Synaptophysin, SV2, PSD95, GAP42, Drebrin, Neurogranin, Rabphilin 3A, and neurotransmitter release (for review, see Murphy and McKernan 2022).

Animal studies using AS knockout (KO) and overexpression models cement further a critical role of the protein in the regulation of dopamine homeostasis. In particular, increased dopamine release and decreased reuptake, low striatal TH and DAT levels and a reduced number of nigral dopaminergic neurons have been reported in AS KO mice (Abeliovich et al. 2000; Chadchankar et al. 2011), although AS deletion was not shown to alter cytosolic dopamine levels (Mosharov et al. 2009). On the other hand, AS aggregates are reported to perturb dopaminergic neurotransmission and induce presynaptic and postsynaptic dysfunction, possibly through interactions with oxidized DA that facilitates further protein aggregation, through the generation of dopamine-modified protein adducts (Conway et al. 2001; Mor et al. 2017). Human AS overexpression evoked dopaminergic terminal loss (Masliah et al. 2000), deficient dopamine release, and altered SV distribution (Janezic et al. 2013), as well as defective DAT function (Lundblad et al. 2012). Exposure of neurons to AS oligomers increases Ca^{2+} intracellular levels resulting in mitochondria and ER stress, reactive oxygen species (ROS) production, and increased DA release, thus initiating a toxic cascade leading to neurodegeneration (Calabresi et al. 2023a,b). Finally, decreased levels of synaptic proteins and alterations in SNARE complex assembly correlating with duration of dementia have been also reported in human PD postmortem material (Mukaetova-Ladinska et al. 2013; Vallortigara et al. 2016), thus underscoring AS-mediated deregulation of synaptic neurotransmission.

Pathogenic Effects of α-Synuclein: Impact on the Nucleus

Following the first observation of AS localization in the nuclear envelope of the Torpedo electric organ (Maroteaux et al. 1988), studies have reported on the presence of AS in the nucleus, even though its function in this compartment is only partially understood. Nuclear AS is implicated in the regulation of gene transcription through direct binding either to naked DNA or to enzymes involved in transcription such as methyltransferases, histone deacetylases,

RNA-interacting proteins, and histones (Somayaji et al. 2021).

Chromatin immunoprecipitation coupled with next-generation sequencing (ChIP-seq) confirmed that the ability of AS to directly bind to supercoiled DNA alters DNA conformation and stability and affects gene expression (Hegde and Rao 2007; Pinho et al. 2019). Further, AS can directly interact with the nucleus-resident DNA methyl transferase 1 forcing its export in the cytosol, thereby causing DNA hypomethylation that increases the expression of various genes, including *SNCA* itself (Desplats et al. 2011). AS can also bind histone 3 (H3) reducing its acetylation, which in turn results in inhibition of gene expression through a disturbance in the balance between histone acetylation and deacetylation (Kontopoulos et al. 2006; Outeiro et al. 2007; Paiva et al. 2017). In addition, AS can interfere with histone methylation; for example, it can enhance histone lysine *N*-methyltransferase 2 (EHMT2) activity, decreasing the expression of genes regulated by the REST complex (Sugeno et al. 2016). Finally, AS can impact nuclear receptor-mediated transcription indirectly via its interaction with retinoic acid (RA). The AS-RA complex translocates to the nucleus, where it activates the RAR and PPAR nuclear receptors and downregulates the orphan receptor, Nurr1, through mobilization of their respective response elements (Yakunin et al. 2012; Volakakis et al. 2015; Davidi et al. 2020). These effects, given the role of Nurr1 in the development and maintenance of dopaminergic neurons, could be detrimental.

Pathogenic Effects of α-Synuclein at the Mitochondria

AS has a cryptic mitochondrial targeting sequence and has been reported to localize to mitochondria and influence mitochondrial dynamics (Devi et al. 2008; Parihar et al. 2008; Nakamura et al. 2011). However, other reports have demonstrated both in vitro and in brains that AS does not directly localize in mitochondria but rather associates with mitochondrial-associated ER membranes (MAMs). This interaction is reduced by pathogenic AS mutations, leading to mitochon-

drial fragmentation (Cooper et al. 2012; Guardia-Laguarta et al. 2014). AS monomers interact with mitochondria and regulate mitophagy events of fusion and fission, as well as transport and degradation of mitochondria (Lurette et al. 2023). In particular, AS promotes mitochondrial fission events and inhibits fusion through the activity of mitofusins. Treatment of isolated brain mitochondria with monomeric AS leads to an increase in ATP production through association of AS and the α-subunit of ATP synthase, suggesting a role of monomeric AS as a mitochondrial bioenergetic regulator (Ludtmann et al. 2018). AS oligomers interact with high affinity with important mitochondrial proteins like VDAC and TOM, proteins required for mitochondrial protein import, leading to their internalization and disruption of mitochondrial function (Guardia-Laguarta et al. 2014; Di Maio et al. 2016). In related research, Bérard et al. (2022) used an optogenetic system to manipulate and monitor AS aggregation in cells. They discovered that AS aggregates transiently and dynamically engage with mitochondria, causing their depolarization, reduced energy production, fragmentation, and destruction through mitophagy, which is dependent on cardiolipin externalization (Bérard et al. 2022). Furthermore, postmortem studies of PD brains showed that aggregated S129-phosphorylated AS preferentially binds to mitochondria (Wang et al. 2019). Treatment of primary neurons with AS fibrils caused the appearance of phosphorylated AS inclusions that appeared associated with the mitochondrial membrane, leading to cytochrome *C* release, and oxidative stress (Prots et al. 2018). Aggregates produced in iPSC-derived neurons bearing AS triplication were shown to promote the opening of osmotic transition pore, causing mitochondrial swelling, ultimately leading to neuronal death (Ludtmann et al. 2016). An additional point is that mitochondrial transport and function may be compromised in the process of the maturation of AS fibrils into LB-like structures (Mahul-Mellier et al. 2020). Cumulatively, these data support the idea that metabolism and function of healthy neurons may depend on the critical interplay between AS and mitochondria (Risiglione et al. 2021). Therefore, targeting the aberrant mitochondrial localization of AS aggregates may prove beneficial for α-synucle-inopathies.

Pathogenic Effects of α-Synuclein: Impact on the ER, Golgi, and Relevant Trafficking

The pathogenic effects of AS on the endoplasmic reticulum (ER) include induction of ER stress and unfolded protein response (UPR), impairment of Ca^{2+} homeostasis, and alterations in the vesicle-dependent protein trafficking. The UPR can be initiated by three different ER-resident stress sensors, inositol-requiring enzyme 1α (IRE1α), PKR-like ER kinase (PERK), and activating transcription factor 6 (ATF6) (Manie et al. 2014). In the absence of ER stress, these proteins remain in an inactive form through binding to BiP, a chaperone protein that acts as a detector of nonproperly folded proteins. Mutated or aggregated AS can penetrate the ER membrane, inducing morphological changes to the ER and binding to BiP (Bellucci et al. 2011; Colla et al. 2012; Gorbatyuk et al. 2012). This promotes the dissociation of BiP from IRE1α, PERK, and ATF6, which then activate a series of cascade reactions directed to preserve cellular proteostasis; however, overreaction may lead to cell death. UPR activation seems to be a major contributor of AS-related cytotoxicity since ER stress markers, such as BiP or p-PERK, are increased in brain material from PD patients (Conn et al. 2004) and genetic or pharmacological targeting of UPR components are beneficial in preclinical models of PD (Colla et al. 2012; Martinez et al. 2019; Siwecka et al. 2023). The fact that BiP overexpression in vivo can resolve ER stress and protect from AS-induced cytotoxicity (Gorbatyuk et al. 2012) further highlights the importance of the BiP–AS interaction in the initiation and maintenance of the UPR and downstream detrimental effects. ER stress conditions can further potentiate AS aggregation feeding a vicious cycle linking AS pathology and ER dysfunction (Jiang et al. 2010; Bellucci et al. 2011).

AS can also interact with vesicular traffic components within the ER affecting ER to Golgi protein transport. The first observation of abnormalities in ER-dependent vesicular traffic came from studies in yeast where overexpression of AS-induced inhibition of vesicle docking to and fusion with the Golgi membrane, which was rescued by overexpression of Rab family members such as Rab1, Rab3A, and Rab8A, suggesting that, except from the ER-Golgi route, AS could impair other steps of the secretory pathway (Cooper et al. 2006; Gitler et al. 2008). Vesicular trafficking is finely orchestrated by intracellular calcium. Aggregated AS can directly bind to SERCA; this interaction distorts ER Ca^{2+} levels, interferes with intracellular Ca^{2+} homeostasis, and compromises vesicle targeting and fusion (Betzer et al. 2018; Kovacs et al. 2021). Importantly, alterations in ER to Golgi and vesicular trafficking occur in the process of LB-like structure maturation and may be responsible for the breakdown of cellular homeostasis (Mahul-Mellier et al. 2020).

Pathogenic Effects of α-Synuclein on the Cytoskeleton

It is notable that in the process of LB-like inclusion formation following AS PFF application and seeding in primary neuronal cultures, neuritic AS aggregates closely apposed to cytoskeletal elements are formed first (Mahul-Mellier et al. 2020). AS may affect the structure and function of the neuronal microtubule cytoskeleton, leading to axonal transport defects (Carnwath et al. 2018; Prots et al. 2018). High concentrations of AS can also alter the actin cytoskeleton when applied to hippocampal neurons, and can subsequently lead to disruption in neuronal functions, including axonal growth and migration. Overexpression of AS in fly neurons increased F-actin levels, promoted mislocalization of the mitochondrial fission proteins, and consequently led to mitochondrial and autophagic-lysosomal dysfunction (Ordonez et al. 2018; Sarkar et al. 2021). Recent data aided by superresolution imaging approaches, revealed the close association of neurofilaments and β-tubulin to pSer129 α-syn in LBs of PD postmortem tissue (Moors et al. 2021). So far, however, there is no evidence to support the hypothesis that AS is a true microtubule associated protein (i.e., a protein able to bind tubulin or microtubules and regulate their behavior). In addition, it is not yet known whether AS binds tubulin directly or via as yet unidentified binding partners.

A Reciprocal Relationship: Pathogenic Effects of α-Synuclein on Protein Degradation Systems

Limiting intracellular AS levels, as well as its pathogenic effects on the function of the intracellular proteolytic machineries represents an obvious therapeutic approach for PD and related α-synucleinopathies. The manner of AS degradation still remains controversial, with both the proteasome (Bennett et al. 1999; Tofaris et al. 2001, 2011; Webb et al. 2003; Shabek et al. 2012) and the lysosome (Paxinou et al. 2001; Webb et al. 2003; Sevlever et al. 2008; Vogiatzi et al. 2008), contributing to AS clearance, in a conformation-, PTM-, cell-type-, and tissue-specific manner (Emmanouilidou et al. 2010a,b; Xilouri et al. 2013a, 2016a). Indicatively, soluble or relatively insoluble (but not fully aggregated) S129-phosphorylated AS appears to be cleared by the proteasome, whereas seeded aggregated AS is cleared by macroautophagy (Pantazopoulou et al. 2021). It has been suggested that AS is degraded via the proteasome under basal conditions in vivo, whereas under conditions where intracellular AS protein load is augmented, the lysosome takes over (Ebrahimi-Fakhari et al. 2011). More recently, a de novo K45, K58, and K60 ubiquitination of AS mediated by NBR1 binding and entry into endosomes in a process that involves ESCRT I–III for subsequent lysosomal degradation was identified, using diverse approaches in living or fixed cells (Zenko et al. 2023). This may be an important pathway for a pool of AS, which is rapidly turning over. We and others have found that WT-soluble AS, but not the A53T and A30P mutants, or phosphorylated or dopamine-modified AS, is degraded, at least partly, via the chaperone-mediated autophagy (CMA) lysosomal pathway (Cuervo et al. 2004; Martinez-Vicente et al. 2008; Vogiatzi et al. 2008; Xilouri et al. 2009; Mak et al. 2010). Further supporting the role of CMA on AS degradation are in vivo findings showing that overexpression of CMA's rate-limiting step, the LAMP2A receptor, concurrent-

ly with human AS in the rat substantia nigra (Xilouri et al. 2013b) and in the *Drosophila* brain (Issa et al. 2018), was capable of mitigating AS levels and alleviating AS-related toxicity. Conversely, CMA deficiency in the rat substantia nigra through LAMP2A down-regulation led to the cytoplasmic accumulation of small AS aggregates, signifying that CMA is responsible for AS turnover within nigral dopaminergic neurons (Xilouri et al. 2016b). Within the lysosome, cathepsin D and cathepsin L have both been reported to clear AS aggregates (Cullen et al. 2009; Bae et al. 2014; McGlinchey and Lee 2015; Prieto et al. 2022).

An interrelated theme to AS degradation is the impact of increased WT- or PD-linked mutant protein load on the function of the proteasome and the lysosome. Initial studies proposed that the PD-linked A30P and A53T mutants evoke proteasomal impairment (Stefanis et al. 2001; Tanaka et al. 2001; Petrucelli et al. 2002; Snyder et al. 2003), although other studies failed to detect such an effect (Martìn-Clemente et al. 2004). In addition to the cell and animal data, reports in human postmortem material also suggest that proteasome function is impaired in sporadic PD patients (for review, see Cook and Petrucelli 2009). Furthermore, many studies highlight a central pathogenic role of aberrant AS on endolysosomal function, focusing mostly on macroautophagy and CMA. Pathological conformations of AS (mutations, oligomeric/aggregated species) have been reported to interfere with different stages of autophagosome formation, maturation, trafficking, and fusion with the lysosome (Xilouri et al. 2016b; Sanchez-Mirasierra et al. 2022). Briefly, impaired macroautophagic activity has been initially reported in in vitro and in vivo synucleinopathy models, in a manner dependent on AS-Rab1a interaction and Atg9 mislocalization (Winslow et al. 2010). Early-stage autophagosome formation was impaired in the presence of both E46K- and A30P PD-linked mutants, in a manner dependent on JNK pathway (Yan et al. 2014; Lei et al. 2019), whereas the A53T mutant protein exerted contradictory effects on mitophagy (Chen et al. 2018; Obergasteiger et al. 2018). Multiple lines of evidence suggest that bulk AS aggregates disrupt endolysosomal trafficking events, including those related to protein secretion partly via exosomes, that impede further AS clearance, thus permitting the persistence of pathological protein species within neurons or glia cells, ultimately leading to cell destruction (Klein and Mazzulli 2018). Intriguingly, it has been proposed that AS secretion via exosomes could act as a protective mechanism against the aberrant effects of AS on lysosomal function (Fussi et al. 2018). In addition, aberrant species or elevated levels of AS can have a detrimental effect on CMA function, likely through excessive binding to Lamp2a, impeding the access of other substrates to this rate-limiting component of the pathway, and thus setting the stage for a vicious cycle of pathogenicity (Cuervo et al. 2004; Xilouri et al. 2009, 2013a,b).

It is interesting to note here that, beyond AS, the aberrant actions of multiple PD-linked genetic defects, such as in LRRK2, VPS35, ATP13A2, and β-glucocerebrosidase (GCase), converge on the lysosome and are often accompanied by AS accumulation due to the ongoing lysosomal impairment (Klein and Mazzulli 2018). In particular, multiple cell- and animal-based studies propose the existence of a bidirectional loop underlying the relationship between *GBA1* mutations, AS, and the lysosome (Smith and Schapira 2022). It has been recently shown that mutant GCase contains a CMA-targeting motif and impairs the formation of the CMA lysosomal translocation complex required to translocate AS into CMA-active lysosomes for degradation, thus providing a new link between AS, CMA, and GCase (Kuo et al. 2022). Interestingly, the protein levels of the CMA markers LAMP2A and HSC70 are decreased in the human substantia nigra and amygdala of PD brains compared to controls (Alvarez-Erviti et al. 2010), and this reduction in LAMP2A levels correlated with increased AS accumulation selectively in regions harboring AS pathology (Murphy et al. 2014, 2015). Altered levels of LAMP2A and/or HSC70 can be detected in peripheral blood mononuclear cells of sporadic PD patients (Wu et al. 2011; Sala et al. 2014; Papagiannakis et al. 2015, 2019), suggesting that a systemic reduction in CMA activity may be present in PD patients.

α-SYNUCLEIN AS A BIOMARKER FOR SYNUCLEINOPATHIES

Many researchers have worked to establish AS-based tools for the evaluation of AS in early stage diagnosis, differential diagnosis of PD from other parkinsonian disorders, and assessment of disease progression (Chopra and Outeiro 2024). This topic has exploded in recent years, and a full accounting is beyond the scope of the current review.

The fact that AS is present in other tissues except the CNS complicates the interpretation of alterations in the levels of the protein. CSF total AS has been quantified using different immunoassays, and the results were controversial ranging from no difference (Mollenhauer et al. 2011; Toledo et al. 2013; Hansson et al. 2014) to a significant decrease observed in PD patients compared to controls (Hall et al. 2012; Kang et al. 2013; Parnetti et al. 2014). This decrease likely reflects the entrapment of soluble interstitial AS within LBs and related aggregates in a manner akin to β-amyloid deposition and related low CSF β-amyloid in AD. The magnitude of the decrease observed is quite low, ~10%–15%, weakening the ability of the assay of total AS in CSF to distinguish PD from controls. The measurement of CSF oligomeric and phosphorylated species has also been pursued. Both conformers were found to be significantly increased in the CSF of PD patients and the oligomer-to-total AS ratio could differentiate PD subjects from controls with higher sensitivity and specificity compared with total AS (Park et al. 2011; Wang et al. 2012; Parnetti et al. 2014; Stewart et al. 2015); yet, these assays have not provided enough separation between PD and controls to be useful as diagnostic biomarkers.

The levels and species of AS have also been examined in other more easily accessible body fluids or tissues. Even though the major source of AS in blood is erythrocytes, most of the work so far has focused on plasma or serum AS. The results from these studies are varied, showing either increased, unaltered, or even decreased levels of AS, highlighting the importance of technical confounders in biological samples with high constituent complexity such as plasma or serum (Duran et al. 2010; Foulds et al.

2013; Shah et al. 2017). In contrast, in studies where aggregated or total AS was quantified in erythrocyte membranes, the results have consistently shown a significant increase in PD samples compared to healthy controls or other related disease groups (Papagiannakis et al. 2018; Abd Elhadi et al. 2019; Li et al. 2021).

Considering that AS conformers are packaged in EVs, which can facilitate disease propagation in the brain, recent studies assessing neuronal cell-adhesion molecule LCAM1-positive EVs isolated from blood or saliva reported significantly higher levels of EV-associated total AS in PD (or even in prodromal PD) compared to controls, a difference that could not be observed in plasma samples from the same groups (Cao et al. 2019; Jiang et al. 2020; Yan et al. 2024). Assessment of pathological aggregated AS in this material may even lead to higher sensitivity and specificity (Kluge et al. 2022).

Another relatively easily accessible material is skin. Skin biopsies have been used mainly as material for immunohistochemistry with antibodies against altered conformations of AS, in particular phosphorylated AS. Such assays have shown high specificity and sensitivity and discriminatory ability compared to controls or non-LB-related parkinsonism (e.g., in Donadio et al. 2014), but results in other studies have been variable, largely likely due to methodological issues. This has culminated in a large multicenter study that clearly established the very high specificity and sensitivity (over 90% for both) of this qualitative assay for differentiating synucleinopathies from controls (Gibbons et al. 2024).

New technological advances have changed our perspective of measuring AS in biofluids. In vitro detection methods in biological material now include conformer-specific immunoassays and electrochemical biosensors (Chen et al. 2022), as well as complex methods based on the addition of recombinant AS and its seeding by relevant biological material, such as CSF. These latter assays that originated as separate methods termed protein-misfolding cyclic amplification (PMCA) (Jung et al. 2017; Nicot et al. 2019), and real-time quaking-induced conversion (RT-QuIC) (Fairfoul et al. 2016; Nakagaki et al. 2021; Huang et al. 2024), have converged on seed amplification assays (SAAs) (Concha-Marambio et al. 2023; Fernandes Gomes et al. 2023; Siderowf et al. 2023), that have tremendously increased the sensitivity and diagnostic accuracy of disease-related AS. Importantly, such assays in the CSF clearly differentiate LB disease-afflicted patients and related prodromal forms from controls and non-LB-related parkinsonism. Beyond the diagnostic utility, such assays have sparked a great degree of enthusiasm as they potentially reflect the ongoing nervous system pathogenetic process. However, the CSF SAA is difficult to implement and requires specialized equipment. At this point, it is qualitative and does not reflect disease progression; it is hoped that refinements of the assay may lead to its further development along these lines. In the meantime, efforts are underway to transfer the success of SAA to more accessible tissues, including skin and serum, among others. An intriguing publication by Okuzumi et al. (2023), in particular, suggests that serum AS SAA may be very promising.

Another landmark in the field would be the development of a PET tracer that could detect abnormal conformations of AS in living patients, as this would provide regional information, and could be followed longitudinally and presumably quantitatively, to provide, among others, meaningful end points for disease-modifying clinical trials. Converging information suggests that there is hope for significant developments in this area in the coming years. Already, Smith et al. (2023) reported specific aberrant AS-targeted PET tracer uptake in the cerebellum of MSA patients, and it is hoped that similar approaches may be successful in PD.

EMERGING CLINICAL TRIALS TARGETING α-SYNUCLEIN IN PARKINSON'S DISEASE

Based on the idea that removal of pathogenic AS species may be beneficial for PD and related synucleinopathies, a number of companies have embarked on clinical trials with the aim to decrease such levels or inhibit the effects of such species through the application of molecular strategies, immunization, or small compounds. Immunization strategies in particular entail both active and passive immunization, with the latter being more advanced in terms of application in large clinical trials. In 2022, two large studies in early PD using different antibodies, with variable specificity toward aggregated AS conformations, failed to show a significant benefit (Lang et al. 2022; Pagano et al. 2022), even though the used antibodies bound completely monomeric AS in the periphery and had shown ameliorating effects in rodent animal models of synucleinopathy. These results raised questions about the validity of the immunization approach, and even the basic principle of targeting pathogenic AS for the treatment of PD (Whone 2022). As argued, however, by Jensen et al. (2023), notwithstanding the rigorous nature of the performed studies, various issues preclude a hasty abandonment of similar approaches: The execution of the animal studies on which the clinical trials were based could be improved, in particular regarding the temporal pattern of treatment, to more closely reflect the human situation; target engagement, in particular of the potentially noxious oligomeric species within neurons, was not demonstrated; it may be difficult to observe motor benefits with neuroprotective strategies even in early disease, as neurodegeneration is already quite advanced, while the progression of the disease is generally quite slow, possibly necessitating studies of longer duration. With these notions in mind, and taking into account the only very recent partial success of similar β-amyloid-targeting strategies in AD, after decades of failures, there is reason to believe that the conduct of further clinical trials targeting AS, possibly in earlier disease stages, such as prodromal or even presymptomatic, as in asymptomatic genetic synucleinopathies, will ultimately demonstrate efficacy (Jensen et al. 2023). In fact, a preordained analysis of a subset of PD patients with characteristics suggesting more rapid progression appears to show a very significant benefit with treatment in one of the two clinical trials mentioned above (Pagano et al. 2024), further reinforcing the notion that anti-AS therapies may prove to be successful in PD and related synucleinopathies.

ACKNOWLEDGMENTS

L.S. has been supported through the grant HFRI-FM17-3013 from the Hellenic Foundation for Research and Innovation (HFRI) and the program Brain Precision TAEDR-0535850, Greece 2.0. K.V. has been supported by the program Brain Precision TAEDR-0535850, Greece 2.0. M.X. has been supported by an HFRI grant for Faculty Members & Researchers (Foundation for Research and Technology-Hellas HFRI-3661), the Brain Precision TAEDR-0535850, Greece 2.0 program, and by the MJFF-024029 grant. E.E. has been supported by a Hellenic Foundation for Research and Innovation (HFRI) grant (581) and the Brain Precision TAEDR-0535850 grant. Partial financial support was received from Special Account for Research grants of NKUA (20131).

REFERENCES

Abd Elhadi S, Grigoletto J, Poli M, Arosio P, Arkadir D, Sharon R. 2019. α-Synuclein in blood cells differentiates Parkinson's disease from healthy controls. *Ann Clin Transl Neurol* **6**: 2426–2436. doi:10.1002/acn3.50944

Abeliovich A, Schmitz Y, Fariñas I, Choi-Lundberg D, Ho WH, Castillo PE, Shinsky N, Verdugo JM, Armanini M, Ryan A, et al. 2000. Mice lacking α-synuclein display functional deficits in the nigrostriatal dopamine system. *Neuron* **25**: 239–252. doi:10.1016/s0896-6273(00)80886-7

Altay MF, Liu AKL, Holton JL, Parkkinen L, Lashuel HA. 2022. Prominent astrocytic α-synuclein pathology with unique post-translational modification signatures unveiled across Lewy body disorders. *Acta Neuropathol Commun* **10**: 163. doi:10.1186/s40478-022-01468-8

Alvarez-Erviti L, Rodriguez-Oroz MC, Cooper JM, Caballero C, Ferrer I, Obeso JA, Schapira AH. 2010. Chaperone-mediated autophagy markers in Parkinson disease brains. *Arch Neurol* **67**: 1464–1472. doi:10.1001/archneurol.2010.198

Alvarez-Erviti L, Seow Y, Schapira AH, Gardiner C, Sargent IL, Wood MJ, Cooper JM. 2011. Lysosomal dysfunction increases exosome-mediated α-synuclein release and transmission. *Neurobiol Dis* **42**: 360–367. doi:10.1016/j.nbd.2011.01.029

Anderson JP, Walker DE, Goldstein JM, de Laat R, Banducci K, Caccavello RJ, Barbour R, Huang J, Kling K, Lee M, et al. 2006. Phosphorylation of Ser-129 Is the dominant pathological modification of α-synuclein in familial and sporadic Lewy body disease. *J Biol Chem* **281**: 29739–29752. doi:10.1074/jbc.M600933200

Andrews SV, Kukkle PL, Menon R, Geetha TS, Goyal V, Kandadai RM, Kumar H, Borgohain R, Mukherjee A, Wadia PM, et al. 2024. The genetic drivers of juvenile, young, and early-onset Parkinson's disease in India. *Mov Disord* **39**: 339–349. doi:10.1002/mds.29676

Appel-Cresswell S, Vilarino-Guell C, Encarnacion M, Sherman H, Yu I, Shah B, Weir D, Thompson C, Szu-Tu C, Trinh J, et al. 2013. α-Synuclein p.H50Q, a novel pathogenic mutation for Parkinson's disease. *Mov Disord* **28**: 811–813. doi:10.1002/mds.25421

Baba M, Nakajo S, Tu PH, Tomita T, Nakaya K, Lee VM, Trojanowski JQ, Iwatsubo T. 1998. Aggregation of α-synuclein in Lewy bodies of sporadic Parkinson's disease and dementia with Lewy bodies. *Am J Pathol* **152**: 879–884.

Bae EJ, Yang NY, Song M, Lee CS, Lee JS, Jung BC, Lee HJ, Kim S, Masliah E, Sardi SP, et al. 2014. Glucocerebrosidase depletion enhances cell-to-cell transmission of α-synuclein. *Nat. Commun* **5**: 4755. doi:10.1038/ncomms 5755

Barrett PJ, Timothy Greenamyre J. 2015. Post-translational modification of α-synuclein in Parkinson's disease. *Brain Res* **1628**: 247–253. doi:10.1016/j.brainres.2015.06.002

Bartels T, Choi JG, Selkoe DJ. 2011. α-Synuclein occurs physiologically as a helically folded tetramer that resists aggregation. *Nature* **477**: 107–110. doi:10.1038/nature 10324

Beach TH, Adler CH, Lue L, Sue LI, Bachalakuri J, Henry-Watson J, Sasse J, Boyer S, Shirohi S, Brooks R, et al. 2009. Unified staging system for Lewy body disorders: correlation with nigrostriatal degeneration, cognitive impairment and motor dysfunction. *Acta Neuropathol* **117**: 613–634. doi:10.1007/s00401-009-0538-8

Bell R, Thrush RJ, Castellana-Cruz M, Oeller M, Staats R, Nene A, Flagmeier P, Xu CK, Satapathy S, Galvagnion C, et al. 2022. N-terminal acetylation of α-synuclein slows down its aggregation process and alters the morphology of the resulting aggregates. *Biochemistry* **61**: 1743–1756. doi:10.1021/acs.biochem.2c00104

Bell R, Castellana-Cruz M, Nene A, Thrush RJ, Xu CK, Kumita JR, Vendruscolo M. 2023. Effects of N-terminal acetylation on the aggregation of disease-related α-synuclein variants. *J Mol Biol* **435**: 167825. doi:10.1016/j.jmb .2022.167825

Bellucci A, Navarria L, Zaltieri M, Falarti E, Bodei S, Sigala S, Battistin L, Spillantini M, Missale C, Spano P. 2011. Induction of the unfolded protein response by α-synuclein in experimental models of Parkinson's disease. *J Neurochem* **116**: 588–605.

Benner EJ, Banerjee R, Reynolds AD, Sherman S, Pisarev VM, Tsiperson V, Nemachek C, Ciborowski P, Przedborski S, Mosley RL, et al. 2008. Nitrated α-synuclein immunity accelerates degeneration of nigral dopaminergic neurons. *PLoS ONE* **3**: e1376. doi:10.1371/journal.pone.000 1376

Bennett MC, Bishop JF, Leng Y, Chock PB, Chase TN, Mouradian MM. 1999. Degradation of α-synuclein by proteasome. *J Biol Chem* **274**: 33855–33858. doi:10 .1074/jbc.274.48.33855

Bérard M, Sheta R, Malvaut S, Rodriguez-Aller R, Teixeira M, Idi W, Turmel R, Alpaugh M, Dubois M, Dahmene M, et al. 2022. A light-inducible protein clustering system for in vivo analysis of α-synuclein aggregation in Parkinson disease. *PLoS Biol* **20**: e3001578. doi:10.1371/journal.pbio .3001578

Betzer C, Movius AJ, Shi M, Gai WP, Zhang J, Jensen PH. 2015. Identification of synaptosomal proteins binding to monomeric and oligomeric α-synuclein. *PLoS ONE* **10**: e0116473. doi:10.1371/journal.pone.0116473

Betzer C, Lassen LB, Olsen A, Kofoed RH, Reimer L, Gregersen E, Zheng J, Calì T, Gai WP, Chen T, et al. 2018.

α-Synuclein aggregates activate calcium pump SERCA leading to calcium dysregulation. *EMBO Rep* **19**: e44617.

Beyer K, Humbert J, Ferrer A, Lao JI, Carrato C, López D, Ferrer I, Ariza A. 2006. Low α-synuclein 126 mRNA levels in dementia with Lewy bodies and Alzheimer disease. *Neuroreport* **17**: 1327–1330. doi:10.1097/01.wnr .0000224773.66904.e7

Binolfi A, Theillet FX, Selenko P. 2012. Bacterial in-cell NMR of human a-synuclein: a disordered monomer by nature? *Biochem Soc Trans* **40**: 950–954. doi:10.1042/BST2012 0096

Borghammer P, Horsager J, Andersen K, Van Den Berge N, Raunio A, Murayama S, Parkkinen L, Myllykangas L. 2021. Neuropathological evidence of body-first vs. brain-first Lewy body disease. *Neurobiol Dis* **161**: 105557. doi:10.1016/j.nbd.2021.105557

Bose A, Petsko GA, Studer L. 2022. Induced pluripotent stem cells: a tool for modeling Parkinson's disease. *Trends Neurosci* **45**: 608–620. doi:10.1016/j.tins.2022.05.001

Braak H, Tredici KD, Rüb U, de Vos RA, Jansen Steur EN, Braak E. 2003. Staging of brain pathology related to sporadic Parkinson's disease. *Neurobiol Aging* **24**: 197–211. doi:10.1016/S0197-4580(02)00065-9

Bu B, Tong X, Li D, Hu Y, He W, Zhao C, Hu R, Li X, Shao Y, Liu C, et al. 2017. N-Terminal acetylation preserves α-synuclein from oligomerization by blocking intermolecular hydrogen bonds. *ACS Chem Neurosci* **8**: 2145–2151. doi:10.1021/acschemneuro.7b00250

Burré J, Sharma M, Tsetsenis T, Buchman V, Etherton MR, Südhof TC. 2010. α-Synuclein promotes SNARE-complex assembly in vivo and in vitro. *Science* **329**: 1663–1667. doi:10.1126/science.1195227

Burré J, Sharma M, Südhof TC. 2012. Systematic mutagenesis of α-synuclein reveals distinct sequence requirements for physiological and pathological activities. *J. Neurosci* **32**: 15227–15242. doi:10.1523/JNEUROSCI.3545-12 .2012

Bussell R, Eliezer D. 2003. A structural and functional role for 11-mer repeats in α-synuclein and other exchangeable lipid binding proteins. *J Mol Biol* **329**: 763–778. doi:10 .1016/S0022-2836(03)00520-5

Butler B, Saha K, Rana T, Becker JP, Sambo D, Davari P, Goodwin JS, Khoshbouei H. 2015. Dopamine transporter activity is modulated by α-synuclein. *J Biol Chem* **290**: 29542–29554. doi:10.1074/jbc.M115.691592

Butler B, Sambo D, Khoshbouei H. 2017. α-Synuclein modulates dopamine neurotransmission. *J Chem Neuroanat* **83-84**: 41–49. doi:10.1016/j.jchemneu.2016.06.001

Calabresi P, Di Lazzaro G, Marino G, Campanelli F, Ghiglieri V. 2023a. Advances in understanding the function of α-synuclein: implications for Parkinson's disease. *Brain* **146**: 3587–3597. doi:10.1093/brain/awad150

Calabresi P, Mechelli A, Natale G, Volpicelli-Daley L, Di Lazzaro G, Ghiglieri V. 2023b. α-Synuclein in Parkinson's disease and other synucleinopathies: from overt neurodegeneration back to early synaptic dysfunction. *Cell Death Dis* **14**: 176. doi:10.1038/s41419-023-05672-9

Cao Z, Wu Y, Liu G, Jiang Y, Wang X, Wang Z, Feng T. 2019. α-Synuclein in salivary extracellular vesicles as a potential biomarker of Parkinson's disease. *Neurosci Lett* **696**: 114–120. doi:10.1016/j.neulet.2018.12.030

Carnwath T, Mohammed R, Tsiang D. 2018. The direct and indirect effects of α-synuclein on microtubule stability in the pathogenesis of Parkinson's disease. *Neuropsychiatr Dis Treat* 14: 1685–1695. doi:10.2147/NDT.S166322

Cenci MA, Björklund A. 2020. Animal models for preclinical Parkinson's research: an update and critical appraisal. *Prog Brain Res* 252: 27–59. doi:10.1016/bs.pbr.2020.02 .003

Chadchankar H, Ihalainen J, Tanila H, Yavich L. 2011. Decreased reuptake of dopamine in the dorsal striatum in the absence of α-synuclein. *Brain Res* 1382: 37–44. doi:10 .1016/j.brainres.2011.01.064

Chartier-Harlin MC, Kachergus J, Roumier C, Mouroux V, Douay X, Lincoln S, Levecque C, Larvor L, Andrieux J, Hulihan M, et al. 2004. α-Synuclein locus duplication as a cause of familial Parkinson's disease. *Lancet* 364: 1167–1169. doi:10.1016/S0140-6736(04)17103-1

Chen J, Ren Y, Gui C, Zhao M, Wu X, Mao K, Li W, Zou F. 2018. Phosphorylation of Parkin at serine 131 by p38 MAPK promotes mitochondrial dysfunction and neuronal death in mutant A53T α-synuclein model of Parkinson's disease. *Cell Death Dis* 9: 700. doi:10.1038/s41419-018-0722-7

Chen R, Gu X, Wang X. 2022. α-Synuclein in Parkinson's disease and advances in detection. *Clin Chim Acta* 529: 76–86. doi:10.1016/j.cca.2022.02.006

Chia R, Sabir MS, Bandres-Ciga S, Saez-Atienzar S, Reynolds RH, Gustavsson E, Walton RL, Ahmed S, Viollet C, Ding J, et al. 2021. Genome sequencing analysis identifies new loci associated with Lewy body dementia and provides insights into its genetic architecture. *Nat Genet* 53: 294–303. doi:10.1038/s41588-021-00785-3

Chopra A, Outeiro TF. 2024. Aggregation and beyond: α-synuclein-based biomarkers in synucleinopathies. *Brain* 147: 81–90. doi:10.1093/brain/awad260

Chung CY, Koprich JB, Siddiqi H, Isacson O. 2009. Dynamic changes in presynaptic and axonal transport proteins combined with striatal neuroinflammation precede dopaminergic neuronal loss in a rat model of AAV α-synucleinopathy. *J Neurosci* 29: 3365–3373. doi:10.1523/JNEUROSCI.5427-08.2009

Colla E, Coune P, Liu Y, Pletnikova O, Troncoso JC, Iwatsubo T, Schneider BL, Lee MK. 2012. Endoplasmic reticulum stress is important for the manifestations of α-synucleinopathy in vivo. *J Neurosci* 32: 3306–3320.

Concha-Marambio L, Pritzkow S, Shahnawaz M, Farris CM, Soto C. 2023. Seed amplification assay for the detection of pathologic α-synuclein aggregates in cerebrospinal fluid. *Nat Protoc* 18: 1179–1196. doi:10.1038/s41596-022-00 787-3

Conn KJ, Gao W, McKee A, Lan MS, Ullman MD, Eisenhauer PB, Fine RE, Wells JM. 2004. Identification of the protein disulfide isomerase family member PDIp in experimental Parkinson's disease and Lewy body pathology. *Brain Res* 1022: 164–172.

Conway KA, Rochet JC, Bieganski RM, Lansbury PT Jr. 2001. Kinetic stabilization of the α-synuclein protofibril by a dopamine-α-synuclein adduct. *Science* 294: 1346–1349.

Cook C, Petrucelli L. 2009. A critical evaluation of the ubiquitin–proteasome system in Parkinson's disease. *Biochim*

Biophys Acta 1792: 664–675. doi:10.1016/j.bbadis.2009 .01.012

Cooper AA, Gitler AD, Cashikar A, Haynes CM, Hill KJ, Bhullar B, Liu K, Xu K, Strathearn KE, Liu F, et al. 2006. α-Synuclein blocks ER-Golgi traffic and Rab1 rescues neuron loss in Parkinson's models. *Science* 313: 324–328.

Cooper O, Seo H, Andrabi S, Guardia-Laguarta C, Graziotto J, Sundberg M, McLean JR, Carrillo-Reid L, Xie Z, Osborn T, et al. 2012. Pharmacological rescue of mitochondrial deficits in iPSC-derived neural cells from patients with familial Parkinson's disease. *Sci Transl Med* 4: 141ra90. doi:10.1126/scitranslmed.3003985

Coughlin DG, Petrovitch H, White LR, Noorigian J, Masaki KH, Ross GW, Duda JE. 2019. Most cases with Lewy pathology in a population-based cohort adhere to the Braak progression pattern but "failure to fit" is highly dependent on staging system applied. *Parkinsonism Relat Disord* 64: 124–131. doi:10.1016/j.parkreldis.2019.03.023

Cremades N, Chen SW, Dobson CM. 2017. Structural characteristics of α-synuclein oligomers. *Int Rev Cell Mol Biol* 329: 79–143. doi:10.1016/bs.ircmb.2016.08.010

Cuervo AM, Stefanis L, Fredenburg R, Lansbury PT, Sulzer D. 2004. Impaired degradation of mutant α-synuclein by chaperone-mediated autophagy. *Science* 305: 1292–1295. doi:10.1126/science.1101738

Cullen V, Lindfors M, Ng J, Paetau A, Swinton E, Kolodziej P, Boston H, Saftig P, Woulfe J, Feany MB, et al. 2009. Cathepsin D expression level affects α-synuclein processing, aggregation, and toxicity in vivo. *Mol Brain* 2: 5. doi:10.1186/1756-6606-2-5

Daida K, Shimonaka S, Shiba-Fukushima K, Ogata J, Yoshino H, Okuzumi A, Hatano T, Motoi Y, Hirunagi T, Katsuno M, et al. 2022. α-Synuclein V15A variant in familial Parkinson's disease exhibits a weaker lipid-binding property. *Mov Disord* 37: 2075–2085. doi:10.1002/mds .29162

Davidi D, Schechter M, Elhadi SA, Matatov A, Nathanson L, Sharon R. 2020. α-Synuclein translocates to the nucleus to activate retinoic-acid-dependent gene transcription. *iScience* 23: 100910.

De Giorgi F, Laferrière F, Zinghirino F, Faggiani E, Lends A, Bertoni M, Yu X, Grélard A, Morvan E, Habenstein B, et al. 2020. Novel self-replicating α-synuclein polymorphs that escape ThT monitoring can spontaneously emerge and acutely spread in neurons. *Sci Adv* 6: eabc4364. doi:10 .1126/sciadv.abc4364

Delenclos M, Trendafilova T, Mahesh D, Baine AM, Moussaud S, Yan IK, Patel T, McLean PJ. 2017. Investigation of endocytic pathways for the internalization of exosome-associated oligomeric α-synuclein. *Front Neurosci* 11: 172.

Delenclos M, Burgess JD, Lamprokostopoulou A, Outeiro TF, Vekrellis K, McLean PJ. 2019. Cellular models of α-synuclein toxicity and aggregation. *J Neurochem* 150: 566–576. doi:10.1111/jnc.14806

Desplats P, Spencer B, Coffee E, Patel P, Michael S, Patrick C, Adame A, Rockenstein E, Masliah E. 2011. α-Synuclein sequesters Dnmt1 from the nucleus: a novel mechanism for epigenetic alterations in Lewy body diseases. *J Biol Chem* 286: 9031–9037.

Devi L, Raghavendran V, Prabhu BM, Avadhani NG, Anandatheerthavarada HK. 2008. Mitochondrial import and accumulation of α-synuclein impair complex I in human dopaminergic neuronal cultures and Parkinson disease brain. *J Biol Chem* 283: 9089–9100. doi:10.1074/jbc.M710012200

Diao J, Burré J, Vivona S, Cipriano DJ, Sharma M, Kyoung M, Südhof TC, Brunger AT. 2013. Native α-synuclein induces clustering of synaptic-vesicle mimics via binding to phospholipids and synaptobrevin-2/VAMP2. *eLife* 2: e00592. doi:10.7554/eLife.00592

Diaw SH, Borsche M, Streubel-Gallasch L, Dulovic-Mahlow M, Hermes J, Lenz I, Seibler P, Klein C, Brüggemann N, Vos M, et al. 2023. Characterization of the pathogenic α-synuclein variant V15A in Parkinson's disease. *NPJ Parkinsons Dis* 9: 148. doi:10.1038/s41531-023-00584-z

Dikiy I, Fauvet B, Jovičić A, Mahul-Mellier AL, Desobry C, El-Turk F, Gitler AD, Lashuel HA, Eliezer D. 2016. Semi-synthetic and in vitro phosphorylation of α-synuclein at y39 promotes functional partly helical membrane-bound states resembling those induced by PD mutations. *ACS Chem Biol* 11: 2428–2437.

Di Maio R, Barrett PJ, Hoffman EK, Barrett CW, Zharikov A, Borah A, Hu X, McCoy J, Chu CT, Burton EA, et al. 2016. α-Synuclein binds to TOM20 and inhibits mitochondrial protein import in Parkinson's disease. *Sci Transl Med* 8: a78. doi:10.1126/scitranslmed.aaf3634

Donadio V, Incensi A, Leta V, Giannoccaro MP, Scaglione C, Martinelli P, Capellari S, Avoni P, Baruzzi A, Liguori R. 2014. Skin nerve α-synuclein deposits: a biomarker for idiopathic Parkinson disease. *Neurology* 82: 1362–1369. doi:10.1212/WNL.0000000000000316

Duran R, Barrero FJ, Morales B, Luna JD, Ramirez M, Vives F. 2010. Plasma α-synuclein in patients with Parkinson's disease with and without treatment. *Mov Disord* 25: 489–493. doi:10.1002/mds.22928

Ebrahimi-Fakhari D, Cantuti-Castelvetri I, Fan Z, Rockenstein E, Masliah E, Hyman BT, et al. 2011. Distinct roles in vivo for the ubiquitin-proteasome system and the autophagy-lysosomal pathway in the degradation of α-synuclein. *J Neurosci* 31: 14508–14520.

Emmanouilidou E, Melachroinou K, Roumeliotis T, Garbis SD, Ntzouni M, Margaritis LH, Stefanis L, Vekrellis K. 2010a. Cell-produced α-synuclein is secreted in a calcium-dependent manner by exosomes and impacts neuronal survival. *J Neurosci* 30: 6838–6851. doi:10.1523/JNEUROSCI.5699-09.2010

Emmanouilidou E, Stefanis L, Vekrellis K. 2010b. Cell-produced α-synuclein oligomers are targeted to, and impair, the 26S proteasome. *Neurobiol Aging* 31: 953–968. doi:10.1016/j.neurobiolaging.2008.07.008

Emmanouilidou E, Minakaki G, Keramioti MV, Xylaki M, Balafas E, Chrysanthou-Piterou M, Kloukina I, Vekrellis K. 2016. GABA transmission via ATP-dependent K$^+$ channels regulate α-synuclein secretion in mouse striatum. *Brain* 139: 871–890. doi:10.1093/brain/awv403

Espay AJ, Vizcarra JA, Marsili L, Lang AE, Simon DK, Merola A, Josephs KA, Fasano A, Morgante F, Savica R, et al. 2019. Revisiting protein aggregation as pathogenic in sporadic Parkinson and Alzheimer diseases. *Neurology* 92: 329–337. doi:10.1212/WNL.0000000000006926

Fairfoul G, McGuire LI, Pal S, Ironside JW, Neumann J, Christie S, Joachim C, Esiri M, Evetts SG, Rolinski M, et al. 2016. α-Synuclein RT-QuIC in the CSF of patients with α-synucleinopathies. *Ann Clin Transl Neurol* 3: 812–818. doi:10.1002/acn3.338

Fares MB, Ait-Bouziad N, Dikiy I, Mbefo MK, Jovicic A, Kiely A, Holton JL, Lee SJ, Gitler AD, Eliezer D, et al. 2014. The novel Parkinson's disease linked mutation G51D attenuates in vitro aggregation and membrane binding of α-synuclein, and enhances its secretion and nuclear localization in cells. *Hum Mol Genet* 23: 4491–4509. doi:10.1093/hmg/ddu165

Fauvet B, Lashuel HA. 2016. Semisynthesis and enzymatic preparation of post-translationally modified α-synuclein. *Methods Mol Biol* 1345: 3–20. doi:10.1007/978-1-4939-2978-8_1

Fauvet B, Fares MB, Samuel F, Dikiy I, Tandon A, Eliezer D, Lashuel HA. 2012. Characterization of semisynthetic and naturally N-acetylated α-synuclein in vitro and in intact cells: implications for aggregation and cellular properties of α-synuclein. *J Biol Chem* 287: 28243–28262. doi:10.1074/jbc.M112.383711

Fernandes Gomes B, Farris CM, Ma Y, Concha-Marambio L, Lebovitz R, Nellgård B, Dalla K, Constantinescu J, Constantinescu R, Gobom J, et al. 2023. α-Synuclein seed amplification assay as a diagnostic tool for parkinsonian disorders. *Parkinsonism Relat Disord* 117: 105807. doi:10.1016/j.parkreldis.2023.105807

Fevga C, Park Y, Lohmann E, Kievit AJ, Breedveld GJ, Ferraro F, de Boer L, van Minkelen R, Hanagasi H, Boon A, et al. 2021. A new α-synuclein missense variant (Thr72Met) in two Turkish families with Parkinson's disease. *Parkinsonism Relat Disord* 89: 63–72. doi:10.1016/j.parkreldis.2021.06.023

Foulds PG, Diggle P, Mitchell JD, Parker A, Hasegawa M, Masuda-Suzukake M, Mann DM, Allsop D. 2013. A longitudinal study on α-synuclein in blood plasma as a biomarker for Parkinson's disease. *Sci Rep* 3: 2540. doi:10.1038/srep02540

Fuchs J, Tichopad A, Golub Y, Munz M, Schweitzer KJ, Wolf B, Berg D, Mueller JC, Gasser T. 2008. Genetic variability in the *SNCA* gene influences α-synuclein levels in the blood and brain. *FASEB J* 22: 1327–1334. doi:10.1096/fj.07-9348com

Fusco G, De Simone A, Gopinath T, Vostrikov V, Vendruscolo M, Dobson CM, Veglia G. 2014. Direct observation of the three regions in α-synuclein that determine its membrane-bound behaviour. *Nat Commun* 5: 3827. doi:10.1038/ncomms4827

Fussi N, Höllerhage M, Chakroun T, Nykänen NP, Rösler TW, Koeglsperger T, Wurst W, Behrends C, Höglinger GU. 2018. Exosomal secretion of α-synuclein as protective mechanism after upstream blockage of macroautophagy. *Cell Death Dis* 9: 757. doi:10.1038/s41419-018-0816-2

Gabrielyan L, Liang H, Minalyan A, Hatami A, John V, Wang L. 2021. Behavioral deficits and brain α-synuclein and phosphorylated serine-129 α-synuclein in male and female mice overexpressing human α-synuclein. *J Alzheimer's Dis* 79: 875–893. doi:10.3233/JAD-200983

Games D, Valera E, Spencer B, Rockenstein E, Mante M, Adame A, Patrick C, Ubhi K, Nuber S, Sacayon P, et al. 2014. Reducing C-terminal-truncated α-synuclein by immunotherapy attenuates neurodegeneration and propagation in Parkinson's disease-like models. *J Neurosci* **34**: 9441–9454. doi:10.1523/JNEUROSCI.5314-13.2014

Gao V, Briano JA, Komer LK, Burré J. 2023. Functional and pathological effects of α-synuclein on synaptic SNARE complexes. *J Mol Biol* **435**: 167714. doi:10.1016/j.jmb.2022.167714

Garcia-Reitboeck P, Anichtchik O, Dalley JW, Ninkina N, Tofaris GK, Buchman VL, Spillantini MG. 2013. Endogenous α-synuclein influences the number of dopaminergic neurons in mouse substantia nigra. *Exp Neurol* **248**: 541–545. doi:10.1016/j.expneurol.2013.07.015

Ghanem SS, Majbour NK, Vaikath NN, Ardah MT, Erskine D, Jensen NM, Fayyad M, Sudhakaran IP, Vasili E, Melachrinou K, et al. 2022. α-Synuclein phosphorylation at serine 129 occurs after initial protein deposition and inhibits seeded fibril formation and toxicity. *Proc Natl Acad Sci* **119**: e2109617119. doi:10.1073/pnas.2109617119

Ghosh D, Sahay S, Ranjan P, Salot S, Mohite GM, Singh PK, Dwivedi S, Carvalho E, Banerjee R, Kumar A, et al. 2014. The newly discovered Parkinson's disease associated Finnish mutation (A53E) attenuates α-synuclein aggregation and membrane binding. *Biochemistry* **53**: 6419–6421. doi:10.1021/bi5010365

Giasson BI, Murray IV, Trojanowski JQ, Lee VMY. 2001. A hydrophobic stretch of 12 amino acid residues in the middle of α-synuclein is essential for filament assembly. *J Biol Chem* **276**: 2380–2386. doi:10.1074/jbc.M008919200

Gibbons CH, Levine T, Adler C, Bellaire B, Wang N, Stohl J, Agarwal P, Aldridge GM, Barboi A, Evidente VGH, et al. 2024. Skin biopsy detection of phosphorylated α-synuclein in patients with synucleinopathies. *JAMA* **331**: 1298–1306. doi:10.1001/jama.2024.0792

Gitler AD, Bevis BJ, Shorter J, Strathearn KE, Hamamichi S, Su LJ, Caldwell KA, Caldwell GA, Rochet JC, McCaffery JM, et al. 2008. The Parkinson's disease protein α-synuclein disrupts cellular Rab homeostasis. *Proc Natl Acad Sci* **105**: 145–150.

Gorbatyuk MS, Shabashvili A, Chen W, Meyers C, Sullivan LF, Salganik M, Lin JH, Lewin AS, Muzyczka N, Gorbatyuk OS. 2012. Glucose regulated protein 78 diminishes α-synuclein neurotoxicity in a rat model of Parkinson disease. *Mol Ther* **20**: 1327–1337.

Grey M, Dunning CJ, Gaspar R, Grey C, Brundin P, Sparr E, Linse S. 2015. Acceleration of α-synuclein aggregation by exosomes. *J Biol Chem* **290**: 2969–2982.

Gribaudo S, Tixador P, Bousset L, Fenyi A, Lino P, Melki R, Peyrin JM, Perrier AL. 2019. Propagation of α-synuclein strains within human reconstructed neuronal network. *Stem Cell Reports* **12**: 230–244. doi:10.1016/j.stemcr.2018.12.007

Guardia-Laguarta C, Area-Gomez E, Rüb C, Liu Y, Magrané J, Becker D, Voos W, Schon EA, Przedborski S. 2014. α-Synuclein is localized to mitochondria-associated ER membranes. *J Neurosci* **34**: 249–259. doi:10.1523/JNEUROSCI.2507-13.2014

Guerreiro R, Ross OA, Kun-Rodrigues C, Hernandez DG, Orme T, Eicher JD, Shepherd CE, Parkkinen L, Darwent L, Heckman MG, et al. 2018. Investigating the genetic architecture of dementia with Lewy bodies: a two-stage genome-wide association study. *Lancet Neurol* **17**: 64–74. doi:10.1016/S1474-4422(17)30400-3

Guo M, Wang J, Zhao Y, Feng Y, Han S, Dong Q, Cui M, Tieu K. 2020. Microglial exosomes facilitate α-synuclein transmission in Parkinson's disease. *Brain* **143**: 1476–1497.

Guo K, Zhang Y, Li L, Zhang J, Rong H, Liu D, Wang J, Jin M, Luo N, Zhang X. 2021. Neuroprotective effect of paeoniflorin in the mouse model of Parkinson's disease through α-synuclein/protein kinase C δ subtype signaling pathway. *Neuroreport* **32**: 1379–1387. doi:10.1097/WNR.0000000000001739

Hall S, Öhrfelt A, Constantinescu R, Andreasson U, Surova Y, Bostrom F, Nilsson C, Widner H, Decraemer H, Någga K, et al. 2012. Accuracy of a panel of 5 cerebrospinal fluid biomarkers in the differential diagnosis of patients with dementia and/or parkinsonian disorders. *Arch Neurol* **69**: 1445–1452. doi:10.1001/archneurol.2012.1654

Hansson O, Hall S, Öhrfelt A, Zetterberg H, Blennow K, Minthon L, Nägga K, Londos E, Varghese S, Majbour NK, et al. 2014. Levels of cerebrospinal fluid α-synuclein oligomers are increased in Parkinson's disease with dementia and dementia with Lewy bodies compared to Alzheimer's disease. *Alzheimers Res Ther* **6**: 25. doi:10.1186/alzrt255

Harischandra DS, Rokad D, Neal ML, Ghaisas S, Manne S, Sarkar S, Panicker N, Zenitsky G, Jin H, Lewis M, et al. 2019. Manganese promotes the aggregation and prion-like cell-to-cell exosomal transmission of α-synuclein. *Sci Signal* **12**: eaau4543.

Hayakawa H, Nakatani R, Ikenaka K, Aguirre C, Choong CJ, Tsuda H, Nagano S, Koike M, Ikeuchi T, Hasegawa M, et al. 2020. Structurally distinct α-synuclein fibrils induce robust parkinsonian pathology. *Mov Disord* **35**: 256–267. doi:10.1002/mds.27887

He Y, Yu Z, Chen S. 2019. α-Synuclein nitration and its implications in Parkinson's disease. *ACS Chem Neurosci* **10**: 777–782. doi:10.1021/acschemneuro.8b00288

Hegde ML, Rao KS. 2007. DNA induces folding in α-synuclein: understanding the mechanism using chaperone property of osmolytes. *Arch Biochem Biophys* **464**: 57–69.

Hong DP, Xiong W, Chang JY, Jiang C. 2011. The role of the C-terminus of human α-synuclein: intra-disulfide bonds between the C-terminus and other regions stabilize non-fibrillar monomeric isomers. *FEBS Lett* **585**: 561–566. doi:10.1016/j.febslet.2011.01.009

Hoyer W, Cherny D, Subramaniam V, Jovin TM. 2004. Impact of the acidic C-terminal region comprising amino acids 109–140 on α-synuclein aggregation in vitro. *Biochemistry* **43**: 16233–16242. doi:10.1021/bi048453u

Huang Y, Liu Z, Li N, Tian C, Yang H, Huo Y, Li Y, Zhang J, Yu Z. 2022. Parkinson's disease derived exosomes aggravate neuropathology in SNCA*A53T mice. *Ann Neurol* **92**: 230–245.

Huang J, Yuan X, Chen L, Hu B, Wang H, Wang Y, Huang W. 2024. Pathological α-synuclein detected by real-time quaking-induced conversion in synucleinopathies. *Exp Gerontol* **187**: 112366. doi:10.1016/j.exger.2024.112366

Issa AR, Sun J, Petitgas C, Mesquita A, Dulac A, Robin M, Mollereau B, Jenny A, Chérif-Zahar B, Birman S. 2018. The lysosomal membrane protein LAMP2A promotes autophagic flux and prevents SNCA-induced Parkinson disease-like symptoms in the *Drosophila* brain. *Autophagy* **14**: 1898–1910. doi:10.1080/15548627.2018.1491489

Cite this article as *Cold Spring Harb Perspect Med* doi: 10.1101/cshperspect.a041645

Iyer A, Roeters SJ, Schilderink N, Hommersom B, Heeren RMA, Woutersen S, Claessens MMAE, Subramaniam V. 2016. The impact of N-terminal acetylation of α-synuclein on phospholipid membrane binding and fibril structure. *J Biol Chem* **291**: 21110–21122. doi:10.1074/jbc.M116.726612

Janezic S, Threlfell S, Dodson PD, Dowie MJ, Taylor TN, Potgieter D, Parkkinen L, Senior SL, Anwar S, Ryan B, et al. 2013. Deficits in dopaminergic transmission precede neuron loss and dysfunction in a new Parkinson model. *Proc Natl Acad Sci* **110**: E4016–E4025. doi:10.1073/pnas.1309143110

Jensen PH, Schlossmacher MG, Stefanis L. 2023. Who ever said it would be easy? Reflecting on two clinical trials targeting α-synuclein. *Mov Disord* **38**: 378–384. doi:10.1002/mds.29318

Jiang P, Gan M, Ebrahim AS, Lin WL, Melrose HL, Yen SH. 2010. ER stress response plays an important role in aggregation of α-synuclein. *Mol Neurodegener* **5**: 56.

Jiang C, Hopfner F, Katsikoudi A, Hein R, Catli C, Evetts S, Huang Y, Wang H, Ryder JW, Kuhlenbaeumer G, et al. 2020. Serum neuronal exosomes predict and differentiate Parkinson's disease from atypical parkinsonism. *J Neurol Neurosurg Psychiatry* **91**: 720–729. doi:10.1136/jnnp-2019-322588

Jin H, Kanthasamy A, Ghosh A, Yang Y, Anantharam V, Kanthasamy AG. 2011. α-Synuclein negatively regulates protein kinase Cδ expression to suppress apoptosis in dopaminergic neurons by reducing p300 histone acetyltransferase activity. *J Neurosci* **31**: 2035–2051. doi:10.1523/JNEUROSCI.5634-10.2011

Jo E, Fuller N, Rand RP, St George-Hyslop P, Fraser PE. 2002. Defective membrane interactions of familial Parkinson's disease mutant A30P α-synuclein. *J Mol Biol* **315**: 799–807. doi:10.1006/jmbi.2001.5269

Jung BC, Lim YJ, Bae EJ, Lee JS, Choi MS, Lee MK, Lee HJ, Kim YS, Lee SJ. 2017. Amplification of distinct α-synuclein fibril conformers through protein misfolding cyclic amplification. *Exp Mol Med* **49**: e314. doi:10.1038/emm.2017.1

Kalia LV, Kalia SK, McLean PJ, Lozano AM, Lang AE. 2013. α-Synuclein oligomers and clinical implications for Parkinson disease. *Ann Neurol* **73**: 155–169. doi:10.1002/ana.23746

Kalia LV, Lang AE, Hazrati LN, Fujioka S, Wszolek ZK, Dickson DW, Ross OA, Van Deerlin VM, Trojanowski JQ, Hurtig HI, et al. 2015. Clinical correlations with Lewy body pathology in *LRRK2*-related Parkinson disease. *JAMA Neurol* **72**: 100–105. doi:10.1001/jamaneurol.2014.2704

Kang JH, Irwin DJ, Chen-Plotkin AS, Siderowf A, Caspell C, Coffey CS, Waligórska T, Taylor P, Pan S, Frasier M, et al. 2013. Association of cerebrospinal fluid β-amyloid 1-42, T-tau, P-tau181, and α-synuclein levels with clinical features of drug-naive patients with early Parkinson disease. *JAMA Neurol* **70**: 1277–1287. doi:10.1001/jamaneurol.2013.3861

Kapasi A, Brosch JR, Nudelman KN, Agrawal S, Foroud TM, Schneider JA. 2020. A novel *SNCA* E83Q mutation in a case of dementia with Lewy bodies and atypical frontotemporal lobar degeneration. *Neuropathology* **40**: 620–626. doi:10.1111/neup.12687

Karampetsou M, Ardah MT, Semitekolou M, Polissidis A, Samiotaki M, Kalomoiri M, Majbour N, Xanthou G, El-Agnaf OMA, Vekrellis K. 2017. Phosphorylated exogenous α-synuclein fibrils exacerbate pathology and induce neuronal dysfunction in mice. *Sci Rep* **7**: 16533. doi:10.1038/s41598-017-15813-8

Kiely AP, Asi YT, Kara E, Limousin P, Ling H, Lewis P, Proukakis C, Quinn N, Lees AJ, Hardy J, et al. 2013. α-Synucleinopathy associated with G51D SNCA mutation: a link between Parkinson's disease and multiple system atrophy? *Acta Neuropathol* **125**: 753–769. doi:10.1007/s00401-013-1096-7

Klein AD, Mazzulli JR. 2018. Is Parkinson's disease a lysosomal disorder? *Brain* **141**: 2255–2262. doi:10.1093/brain/awy147

Kluge A, Bunk J, Schaeffer E, Drobny A, Xiang W, Knacke H, Bub S, Lückstädt W, Arnold P, Lucius R, et al. 2022. Detection of neuron-derived pathological α-synuclein in blood. *Brain* **145**: 3058–3071. doi:10.1093/brain/awac115

Kontopoulos E, Parvin JD, Feany MB. 2006. α-Synuclein acts in the nucleus to inhibit histone acetylation and promote neurotoxicity. *Hum Mol Genet* **15**: 3012–3023.

Kovacs G, Reimer L, Jensen PH. 2021. Endoplasmic reticulum-based calcium dysfunctions in synucleinopathies. *Front Neurol* **12**: 742625.

Krohn L, Heilbron K, Blauwendraat C, Reynolds RH, Yu E, Senkevich K, Rudakou U, Estiar MA, Gustavsson EK, Brolin K, et al. 2022. Genome-wide association study of REM sleep behavior disorder identifies polygenic risk and brain expression effects. *Nat Commun* **13**: 7496. doi:10.1038/s41467-022-34732-5

Krüger R, Kuhn W, Müller T, Woitalla D, Graeber M, Kösel S, Przuntek H, Epplen JT, Schöls L, Riess O. 1998. Ala30Pro mutation in the gene encoding α-synuclein in Parkinson's disease. *Nat Genet* **18**: 106–108. doi:10.1038/ng0298-106

Kunadt M, Eckermann K, Stuendl A, Gong J, Russo B, Strauss K, Rai S, Kügler S, Falomir Lockhart L, Schwalbe M, et al. 2015. Extracellular vesicle sorting of α-synuclein is regulated by sumoylation. *Acta Neuropathol* **129**: 695–713. doi:10.1007/s00401-015-1408-1

Kuo SH, Tasset I, Cheng MM, Diaz A, Pan MK, Lieberman OJ, Hutten SJ, Alcalay RN, Kim S, Ximénez-Embún P, et al. 2022. Mutant glucocerebrosidase impairs α-synuclein degradation by blockade of chaperone-mediated autophagy. *Sci Adv* **8**: eabm6393. doi:10.1126/sciadv.abm6393

Lang AE, Siderowf AD, Macklin EA, Poewe W, Brooks DJ, Fernandez HH, Rascol O, Giladi N, Stocchi F, Tanner CM, et al. 2022. Trial of cinpanemab in early Parkinson's disease. *N Engl J Med* **387**: 408–420. doi:10.1056/NEJMoa2203395

Lau A, So RWL, Lau HHC, Sang JC, Ruiz-Riquelme A, Fleck SC, Stuart E, Menon S, Visanji NP, Meisl G, et al. 2020. α-Synuclein strains target distinct brain regions and cell types. *Nat Neurosci* **23**: 21–31. doi:10.1038/s41593-019-0541-x

Lee HJ, Choi C, Lee SJ. 2002. Membrane-bound α-synuclein has a high aggregation propensity and the ability to seed the aggregation of the cytosolic form. *J Biol Chem* **277**: 671–678. doi:10.1074/jbc.M107045200

Lee HJ, Patel S, Lee SJ. 2005. Intravesicular localization and exocytosis of α-synuclein and its aggregates. *J Neurosci* **25:** 6016–6024. doi:10.1523/JNEUROSCI.0692-05.2005

Lee HJ, Suk JE, Bae EJ, Lee JH, Paik SR, Lee SJ. 2008. Assembly-dependent endocytosis and clearance of extracellular α-synuclein. *Int J Biochem Cell Biol* **40:** 1835–1849. doi:10.1016/j.biocel.2008.01.017

Lee SJ, Lim HS, Masliah E, Lee HJ. 2011. Protein aggregate spreading in neurodegenerative diseases: problems and perspectives. *Neurosci Res* **70:** 339–348. doi:10.1016/j.neures.2011.05.008

Lee JG, Takahama S, Zhang G, Tomarev SI, Ye Y. 2016. Unconventional secretion of misfolded proteins promotes adaptation to proteasome dysfunction in mammalian cells. *Nat Cell Biol* **18:** 765–776. doi:10.1038/ncb3372

Lee BE, Kim HY, Kim HJ, Jeong H, Kim BG, Lee HE, Lee J, Kim HB, Lee SE, Yang YR, et al. 2020. O-GlcNAcylation regulates dopamine neuron function, survival and degeneration in Parkinson disease. *Brain* **143:** 3699–3716. doi:10.1093/brain/awaa320

Lei Z, Cao G, Wei G. 2019. A30p mutant α-synuclein impairs autophagic flux by inactivating JNK signaling to enhance ZKSCAN3 activity in midbrain dopaminergic neurons. *Cell Death Dis* **10:** 133. doi:10.1038/s41419-019-1364-0

Lesage S, Anheim M, Letournel F, Bousset L, Honoré A, Rozas N, Pieri L, Madiona K, Dürr A, Melki R, et al. 2013. G51d α-synuclein mutation causes a novel parkinsonian-pyramidal syndrome. *Ann Neurol* **73:** 459–471. doi:10.1002/ana.23894

Levine PM, De Leon CA, Galesic A, Balana A, Marotta NP, Lewis YE, Pratt MR. 2017. O-GlcNAc modification inhibits the calpain-mediated cleavage of α-synuclein. *Bioorg Med Chem* **25:** 4977–4982. doi:10.1016/j.bmc.2017.04.038

Li W, West N, Colla E, Pletnikova O, Troncoso JC, Marsh L, Dawson TM, Jäkälä P, Hartmann T, Price DL, et al. 2005. Aggregation promoting C-terminal truncation of α-synuclein is a normal cellular process and is enhanced by the familial Parkinson's disease-linked mutations. *PNAS* **102:** 2162–2167. doi:10.1073/pnas.0406976102

Li XY, Li W, Li X, Li XR, Sun L, Yang W, Cai Y, Chen Z, Wu J, Wang C, et al. 2021. Alterations of erythrocytic phosphorylated α-synuclein in different subtypes and stages of Parkinson's disease. *Front Aging Neurosci* **13:** 623977. doi:10.3389/fnagi.2021.623977

Liani E, Eyal A, Avraham E, Shemer R, Szargel R, Berg D, Bornemann A, Riess O, Ross CA, Rott R, et al. 2004. Ubiquitylation of synphilin-1 and α-synuclein by SIAH and its presence in cellular inclusions and Lewy bodies imply a role in Parkinson's disease. *Proc Natl Acad Sci* **101:** 5500–5505. doi:10.1073/pnas.0401081101

Limbocker R, Staats R, Chia S, Ruggeri FS, Mannini B, Xu CK, Perni M, Cascella R, Bigi A, Sasser LR, et al. 2021. Squalamine and its derivatives modulate the aggregation of amyloid-β and α-synuclein and suppress the toxicity of their oligomers. *Front Neurosci* **15:** 680026. doi:10.3389/fnins.2021.680026

Liu H, Koros C, Strohäker T, Schulte C, Bozi M, Varvaresos S, Ibáñez de Opakua A, Simitsi AM, Bougea A, Voumvourakis K, et al. 2021. A novel SNCA A30G mutation causes familial Parkinson's disease. *Mov Disord* **36:** 1624–1633. doi:10.1002/mds.28534

Logan T, Bendor J, Toupin C, Thorn K, Edwards RH. 2017. α-Synuclein promotes dilation of the exocytotic fusion pore. *Nat Neurosci* **20:** 681–689. doi:10.1038/nn.4529

Ludtmann MHR, Angelova PR, Ninkina NN, Gandhi S, Buchman VL, Abramov AY. 2016. Monomeric α-synuclein exerts a physiological role on brain ATP synthase. *J Neurosci* **36:** 10510–10521. doi:10.1523/JNEUROSCI.1659-16.2016

Ludtmann MHR, Angelova PR, Horrocks MH, Choi ML, Rodrigues M, Baev AY, Berezhnov AV, Yao Z, Little D, Banushi B, et al. 2018. α-Synuclein oligomers interact with ATP synthase and open the permeability transition pore in Parkinson's disease. *Nat Commun* **9:** 2293. doi:10.1038/s41467-018-04422-2

Lundblad M, Decressac M, Mattsson B, Björklund A. 2012. Impaired neurotransmission caused by overexpression of α-synuclein in nigral dopamine neurons. *Proc Natl Acad Sci* **109:** 3213–3219. doi:10.1073/pnas.1200575109

Lurette O, Martín-Jiménez R, Khan M, Sheta R, Jean S, Schofield M, Teixeira M, Rodriguez-Aller R, Perron I, Oueslati A, et al. 2023. Aggregation of α-synuclein disrupts mitochondrial metabolism and induce mitophagy via cardiolipin externalization. *Cell Death Dis* **14:** 729. doi:10.1038/s41419-023-06251-8

Ma MR, Hu ZW, Zhao YF, Chen YX, Li YM. 2016. Phosphorylation induces distinct α-synuclein strain formation. *Sci Rep* **6:** 37130. doi:10.1038/srep37130

Madsen DA, Schmidt SI, Blaabjerg M, Meyer M. 2021. Interaction between Parkin and α-synuclein in PARK2-mediated Parkinson's disease. *Cells* **10:** 283. doi:10.3390/cells10020283

Magalhães P, Lashuel HA. 2022. Opportunities and challenges of α-synuclein as a potential biomarker for Parkinson's disease and other synucleinopathies. *NPJ Park Dis* **8:** 93. doi:10.1038/s41531-022-00357-0

Mahul-Mellier AL, Burtscher J, Maharjan N, Weerens L, Croisier M, Kuttler F, Leleu M, Knott GW, Lashuel HA. 2020. The process of Lewy body formation, rather than simply α-synuclein fibrillization, is one of the major drivers of neurodegeneration. *Proc Natl Acad Sci* **117:** 4971–4982. doi:10.1073/pnas.1913904117

Mak SK, McCormack AL, Manning-Boğ AB, Cuervo AM, Di Monte DA. 2010. Lysosomal degradation of α-synuclein in vivo. *J Biol Chem* **285:** 13621–13629. doi:10.1074/jbc.M109.074617

Manzanza NDO, Sedlackova L, Kalaria RN. 2021. α-Synuclein post-translational modifications: implications for pathogenesis of Lewy body disorders. *Front Aging Neurosci* **13:** 690293. doi:10.3389/fnagi.2021.690293

Maroteaux L, Campanelli JT, Scheller RH. 1988. Synuclein: a neuron-specific protein localized to the nucleus and presynaptic nerve terminal. *J Neurosci* **8:** 2804–2815.

Marotta NP, Lin YH, Lewis YE, Ambroso MR, Zaro BW, Roth MT, Arnold DB, Langen R, Pratt MR. 2015. O-GlcNAc modification blocks the aggregation and toxicity of the protein α-synuclein associated with Parkinson's disease. *Nat Chem* **7:** 913–920. doi:10.1038/nchem.2361

Martìn-Clemente B, Alvarez-Castelao B, Mayo I, Sierra AB, Díaz V, Milán M, Fariñas I, Gómez-Isla T, Ferrer I, Castaño JG. 2004. α-Synuclein expression levels do not significantly affect proteasome function and expression in mice and stably transfected PC12 cell lines. *J Biol Chem* **279:** 52984–52990. doi:10.1074/jbc.M409028200

Martinez J, Moeller I, Erdjument-Bromage H, Tempst P, Lauring B. 2003. Parkinson's disease-associated α-synuclein is a calmodulin substrate. *J Biol Chem* **278:** 17379–17387. doi:10.1074/jbc.M209020200

Martinez A, Lopez N, Gonzalez C, Hetz C. 2019. Targeting of the unfolded protein response (UPR) as therapy for Parkinson's disease. *Biol Cell* **111:** 161–168.

Martinez-Vicente M, Talloczy Z, Kaushik S, Massey AC, Mazzulli J, Mosharov EV, Hodara R, Fredenburg R, Wu DC, Follenzi A. 2008. Dopamine-modified α-synuclein blocks chaperone-mediated autophagy. *J Clin Invest* **118:** 777–788. doi:10.1172/JCI32806

Masliah E, Rockenstein E, Veinbergs I, Mallory M, Hashimoto M, Takeda A, Sagara Y, Sisk A, Mucke L. 2000. Dopaminergic loss and inclusion body formation in α-synuclein mice: implications for neurodegenerative disorders. *Science* **287:** 1265–1269. doi:10.1126/science.287.5456.1265

McGlinchey RP, Lee JC. 2015. Cysteine cathepsins are essential in lysosomal degradation of α-synuclein. *Proc Natl Acad Sci* **112:** 9322–9327. doi:10.1073/pnas.1500937112

McLean JR, Hallett PJ, Cooper O, Stanley M, Isacson O. 2012. Transcript expression levels of full-length α-synuclein and its three alternatively spliced variants in Parkinson's disease brain regions and in a transgenic mouse model of α-synuclein overexpression. *Mol Cell Neurosci* **49:** 230–239. doi:10.1016/j.mcn.2011.11.006

Melachroinou K, Divolis G, Tsafaras G, Karampetsou M, Fortis S, Stratoulias Y, Papadopoulou G, Kriebardis AG, Samiotaki M, Vekrellis K. 2024. Endogenous α-synuclein is essential for the transfer of pathology by exosome-enriched extracellular vesicles, following inoculation with preformed fibrils in vivo. *Aging Dis* **15:** 869–892.

Miraglia F, Ricci A, Rota L, Colla E. 2018. Subcellular localization of α-synuclein aggregates and their interaction with membranes. *Neural Regen Res* **13:** 1136–1144. doi:10.4103/1673-5374.235013

Mohamed NV, Larroquette F, Beitel LK, Fon EA, Durcan TM. 2019. One step into the future: new iPSC tools to advance research in Parkinson's disease and neurological disorders. *J Parkinsons Dis* **9:** 265–281. doi:10.3233/JPD-181515

Mollenhauer B, Locascio JJ, Schulz-Schaeffer W, Sixel-Döring F, Trenkwalder C, Schlossmacher MG. 2011. α-Synuclein and tau concentrations in cerebrospinal fluid of patients presenting with parkinsonism: a cohort study. *Lancet Neurol* **10:** 230–240. doi:10.1016/S1474-4422(11)70014-X

Moon SP, Balana AT, Galesic A, Rakshit A, Pratt MR. 2020. Ubiquitination can change the structure of the α-synuclein amyloid fiber in a site selective fashion. *J Org Chem* **85:** 1548–1555. doi:10.1021/acs.joc.9b02641

Moors TE, Maat CA, Niedieker D, Mona D, Petersen D, Timmermans-Huisman E, Kole J, El-Mashtoly SF, Spycher L, Zago W, et al. 2021. The subcellular arrangement of α-synuclein proteoforms in the Parkinson's disease brain as revealed by multicolor STED microscopy. *Acta Neuropathol* **142:** 423–448. doi:10.1007/s00401-021-02329-9

Mor DE, Tsika E, Mazzulli JR, Gould NS, Kim H, Daniels MJ, Doshi S, Gupta P, Grossman JL, Tan VX, et al. 2017. Dopamine induces soluble α-synuclein oligomers and nigrostriatal degeneration. *Nat Neurosci* **20:** 1560–1568.

Mosharov EV, Larsen KE, Kanter E, Phillips KA, Wilson K, Schmitz Y, Krantz DE, Kobayashi K, Edwards RH, Sulzer D. 2009. Interplay between cytosolic dopamine, calcium, and α-synuclein causes selective death of substantia nigra neurons. *Neuron* **62:** 218–229. doi:10.1016/j.neuron.2009.01.033

Mueller JC, Fuchs J, Hofer A, Zimprich A, Lichtner P, Illig T, Berg D, Wüllner U, Meitinger T, Gasser T. 2005. Multiple regions of α-synuclein are associated with Parkinson's disease. *Ann Neurol* **57:** 535–541. doi:10.1002/ana.20438

Mukaetova-Ladinska EB, Andras A, Milne J, Abdel-All Z, Borr I, Jaros E, Perry RH, Honer WG, Cleghorn A, Doherty J, et al. 2013. Synaptic proteins and choline acetyltransferase loss in visual cortex in dementia with Lewy bodies. *J Neuropathol Exp Neurol* **72:** 53–60. doi:10.1097/NEN.0b013e31827c5710

Murphy J, McKernan DP. 2022. The effect of aggregated α synuclein on synaptic and axonal proteins in Parkinson's disease—a systematic review. *Biomolecules* **12:** 1199. doi:10.3390/biom12091199

Murphy KE, Gysbers AM, Abbott SK, Tayebi N, Kim WS, Sidransky E, Cooper A, Garner B, Halliday GM. 2014. Reduced glucocerebrosidase is associated with increased α-synuclein in sporadic Parkinson's disease. *Brain* **137:** 834–848.

Murphy KE, Gysbers AM, Abbott SK, Spiro AS, Furuta A, Cooper A, Garner B, Kabuta T, Halliday GM. 2015. Lysosomal-associated membrane protein 2 isoforms are differentially affected in early Parkinson's disease. *Mov Disord* **30:** 1639–1647.

Musgrove RE, Helwig M, Bae EJ, Aboutalebi H, Lee SJ, Ulusoy A, Di Monte DA. 2019. Oxidative stress in vagal neurons promotes parkinsonian pathology and intercellular α-synuclein transfer. *J Clin Invest* **129:** 3738–3753. doi:10.1172/JCI127330

Nakagaki T, Nishida N, Satoh K. 2021. Development of α-synuclein real-time quaking-induced conversion as a diagnostic method for α-synucleinopathies. *Front Aging Neurosci* **13:** 703984. doi:10.3389/fnagi.2021.703984

Nakamura K, Nemani VM, Azarbal F, Skibinski G, Levy JM, Egami K, Munishkina L, Zhang J, Gardner B, Wakabayashi J, et al. 2011. Direct membrane association drives mitochondrial fission by the Parkinson disease-associated protein α-synuclein. *J Biol Chem* **286:** 20710–20726. doi:10.1074/jbc.M110.213538

Nalls MA, Blauwendraat C, Vallerga CL, Heilbron K, Bandres-Ciga S, Chang D, Tan M, Kia DA, Noyce AJ, Xue A, et al. 2019. Identification of novel risk loci, causal insights, and heritable risk for Parkinson's disease: a meta-analysis of genome-wide association studies. *Lancet Neurol* **18:** 1091–1102. doi:10.1016/S1474-4422(19)30320-5

Nellikka RK, Bhaskar BR, Sanghrajka K, Patil SS, Das D. 2021. α-Synuclein kinetically regulates the nascent fusion pore dynamics. *Proc Natl Acad Sci* **118:** e2021742118. doi:10.1073/pnas.2021742118

Nemani VM, Lu W, Berge V, Nakamura K, Onoa B, Lee MK, Chaudhry FA, Nicoll RA, Edwards RH. 2010. Increased expression of α-synuclein reduces neurotransmitter release by inhibiting synaptic vesicle reclustering after endocytosis. *Neuron* **65**: 66–79. doi:10.1016/j.neuron.2009.12.023

Nicot S, Verchère J, Bélondrade M, Mayran C, Bétemps D, Bougard D, Baron T. 2019. Seeded propagation of α-synuclein aggregation in mouse brain using protein misfolding cyclic amplification. *FASEB J* **33**: 12073–12086. doi:10.1096/fj.201900354R

Obergasteiger J, Frapporti G, Pramstaller PP, Hicks AA, Volta M. 2018. A new hypothesis for Parkinson's disease pathogenesis: GTPase-p38 MAPK signaling and autophagy as convergence points of etiology and genomics. *Mol Neurodegener* **13**: 40. doi:10.1186/s13024-018-0273-5

Ohgita T, Namba N, Kono H, Shimanouchi T, Saito H. 2022. Mechanisms of enhanced aggregation and fibril formation of Parkinson's disease-related variants of α-synuclein. *Sci Rep* **12**: 6770. doi:10.1038/s41598-022-10789-6

Okuzumi A, Hatano T, Matsumoto G, Nojiri S, Ueno SI, Imamichi-Tatano Y, Kimura H, Kakuta S, Kondo A, Fukuhara T, et al. 2023. Propagative α-synuclein seeds as serum biomarkers for synucleinopathies. *Nat Med* **29**: 1448–1455. doi:10.1038/s41591-023-02358-9

Ordonez DG, Lee MK, Feany MB. 2018. α-Synuclein induces mitochondrial dysfunction through spectrin and the actin cytoskeleton. *Neuron* **97**: 108–124.e6. doi:10.1016/j.neuron.2017.11.036

Ostrerova N, Petrucelli L, Farrer M, Mehta N, Alexander P, Choi P, Palacino J, Hardy J, Wolozin B. 1999. α-Synuclein shares physical and functional homology with 14-3-3 proteins. *J Neurosci* **19**: 5782–5791. doi:10.1523/JNEUROSCI.19-14-05782.1999

Oueslati A, Fournier M, Lashuel HA. 2010. Role of post translational modifications in modulating the structure, function and toxicity of α-synuclein: implications for Parkinson's disease pathogenesis and therapies. *Prog Brain Res* **183**: 115–145. doi:10.1016/S0079-6123(10)83007-9

Oueslati A, Paleologou KE, Schneider BL, Aebischer P, Lashuel HA. 2012. Mimicking phosphorylation at serine 87 inhibits the aggregation of human α-synuclein and protects against its toxicity in a rat model of Parkinson's disease. *J Neurosci* **32**: 1536–1544. doi:10.1523/JNEUROSCI.3784-11.2012

Outeiro TF, Lindquist S. 2003. Yeast cells provide insight into α-synuclein biology and pathobiology. *Science* **302**: 1772–1775.

Outeiro TF, Kontopoulos E, Altmann SM, Kufareva I, Strathearn KE, Amore AM, Volk CB, Maxwell MM, Rochet JC, McLean PJ, et al. 2007. Sirtuin 2 inhibitors rescue α-synuclein-mediated toxicity in models of Parkinson's disease. *Science* **317**: 516–519.

Pagano G, Taylor KI, Anzures-Cabrera J, Marchesi M, Simuni T, Marek K, Postuma RB, Pavese N, Stocchi F, Azulay JP, et al. 2022. Trial of prasinezumab in early-stage Parkinson's disease. *N Engl J Med* **387**: 421–432. doi:10.1056/NEJMoa2202867

Pagano G, Taylor KI, Anzures Cabrera J, Simuni T, Marek K, Postuma RB, Pavese N, Stocchi F, Brockmann K, Svoboda H, et al. 2024. Prasinezumab slows motor progression in rapidly progressing early-stage Parkinson's disease. *Nat Med* **30**: 1096–1103. doi:10.1038/s41591-024-02886-y

Paiva I, Pinho R, Pavlou MA, Hennion M, Wales P, Schütz AL, Rajput A, Szego ÉM, Kerimoglu C, Gerhardt E, et al.

2017. Sodium butyrate rescues dopaminergic cells from α-synuclein-induced transcriptional deregulation and DNA damage. *Hum Mol Genet* **26**: 2231–2246.

Pan H, Liu Z, Ma J, Li Y, Zhao Y, Zhou X, Xiang Y, Wang Y, Zhou X, He R, et al. 2023. Genome-wide association study using whole-genome sequencing identifies risk loci for Parkinson's disease in Chinese population. *NPJ Parkinsons Dis* **9**: 22. doi:10.1038/s41531-023-00456-6

Pantazopoulou M, Brembati V, Kanellidi A, Bousset L, Melki R, Stefanis L. 2021. Distinct α-synuclein species induced by seeding are selectively cleared by the lysosome or the proteasome in neuronally differentiated SH-SY5Y cells. *J Neurochem* **156**: 880–896. doi:10.1111/jnc.15174

Papadimitriou D, Antonelou R, Miligkos M, Maniati M, Papagiannakis N, Bostantjopoulou S, Leonardos A, Koros C, Simitsi A, Papageorgiou SG, et al. 2016. Motor and nonmotor features of carriers of the p.A53T α-synuclein mutation: a longitudinal study. *Mov Disord* **31**: 1226–1230. doi:10.1002/mds.26615

Papagiannakis N, Xilouri M, Koros C, Stamelou M, Antonelou R, Maniati M, Papadimitriou D, Moraitou M, et al. 2015. Lysosomal alterations in peripheral blood mononuclear cells of Parkinson's disease patients. *Mov Disord* **30**: 1830–1834. doi:10.1002/mds.26433

Papagiannakis N, Koros C, Stamelou M, Simitsi AM, Maniati M, Antonelou R, Papadimitriou D, Dermentzaki G, Moraitou M, Michelakakis H, et al. 2018. α-Synuclein dimerization in erythrocytes of patients with genetic and non-genetic forms of Parkinson's disease. *Neurosci Lett* **672**: 145–149. doi:10.1016/j.neulet.2017.11.012

Papagiannakis N, Xilouri M, Koros C, Simitsi AM, Stamelou M, Maniati M, Stefanis L. 2019. Autophagy dysfunction in peripheral blood mononuclear cells of Parkinson's disease patients. *Neurosci Lett* **704**: 112–115. doi:10.1016/j.neulet.2019.04.003

Parihar MS, Parihar A, Fujita M, Hashimoto M, Ghafourifar P. 2008. Mitochondrial association of α-synuclein causes oxidative stress. *Cell Mol life Sci* **65**: 1272–1284. doi:10.1007/s00018-008-7589-1

Park MJ, Cheon SM, Bae HR, Kim SH, Kim JW. 2011. Elevated levels of α-synuclein oligomer in the cerebrospinal fluid of drug-naïve patients with Parkinson's disease. *J Clin Neurol* **7**: 215–222. doi:10.3988/jcn.2011.7.4.215

Parkkinen L, Pirttilä T, Alafuzoff I. 2008. Applicability of current staging/categorization of α-synuclein pathology and their clinical relevance. *Acta Neuropathol* **115**: 399–407. doi:10.1007/s00401-008-0346-6

Parnetti L, Chiasserini D, Persichetti E, Eusebi P, Varghese S, Qureshi MM, Dardis A, Deganuto M, De Carlo C, Castrioto A, et al. 2014. Cerebrospinal fluid lysosomal enzymes and α-synuclein in Parkinson's disease. *Mov Disord* **29**: 1019–1027. doi:10.1002/mds.25772

Parra-Rivas LA, Madhivanan K, Aulston BD, Wang L, Prakashchand DD, Boyer NP, Saia-Cereda VM, Branes-Guerrero K, Pizzo DP, Bagchi P, et al. 2023. Serine-129 phosphorylation of α-synuclein is an activity-dependent trigger for physiologic protein-protein interactions and synaptic function. *Neuron* **111**: 4006–4023.e10. doi:10.1016/j.neuron.2023.11.020

Pasanen P, Myllykangas L, Siitonen M, Raunio A, Kaakkola S, Lyytinen J, Tienari PJ, Pöyhönen M, Paetau A. 2014. Novel α-synuclein mutation A53E associated with atypical multiple system atrophy and Parkinson's disease-type pathology. *Neurobiol Aging* **35**: 2180.e1-5. doi:10.1016/j.neurobiolaging.2014.03.024

Paxinou E, Chen Q, Weisse M, Giasson BI, Norris EH, Rueter SM, Trojanowski JQ, Lee VM, Ischiropoulos H. 2001. Induction of α-synuclein aggregation by intracellular nitrative insult. *J Neurosci* **21**: 8053–8061. doi:10.1523/JNEUROSCI.21-20-08053.2001

Peng C, Gathagan RJ, Covell DJ, Medellin C, Stieber A, Robinson JL, Zhang B, Pitkin RM, Olufemi MF, Luk KC, et al. 2018. Cellular milieu imparts distinct pathological α-synuclein strains in α-synucleinopathies. *Nature* **557**: 558–563. doi:10.1038/s41586-018-0104-4

Permanne B, Sand A, Ousson S, Nény M, Hantson J, Schubert R, Wiessner C, Quattropani A, Beher D. 2022. O-GlcNAcase inhibitor ASN90 is a multimodal drug candidate for tau and α-synuclein proteinopathies. *ACS Chem Neurosci* **13**: 1296–1314. doi:10.1021/acschemneuro.2c00057

Perni M, Galvagnion C, Maltsev A, Meisl G, Müller MB, Challa PK, Kirkegaard JB, Flagmeier P, Cohen SI, Cascella R, et al. 2017. A natural product inhibits the initiation of α-synuclein aggregation and suppresses its toxicity. *Proc Natl Acad Sci* **114**: E1009–E1017. doi:10.1073/pnas.1616191114

Petrucelli L, O'Farrell C, Lockhart PJ, Baptista M, Kehoe K, Vink L, Choi P, Wolozin B, Farrer M, Hardy J, et al. 2002. Parkin protects against the toxicity associated with mutant α-synuclein: proteasome dysfunction selectively affects catecholaminergic neurons. *Neuron* **36**: 1007–1019. doi:10.1016/S0896-6273(02)01125-X

Pinho R, Paiva I, Jercic KG, Fonseca-Ornelas L, Gerhardt E, Fahlbusch C, Garcia-Esparcia P, Kerimoglu C, Pavlou MAS, Villar-Piqué A, et al. 2019. Nuclear localization and phosphorylation modulate pathological effects of α-synuclein. *Hum Mol Genet* **28**: 31–50.

Polymeropoulos MH, Lavedan C, Leroy E, Ide SE, Dehejia A, Dutra A, Pike B, Root H, Rubenstein J, Boyer R, et al. 1997. Mutation in the α-synuclein gene identified in families with Parkinson's disease. *Science* **276**: 2045–2047. doi:10.1126/science.276.5321.2045

Prieto S, Drobny HA, Marques ARA, Di Spiezio A, Dobert JP, Balta D, Werner C, Rizo T, Gallwitz L, Bub S, et al. 2022. Recombinant pro-CTSD (cathepsin D) enhances SNCA/α-synuclein degradation in α-Synucleinopathy models. *Autophagy* **18**: 1127–1151. doi:10.1080/15548627.2022.2045534

Prigione A, Piazza F, Brighina L, Begni B, Galbussera A, DiFrancesco JC, Andreoni S, Piolti R, Ferrarese C. 2010. α-Synuclein nitration and autophagy response are induced in peripheral blood cells from patients with Parkinson disease. *Neurosci Lett* **477**: 6–10. doi:10.1016/j.neulet.2010.04.022

Prots I, Grosch J, Brazdis R-M, Simmnacher K, Veber V, Havlicek S, Hannappel C, Krach F, Krumbiegel M, Schütz O, et al. 2018. α-Synuclein oligomers induce early axonal dysfunction in human iPSC-based models of Synucleinopathies. *Proc Natl Acad Sci* **115**: 7813–7818. doi:10.1073/pnas.1713129115

Proukakis C, Dudzik CG, Brier T, MacKay DS, Cooper JM, Millhauser GL, Houlden H, Schapira AH. 2013. A novel α-synuclein missense mutation in Parkinson disease.

Neurology **80**: 1062–1064. doi:10.1212/WNL.0b013e31828727ba

Przedborski S, Chen Q, Vila M, Giasson BI, Djaldatti R, Vukosavic S, Souza JM, Jackson-Lewis V, Lee VM, Ischiropoulos H. 2001. Oxidative post-translational modifications of α-synuclein in the 1-methyl-4-phenyl-1,2,3,6-tetrahydropyridine (MPTP) mouse model of Parkinson's disease. *J Neurochem* **76**: 637–640. doi:10.1046/j.1471-4159.2001.00174.x

Recasens A, Ulusoy A, Kahle PJ, Di Monte DA, Dehay B. 2018. In vivo models of α-synuclein transmission and propagation. *Cell Tissue Res* **373**: 183–193. doi:10.1007/s00441-017-2730-9

Reimer L, Gram H, Jensen NM, Betzer C, Yang L, Jin L, Shi M, Boudeffa D, Fusco G, De Simone A, et al. 2022. Protein kinase R dependent phosphorylation of α-synuclein regulates its membrane binding and aggregation. *PNAS Nexus* **1**: pgac259.

Rey NL, George S, Steiner JA, Madaj Z, Luk KC, Trojanowski JQ, Lee VMY, Brundin P. 2018. Spread of aggregates after olfactory bulb injection of α-synuclein fibrils is associated with early neuronal loss and is reduced long term. *Acta Neuropathol* **135**: 65–83. doi:10.1007/s00401-017-1792-9

Risiglione P, Zinghirino F, Di Rosa MC, Magrì A, Messina A. 2021. α-Synuclein and mitochondrial dysfunction in Parkinson's disease: the emerging role of VDAC. *Biomolecules* **11**: 718. doi:10.3390/biom11050718

Roberts RF, Wade-Martins R, Alegre-Abarrategui J. 2015. Direct visualization of α-synuclein oligomers reveals previously undetected pathology in Parkinson's disease brain. *Brain* **138**: 1642–1657. doi:10.1093/brain/awv040

Rodriguez-Araujo G, Nakagami H, Hayashi H, Mori M, Shiuchi T, Minokoshi Y, Nakaoka Y, Takami Y, Komuro I, Morishita R, et al. 2013. α-Synuclein elicits glucose uptake and utilization in adipocytes through the Gab1/PI3K/Akt transduction pathway. *Cell Mol Life Sci* **70**: 1123–1133. doi:10.1007/s00018-012-1198-8

Rott R, Szargel R, Shani V, Hamza H, Savyon M, Abd Elghani F, Bandopadhyay R, Engelender S. 2017. SUMOylation and ubiquitination reciprocally regulate α-synuclein degradation and pathological aggregation. *Proc Natl Acad Sci* **114**: 13176–13181. doi:10.1073/pnas.1704351114

Rousseaux MW, Revelli JP, Vázquez-Vélez GE, Kim JY, Craigen E, Gonzales K, Beckinghausen J, Zoghbi HY. 2018. Depleting Trim28 in adult mice is well tolerated and reduces levels of α-synuclein and tau. *Elife* **7**: e36768. doi:10.7554/eLife.36768

Rutherford NJ, Moore BD, Golde TE, Giasson BI. 2014. Divergent effects of the H50Q and G51D SNCA mutations on the aggregation of α-synuclein. *J Neurochem* **131**: 859–867. doi:10.1111/jnc.12806

Sala G, Stefanoni G, Arosio A, Riva C, Melchionda L, Saracchi E, Fermi S, Brighina L, Ferrarese C. 2014. Reduced expression of the chaperone-mediated autophagy carrier hsc70 protein in lymphomonocytes of patients with Parkinson's disease. *Brain Res* **1546**: 46–52. doi:10.1016/j.brainres.2013.12.017

Sanchez-Mirasierra I, Ghimire S, Hernandez-Diaz S, Soukup SF. 2022. Targeting macroautophagy as a therapeutic opportunity to treat Parkinson's disease. *Front Cell Dev Biol* **10**: 921314. doi:10.3389/fcell.2022.921314

Sarkar S, Olsen AL, Sygnecka K, Lohr KM, Feany MB. 2021. α-Synuclein impairs autophagosome maturation through abnormal actin stabilization. *PLoS Genet* **17:** e1009359. doi:10.1371/journal.pgen.1009359

Scott D, Roy S. 2012. α-Synuclein inhibits intersynaptic vesicle mobility and maintains recycling-pool homeostasis. *J Neurosci* **32:** 10129–10135. doi:10.1523/JNEUROSCI.0535-12.2012

Scott DA, Tabarean I, Tang Y, Cartier A, Masliah E, Roy S. 2010. A pathologic cascade leading to synaptic dysfunction in α-synuclein-induced neurodegeneration. *J Neurosci* **30:** 8083–8095. doi:10.1523/JNEUROSCI.1091-10.2010

Sekiya H, Tsuji A, Hashimoto Y, Takata M, Koga S, Nishida K, Futamura N, Kawamoto M, Kohara N, Dickson DW, et al. 2022. Discrepancy between distribution of α-synuclein oligomers and Lewy-related pathology in Parkinson's disease. *Acta Neuropathol Commun* **10:** 133. doi:10.1186/s40478-022-01440-6

Sevlever D, Jiang P, Yen SH. 2008. Cathepsin D is the main lysosomal enzyme involved in the degradation of α-synuclein and generation of its carboxy-terminally truncated species. *Biochemistry* **47:** 9678–9687.

Shabek N, Herman-Bachinsky Y, Buchsbaum C, Lewinson O, Haj-Yahya M, Hejjaoui M, Lashuel HA, Sommer T, Brik A, Ciechanover A. 2012. The size of the proteasomal substrate determines whether its degradation will be mediated by mono- or polyubiquitylation. *Mol Cell* **48:** 87–97. doi:10.1016/j.molcel.2012.07.011

Shah AN, Hiew KW, Han P, Parsons RB, Chang RCC, Legido-Quigley C. 2017. α-Synuclein in bio fluids and tissues as a potential biomarker for Parkinson's disease. *JADP* **2:** 013.

Shahmoradian SH, Lewis AJ, Genoud C, Hench J, Moors TE, Navarro PP, Castaño-Díez D, Schweighauser G, Graff-Meyer A, Goldie KN, et al. 2019. Lewy pathology in Parkinson's disease consists of crowded organelles and lipid membranes. *Nat Neurosci* **22:** 1099–1109. doi:10.1038/s41593-019-0423-2

Siderowf A, Concha-Marambio L, Lafontant DE, Farris CM, Ma Y, Urenia PA, Nguyen H, Alcalay RN, Chahine LM, Foroud T, et al. 2023. Assessment of heterogeneity among participants in the Parkinson's progression markers initiative cohort using α-synuclein seed amplification: a cross-sectional study. *Lancet Neurol* **22:** 407–417. doi:10.1016/S1474-4422(23)00109-6

Singleton AB, Farrer M, Johnson J, Singleton A, Hague S, Kachergus J, Hulihan M, Peuralinna T, Dutra A, Nussbaum R, et al. 2003. α-Synuclein locus triplication causes Parkinson's disease. *Science* **302:** 841. doi:10.1126/science.1090278

Sivakumar P, Nagashanmugam KB, Priyatharshni S, Lavanya R, Prabhu N, Ponnusamy S. 2023. Review on the interactions between dopamine metabolites and α-synuclein in causing Parkinson's disease. *Neurochem Int* **162:** 105461. doi:10.1016/j.neuint.2022.105461

Siwecka N, Saramowicz K, Galita G, Rozpędek-Kamińska W, Majsterek I. 2023. Inhibition of protein aggregation and endoplasmic reticulum stress as a targeted therapy for α-synucleinopathy. *Pharmaceutics* **15:** 2051.

Smith L, Schapira AHV. 2022. GBA variants and Parkinson disease: mechanisms and treatments. *Cells* **11:** 1261. doi:10.3390/cells11081261

Snyder H, Mensah K, Theisler C, Lee J, Matouschek A, Wolozin B. 2003. Aggregated and monomeric α-synuclein bind to the S6' proteasomal protein and inhibit proteasomal function. *J Biol Chem* **278:** 11753–11759. doi:10.1074/jbc.M208641200

Sonustun B, Altay MF, Strand C, Ebanks K, Hondhamuni G, Warner TT, Lashuel HA, Bandopadhyay R. 2022. Pathological relevance of post-translationally modified α-synuclein (pSer87, pSer129, nTyr39) in idiopathic Parkinson's disease and multiple system atrophy. *Cells* **11:** 906. doi:10.3390/cells11050906

Sorrentino ZA, Giasson BI. 2020. The emerging role of α-synuclein truncation in aggregation and disease. *J Biol Chem* **295:** 10224–10244. doi:10.1074/jbc.REV120.011743

Spillantini MG, Schmidt ML, Lee VM, Trojanowski JQ, Jakes R, Goedert M. 1997. α-Synuclein in Lewy bodies. *Nature* **388:** 839–840. doi:10.1038/42166

Spillantini MG, Crowther RA, Jakes R, Hasegawa M, Goedert M. 1998. α-Synuclein in filamentous inclusions of Lewy bodies from Parkinson's disease and dementia with Lewy bodies. *Proc Natl Acad Sci* **95:** 6469–6473. doi:10.1073/pnas.95.11.6469

Stefanis L. 2012. α-Synuclein in Parkinson's disease. *Cold Spring Harb Perspect Med* **2:** a009399. doi:10.1101/cshperspect.a009399

Stefanis L, Larsen KE, Rideout HJ, Sulzer D, Greene LA. 2001. Expression of A53T mutant but not wild-type α-synuclein in PC12 cells induces alterations of the ubiquitin-dependent degradation system, loss of dopamine release, and autophagic cell death. *J Neurosci* **21:** 9549–9560. doi:10.1523/JNEUROSCI.21-24-09549.2001

Stewart T, Sossi V, Aasly JO, Wszolek ZK, Uitti RJ, Hasegawa K, Yokoyama T, Zabetian CP, Leverenz JB, Stoessl AJ, et al. 2015. Phosphorylated α-synuclein in Parkinson's disease: correlation depends on disease severity. *Acta Neuropathol Commun* **3:** 7. doi:10.1186/s40478-015-0185-3

Strohäker T, Jung BC, Liou SH, Fernandez CO, Riedel D, Becker S, Halliday GM, Bennati M, Kim WS, Lee SJ, et al. 2019. Structural heterogeneity of α-synuclein fibrils amplified from patient brain extracts. *Nat Commun* **10:** 5535. doi:10.1038/s41467-019-13564-w

Stuendl A, Kunadt M, Kruse N, Bartels C, Moebius W, Danzer KM, Mollenhauer B, Schneider A. 2016. Induction of α-synuclein aggregate formation by CSF exosomes from patients with Parkinson's disease and dementia with Lewy bodies. *Brain* **139:** 481–494. doi:10.1093/brain/awv346

Sun J, Wang L, Bao H, Roy S. 2019. Functional cooperation of α-synuclein and VAMP2 in synaptic vesicle recycling. *Proc Natl Acad Sci* **116:** 11113–11115. doi:10.1073/pnas.1903049116

Sun Y, Long H, Xia W, Wang K, Zhang X, Sun B, Cao Q, Zhang Y, Dai B, Li D, et al. 2021. The hereditary mutation G51D unlocks a distinct fibril strain transmissible to wild-type α-synuclein. *Nat Commun* **12:** 6252. doi:10.1038/s41467-021-26433-2

Surguchov A. 2024. Nuclear α-synuclein regulates transcription and affects neuronal differentiation. *FEBS J* **291**: 1886–1888. doi:10.1111/febs.17072

Surmeier DJ, Schumacker PT. 2013. Calcium, bioenergetics, and neuronal vulnerability in Parkinson's disease. *J Biol Chem* **288**: 10736–10741. doi:10.1074/jbc.R112.410530

Swant J, Goodwin JS, North A, Ali AA, Gamble-George J, Chirwa S, Khoshbouei H. 2011. α-Synuclein stimulates a dopamine transporter-dependent chloride current and modulates the activity of the transporter. *J Biol Chem* **286**: 43933–43943. doi:10.1074/jbc.M111.241232

Tanaka Y, Engelender S, Igarashi S, Rao RK, Wanner T, Tanzi RE, Sawa A, Dawson VL, Dawson TM, Ross CA. 2001. Inducible expression of mutant α-synuclein decreases proteasome activity and increases sensitivity to mitochondria-dependent apoptosis. *Hum Mol Genet* **10**: 919–926. doi:10.1093/hmg/10.9.919

Tavassoly O, Yue J, Vocadlo DJ. 2021. Pharmacological inhibition and knockdown of *O*-GlcNAcase reduces cellular internalization of α-synuclein preformed fibrils. *FEBS J* **288**: 452–470. doi:10.1111/febs.15349

Tehranian R, Montoya SE, Van Laar AD, Hastings TG, Perez RG. 2006. α-Synuclein inhibits aromatic amino acid decarboxylase activity in dopaminergic cells. *J Neurochem* **99**: 1188–1196. doi:10.1111/j.1471-4159.2006.04146.x

Theillet FX, Binolfi A, Bekei B, Martorana A, Rose HM, Stuiver M, Verzini S, Lorenz D, van Rossum M, Goldfarb D, et al. 2016. Structural disorder of monomeric α-synuclein persists in mammalian cells. *Nature* **530**: 45–50. doi:10.1038/nature16531

Tofaris GK, Layfield R, Spillantini MG. 2001. α-Synuclein metabolism and aggregation is linked to ubiquitin-independent degradation by the proteasome. *FEBS Lett* **509**: 22–26. doi:10.1016/S0014-5793(01)03115-5

Tofaris GK, Kim HT, Hourez R, Jung JW, Kim KP, Goldberg AL. 2011. Ubiquitin ligase Nedd4 promotes α-synuclein degradation by the endosomal-lysosomal pathway. *Proc Natl Acad Sci* **108**: 17004–17009. doi:10.1073/pnas.1109356108

Toledo JB, Korff A, Shaw LM, Trojanowski JQ, Zhang J. 2013. CSF α-synuclein improves diagnostic and prognostic performance of CSF tau and Aβ in Alzheimer's disease. *Acta Neuropathol* **126**: 683–697. doi:10.1007/s00401-013-1148-z

Ueda J, Uemura N, Ishimoto T, Taguchi T, Sawamura M, Nakanishi E, Ikuno M, Matsuzawa S, Yamakado H, Takahashi R. 2023. Ca^{2+}–calmodulin–calcineurin signaling modulates α-synuclein transmission. *Mov Disord* **38**: 1056–1067. doi:10.1002/mds.29401

Uversky VN. 2003. A protein-chameleon: conformational plasticity of α-synuclein, a disordered protein involved in neurodegenerative disorders. *J Biomol Struct Dyn* **21**: 211–234. doi:10.1080/07391102.2003.10506918

Uversky VN, Li J, Fink AL. 2001. Evidence for a partially folded intermediate in α-synuclein fibril formation. *J Biol Chem* **276**: 10737–10744. doi:10.1074/jbc.M010907200

Vallortigara J, Whitfield D, Quelch W, Alghamdi A, Howlett D, Hortobágyi T, Johnson M, Attems J, O'Brien JT, Thomas A, et al. 2016. Decreased levels of VAMP2 and monomeric α-synuclein correlate with duration of dementia. *J Alzheimers Dis* **50**: 101–110. doi:10.3233/JAD-150707

Verma DK, Ghosh A, Ruggiero L, Cartier E, Janezic E, Williams D, Jung EG, Moore M, Seo JB, Kim YH. 2020. The SUMO conjugase Ubc9 protects dopaminergic cells from cytotoxicity and enhances the stability of α-synuclein in Parkinson's disease models. *eNeuro* **7**: ENEURO.0134-20.2020. doi:10.1523/ENEURO.0134-20.2020

Vogiatzi T, Xilouri M, Vekrellis K, Stefanis L. 2008. Wild type α-synuclein is degraded by chaperone-mediated autophagy and macroautophagy in neuronal cells. *J Biol Chem* **283**: 23542–23556. doi:10.1074/jbc.M801992200

Volakakis N, Tiklova K, Decressac M, Papathanou M, Mattsson B, Gillberg L, Nobre A, Björklund A, Perlmann T. 2015. Nurr1 and retinoid X receptor ligands stimulate ret signaling in dopamine neurons and can alleviate α-synuclein disrupted gene expression. *J Neurosci* **35**: 14370–14385.

Wakabayashi K. 2020. Where and how α-synuclein pathology spreads in Parkinson's disease. *Neuropathology* **40**: 415–425. doi:10.1111/neup.12691

Wang Y, Shi M, Chung KA, Zabetian CP, Leverenz JB, Berg D, Srulijes K, Trojanowski JQ, Lee VM, Siderowf AD, et al. 2012. Phosphorylated α-synuclein in Parkinson's disease. *Sci Transl Med* **4**: 121ra20.

Wang L, Das U, Scott DA, Tang Y, McLean PJ, Roy S. 2014. α-Synuclein multimers cluster synaptic vesicles and attenuate recycling. *Curr Biol* **24**: 2319–2326. doi:10.1016/j.cub.2014.08.027

Wang X, Becker K, Levine N, Zhang M, Lieberman AP, Moore DJ, Ma J. 2019. Pathogenic α-synuclein aggregates preferentially bind to mitochondria and affect cellular respiration. *Acta Neuropathol Commun* **7**: 41. doi:10.1186/s40478-019-0696-4

Webb JL, Ravikumar B, Atkins J, Skepper JN, Rubinsztein DC. 2003. α-Synuclein is degraded by both autophagy and the proteasome. *J Biol Chem* **278**: 25009–25013. doi:10.1074/jbc.M300227200

Weston LJ, Cook ZT, Stackhouse TL, Sal MK, Schultz BI, Tobias ZJC, Osterberg VR, Brockway NL, Pizano S, Glover G, et al. 2021. In vivo aggregation of presynaptic α-synuclein is not influenced by its phosphorylation at serine-129. *Neurobiol Dis* **152**: 105291. doi:10.1016/j.nbd.2021.105291

Whone A. 2022. Monoclonal antibody therapy in Parkinson's disease—the end? *N Engl J Med* **387**: 466–467. doi:10.1056/NEJMe2207681

Wijesekara N, Ahrens R, Wu L, Langman T, Tandon A, Fraser PE. 2021. α-Synuclein regulates peripheral insulin secretion and glucose transport. *Front Aging Neurosci* **13**: 665348. doi:10.3389/fnagi.2021.665348

Williams DM, Thorn DC, Dobson CM, Meehan S, Jackson SE, Woodcock JM, Carver JA. 2021. The amyloid fibril-forming β-sheet regions of amyloid β and α-synuclein preferentially interact with the molecular chaperone 14-3-3ζ. *Molecules* **26**: 6120. doi:10.3390/molecules26206120

Winslow AR, Chen CW, Corrochano S, Acevedo-Arozena A, Gordon DE, Peden AA, Lichtenberg M, Menzies FM, Ravikumar B, Imarisio S, et al. 2010. α-Synuclein impairs macroautophagy: implications for Parkinson's disease. *J Cell Biol* **190**: 1023–1037. doi:10.1083/jcb.201003122

Woerman AL. 2021. Strain diversity in neurodegenerative disease: an argument for a personalized medicine approach to diagnosis and treatment. *Acta Neuropathol* **142:** 1–3. doi:10.1007/s00401-021-02311-5

Wu G, Wang X, Feng X, Zhang A, Li J, Gu K, Huang J, Pang S, Dong H, Gao H, et al. 2011. Altered expression of autophagic genes in the peripheral leukocytes of patients with sporadic Parkinson's disease. *Brain Res* **1394:** 105–111. doi:10.1016/j.brainres.2011.04.013

Wu S, Hernandez Villegas NC, Sirkis DW, Thomas-Wright I, Wade-Martins R, Schekman R. 2023. Unconventional secretion of α-synuclein mediated by palmitoylated DNAJC5 oligomers. *Elife* **12:** e85837. doi:10.7554/eLife .85837

Xie YX, Naseri NN, Fels J, Kharel P, Na Y, Lane D, Burré J, Sharma M. 2022. Lysosomal exocytosis releases pathogenic α-synuclein species from neurons in synucleinopathy models. *Nat Commun* **13:** 4918. doi:10.1038/s41467-022-32625-1

Xilouri M, Vogiatzi T, Vekrellis K, Park D, Stefanis L. 2009. Abberant α-synuclein confers toxicity to neurons in part through inhibition of chaperone-mediated autophagy. *PLoS ONE* **4:** e5515. doi:10.1371/journal.pone.0005515

Xilouri M, Brekk OR, Stefanis L. 2013a. α-Synuclein and protein degradation systems: a reciprocal relationship. *Mol Neurobiol* **47:** 537–551. doi:10.1007/s12035-012-8341-2

Xilouri M, Brekk OR, Landeck N, Pitychoutis PM, Papasilekas T, Papadopoulou-Daifoti Z, Kirik D, Stefanis L. 2013b. Boosting chaperone-mediated autophagy in vivo mitigates α-synuclein-induced neurodegeneration. *Brain* **136:** 2130–2146. doi:10.1093/brain/awt131

Xilouri M, Brekk OR, Polissidis A, Chrysanthou-Piterou M, Kloukina I, Stefanis L. 2016a. Impairment of chaperone-mediated autophagy induces dopaminergic neurodegeneration in rats. *Autophagy* **12:** 2230–2247. doi:10.1080/15548627.2016.1214777

Xilouri M, Brekk OR, Stefanis L. 2016b. Autophagy and α-synuclein: relevance to Parkinson's disease and related synucleopathies. *Mov Disord* **31:** 178–192. doi:10.1002/mds.26477

Yakunin E, Loeb V, Kisos H, Biala Y, Yehuda S, Yaari Y, Selkoe DJ, Sharon R. 2012. α-Synuclein neuropathology is controlled by nuclear hormone receptors and enhanced by docosahexaenoic acid in a mouse model for Parkinson's disease. *Brain Pathol* **22:** 280–294.

Yamada K, Iwatsubo T. 2018. Extracellular α-synuclein levels are regulated by neuronal activity. *Mol Neurodegener* **13:** 9. doi:10.1186/s13024-018-0241-0

Yan JQ, Yuan YH, Gao YN, Huang JY, Ma KL, Gao Y, Zhang WQ, Guo XF, Chen NH. 2014. Overexpression of human E46K mutant α-synuclein impairs macroautophagy via inactivation of JNK1-Bcl-2 pathway. *Mol Neurobiol* **50:** 685–701. doi:10.1007/s12035-014-8738-1

Yan S, Jiang C, Janzen A, Barber TR, Seger A, Sommerauer M, Davis JJ, Marek K, Hu MT, Oertel WH, et al. 2024. Neuronally derived extracellular vesicle α-synuclein as a serum biomarker for individuals at risk of developing Parkinson disease. *JAMA Neurol* **81:** 59–68. doi:10 .1001/jamaneurol.2023.4398

Yang Y, Shi Y, Schweighauser M, Zhang X, Kotecha A, Murzin AG, Garringer HJ, Cullinane PW, Saito Y, Foroud T, et al. 2022. Structures of α-synuclein filaments from human brains with Lewy pathology. *Nature* **610:** 791–795. doi:10 .1038/s41586-022-05319-3

Yang Y, Garringer HJ, Shi Y, Lövestam S, Peak-Chew S, Zhang X, Kotecha A, Bacioglu M, Koto A, Takao M, et al. 2023. New SNCA mutation and structures of α-synuclein filaments from juvenile-onset synucleinopathy. *Acta Neuropathol* **145:** 561–572. doi:10.1007/s00401-023-02 550-8

Zarranz JJ, Alegre J, Gómez-Esteban JC, Lezcano E, Ros R, Ampuero I, Vidal L, Hoenicka J, Rodriguez O, Atarés B, et al. 2004. The new mutation, E46K, of α-synuclein causes Parkinson and Lewy body dementia. *Ann Neurol* **55:** 164–173. doi:10.1002/ana.10795

Zenko D, Marsh J, Castle AR, Lewin R, Fischer R, Tofaris GK. 2023. Monitoring α-synuclein ubiquitination dynamics reveals key endosomal effectors mediating its trafficking and degradation. *Sci Adv* **9:** eadd8910. doi:10 .1126/sciadv.add8910

Zhang Z, Kang SS, Liu X, Ahn EH, Zhang Z, He L, Iuvone PM, Duong DM, Seyfried NT, Benskey MJ, et al. 2017. Asparagine endopeptidase cleaves α-synuclein and mediates pathologic activities in Parkinson's disease. *Nat Struct Mol Biol* **24:** 632–642. doi:10.1038/nsmb.3433

Leucine-Rich Repeat Kinase 2: Pathways to Parkinson's Disease

Suzanne R. Pfeffer[1,2,3] and Dario R. Alessi[2,3,4]

[1]Department of Biochemistry, Stanford University School of Medicine, Stanford, California 94305-5307, USA

[2]Aligning Science Across Parkinson's (ASAP) Collaborative Research Network, Chevy Chase, Maryland 20815, USA

[3]LRRK2 Investigative Therapeutics Exchange (LITE), New York, New York 10120, USA

[4]MRC Protein Phosphorylation and Ubiquitylation Unit, Sir James Black Centre, School of Life Sciences, University of Dundee, Dundee DD1 5EH, United Kingdom

Correspondence: pfeffer@stanford.edu

The past 10 years have seen tremendous progress in our understanding of leucine-rich repeat kinase 2 (LRRK2) and how mutations activate the kinase and trigger downstream pathology, contributing to Parkinson's disease. A breakthrough came from the identification of key LRRK2 substrates—a subset of small guanosine triphosphatases (GTPases) called Rab proteins. Cryoelectron microscopy has revealed structures of LRRK2 and showed how inhibitors engage and inhibit the kinase. Biochemical experiments have revealed how LRRK2 is recruited to membranes to phosphorylate Rab substrates. LRRK2 activation during lysosomal stress triggers Rab phosphorylation, altering the repertoire of Rab-binding partners. Resulting phospho-Rab-effector complexes have prominent effects in specific cell types, disrupting primary cilia and impairing Hedgehog signaling—effects that can be reversed by LRRK2 inhibitors. This disruption in Hedgehog signaling represents a convergence point linking genetic and idiopathic forms of Parkinson's. Together, these findings support the therapeutic potential of LRRK2 inhibitors in Parkinson's disease.

In most instances, the cause of Parkinson's disease (PD) remains unknown. However, ~15% of patients carry genetic variants linked to the disease, with mutations in the leucine-rich repeat kinase 2 (LRRK2) gene being among the most prevalent (Westenberger et al. 2024; Krüger et al. 2025). Given the prominence of LRRK2 in both familial and sporadic PD, it is worth dedicating a full article to this kinase and its role in disease mechanisms and therapeutic targeting.

In 2004, mutations in *LRRK2* were identified as the cause of PARK8 PD (Paisán-Ruíz et al. 2004; Zimprich et al. 2004). LRRK2 encodes a protein of 288 kDa that is comprised of multiple domains including (from the amino terminus) an armadillo (ARM) domain, ankyrin (ANK) repeats, leucine-rich repeats (LRRs), a short (19-residue) hinge-helix, an atypical Roco family guanosine triphosphatase (GTPase) domain consisting of a GTP-binding Ras of complex proteins (ROC) coupled to two tandem–fold

Copyright © 2025 Cold Spring Harbor Laboratory Press; all rights reserved
Cite this article as *Cold Spring Harb Perspect Med* doi: 10.1101/cshperspect.a041620

carboxy terminal of ROC (COR-A and COR-B), a catalytic kinase (KIN) domain, a β-propeller (WD40) domain, and an extended (29-residue) carboxy-terminal αC-helix (Fig. 1). Four human proteins contain a ROC–COR GTPase domain and form the ROCO protein family: the homologous kinases LRRK1 and LRRK2; death-associated protein kinase 1 (DAPK1), which has an unrelated kinase domain located at its amino terminus; and malignant fibrous histiocytoma amplified sequence 1 (MASL1), which lacks a kinase domain entirely (Dihanich et al. 2014).

We and others have recently reviewed the biochemistry of LRRK2, its interaction with ki-nase inhibitors, and recruitment to membranes and activation (Alessi and Pfeffer 2024). Here we will focus on the cell-type-specific functions of LRRK2 and how hyperactivation in those cell types may contribute to PD. Many aspects of LRRK2 function point to the lysosome as a key player in PD pathogenesis. Nevertheless, as discussed herein, LRRK2 hyperactivation—driven either by PD-associated mutations or by endo-lysosomal dysfunction—disrupts multiple disease-relevant cellular pathways. These findings underscore the therapeutic potential and importance of developing effective inhibitors to counteract LRRK2 kinase hyperactivation for disease-modifying treatment of PD.

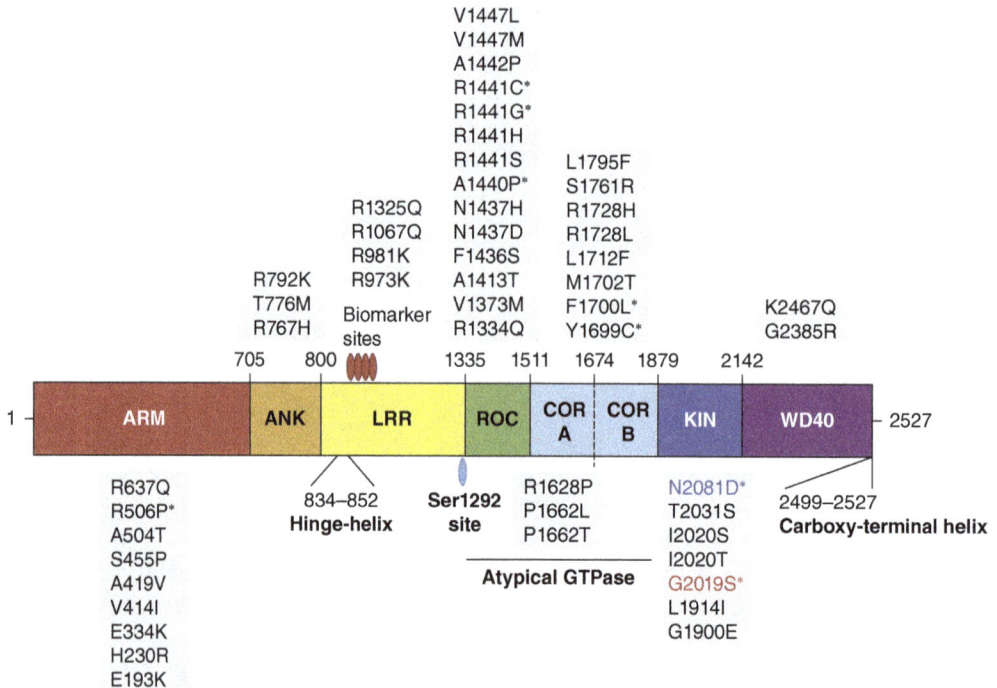

Figure 1. Leucine-rich repeat kinase 2 (LRRK2) domain structure and pathogenic mutations. LRRK2 domain structure highlighting the location of 45 variants that have been reported to promote LRRK2 activity at least 1.5-fold, as assessed in a cellular overexpression assay. Variants marked with an asterisk have been studied in patient-derived neutrophils and/or monocytes. The G2019S variant that is found in 80% of people with LRRK2 mutations is highlighted in red. (ARM) Armadillo, (ANK) ankyrin, (LRRs) leucine-rich repeats, (ROC) Ras of complex proteins, (COR-A, COR-B) carboxy-terminal of ROC A and B (COR-A, COR-B), (KIN) kinase, and WD40 domains. Biomarker and Ser1292 phosphorylation sites and other details are discussed in Alessi and Pfeffer (2024). (Reprinted from Alessi and Pfeffer 2024, © The Authors, under a Creative Commons Attribution 4.0 International License.)

LRRK2 MUTATIONS

The most extensively studied pathogenic mutation in LRRK2 is G2019S, located within the kinase domain's conserved Asp-Tyr-Gly magnesium-binding motif, resulting in approximately a twofold increase in kinase activity (Greggio et al. 2006; Jaleel et al. 2007; West et al. 2007). This variant is particularly prevalent in certain populations: in Ashkenazi Jewish individuals, it accounts for 15%–20% of all PD cases and up to 30% of familial PD cases, while among North African populations, G2019S is responsible for 30%–40% of familial cases and up to 40% of all PD cases (Dächsel and Farrer 2010). Less frequent mutations in other LRRK2 domains also significantly impact kinase function. The R1441G mutation in the ROC GTPase domain disrupts GTPase function and is notably common in the Basque population, contributing to ~15%–20% of familial PD cases and 3%–5% of sporadic cases (Gaig et al. 2006; Correia Guedes et al. 2010). Additionally, in East Asian populations, the prevalent pathogenic mutation R1628P, situated in the COR domain, increases PD risk (Ross et al. 2008). Another common East Asian variant, G2385R, resides within the WD40 domain and is a significant risk factor for late-onset PD characterized by classic motor symptoms (Farrer et al. 2007). The R1441G variant enhances LRRK2-mediated Rab10 phosphorylation in cells approximately fourfold, while the R1628P and G2385R variants activate LRRK2 more modestly, approximately twofold.

A recent comprehensive analysis cataloged 283 distinct LRRK2 variants from 4660 individuals in the Movement Disorders Society Genetic Mutation Database (Krüger et al. 2025). Since the initiation of this database in 2017, the number of cataloged variants and patients has grown ~10-fold and fourfold, respectively. This rapid expansion underscores the increasing complexity and significance of understanding how specific LRRK2 variants contribute to PD pathology.

Systematic mutational analysis of 100 PD-linked LRRK2 variants identified 23 mutations in all domains that increase kinase activity (Kalogeropulou et al. 2022). Variants located within the ROC–COR domains display the strongest

activating effects on Rab10 phosphorylation in cells (Kalogeropulou et al. 2022). Although it was initially proposed that ROC–COR domain variants primarily affect LRRK2 activity through altered GTPase function, recent findings suggest these PD-associated mutations predominantly occur at residues normally involved in stabilizing LRRK2's inactive conformation (Deniston et al. 2020; Myasnikov et al. 2021; Kalogeropulou et al. 2022). Thus, pathogenic variants disrupt these interfaces, destabilizing the inactive state and consequently increasing kinase activity. It is important to note that the effects of LRRK2 mutations differ significantly when comparing in vitro kinase assays to cellular assays.

Only a subset of activating mutations, specifically those in the COR-B (Y1699C, R1728H/L, S1761R) and kinase domains (G2019S, I2020T, T2031S), consistently increase kinase activity in vitro with immunoprecipitated LRRK2 (Kalogeropulou et al. 2022). Discrepancies between in vitro and cellular findings likely result from differences in LRRK2's membrane localization and substrate accessibility. Finally, the reason why specific activating mutations, such as N2081D, are uniquely associated with Crohn's disease (despite similar kinase activation to other variants without this association) remains unclear (Hui et al. 2018). Clarifying these differences is crucial for understanding clinical implications and identifying targeted therapeutic opportunities for distinct LRRK2 variants.

RAB GTPASES ARE LRRK2 SUBSTRATES

A state-of-the-art phosphoproteomics strategy identified a subset of Rab GTPases as bona fide LRRK2 substrates (Steger et al. 2016; Willett et al. 2017). Specifically, ~10 out of the ~65 human Rabs (Rab3A/B/C/D, Rab8A/B, Rab10, Rab12, Rab29, Rab35, and Rab43) are endogenous LRRK2 substrates, phosphorylated at a single conserved threonine (or serine) residue at position 73 (or its equivalent) (Steger et al. 2016, 2017) that lies in the middle of the Rab effector binding, Switch II motif. Note that some of these Rabs are ubiquitously expressed (e.g.,

Rab8A, 10, 12), whereas others are more cell-type-specific (cf. Rabs 29, 35, 43). How this impacts PD is not yet clear, but as discussed below, it suggests that mechanisms of LRRK2 membrane recruitment and activation may rely on different Rabs in specific cell types in which these Rabs are especially expressed.

Rab proteins are key regulators of cellular secretory and endocytic pathways (Pfeffer 2017). Most are stably modified at their carboxyl termini by the attachment of two, 20 carbon, branched geranylgeranyl groups that anchor Rabs to membranes (Seabra 1998). Rabs localize to distinct membrane surfaces, where they recruit effector proteins, including cytoskeletal motor proteins and motor protein adaptors, vesicle tethering proteins, and other membrane trafficking factors. These interactions drive transport vesicle formation, motility, docking, and fusion with target membranes (Pfeffer 2017).

Rabs bind to effectors only when they have GTP bound; they are inactivated by GTPase-activating proteins (GAPs), which enhance GTP hydrolysis, converting them to their inactive guanosine diphosphate (GDP)-bound state. Two abundant cytosolic proteins, GDP dissociation inhibitors (GDIs), recognize GDP-bound Rabs and extract them from membranes by binding both the Rab and its prenyl moieties (Rak et al. 2003). GDI-Rab complexes then deliver Rabs back to their membranes of origin, where guanine nucleotide exchange factors (GEFs) facilitate GDP release, allowing cytosolic GTP to occupy the nucleotide-binding pocket.

The LRRK2-phosphorylated T73 residue lies within the Rab switch II region, a critical site for effector binding that is sensitive to the bound nucleotide (GDP or GTP; Stroupe and Brunger 2000). Phosphorylation by LRRK2 at this site disrupts Rab interactions with a wide array of effectors, including GDIs, which are essential for Rab recycling (Steger et al. 2017). Specifically, phosphorylation of Rab8A interferes with its binding to its GEF (Steger et al. 2016) and also blocks interactions with many of its effectors (Steger et al. 2016, 2017). As a result, phosphorylated Rabs become functionally unavailable because they cannot bind their normal effectors, and they become trapped on

the membranes upon which they are phosphorylated because they cannot be retrieved by GDI.

It is important to note that at steady state, only ~1% of total Rab10 is phosphorylated (Ito et al. 2016; Karayel et al. 2020), and this low phosphorylation stoichiometry is likely to be similar or even lower for the other LRRK2-phosphorylated Rab proteins. Rab proteins are phosphorylated only on specific membranes where LRRK2 is locally recruited. This spatially restricted and limited phosphorylation is critical, as a global blockade of Rab-effector interactions would likely have severe consequences for cellular function. As discussed below, cellular phosphatases actively dephosphorylate Rab proteins to maintain low overall levels of phosphorylation. Nevertheless, even the modification of a small fraction of a highly abundant Rab protein, when confined to a specific subcellular region, can exert significant physiological effects. As discussed below, LRRK2-phosphorylated Rab proteins bind preferentially to a group of phospho-Rab-specific effector proteins, such as RILPL1, which exhibit stronger interactions with phosphorylated Rabs compared with their nonphosphorylated counterparts.

Most LRRK2 (~90%) is cytosolic (Biskup et al. 2006; Berger et al. 2010; Purlyte et al. 2018; Gomez et al. 2019); only a small fraction of LRRK2 is membrane-associated, and it appears that only this membrane-localized pool is catalytically active (Gomez et al. 2019). LRRK2 may appear to be inactive in cytosol because most of its substrates are bound to GDI and thus unavailable for modification. In addition to serving as substrates, Rabs also recruit LRRK2 to membrane domains that are rich in particular Rab substrates; they hold Rabs on those membranes to increase phosphorylation efficiency.

LRRK2 contains three distinct Rab-binding sites on the LRRK2 ARM domain (Fig. 1; for review, see Alessi and Pfeffer 2024). In addition to Rab8A (but less so, Rab10; Vides et al. 2022), site 1 interacts with Rab29 and its close homologs Rab32 and Rab38 (Fig. 2; McGrath et al. 2021). Notably, the Rab32 S71R variant, which has been linked to autosomal-dominant PD in rare families, enhances LRRK2 activation more strongly when overexpressed in cells (Gustavsson et al. 2024; Hop et al. 2024). Site 3 binds to Rab12 (Dhekne et al. 2023; Wang et al. 2023; Li et al. 2024), and disruption of this site significantly reduces the steady-state level of LRRK2 in cells—more so than disruption of site 1. Site 2 binds specifically to LRRK2-phosphorylated Rabs, forming a feedforward loop that promotes further LRRK2 recruitment and activation at membrane surfaces (Vides et al. 2022). The ARM domain can bind multiple Rabs simultaneously, and cooperative engagement at these sites enhances LRRK2's membrane association (Vides et al. 2022). Multivalent Rab binding is thought to be a key mechanism by which LRRK2 is selectively recruited to specific subcellular membranes (Fig. 2).

As discussed below, in macrophages, LRRK2 is activated and recruited to lysosomes following activation of the STING pathway, subsequent to lipidation of GABARAP, a member of the ATG8 protein family. Lysosome-decorating GABARAP binds to the ARM domain of LRRK2, facilitating its recruitment to lysosomal surfaces where LRRK2 becomes activated (Bentley-DeSousa et al. 2025). Thus, multiple mechanisms have evolved to control the recruitment of LRRK2 to specific membranes in cells.

Several groups have used membrane relocalization strategies to test whether LRRK2 can remain functional when redirected to nonnative cellular compartments (Gomez et al. 2019; Kluss et al. 2022). In one key experiment, Rab29 was artificially anchored to mitochondria, and it successfully recruited LRRK2 to the mitochondrial surface. LRRK2 appeared to become activated at this site, as unexpectedly phospho-Rab10 also accumulated on mitochondria in this setup (Gomez et al. 2019). Normally, Rab proteins localize to membranes where they encounter their cognate GEFs—which are absent from mitochondria—and are stabilized through GTP binding and effector interactions (Aivazian et al. 2006; Barr 2013). These results revealed that GDI, which is responsible for recycling Rabs, can promiscuously deliver Rabs to nonnative membranes. As discussed earlier, once a Rab is phosphorylated by LRRK2, it can no longer be retrieved by GDI (Steger et al. 2016),

leading to its entrapment at these inappropriate locations. Such mislocalization is especially relevant in the context of lysosome damage and repair. Under normal conditions, lysosomes lack Rab proteins on their surface. However, in cells undergoing lysosomal stress, LRRK2 and phospho-Rabs are frequently detected at lysosomes, suggesting a similar mechanism of mistargeted Rab trapping may be at play (discussed below).

Researchers often rely on phosphomimetic and/or nonphosphorylatable mutants of phosphoproteins to understand the consequences of a particular phosphorylation process. Using phosphomimetic (T73E) and nonphosphorylatable (T73A) Rab10 mutants, Dhekne et al. (2018) showed that both variants were nonfunctional. The TA mutant failed to localize to membranes and was poorly prenylated, and the T73E mutant that should behave like phospho-Rab10 protein does not effectively bind to its phospho-Rab effector, RILPL1, indicating that glutamate substitution does not mimic true phosphorylation. Similar observations were made with Rab8A: the TA mutant and T72E mutants were incorrectly localized and the hoped-for phosphomimetic TE form only poorly bound its phospho-Rab-specific binding partner. These findings underscore the perils of using phosphorylation mimicking mutations to study Rab GTPase phosphorylation.

REVERSAL OF LRRK2 ACTION BY PPM FAMILY PHOSPHATASES

Berndsen et al. (2019) carried out a phosphatome-wide small interfering RNA (siRNA) screen in human A549 cells, targeting 264 phosphatase catalytic subunits and 56 regulatory subunits to identify phosphatases that reverse LRRK2 Rab10 phosphorylation. This study made use of a highly specific anti-pRab10 antibody and included incubating cells for 5 min with MLi-2 LRRK2 kinase inhibitor to block Rab rephosphorylation and amplify detection of a Rab-specific phosphatase. The top hit from this screen was PPM1H, which is a member of the Mg^{2+}/Mn^{2+}-dependent, PPM family of protein phosphatases (Kamada et al. 2020).

Based on a substrate trapping form of PPM1A, Berndsen et al. (2019) created a substrate trapping PPM1H mutant protein. The substrate trapping PPM1H D288A specifically trapped only Rab8A and Rab10 as determined by mass spectrometry of immunoprecipitates, confirming the remarkable specificity of this phosphatase for pRab proteins. As described below, knockout (KO) of PPM1H showed the same phenotypes as hyperactive LRRK2 in cultured cells and in the brain.

PPM1H adopts a conserved phosphatase fold and features a unique 110-residue "flap domain" that contributes to its Rab substrate specificity (Waschbüsch et al. 2021). Golgi localization of PPM1H is mediated by an amino-terminal, 37-residue amphipathic helix, positioning it to encounter its primary targets: phosphorylated Rab8, Rab10, and Rab29 (Yeshaw et al. 2023). Under conditions of mitochondrial depolarization, a small fraction of PPM1H also localizes to mitochondrial surfaces (Berndsen et al. 2019; Bagnoli et al. 2025). In vitro reconstitution experiments demonstrated that this amphipathic helix alone is sufficient to target PPM1H to liposomes (Yeshaw et al. 2023). Moreover, PPM1H activity is enhanced in the presence of highly curved 50 nm liposomes, and this activation is dependent on the engagement of the amino-terminal helix with the liposome surface (Yeshaw et al. 2023). Although purified PPM1H is capable of dephosphorylating pRab12 in vitro, it does not efficiently dephosphorylate pRab12 in cells (Berndsen et al. 2019; Chiang et al. 2025). This discrepancy may be due to the limited colocalization of Rab12 with PPM1H under normal cellular conditions (Yeshaw et al. 2023). Supporting this idea, forced relocalization of both pRab12 and PPM1H to mitochondrial membranes in cells resulted in efficient dephosphorylation of pRab12 (Yeshaw et al. 2023).

Although PPM1H is highly specific for pRab8A and pRab10, close inspection of phospho-Rab dephosphorylation in PPM1H KO cells indicates that additional phosphatases also contribute to counteracting LRRK2 action, as pRab10 levels decline sharply even in the absence of PPM1H (Berndsen et al. 2019; Chiang et al. 2025). A subsequent phosphatome-wide screen in mouse 3T3 cells has revealed that the related PPM1M phosphatase shows preference for LRRK2-phosphorylated Rab12 (Chiang et al. 2025). PPM1M may not have been picked up in the original screen as it is expressed at barely detectable levels in A549 cells used in that study. Interestingly, loss of either PPM1H (Berndsen et al. 2019; Khan et al. 2021) or PPM1M (Chiang et al. 2025) influences phospho-Rab levels in cells and in mouse tissues, suggesting that both contribute and are not entirely redundant enzymes. Importantly, loss-of-function mutations in PPM1H and PPM1M may eventually be linked to PD as they are seen in PD cases (Khan et al. 2021; Chiang et al. 2025). Given their roles in counteracting LRRK2 signaling, these phosphatases represent potential therapeutic targets, provided agents can be developed to enhance their activity.

LRRK2 MUTATIONS ARE GAIN-OF-FUNCTION

Multiple lines of evidence support the conclusion that PD-associated LRRK2 mutations act through a gain-of-function mechanism. All pathogenic variants, including G2019S and R1441C/G, increase LRRK2 kinase activity, leading to excessive phosphorylation of Rab GTPases and disruption of critical cellular processes described below. Rather than impairing LRRK2's normal function, these mutations enhance its enzymatic activity, driving cellular dysfunction. Genetic studies reinforce this interpretation: KO of the phosphatases PPM1H or PPM1M, both of which normally counteract LRRK2 signaling, phenocopies the effects of hyperactive LRRK2 mutants in the nigrostriatal circuit. This suggests that the observed phenotypes are due to elevated levels of phosphorylated Rab proteins. Further supporting this model, a recent study showed that 3 months of treatment with LRRK2 inhibitors in mice carrying hyperactive LRRK2 mutations reversed all associated brain phenotypes (Jaimon et al. 2025). These findings provide strong rationale for the continued development of LRRK2 kinase inhibitors as promising disease-modifying therapies for PD.

PHOSPHO-RAB-SPECIFIC EFFECTORS

Mass spectrometry analyses of Rab GTPase immunoprecipitates, performed with and without LRRK2 kinase inhibition, identified a group of proteins that preferentially bind phosphorylated Rabs over their nonphosphorylated counterparts (Steger et al. 2017). This approach revealed RILPL1, RILPL2, EHBP1L1, and SPAG9/JIP4 as binding partners of phospho-Rab8A; notably, phospho-Rab10 and phospho-Rab12 also interact with RILPL1 and RILPL2. Earlier work by Matsui et al. (2012) characterized a Rab-interacting lysosomal protein (RILP) homology domain (RH2) shared by RILPL1, RILPL2, JNK-interacting protein 3 (JIP3), and JNK-interacting protein 4 (JIP4). JIP3 and JIP4 serve as adaptors for cargo transport via microtubule-based dynein/dynactin and kinesin-1 motors, which drive vesicle trafficking in opposite directions. These RH2-containing proteins form dimers and bind to LRRK2-phosphorylated Rab8 and Rab10 via conserved basic residues within the RH2 motif that interact specifically with the phosphate group on the modified Rab (Waschbüsch et al. 2020).

RILPL2 is also a known interacting partner of Myosin Va (MyoVa). It binds the globular tail domain of MyoVa through its amino-terminal RH1 domain (Lisé et al. 2009) and engages phosphorylated Rab8 or Rab10 through its carboxy-terminal RH2 domain (residues 129–165; Waschbüsch et al. 2020). Additionally, RILPL2 has been shown to regulate the motor activity of MyoVa (Cao et al. 2019). This dual binding ability allows RILPL2 to act as an adaptor linking MyoVa to phospho-Rabs anchored on membranes, thereby coordinating actin-based vesicle transport. MyoVa itself directly binds phospho-Rab10 with high affinity (475 nM), in contrast to its much weaker binding to phospho-Rab8A (6.6 μM; Dhekne et al. 2021). This strong interaction results in sequestration of MyoVa at the mother centriole in cells expressing hyperactive LRRK2 (Dhekne et al. 2021). While more phospho-Rab effectors will likely be identified, those known so far clearly function in the regulation of cytoskeletal motor proteins and organelle motility, underscoring the central role of phospho-Rab signaling in intracellular transport dynamics.

CELL-TYPE-SPECIFIC ROLES FOR LRRK2

As discussed below, recent studies highlight two major categories of LRRK2 activation consequences. Especially in lysosome-rich macrophages that express high levels of LRRK2, Rab29, and Rab32 (but also in other cell types), LRRK2 is recruited to damaged lysosomes (Eguchi et al. 2018; Bonet-Ponce et al. 2020; Herbst et al. 2020; Kuwahara and Iwatsubo 2024) and likely contributes to lysosome and lysosome-related organelle (LRO) exocytosis (Bogacki et al. 2025). In addition, in certain classes of neurons and astrocytes, LRRK2 hyperactivation inhibits primary ciliogenesis (Dhekne et al. 2018; Lara Ordóñez et al. 2019; Brahmia et al. 2024), disrupting downstream Sonic Hedgehog (Shh) signaling pathways that require signaling-competent cilia.

Primary Cilia Loss

When a subset of Rab family LRRK2 substrates was first identified (Steger et al. 2016, 2017), several were noted to share a common link to the process of primary cilia formation. Primary cilia are solitary, nonmotile organelles that extend from the surface of most mammalian cell types, including neurons and astrocytes (Fig. 3). These microtubule-based structures originate from the mother centriole and typically measure 3–10 μm in length and ~0.2 μm in diameter. Cilia form during the G_0 or G_1 phase of the cell cycle in nondividing cells. They function as cellular antennae, playing crucial roles in development, tissue homeostasis, and organ function by mediating signaling pathways such as Shh (Mill et al. 2023; Hilgendorf et al. 2024). It was recently discovered that cilia may contact as many as 40 different neighboring neurons (Wu et al. 2024), and these interactions appear to involve functional synapses, explaining the presence of multiple neuropeptide and neurotransmitter receptors on the surfaces of primary cilia (Sheu et al. 2022; Wu et al. 2024).

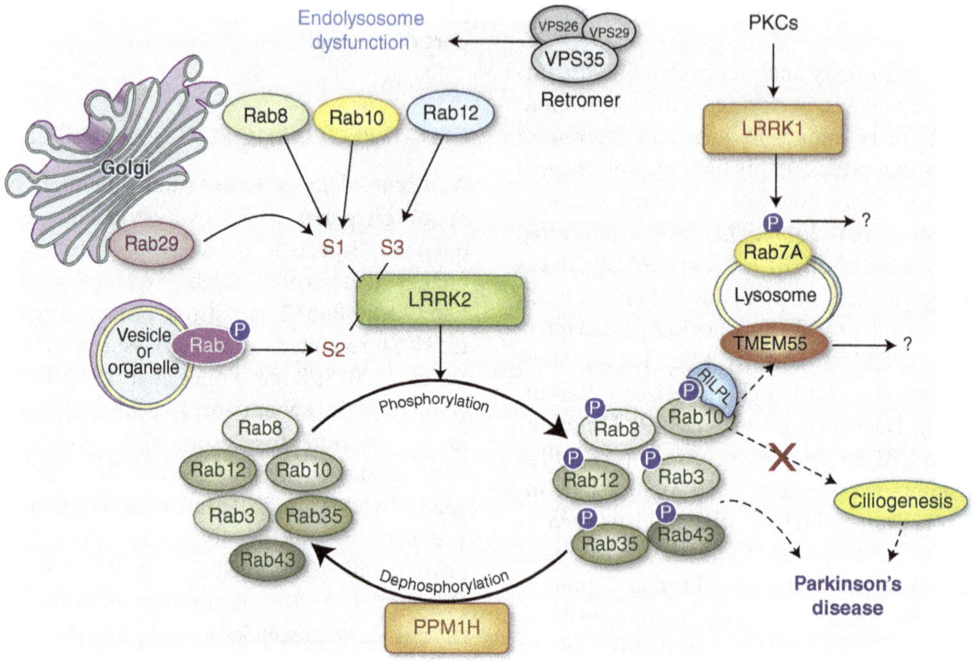

Figure 2. A summary of our knowledge of the biochemistry of the leucine-rich repeat kinase 2 (LRRK2) pathway. S1, S2, and S3 are Rab or phospho-Rab (site 2) binding sites in LRRK2. LRRK2 substrate Rabs are shown; endolysosomal Rab7A is a LRRK1 substrate. Lysosome stress mislocalizes LRRK2 and phospho-Rabs to lysosome surfaces. Protein kinase C (PKC) activates LRRK1. (For further details, see Alessi and Pfeffer 2024; reprinted from Alessi and Pfeffer 2024, © The Authors, under a Creative Commons Attribution 4.0 International License.)

Rab phosphorylation by LRRK2 kinase blocks the process of ciliogenesis (Steger et al. 2017; Dhekne et al. 2018; Lara Ordónez et al. 2019; Brahmia et al. 2024). Inhibition requires pRab10 and its phospho-Rab effector, RILPL1 (but not RILPL2; see also Li et al. 2024). Inhibition has been reported in multiple cultured cell lines and human LRRK2 G2019S-induced pluripotent stem cells (iPSCs), but only in certain cell types in the brains of *LRRK2* mutant mice. Importantly, loss of cilia in cultured cells is directly proportional to the loss of Shh signaling (Dhekne et al. 2018) that is known to require primary cilia (Bangs and Anderson 2017). Proof for a role for wild-type LRRK2 in regulating ciliogenesis comes from the finding that loss of the LRRK2-counteracting PPM1H or PPM1M Rab phosphatases decreases ciliation in wild-type mouse embryo fibroblasts (Berndsen et al. 2019; Chiang et al. 2025) and in specific cell types in the brain (Khan et al. 2024; Chiang et al. 2025).

How LRRK2 blocks primary cilia formation remains poorly understood. LRRK2-mediated Rab10 phosphorylation blocks ciliogenesis at the earliest stages: RILPL1-phospho-Rab10 complexes interfere with the recruitment of TTBK2 and the removal of CP110 (Sobu et al. 2021), a protein that caps the mother centriole and must be displaced for primary cilia formation (Goetz et al. 2012). By blocking this step, LRRK2 effectively prevents ciliogenesis, impairing Shh signaling. Live cell microscopy revealed that LRRK2 also increases the rate of cilia loss in a Rab10- and RILPL1-independent manner, possibly by phosphorylation of another key Rab such as Rab8A. In addition to disrupting primary cilia, in certain cell types, phospho-Rab10-RILPL1 complexes also disrupt centriolar cohesion (Madero-Pérez et al. 2018), the tight association between the two centrioles by interfering with the proper centrosomal localization of CDK5RAP2 (Fdez et al. 2022).

Ciliary Signaling in the Nigrostriatal Circuit

How would cilia loss link to PD? A hallmark of PD is the degeneration of dopaminergic neurons in the substantia nigra that project to the striatum. This has drawn attention to the importance of Shh signaling within the nigrostriatal circuit. In 2012, Kottmann and colleagues demonstrated that dopaminergic neurons depend on Shh signaling for survival, despite the fact that they do not express Shh receptors (Fig. 4; Gonzalez-Reyes et al. 2012). Instead, they project to the dorsal striatum and secrete Shh onto cholinergic and parvalbumin interneurons. In response, these interneurons produce two potent neuroprotective factors: glial cell–derived neurotrophic factor (GDNF) from cholinergic neurons and Neurturin (NRTN) from parvalbumin neurons (Fig. 4; Gonzalez-Reyes et al. 2012). Disruption of Shh expression in dopaminergic neurons leads to progressive degeneration of both dopaminergic and striatal cholinergic neurons, resulting in motor symptoms reminiscent of PD (Gonzalez-Reyes et al. 2012). This reciprocal signaling loop is essential for maintaining structural and neurochemical homeostasis in the nigrostriatal pathway.

In both LRRK2 pathway mutant mice and patients with idiopathic PD, it is specifically the Shh signal-receiving cholinergic and parvalbumin interneurons in the striatum that lose their primary cilia, rendering them unresponsive to Shh (Dhekne et al. 2018; Khan et al. 2021, 2024; Lin et al. 2025). Curiously, the much more abundant medium spiny neurons that also express LRRK2 at high levels retain their cilia in both LRRK2 mutant mice and postmortem human PD brain (Dhekne et al. 2018; Khan et al. 2021, 2024; Lin et al. 2025). However, within the populations of cholinergic and parvalbumin interneurons, those with the highest LRRK2 expression levels show the most pronounced cilia loss. Thus, factors beyond LRRK2 expression levels influence a cell's vulnerability to ciliary disruption.

Single-nucleus RNA sequencing and fluorescence in situ hybridization analyses revealed that cilia-deficient cholinergic and parvalbumin interneurons show reduced expression of GDNF and NRTN, respectively, two neurotrophic factors essential for dopaminergic neuron support (Khan et al. 2024; Lin et al. 2025; Nair et al. 2025). Additionally, LRRK2 mutations were found to reduce brain-derived neurotrophic factor (BDNF) production by striatal astrocytes (Nair et al. 2025). These observations emphasize the pivotal role of primary cilia in sustaining neuroprotective signaling in the nigrostriatal circuit and suggest that cilia dysfunction in specific striatal cell types may contribute to dopaminergic neuron vulnerability in PD (Fig. 4).

In summary, LRRK2 mutations disrupt cilia formation in cholinergic and parvalbumin interneurons and astrocytes in the dorsolateral striatum. This ciliary loss leads to decreased secretion of GDNF, NRTN, and BDNF, resulting in cell loss, particularly among the interneuron populations, in both mice and humans. These findings directly support the model proposed by Gonzalez-Reyes et al. (2012), describing a non-cell-autonomous circuit wherein Shh signaling from dopaminergic neurons promotes reciprocal neurotrophic factor production by striatal interneurons.

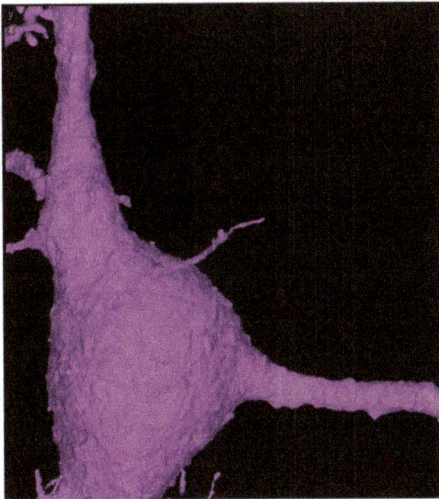

Figure 3. Primary cilia on brain cortical neurons. Shown is a primary cilium extending from a layer 2 projection neuron. Axons from as many as 40 other neurons contact the primary cilium and as many as half for synapse-like connections. (Reprinted, with permission, from Wu et al. 2024, © Elsevier.)

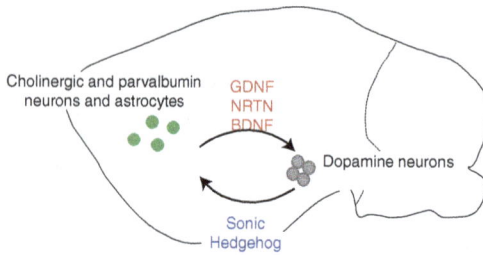

Figure 4. Sonic Hedgehog signaling in the nigrostriatal circuit. See text for details. (GDNF) Glial cell-derived neurotrophic factor, (NRTN) Neurturin, (BDNF) brain-derived neurotrophic factor. (Figure based on data in Gonzalez-Reyes et al. 2012.)

Further work by Kottmann and colleagues explored the impact of diminished Shh signaling, resulting from midbrain dopaminergic neuron degeneration, on the development of levodopa-induced dyskinesia (LID), a common side effect of long-term levodopa (L-DOPA) treatment (Malave et al. 2021). They showed that pharmacological activation of Shh signaling reduced involuntary movements in both neurotoxic and genetic models of PD. Conversely, suppressing Shh secretion from dopaminergic neurons or blocking its reception by cholinergic interneurons exacerbated LID. Remarkably, enhancing Shh signaling in cholinergic interneurons rendered sensitized models resistant to LID. These findings suggest that restoring or augmenting Shh signaling may represent a promising therapeutic strategy to reduce dyskinesia in Parkinson's patients undergoing L-DOPA therapy.

OTHER CELL-TYPE-SPECIFIC ROLES FOR LRRK2: LYSOSOME-RELATED ORGANELLE RELEASE

Lysosome-Related Organelles and LRRK2

Lysosome-related organelles (LROs) are specialized organelles that share structural and biochemical features with conventional lysosomes but are adapted to perform cell-type-specific functions. Examples of LROs include the following: melanosomes in melanocytes, which transport melanin for pigmentation; secretory gran-

ules in cytotoxic T lymphocytes, which release perforin and granzymes; platelet dense granules, which store factors essential for blood clotting; lamellar bodies in lung type II pneumocytes, which store and secrete surfactant to reduce alveolar surface tension; and maturing phagosomes in macrophages and neutrophils, which are involved in microbial degradation and immune defense (Marks et al. 2013). LRRK2 has been implicated in the regulation of secretory events involving LROs in at least some of these specialized cell types. Its specific role is not yet clear but could reflect a role in vesicle trafficking, membrane remodeling, and/or regulated exocytosis, all critical to proper LRO function.

Early *LRRK2* KO mouse models revealed a striking phenotype by 6 weeks of age, characterized by a marked increase in both the number and size of secondary lysosomes in kidney proximal tubule cells, as well as lamellar bodies in lung type II pneumocytes (Tong et al. 2010; Herzig et al. 2011). A critical role for LRRK2 in lung function was further supported by studies in nonhuman primates treated with LRRK2 inhibitors (Fuji et al. 2015; Baptista et al. 2020; Miller et al. 2023). These animals developed early signs of lung fibrosis, attributed to the accumulation of pulmonary surfactant within type II pneumocytes. Similar pathological changes were observed in LRRK2 inhibitor–treated mice (Bryce et al. 2021), underscoring the importance of LRRK2 activity in regulating surfactant processing and secretion in the lung.

Administration of the LRRK2 inhibitor DNL201 in phase 1 and 1b clinical trials led to a marked decrease in the urinary levels of the lysosomal lipid bis(monoacylglycerol)phosphate (BMP) in both healthy volunteers and PD patients (Jennings et al. 2022). This observation suggests that LRRK2 inhibition impairs the exocytic release of lysosome-derived vesicles from kidney cells into the urine (Fuji et al. 2015). Supporting this model, BMP levels are elevated in the urine of both manifesting and nonmanifesting carriers of LRRK2 pathogenic mutations, consistent with increased lysosomal exocytosis driven by LRRK2 hyperactivity (Alcalay et al. 2020).

The precise role of LRRK2 in LRO exocytosis remains incompletely understood. Notably, as discussed above, Rab29 and its closely related family members Rab32 and Rab38 bind to the LRRK2 ARM domain (Waschbüsch et al. 2014; McGrath et al. 2021). These Rab proteins exhibit tissue-specific expression and are localized to LRO surfaces in particular cell types. This raises the possibility that tissue-specific Rabs may recruit LRRK2 to the surfaces of secretory granules or other LROs, where LRRK2 could phosphorylate additional Rab substrates. These phospho-Rabs could then act as docking platforms for motor protein adaptors such as RILPL2, which binds both phospho-Rabs and the actin-based motor protein MyoVa (Dhekne et al. 2021). The possible relevance of this model is underscored by evidence that MyoVa plays a role in secretory vesicle release in chromaffin cells (Desnos et al. 2007). In this scenario, LRRK2-mediated Rab phosphorylation would enhance MyoVa recruitment to the vesicle surface, thereby facilitating vesicle docking and subsequent exocytosis.

Special Roles for LRRK2 in Macrophages

LRRK2 plays distinct roles in macrophages, where both LRRK2 and its Rab29 activator are highly expressed. In macrophages and microglia, the LRRK2 G2019S mutation suppresses lysosomal degradative function by inhibiting the expression and nuclear localization of the MiT-TFE family of transcription factors, which regulate numerous lysosomal genes (Yadavalli and Ferguson 2023). These findings suggest that LRRK2 hyperactivation may impair lysosomal homeostasis, thereby increasing the risk of PD.

In macrophages, LRRK2 is recruited to lysosomes and activated in response to lysosomotropic agents such as chloroquine (Eguchi et al. 2018). It is also activated upon infection with pathogens like *Mycobacterium tuberculosis*, *Listeria monocytogenes*, and *Candida albicans*, which trigger endomembrane damage (Herbst et al. 2020). How this impacts patients is not yet clear.

Lysosomes routinely experience low-level membrane damage, repaired through mechanisms similar to those used for plasma membrane repair. One rapid repair response involves the ESCRT machinery, which accumulates on lysosomal membranes within seconds of LLOME-induced rupture in a calcium-dependent manner (Scheffer et al. 2014; Radulovic et al. 2018; Skowyra et al. 2018). In parallel, calcium-induced sphingomyelin scrambling helps reseal minor lesions. Another pathway involves recruitment of PI4K2A to damaged lysosomes, where it generates phosphatidylinositol 4-phosphate. This lipid recruits oxysterol-binding proteins such as OSBP and ORP1L to mediate ER-to-lysosome transfer of phosphatidylserine and cholesterol, aiding membrane repair (Tan and Finkel 2022). Additionally, the lipid transfer protein ATG2 and VPS13C (Wang et al. 2025) contribute by catalyzing lipid transfer from the ER to damaged lysosomes.

LRRK2 is recruited to damaged lysosomes but more slowly than VPS13C and OSBP (Wang et al. 2025), in association with activation of the conjugation of ATG8 to single membranes (CASM) pathway. CASM is a noncanonical autophagy-related process in which ATG8 proteins such as LC3 and GABARAP are lipidated and conjugated to the cytoplasmic surface of endolysosomal membranes. CASM is activated by lysosome damage, STING pathway signaling, and agonists of the lysosomal Ca^{2+} channel TRPML1—all of which perturb lysosomal ion and pH homeostasis (Durgan and Florey 2022).

Recent findings from Ferguson's group and others (Kuwahara and Iwatsubo 2024; Bentley-DeSousa et al. 2025; Huang et al. 2025) demonstrate that CASM promotes the recruitment and activation of LRRK2 at lysosomes. This process appears to be mediated through a direct interaction between the lipidated form of the ATG8 family member, GABARAP, and the ARM domain of LRRK2 (Bentley-DeSousa et al. 2025). Following this recruitment, phospho-Rabs accumulate on lysosomes, likely because once Rabs are phosphorylated by LRRK2, they can no longer be retrieved by GDIs. The presence of phospho-Rabs at the lysosome thus serves as a sign of LRRK2 activation at this compartment.

The physiological roles of lysosome-localized LRRK2 and phospho-Rabs are not yet fully understood. One study has shown that the PD-linked VPS35[D620N] mutation, which impacts the function of the endosome-Golgi and endosome-plasma membrane cargo sorting retromer complex, also causes lysosomal dysfunction and stimulates LRRK2 recruitment and Rab phosphorylation at the lysosome (Pal et al. 2023). This, in turn, leads to the recruitment of the phospho-Rab effector RILPL1 to the lysosomal surface, where it binds to TMEM55B—a lysosomal integral membrane protein (Pal et al. 2023). TMEM55B also interacts with other LRRK2 pathway–associated proteins implicated in PD, including JIP3/4 and VPS13C (Willett et al. 2017).

The consequences of these molecular events for lysosome function remain unclear. One possibility is that phospho-Rabs on damaged lysosomes promote exocytic clearance, analogous to their proposed role on LROs. Supporting this hypothesis, Eguchi et al. (2018) reported LRRK2-dependent changes in extracellular lysosomal enzyme levels following chloroquine treatment. Further work is needed to define the mechanisms and functional outcomes of LRRK2 and phospho-Rab accumulation at damaged lysosomes.

CONVERGENT PATHWAYS TO PD

Current genome-wide association studies have identified a large number of genetic variants associated with PD, many of which converge on common cellular pathways. Here, we highlight two key PD risk factors, mutations in glucosylceramidase β1 (GBA1) and PTEN-induced kinase 1 (PINK1), that impact a shared pathway: Shh signaling in the nigrostriatal circuit (Fig. 5).

GBA1, is among the most common genetic risk factors for PD. Up to 10% of PD patients carry heterozygous loss-of-function mutations in *GBA1* (Westenberger et al. 2024). These mutations impair GBA1 activity, leading to lysosomal accumulation of undegraded substrates such as glucosylceramide and lysosomal dysfunction. This disrupts autophagy and other cellular clearance pathways, contributing to the accumulation of α-synuclein, a pathological hallmark of PD. Importantly, while *GBA1* mutations do not directly cause PD, they increase disease risk and accelerate progression, often leading to earlier onset and more rapid cognitive decline.

Although impaired lysosomal function provides a plausible explanation for the link between *GBA1* mutations and PD, it does not explain why mutations in other lysosomal enzymes that lead to lysosome dysfunction are not similarly associated with PD risk. Recent studies suggest an alternative mechanism involving cholesterol homeostasis. The plasma membrane contains three cholesterol pools: one bound to proteins or sphingomyelin, one critical for membrane integrity, and a third, "accessible" pool that is chemically active and detectable by cholesterol-sensitive probes. These pools are interconvertible, depending on lipid composition (Das et al. 2014; Radhakrishnan et al. 2020).

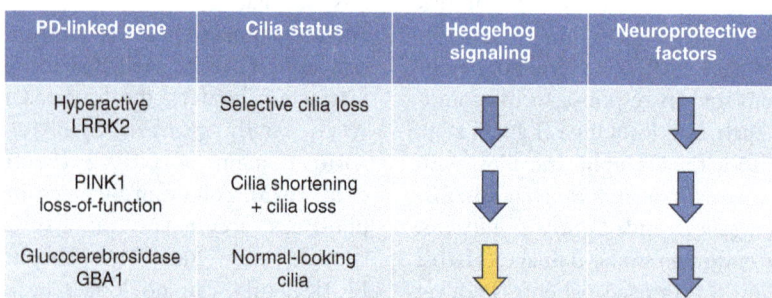

PD-linked gene	Cilia status	Hedgehog signaling	Neuroprotective factors
Hyperactive LRRK2	Selective cilia loss	⬇	⬇
PINK1 loss-of-function	Cilia shortening + cilia loss	⬇	⬇
Glucocerebrosidase GBA1	Normal-looking cilia	⬇	⬇

Figure 5. Convergent pathways to Parkinson's disease (PD). (For details, see text and Dhekne et al. 2021; Khan et al. 2021, 2024; Bagnoli et al. 2025; Nair et al. 2025.)

Relevant to PD, Rohatgi and colleagues demonstrated that accessible cholesterol at the ciliary membrane is essential for activating the Shh signaling pathway (Radhakrishnan et al. 2020; Kinnebrew et al. 2021, 2022).

Nair et al. (2025) demonstrated that in multiple patient-derived, heterozygous or compound heterozygous fibroblasts carrying *GBA1* mutations, substrate accumulation leads to a buildup of accessible cholesterol within lysosomes. This depletes accessible cholesterol from the plasma membrane and ciliary membrane. As a result, Shh signaling is impaired, despite the presence of structurally intact cilia. In both cultured cells and the striatum of *Gba1* mutant mice, this disruption leads to reduced expression of neuroprotective factors and fewer fine dopaminergic processes, phenotypes that closely mirror those seen in *LRRK2* mutants. These findings suggest a convergent mechanism by which *GBA1* and *LRRK2* mutations disrupt Shh signaling and contribute to PD pathogenesis, offering new insights into the impact of lysosomal dysfunction.

PINK1 mutations are a known cause of autosomal-recessive early-onset PD. Bagnoli et al. (2025) reported that *PINK1* KO mice exhibit defective ciliogenesis in striatal cholinergic interneurons and astrocytes, impairing Shh-dependent induction of GDNF transcription. The resulting loss of cilia led to reduced GDNF expression. Importantly, this phenotype was not worsened in double-mutant *LRRK2* and *PINK1* mice, suggesting that PINK1 affects ciliogenesis via a mechanism distinct from LRRK2. However, both mutations converge on the same outcome: impaired Shh signaling.

Together, these findings point to a critical neuroprotective circuit involving Shh signaling in the nigrostriatal pathway. It is likely that additional PD risk genes will be found to converge on this pathway, reinforcing its importance in disease pathogenesis.

A distinctive feature of *LRRK2*-associated PD is that 30% of individuals carrying the G2019S mutation and 70% of carriers of other mutations (including I2020T, R1441C, R1441G, and Y1699C) do not exhibit Lewy body pathology, despite clear evidence of dopaminergic

neuron degeneration (Kalia et al. 2015). Additionally, patients with LRRK2 PD show a lower rate of positive α-synuclein seed amplification assay results (Siderowf et al. 2023). These findings suggest that a critical driver of LRRK2 PD may be the selective loss of neuroprotection within the nigrostriatal circuit, which is essential for dopamine neuron survival. Interestingly, even in the apparent absence of synuclein aggregation in some LRRK2 mutation carriers, a recent study reported TDP-43 aggregates in 73% of LRRK2 PD brains compared to only 18% in non-LRRK2 PD cases (Agin-Liebes et al. 2023), highlighting lysosomal dysfunction as another important area for future investigation related to LRRK2 PD pathogenesis.

SUMMARY AND FUTURE PERSPECTIVES

The past decade has brought transformative progress in our understanding of LRRK2 structure and function, and how Rabs lie at the nexus of the cellular pathways LRRK2 influences. Structural studies have revealed key insights into how pathogenic mutations enhance kinase activity by destabilizing the inactive conformation, while mechanistic investigations have begun to show how LRRK2 is activated, recruited to membranes, and targeted to specific subcellular compartments. It is now evident that the consequences of LRRK2 activity are cell-type-specific. For example, disease-linked *LRRK2* mutations selectively disrupt Shh signaling in rare populations of striatal cholinergic and parvalbumin interneurons by causing the loss of primary cilia—an effect not seen in the more abundant medium spiny neurons. In (nonciliated) macrophages that contain high levels of LRRK2 and Rab29 and Rab32, a significant fraction of LRRK2 becomes localizes to lysosomes and is activated there upon lysosome damage. This striking selectivity underscores the importance of cellular context in determining LRRK2's impact and raises several pressing questions for the field: How does organelle dysfunction, particularly at the lysosome, lead to LRRK2 recruitment and activation? What are the precise molecular events that activate LRRK2 on membranes? How does LRRK2 negatively regulate GBA1 activity, and how is this mechanistically linked to PD pathology? What role does LRRK2 play in lysosome and LRO exocytosis, and how does this intersect with its ability to respond to lysosomal stress or damage? How does chronic LRRK2 activation contribute to lysosomal dysfunction, and what are the long-term consequences for macro-autophagy and selective pathways such as mitophagy? Are mitochondrial defects driven by impaired mitophagy or via other indirect pathways downstream from LRRK2 hyperactivation? What are the molecular mechanisms by which LRRK2 regulates ciliogenesis, and which ciliated cell types are most relevant to PD pathology? Does LRRK2 activity in peripheral immune cells and microglia contribute to neuroinflammatory processes that drive PD pathogenesis? Do certain individuals with nonfamilial PD exhibit underlying LRRK2 pathway dysfunction, and how can these patients be identified and stratified to determine whether they would benefit most from LRRK2-targeted therapies? Could administering LRRK2 inhibitor therapy to individuals with hyperactivating LRRK2 variants prior to symptom onset help prevent or delay the development of PD? This will require the development of robust biomarkers and stratification tools to identify patients with LRRK2-driven disease. Continued progress in answering these questions will deepen our understanding of LRRK2 biology and enable more precise therapeutic strategies aimed at halting or reversing disease progression in Parkinson's patients.

COMPETING INTEREST STATEMENT

The authors declare no competing interests.

ACKNOWLEDGMENTS

We apologize to any authors whose relevant work was not cited due to space limitations. Current work in our laboratories is funded by the joint efforts of The Michael J. Fox Foundation for Parkinson's Research (MJFF) and Aligning Science Across Parkinson's (ASAP) initiative and the LRRK2 Investigative Therapeutics Exchange (LITE) initiative. MJFF administers the grant (ASAP-000463) on behalf of ASAP and itself (to D.R.A. and S.R.P.). The University of Dundee administered the grant (Grant ID: MJFF-025924) on behalf of LITE and itself. Funds were also provided by the Medical Research Council (grant no. MC_UU_00018/1 to D.R.A.).

REFERENCES

Agin-Liebes J, Hickman RA, Vonsattel JP, Faust PL, Flowers X, Utkina Sosunova I, Ntiri J, Mayeux R, Surface M, Marder K, et al. 2023. Patterns of TDP-43 deposition in brains with *LRRK2 G2019S* mutations. *Mov Disord* **38:** 1541–1545. doi:10.1002/mds.29449

Aivazian D, Serrano RL, Pfeffer S. 2006. TIP47 is a key effector for Rab9 localization. *J Cell Biol* **173:** 917–926. doi:10.1083/jcb.200510010

Alcalay RN, Hsieh F, Tengstrand E, Padmanabhan S, Baptista M, Kehoe C, Narayan S, Boehme AK, Merchant K. 2020. Higher urine bis(monoacylglycerol)phosphate levels in LRRK2 G2019S mutation carriers: implications for therapeutic development. *Mov Disord* **35:** 134–141. doi:10.1002/mds.27818

Alessi DR, Pfeffer SR. 2024. Leucine-rich repeat kinases. *Annu Rev Biochem* **93:** 261–287. doi:10.1146/annurev-biochem-030122-051144

Bagnoli E, Lin YE, Burel S, Jaimon E, Antico O, Themistokleous C, Nikoloff JM, Squires S, Morella I, Watzlawik JO, et al. 2025. Endogenous LRRK2 and PINK1 function in a convergent neuroprotective ciliogenesis pathway in the brain. *Proc Natl Acad Sci* **122:** e2412029122. doi:10.1073/pnas.2412029122

Bangs F, Anderson KV. 2017. Primary cilia and mammalian hedgehog signaling. *Cold Spring Harb Perspect Biol* **9:** a028175. doi:10.1101/cshperspect.a028175

Baptista MAS, Merchant K, Barrett T, Bhargava S, Bryce DK, Ellis JM, Estrada AA, Fell MJ, Fiske BK, Fuji RN, et al. 2020. LRRK2 inhibitors induce reversible changes in nonhuman primate lungs without measurable pulmonary deficits. *Sci Transl Med* **12:** eaav0820. doi:10.1126/scitranslmed.aav0820

Barr FA. 2013. Review series: Rab GTPases and membrane identity: causal or inconsequential? *J Cell Biol* **202:** 191–199. doi:10.1083/jcb.201306010

Bentley-DeSousa A, Roczniak-Ferguson A, Ferguson SM. 2025. A STING–CASM–GABARAP pathway activates LRRK2 at lysosomes. *J Cell Biol* **224:** e202310150. doi:10.1083/jcb.202310150

Berger Z, Smith KA, Lavoie MJ. 2010. Membrane localization of LRRK2 is associated with increased formation of the highly active LRRK2 dimer and changes in its phosphorylation. *Biochemistry* **49:** 5511–5523. doi:10.1021/bi100157u

Berndsen K, Lis P, Yeshaw WM, Wawro PS, Nirujogi RS, Wightman M, Macartney T, Dorward M, Knebel A, Tonelli F, et al. 2019. PPM1H phosphatase counteracts LRRK2 signaling by selectively dephosphorylating Rab proteins. *eLife* **8:** e50416. doi:10.7554/eLife.50416

Cite this article as *Cold Spring Harb Perspect Med* doi: 10.1101/cshperspect.a041620

Biskup S, Moore DJ, Celsi F, Higashi S, West AB, Andrabi SA, Kurkinen K, Yu SW, Savitt JM, Waldvogel HJ, et al. 2006. Localization of LRRK2 to membranous and vesicular structures in mammalian brain. *Ann Neurol* **60:** 557–569. doi:10.1002/ana.21019

Bogacki EC, Longmore G, Lewis PA, Herbst S. 2025. GPNMB is a biomarker for lysosomal dysfunction and is secreted via LRRK2-modulated lysosomal exocytosis. bioRxiv doi:10.1101/2025.01.01.630988

Bonet-Ponce L, Beilina A, Williamson CD, Lindberg E, Kluss JH, Saez-Atienzar S, Landeck N, Kumaran R, Mamais A, Bleck CKE, et al. 2020. LRRK2 mediates tubulation and vesicle sorting from lysosomes. *Sci Adv* **6:** eabb2454. doi:10.1126/sciadv.abb2454

Brahmia B, Naaldijk Y, Sarkar P, Parisiadou L, Hilfiker S. 2024. Pathogenic LRRK2 causes age-dependent and region-specific deficits in ciliation, innervation and viability of cholinergic neurons. *eLife* **13:** RP101135. doi:10.7554/eLife.101135.1

Bryce DK, Ware CM, Woodhouse JD, Ciaccio PJ, Ellis JM, Hegde LG, Kuruvilla S, Maddess ML, Markgraf CG, Otte KM, et al. 2021. Characterization of the onset, progression, and reversibility of morphological changes in mouse lung after pharmacological inhibition of leucine-rich kinase 2 kinase activity. *J Pharmacol Exp Ther* **377:** 11–19. doi:10.1124/jpet.120.000217

Cao QJ, Zhang N, Zhou R, Yao LL, Li XD. 2019. The cargo adaptor proteins RILPL2 and melanophilin co-regulate myosin-5a motor activity. *J Biol Chem* **294:** 11333–11341. doi:10.1074/jbc.RA119.007384

Chiang CY, Pratuseviciute N, Lin YE, Adhikari A, Yeshaw WM, Flitton C, Sherpa PL, Tonelli F, Rektorova I, Lynch T, et al. 2025. PPM1M, a LRRK2-counteracting, phosphoRab12-preferring phosphatase with potential link to Parkinson's disease. bioRxiv doi:10.1101/2025.03.19.644182

Correia Guedes L, Ferreira JJ, Rosa MM, Coelho M, Bonifati V, Sampaio C. 2010. Worldwide frequency of G2019S LRRK2 mutation in Parkinson's disease: a systematic review. *Parkinsonism Relat Disord* **16:** 237–242. doi:10.1016/j.parkreldis.2009.11.004

Dächsel JC, Farrer MJ. 2010. LRRK2 and Parkinson disease. *Arch Neurol* **67:** 542–547. doi:10.1001/archneurol.2010.79

Das A, Brown MS, Anderson DD, Goldstein JL, Radhakrishnan A. 2014. Three pools of plasma membrane cholesterol and their relation to cholesterol homeostasis. *eLife* **3:** e02882. doi:10.7554/eLife.02882

Deniston CK, Salogiannis J, Mathea S, Snead DM, Lahiri I, Matyszewski M, Donosa O, Watanabe R, Böhning J, Shiau AK, et al. 2020. Structure of LRRK2 in Parkinson's disease and model for microtubule interaction. *Nature* **588:** 344–349. doi:10.1038/s41586-020-2673-2

Desnos C, Huet S, Fanget I, Chapuis C, Böttiger C, Racine V, Sibarita JB, Henry JP, Darchen F. 2007. Myosin Va mediates docking of secretory granules at the plasma membrane. *J Neurosci* **27:** 10636–10645. doi:10.1523/JNEUROSCI.1228-07.2007

Dhekne HS, Yanatori I, Gomez RC, Tonelli F, Diez F, Schüle B, Steger M, Alessi DR, Pfeffer SR. 2018. A pathway for Parkinson's disease LRRK2 kinase to block primary cilia

and Sonic hedgehog signaling in the brain. *eLife* **7:** e40202. doi:10.7554/eLife.40202

Dhekne HS, Yanatori I, Vides EG, Sobu Y, Diez F, Tonelli F, Pfeffer SR. 2021. LRRK2-phosphorylated rab10 sequesters Myosin Va with RILPL2 during ciliogenesis blockade. *Life Sci Alliance* **4:** e202101050. doi:10.26508/lsa.202101050

Dhekne HS, Tonelli F, Yeshaw WM, Chiang CY, Limouse C, Jaimon E, Purlyte E, Alessi DR, Pfeffer SR. 2023. Genome-wide screen reveals Rab12 GTPase as a critical activator of Parkinson's disease-linked LRRK2 kinase. *eLife* **12:** e87098. doi:10.7554/eLife.87098

Dihanich S, Civiero L, Manzoni C, Mamais A, Bandopadhyay R, Greggio E, Lewis PA. 2014. GTP binding controls complex formation by the human ROCO protein MASL1. *FEBS J* **281:** 261–274. doi:10.1111/febs.12593

Durgan J, Florey O. 2022. Many roads lead to CASM: diverse stimuli of noncanonical autophagy share a unifying molecular mechanism. *Sci Adv* **8:** eabo1274. doi:10.1126/sciadv.abo1274

Eguchi T, Kuwahara T, Sakurai M, Komori T, Fujimoto T, Ito G, Yoshimura SI, Harada A, Fukuda M, Koike M, et al. 2018. LRRK2 and its substrate Rab GTPases are sequentially targeted onto stressed lysosomes and maintain their homeostasis. *Proc Natl Acad Sci* **115:** E9115–E9124. doi:10.1073/pnas.1812196115

Farrer MJ, Stone JT, Lin CH, Dächsel JC, Hulihan MM, Haugarvoll K, Ross OA, Wu RM. 2007. Lrrk2 G2385R is an ancestral risk factor for Parkinson's disease in Asia. *Parkinsonism Relat Disord* **13:** 89–92. doi:10.1016/j.parkreldis.2006.12.001

Fdez E, Madero-Pérez J, Lara Ordóñez AJ, Naaldijk Y, Fasiczka R, Aiastui A, Ruiz-Martínez J, López de Munain A, Cowley SA, Wade-Martins R, et al. 2022. Pathogenic LRRK2 regulates centrosome cohesion via Rab10/RILPL1-mediated CDK5RAP2 displacement. *iScience* **25:** 104476. doi:10.1016/j.isci.2022.104476

Fuji RN, Flagella M, Baca M, Baptista MA, Brodbeck J, Chan BK, Fiske BK, Honigberg L, Jubb AM, Katavolos P, et al. 2015. Effect of selective LRRK2 kinase inhibition on non-human primate lung. *Sci Transl Med* **7:** 273ra15. doi:10.1126/scitranslmed.aaa3634

Gaig C, Ezquerra M, Marti MJ, Muñoz E, Valldeoriola F, Tolosa E. 2006. LRRK2 mutations in Spanish patients with Parkinson disease: frequency, clinical features, and incomplete penetrance. *Arch Neurol* **63:** 377–382. doi:10.1001/archneur.63.3.377

Goetz SC, Liem KF Jr, Anderson KV. 2012. The spinocerebellar ataxia-associated gene Tau tubulin kinase 2 controls the initiation of ciliogenesis. *Cell* **151:** 847–858. doi:10.1016/j.cell.2012.10.010

Gomez RC, Wawro P, Lis P, Alessi DR, Pfeffer SR. 2019. Membrane association but not identity is required for LRRK2 activation and phosphorylation of Rab GTPases. *J Cell Biol* **218:** 4157–4170. doi:10.1083/jcb.201902184

Gonzalez-Reyes LE, Verbitsky M, Blesa J, Jackson-Lewis V, Paredes D, Tillack K, Phani S, Kramer ER, Przedborski S, Kottmann AH. 2012. Sonic hedgehog maintains cellular and neurochemical homeostasis in the adult nigrostriatal circuit. *Neuron* **75:** 306–319. doi:10.1016/j.neuron.2012.05.018

Greggio E, Jain S, Kingsbury A, Bandopadhyay R, Lewis P, Kaganovich A, van der Brug MP, Beilina A, Blackinton J, Thomas KJ, et al. 2006. Kinase activity is required for the toxic effects of mutant LRRK2/dardarin. *Neurobiol Dis* **23:** 329–341. doi:10.1016/j.nbd.2006.04.001

Gustavsson EK, Follett J, Trinh J, Barodia SK, Real R, Liu Z, Grant-Peters M, Fox JD, Appel-Cresswell S, Stoessl AJ, et al. 2024. RAB32 ser71arg in autosomal dominant Parkinson's disease: linkage, association, and functional analyses. *Lancet Neurol* **23:** 603–614. doi:10.1016/S1474-4422 (24)00121-2

Herbst S, Campbell P, Harvey J, Bernard EM, Papayannopoulos V, Wood NW, Morris HR, Gutierrez MG. 2020. LRRK2 activation controls the repair of damaged endomembranes in macrophages. *EMBO J* **39:** e104494. doi:10 .15252/embj.2020104494

Herzig MC, Kolly C, Persohn E, Theil D, Schweizer T, Hafner T, Stemmelen C, Troxler TJ, Schmid P, Danner S, et al. 2011. LRRK2 protein levels are determined by kinase function and are crucial for kidney and lung homeostasis in mice. *Hum Mol Genet* **20:** 4209–4223. doi:10.1093/ hmg/ddr348

Hilgendorf KI, Myers BR, Reiter JF. 2024. Emerging mechanistic understanding of cilia function in cellular signalling. *Nat Rev Mol Cell Biol* **25:** 555–573. doi:10.1038/ s41580-023-00698-5

Hop PJ, Lai D, Keagle PJ, Baron DM, Kenna BJ, Kooyman M, Shankaracharya, Halter C, Straniero L, Asselta R, et al. 2024. Systematic rare variant analyses identify RAB32 as a susceptibility gene for familial Parkinson's disease. *Nat Genet* **56:** 1371–1376. doi:10.1038/s41588-024-01787-7

Huang T, Sun C, Du F, Chen ZJ. 2025. STING-induced noncanonical autophagy regulates endolysosomal homeostasis. *Proc Natl Acad Sci* **122:** e2415422122. doi:10 .1073/pnas.2415422122

Hui KY, Fernandez-Hernandez H, Hu J, Schaffner A, Pankratz N, Hsu NY, Chuang LS, Carmi S, Villaverde N, Li X, et al. 2018. Functional variants in the LRRK2 gene confer shared effects on risk for Crohn's disease and Parkinson's disease. *Sci Transl Med* **10:** eaai7795. doi:10.1126/sci translmed.aai7795

Ito G, Katsemonova K, Tonelli F, Lis P, Baptista MA, Shpiro N, Duddy G, Wilson S, Ho PW, Ho SL, et al. 2016. Phostag analysis of Rab10 phosphorylation by LRRK2: a powerful assay for assessing kinase function and inhibitors. *Biochem J* **473:** 2671–2685. doi:10.1042/BCJ20160557

Jaimon E, Lin YE, Tonelli F, Antico O, Alessi DR, Pfeffer SR. 2025. Restoration of striatal neuroprotective pathways by kinase inhibitor treatment of Parkinson's linked-LRRK2 mutant mice. *Sci Signal* doi:10.1126/scisignal.ads5761

Jaleel M, Nichols RJ, Deak M, Campbell DG, Gillardon F, Knebel A, Alessi DR. 2007. LRRK2 phosphorylates moesin at threonine-558: characterization of how Parkinson's disease mutants affect kinase activity. *Biochem J* **405:** 307–317. doi:10.1042/BJ20070209

Jennings D, Huntwork-Rodriguez S, Henry AG, Sasaki JC, Meisner R, Diaz D, Solanoy H, Wang X, Negrou E, Bondar VV, et al. 2022. Preclinical and clinical evaluation of the LRRK2 inhibitor DNL201 for Parkinson's disease. *Sci Transl Med* **14:** eabj2658. doi:10.1126/scitranslmed .abj2658

Kalia LV, Lang AE, Hazrati LN, Fujioka S, Wszolek ZK, Dickson DW, Ross OA, Van Deerlin VM, Trojanowski JQ, Hurtig HI, et al. 2015. Clinical correlations with Lewy body pathology in *LRRK2*-related Parkinson disease. *JAMA Neurol* **72:** 100–105. doi:10.1001/jamaneurol.2014 .2704

Kalogeropulou AF, Purlyte E, Tonelli F, Lange SM, Wightman M, Prescott AR, Padmanabhan S, Sammler E, Alessi DR. 2022. Impact of 100 LRRK2 variants linked to Parkinson's disease on kinase activity and microtubule binding. *Biochem J* **479:** 1759–1783. doi:10.1042/BCJ2022 0161

Kamada R, Kudoh F, Ito S, Tani I, Janairo JIB, Omichinski JG, Sakaguchi K. 2020. Metal-dependent Ser/Thr protein phosphatase PPM family: evolution, structures, diseases and inhibitors. *Pharmacol Ther* **215:** 107622. doi:10.1016/ j.pharmthera.2020.107622

Karayel Ö, Tonelli F, Virreira Winter S, Geyer PE, Fan Y, Sammler EM, Alessi DR, Steger M, Mann M. 2020. Accurate MS-based Rab10 phosphorylation stoichiometry determination as readout for LRRK2 activity in Parkinson's disease. *Mol Cell Proteomics* **19:** 1546–1560. doi:10 .1074/mcp.RA120.002055

Khan SS, Sobu Y, Dhekne HS, Tonelli F, Berndsen K, Alessi DR, Pfeffer SR. 2021. Pathogenic LRRK2 control of primary cilia and Hedgehog signaling in neurons and astrocytes of mouse brain. *eLife* **10:** e67900. doi:10.7554/eLife .67900

Khan SS, Jaimon E, Lin YE, Nikoloff J, Tonelli F, Alessi DR, Pfeffer SR. 2024. Loss of primary cilia and dopaminergic neuroprotection in pathogenic LRRK2-driven and idiopathic Parkinson's disease. *Proc Natl Acad Sci* **121:** e2402206121. doi:10.1073/pnas.2402206121

Kinnebrew M, Luchetti G, Sircar R, Frigui S, Viti LV, Naito T, Beckert F, Saheki Y, Siebold C, Radhakrishnan A, et al. 2021. Patched 1 reduces the accessibility of cholesterol in the outer leaflet of membranes. *eLife* **10:** e70504. doi:10 .7554/eLife.70504

Kinnebrew M, Woolley RE, Ansell TB, Byrne EFX, Frigui S, Luchetti G, Sircar R, Nachtergaele S, Mydock-McGrane L, Krishnan K, et al. 2022. Patched 1 regulates smoothened by controlling sterol binding to its extracellular cysteine-rich domain. *Sci Adv* **8:** eabm5563. doi:10.1126/sciadv .abm5563

Kluss JH, Bonet-Ponce L, Lewis PA, Cookson MR. 2022. Directing LRRK2 to membranes of the endolysosomal pathway triggers RAB phosphorylation and JIP4 recruitment. *Neurobiol Dis* **170:** 105769. doi:10.1016/j.nbd.2022 .105769

Krüger C, Lim SY, Buhrmann A, Fahrig FL, Gabbert C, Bahr N, Madoev H, Marras C, Klein C, Lohmann K. 2025. Updated MDSGene review on the clinical and genetic spectrum of LRRK2 variants in Parkinson's disease. *NPJ Parkinsons Dis* **11:** 30. doi:10.1038/s41531-025-00881-9

Kuwahara T, Iwatsubo T. 2024. CASM mediates LRRK2 recruitment and activation under lysosomal stress. *Autophagy* **20:** 1692–1693. doi:10.1080/15548627.2024.2330032

Lara Ordóñez AJ, Fernández B, Fdez E, Romo-Lozano M, Madero-Pérez J, Lobbestael E, Baekelandt V, Aiastui A, López de Munaín A, Melrose HL, et al. 2019. RAB8, RAB10 and RILPL1 contribute to both LRRK2 kinase–mediated centrosomal cohesion and ciliogenesis deficits.

Hum Mol Genet **28:** 3552–3568. doi:10.1093/hmg/ddz 201

Li X, Zhu H, Huang BT, Li X, Kim H, Tan H, Zhang Y, Choi I, Peng J, Xu P, et al. 2024. RAB12-LRRK2 complex suppresses primary ciliogenesis and regulates centrosome homeostasis in astrocytes. *Nat Commun* **15:** 8434. doi:10.1038/s41467-024-52723-6

Lin YE, Jaimon E, Tonelli F, Pfeffer SR. 2025. Pathogenic LRRK2 mutations cause loss of primary cilia and Neurturin in striatal parvalbumin interneurons. *Life Sci Alliance* **8:** e202402922. doi:10.26508/lsa.202402922

Lisé MF, Srivastava DP, Arstikaitis P, Lett RL, Sheta R, Viswanathan V, Penzes P, O'Connor TP, El-Husseini A. 2009. Myosin-Va-interacting protein, RILPL2, controls cell shape and neuronal morphogenesis via Rac signaling. *J Cell Sci* **122:** 3810–3821. doi:10.1242/jcs.050344

Madero-Pérez J, Fernández B, Lara Ordóñez AJ, Fdez E, Lobbestael E, Baekelandt V, Hilfiker S. 2018. RAB7L1-mediated relocalization of LRRK2 to the Golgi complex causes centrosomal deficits via RAB8A. *Front Mol Neurosci* **11:** 417. doi:10.3389/fnmol.2018.00417

Malave L, Zuelke DR, Uribe-Cano S, Starikov L, Rebholz H, Friedman E, Qin C, Li Q, Bezard E, Kottmann AH. 2021. Dopaminergic co-transmission with sonic hedgehog inhibits abnormal involuntary movements in models of Parkinson's disease and L-Dopa induced dyskinesia. *Commun Biol* **4:** 1071. doi:10.1038/s42003-021-02567-3

Marks MS, Heijnen HF, Raposo G. 2013. Lysosome-related organelles: unusual compartments become mainstream. *Curr Opin Cell Biol* **25:** 495–505. doi:10.1016/j.ceb.2013.04.008

Matsui T, Ohbayashi N, Fukuda M. 2012. The Rab interacting lysosomal protein (RILP) homology domain functions as a novel effector domain for small GTPase Rab36: Rab36 regulates retrograde melanosome transport in melanocytes. *J Biol Chem* **287:** 28619–28631. doi:10.1074/jbc.M112.370544

McGrath E, Waschbüsch D, Baker BM, Khan AR. 2021. LRRK2 binds to the Rab32 subfamily in a GTP-dependent manner via its armadillo domain. *Small GTPases* **12:** 133–146. doi:10.1080/21541248.2019.1666623

Mill P, Christensen ST, Pedersen LB. 2023. Primary cilia as dynamic and diverse signalling hubs in development and disease. *Nat Rev Genet* **24:** 421–441. doi:10.1038/s41576-023-00587-9

Miller GK, Kuruvilla S, Jacob B, LaFranco-Scheuch L, Bakthavatchalu V, Flor J, Flor K, Ziegler J, Reichard C, Manfre P, et al. 2023. Effects of LRRK2 inhibitors in nonhuman primates. *Toxicol Pathol* **51:** 232–245. doi:10.1177/01926233231205895

Myasnikov A, Zhu H, Hixson P, Xie B, Yu K, Pitre A, Peng J, Sun J. 2021. Structural analysis of the full-length human LRRK2. *Cell* **184:** 3519–3527.e10. doi:10.1016/j.cell.2021.05.004

Nair SV, Jaimon E, Adhikari A, Nikoloff J, Pfeffer SR. 2025. Lysosomal glucocerebrosidase is needed for ciliary Hedgehog signaling: a convergent pathway to Parkinson's disease. bioRxiv doi:10.1101/2025.01.20.633968

Paisán-Ruíz C, Jain S, Evans EW, Gilks WP, Simón J, van der Brug M, López de Munain A, Aparicio S, Gil AM, Khan N, et al. 2004. Cloning of the gene containing mutations that cause PARK8-linked Parkinson's disease. *Neuron* **44:** 595–600. doi:10.1016/j.neuron.2004.10.023

Pal P, Taylor M, Lam PY, Tonelli F, Hecht CA, Lis P, Nirujogi RS, Phung TK, Yeshaw WM, Jaimon E, et al. 2023. Parkinson's VPS35[D620N] mutation induces LRRK2-mediated lysosomal association of RILPL1 and TMEM55B. *Sci Adv* **9:** eadj1205. doi:10.1126/sciadv.adj1205

Pfeffer SR. 2017. Rab GTPases: master regulators that establish the secretory and endocytic pathways. *Mol Biol Cell* **28:** 712–715. doi:10.1091/mbc.e16-10-0737

Purlyte E, Dhekne HS, Sarhan AR, Gomez R, Lis P, Wightman M, Martinez TN, Tonelli F, Pfeffer SR, Alessi DR. 2018. Rab29 activation of the Parkinson's disease-associated LRRK2 kinase. *EMBO J* **37:** 1–18. doi:10.15252/embj.201798099

Radhakrishnan A, Rohatgi R, Siebold C. 2020. Cholesterol access in cellular membranes controls Hedgehog signaling. *Nat Chem Biol* **16:** 1303–1313. doi:10.1038/s41589-020-00678-2

Radulovic M, Schink KO, Wenzel EM, Nähse V, Bongiovanni A, Lafont F, Stenmark H. 2018. ESCRT-mediated lysosome repair precedes lysophagy and promotes cell survival. *EMBO J* **37:** e99753. doi:10.15252/embj.201899753

Rak A, Pylypenko O, Durek T, Watzke A, Kushnir S, Brunsveld L, Waldmann H, Goody RS, Alexandrov K. 2003. Structure of Rab GDP-dissociation inhibitor in complex with prenylated YPT1 GTPase. *Science* **302:** 646–650. doi:10.1126/science.1087761

Ross OA, Wu YR, Lee MC, Funayama M, Chen ML, Soto AI, Mata IF, Lee-Chen GJ, Chen CM, Tang M, et al. 2008. Analysis of Lrrk2 R1628P as a risk factor for Parkinson's disease. *Ann Neurol* **64:** 88–92. doi:10.1002/ana.21405

Scheffer LL, Sreetama SC, Sharma N, Medikayala S, Brown KJ, Defour A, Jaiswal JK. 2014. Mechanism of Ca^{2+}-triggered ESCRT assembly and regulation of cell membrane repair. *Nat Commun* **5:** 5646. doi:10.1038/ncomms6646

Seabra MC. 1998. Membrane association and targeting of prenylated Ras-like GTPases. *Cell Signal* **10:** 167–172. doi:10.1016/S0898-6568(97)00120-4

Sheu SH, Upadhyayula S, Dupuy V, Pang S, Deng F, Wan J, Walpita D, Pasolli HA, Houser J, Sanchez-Martinez S, et al. 2022. A serotonergic axon-cilium synapse drives nuclear signaling to alter chromatin accessibility. *Cell* **185:** 3390–3407.e18. doi:10.1016/j.cell.2022.07.026

Siderowf A, Concha-Marambio L, Lafontant DE, Farris CM, Ma Y, Urenia PA, Nguyen H, Alcalay RN, Chahine LM, Foroud T, et al. 2023. Assessment of heterogeneity among participants in the Parkinson's Progression Markers Initiative cohort using α-synuclein seed amplification: a cross-sectional study. *Lancet Neurol* **22:** 407–417. doi:10.1016/S1474-4422(23)00109-6

Skowyra ML, Schlesinger PH, Naismith TV, Hanson PI. 2018. Triggered recruitment of ESCRT machinery promotes endolysosomal repair. *Science* **360:** eaar5078. doi:10.1126/science.aar5078

Sobu Y, Wawro PS, Dhekne HS, Yeshaw WM, Pfeffer SR. 2021. Pathogenic LRRK2 regulates ciliation probability upstream of tau tubulin kinase 2 via Rab10 and RILPL1 proteins. *Proc Natl Acad Sci* **118:** e2005894118. doi:10.1073/pnas.2005894118

Steger M, Tonelli F, Ito G, Davies P, Trost M, Vetter M, Wachter S, Lorentzen E, Duddy G, Wilson S, et al. 2016. Phosphoproteomics reveals that Parkinson's disease kinase LRRK2 regulates a subset of Rab GTPases. *eLife* **5**: e12813. doi:10.7554/eLife.12813

Steger M, Diez F, Dhekne HS, Lis P, Nirujogi RS, Karayel O, Tonelli F, Martinez TN, Lorentzen E, Pfeffer SR, et al. 2017. Systematic proteomic analysis of LRRK2-mediated Rab GTPase phosphorylation establishes a connection to ciliogenesis. *eLife* **6**: e31012. doi:10.7554/eLife.31012

Stroupe C, Brunger AT. 2000. Crystal structures of a Rab protein in its inactive and active conformations. *J Mol Biol* **304**: 585–598. doi:10.1006/jmbi.2000.4236

Tan JX, Finkel T. 2022. A phosphoinositide signalling pathway mediates rapid lysosomal repair. *Nature* **609**: 815–821. doi:10.1038/s41586-022-05164-4

Tong Y, Yamaguchi H, Giaime E, Boyle S, Kopan R, Kelleher RJ, Shen J. 2010. Loss of leucine-rich repeat kinase 2 causes impairment of protein degradation pathways, accumulation of α-synuclein, and apoptotic cell death in aged mice. *Proc Natl Acad Sci* **107**: 9879–9884. doi:10.1073/pnas.1004676107

Vides EG, Adhikari A, Chiang CY, Lis P, Purlyte E, Limouse C, Shumate JL, Spínola-Lasso E, Dhekne HS, Alessi DR, et al. 2022. A feed-forward pathway drives LRRK2 kinase membrane recruitment and activation. *eLife* **11**: e79771. doi:10.7554/eLife.79771

Wang X, Bondar VV, Davis OB, Maloney MT, Agam M, Chin MY, Cheuk-Nga Ho A, Ghosh R, Leto DE, Joy D, et al. 2023. Rab12 is a regulator of LRRK2 and its activation by damaged lysosomes. *eLife* **12**: e87255. doi:10.7554/eLife.87255

Wang X, Xu P, Bentley-DeSousa A, Hancock-Cerutti W, Cai S, Johnson BT, Tonelli F, Shao L, Talaia G, Alessi DR, et al. 2025. Lysosome damage triggers acute formation of ER to lysosomes membrane tethers mediated by the bridge-like lipid transport protein VPS13C. bioRxiv doi:10.1101/2024.06.08.598070

Waschbüsch D, Michels H, Strassheim S, Ossendorf E, Kessler D, Gloeckner CJ, Barnekow A. 2014. LRRK2 transport is regulated by its novel interacting partner Rab32. *PLoS One* **9**: e111632. doi:10.1371/journal.pone.0111632

Waschbüsch D, Purlyte E, Pal P, McGrath E, Alessi DR, Khan AR. 2020. Structural basis for Rab8a recruitment of RILPL2 via LRRK2 phosphorylation of Switch 2. *Structure* **28**: 406–417.e6. doi:10.1016/j.str.2020.01.005

Waschbüsch D, Berndsen K, Lis P, Knebel A, Lam YP, Alessi DR, Khan AR. 2021. Structural basis for the specificity of PPM1H phosphatase for Rab GTPases. *EMBO Rep* **22**: e52675. doi:10.15252/embr.202152675

West AB, Moore DJ, Choi C, Andrabi SA, Li X, Dikeman D, Biskup S, Zhang Z, Lim KL, Dawson VL, et al. 2007. Parkinson's disease-associated mutations in LRRK2 link enhanced GTP-binding and kinase activities to neuronal toxicity. *Hum Mol Genet* **16**: 223–232. doi:10.1093/hmg/ddl471

Westenberger A, Skrahina V, Usnich T, Beetz C, Vollstedt EJ, Laabs BH, Paul JJ, Curado F, Skobalj S, Gaber H, et al. 2024. Relevance of genetic testing in the gene-targeted trial era: the Rostock Parkinson's disease study. *Brain* **147**: 2652–2667. doi:10.1093/brain/awae188

Willett R, Martina JA, Zewe JP, Wills R, Hammond GRV, Puertollano R. 2017. TFEB regulates lysosomal positioning by modulating TMEM55B expression and JIP4 recruitment to lysosomes. *Nat Commun* **8**: 1580. doi:10.1038/s41467-017-01871-z

Wu JY, Cho SJ, Descant K, Li PH, Shapson-Coe A, Januszewski M, Berger DR, Meyer C, Casingal C, Huda A, et al. 2024. Mapping of neuronal and glial primary cilia contactome and connectome in the human cerebral cortex. *Neuron* **112**: 41–55.e3. doi:10.1016/j.neuron.2023.09.032

Yadavalli N, Ferguson SM. 2023. LRRK2 suppresses lysosome degradative activity in macrophages and microglia through MiT-TFE transcription factor inhibition. *Proc Natl Acad Sci* **120**: e2303789120. doi:10.1073/pnas.2303789120

Yeshaw WM, Adhikari A, Chiang CY, Dhekne HS, Wawro PS, Pfeffer SR. 2023. Localization of PPM1H phosphatase tunes Parkinson's disease-linked LRRK2 kinase-mediated Rab GTPase phosphorylation and ciliogenesis. *Proc Natl Acad Sci* **120**: e2315171120. doi:10.1073/pnas.2315171120

Zimprich A, Biskup S, Leitner P, Lichtner P, Farrer M, Lincoln S, Kachergus J, Hulihan M, Uitti RJ, Calne DB, et al. 2004. Mutations in LRRK2 cause autosomal-dominant parkinsonism with pleomorphic pathology. *Neuron* **44**: 601–607. doi:10.1016/j.neuron.2004.11.005

Cite this article as *Cold Spring Harb Perspect Med* doi: 10.1101/cshperspect.a041620

Exploring Parkinson's through the Lens of Genomics and Bioinformatics

Vilas Menon

Center for Translational and Computational Neuroimmunology, Department of Neurology, Columbia University Irving Medical Center, New York, New York 10035, USA

Correspondence: vm2545@cumc.columbia.edu

Within the last three decades, revolutions in genomics data generation and bioinformatics analysis techniques have profoundly impacted our understanding of the molecular mechanisms of Parkinson's disease (PD). From the description of the first PD-associated risk gene in 1997 through today, new technologies have revolutionized approaches to identify genetic and molecular mechanisms implicated in human health and disease. Spurred by the dramatically decreasing costs for genotyping, genome sequencing, and transcriptomics approaches, the ability to profile large cohorts of human populations or model organisms has accelerated the understanding of disease susceptibility, pathways, and genes. Thus far, ~30 genetic loci have been unequivocally linked to the pathogenesis of PD, highlighting essential molecular pathways underlying this common disorder. More recently, the advent of single-cell transcriptomics techniques applied to human brain tissue has implicated cell-type-specific dysregulation and vulnerability (beyond the loss of dopaminergic neurons) in the disease. Herein, we discuss how neurogenomics and bioinformatics are applied to dissect the nature of this complex disease with the overall aim of identifying new targets for therapeutic interventions.

The pathogenesis of Parkinson's disease (PD) as we understand it today includes a broad spectrum of metabolic pathways, including oxidative stress caused by mitochondrial dysfunction, inflammation, abnormal protein metabolism, and aging (reviewed in Dauer and Przedborski 2003 and Dawson and Dawson 2003).[1] Most of these pathophysiological links with PD result from studies of gene mutations implicated in familial, highly penetrant forms of PD. Identifying risk loci and variants associated with the more common nonfamilial form of PD requires larger sample sizes in terms of genetics and molecular studies, given the smaller effect sizes of these "sporadic" variants. With the increasing scale of modern genomics technologies, numerous genetic risk loci involved in nonfamilial PD are being uncovered, raising new research questions and hopes to better understand and treat this neurodegenerative condition more comprehensively. In addition, the scaling up of transcriptomics and proteomics

---

[1]This is an update to a previous article published in *Cold Spring Harbor Perspectives in Medicine* [Scholz et al. (2012). *Cold Spring Harb Perspect Med* **2**: a009449. doi:10.1101/cshperspect.a009449].

Copyright © 2025 Cold Spring Harbor Laboratory Press; all rights reserved
Cite this article as *Cold Spring Harb Perspect Med* doi: 10.1101/cshperspect.a041621

efforts on human brain tissue has led to the identification of cell-type-specific pathways that may be disrupted by PD variants. In this work, we will introduce some of the main concepts and discussions in PD genetics, outline the current status of PD genomics and bioinformatics, discuss the impact of new sequencing technologies, and chart the necessary steps for translating these findings into targeted therapeutics.

"MENDELIAN" VERSUS COMPLEX DISEASE: SIMILAR IDEAS, DIFFERENT CONCEPTS

For simplicity, geneticists have separated genetic diseases into two main categories. The first one refers to "Mendelian" diseases such as Huntington's disease, muscular dystrophy, and cystic fibrosis. "Mendelian" diseases are defined as the classical familial forms of disease in which the underlying genetic defect manifests in disease symptoms in a large proportion of mutation carriers; from a discovery point of view, the high penetrance of variants in Mendelian diseases allows for inference of disease loci through typical inheritance patterns can be inferred. Initially, this class of disease was the mainstay of genetic research; this is primarily because, until recently, the standard genetic approach for disease gene discovery was linkage study, a technique that relies on targeted examination of large families with multiple affected individuals. By focusing on families with these risk variants, limited bandwidth for sequencing and genotyping could be organized effectively to identify key genetic signals. For the most part, this approach was very successful, with nearly 3000 Mendelian disorders deciphered to date (Lander 2011). However, linkage study design is only applicable to a small proportion of human ailments that are typically rather rare in the general population. Only a small subset of PD patients reports a positive family history of the disease, and thus it is not surprising that the first mutations identified as a cause for PD were identified in this small subset of patients using linkage studies. Studying the genetics of complex diseases, which constitute the second disease category, has proven to be more challenging. Indeed, the most prevalent neurodegenerative diseases—

such as PD and Alzheimer's disease—are good examples of complex diseases that can present in familial or nonfamilial "varieties." Specifically, in the vast majority of patients, a family history of PD is absent, and the etiology is less clear. It is in these patients that most of the research in the last few years has been focused.

The differences in studying complex disease as opposed to familial disease lie within the study designs, the techniques used, the degree to which a particular genetic variant confers risk to disease, and the mechanistic hypotheses of disease pathogenesis. These concepts require precise technical definitions, as outlined below:

1. **Distinction between association and causation.** When a genetic variant is investigated for a potential contribution to disease, a number of questions need to be addressed. Is the variant commonly present in the population? If so, is the frequency of this variant significantly different in cases versus controls? If this observation is statistically robust, then significant association with a particular phenotype has been determined. However, it is important to emphasize that establishing significant association should not be misconstrued as drawing inferences about causation. For example, the APOE 14 allele on chromosome 19 has been consistently associated with increased risk for developing Alzheimer's disease, but carrying this risk allele is neither necessary nor sufficient to cause disease. Similarly, variants in the GBA1 locus have been implicated in PD, but not in a fully deterministic fashion. Although an association signal sometimes implies genes or genetic regions that play a causative role in the pathogenesis of disease, it is not appropriate to assume that this applies to all instances in which associations are observed.

2. **"Common disease–common variant hypothesis" versus "common disease–rare variants hypothesis."** In contrast to monogenic diseases in which mutations in a single gene is sufficient to cause disease, complex diseases are thought to be result from a combination of multiple genetic, environmental, and stochastic factors. Two distinctive concepts have

Cite this article as *Cold Spring Harb Perspect Med* doi: 10.1101/cshperspect.a041621

been postulated for the detection of genetic variants underlying common, complex diseases. The "common disease–common variant hypothesis" posits that multiple, common small-risk variants of small effect size interact to cause common disease (Reich and Lander 2001). This hypothesis is the core basis for genome-wide association studies (GWAS), a study design that relies on testing several hundred thousand common genetic variants throughout the human genome in large case-control cohorts. Over the past few years, hundreds of new gene loci and pathways, including 16 PD loci (Fig. 1; Simón-Sánchez et al. 2009; International Parkinson Disease Genomics Consortium et al. 2011), have been implicated with various human disease traits using a GWAS design (an updated catalog of identified genetic risk loci can be found at www.genome.gov/gwastudies). However, heritability estimates have shown that large proportions of genetic risk underlying complex disease have not yet been explained (Manolio et al. 2009). In PD, for example, only ~60% of heritability is understood, depending on the population studied (International Parkinson Disease Genomics Consortium et al. 2011). This "missing heritability" is at the center of much of the current debate, and possible explanations that have been brought forward include lack of power to detect common low-risk variants, rare variants, gene–gene interactions, gene–environment interactions, structural variants such as deletions or duplications, and inversions (Manolio et al. 2009). In particular, the "common disease–rare variants hypothesis" has gained much traction, mainly because of the introduction of advanced sequencing technologies that allow cost-effective sequencing of entire genomes. An early lesson learned from these next-generation sequencing technologies is that there are numerous rare variants in the human genome, which have not yet been systematically explored and may account for a portion of the missing heritability estimates. As genetic association studies become larger, more insights will be gained into the pathogenic relevance of rare genetic variability.

DISSECTING THE GENETICS OF COMPLEX DISEASE: THE REVOLUTION OF GENOMICS AND BIOINFORMATICS

Genomic research in the past few years has been defined by the rapid integration of technological advances in sequencing and molecular biology, deriving ultimately from the human genome project. To reconstruct the developments, successes, and challenges in genomic PD research, we will point out some of the past milestones of the genomics revolution, and discuss the events and developments achieved so far (Fig. 2 shows the selected landmark discoveries in genomic PD research). The year 1997 marks the starting point for PD genomics, whereby disease associations across the entire genome could be measured. Using a linkage study approach, Polymeropoulos et al. (1997) reported the discovery of missense mutations in the SNCA gene, coding for α-synuclein, to underlie a rare familial form of PD. This finding was crucial in that it provided clear evidence that there are genetic forms of disease, a view of which was evolving from prior concepts that PD was a nongenetically driven disease. The identification of a disease-causing gene allowed the generation of cell- and animal-based model systems for studying the functional mechanisms in disease pathogenesis (Feany and Bender 2000; Masliah et al. 2000). Moreover, subsequent screening studies showed that variability at the SNCA locus not only plays a role in this familial form of PD, but is also associated with risk for disease in sporadic cases (Farrer et al. 2001; Maraganore et al. 2006). Soon after the discovery of SNCA, other "Mendelian" PD genes (Parkin, PINK1, DJ1, and LRRK2) were revealed using a linkage mapping design (KItada et al. 1998; Valente et al. 2001; Bonifati et al. 2003; Paisán-Ruíz et al. 2004; Zimprich et al. 2004; refer to Westenberger et al. 2024 for a more detailed discussion on PD genetics). In contrast to the advances in dissecting the genetics of rare familial forms of PD, the genetic factors influencing common sporadic cases remained enigmatic. The publication of the human genome sequence in 2001 marked an exciting turning point (Lander et al. 2001; Venter et al. 2001). For the first time, researchers had

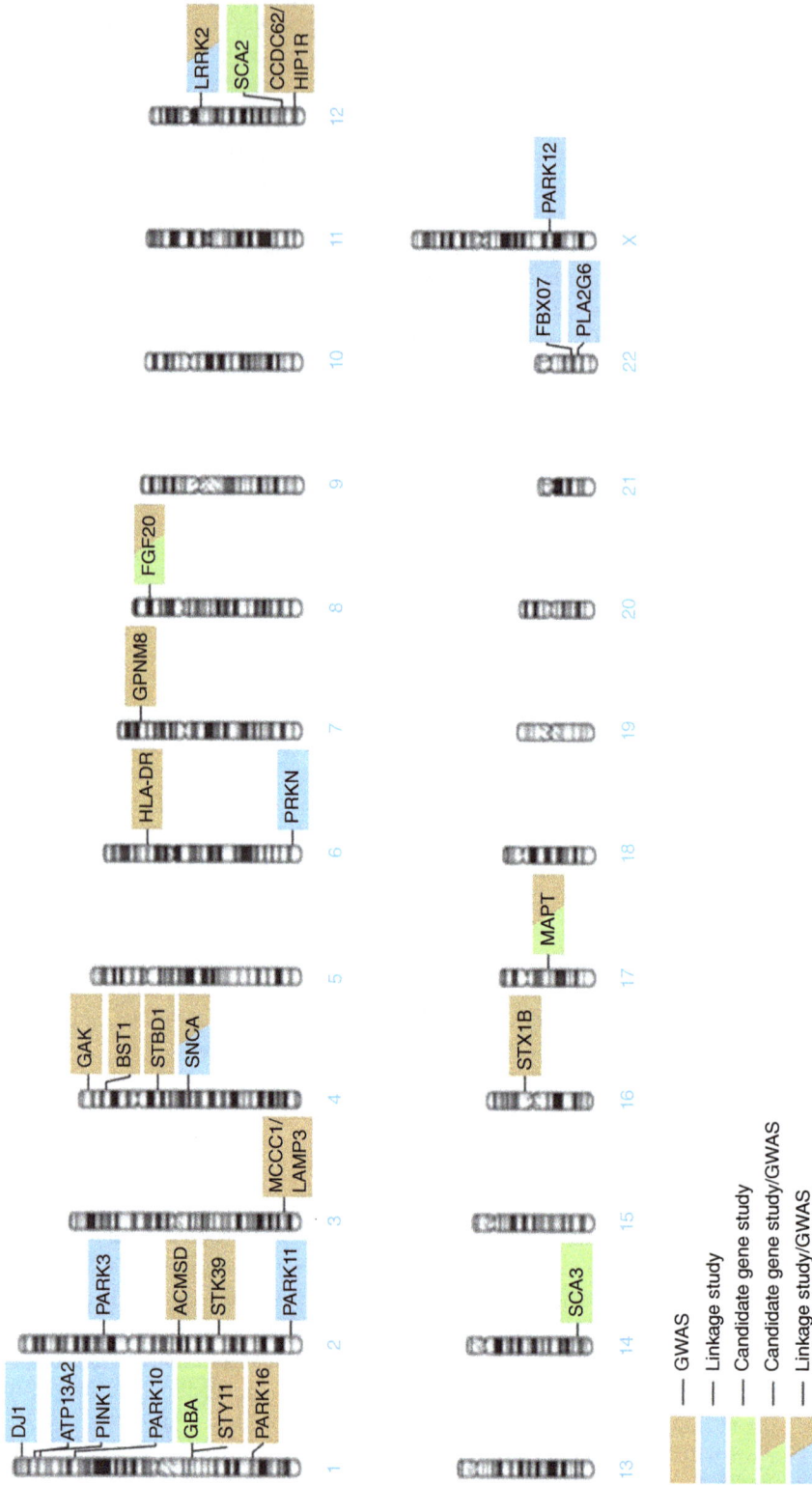

Figure 1. An overview of the genetic loci implicated in the pathogenesis of Parkinson's disease (PD). The position of each locus relative to the ideogram of each chromosome is depicted. The background color of each box indicates the method that was used to identify this locus. (GWAS) Genome-wide association study. (Figure and figure legend reprinted from Scholz et al. 2012, © Cold Spring Harbor Laboratory Press.)

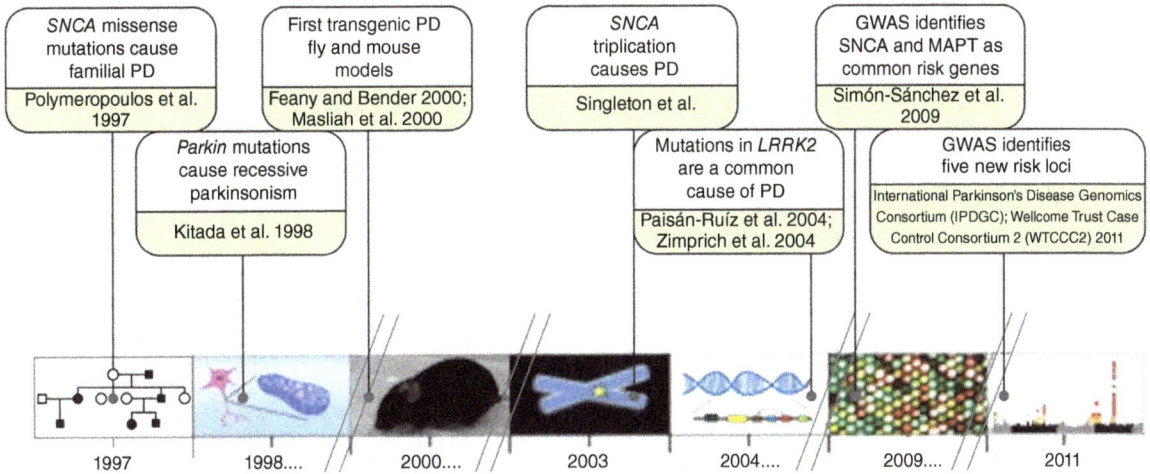

Figure 2. Highlights of key genomic discoveries in Parkinson's disease (PD) over the past decade and a half. (Figure and figure legend reprinted from Scholz et al. 2012, © Cold Spring Harbor Laboratory Press.)

the opportunity to examine the sequence of an entire human genome and, 4 years later, a detailed catalog of common genetic variants (the HapMap) became available (www.hap.map .ncbi.nlm.nih.gov). This catalog was the starting point for studying the genomics of complex diseases. Soon microarray platforms for genotyping hundreds of thousands of these common variants throughout the human genome were developed and genome-wide association testing became a reality. The revolutionary aspect of this new technology was that it provided a cost-effective tool to rapidly scan hundreds of thousands of variants in the genome in thousands of individuals. In PD, genome-wide association strategies have been remarkably successful; because of this new technology, the number of risk loci implicated in PD pathogenesis has doubled over the last 3 years (Fig. 2).

As new sequencing technologies for sequencing entire genomes emerge, the era of GWAS is diminishing in importance. In the same way, as GWAS was the standard technology used to search for common risk variants, next-generation sequencing technologies are going to determine the exploration of rare genetic variability and of structural genomic rearrangements. With costs of sequencing dropping and the speed of genomic data generation reaching unprecedented scales, data handling and analysis have become big chal-

lenges in modern PD research. In fact, the generation of genomics data, and the corresponding identification of disease-associated variants across the genome, has accelerated so rapidly that the amount of data produced exceeds the exponential growth of computer processing speed known as Moore's law. In other words, the bottleneck for genomic discovery is no longer generating extensive data sets, but rather storing and analyzing the information; this problem has even generated interest in the lay press (Pollack 2011). To put the advances of genomics into perspective, it took several thousand researchers 13 years to sequence the first human genome at a cost of $3 billion (www.genome.gov); today, one technician can sequence an entire genome in a few days for ~$500. This sequencing revolution has also accelerated the pace of transcriptomics, with the development of bulk and single-cell RNA-seq, whereby RNA in tissues and cells can be reverse transcribed, amplified, and sequenced in the same way as genome data. As technical innovation in genomics and other -omics methods increase at a rapid pace, the development of computational medicine and its diverse applications, commonly referred to as bioinformatics, is confronted with challenging demands to develop user-friendly methods to parse, analyze, and share genomic data. Automation of sequence data filtering, alignment with reference genomes, variant calling and statistical analyses across various platforms, even a decade later, is still not fully harmonized. Moreover, in-depth analyses of epistatic effects, or gene–gene interactions, as well as investigations of gene–environment interactions, which are at the center of research ambitions aimed at resolving the complete genomic architecture of complex diseases, depend on sophisticated bioinformatics algorithms.

INFORMATICS OF GENOMICS, TRANSCRIPTOMICS, PROTEOMICS, AND METABOLOMICS

Developing comprehensive informatics tools requires, concurrent with the development of genome-centric algorithms, a deeper understanding of the functional consequences of genetic variability on disease vulnerability. In other words, how do we translate the relatively invariant nature of our genomes into what is surely a dynamic, evolving risk of disease? This understanding involves connecting genetic information onto the full molecular space, which includes genetics, epigenetics (i.e., noncoding changes such DNA methylation, chromatin conformation, noncoding RNAs—all affecting the readout of the genome), gene expression, protein expression, and the resultant metabolic output of all of these processes. So far, technological advances in gene expression analyses (RNA expression or transcriptomics) have paralleled those in genome analyses, as similar chemistries allow for the rapid, high-throughput ascertainment of both DNA and RNA. For example, studies from a decade ago linked data from PD GWAS to specific gene expression and epigenetic changes in brain tissue, as was reported by the International Parkinson's Disease Genomics Consortium (IPDGC), Wellcome Trust Case Control Consortium 2 (WTCCC2) (2011). However, the more complex understanding of the relationships between the static genome, the epigenome, and RNA expression, including regulatory RNAs such as small interfering RNA (siRNA) and microRNA remain underexplored. Additionally, the technologies for detecting and quantifying proteins (proteomics) and metabolites (metabolomics) have been rapidly advancing, as have techniques to localize mo-

lecular signatures in space in intact tissue. Thus, the next step forward is coupling the information in the nucleic acid space to the proteomic (both at the transcriptional and posttranscriptional levels) and the metabolomic spaces, taking into consideration the spatial distribution of these modalities. Collectively, we refer to the science and technology of measuring these biological molecules as "-omics"—genomics, epigenomics, transcriptomics, proteomics, and metabolomics. The ongoing revolution in "-omics" technologies, coupled with advances in bioinformatics, has greatly expanded our ability to link genetic architecture to its functional output, namely, the expression of genes, proteins, and metabolites. Thus, a key challenge to the field is to detect and quantitatively measure the full complement of biological molecules and to integrate these into a meaningful understanding of both normal function and dysfunction (e.g., disease).

Although "-omics"-driven research holds great promise in understanding disease biology, enthusiasm should be tempered. The current state-of-the-art technologically to meaningfully interpret large and diverse data sets is limiting. There is a snowballing complexity of analyses as we integrate data across multiple domains, which includes not just those from the molecular space (e.g., RNA, protein, and metabolite expression), but also the clinical space, which includes neuroimaging among other sources. For example, how does genetic risk translate from the subcellular organelle (e.g., mitochondrion), cellular (e.g., midbrain dopaminergic neuron), circuit (e.g., nigrostriatal pathway), and organ (central nervous system) levels to an individual's risk for developing PD? This is further complicated by the complex interactions we as individuals have with other organisms, be they the microbiome of our gastrointestinal systems to the human interactions that create our social and environmental communities. On examining human postmortem tissues, Heiko Braak proposed a new framework for PD wherein the disease process begins outside the midbrain, the traditional locus of vulnerable nigrostriatal dopaminergic neuronal cell bodies that are the uniformly affected in PD (Braak et al. 2002, 2003a). One concept that followed from this is that PD could be initiated in the enteric nervous system with ascending pathology to involve the dorsal motor nucleus of the vagus (Braak et al. 2003b; reviewed in Hawkes et al. 2010). This hypothesis has spawned a new field examining the gut–brain axis in PD, but large-scale data generation on the microbiome linked with human molecular data is still relatively nascent (see Gao et al. 2024 for a more detailed discussion on this topic). As we test new hypotheses of pathogenesis such as these, the informatic complexity increases substantially. One possibility is that the informatic challenges will require the integration of -omics data with other data to derive empirically testable biological networks—those that explain an individual's risk for PD. Once risk is stratified, we may then be positioned to consider earlier interventions that could alter the natural history of PD.

IDENTIFICATION OF ABERRANT NETWORK ACTIVITIES

As discussed above, increasing evidence indicates that complex diseases such as PD are associated with multiple genetic polymorphisms that are postulated to affect biological networks (Schadt 2009; Tan et al. 2009; Meyerson et al. 2010). Biological networks represent a series of actions among molecules that lead to a certain specific change in a cell (Croft et al. 2011). These networks may subserve many categories of cellular activities (Chang et al. 2009; Wang et al. 2009), including cell signaling networks, protein–protein interactions, and metabolic networks (Kim et al. 2010). Identifying aberrancies involving small numbers of genes in specific biological networks may lead to more precise diagnosis and treatment of a disease. However, because tens or hundreds of genes are often involved, conventional experimental systems developed for identifying aberrant gene(s) are inadequate for deciphering aberrancies in network activity. High-throughput technologies enable the simultaneous detection of a large number of alterations in molecular components, or nodes in network parlance. As noted above, these high-throughput technologies were developed to fill this gap and the first successful use of

technology was in DNA sequencing (Sanger et al. 1977), which later developed into a gambit of genomics. As other -omics technologies developed and enabled the simultaneous detection of a large number of alterations in protein expression and metabolites, the correlations and dependencies between molecular components have become complex. Multiple algorithms to infer cell–cell communication disruption in the context of disease are now being applied to these questions (Browaeys et al. 2020; Jin et al. 2021). These studies and approaches, when understood in a network context, offer a systems level perspective on processes underling disease initiation and progression.

Analyses of high-dimensional data using robust new informatics tools allows for the integration of these different data sources to yield new information about network functions and dysfunctions. Figure 3 depicts the workflow of a typical study involving high-dimensional -omics data to identify aberrant network activities. Such data analyses often reveal that our current understanding of molecular and chemical biology underlying cellular functions remains incomplete (Fig. 3; Ochs et al. 2011). However, an increasingly greater number of open-access databases complement the analysis of aberrant network activities from these high-dimensional data (Peri et al. 2003; Kanehisa et al. 2004; Schaefer et al. 2009; Browaeys et al. 2020; Jin et al. 2021). In addition, numerous tools have been developed for visually exploring and analyzing biological networks, including Cytoscape (Smoot et al. 2011), VisANT (Hu et al. 2009), GeneGO (www.genego.com), Ingenuity (www.ingenuity.com), and Pathway Studio (Nikitin et al. 2003).

In recent years, questions have been raised as to how genetic differences between individuals lead to differences in disease networks and, thus, to differences in phenotypes (Taylor et al. 2009; Vaske et al. 2010). Data-driven computational models, such as PARADIGM (Vaske et al. 2010) and ResponseNet (Yeger-Lotem et al. 2009) among others (Amit et al. 2009; Tan et al. 2009; Kreeger and Lauffenburger 2010; Chien et al. 2011), have been especially useful in this area. For example, progress has been made in applying information about aberrant networks

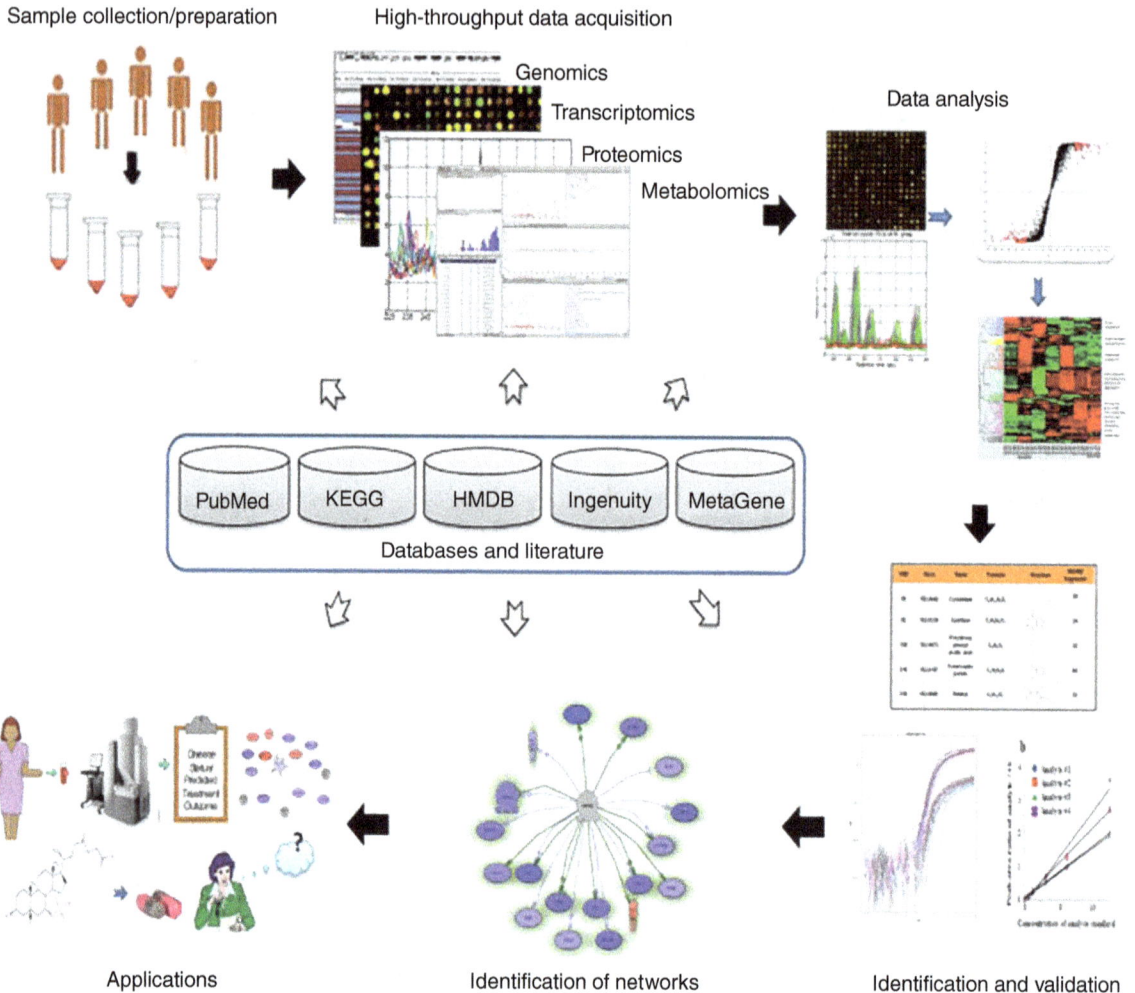

Figure 3. A hypothetical systems-based approach to identify aberrant networks of disease. Data (including biological, clinical, imaging) and samples are collected from a population. High-dimensional -omics data are acquired, integrated with clinical data, analyzed, and validated to identify networks involved in disease. Genomics will predict aberrant networks, whereas transcriptomics, proteomics, and metabolomics will report the outcomes of these networks. In turn, these networks and the aberrant nodes that are perturbed in disease can then be used to develop biomarkers, prognostic markers (e.g., markers that report disease progress or therapeutic efficacy), and rational therapeutics. The process is not inherently unidirectional nor is it intended to be single pass. Instead, as technologies improve, the process can be employed in an iterative fashion to refine nodes within aberrant disease networks and to generate better biomarkers, targets, and therapies. The approach is predicated on robust bioinformatics, and analytics that are critical to our abilities translate high-dimensional data to our understanding of disease and its treatment. (Image is from Wang et al. 2012. Figure and figure legend from Scholz et al. 2012, © Cold Spring Harbor Laboratory Press.)

to the identification of functional modules; that is, groups of biological entities (e.g., gene, protein) that perform biological tasks (e.g., protein degradation) that are dependent on each con-stituent part (Qiu et al. 2009; Wu et al. 2010). However, sequenced mutations, copy number alterations, gene fusion events, or epigenetic changes are not well represented in these models. Nevertheless, information derived from putative aberrant network activities will facilitate the detection of biomarkers (Singh et al. 2009), the identification of novel drug targets (Andre et al. 2009), the classification of disease types (Gatza et al. 2010), and the prediction of clinical outcomes (Taylor et al. 2009; Cerami et al. 2010; Chen et al. 2010; Gatza et al. 2010). Increasing use of -omics data-driven methodologies for identifying biomarkers and therapeutic targets will also include machine learning methods (Andre et al. 2009; Singh et al. 2009), graph theory (Taylor et al. 2009), and statistical methods (Singh et al. 2009).

Obviously, understanding of the aberrancies in biological networks responsible for complex diseases is far from complete and the use of high-dimensional data together with large-scale biological databases (e.g., protein–protein interaction and pathway databases) will be crucial for uncovering aberrant biological processes. However, the challenge is to accommodate the large volume of -omics data, which is growing exponentially. Additionally, much work needs to be done to identify the subnetworks of metabolic reactions associated with diseases such as PD. Moreover, a reliable computational approach to identify subnetwork-associated disease processes is currently limited by the incompleteness of the available interactome maps and limitations of the existing tools.

As widely acknowledged, gaining an integrated understanding of the interactions among the genome, proteome, metabolome, and environment as mediated by the underlying cellular network, may offer a basis for future advances. However, some of the most difficult problems in this area include discovering dynamic (rather than static) processes in cells, connecting molecular-level network activities to functional behavior at the cellular level, developing data-driven computational models that reflect the causal relationships between molecules (including those designated as drug targets), biomarkers (as potential readouts of network modulation as would be the case with a disease-modifying therapy), measuring the consequent changes that occur in cellular dynamic processes, and predicting the impact of an intervention on

the system to effect a change consistent with a beneficial outcome. An especially challenging focus of importance is to find ways to model changes in biological entities that could affect the dynamics of the biological process using large-scale and diverse -omics data. To accommodate these challenges, network-based research is shifting toward integrated multiple networks or networks composed of heterogeneous large-scale data elements. To process large-scale -omics data, we need to develop next-generation algorithms and tools to study the relationships between aberrant human genes, proteins, and interactome networks. It is in this context that we delineate new disease-associated biological molecules in relation to disease-specific networks, that we understand how network perturbations can lead to disease, and that we use this knowledge to develop better diagnostics and therapeutics for disease. It is in the integration of these various data sets both temporally and spatially that we begin to see the emergent properties of disease-specific networks and that we then use this knowledge to modify disease natural history (Fig. 4).

INFORMATICS IN BIOMARKER DISCOVERY

In the era of -omics, moving forward in the clinical management of PD will require the development of specific and sensitive biomarkers that could detect disease early, define a prognostic trajectory of disease, and/or provide an indicator of response to therapy. Ideally, these biomarkers would be present early enough in the disease course that interventions could halt progression or even reverse damage. Risk alleles in and of themselves are biomarkers, measurable entities that denote risk for disease; however, they do so in a nonspecific and nonsensitive manner. The technologies outlined above provide a way forward as we link genetic information to the other molecular and organismal levels, translating disease risk to disease-specific networks. In fact, a number of studies have used various -omics approaches to identify potential PD biomarkers (reviewed in Caudle et al. 2010). As we increase

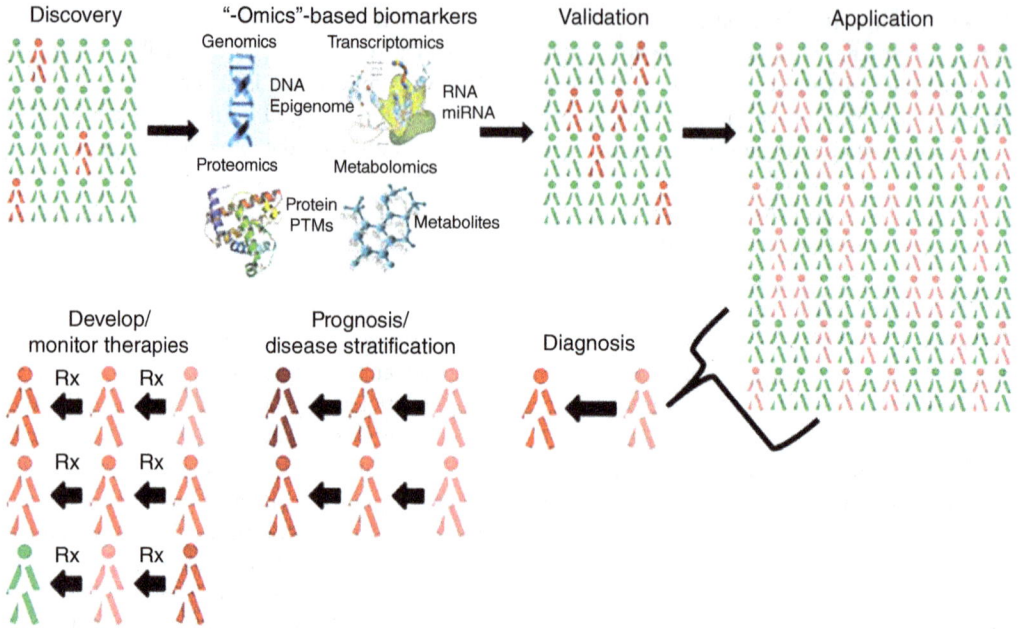

Figure 4. Application of "-omics"-based biomarker strategy to discover, validate, and apply molecular profiles to disease diagnosis, prognosis, and therapeutic development. Biomarker discovery efforts follow a predictable model. A discovery cohort of case (red) and control (green) subjects is amassed. Biological samples along with clinical, demographic, and other data are collected. High-dimensional genomic, transcriptomic, proteomic, and metabolomic data are generated and integrated with clinical data to elucidate the dynamic networks and their critical nodes that contribute to risk and evolution of disease. These pathways are then validated in a separate cohort of cases and controls. Robust, sensitive, and specific profiles can then be applied on a population scale to provide readouts for individuals' risk (pink) for disease. These profiles can be informative to disease diagnosis (pink to red), to prognosis and disease stratification (affected individuals with different temporal progression and severity), and in developing and monitoring therapies that slow, halt, or reverse disease progression. Additional historical (environment, lifestyle), clinical, and imaging data (e.g., PET, SPECT) will be integrated with molecular pathway data that will also be informative in disease diagnosis, stratification, and therapies. (PTMs) Posttranslational modifications. Images of myoglobin structure (www.en.wikipedia.org/wiki/File: Myoglobin.png) and ribosome/mRNA translation (www.en.wikipedia.org/wiki/File:Ribosome_mRNA_translation_en.svg) have been released to the public domain. (Figure and figure legend reprinted from Scholz et al. 2012, © Cold Spring Harbor Laboratory Press.)

our understanding of these disease-specific networks, we can begin to identify other tissues that may be affected and that could be used as surrogates for ongoing monitoring of disease processes within the CNS. For example, induced pluripotent stem (iPS) cells derived from patients may become an important resource not only for understanding disease biology (Park et al. 2008) and fueling biomarker elucidation, but also as a strategic part of therapeutics development (Wernig et al. 2008; Cooper et al. 2010). A more immediate application of recent advances in genomics and informatics to bio-

marker development is as an initial screen for risk. As more risk loci are identified and as the costs of screening decline, more widespread screening of populations of individuals will become feasible. Although this would not be a particularly sensitive or specific screen, it would provide for an economical triaging of high-risk subjects for more in-depth screening. It is also important to note that modalities other than molecular markers will form an important component of our biomarker armamentarium and that these modalities will need to be integrated within the network-centric approach. These modalities range from the relatively inexpensive and insensitive (history of constipation; hyposmia) to the expensive and sensitive (PET or SPECT neuroimaging).

As noted above, GWAS studies have been performed to reveal the genetic association of disease with SNPs. Despite the success of GWAS, the heritability of common disorders such as PD cannot be fully explained by the genes that have been discovered. A similar pattern is seen when we expand our search for biomarkers of disease risk to the other -omics. A number of reasons can be ascribed to the limitations of these studies for common diseases with complex traits. One explanation is that the current biostatistical analyses are agnostic or unbiased and thus ignore what is known about disease pathology. In addition, the linear modeling in GWAS analyses usually considers only one SNP at a time, whereas ignoring the genomic and epigenomic factors of each SNP. However, recently, there has been a shift away from this approach toward a more holistic one that recognizes the complexity of the genotype–phenotype relationship that is likely characterized by significant genetic heterogeneity and gene–gene and gene–environment interactions. Furthermore, strategies have been used to iteratively mine GWAS data, including the identification of potential PD targets and pathways using meta-analyses of previous GWAS and neuronal transcriptomic profiling studies (Zheng et al. 2010; Edwards et al. 2011).

The limitations of a linear model and other parametric statistical approaches have motivated the development of data mining and machine learning methods (Hastie et al. 2009). The advantage of these computational approaches is that they make fewer assumptions about the functional form of the model and the effects being modeled. In effect, data mining and machine learning methods are much more consistent with the notion of having the data direct the model, rather than forcing the data to fit a predetermined model. Several reviews highlight the need for these newer methods, including machine learning approaches such as random forests (RFs) and multifactor dimensionality reduction (MDR), which have been developed

to address some of these issues (Tarca et al. 2007; Ressom et al. 2008; Moore 2010; Sun 2010). It is also clear that evolving informatics approaches will play an important role in addressing the complexity of the underlying molecular basis of many common human diseases. These methodologies have the potential to identify other molecular species (proteins, metabolites), which could serve as disease biomarkers in addition to the genome and interactome. In identifying nodal proteins, proteins that are present at the nodes of a network, a proteomics approach provides a great platform, which typically assesses proteins in an unbiased fashion and provides the means to study the proteomic profile of a complex biological system on a large scale. Several technologies continue to evolve that are composed of integrated technical components, including separation technology, mass spectrometry (MS), and bioinformatics data processing. With advances in analytical technology and statistical analyses, several studies have set out to develop proteomic "molecular profiles" of PD tissues, including blood, cerebrospinal fluid (CSF), and postmortem brain (reviewed in Caudle et al. 2010). Similar methodologies and analytical approaches are being used in the search for "metabolic profiles" of PD, which include compounds such as lipids, amino acids, fatty acids, amines, alcohols, sugars, organic phosphates, hydroxyl acids, aromatics, purines, and other high abundance or clinically important molecules; however, these databases are incomplete for secondary metabolites, drugs, and environmental compounds.

As technology and analytic methods improve, we will generate more complete annotations of the genomic, transcriptomic, proteomic, and metabolic spaces, which would greatly enhance the analysis of specific pathways and molecules involved in PD and would yield additional insight into the pathogenesis of the disease. In addition, it would provide a platform for future meta-analytic studies, which could assuage much of the between-study variability currently encountered when analyzing multiple studies. However, despite the advance in -omics, there are still several issues that need to be addressed and resolved. Most importantly, issues around data integration, analysis, and interpretation pose a great challenge, especially in the context of ever-expanding data being generated.

CELL-TYPE DYSREGULTION AND TARGETS IDENTIFIED BY HIGH-RESOLUTION TRNSCRIPTOMICS

Whereas every brain cell harbors the same genome (somatic mutations and mosaicism notwithstanding), the transcriptomic profiles of these cells vary substantially across broad functional classes, brain regions, age, and disease. Advancements in single-cell transcriptomics technologies over the last decade have allowed for the profiling of multiple brain regions across species in health and disease; indeed, large-scale efforts such as the NIH's BRAIN Initiative and the disease-focused Accelerating Medicines Partnerships (for both Alzheimer's disease and PD) have resulted in data sets from hundreds of individuals profiled at the single-cell level. Importantly, these approaches have implicated a multitude of cell types dysregulated in diseases such as PD. Initial single-cell research efforts have focused on identifying drivers of neuronal death in PD (Kamath et al. 2022), identifying particular subsets of dopaminergic neurons in the substantia nigra that appeared to be more susceptible than others in human tissue obtained from PD patients. This heterogeneity among dopaminergic neurons suggests intriguing mechanisms by which certain neuronal subsets may be able to withstand the degenerative process, potentially nominating pathways or targets that could promote neuronal resilience.

Beyond the neuronal space, elucidating the full complement of cell types associated with neuronal resilience and risk in PD may yield insight into PD pathology and progression, ultimately leading to potential cell-type-specific targets for new disease-modifying therapies. For example, whereas neuronal susceptibility has been studied from a cell-autonomous perspective, but mounting evidence suggests that CNS-resident and peripheral immune cells also contribute to neurodegeneration (Garden

and La Spada 2012). Moreover, genes implicated in the regulation of leukocyte/lymphocyte activity and cytokine-mediated signaling as well as genetic variants for the microglial receptor *TREM2* have been linked to increased risk for several neurodegenerative disorders, including PD (Holmans et al. 2013; Heavener and Bradshaw 2022). Ongoing efforts from the NIH's Accelerating Medicines Partnership—Parkinson's Disease, among others, are profiling the full array of cell types in the human brain to gather a more complete picture of cell-type-specific dysregulation in various stages of the disease.

TRANSLATING GENOMICS INTO RATIONAL THERAPEUTICS

Drug discovery is the process by which new drugs are identified. The traditional method relies on trial-and-error in testing chemical substances against purified molecules and cultured cells, and subsequently examining their effects on a model of disease before taking such a candidate into clinical studies. However, in the last few decades, a new approach, termed rational drug discovery (RDD) has been adopted, which relies on characterized molecular mechanisms of disease. RDD posits that modulation of a specific target, putatively causal in disease pathogenesis, will have therapeutic value. This raises two fundamental questions: What is a specific target and how is a modulator of this target found? The first question is at the core of drug discovery.

Current drug discovery assumes that diseases can be characterized by a faulty protein structure or aberrant expression of a protein encoded by a variant or differentially expressed gene, and that identification of candidate drugs that modulate the activity of the proteins will have an effect on phenotype and/or disease outcome. This view is exemplified by expectations heralded with human genome sequencing technology, in which it was estimated that ∼8000 genes would be available as drug targets (Imming et al. 2006). Currently, the number of drug targets correlated with genetic variation or

Cite this article as *Cold Spring Harb Perspect Med* doi: 10.1101/cshperspect.a041621

polymorphisms is ~220 (Russ and Lampel 2005). As we move from monogenic diseases to complex diseases, there is no consensus as to how genetics will inform drug discovery. However, it is expected that the discovery of aberrant genetic networks, populated by aberrant proteomic and metabolic nodes, will provide target candidates for therapeutics development. It is in this arena that disease-specific networks and nodes can be used for the rational design of common and even personalized therapeutics.

From the above, it should be clear that defining a disease target is not formulaic. However, once a target is chosen, its manipulation unfolds in one of two ways: target- or ligand-based drug discovery. Target-based drug discovery starts with the 3D structure of a target. If a 3D structure of the target is not available, it may be empirically determined by crystallography or generated informatically via homology modeling using proteins with similar domains as a template.

Alternatively, ligand-based drug discovery starts with structural information of a known or predicted ligand. In either case, a pharmacophore is designed as bait. A pharmacophore is an abstract description of the molecular features that are required for interaction between a ligand and a target. More specifically, the IUPAC defines a pharmacophore to be "an ensemble of steric and electronic features that are necessary to ensure the optimal supramolecular interactions with a specific biological target and to trigger (or block) its biological response" (Wermuth et al. 1998). Because target/ligand interactions are "polar positive," "polar negative," or "hydrophobic," typical features considered in designing a pharmacophore are hydrophobic, aromatic, a hydrogen bond acceptor, a hydrogen bond donor, cation, or anion moieties. Because ligand-based drug discovery relies on knowledge of other molecules known to bind a biological target, the minimum necessary structural characteristics derived from these molecules are used in designing the pharmacophore, which then can be used to identify similar compounds via screening of chemical libraries or through de novo synthesis. The basic principle is that similar molecules behave similarly. In other words, similar chemical groups and entities will have similar biological effects. This gives rise to the concept of the structure–activity relationship, or

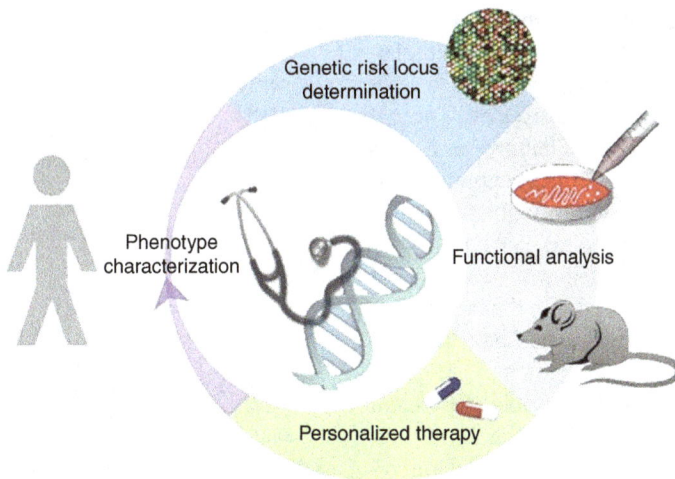

Figure 5. Genomics will play an integral role in the development of personalized therapeutics. The availability of detailed phenotype data from large patient/control cohorts is an important prerequisite for high-throughput genetic screening studies, including genome-wide association studies and genomic sequencing. After genetic risk loci have been dissected, in silico, in vitro, and in vivo analyses establish the underlying functional pathways and help to posit targets for rational, personalized therapies. (Figure and figure legend reprinted from Scholz et al. 2012, © Cold Spring Harbor Laboratory Press.)

SAR. As 3D structures of biological targets increase, and detailed information about molecular interactions between ligand and target becomes available, the application of SAR has become more complex. Drug discovery has evolved with the use of high-performance computing to enable computer-aided drug design (CADD) and sophisticated statistical algorithms and molecular dynamic tools to provide quantitative methodologies for SAR (QSAR) to rank order the potential potency of a number of biologically similar compounds. Once a series of compounds is identified using the above approach, they are labeled as "hits," which are ready to be tested in biological assay systems. As shown by the analytical bottlenecks in PD genomics and biomarker discovery, the evolution of bioinformatics will be critical to the successful development of improved PD therapeutics. The role of bioinformatics in connecting aberrant networks and nodes fundamental to PD pathogenesis with validated targets developed through computational chemistry and target modeling is anticipated to accelerate the process of drug discovery.

THE ROAD AHEAD

The main aim of genomic research is to identify pathways that are suitable for targeted therapeutic interventions to prevent, slow, halt, or reverse neurodegenerative disease processes. To that end, the success of translational research rests on the resolution of the complex genomic architecture of human disease, translating this to understanding aberrant networks, cell types, pathways, and nodes associated with disease, and implementing this knowledge in the rational design of therapeutics, which could be tailored to the individual (personalized therapy) (Fig. 5). However, this success is not only dependent on the advancement of technologies and their applications. Success will also depend on regional, national, and even international collaborative efforts. In much the same way that the neurogenetics field has evolved, moving forward will require the collective efforts of scientists, clinicians, healthcare providers, policymakers, and, importantly, patients. The impact of these efforts will also go beyond translational research and therapeutics development. Given the potential to revolutionize medicine, a host of societal issues will need to be addressed, including socioeconomic, ethical, clinical acceptance, medical education, cost-effectiveness, and regulatory considerations.

REFERENCES

*Reference is also in this subject collection.

Amit I, Garber M, Chevrier N, Leite AP, Donner Y, Eisenhaure T, Guttman M, Grenier JK, Li W, Zuk O, et al. 2009. Unbiased reconstruction of a mammalian transcriptional network mediating pathogen responses. *Science* **326:** 257–263. doi:10.1126/science.1179050

Andre F, Job B, Dessen P, Tordai A, Michiels S, Liedtke C, Richon C, Yan K, Wang B, Vassal G, et al. 2009. Molecular characterization of breast cancer with high-resolution oligonucleotide comparative genomic hybridization array. *Clin Cancer Res* **15:** 441–451. doi:10.1158/1078-0432 .CCR-08-1791

Bonifati V, Rizzu P, van Baren MJ, Schaap O, Breedveld GJ, Krieger E, Dekker MC, Squitieri F, Ibanez P, Joosse M, et al. 2003. Mutations in the *DJ-1* gene associated with autosomal recessive early-onset parkinsonism. *Science* **299:** 256–259. doi:10.1126/science.1077209

Braak H, Del Tredici K, Bratzke H, Hamm-Clement J, Sandmann-Keil D, Rüb U. 2002. Staging of the intracerebral inclusion body pathology associated with idiopathic Parkinson's disease (preclinical and clinical stages). *J Neurol* **249:** iii1–iii5. doi:10.1007/s00415-002-1301-4

Braak H, Del Tredici K, Rüb U, de Vos RA, Jansen Steur EN, Braak E. 2003a. Staging of brain pathology related to sporadic Parkinson's disease. *Neurobiol Aging* **24:** 197–211. doi:10.1016/S0197-4580(02)00065-9

Braak H, Rüb U, Gai WP, Del Tredici K. 2003b. Idiopathic Parkinson's disease: possible routes by which vulnerable neuronal types may be subject to neuroinvasion by an unknown pathogen. *J Neural Transm* **110:** 517–536. doi:10.1007/s00702-002-0808-2

Browaeys R, Saelens W, Saeys Y. 2020. Nichenet: modeling intercellular communication by linking ligands to target genes. *Nat Methods* **17:** 159–162. doi:10.1038/s41592-019-0667-5

Caudle WM, Bammler TK, Lin Y, Pan S, Zhang J. 2010. Using "omics" to define pathogenesis and biomarkers of Parkinson's disease. *Expert Rev Neurother* **10:** 925–942. doi:10.1586/ern.10.54

Cerami E, Demir E, Schultz N, Taylor BS, Sander C. 2010. Automated network analysis identifies core pathways in glioblastoma. *PLoS ONE* **5:** e8918. doi:10.1371/journal .pone.0008918

Chang JT, Carvalho C, Mori S, Bild AH, Gatza ML, Wang Q, Lucas JE, Potti A, Febbo PG, West M, et al. 2009. A genomic strategy to elucidate modules of oncogenic pathway signaling networks. *Mol Cell* **34:** 104–114. doi:10 .1016/j.molcel.2009.02.030

Chen J, Sam L, Huang Y, Lee Y, Li J, Liu Y, Xing HR, Lussier YA. 2010. Protein interaction network underpins concordant prognosis among heterogeneous breast cancer signatures. *J Biomed Inform* **43**: 385–396. doi:10.1016/j.jbi.2010.03.009

Chien CH, Sun YM, Chang WC, Chiang-Hsieh PY, Lee TY, Tsai WC, Horng JT, Tsou AP, Huang HD. 2011. Identifying transcriptional start sites of human microRNAs based on high-throughput sequencing data. *Nucleic Acids Res* **392**: 9345–9356. doi:10.1093/nar/gkr604

Cooper O, Hargus G, Deleidi M, Blak A, Osborn T, Marlow E, Lee K, Levy A, Perez-Torres E, Yow A, et al. 2010. Differentiation of human ES and Parkinson's disease iPS cells into ventral midbrain dopaminergic neurons requires a high activity form of SHH, FGF8a and specific regionalization by retinoic acid. *Mol Cell Neurosci* **45**: 258–266. doi:10.1016/j.mcn.2010.06.017

Croft D, O'Kelly G, Wu G, Haw R, Gillespie M, Matthews L, Caudy M, Garapati P, Gopinath G, Jassal B, et al. 2011. Reactome: a database of reactions, pathways and biological processes. *Nucleic Acids Res* **39**: D691–D697. doi:10.1093/nar/gkq1018

Dauer W, Przedborski S. 2003. Parkinson's disease: mechanisms and models. *Neuron* **39**: 889–909. doi:10.1016/S0896-6273(03)00568-3

Dawson TM, Dawson VL. 2003. Molecular pathways of neurodegeneration in Parkinson's disease. *Science* **302**: 819–822. doi:10.1126/science.1087753

Edwards YJ, Beecham GW, Scott WK, Khuri S, Bademci G, Tekin D, Martin ER, Jiang Z, Mash DC, Ffrench-Mullen J, et al. 2011. Identifying consensus disease pathways in Parkinson's disease using an integrative systems biology approach. *PLoS ONE* **6**: e16917. doi:10.1371/journal.pone.0016917

Farrer M, Maraganore DM, Lockhart P, Singleton A, Lesnick TG, de Andrade M, West A, de Silva R, Hardy J, Hernandez D. 2001. α-synuclein gene haplotypes are associated with Parkinson's disease. *Hum Mol Genet* **10**: 1847–1851. doi:10.1093/hmg/10.17.1847

Feany MB, Bender WW. 2000. A *Drosophila* model of Parkinson's disease. *Nature* **404**: 394–398. doi:10.1038/35006074

* Gao V, Crawford CV, Burré J. 2024. The gut–brain axis in Parkinson's disease. *Cold Spring Harb Perspect Med* doi:10.1101/cshperspect.a041618

Garden GA, La Spada AR. 2012. Intercellular (mis)communication in neurodegenerative disease. *Neuron* **73**: 886–901. doi:10.1016/j.neuron.2012.02.017

Gatza ML, Lucas JE, Barry WT, Kim JW, Wang Q, Crawford MD, Datto MB, Kelley M, Mathey-Prevot B, Potti A, et al. 2010. A pathway-based classification of human breast cancer. *Proc Natl Acad Sci* **107**: 6994–6999. doi:10.1073/pnas.0912708107

Hastie T, Tibshirani R, Friedman J. 2009. *The elements of statistical learning: data mining, inference, and prediction*, 2nd ed. Springer, New York.

Hawkes CH, Del Tredici K, Braak H. 2010. A timeline for Parkinson's disease. *Parkinsonism Relat Disord* **16**: 79–84. doi:10.1016/j.parkreldis.2009.08.007

Heavener KS, Bradshaw EM. 2022. The aging immune system in Alzheimer's and Parkinson's diseases. *Semin Immunopathol* **44**: 649–657. doi:10.1007/s00281-022-00944-6

Holmans P, Moskvina V, Jones L, Sharma M; International Parkinson's Disease Genomics Consortium; Vedernikov A, Buchel F, Saad M, Bras J, Bettella F, et al. 2013. A pathway-based analysis provides additional support for an immune-related genetic susceptibility to Parkinson's disease. *Hum Mol Genet* **22**: 1039–1049. doi:10.1093/hmg/dds492

Hu Z, Hung JH, Wang Y, Chang YC, Huang CL, Huyck M, DeLisi C. 2009. VisANT 3.5: multi-scale network visualization, analysis and inference based on the gene ontology. *Nucleic Acids Res* **37**: W115–W121. doi:10.1093/nar/gkp406

Imming P, Sinning C, Meyer A. 2006. Drugs, their targets and the nature and number of drug targets. *Nat Rev Drug Discov* **5**: 821–834. doi:10.1038/nrd2132

International HapMap Consortium. 2005. Haplotype map of the human genome. *Nature* **437**: 1299–1320. doi:10.1038/nature04226

International Parkinson Disease Genomics Consortium; Nalls MA, Plagnol V, Hernandez DG, Sharma M, Sheerin UM, Saad M, Simón-Sánchez J, Schulte C, Lesage S, et al. 2011. Imputation of sequence variants for identification of genetic risks for Parkinson's disease: a meta-analysis of genome-wide association studies. *Lancet* **377**: 641–649. doi:10.1016/S0140-6736(10)62345-8

International Parkinson's Disease Genomics Consortium (IPDGC); Wellcome Trust Case Control Consortium 2 (WTCCC2). 2011. A two-stage meta-analysis identifies several new loci for Parkinson's disease. *PLoS Genet* **7**: e1002142. doi:10.1371/journal.pgen.1002142

Jin S, Guerrero-Juarez CF, Zhang L, Chang I, Ramos R, Kuan CH, Myung P, Plikus MV, Nie Q. 2021. Inference and analysis of cell–cell communication using CellChat. *Nat Commun* **12**: 1088. doi:10.1038/s41467-021-21246-9

Kamath T, Abdulraouf A, Burris SJ, Langlieb J, Gazestani V, Nadaf NM, Balderrama K, Vanderburg C, Macosko EZ. 2022. Single-cell genomic profiling of human dopamine neurons identifies a population that selectively degenerates in Parkinson's disease. *Nat Neurosci* **25**: 588–595. doi:10.1038/s41593-022-01061-1

Kanehisa M, Goto S, Kawashima S, Okuno Y, Hattori M. 2004. The KEGG resource for deciphering the genome. *Nucleic Acids Res* **32**: D277–D280. doi:10.1093/nar/gkh063

Kim TY, Kim HU, Lee SY. 2010. Data integration and analysis of biological networks. *Curr Opin Biotechnol* **21**: 78–84. doi:10.1016/j.copbio.2010.01.003

Kitada T, Asakawa S, Hattori N, Matsumine H, Yamamura Y, Minoshima S, Yokochi M, Mizuno Y, Shimizu N. 1998. Mutations in the parkin gene cause autosomal recessive juvenile parkinsonism. *Nature* **392**: 605–608. doi:10.1038/33416

Kreeger PK, Lauffenburger DA. 2010. Cancer systems biology: a network modeling perspective. *Carcinogenesis* **31**: 2–8. doi:10.1093/carcin/bgp261

Lander ES. 2011. Initial impact of the sequencing of the human genome. *Nature* **470**: 187–197. doi:10.1038/nature09792

Lander ES, Linton LM, Birren B, Nusbaum C, Zody MC, Baldwin J, Devon K, Dewar K, Doyle M, FitzHugh W, et al. 2001. Initial sequencing and analysis of the human genome. *Nature* **409:** 860–921. doi:10.1038/35057062

Manolio TA, Collins FS, Cox NJ, Goldstein DB, Hindorff LA, Hunter DJ, McCarthy MI, Ramos EM, Cardon LR, Chakravarti A, et al. 2009. Finding the missing heritability of complex diseases. *Nature* **461:** 747–753. doi:10.1038/nature08494

Maraganore DM, de Andrade M, Elbaz A, Farrer MJ, Ioannidis JP, Kruger R, Rocca WA, Schneider NK, Lesnick TG, Lincoln SJ, et al. 2006. Collaborative analysis of α-synuclein gene promoter variability and Parkinson disease. *JAMA* **296:** 661–670. doi:10.1001/jama.296.6.661

Mardis ER. 2006. Anticipating the 1,000 dollar genome. *Genome Biol* **7:** 112. doi:10.1186/gb-2006-7-7-112

Masliah E, Rockenstein E, Veinbergs I, Mallory M, Hashimoto M, Takeda A, Sagara Y, Sisk A, Mucke L. 2000. Dopaminergic loss and inclusion body formation in α-synuclein mice: implications for neurodegenerative disorders. *Science* **287:** 1265–1269. doi:10.1126/science.287.5456.1265

Meyerson M, Gabriel S, Getz G. 2010. Advances in understanding cancer genomes through second-generation sequencing. *Nat Rev Genet* **11:** 685–696. doi:10.1038/nrg2841

Moore JH. 2010. Detecting, characterizing, and interpreting nonlinear gene–gene interactions using multifactor dimensionality reduction. *Adv Genet* **72:** 101–116. doi:10.1016/B978-0-12-380862-2.00005-9

Nikitin A, Egorov S, Daraselia N, Mazo I. 2003. Pathway studio—the analysis and navigation of molecular networks. *Bioinformatics* **19:** 2155–2157. doi:10.1093/bioinformatics/btg290

Ochs MF, Karchin R, Ressom H, Gentleman R. 2011. Identification of aberrant pathway and network activity from high-throughput data—workshop introduction. *Pac Symp Biocomput* **2011:** 364–368.

Paisán-Ruíz C, Jain S, Evans EW, Gilks WP, Simón J, van der Brug M, de Munain López A, Aparicio S, Gil AM, Khan N, et al. 2004. Cloning of the gene containing mutations that cause PARK8-linked Parkinson's disease. *Neuron* **44:** 595–600. doi:10.1016/j.neuron.2004.10.023

Park IH, Arora N, Huo H, Maherali N, Ahfeldt T, Shimamura A, Lensch MW, Cowan C, Hochedlinger K, Daley GQ. 2008. Disease-specific induced pluripotent stem cells. *Cell* **134:** 877–886. doi:10.1016/j.cell.2008.07.041

Peri S, Navarro JD, Amanchy R, Kristiansen TZ, Jonnalagadda CK, Surendranath V, Niranjan V, Muthusamy B, Gandhi TK, Gronborg M, et al. 2003. Development of human protein reference database as an initial platform for approaching systems biology in humans. *Genome Res* **13:** 2363–2371. doi:10.1101/gr.1680803

Pollack A. 2011. DNA sequencing caught in deluge of data. *The New York Times*, November 30.

Polymeropoulos MH, Lavedan C, Leroy E, Ide SE, Dehejia A, Dutra A, Pike B, Root H, Rubenstein J, Boyer R, et al. 1997. Mutation in the α-synuclein gene identified in families with Parkinson's disease. *Science* **276:** 2045–2047. doi:10.1126/science.276.5321.2045

Qiu YQ, Zhang S, Zhang XS, Chen L. 2009. Identifying differentially expressed pathways via a mixed integer linear programming model. *IET Syst Biol* **3:** 475–486. doi:10.1049/iet-syb.2008.0155

Reich DE, Lander ES. 2001. On the allelic spectrum of human disease. *Trends Genet* **17:** 502–510. doi:10.1016/S0168-9525(01)02410-6

Ressom HW, Varghese RS, Zhang Z, Xuan J, Clarke R. 2008. Classification algorithms for phenotype prediction in genomics and proteomics. *Front Biosci* **13:** 691–708. doi:10.2741/2712

Russ AP, Lampel S. 2005. The druggable genome: an update. *Drug Discov Today* **10:** 1607–1610. doi:10.1016/S1359-6446(05)03666-4

Sanger F, Air GM, Barrell BG, Brown NL, Coulson AR, Fiddes CA, Hutchison CA, Slocombe PM, Smith M. 1977. Nucleotide sequence of bacteriophage ϕX174 DNA. *Nature* **265:** 687–695. doi:10.1038/265687a0

Schadt EE. 2009. Molecular networks as sensors and drivers of common human diseases. *Nature* **461:** 218–223. doi:10.1038/nature08454

Schaefer CF, Anthony K, Krupa S, Buchoff J, Day M, Hannay T, Buetow KH. 2009. PID: the pathway interaction database. *Nucleic Acids Res* **37:** D674–D679. doi:10.1093/nar/gkn653

Scholz SW, Mhyre T, Ressom H, Shah S, Federoff HJ. 2012. Genomics and bioinformatics of Parkinson's disease. *Cold Spring Harb Perspect Med* **2:** a009449. doi:10.1101/cshperspect.a009449

Simón-Sánchez J, Schulte C, Bras JM, Sharma M, Gibbs JR, Berg D, Paisan-Ruiz C, Lichtner P, Scholz SW, Hernandez DG, et al. 2009. Genome-wide association study reveals genetic risk underlying Parkinson's disease. *Nat Genet* **15:** 15. doi:10.1038/nm0109-15

Singh A, Greninger P, Rhodes D, Koopman L, Violette S, Bardeesy N, Settleman J. 2009. A gene expression signature associated with "K-Ras addiction" reveals regulators of EMT and tumor cell survival. *Cancer Cell* **15:** 489–500. doi:10.1016/j.ccr.2009.03.022

Singleton AB, Farrer M, Johnson J, Singleton A, Hague S, Kachergus J, Hulihan M, Peuralinna T, Dutra A, Nussbaum R, et al. 2003. α-Synuclein locus triplication causes Parkinson's disease. *Science* **302:** 841. doi:10.1126/science/1090278.

Smoot ME, Ono K, Ruscheinski J, Wang PL, Ideker T. 2011. Cytoscape 2.8: new features for data integration and network visualization. *Bioinformatics* **27:** 431–432. doi:10.1093/bioinformatics/btq675

Sun YV. 2010. Multigenic modeling of complex disease by random forests. *Adv Genet* **72:** 73–99. doi:10.1016/B978-0-12-380862-2.00004-7

Tan CS, Bodenmiller B, Pasculescu A, Jovanovic M, Hengartner MO, Jorgensen C, Bader GD, Aebersold R, Pawson T, Linding R. 2009. Comparative analysis reveals conserved protein phosphorylation networks implicated in multiple diseases. *Sci Signal* **2:** ra39. doi:10.1126/scisignal.2000316

Tarca AL, Carey VJ, Chen XW, Romero R, Drăghici S. 2007. Machine learning and its applications to biology. *PLoS Comput Biol* **3:** e116. doi:10.1371/journal.pcbi.0030116

Taylor IW, Linding R, Warde-Farley D, Liu Y, Pesquita C, Faria D, Bull S, Pawson T, Morris Q, Wrana JL. 2009. Dynamic modularity in protein interaction networks predicts breast cancer outcome. *Nat Biotechnol* **27:** 199–204. doi:10.1038/nbt.1522

Valente EM, Bentivoglio AR, Dixon PH, Ferraris A, Ialongo T, Frontali M, Albanese A, Wood NW. 2001. Localization of a novel locus for autosomal recessive early-onset Parkinsonism, PARK6, on human chromosome 1p35–p36. *Am J Hum Genet* **68:** 895–900. doi:10.1086/319522

Vaske CJ, Benz SC, Sanborn JZ, Earl D, Szeto C, Zhu J, Haussler D, Stuart JM. 2010. Inference of patient-specific pathway activities from multi-dimensional cancer genomics data using PARADIGM. *Bioinformatics* **26:** i237–i245. doi:10.1093/bioinformatics/btq182

Venter JC, Adams MD, Myers EW, Li PW, Mural RJ, Sutton GG, Smith HO, Yandell M, Evans CA, Holt RA, et al. 2001. The sequence of the human genome. *Science* **291:** 1304–1351. doi:10.1126/science.1058040

Wang CC, Cirit M, Haugh JM. 2009. PI3K-dependent cross-talk interactions converge with Ras as quantifiable inputs integrated by Erk. *Mol Syst Biol* **5:** 246. doi:10.1038/msb .2009.4

Wang J, Zhang Y, Marian C, Ressom HW. 2012. Identification of aberrant pathways and network activities from high-throughput data. *Brief Bioinform* **13:** 406–419. doi:10.1093/bib/bbs001

Wermuth CG, Ganellin CR, Lindberg P, Mitscher LA. 1998. Glossary of terms used in medicinal chemistry (IUPAC Genomics and Bioinformatics of Parkinson's Disease Recommendations 1998). *Pure Appl Chem* **70:** 1129–1143. doi:10.1351/pac199870051129

Wernig M, Zhao JP, Pruszak J, Hedlund E, Fu D, Soldner F, Broccoli V, Constantine-Paton M, Isacson O, Jaenisch R. 2008. Neurons derived from reprogrammed fibroblasts functionally integrate into the fetal brain and improve symptoms of rats with Parkinson's disease. *Proc Natl Acad Sci* **105:** 5856–5861. doi:10.1073/pnas.0801677105

* Westenberger A, Brüggemann N, Klein C. 2024. Genetics of Parkinson's disease: from causes to treatment. *Cold Spring Harb Perspect Med* **12:** a041774. doi:10.1101/cshperspect .a041774

Wu G, Feng X, Stein L. 2010. A human functional protein interaction network and its application to cancer data analysis. *Genome Biol* **11:** R53. doi:10.1186/gb-2010-11-5-r53

Yeger-Lotem E, Riva L, Su LJ, Gitler AD, Cashikar AG, King OD, Auluck PK, Geddie ML, Valastyan JS, Karger DR, et al. 2009. Bridging high-throughput genetic and transcriptional data reveals cellular responses to α-synuclein toxicity. *Nat Genet* **41:** 316–323. doi:10.1038/ng .337

Zheng B, Liao Z, Locascio JJ, Lesniak KA, Roderick SS, Watt ML, Eklund AC, Zhang-James Y, Kim PD, Hauser MA, et al. 2010. *PGC-1α*, a potential therapeutic target for early intervention in Parkinson's disease. *Sci Transl Med* **2:** 52ra73. doi:10.1126/scitranslmed.3001059

Zimprich A, Biskup S, Leitner P, Lichtner P, Farrer M, Lincoln S, Kachergus J, Hulihan M, Uitti RJ, Calne DB, et al. 2004. Mutations in LRRK2 cause autosomal-dominant parkinsonism with pleomorphic pathology. *Neuron* **44:** 601–607. doi:10.1016/j.neuron.2004.11.005

Functional Neuroanatomy of the Normal and Pathological Basal Ganglia

José L. Lanciego[1,2,3] and José A. Obeso[2,3,4]

[1]CNS Gene Therapy Program, Center for Applied Medical Research (CIMA), University of Navarra Medical School, 31008 Pamplona, Spain

[2]Centro de Investigación Biomédica en Red de Enfermedades Neurodegenerativas (CiberNed-ISCIII), 28029 Madrid, Spain

[3]Aligning Science Across Parkinson's (ASAP) Collaborative Research Network, Chevy Chase, Maryland 20815, USA

[4]HM CINAC (Centro Integral de Neurociencias Abarca Campal), Hospital Universitario HM Puerta del Sur, HM Hospitales, 28938 Madrid, Spain

Correspondence: jlanciego@unav.es; jobeso.hmcinac@hmhospitales.com

The term "basal ganglia" refers to a group of interconnected subcortical nuclei engaged in motor planning and movement initiation, executive functions, behaviors, and emotions. Dopamine released from the substantia nigra is the underlying driving force keeping the basal ganglia network under proper equilibrium and, indeed, reduction of dopamine levels triggers basal ganglia dysfunction, setting the groundwork for several movement disorders. The canonical basal ganglia model has been instrumental for most of our current understanding of the normal and pathological functioning of this subcortical network. This model explains how cortical information flows through the basal ganglia nuclei back to the cortex by going through two pathways with opposing effects that together lead to the proper execution of a given movement. The basal ganglia model has paved the way for the standard clinical management of Parkinson's disease, where pharmacological and neurosurgical treatments in place collectively afford an impressive symptomatic alleviation. Although much of the model has remained, the canonical model has been enriched with new arrivals gathered from evidence provided in the last three decades. Here, we sought to provide a comprehensive review of the basal ganglia network, with emphasis on structure, connectivity patterns, and basic operational principles, both in normal and pathological conditions.

The basal ganglia are made up of a variety of subcortical nuclei engaged primarily in movement initiation and planning, together with few other roles such as motor learning, executive functions, behavior, reward, and emotions.[5] The term "basal ganglia" in the strictest sense refers to subcortical nuclei located deep in the brain hemispheres (comprising the striatum

---

[5]This is an update to a previous article published in *Cold Spring Harbor Perspectives in Medicine* [Lanciego et al. (2012). *Cold Spring Harb Perspect Biol* **2:** a009621. doi:10.1101/cshperspect.a009621].

Copyright © 2025 Cold Spring Harbor Laboratory Press; all rights reserved
Cite this article as *Cold Spring Harb Perspect Med* doi: 10.1101/cshperspect.a041617

or caudate, putamen, and globus pallidus nuclei), whereas the so-called "basal ganglia-related nuclei" are made of structures located in the diencephalon (subthalamic nucleus [STN]), mesencephalon (substantia nigra), and pons (pedunculopontine nucleus [PPN]). Clinical observations made in the early years of the twentieth century settled the conceptual framework of basal ganglia function, by showing that lesions of the lenticular nucleus and the STN triggered the appearance of parkinsonian signs, dystonia, and hemiballismus (Wilson 1925; Purdon-Martin 1927). These observations led to the distinction between extrapyramidal and pyramidal systems (related to basal ganglia and corticospinal systems, respectively). Although often used in the neurological clinical practice over many years, at present these terms are considered obsolete and thus are no longer in use.

The basal ganglia and related nuclei can broadly be categorized as input, output, and intrinsic nuclei. The caudate (CN), putamen (Put), and accumbens (Acb) nuclei are the input nuclei, receiving incoming information from different sources, such as cortical (ipsi- and contralateral), thalamic, and nigral. Regarding neurotransmission, corticostriatal and thalamostriatal projections are glutamatergic and of an excitatory nature, whereas nigrostriatal projections are dopaminergic and considered modulatory, since they can trigger either inhibition or excitation as a function of the subtype of dopamine receptors located in postsynaptic striatal neurons. The basal ganglia output nuclei are the structures sending information to the thalamus and comprise the internal segment of the globus pallidus (GPi) and the substantia nigra pars reticulate (SNr). Basal ganglia output pathways (pallidothalamic and nigrothalamic) are of an inhibitory nature, using GABA as a neurotransmitter. Finally, the so-called intrinsic nuclei are located between the input and output nuclei and comprise the external segment of the globus pallidus (GPe), the STN, and the substantia nigra pars compacta (SNc). The output of the GPe is GABA-mediated (inhibitory), STN efferent projections are glutamatergic (excitatory), whereas SNc dopaminergic output is considered as modulatory. Information coming from cortical and thalamic sources reaches the input nuclei and it is further processed within the basal ganglia system before being sent to the thalamus, which, in turn, sends information back to the cerebral cortex. The proper function of the basal ganglia requires dopamine to be released in the input nuclei to further keep an adequate balance between excitation and inhibition within the network. Basal ganglia components are shown in Figure 1. While this paper and the one by Wichmann (2024) share some common information about the functional neuroanatomy of the basal ganglia, their primary focus differs. This work centers on the anatomy, with only brief mentions of Parkinson's disease and dyskinesia. In contrast, Wichmann (2024) emphasizes the pathophysiology of Parkinson's disease, with anatomical details provided mainly as background.

BASAL GANGLIA AND RELATED NUCLEI

Input Nuclei: the Striatum

The striatum is by far the largest subcortical brain structure, with an estimated volume of 10 cm^3 in the human brain (Schröeder et al. 1975). It is a heterogeneous structure made of several types of neuronal populations (projection neurons and interneurons) that receive cortical and subcortical inputs from different sources and projects downstream to basal ganglia nuclei.

Striatal Neurons: Projection Neurons and Interneurons

Striatal neurons can be divided into projection neurons and interneurons, roughly accounting for 90% and 10% of striatal neurons in the human brain. Projection neurons (striatofugal neurons) are more often termed as medium-sized spiny neurons (MSNs) because have small-to-medium cellular somata (20 μm in diameter) and highly arborized multipolar dendritic processes covered by dendritic spines (Fig. 2). All striatal neurons are inhibitory and use GABA as neurotransmitter. According to their projection patterns, striatofugal MSNs

Cite this article as *Cold Spring Harb Perspect Med* doi: 10.1101/cshperspect.a041617

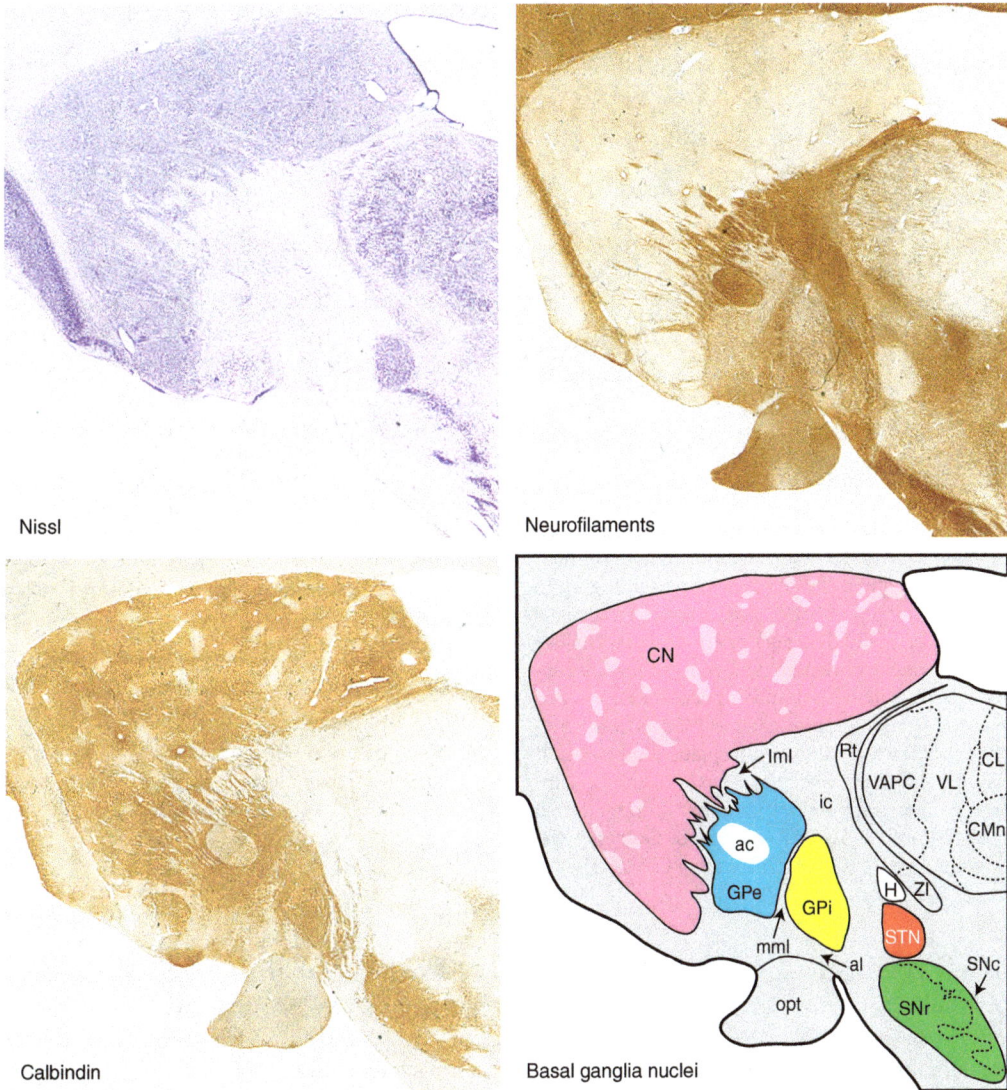

Figure 1. The basal ganglia and related nuclei. Cyto- and chemo-architectural stains performed in parasagittal sections of the nonhuman primate brain showing the location and boundaries of all components of the basal ganglia network (Lanciego and Vázquez 2012, © 2011, The Author(s)). (CN) Caudate nucleus, (GPe) external division of the globus pallidus, (GPi) internal division of the globus pallidus, (STN) subthalamic nucleus, (SNc) substantia nigra pars compacta, (SNr) substantia nigra pars reticulate, (Rt) thalamic reticular nucleus, (VAPC) ventral anterior thalamic nucleus, parvicellular part, (VL) ventral lateral thalamic nucleus, (CL) central lateral nucleus, (CMn) centromedian nucleus, (H) Forel's field H, (ZI) zona incerta, (ic) internal capsule, (opc) optic tract, (al) ansa lenticularis, (lml) lateral medullary lamina, (mml) medial medullary lamina.

could be divided further into those innervating the GPe and those projecting to the basal ganglia output nuclei GPi and SNr. Another differential feature is represented by the type of dopamine receptor expressed by striatofugal MSNs, where those targeting the GPe contain the dopamine receptor type 2 (D_2R) and those projecting to the basal ganglia output nuclei express the dopamine receptor type 1 (D_1R). Moreover, D_2R MSNs contain the neuropeptide enkephalin,

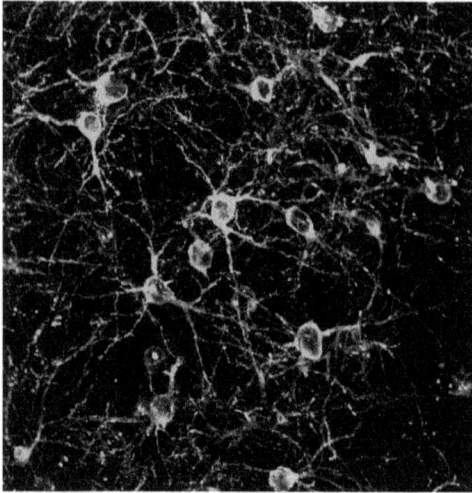

Figure 2. Striatal medium-sized spiny neurons (MSNs). Striatal MSNs are the most abundant neuronal phenotype in the striatum (representing up to 95% of the total number of striatal neurons). Striatal MSNs (20 μm in diameter) are multipolar stellate cells with radially oriented dendrites, which are covered with small postsynaptic specializations called dendritic spines. Photomicrograph taken from striatonigral-projecting MSNs, retrogradely labeled following the delivery of rabies virus into the rat substantia nigra pars reticulata (work conducted in collaboration with Lydia Kerkerian-Le Goff and Pascal Salin, Aix-Marseille University, CNRS, IBDM, Marseille, France).

whereas D_1R MSNs express substance P and dynorphin.

Besides projection neurons, the striatum also contains several different types of local-circuit neurons with smooth dendrites (e.g., interneurons). Striatal interneurons are typically categorized into four main groups depending on their neurochemical phenotypes and morphological characteristics (for review, see Kawaguchi et al. 1995). The most abundant group of interneurons is made of large-sized aspiny cholinergic neurons that use acetylcholine as neurotransmitter. These neurons are characterized by a continuous constant firing pattern and, therefore, are also known as tonically active neurons (TANs). Another major group of striatal interneurons is represented by GABAergic neurons expressing the calcium-binding protein parvalbumin. These interneurons have their own characteristic electrophysiological fingerprint and are also termed as fast-spiking interneurons (FSIs). The remaining two subtypes of striatal interneurons are both of GABAergic phenotype, one containing the calcium-binding protein calretinin, the other one using nitric oxide as neurotransmitter. Together with projection MSNs, interneurons form complex intrastriatal microcircuits, where both TANs and FSIs modulate the dopamine-driven output activity of MSNs, whereas calretinin and nitrergic interneurons innervate TANs and FSIs (Salin et al. 2009). All major four types of striatal interneurons are shown in Figure 3.

Striatal Compartments: Striosomes and Matrix

Although the caudate and putamen nuclei appear as rather homogeneous structures when viewed with cytoarchitectural techniques as well as in terms of dopaminergic innervation, the use of some specific markers highlights the presence of two subdivisions, known as striosomes and matrix compartments (Fig. 1). Initial evidence for this compartmental organization was provided in the late 1970s by Graybiel and Ragsdale (1978) using the histochemical silver staining of the enzyme acetylcholinesterase (AChE). A characteristic staining pattern was observed, with patchy-like areas showing very weak AChE signal (striosomes), embedded in a background area showing strong AChE stain (matrix). Moreover, striosomes and matrix compartments also exhibited a differential expression of the μ opioid receptor, highly enriched in striosomes and weakly expressed in the matrix (Perth et al. 1976; Desban et al. 1995). Furthermore, striosomes and matrix compartments also have differential neurochemical signatures, with strong immunoreactivity against enkephalin, substance P, GABA, and neurotensin observed in the striosomal compartment and an enriched expression for calcium-binding proteins such as calbindin and parvalbumin characterizing the matrix compartment (Gerfen et al. 1985; Prensa et al. 1999).

Besides neurochemical signatures disclosing striosomal and matrix compartments, another

Cite this article as *Cold Spring Harb Perspect Med* doi: 10.1101/cshperspect.a041617

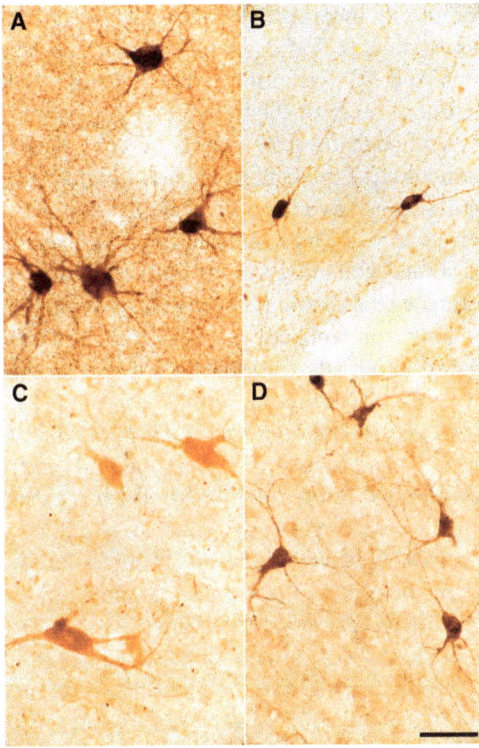

Figure 3. Striatal interneurons. Striatal interneurons account for 5%–10% of striatal neurons and comprise up to four main types of local-circuit neurons. Cholinergic interneurons (*A*) have a large soma size, use acetylcholine as a neurotransmitter, and are also known as tonically active neurons (TANs). Parvalbumin neurons (*B*) are GABAergic interneurons. Because of its own electrophysiological characteristic fingerprint, these interneurons are also termed as fast-spiking interneurons (FSIs). Both TANs and FSIs exert a modulatory role toward striatal MSNs. Other subtypes of striatal interneurons are represented by GABAergic interneurons expressing the calcium-binding protein calretinin (CR$^+$; *C*) and by nNOS$^+$ inhibitory interneurons using nitric oxide as neurotransmitter (*D*). Both CR$^+$ and nNOS$^+$ interneurons modulate the activity of TANs and FSIs; therefore, striatal interneurons altogether create sophisticated striatal microcircuits finally shaping striatal output (Salin et al. 2009). (Images taken from human postmortem brain samples at the level of the postcommissural putamen.) Scale bar, 50 μm.

layer of complexity is represented by the differential connectivity patterns that characterize both compartments. Despite the fact that striatal MSNs have extensive dendritic arborizations

and profuse axon collaterals, these processes remain restricted within the striatal compartment in which they are located. In other words, in terms of MSN arborizations, striosomal MSNs remained confined within the boundaries of the striosome where they are located, and they never entered the matrix compartment. The same also applies to matrix MSNs, with neuronal arborizations spreading over the matrix compartment without entering the striosomes. The characteristic compartmental segregation of striatal MSNs was first reported in the late 1980s (Penny et al. 1988; Kawaguchi et al. 1989) and later confirmed by Fujiyama et al. (2011). To what extent striosomes and matrix can be viewed as isolated compartments is also reflected by their own pattern of efferent and afferent connections. Regarding striatofugal projections, matrix MSNs innervate the GPe, GPi, and SNr, whereas striosomal MSNs preferentially target SNc, with axon collaterals reaching GPe, GPi, and SNr (Gerfen et al. 1985; Bolam et al. 1988; Kawaguchi et al. 1989; Giménez-Amaya and Graybiel 1990; Fujiyama et al. 2011). Glutamatergic inputs originated from the cerebral cortex, thalamus, and amygdala, as well as dopaminergic afferents arising from the SNc also show a heterogeneous distribution throughout the striatal compartments. The matrix compartment is mainly innervated by striatal glutamatergic afferents coming from cortical motor and sensory areas and thalamostriatal projections originated in the caudal intralaminar nuclei, as well as dopaminergic nigrostriatal projections. In contrast, cortical limbic areas, amygdala, and parts of the SNc preferentially project to striosomes (Graybiel 1984, 1990; Donoghue and Herkenham 1986; Ragsdale and Graybiel 1988; Gerfen 1992; Sadikot et al. 1992a,b; Kincaid and Wilson 1996). Despite long-existing evidence and extensive characterization of striatal compartments across many different animal species, an accurate explanation of the functional correlate underlying this organization is still lacking. In addition to the canonical matrix D$_1$-MSNs and D$_2$-MSNs direct and indirect basal ganglia pathways, recent evidence supported a conceptually equivalent parallel system arising from the striosomal compartment and targeting dopaminergic neurons of the SNc, where striosomal D1-MSNs and D2-MSNs oppositely modulate striatal dopamine release (Lazaridis et al. 2024).

Striatal Afferent Systems: Cerebral Cortex, Thalamus, Substantia Nigra Pars Compacta, and Raphe

Corticostriatal Pathway. The entire cerebral cortex projects to the striatum through glutamatergic projections that make asymmetric synaptic contacts with dendritic spines of striatal MSNs. The bulk of corticostriatal projections innervate the matrix compartment in keeping with a well-characterized topographical pattern (Selemon and Goldman-Rakic 1985; Haber et al. 2000; McFarland and Haber 2000; Borra et al. 2021). As indicated above, striosomes receive afferent projections arising from limbic cortical areas (Gerfen 1984; Donoghue and Herkenham 1986). Layer V pyramidal neurons are the main sources of corticostriatal projections reaching the matrix and striosomes (upper and lower layer V neurons, respectively), although layer III pyramidal neurons are another moderate source of inputs targeting the matrix compartment. Regarding corticostriatal layer V pyramidal neurons, up to two distinct types have been deeply characterized, comprising the so-called pyramidal tract neurons (PT-type) and intratelencephalic neurons (IT-type). Neurons of the PT-type are the origin of the cortico-pyramidal tract, innervating the striatum via axon collaterals. In contrast, IT-type neurons directly innervate the striatum without giving rise to cortico-pyramidal projections (Reiner et al. 2003, 2010; Deng et al. 2015). Although IT-type neurons provide bilateral projections to the striatum, PT-type neurons only innervate the ipsilateral striatum (Reiner et al. 2003). Furthermore, IT- and PT-type neurons also differ in terms of targeted postsynaptic striatal neurons. In this regard, whereas D_1R MSNs are the main recipient for IT-type afferents, PT-type neurons mainly contact D_2R MSNs (Lei et al. 2004). This statement has been later challenged by Baillon et al. (2008) who reported that IT-type neurons equally excite both types of striatofugal neurons. However, it is worth noting that this is not completely inconsistent with a preferential innervation of D_1R MSNs by IT-type cortical neurons and PT-type innervation reaching D_2R neurons (Lei et al. 2004).

Although the role played by contralateral corticostriatal projections in basal ganglia pathophysiology has often been neglected, this pathway collectively represents an important source of striatal glutamatergic afferents. It has been estimated that crossed corticostriatal projections roughly represent between 18% and 30% of total glutamatergic afferents to the striatum from cortical sources (Borra et al. 2021; Rico et al. 2024). The main contributors to contralateral corticostriatal projections are layer V pyramidal neurons located in the superior frontal, precentral, and anterior cingulate gyri (Rico et al. 2024).

Thalamostriatal Pathway. Another major source of excitatory glutamatergic inputs to the striatum is represented by thalamostriatal projections. Initial descriptions of the thalamostriatal system were made in the mid-twentieth century (Vogt and Vogt 1941; Powell and Cowan 1956). Although this pathway is well preserved across animal species and properly characterized anatomically, its pathophysiological role remains to be elucidated and indeed the thalamostriatal system has been often neglected in classical basal ganglia models. Neuroanatomical tract-tracing studies revealed that thalamostriatal afferents contact both types of striatal MSNs (Dubé et al. 1988; Sidibé and Smith 1999; Gonzalo et al. 2002; Bacci et al. 2004; Lanciego et al. 2004; Smith et al. 2004; Castle et al. 2005) as well as different types of striatal local-circuit interneurons (Meredith and Wouterlood 1990; Lapper and Bolam 1992; Lapper et al. 1992; Bennet and Bolam 1993; Rudkin and Sadikot 1999; Sidibé and Smith 1999; Thomas et al. 2000). In terms of postsynaptic targets for thalamostriatal projections, it has been broadly considered that thalamostriatal axon terminals preferentially contact dendritic shafts of striatal MSNs (Sadikot et al. 1990; Smith and Bolam 1990; Sidibé and Smith 1996; Smith et al. 2004), whereas dendritic spines were contacted by corticostriatal terminals. Such a segregation regarding postsynaptic structures for corticostriatal and thalamostriatal axon terminals has been revisited upon the demonstration that

Cite this article as *Cold Spring Harb Perspect Med* doi: 10.1101/cshperspect.a041617

thalamostriatal axons arising from the centromedian–parafascicular thalamic (CM–Pf) complex only reach dendritic shafts of MSNs, whereas thalamic inputs from nuclei other than the CM–Pf contact dendritic spines (Raju et al. 2006). Furthermore, although the midline and intralaminar nuclei are known to be the main origin for thalamostriatal projections (Berendse and Groenewegen 1990; Groenewegen and Berendse 1994), the ventral thalamic motor nuclei are also minor contributors to the thalamostriatal system (McFarland and Haber 2001).

Besides differential connectivity patterns for corticostriatal and thalamostriatal projections, another key difference between both glutamatergic sources of inputs to the striatum is represented by the type of glutamate transporter expressed by neurons giving rise to these systems. While the isoform 1 of the vesicular glutamate transporter (vGlut1) identifies corticostriatal projections, thalamostriatal terminals contain the isoform 2 of vGlut (vGlut2) (Kaneko et al. 2002; Fujiyama et al. 2004, 2006; Raju and Smith 2005; Raju et al. 2006). However, it is worth noting that vGlut1 transcripts were found in neurons located in ventral relay and associative thalamic nuclei (Barroso-Chinea et al. 2007); these nuclei also contributed substantially to the thalamostriatal system (Beckstead 1984a,b; Smith and Parent 1986; Berendse and Groenewegen 1990; Erro et al. 2001, 2002; McFarland and Haber 2001; Van der Werf et al. 2002). Furthermore, coexpression of vGlut1 and vGlut2 transcripts within single thalamostriatal-projecting neurons located in thalamic territories other than the midline and intralaminar nuclei has been reported elsewhere (Barroso-Chinea et al. 2008). In summary, the thalamostriatal system can be regarded as a dual system, made of a main pathway originating from the midline and intralaminar nuclei that express vGlut2, together with another minor contributor represented by projections arising from ventral and associative thalamic nuclei expressing both vGlut isoforms.

Nigrostriatal Pathway. Nigrostriatal projections arising from the SNc (A9 group from Dahlström and Fuxe [1964]) represent the major source of dopaminergic innervation of the striatum. The retrorubral field (A8) and the ventral tegmental area (VTA) are the origins of dopaminergic mesostriatal and mesolimbic projections. Dopaminergic neurons of the SNc are organized in dorsal and ventral tiers (SNcD and SNcV, respectively). Dopaminergic terminals densely innervate the matrix compartment, whereas striosomes exhibit different densities of nigrostriatal axon terminals (Prensa and Parent 2001). Nigrostriatal projections target both types of striatal MSNs. One of the main foundations of the canonical model of the basal ganglia is based on a dual effect of dopamine when targeting either D_1R- or D_2R-expressing postsynaptic striatal MSNs. It has been broadly accepted that dopamine exerts and excitatory effect when targeting postsynaptic D_1Rs and an inhibitory role in D_2R-expressing MSNs. Dopaminergic cells exhibit extensive axonal arborizations, and indeed it has been estimated in rats that a single dopaminergic axon innervates up to 75,000 striatal MSNs and that each individual MSN is under the influence of between 95 and 194 dopaminergic neurons on average (Matsuda et al. 2009). These figures are calculated to be much higher in primates, where one dopaminergic neuron may lead to 1,000,000 synaptic contacts onto striatal MSNs. Such extensive dopaminergic connectivity paved the ground to postulate the hypothesis of "too many mouths to feed" (Bolam and Pissadaki 2012) where the unique cellular architecture of dopaminergic axons is pushing nigral cells to a huge metabolic and energetic stress that might account for the high vulnerability of dopaminergic neurons to degenerate. Although this is viewed as an appealing hypothesis, some concerns still remain. On one hand, terminal axonal arborizations were reconstructed in a reduced number of nigrostriatal neurons ($n = 8$) by taking advantage of the Sindbis virus vector with a membrane-targeted green fluorescent protein, and such a viral vector is indeed known to induce some degree of sprouting in unmyelinated axons (Gauthier-Campbell et al. 2004; Kuramoto 2019). On the other hand, earlier work made by reconstructing single nigrostriatal axonal arborizations with the microiontopheric intracellular delivery of biotinylated dextran amine in rats reported a more restrictive pattern of axonal arborization (Prensa and Parent 2001). Finally, unpublished work from different research teams analyzing connectivity patterns of nigrostriatal axon terminals in nonhuman primates failed to reproduce these highly elaborated axonal arborizations formerly reported in rodents.

Additional Sources of Striatal Afferent Projections. Besides the main striatal afferent systems listed above, two more additional pathways deserve to be considered, namely, glutamatergic projections arising from the amygdaloid complex that preferentially innervate the striosome compartment (Ragsdale and Graybiel 1988) and serotoninergic projections from the dorsal raphe nuclei, the latter that although well-characterized some time ago (Andén et al. 1966; Bobillier et al. 1976; Szabo 1980), their functional correlate still remains to be properly disclosed. Studies carried out in neurotoxin-based rodent models of Parkinson's disease revealed intense sprouting of serotoninergic fibers reaching the striatum in MPTP-treated mice, this hyperinnervation being more intense in animals showing the highest depletion of dopaminergic nigrostriatal neurons (Rozas et al. 1998). Another important piece of evidence is related to the paradoxical effect of levodopa medication (L-dopa) when administered to patients in the late stages of parkinsonism. Exogenous L-dopa is expected to be converted to dopamine by the enzyme dopadecarboxylase within dopaminergic axon terminals, although in patients with severe dopaminergic depletion, there is a huge loss of dopaminergic terminals while the beneficial symptomatic effect of this medication in largely preserved. Experiments conducted in hemi-lesioned rats revealed the ability of serotoninergic terminals for converting L-dopa to dopamine (López et al. 2001). Later on, serotoninergic striatal afferents have gained increased interest because of their putative engagement in underlying graft-induced dyskinesia (López et al. 2001; López-Real et al. 2003; Carta et al. 2007, 2008a,b; 2010).

Striatal Efferent Systems: Striatum-GPe and Striatum-GPI/SNr

Striatofugal projections can be broadly categorized as D_2R MSNs innervating the GPe and D_1R MSNs projecting downstream to the GPi and SNr, each MSN subtype representing the first neuron of the indirect and direct basal ganglia pathways, respectively. Although it has been initially considered that striatal efferents reaching GPe and GPi nuclei were collaterals of striatonigral projections (Fox and Rafols 1975), in the current scenario up to three distinctive types of striatofugal MSNs are described, innervating either the GPe or the GPi of the SNr nuclei (Beckstead and Cruz 1986). That said, it is also widely accepted that efferent striatal MSNs interact with different target structures through axon collaterals (for review, see Parent et al. 2000) and indeed the presence of a substantial number of striatonigral MSNs that also innervate the GPe in rats through axonal collateralization has been characterized in the rat brain (Castle et al. 2005). Furthermore, although accumulated evidence (anatomical, electrophysiological, and neurochemical) strongly supports a strict complementary distribution of D_1Rs and D_2Rs within different types of striatofugal MSNs (Beckstead 1988; Gerfen et al. 1990; Bertrán-González et al. 2010), such a sharp segregated expression of dopaminergic postsynaptic receptors still is a matter of debate. In this regard, studies carried out in BAC-transgenic mice revealed the presence of a small number of striatal MSNs (<6% of total MSNs) coexpressing both types of dopaminergic receptors (Bertrán-González et al. 2008). Similar data were reported in nonhuman primates where small percentages of striatal MSNs of the caudate and putamen nuclei innervating either the GPe and GPi nuclei express G protein–coupled receptor heteromers made of D_1Rs and D_2Rs (Rico et al. 2017). Data gathered from juxtacellular tract-tracing studies followed by single-axon reconstruction revealed that approximately half of striatofugal MSNs innervating the GPi and SNr output nuclei also send axon collaterals reaching the GPe, whereas most striatal MSNs directly innervate the GPe (Kawaguchi et al. 1990; Parent et al. 2000; Wu et al. 2000). However, single-axon tracing experiments are often limited by the rather low number of reconstructed axons and like any other anatomical study only reflect a static picture, without providing information on the relative functional importance for each reconstructed axon collateral when compared with the main axonal terminal arborization.

Output Nuclei: Internal Segment of the Globus Pallidus and Substantia Nigra Pars Reticulata

The GPi (sometimes also referred as the medial division of the globus pallidus) and the SNr are often grouped together as the basal ganglia output nuclei because they share a number of cyto- and chemo-architectural characteristics and to some extent similar sources of afferent and efferent systems. Both nuclei are made of inhibitory GABAergic neurons with a high rate of discharge that fire tonically to inhibit their target structures. The term "pallidus" refers to its smaller cellular density compared to the caudate and putamen nuclei. This feature, together with the presence of a large number of myelinated striatofugal fiber bundles crossing the GPi, finally leads to a paler appearance when compared to the striatum.

The GPi and the SNr receive two different types of inputs. Striatal D_1R MSNs are the major source of inhibition to GPi and SNr through the basal ganglia direct pathway. In contrast, vGlut2$^+$ glutamatergic STN efferent neurons provide excitatory inputs to GPi and SNr, as part of the indirect pathway. In summary, both GABAergic and glutamatergic inputs converge onto basal ganglia output nuclei, which, in turn, innervate thalamic targets and brainstem nuclei such as the inferior colliculus and the tegmental PPN. Thalamic-recipient areas of basal ganglia output information are slightly different when considering pallidothalamic and nigrothalamic projections, and these differences enabled an accurate parcellation of the ventral thalamic nuclei according to hodological criteria (Ilinsky and Kultas-Ilinsky 1987; Percheron et al. 1996). Pallidothalamic projections mainly innervate densicellular and parvicellular territories of the ventral anterior motor thalamic nucleus (VAdc and VApc, respectively), whereas nigrothalamic projections mainly target the magnocellular division of the ventral anterior

motor thalamic nucleus (VAmc). Furthermore, the caudal intralaminar nuclei (CM–Pf complex) are also a major source of basal ganglia output projections. Pallidal neurons innervate the thalamus through two different pathways, as follows: Neurons located in lateral GPi territories give rise to the so-called "ansa lenticularis," that course around the internal capsule, finally reaching the prerubral field (field H of Forel). In contrast, axons from neurons located in medial GPi territories travel through the internal capsule to form the "lenticular fasciculus" (field H2 of Forel), located between the STN and the zona incerta. The ansa lenticularis and the lenticular fasciculus merge together at the level of field H of Forel giving rise to the "thalamic fasciculus" (field H1 of Forel), and finally reaching ventral and intralaminar thalamic nuclei.

Intrinsic Nuclei

Globus Pallidus, External Segment

The GPe (also known as the lateral division of the globus pallidus) is part of the pallidal complex together with the GPi and the ventral pallidus. The GPe is surrounded laterally by the putamen, which is separated by the lateral medullary lamina. Another thin layer of myelinated fibers named as the medial medullary lamina represents the boundary between the GPe and the GPi (Fig. 1). Both pallidal subdivisions share several cyto- and chemo-architectural similarities since both nuclei are made of sparsely distributed GABAergic neurons with large soma sizes; these neurons are characterized by an enriched expression of the calcium-binding protein parvalbumin.

GPe neurons receive a GABAergic projection from D_2R-containing striatal MSNs, an inhibitory input representing the first synaptic relay station for the basal ganglia indirect pathway (see the canonical basal ganglia model below). Striatal MSNs projecting to the GPe nucleus also show an enriched expression of adenosine type 2A receptors ($A_{2A}Rs$), and therefore antibodies against $A_{2A}Rs$ have been used to properly delineate GPe boundaries (Rosin et al. 1998; Bogenpohl et al. 2012). GPe neurons are also reciprocally connected with STN neurons from which they receive strong vGlut2$^+$ glutamatergic projections. Such a close interrelationship between the GPe and the STN nuclei has paved the way for no longer considering the GPe as a simply relay station between the striatum and the STN (Shink et al. 1996; Joel and Weiner 1997; Nambu 2004; Nambu and Chiken 2024). Furthermore, both segments of the globus pallidus are known to be reciprocally connected. Finally, an additional minor source of glutamatergic afferents to the GPe is represented by inputs arising from the caudal intralaminar nuclei (Kincaid et al. 1991; Marini et al. 1999; Yashukawa et al. 2004). More recently, another layer of complexity was added upon the identification of a subpopulation of GPe neurons projecting back to both types of striatal MSNs as well as striatal GABAergic interneurons (Mallet et al. 2012, 2016; Abdi et al. 2015; Glajch et al. 2016). These neurons, known as arkypallidal neurons, are considered as a feedback circuit interconnecting the striatum and the GPe in a bidirectional fashion (for reviews, see Fang and Creed 2024; Guilhemsang and Mallet 2024).

Subthalamic Nucleus

The STN (of Luys) has been the focus of extensive research and currently represents the gold-standard neurosurgical target for deep brain stimulation (for review, see Guridi et al. 2018). The STN is a small subcortical nucleus made of densely packed glutamatergic excitatory neurons and located in a crossroad area immediately ventral to the zona incerta, rostral to the substantia nigra, and the red nucleus, and surrounded by a number of fiber bundles such as the ansa lenticularis, lenticular fasciculus, and thalamic fasciculus (Fig. 4). The canonical basal ganglia model (Albin et al. 1989; DeLong 1990) provided an oversimplified view of STN afferent and efferent systems, where the STN receives a GABAergic projection from the GPe nucleus and in turn, STN output reaches the basal ganglia output nuclei. At present, STN connectivity patterns have been enriched upon the introduction of the so-called "hyperdirect pathway," represented by direct glutamatergic projections re-

ceived from the cerebral cortex (Nambu et al. 2002, 2023; Nambu 2004). Furthermore, another source of glutamatergic afferents to the STN arises from the ipsilateral caudal intralaminar nuclei (Sugimoto and Hattori 1983; Sugimoto et al. 1983; Royce and Mourey 1985; Sadikot et al. 1992a,b; Sidibé and Smith 1996; Marini et al. 1999; Gonzalo et al. 2002; Lanciego et al. 2004, 2009; Castle et al. 2005; Gonzalo-Martín et al. 2024), together with a weaker projection from the contralateral intralaminar nuclei (Gerfen et al. 1982; Castle et al. 2005). Finally, the STN receives a sparse dopaminergic projection from the SNc, as a part of the so-called nigrostriatal system (for review, see Rommelfanger and Wichmann 2010). This projection likely plays an important role in terms of basal ganglia pathophysiology.

The cerebral cortex sends direct excitatory projections to the STN, which, in turn, innervates the basal ganglia output nuclei; therefore, providing a way for motor-related cortical areas to gain direct access to the output nuclei by bypassing the input nuclei (this is where the term hyperdirect pathway comes from). This pathway is characterized by shorter conduction times than those funneled through direct and indirect basal ganglia pathways, therefore, capable of exerting a quick and powerful excitation to basal ganglia output centers. Furthermore, STN hyperdirect afferent systems arising from the primary motor area, supplementary motor cortex, premotor cortices, frontal, and supplementary eye field areas are somatotopically organized at the level of the STN (Nambu et al. 1996, 2002). Thalamo-subthalamic projections have been described in rodents, cats, and in nonhuman primates are also topographically organized (Sadikot et al. 1992a,b; Lanciego et al. 2004; Gonzalo-Martín et al. 2024) and reach STN efferent neurons innervating the GPe and GPi/SNr (Castle et al. 2005).

STN output is mainly directed toward the basal ganglia output nuclei, a projection typically characterized by its highly branched axonal processes. Most STN efferent neurons simultaneously innervate the GPi, GPe, and SNr through axon collaterals (Van der Kooy and Hattori 1980; Kita et al. 1983; Kita and Kitai

1987; Plenz and Kitai 1999; Castle et al. 2005). According to their pattern of axonal collateralization, up to five distinct subtypes of STN projection neurons have been characterized (Parent et al. 2000; Sato et al. 2000). Finally, a minor STN projection reaching the ventral thalamic motor nuclei (Nauta and Cole 1978; Rico et al. 2010), this projection arising from STN neurons other than those innervating the output nuclei (Rico et al. 2010).

Substantia Nigra Pars Compacta

Up to three distinctive types of dopaminergic centers can be found in the ventral midbrain, namely, the VTA, the SNc, and the retrorubral field (RRF); these nuclei also named as A10, A9, and A8, respectively, according to Dahlström and Fuxe (1964) and are most often identified by its expression of the enzyme tyrosine hydroxylase (Fig. 5). SNc dopaminergic neurons can be subdivided into several clusters or cell groups, with a preferential distribution within the SNc and expressing different molecular signatures. Earlier experiments performed in neurotoxin-based animal models of Parkinson's disease revealed two SNc cell groups showing a differential vulnerability to degenerate; these groups were composed by calbindin-positive neurons of the so-called "dorsal tier" of the SNc (SNcD; more resilient neurons) (Dopeso-Reyes et al. 2014) and more vulnerable dopaminergic neurons located in the "ventral tier" (SNcV) that are calbindin-negative and positive for aldehyde dehydrogenase type 1a1 (Aldh1a1) and for Girk2 (German et al. 1989; Damier et al. 1999a,b). More recently, a specific subpopulation of Aldh1a1 neurons in the SNcV expressing the angiotensin receptor type 1 (AGTR1) has been identified as the dopaminergic neuron subtype most susceptible to neurodegeneration in parkinsonian patients (Kamath et al. 2022).

Neuromelanin pigmentation is a characteristic feature for dopaminergic neurons in primates (human and nonhuman primates). In these species, pigmentation is restricted to the SNc, VTA, locus coeruleus, and the dorsal motor nucleus of the vagus nerve (the latter far less pigmented and not visible to the naked eye). Of particular importance, it is worth noting that pigmented centers in the human brain are all key relay stations for the so-called Braak hypothesis (related to the prion-like spread of α-synuclein as the underlying mechanism supporting disease progression) (see Braak et al. 2002, 2003). Neuromelanin is a dark pigment that accumulates progressively with increased age resulting from physiological dopamine oxidation (Fedorow et al. 2005, 2006). Interestingly, pigmentation of dopaminergic neurons is preserved in human albinism, because the gene coding for cutaneous melanin is different than the one coding for neuromelanin. The potential role for neuromelanin in the pathophysiology of Parkinson's disease has long been neglected since most commonly used laboratory animals such as rodents lack neuromelanin pigmentation (Marsden 1961). However, neuromelanin has gained momentum very recently upon the introduction of rodent and nonhuman primate models of Parkinson's disease based on the viral vector-mediated enhancement of human tyrosinase, the rate-limiting enzyme of melanin synthesis (Carballo-Carbajal et al. 2019; Chocarro et al. 2023). Under parkinsonian conditions, there is a time-dependent progressive loss of dopaminergic neurons leading to the appearance of the cardinal symptoms once neuron degeneration goes beyond 50% of cell loss (Fig. 5). The main neuropathological hallmark of Parkinson's disease is represented by the presence of intracytoplasmic aggregates of misfolded α-synuclein known as Lewy bodies (LBs) (Spillantini et al. 1997). Intracytoplasmic LBs are ring-shaped inclusions and are also often observed as multivesicular LBs (Fig. 5). Marinesco bodies are another type of intracellular inclusions that can be found in the nucleus of dopaminergic neurons from aged patients. To what extent the presence of LBs can be considered as the pathological mechanism leading to dopaminergic cell death still remains controversial (Milber et al. 2012). Furthermore, intracellular accumulation of misfolded tau protein (known as tangles) is also often found in advanced stages of Parkinson's disease, as a good example of existing copathologies (Fig. 5). In this regard, the potential role of tau aggregates in trig-

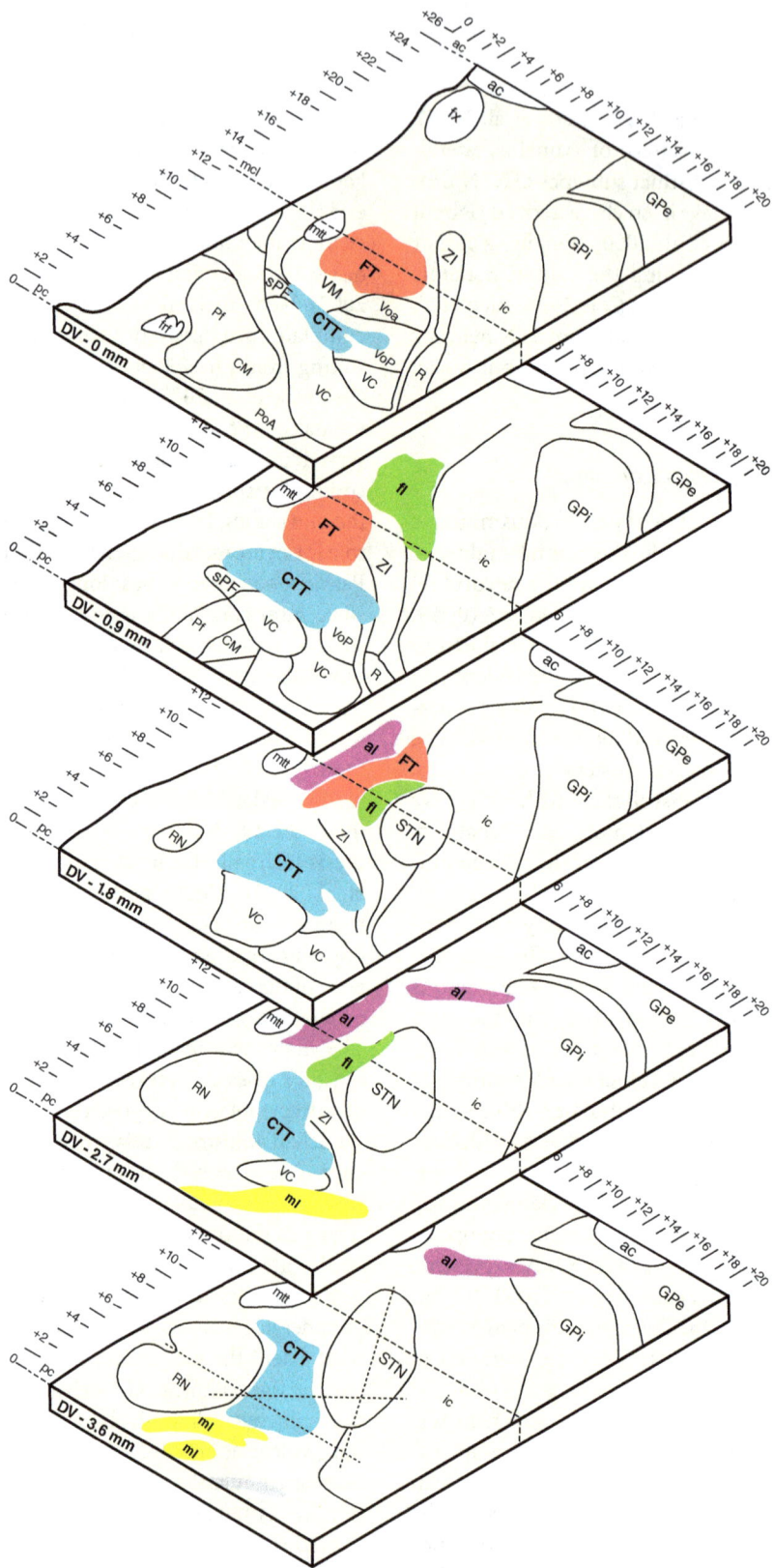

Figure 4. (*See following page for legend.*)

gering dopaminergic degeneration currently is at the center of an ongoing, heated debate (Chu et al. 2024; Espay and Lees 2024; Li and Li 2024).

BASAL GANGLIA FUNCTION AND DYSFUNCTION: THE BASAL GANGLIA MODEL

The Canonical Basal Ganglia Model

The classic model of the functional organization of the basal ganglia was formulated in the 1980s and was based on the concept that cortical information flows to the striatum to the output nuclei and finally reaches back to the cortex via a thalamic relay. Information is funneled this way through parallel corticobasal ganglia-thalamo-cortical loops. Initially intended to address the pathophysiology of movement, the presence of parallel circuits dealing with oculomotor control, executive functions, and emotions was also recognized.

The basis of the model stands of a dichotomy role for dopaminergic nigrostriatal projections when reaching striatopallidal neurons giving rise to the direct and indirect pathways (Fig. 6), each pathway with an opposite functional effect on basal ganglia output nuclei (Albin et al. 1989; DeLong 1990). Dopamine excites GABAergic transmission of D_1R striatal MSNs giving rise to the direct pathway that inhibits tonic activity of the output nuclei leading to a pause in neuronal firing. Simultaneously, dopamine inhibits GABAergic D_2R striatal MSNs of the indirect pathway, therefore, releasing inhibition of GPe neurons and further enhancing the inhibition received by STN neurons, which, in turn, reduces excitatory inputs to the GPi/SNr (Gerfen et al. 1990). Accordingly, basal ganglia direct and indirect pathways are viewed as opposite functional projection systems that facilitate and inhibit movements and behaviors, respectively. Under conditions of dopamine depletion, there is a reduced excitation of D_1R striatal MSNs together with increased inhibition of D_2R striatal MSNs. The synergistic effect between reduced GABAergic transmission in MSNs of the direct pathway and activation of

---

Figure 4. Subthalamic nucleus and neighboring fiber tracts. Diagrams are modified from Morel's stereotaxic atlas (Morel 2007) showing different dorsoventral axial planes at the level of the subthalamic area. The Forel's field is a bottleneck area occupying an anatomical position between the subthalamic nucleus (STN) and the internal capsule at the lateral border, and the red nucleus and the mamillothalamic tract at the medial border. Forel's field is traversed from its anterior to the posterior axis by the lenticular fasciculus (Forel's field H2; green) and the thalamic fasciculus (Forel's field H1; red) forming a bundle anterior and medial to the STN. A subset of the internal segment of the globus pallidus (GPi) efferent projections form the ansa lenticularis (purple), located anterior and medial to the STN. The ansa lenticularis and the lenticular fasciculus merge together to form the thalamic fasciculus. The anatomical position of the mamillothalamic tract (white color) corresponds to the midpoint of the intercommissural line (a line linking the anterior and posterior commissures), and it is a reference of particular importance for stereotaxic surgeries targeting the STN. The cerebello-rubro-thalamic tract (blue) is located in a more posterior position to the STN and close to the red nucleus. The zona incerta is located dorsal and posterior to the STN. This area contains GABAergic neurons receiving inputs from the cortex and spinal cord and projects to the thalamus. The medial lemniscus (yellow) is a fiber bundle arising from the spinal cord and brainstem that traverses the Forel's field at the level of the posterior commissure and finally reaches thalamic targets. Finally, nigrostriatal projections go through the STN and the zona incerta, crossing the internal capsule parallel to Forel's field H2. The functional importance of Forel's field was described in classic surgeries for movement disorders conducted during the 1960s. This surgery targeting Forel's field was called campectomy and improved tremor and rigidity. Lesions placed at the level of the zona incerta and in Forel's field H1 and H2 abolished dyskinesia. (al) Ansa lenticularis, (fl) lenticular fasciculus, (FT) thalamic fasciculus, (CTT) cerebello-rubro-thalamic tract, (mtt) mamillothalamic tract, (ml) medial lemniscus, (ac) anterior commissure, posterior commissure (pc), (ic) internal capsule, (frf) fasciculus retroflexus, (fx) fornix, (GPe) external division of the globus pallidus, (GPi) internal division of the globus pallidus, (ZI) zona incerta, (R) reticular nucleus, (VM) ventromedial thalamic nucleus, (Voa) ventral oralis anterior thalamic nucleus, (Vop) ventral oralis posterior thalamic nucleus, (VC) ventrocaudal thalamic nucleus, (Pf) parafascicular nucleus, (CM) centromedian nucleus, (RN) red nucleus, (STN) subthalamic nucleus.

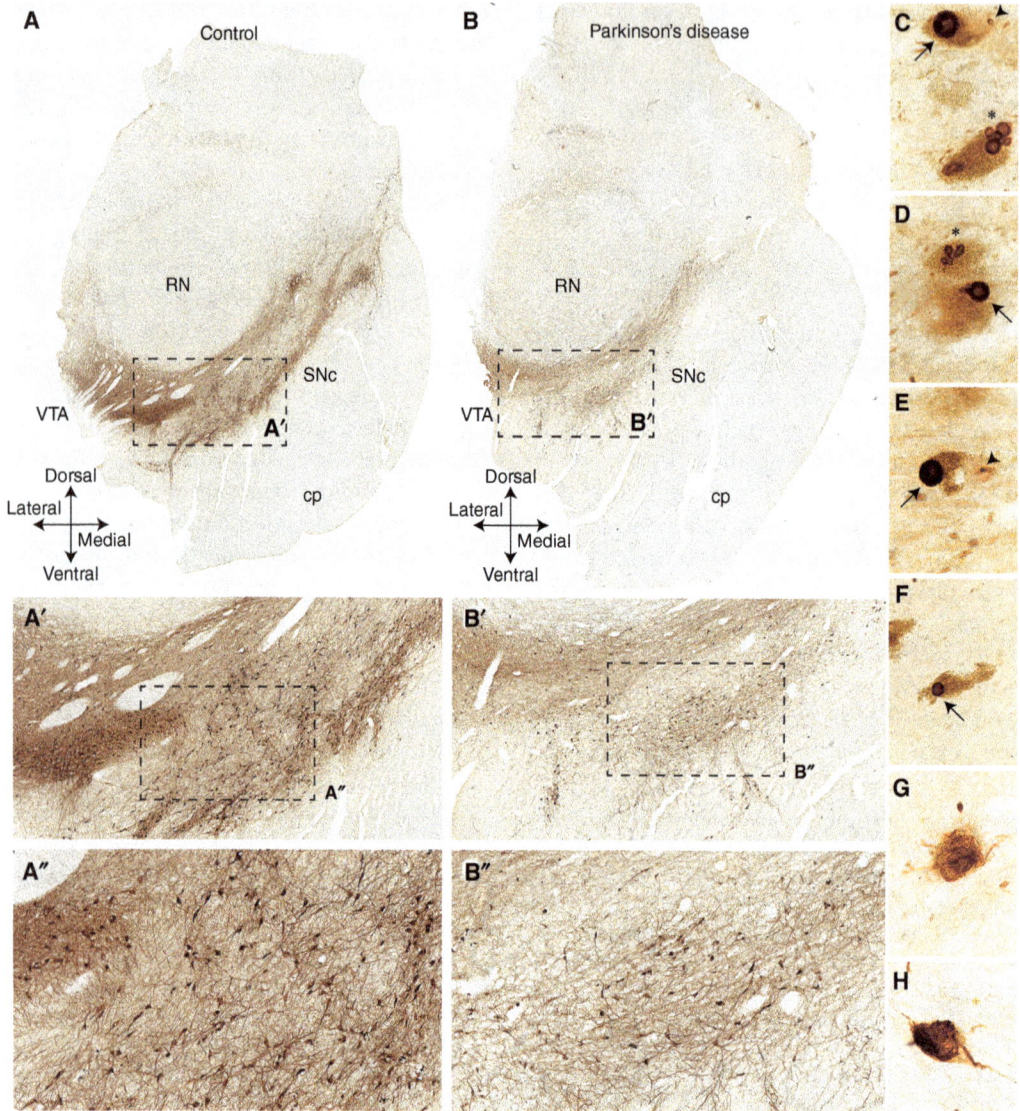

Figure 5. The substantia nigra pars compacta (SNc) in healthy and diseased conditions. The SNc is located ventral to the red nucleus and dorsal to the cerebral peduncle. Basal ganglia intrinsic nuclei contain dopaminergic neurons projecting to the putamen. Tyrosine hydroxylase (TH) is the rate-limiting enzyme for dopamine synthesis most often used as the marker of choice for disclosing dopaminergic neurons. Here, postmortem brain samples were used for comparing the number of TH$^+$ neurons in the substantia nigra of a healthy, control individual (*A–A″*) and the degree of dopaminergic cell loss observed in a parkinsonian patient in an advanced disease stage (Hoehn and Yahr stage of 5; *B–B″*). The main neuropathological hallmark of Parkinson's disease is represented by intracytoplasmic aggregates of misfolded α-synuclein known as Lewy bodies (LBs; *C–F*). Both single and multivesicular LBs (arrows and asterisks, respectively) can be found in pigmented dopaminergic neurons. Intranuclear aggregates called Marinesco bodies (arrowheads) can also be often found in these neurons, and are taken as indirect indicators for advanced aging. Finally, the potential role of copathologies in triggering dopaminergic cell death cannot be ruled out. Intracellular aggregates of misfolded tau protein (known as tangles) are also frequently observed in the cytoplasm of dopaminergic neurons (*G,H*).

indirect circuit neurons leading to increased STn activity overactivates GABAergic neurons of the GPi/SNr projecting to the thalamus, ultimately reducing thalamocortical transmission, thus impeding movement initiation and execution. Seminal studies carried out in macaques by Crossman (1987) lead to a proper understanding of basal ganglia circuits in dyskinetic states (e.g., hemichorea-ballism and levodopa-induced dyskinesia), resulting from decreased basal ganglia output and, therefore, functionally opposite to the parkinsonian state (Obeso et al. 2000, 2008).

The dual concept of the canonical basal ganglia model (sometimes also referred as the "boxes and arrows model") has paved the way for most of our current understanding of basal ganglia function and dysfunction and indeed gained strong support from many different experimental and clinical sources. Of particular importance, this model has been instrumental in setting up therapeutic interventions such as deep brain stimulation (Rosenow et al. 2004) and repeated apomorphine administration (Luquin et al. 1993), to mention just a few. That said,

since the introduction of the canonical model some difficulties were found when trying to accommodate new findings within this model, creating paradoxes that have been hotly discussed (Crossman and Obeso 2016). For instance, the current model predicts that lesion of basal ganglia output nuclei should induce a reduced inhibition of thalamic neurons innervating the cortex, finally leading to movement disinhibition. However, in macaques and parkinsonian patients, pallidotomy not only does not induce involuntary movements but, on the contrary, it abolishes dyskinesia (Carpenter et al. 1950; Lozano et al. 1995). Another classical example is represented by the fact that a lesion of either the STN, GPi, or motor thalamus does not deteriorate motor function (Marsden and Obeso 1994), but it is actually associated with substantial clinical benefit in parkinsonian patients (Obeso et al. 2009). Indeed, activities of daily life were not impaired in a patient treated with dual unilateral pallidotomy and subthalamotomy, as reported elsewhere (Obeso et al. 2009).

Since its introduction in the 1980s and despite existing limitations for the classic basal

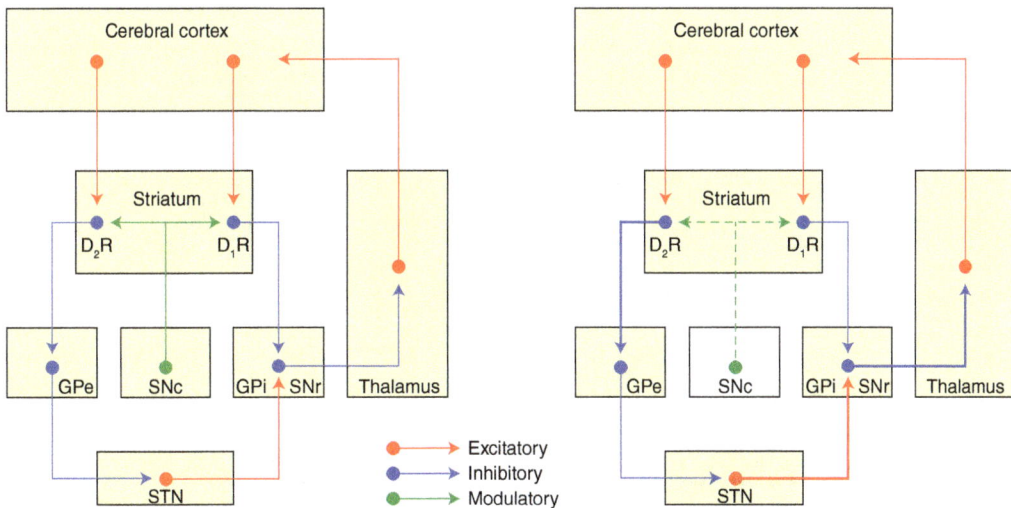

Figure 6. The canonical basal ganglia model. The motor circuit is composed of a corticostriatal (putaminal) projection, two major striatofugal projection systems giving rise to the direct and indirect pathways, and efferent pallido-thalamo-cortical projections closing the motor loop. The thickness of the arrows represents the functional state of a given circuit (thicker arrows show hyperactive pathways; thinner arrows represent hypoactive circuits). Excitatory pathways are color-coded in red, GABAergic inhibitory circuits are color-coded in blue, whereas modulatory nigrostriatal dopaminergic projections are drawn in green.

J.L. Lanciego and J.A. Obeso

ganglia model, most of the model has remained well preserved and it still stimulates considerable advances in our understanding of basal ganglia organization. In this regard, strong support for the classic model has been gathered from studies in rats with optogenetics. In the first study, the selective stimulation of D_2R MSNs (indirect pathway) resulted in movement arrest, whereas activation of D_1R MSNs leads to movement activation (Kravitz et al. 2010); therefore, confirming the model's predictions assuming that direct and indirect basal ganglia pathways are in functional equilibrium, facilitating, and inhibiting movement, respectively. In parallel, evidence was provided showing that loss of DARP-32 in the direct striatonigral-projecting neurons abolished levodopa-induced dyskinesia, whereas in striatopallidal neurons (indirect pathway) resulted in increased locomotor activity together with a reduced haloperidol-induced catalepsy (Bateup et al. 2010).

Current Concepts

The original formulation of the basal ganglia model was viewed as a straightforward motor loop, where cortical information enters the input nuclei; it is processed within the basal ganglia network and returned to the cortex after a thalamic relay. Basic and clinical research con-

ducted in the past few decades has added several extra layers of complexity to the classic model, currently enriched with cortical and subcortical projections interacting with internal reentry loops (Fig. 7), properly designed for selecting and inhibiting a variety of simultaneously occurring events and signals.

Functional Subdivisions: Basal Ganglia Domains

The basal ganglia are functionally subdivided into motor, associative, and limbic/emotional domains based on their relationship with relevant cortical projection areas and the engagement of these regions. Thus, in addition to the well-known role of the basal ganglia in motor control, there is now a better appreciation for other functions such as attention, time estimation, implicit learning, habit formation, reward-related behavior, and emotions, all of which are associated with the activation of cortical loops connecting to the caudate nucleus, the anterior putamen or the ventral putamen (Hikosaka et al. 2002; Buhusi and Meck 2005; Yin and Knowlton 2006). Evidence for this concept has arisen from studies in human and nonhuman primates.

In macaques, bicuculline injections in the motor putamen elicited focal, brief muscle jerks (tic-like) together with increased neuronal activ-

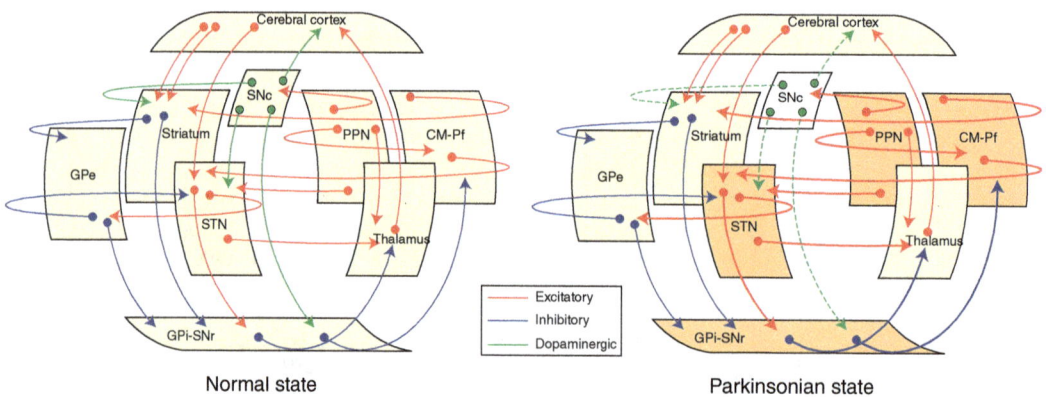

Figure 7. Updated views of the classic basal model. Besides traditional cortico-basal ganglia-thalamocortical circuits, several transverse loops have been described in the last few years, most of them with a putative modulatory role. Dopamine depletion leads to both enhanced inhibitions and excitations within the basal ganglia network. Several basal ganglia nuclei are hyperactive, most of them appointed as surgical target candidates for deep brain stimulation.

Cite this article as *Cold Spring Harb Perspect Med* doi: 10.1101/cshperspect.a041617

ity in the GPe, marked reduction in the GPi, and increased firing in the primary motor cortex, time-locked to the jerks (McCairn et al. 2009). This is in support that movement release is punctually associated with a reduction of GPi output activity. Moreover, focal bicuculline injections, blocking specific subregions of the striatum, GPe, and STN, produced dyskinesia, stereotypies, and hyperactivity (among other behavioral abnormalities) when injected into the posterolateral segment (motor-related), associative, and limbic regions (François et al. 2004; Karachi et al. 2009). Injections in the ventral striatum elicited sexual behaviors (erection and ejaculation) and vomiting (Worbe et al. 2009).

Human activation studies using fMRI and PET showed topographical segregation according to the requested task and underlying functions. Thus, activation in the posterior putamen (sensorimotor territory) was consistently encountered for movements and presented a somatotopical organization, with the leg representation lying dorsal, face ventral, and arm in between, as expected (Lehéricy et al. 2005). Preparatory activation as well as finger movement sequencing was located more rostrally in the anterior putamen, whereas activation of the associative territory was observed during tasks such as motor internal representation selection and planning of sequential emotions (Monchi et al. 2006). In contrast, tasks evoking emotional responses typically activate the accumbens nucleus at the level of the ventral striatum.

Functional Neuroanatomy of the Basal Ganglia

Corticostriatal Connections. GABAergic striatal MSNs receive profuse glutamatergic inputs from most cortical regions (ipsi- and contralateral cortex), the thalamus, and the amygdala (the latter to a lower extent). Cortical axons form asymmetric synapses with the heads of dendritic spines, whereas thalamic inputs mainly reach dendritic shafts of striatal MSNs. Additional inputs to the MSNs arise from dopaminergic neurons through the nigrostriatal pathway, as well as cholinergic and GABAergic striatal interneurons, which tend to form sym-

metric synaptic contacts. In the rat, it has been estimated that up to 5000 glutamatergic afferents reach each striatal MSN, whereas up to 100 striatal MSNs innervate each pallidal neuron. This arrangement is likely setting the anatomical basis for striatal function, in particular filtering incoming/outgoing signals. The capacity of the striatum to filter massive incoming signals is facilitated by several mechanisms, among which the nigrostriatal dopaminergic projection plays a fundamental role. Activation of selected nigrostriatal projections leads to increased dopamine levels in innervated striatal regions in parallel with a marked dopamine reduction in adjacent zones. This is supported by experimental evidence indicating that dopamine enhances synchronous corticostriatal afferent volleys while simultaneously inhibiting other inputs (Bamford et al. 2004). The establishment of a gradient or salient pattern of activity is thought to govern dopaminergic signaling, likely underlying the responses to reward and addiction, habit formation, and motor learning. Another important component of striatal internal functioning is provided by striatal cholinergic TNAs and GABAergic FISs interneurons. Phasic release of dopamine induces pauses in TANs facilitating corticostriatal transmission (Bonsi et al. 2011). Moreover, GABAergic interneurons, also modulated by dopamine and by axon collaterals of MSNs, mediate a powerful intrastriatal inhibition (Tepper et al. 2004). In summary, interneurons modulate striatal excitability by providing a mechanism to facilitate a given pool of MSNs excited from the cortex, while competing stimuli are canceled.

Corticosubthalamic Connections. The STN nucleus is subdivided into motor, associative, and limbic regions. The dorsolateral part of the STN corresponds to the motor region. In healthy macaques, STN neurons fire at 20–30 Hz and provide tonic excitatory input to both segments of the globus pallidus. Lesion or blockade of the STN is associated with a significant reduction in the mean firing rate of pallidal neurons.

Initial studies recording STN activity in awake macaques performing a task indicated that neuronal firing occurred after or coincident with movement initiation (DeLong et al. 1985), suggesting that the STN was not involved in the onset of movement, in keeping with the prevailing idea that the STN was mainly related to movement inhibition. However, later studies recording single-cell activity and local field potentials in parkinsonian patients provided clear evidence for STN activation up to 1 second before movement initiation (Rodríguez-Oroz et al. 2001; Alegre et al. 2005). Accordingly, when a voluntary movement is about to be performed, the same cortical volley arriving to the putamen is also reaching the STN. Indeed, studies in macaques have shown that the cortico-STN-GPe/GPi projection is a fast-conducting system, well placed to modulate basal ganglia output (Nambu 2004).

Subcortical Connections: Thalamobasal Ganglia Loops and Brainstem Circuits. The role and importance of projections and feedback loops between the basal ganglia and several subcortical regions have often been neglected and not included in initial versions of the classic basal ganglia model. Loops engaging the cerebral cortex and the basal ganglia are not the first evolutionary example of closed-loop circuits involving the basal ganglia. Phylogenetically older subcortical loops between the basal ganglia and brainstem motor structures have been extensively documented, including the superior and inferior colliculi, periaqueductal gray, PPN, cuneiform area, parabrachial complex, and various pontine and medullary reticular nuclei (for reviews, see Redgrave et al. 2010, 2011; Smith et al. 2011).

Pathophysiology of the Basal Ganglia

Movement disorders comprise a variety of motor problems, not all of which are associated with basal ganglia dysfunction. Those that have a clearly established pathological basis and are caused by pathophysiological mechanisms directly engaging basal ganglia can be categorized as (1) the parkinsonian syndrome, composed of rigidity, akinesia/bradykinesia, and resting tremor; (2) dystonia, characterized by prolonged muscular spasms and abnormal postures; and (3) chorea-ballism, in which fragments of movements flow irregularly from one

body segment to another, to cause a dance-like appearance. When the amplitude of the movement is very large, the term "ballism" is applied. From a practical point of view, levodopa-induced dyskinesia in parkinsonian patients, which are by large choreic in nature, represent the most frequent cause of choreic dyskinesia in the general population.

The Parkinsonian Syndrome

The best-known example and cause of the parkinsonian syndrome is Parkinson's disease, characterized by poverty of spontaneous movements (hypokinesia) and slowness of voluntary movements (bradykinesia), together with increased muscular tone (rigidity) and resting tremor. Dopamine striatal depletion resulting in dopaminergic cell loss in the SNc is the cause of the major clinical features of Parkinson's disease. A large bulk of accumulated evidence gathered from both experimental animal models and patients with Parkinson's disease showed that dopamine depletion shifts the balance of basal ganglia activity toward the indirect circuit, leading to excessive activity of the STN that overstimulates the GPi/SNr. Increased basal ganglia output overinhibits thalamocortical projections, reducing cortical neuronal activity associated with movement initiation (DeLong 1990; Obeso et al. 2000).

The way in which the basal ganglia modulate and respond to external input is also distorted in the parkinsonian state. The putamen, STN, and GPi show augmented neuronal responses to peripheral stimulation, which impairs the physiological selection or filtering of incoming signals discussed above. Overall, the basal ganglia are shifted toward inhibiting cortically aided movements by an increased activation of the STN-GPi network and reduced excitability in the direct cortico-putaminal-GPi projection (Obeso et al. 2008).

It is also worth noting that our current understanding of basal ganglia pathophysiology does not provide an adequate explanation for two more cardinal features of Parkinson's disease, namely, rigidity and tremor. Regarding tremor, studies carried out in parkinsonian patients during STN-DBS as well as in MPTP-treated macaques have started to unravel some interesting features (Heimer et al. 2002). Nevertheless, it is fascinating to realize that when nearly reaching 200 years after James Parkinson described the "Shaking Palsy" as the parkinsonian tremor, its origin remains a mystery.

Pathophysiology of Dyskinesia

Chorea-ballism is the abnormal release of fragments of normal movement. In simple terms, the pathophysiology of dyskinetic movements may be seen as the opposite of parkinsonism. In most cases, chorea is drug-related, such as levodopa in Parkinson's disease, in neurodegenerative disorders such as Huntington's disease, and secondary to systemic conditions like either hyperthyroidism or lupus erythematosus.

According to the basal ganglia canonical model, dyskinesia (e.g., either chorea or ballism) results from reduced activity in the STN-GPi projection leading to decreased firing in the GPi output (Crossman 1987; DeLong 1990). Choreatic or ballistic movements can occur secondary to focal lesion in the basal ganglia, such as the STN. In the 1980s, seminal work performed in naive macaques showed that local administration of bicuculline to block the striato-GPe projection induced chorea-ballism in the contralateral hemibody. This was associated with increased 2-deoxyglucose uptake in the STN as seen with autoradiography (Mitchell et al. 1989), indicating excessive inhibition from the GPe onto the STN, consequently decreasing glutamatergic excitatory STN activity directed to the GPi. Neuronal recording of extracellular activity in the GPi of choreic macaques (induced by bicuculline injections into the GPe) showed a reduction in neuronal firing rate and an abnormal pattern of neuronal synchronization (Filion 2000). This initial evidence supported a preponderant role for the indirect pathway in the induction of dyskinesia. Subsequently, recordings of neuronal activity in the GPe and GPi in MPTP-treated macaques and parkinsonian patients during STN-DBS, further showed the increase and reduction of GPe and GPi normal activity, respectively, as a basic pathophysiolog-ical feature underlying choreic dyskinesia (Papa et al. 1999). Later on, evidence was provided supporting increased and extrasynaptic shifting of N-methyl-D-aspartate (NMDA) glutamatergic receptors (Gardoni et al. 2006) and abnormal D1R receptor activation in the striatum of dyskinetic animals (Berthet et al. 2009), suggesting a pivotal role of the direct pathway as a mediator of levodopa-induced dyskinesia (Bateup et al. 2010).

CONCLUDING REMARKS

The basal ganglia and related nuclei are a network made of multiple parallel loops and re-entering circuits whereby motor, associative, and limbic territories are mainly engaged in movement control, behavior, and emotions. This is likely to be sustained by the same basic architectural and functional organization, differentially applied to (1) selection and facilitation of prefrontal striatopallidal activity during the performance and acquisition of new activities and tasks (goal-directed system); (2) reinforcement learning to create habitual responses automatically performed by the motor circuits (habit system); and (3) stopping and ongoing activity and switching to a new one if needed, which is mainly mediated by the inferior frontal cortex/STN-cortical circuits. Abnormalities in these domains and functions lead to movement disorders of basal ganglia origin, such as parkinsonism, dyskinesia, obsessive-compulsive disorders, and mood alterations (e.g., apathy and euphoria).

ACKNOWLEDGMENTS

In loving memory of Mahlon DeLong, MD (1938–2024), a friend, a mentor, and a continuing source of inspiration for anyone engaged in basal ganglia research and movement disorders. This research was funded in whole or in part by Aligning Science Across Parkinson's (Grant No. ASAP-020505) through the Michael J. Fox Foundation for Parkinson's Research (MJFF). Conducted work was also funded by MICIN/AIE/10.12039/5011000011033 (Grant No. PID 2020-120308RB), by CiberNed Intramural Collaborative Projects (Grant No. PI2020/09), and by the Spanish Fundación Mutua Madrileña de Investigación Médica (Grant No. FMM2020).

REFERENCES

*Reference is in this subject collection.

Abdi A, Mallet N, Mohamed FY, Sharott A, Dodson PD, Nakamura KC, Suri S, Avery SV, Larvin JT, Garas FN, et al. 2015. Prototypic and arkypallidal neurons in the dopamine-intact external globus pallidus. *J Neurosci* **35**: 6667–6688. doi:10.1523/JNEUROSCI.4662-14.2015

Albin RL, Young AB, Penney JB. 1989. The functional anatomy of basal ganglia disorders. *Trends Neurosci* **12**: 366–375. doi:10.1016/0166-2236(89)90074-X

Alegre M, Alonso-Frech F, Rodríguez-Oroz MC, Guridi J, Zamarbide I, Valencia M, Manrique M, Obeso JA, Artieda J. 2005. Movement-related changes in oscillatory activity in the human subthalamic nucleus: ipsilateral vs. contralateral movements. *Eur J Neurosci* **22**: 2315–2324. doi:10.1111/j.1460-9568.2005.04409.x

Andén NE, Dahlström A, Fuxe K, Olson L, Ungerstedt U. 1966. Ascending noradrenaline neurons from the pons and the medulla oblongata. *Experientia* **22**: 44–45. doi:10.1007/BF01897761

Bacci JJ, Kachidian P, Kerkerian-Le Goff L, Salin P. 2004. Intralaminar thalamic nuclei lesions: widespread impact on dopamine denervation-mediated cellular defects in the rat basal ganglia. *J Neuropathol Exp Neurol* **63**: 20–31. doi:10.1093/jnen/63.1.20

Baillon B, Mallet N, Bézard E, Lanciego JL, Gonon F. 2008. Intratelencephalic corticostriatal neurons equally excite striatonigral and striatopallidal neurons and their discharge activity is selectively reduced in experimental parkinsonism. *Eur J Neurosci* **27**: 2313–2321. doi:10.1111/j.1460-9568.2008.06192.x

Bamford NS, Zhang H, Schmitz Y, Wu NP, Cepeda C, Levine MS, Schmauss C, Zakharenko SS, Zablow L, Sulzer D. 2004. Heterosynaptic dopamine neurotransmission selects sets of corticostriatal terminals. *Neuron* **42**: 653–663. doi:10.1016/S0896-6273(04)00265-X

Barroso-Chinea P, Castle M, Aymerich MS, Pérez-Manso M, Erro E, Tuñón T, Lanciego JL. 2007. Expression of the mRNAs encoding for the vesicular glutamate transporters 1 and 2 in the rat thalamus. *J Comp Neurol* **501**: 703–715. doi:10.1002/cne.21265

Barroso-Chinea P, Castle M, Aymerich MS, Lanciego JL. 2008. Expression of vesicular glutamate transporters 1 and 2 in the cells of origin of the rat thalamostriatal pathway. *J Chem Neuroanat* **35**: 101–107. doi:10.1016/j.jchemneu.2007.08.001

Bateup HS, Santini E, Shen W, Birnbaum S, Valjent E, Surmeier DJ, Fisone G, Nestler EJ, Greengard P. 2010. Distinct subclasses of medium spiny neurons differentially regulate striatal motor behaviors. *Proc Natl Acad Sci* **107**: 14845–14850. doi:10.1073/pnas.1009874107

Beckstead RM. 1984a. The thalamostriatal projection in the cat. *J Comp Neurol* **223**: 313–346. doi:10.1002/cne.902230302

Beckstead RM. 1984b. A projection to the striatum from the medial subdivision of the posterior group of the thalamus in the cat. *Brain Res* **300**: 351–356. doi:10.1016/0006-8993(84)90845-X

Beckstead RM. 1988. Association of dopamine D1 and D2 receptors with specific cellular elements in the basal ganglia of the cat: the uneven topography of dopamine receptors in the striatum is determined by intrinsic striatal cells, not nigrostriatal axons. *Neuroscience* **27**: 851–863. doi:10.1016/0306-4522(88)90188-1

Beckstead RM, Cruz CJ. 1986. Striatal axons to the globus pallidus, entopeduncular nucleus and substantia nigra come mainly from separate cell populations in the cat. *Neuroscience* **19**: 147–158. doi:10.1016/0306-4522(86)90012-6

Bennet BD, Bolam JP. 1993. Characterization of calretinin-immunoreactive structures in the striatum of the rat. *Brain Res* **609**: 137–148. doi:10.1016/0006-8993(93)90866-L

Berendse HW, Groenewegen HJ. 1990. Organization of the thalamostriatal projections in the rat, with special emphasis on the ventral striatum. *J Comp Neurol* **299**: 187–228. doi:10.1002/cne.902990206

Berthet A, Porras G, Doudnikoff E, Stark H, Cador M, Bezard E, Bloch B. 2009. Pharmacological analysis demonstrates dramatic alteration of D_1 dopamine receptor neuronal distribution in the rat analog of L-DOPA-induced dyskinesia. *J Neurosci* **29**: 4829–4835. doi:10.1523/JNEUROSCI.5884-08.2009

Bertrán-González J, Bosch C, Maroteaux M, Matamales M, Hervé D, Valjent E, Girault JA. 2008. Opposing patterns of signaling activation in dopamine D_1 and D_2 receptor-expressing striatal neurons in response to cocaine and haloperidol. *J Neurosci* **28**: 5671–5685. doi:10.1523/JNEUROSCI.1039-08.2008

Bertrán-González J, Hervé D, Girault JA, Valjent E. 2010. What is the degree of segregation between striatonigral and striatopallidal projections? *Front Neuroanat* **4**: 136.

Bobillier R, Seguin S, Petitjean F, Slavert D, Turet M, Jouvet M. 1976. What is the degree of segregation between striatonigral and striatopallidal projections? *Front Neuroanat* **4**: 136.

Bogenpohl JW, Ritter SL, Hall RA, Smith Y. 2012. Adenosine A_2A receptor in the monkey basal ganglia: ultrastructural localization and colocalization with the metabotropic glutamate receptor 5 in the striatum. *J Comp Neurol* **520**: 570–589. doi:10.1002/cne.22751

Bolam JP, Pissadaki EK. 2012. Living on the edge with too many mouths to feed: why dopamine neurons die. *Mov Disord* **27**: 1478–1483. doi:10.1002/mds.25135

Bolam JP, Izzo PN, Graybiel AM. 1988. Cellular substrates of the histochemically defined striosome/matrix system of the caudate nucleus: a combined Golgi and immunocytochemical study in cat and ferret. *Neuroscience* **24**: 853–875. doi:10.1016/0306-4522(88)90073-5

Bonsi P, Cuomo D, Martella G, Madeo G, Schirinzi T, Puglisi F, Ponteiro G, Pisani A. 2011. Centrality of striatal cholinergic transmission in basal ganglia function. *Front Neuroanat* **5**: 6. doi:10.3389/fnana.2011.00006

Borra E, Rizzo M, Gerbella M, Rozzi S, Luppino G. 2021. Laminar origin of corticostriatal projections to the motor putamen in the macaque brain. *J Neurosci* **41**: 1455–1469. doi:10.1523/JNEUROSCI.1475-20.2020

Braak H, Del Tredici K, Bratzke H, Hamm-Clement J, Sandmann-Keil D, Rüb U. 2002. Staging of the intracerebral inclusion body pathology associated with idiopathic Parkinson's disease (preclinical and clinical stages). *J Neurol* **249**(Suppl 3): III/1–III/5. doi:10.1007/s00415-002-1301-4

Braak H, Del Tredici K, Rüb U, de Vos RAI, Jansen-Steur ENH, Braak E. 2003. Staging of brain pathology related to sporadic Parkinson's disease. *Neurobiol Aging* **24**: 197–211. doi:10.1016/S0197-4580(02)00065-9

Buhusi CV, Meck WH. 2005. What makes us tick? Functional and neural mechanisms of interval timing. *Nat Rev Neurosci* **6**: 755–765. doi:10.1038/nrn1764

Carballo-Carbajal I, Laguna A, Romero-Giménez J, Cuadros T, Bové J, Martinez-Vicente M, Parent A, Gonzalez-Sepulveda M, Peñuelas N, Torra A, et al. 2019. Brain tyrosinase overexpression implicates age-dependent neuromelanin production in Parkinson's disease pathogenesis. *Nat Commun* **10**: 973. doi:10.1038/s41467-019-08858-y

Carpenter MB, Whittier JR, Mettler FA. 1950. Analysis of choreoid hyperkinesia in the Rhesus monkey; surgical and pharmacological analysis of hyperkinesia resulting from lesions in the subthalamic nucleus of Luys. *J Comp Neurol* **92**: 293–331. doi:10.1002/cne.900920303

Carta M, Carlsson T, Kirik D, Björklund A. 2007. Dopamine released from 5-HT terminals is the cause of L-DOPA-induced dyskinesia in parkinsonian rats. *Brain* **130**: 1819–1833. doi:10.1093/brain/awm082

Carta M, Carlsson T, Muñoz A, Kirik D, Björklund A. 2008a. Involvement of the serotonin system in L-dopa-induced dyskinesias. *Parkinsonism Relat Disord* **14**: S154–S158. doi:10.1016/j.parkreldis.2008.04.021

Carta M, Carlsson T, Muñoz A, Kirik D, Björklund A. 2008b. Serotonin-dopamine interaction and maintenance of L-DOPA-induced dyskinesias. *Progr Brain Res* **172**: 465–478. doi:10.1016/S0079-6123(08)00922-9

Carta M, Carlsson T, Muñoz A, Kirik D, Björklund A. 2010. Role of serotonin neurons in the induction of levodopa- and graft-induced dyskinesias in Parkinson's disease. *Mov Disord* **25**: S174–S179. doi:10.1002/mds.22792

Castle M, Aymerich MS, Sánchez-Escobar C, Gonzalo N, Obeso JA, Lanciego JL. 2005. Thalamic innervation of the direct and indirect basal ganglia pathways in the rat: ipsi- and contralateral projections. *J Comp Neurol* **483**: 143–153. doi:10.1002/cne.20421

Chocarro J, Rico AJ, Ariznabarreta G, Roda E, Honrubia A, Collantes M, Peñuelas I, Vázquez A, Rodríguez-Pérez AI, Labandeira-García JL, et al. 2023. Neuromelanin accumulation drives endogenous synucleinopathy in non-human primates. *Brain* **146**: 5000–5014. doi:10.1093/brain/awad331

Chu Y, Hirst WD, Federoff HJ, Harms AS, Stoessl AJ, Kordower JH. 2024. Nigrostriatal tau pathology in parkinsonism and Parkinson's disease. *Brain* **147**: 444–457. doi:10.1093/brain/awad388

Crossman AR. 1987. Primate models of dyskinesia: the experimental approach to the study of basal ganglia-related involuntary movement disorders. *Neuroscience* **21**: 1–40. doi:10.1016/0306-4522(87)90322-8

Crossman AR, Obeso JA. 2016. Functions of the basal ganglia—paradox or no paradox? *Mov Disord* **31**: 1120–1121. doi:10.1002/mds.26745

Dahlström A, Fuxe K. 1964. Localization of monoamines in the lower brain stem. *Experientia* **20**: 398–399. doi:10.1007/BF02147990

Damier P, Hirsch EC, Agid Y, Graybiel AM. 1999a. The substantia nigra of the human brain. I: Nigrosomes and the nigral matrix, a compartmental organization based on calbindin D(28K) immunohistochemistry. *Brain* **122**: 1421–1436. doi:10.1093/brain/122.8.1421

Damier P, Hirsch EC, Agid Y, Graybiel AM. 1999b. The substantia nigra of the human brain. II: Patterns of loss of dopamine-containing neurons in Parkinson's disease. *Brain* **122**: 1437–1448. doi:10.1093/brain/122.8.1437

DeLong MR. 1990. Primate models of movement disorders of basal ganglia origin. *Trends Neurosci* **13**: 281–285. doi:10.1016/0166-2236(90)90110-V

DeLong MR, Crutcher MD, Georgopoulus AP. 1985. Primate globus pallidus and subthalamic nucleus: functional organization. *J Neurophysiol* **53**: 530–543. doi:10.1152/jn.1985.53.2.530

Deng Y, Lanciego J, Kerkerian-Le Goff L, Coulon P, Salin P, Kachidian P, Lei W, Del Mar N, Reiner A. 2015. Differential organization of cortical inputs to striatal projection neurons of the matrix compartment in rats. *Front Sys Neurosci* **9**: 51. doi:10.3389/fnsys.2015.00051

Desban M, Gauchy C, Glowinski J, Kemel ML. 1995. Heterogeneous topographical distribution of the striatonigral and striatopallidal neurons in the matrix compartment of the cat caudate nucleus. *J Comp Neurol* **352**: 117–133. doi:10.1002/cne.903520109

Donoghue JP, Herkenham M. 1986. Neostriatal projections from individual cortical fields conform to histochemically distinct striatal compartments in the rat. *Brain Res* **365**: 397–403. doi:10.1016/0006-8993(86)91658-6

Dopeso-Reyes IG, Rico AJ, Roda E, Sierra S, Pignataro D, Lanz M, Sucunza D, Chang-Azancot L, Lanciego JL. 2014. Calbindin content and differential vulnerability of midbrain efferent dopaminergic neurons in macaques. *Front Neuroanat* **8**: 146. doi:10.3389/fnana.2014.00146

Dubé L, Smith AD, Bolam JP. 1988. Identification of synaptic terminals of thalamic or cortical origin in contact with distinct medium-sized spiny neurons in the rat neostriatum. *J Comp Neurol* **267**: 455–471. doi:10.1002/cne.902670402

Erro E, Lanciego JL, Arribas J, Giménez-Amaya JM. 2001. Striatal input from the ventrobasal complex of the rat thalamus. *Histochem Cell Biol* **115**: 447–454. doi:10.1007/s004180100273

Erro ME, Lanciego JL, Giménez-Amaya JM. 2002. Re-examination of the thalamostriatal projections in the rat with retrograde tracers. *Neurosci Res* **42**: 45–55. doi:10.1016/S0168-0102(01)00302-9

Espay AJ, Lees AJ. 2024. Are we entering the "*Tau*-lemaic" era of Parkinson's disease? *Brain* **147**: 330–332. doi:10.1093/brain/awae002

Fang LZ, Creed MC. 2024. Updating the striatal-pallidal wiring diagram. *Nat Neurosci* **27**: 15–27. doi:10.1038/s41593-023-01518-x

Fedorow H, Tribl F, Halliday G, Gerlanch M, Riederer P, Double KL. 2005. Neuromelanin in human dopamine neurons: comparison with peripheral melanins and relevance to Parkinson's disease. *Progr Neurobiol* **75**: 109–124. doi:10.1016/j.pneurobio.2005.02.001

Fedorow H, Halliday GM, Rickert CH, Gerlach M, Riederer P, Double KL. 2006. Evidence for specific phases in the development of human neuromelanin. *Neurobiol Aging* **27**: 506–512. doi:10.1016/j.neurobiolaging.2005.02.015

Filion M. 2000. Physiological basis of dyskinesia. *Ann Neurol* **47**: S35–S40.

Fox CA, Rafols JA. 1975. The radial fibers in the globus pallidus. *J Comp Neurol* **159**: 177–199. doi:10.1002/cne .901590203

François C, Grabli D, McCairn K, Jan C, Karachi C, Hirsch EC, Féger J, Tremblay L. 2004. Behavioural disorders induced by external globus pallidus dysfunction in primates. II: Anatomical study. *Brain* **127**: 2055–2070. doi:10.1093/brain/awh239

Fujiyama F, Kuramoto E, Okamoto K, Hioki H, Furuta T, Zhou L, Nomura S, Kaneko T. 2004. Presynaptic localization of an AMPA-type glutamate receptor in corticostriatal and thalamostriatal axon terminals. *Eur J Neurosci* **20**: 3322–3330. doi:10.1111/j.1460-9568.2004.03807.x

Fujiyama F, Unzai T, Nakamura K, Nomura S, Kaneko T. 2006. Difference in organization of corticostriatal and thalamostriatal synapses between patch and matrix compartments of rat neostriatum. *Eur J Neurosci* **24**: 2813–2824. doi:10.1111/j.1460-9568.2006.05177.x

Fujiyama F, Sohn J, Nakano T, Furuta T, Nakamura KC, Matsuda W, Kaneko T. 2011. Exclusive and common targets of neostriatofugal projections of rat striosome neurons: a single neuron-tracing study using a viral vector. *Eur J Neurosci* **33**: 668–677. doi:10.1111/j.1460-9568 .2010.07564.x

Gardoni F, Picconi B, Ghiglieri V, Pilli F, Bagetta V, Bernardi G, Cattabeni F, Di Luca M, Calabresi P. 2006. A critical interaction between NR2B and MAGUK in L-DOPA induced dyskinesia. *J Neurosci* **26**: 2914–2922. doi:10.1523/ JNEUROSCI.5326-05.2006

Gauthier-Campbell C, Bredt DS, Murphy TH, El-Husseini AED. 2004. Regulation of dendritic branching and filopodia formation in hippocampal neurons by specific acylated protein motifs. *Mol Biol Cell* **15**: 2205–2217. doi:10 .1091/mbc.e03-07-0493

Gerfen CR. 1984. The neostriatal mosaic: compartmentalization of corticostriatal input and striatonigral output systems. *Nature* **311**: 461–464. doi:10.1038/311461a0

Gerfen CR. 1992. The neostriatal mosaic: multiple levels of compartmental organization. *Trends Neurosci* **15**: 133–139. doi:10.1016/0166-2236(92)90355-C

Gerfen CR, Stained WA, Arbuthnott GW, Fibiger HC. 1982. Crossed connections of the substantia nigra in the rat. *J Comp Neurol* **207**: 283–303. doi:10.1002/cne.902070308

Gerfen CR, Baimbridge KG, Miller JJ. 1985. The neostriatal mosaic: compartmental distribution of calcium-binding protein and parvalbumin in the basal ganglia of the rat and monkey. *Proc Natl Acad Sci* **82**: 8780–8784. doi:10 .1073/pnas.82.24.8780

Gerfen CR, Engber T, Mahan L, Susel Z, Chase T, Monsma F, Sibley D. 1990. D_1 and D_2 dopamine receptor-regulated gene expression of striatonigral and striatopallidal neu-

rons. *Science* **250**: 1429–1432. doi:10.1126/science.2147 780

German DC, Manaye K, Smith WK, Woodward DJ, Saper CB. 1989. Midbrain dopaminergic cell loss in Parkinson's disease: computer visualization. *Ann Neurol* **26**: 507–514. doi:10.1002/ana.410260403

Giménez-Amaya JM, Graybiel AM. 1990. Compartmental origins of the striatopallidal projection in the primate. *Neuroscience* **34**: 111–126. doi:10.1016/0306-4522(90) 90306-O

Glajch KE, Kelver DA, Hegeman DJ, Cui Q, Xenias HS, Agustine EC, Hernández VM, Verma N, Huang TY, Luo M, et al. 2016. Npas1+ pallidal neurons target striatal projection neurons. *J Neurosci* **36**: 5472–5488. doi:10 .1523/JNEUROSCI.1720-15.2016

Gonzalo N, Lanciego JL, Castle M, Vázquez A, Erro E, Obeso JA. 2002. The parafascicular thalamic complex and basal ganglia circuitry: further complexity to the basal ganglia model. *Thal Rel Sys* **1**: 341–348. doi:10.1017/S1472928 802000079

Gonzalo-Marín E, Alonso-Martinez C, Prensa Sepúlveda L, Clasca F. 2024. Micropopulation mapping of the mouse parafascicular nucleus connections reveals diverse input-output motifs. *Front Neuroanat* **17**: 1305500. doi:10 .3389/fnana.2023.1305500

Graybiel AM. 1984. Correspondence between the dopamine islands and striosomes of the mammalian striatum. *Neuroscience* **13**: 1157–1187. doi:10.1016/0306-4522(84) 90293-8

Graybiel AM. 1990. Neurotransmitters and neuromodulators in the basal ganglia. *Trends Neurosci* **13**: 244–254. doi:10.1016/0166-2236(90)90104-I

Graybiel AM, Ragsdale CW. 1978. Histochemically distinct compartments in the striatum of human, monkeys, and cat demonstrated by acetylthiocholinesterase staining. *Proc Natl Acad Sci* **75**: 5723–5726. doi:10.1073/pnas.75 .11.5723

Groenewegen HJ, Berendse HW. 1994. The specificity of the "nonspecific" midline and intralaminar thalamic nuclei. *Trends Neurosci* **17**: 52–57. doi:10.1016/0166-2236(94) 90074-4

Guilhemsang L, Mallet NP. 2024. Arkypallidal neurons in basal ganglia circuits: unveiling novel pallidostriatal loops? *Curr Opin Neurobiol* **84**: 102814. doi:10.1016/j .conb.2023.102814

Guridi J, Rodriguez-Rojas R, Carmona-Abellán M, Parras O, Becerra V, Lanciego JL. 2018. History and future challenges of the subthalamic nucleus as surgical target: review article. *Mov Disord* **33**: 1540–1550. doi:10.1002/mds .92

Haber SN, Fudge JL, McFarland NR. 2000. Striatonigrostriatal pathways in primates form an ascending spiral from the shell to the dorsolateral striatum. *J Neurosci* **20**: 2369–2382. doi:10.1523/JNEUROSCI.20-06-02369.2000

Heimer G, Bar-Gad I, Goldberg JA, Bergman H. 2002. Dopamine replacement therapy reverses abnormal synchronization of pallidal neurons in the 1-methyl-4-phenyl-1,2,3,6-tetrahydropyridine primate model of Parkinsonism. *J Neurosci* **22**: 7850–7855. doi:10.1523/JNEUROSCI .22-18-07850.2002

Cite this article as *Cold Spring Harb Perspect Med* doi: 10.1101/cshperspect.a041617

Hikosaka O, Nakamura K, Sakai K, Nakahara H. 2002. Central mechanisms of motor skill learning. *Curr Opin Neurobiol* 12: 217–222. doi:10.1016/S0959-4388(02)00307-0

Ilinsky JA, Kultas-Ilinsky K. 1987. Saggital cytoarchitectonic maps of *Macaca mulatta* thalamus with a revised nomenclature of the motor-related nuclei validated by observations on their connectivity. *J Comp Neurol* 262: 331–364. doi:10.1002/cne.902620303

Joel D, Weiner I. 1997. The connections of the primate subthalamic nucleus: indirect pathways and the open-interconnected scheme of basal ganglia-thalamocortical circuitry. *Brain Res Rev* 23: 62–78. doi:10.1016/S0165-0173(96)00018-5

Kamath T, Abdulraouf A, Burris SJ, Langlieg J, Gazestani V, Nadaf NM, Balderrama K, Vanderburg C, Macosko EZ. 2022. Single-cell genomic profiling of human dopamine neurons identifies a population that selectively degenerates in Parkinson's disease. *Nat Neurosci* 25: 588–595. doi:10.1038/s41593-022-01061-1

Kaneko T, Fujiyama F, Hioki H. 2002. Immunohistochemical localization of candidates for vesicular glutamate transporters in the rat brain. *J Comp Neurol* 444: 39–62. doi:10.1002/cne.10129

Karachi C, Grabli D, Baup N, Mounayar S, Tandé D, François C, Hirsch EC. 2009. Dysfunction of the subthalamic nucleus induces behavioral and movement disorders in monkeys. *Mov Disord* 24: 1183–1192. doi:10.1002/mds.22547

Kawaguchi Y, Wilson CJ, Emson PC. 1989. Intracellular recording of identified neostriatal patch and matrix spiny cells in a slice preparation preserving cortical inputs. *J Neurophysiol* 62: 1052–1068. doi:10.1152/jn.1989.62.5.1052

Kawaguchi Y, Wilson CJ, Emson PC. 1990. Projection subtypes of rat neostriatal matrix cells revealed by intracellular injection of biocytin. *J Neurosci* 10: 3421–3438. doi:10.1523/JNEUROSCI.10-10-03421.1990

Kawaguchi Y, Wilson CJ, Augood SJ, Emson PC. 1995. Striatal interneurons: chemical, physiological and morphological characterization. *Trends Neurosci* 18: 527–535. doi:10.1016/0166-2236(95)98374-8

Kincaid AE, Wilson CJ. 1996. Corticostriatal innervation of the patch and matrix in the rat neostriatum. *J Comp Neurol* 374: 578–592. doi:10.1002/(SICI)1096-9861(19961028)374:4<578::AID-CNE7>3.0.CO;2-Z

Kincaid Y, Wilson CJ, Emson PC. 1991. Projection subtypes of rat neostriatal matrix cells revealed by intracellular injection of biocytin. *J Neurosci* 10: 3241–3438.

Kita H, Kitai ST. 1987. Efferent projections of the subthalamic nucleus in the rat: light and electron microscopic analysis with the PHA-L method. *J Comp Neurol* 260: 435–452. doi:10.1002/cne.902600309

Kita H, Chang HT, Kitai ST. 1983. The morphology of intracellularly labeled rat subthalamic neurons: a light microscopic analysis. *J Comp Neurol* 215: 245–257. doi:10.1002/cne.902150302

Kravitz AV, Freeze BS, Parker PR, Kay K, Thwin MT, Deisseroth K, Kreitzer AC. 2010. Regulation of parkinsonian motor behaviours by optogenetic control of basal ganglia circuitry. *Nature* 466: 622–626. doi:10.1038/nature09159

Kuramoto E. 2019. Method for labeling and reconstruction of single neurons using Sindbis virus vectors. *J Chem Neuroanat* 100: 101648. doi:10.1016/j.jchemneu.2019.05.002

Lanciego JL, Vázquez A. 2012. The basal ganglia and thalamus of the long-tailed macaque in stereotaxic coordinates. A template atlas based on coronal, sagittal and horizontal brain sections. *Brain Struct Funct* 217: 613–666. doi:10.1007/s00429-011-0370-5

Lanciego JL, Gonzalo N, Castle M, Sánchez-Escobar C, Aymerich MS, Obeso JA. 2004. Thalamic innervation of striatal and subthalamic neurons projecting to the rat entopeduncular nucleus. *Eur J Neurosci* 19: 1267–1277. doi:10.1111/j.1460-9568.2004.03244.x

Lanciego JL, López IP, Rico AJ, Aymerich MS, Pérez-Manso M, Conte L, Combarro C, Roda E, Molina C, Gonzalo N, et al. 2009. The search for a role of the caudal intralaminar nuclei in the pathophysiology of Parkinson's disease. *Brain Res Bull* 78: 55–59. doi:10.1016/j.brainresbull.2008.08.008

Lapper SR, Bolam JP. 1992. Input from the frontal cortex and the parafascicular nucleus to cholinergic interneurons in the dorsal striatum of the rat. *Neuroscience* 51: 533–545. doi:10.1016/0306-4522(92)90293-B

Lapper SR, Smith Y, Sadikot AF, Parent A, Bolam JP. 1992. Cortical input to parvalbumin-immunoreactive neurones in the putamen of the squirrel monkey. *Brain Res* 580: 215–224. doi:10.1016/0006-8993(92)90947-8

Lazaridis I, Crittenden JR, Ahn G, Hirokane K, Wickersham IR, Yoshida T, Mahar A, Skara V, Loftus JH, Parvataneni K, et al. 2024. Striosomes control dopamine via dual pathways paralleling canonical basal ganglia circuits. *Curr Biol* S0960-9822(24)01338-1. doi:10.1016/j.cub.2024.09.070

Lehéricy S, Benali H, Van de Moortele PF, Pélégrini-Issac M, Waechter T, Ugurbil K, Doyon J. 2005. Distinct basal ganglia territories are engaged in early and advanced motor learning sequence. *Proc Natl Acad Sci* 102: 12566–12571. doi:10.1073/pnas.0502762102

Lei W, Jiao Y, Del Mar N, Reiner A. 2004. Evidence for differential cortical input to direct pathway versus indirect pathway striatal projection neurons in rats. *J Neurosci* 24: 8289–8299. doi:10.1523/JNEUROSCI.1990-04.2004

Li W, Li JY. 2024. Overlaps and divergences between tauopathies and synucleinopathies: a duet of neurodegeneration. *Translat Neurodegen* 13: 16. doi:10.1186/s40035-024-00407-y

López A, Muñoz A, Guerra MJ, Labandeira-García JL. 2001. Mechanisms of the effects of exogenous levodopa on the dopamine-denervated striatum. *Neuroscience* 103: 639–651. doi:10.1016/S0306-4522(00)00588-1

López-Real A, Rodríguez-Pallares J, Guerra MJ, Labandeiragarcía JL. 2003. Localization and functional significance of striatal neurons immunoreactive to aromatic L-amino acid decarboxylase or tyrosine hydroxylase in rat parkinsonian models. *Brain Res* 969: 135–146. doi:10.1016/S0006-8993(03)02291-1

Lozano AM, Lang AE, Galvez-Jimenez N, Miyasaki J, Duff J, Hutchinson WD, Dostrovsky JO. 1995. The effect of GPi pallidotomy on motor function in Parkinson's disease. *Lancet* 346: 1383–1387. doi:10.1016/S0140-6736(95)92404-3

Luquin MR, Laguna J, Herrero MT, Obeso JA. 1993. Behavioral tolerance to repeated apomorphine administration in parkinsonian monkeys. *J Neurol Sci* **114**: 40–44. doi:10 .1016/0022-510X(93)90046-2

Mallet N, Micklem BR, Henny P, Brown MT, Williams C, Bolam JP, Nakamura KC, Magill PJ. 2012. Dichotomous organization of the external globus pallidus. *Neuron* **74**: 1075–1086. doi:10.1016/j.neuron.2012.04.027

Mallet N, Schmidt R, Leventhal D, Chen F, Amer N, Boraud T, Berke JL. 2016. Arkypallidal cells send a stop signal to striatum. *Neuron* **89**: 308–316. doi:10.1016/j.neuron.2015 .12.017

Marini G, Pianca L, Tredici G. 1999. Descending projections arising from the parafascicular nucleus in rats: trajectory of fibers, projection pattern and mapping of termina-tions. *Somatosens Motor Res* **16**: 207–222. doi:10.1080/ 08990229970465

Marsden CD. 1961. Pigmentation in the nucleus substantie nigrae of mammals. *J Anat* **95**: 256–261.

Marsden CD, Obeso JA. 1994. The functions of the basal ganglia and the paradox of stereotaxic surgery in Parkin-son's disease. *Brain* **117**: 877–897. doi:10.1093/brain/117 .4.877

Matsuda W, Furuta T, Nakamura KC, Hioki H, Fujiyama E, Arai R, Kaneko T. 2009. Single nigrostriatal dopaminergic neurons form widely spread and highly dense axonal ar-borizations in the neostriatum. *J Neurosci* **29**: 444–453. doi:10.1523/JNEUROSCI.4029-08.2009

McCairn KW, Bronfeld M, Belelovsky K, Bar-Gad I. 2009. The neurophysiological correlates of motor tics following focal striatal disinhibition. *Brain* **132**: 2125–2138. doi:10 .1093/brain/awp142

McFarland NR, Haber SN. 2000. Convergent inputs from thalamic motor nuclei and frontal cortical areas to the dorsal striatum in the primate. *J Neurosci* **20**: 3798–3813. doi:10.1523/JNEUROSCI.20-10-03798.2000

McFarland NR, Haber SN. 2001. Organization of thalamos-triatal terminals from the ventral motor nuclei in the macaque. *J Comp Neurol* **429**: 321–336. doi:10.1002/ 1096-9861(20000108)429:2<321::AID-CNE11>3.0.CO ;2-A

Meredith GE, Wouterlood FG. 1990. Hippocampal and mid-line thalamic fibers and terminals in relation to the cho-line acetyltransferase-immunoreactive neurons in nucle-us accumbens of the rat: a light and electron microscopic study. *J Comp Neurol* **296**: 204–221. doi:10.1002/cne .902960203

Milber JM, Noorigian JV, Morley JF, Petrovitch H, White L, Ross GW, Duda JE. 2012. Lewy pathology is not the first sign of degeneration in vulnerable neurons in Parkinson disease. *Neurology* **79**: 2307–2314. doi:10.1212/WNL .0b013e318278fe32

Mitchell JJ, Clarke CE, Boyce S, Robertson RG, Peggs D, Sambrook MA, Crossman AR. 1989. Neural mechanisms underlying parkinsonian symptoms based upon regional uptake of 2-deoxyglucose in monkeys exposed to 1-meth-yl-4-phenyl-1,2,3,6-tetrahydropyridine. *Neuroscience* **32**: 213–226. doi:10.1016/0306-4522(89)90120-6

Monchi O, Petrides M, Strafella AP, Worsley KJ, Doyon J. 2006. Functional role of the basal ganglia in the planning and execution of actions. *Ann Neurol* **59**: 257–264. doi:10 .1002/ana.20742

Morel A. 2007. *Stereotactic atlas of the human thalamus and basal ganglia*. CRC, Boca Raton, FL.

Nambu A. 2004. A new dynamic model of the cortico-basal ganglia loop. *Progr Brain Res* **143**: 461–466. doi:10.1016/ S0079-6123(03)43043-4

Nambu A, Chiken S. 2024. External segment of the globus pallidus in health and disease: its interactions with the striatum and subthalamic nucleus. *Neurobiol Dis* **190**: 106362. doi:10.1016/j.nbd.2023.106362

Nambu A, Takada M, Inase M, Tokuno H. 1996. Dual so-matotopical representations in the primate subthalamic nucleus: evidence for ordered but reversed body-map transformations from the primary motor cortex and the supplementary motor area. *J Neurosci* **16**: 2671–2683. doi:10.1523/JNEUROSCI.16-08-02671.1996

Nambu A, Tokuno H, Takada M. 2002. Functional signifi-cance of the cortico-subthalamo-pallidal "hyperdirect" pathway. *Neurosci Res* **43**: 111–117. doi:10.1016/S0168-0102(02)00027-5

Nambu A, Chiken S, Sano H, Hatanaka N, Obeso JA. 2023. Dynamic activity model of movement disorders: the fun-damental role of the *hyperdirect* pathway. *Mov Disord* **38**: 2145–2150. doi:10.1002/mds.29646

Nauta HJW, Cole M. 1978. Efferent projections of the sub-thalamic nucleus: an autoradiographic study in monkey and cat. *J Comp Neurol* **180**: 1–16.

Obeso JA, Rodriguez-Oroz MC, Rodriguez M, Lanciego JL, Artieda J, Gonzalo N, Olanow CW. 2000. Pathophysiol-ogy of the basal ganglia in Parkinson's disease. *Trends Neurosci* **23**: S8–S19. doi:10.1016/S1471-1931(00)000 28-8

Obeso JA, Rodríguez-Oroz MC, Benitez-Termino B, Blesa FJ, Guridi J, Marin C, Rodriguez M. 2008. Functional organization of the basal ganglia: therapeutic implications for Parkinson's disease. *Mov Disord* **23**: S548–S559. doi:10.1002/mds.22062

Obeso JA, Jahanshahi M, Alvarez L, Macias R, Pedroso I, Wilkinson L, Pavon N, Day B, Pinto S, Rodriguez-Oroz MC, et al. 2009. What can man do without basal ganglia motor output? The effect of combined unilateral sub-thalamotomy and pallidotomy in a patient with Parkin-son's disease. *Exp Neurol* **220**: 283–292. doi:10.1016/j .expneurol.2009.08.030

Papa SM, Desimone R, Fiorani M, Oldfield EH. 1999. Inter-nal globus pallidus discharge is nearly suppressed during levodopa-induced dyskinesias. *Ann Neurol* **46**: 732–738. doi:10.1002/1531-8249(199911)46:5<732::AID-ANA8>3.0 .CO;2-Q

Parent A, Sato F, Wu Y, Gauthier J, Lévesque M, Parent M. 2000. Organization of the basal ganglia: the importance of axonal collateralization. *Trends Neurosci* **23**: S20–S27. doi:10.1016/S1471-1931(00)00022-7

Penny GR, Wilson CJ, Kitai ST. 1988. Relationship of the axonal and dendritic geometry of spiny projection neu-rons to the compartmental organization of the neostria-tum. *J Comp Neurol* **269**: 275–289. doi:10.1002/cne .902690211

Percheron G, François C, Talbi B, Yelnik J, Fénelon G. 1996. The primate motor thalamus. *Brain Res Rev* **22**: 93–181. doi:10.1016/0165-0173(96)00003-3

Perth CB, Kuhar MI, Snyder SH. 1976. Opiate receptor: autoradiographic localization in rat brain. *Proc Natl Acad Sci* **73**: 3729–3733. doi:10.1073/pnas.73.10.3729

Cite this article as *Cold Spring Harb Perspect Med* doi: 10.1101/cshperspect.a041617

Plenz D, Kitai ST. 1999. A basal ganglia pacemaker formed by the subthalamic nucleus and external globus pallidus. *Nature* **400:** 677–682. doi:10.1038/23281

Powell TPS, Cowan WM. 1956. A study of thalamo-striate relations in the monkey. *Brain* **79:** 364–366. doi:10.1093/brain/79.2.364

Prensa L, Parent A. 2001. The nigrostriatal pathway in the rat: a single-axon study of the relationship between dorsal and ventral tier nigral neurons and the striosome/matrix striatal compartments. *J Neurosci* **21:** 7247–7260. doi:10.1523/JNEUROSCI.21-18-07247.2001

Prensa L, Giménez-Amaya JM, Parent A. 1999. Chemical heterogeneity of the striosomal compartment in the human striatum. *J Comp Neurol* **413:** 603–618. doi:10.1002/(SICI)1096-9861(19991101)413:4<603::AID-CNE9>3.0.CO;2-K

Purdon-Martin J. 1927. Hemichorea resulting from a local lesion of the brain (the syndrome of the body of Luys). *Brain* **50:** 637–649. doi:10.1093/brain/50.3-4.637

Ragsdale CW, Graybiel AM. 1988. Fibers from the basolateral nucleus of the amygdala selectively innervate striosomes in the caudate nucleus of the cat. *J Comp Neurol* **269:** 506–522. doi:10.1002/cne.902690404

Raju DV, Smith Y. 2005. Differential localization of vesicular glutamate transporters 1 and 2 in the rat striatum. In *The basal ganglia VIII* (ed. Bolam JP, et al.), pp. 601–610. Springer, New York.

Raju DV, Shah DJ, Wright TM, Hall RA, Smith Y. 2006. Differential synaptology of vGlut2-containing thalamostriatal afferents between the patch and matrix compartments in rats. *J Comp Neurol* **499:** 231–243. doi:10.1002/cne.21099

Redgrave P, Coizet V, Comoli E, McHaffie JG, Leriche M, Vautrelle N, Haves LM, Overton P. 2010. Interactions between the midbrain superior colliculus and the basal ganglia. *Front Neuroanat* **4:** 132. doi:10.3389/fnana.2010.00132

Redgrave P, Vautrelle N, Reynolds JN. 2011. Functional properties of the basal ganglia's re-entrant loop architecture: selection and reinforcement. *Neuroscience* **198:** 138–151. doi:10.1016/j.neuroscience.2011.07.060

Reiner A, Jiao Y, Del Mar N, Laverghetta AV, Lei WL. 2003. Differential morphology of pyramidal tract-type and intratelencephalically projecting-type corticostriatal neurons and their intrastriatal terminals in rats. *J Comp Neurol* **457:** 420–440. doi:10.1002/cne.10541

Reiner A, Hart NM, Lei W, Deng Y. 2010. Corticostriatal projection neurons—dichotomous types and dichotomous functions. *Front Neuroanat* **4:** 142. doi:10.3389/fnana.2010.00142

Rico AJ, Barroso-Chinea P, Conte-Perales L, Roda E, Gómez-Bautista V, Gendive M, Obeso JA, Lanciego JL. 2010. A direct projection from the subthalamic nucleus to the ventral thalamus in monkeys. *Neurobiol Dis* **39:** 381–392. doi:10.1016/j.nbd.2010.05.004

Rico AJ, Dopeso-Reyes IG, Martínez-Pinilla E, Sucunza D, Pignataro D, Roda E, Marín-Ramos D, Labandeira-García JL, George SR, Franco R, et al. 2017. Neurochemical evidence supporting dopamine D1-D2 receptor heteromers in the striatum of the long-tailed macaque: changes following dopaminergic manipulation. *Brain Struct Funct* **222:** 1767–1784. doi:10.1007/s00429-016-1306-x

Rico AJ, Corcho A, Chocarro J, Ariznabarreta G, Roda E, Honrubia A, Arnaiz P, Lanciego JL. 2024. Development and characterization of a non-human primate model of disseminated synucleinopathy. *Front Neuroanat* **18:** 1355940. doi:10.3389/fnana.2024.1355940

Rodríguez-Oroz MC, Rodriguez M, Guridi J, Mewes K, Chockkman V, Vitek J, DeLong MR, Obeso JA. 2001. The subthalamic nucleus in Parkinson's disease: somatotopic organization and physiological characteristics. *Brain* **124:** 1777–1790. doi:10.1093/brain/124.9.1777

Rommelfanger KS, Wichmann T. 2010. Extrastriatal dopaminergic circuits of the basal ganglia. *Front Neuroanat* **4:** 139. doi:10.3389/fnana.2010.00139

Rosenow JM, Mogilnert AY, Ahmed A, Rezai AR. 2004. Deep brain stimulation for movement disorders. *Neurol Res* **26:** 9–20. doi:10.1179/016164104773026480

Rosin DL, Robeya A, Woodard RL, Guyenet PG, Linden J. 1998. Immunohistochemical localization of adenosine A_{2A} receptors in the rat central nervous system. *J Comp Neurol* **401:** 163–186. doi:10.1002/(SICI)1096-9861(19981116)401:2<163::AID-CNE2>3.0.CO;2-D

Royce GJ, Mourey RJ. 1985. Efferent connections of the centromedian and parafascicular thalamic nuclei: an autoradiographic investigation in the cat. *J Comp Neurol* **235:** 277–300. doi:10.1002/cne.902350302

Rozas G, Liste I, Guerra MJ, Labandeira-Garcia JL. 1998. Sprouting of the serotonergic afferents into striatum after selective lesion of the dopaminergic system by MPTP in adult mice. *Neurosci Lett* **245:** 151–154. doi:10.1016/S0304-3940(98)00198-0

Rudkin TM, Sadikot AE. 1999. Thalamic input to parvalbumin-immunoreactive GABAergic interneurons: organization in normal striatum and effect of neonatal decortication. *Neuroscience* **88:** 1165–1175. doi:10.1016/S0306-4522(98)00265-6

Sadikot AF, Parent A, François C. 1990. The centre median and parafascicular thalamic nuclei project respectively to the sensorimotor and associative-limbic striatal territories in the squirrel monkey. *Brain Res* **510:** 161–165. doi:10.1016/0006-8993(90)90746-X

Sadikot AF, Parent A, François C. 1992a. Efferent connections of the centromedian and parafascicular thalamic nuclei in the squirrel monkey: a PHA-L study of subcortical projections. *J Comp Neurol* **315:** 137–159. doi:10.1002/cne.903150203

Sadikot AF, Parent A, Smith Y, Bolam JP. 1992b. Efferent connections of the centromedian and parafascicular thalamic nuclei in the squirrel monkey: a light and electron microscopic study of the thalamostriatal projection in relation to striatal heterogeneity. *J Comp Neurol* **320:** 228–242. doi:10.1002/cne.903200207

Salin P, López IP, Kachidian P, Barroso-Chinea P, Rico AJ, Gómez-Bautista V, Coulon P, Kerkerian-Le Goff L, Lanciego JL. 2009. Changes to interneuron-driven striatal microcircuits in a rat model of Parkinson's disease. *Neurobiol Dis* **34:** 545–552. doi:10.1016/j.nbd.2009.03.006

Sato F, Parent M, Lévesque M, Parent A. 2000. Axonal branching pattern of neurons of the subthalamic nucleus in primates. *J Comp Neurol* **424:** 142–152. doi:10.1002/1096-9861(20000814)424:1<142::AID-CNE10>3.0.CO;2-8

Schröeder KF, Hopf A, Lange H, Thörner G. 1975. Struktur-analysen des striatum, pallidum und nucleus subthalami-cus beim menschen. I: Striatum. *J Hirnforschung* **16**: 333–350.

Selemon LD, Goldman-Rakic PS. 1985. Longitudinal topography and interdigitation of corticostriatal projections in the rhesus monkey. *J Neurosci* **5**: 776–794. doi:10.1523/JNEUROSCI.05-03-00776.1985

Shink E, Bevan MD, Bolam JP, Smith Y. 1996. The subthalamic nucleus and the external pallidum: two tightly interconnected structures that control the output of the basal ganglia in the monkey. *Neuroscience* **73**: 335–357. doi:10.1016/0306-4522(96)00022-X

Sidibé M, Smith Y. 1996. Differential synaptic innervation of striatofugal neurones projecting to the internal or external segments of the globus pallidus by thalamic afferents in the squirrel monkey. *J Comp Neurol* **365**: 445–465. doi:10.1002/(SICI)1096-9861(19960212)365:3<445::AID-CNE8>3.0.CO;2-4

Sidibé M, Smith Y. 1999. Thalamic inputs to striatal interneurons in monkeys: synaptic organization and co-localization of calcium-binding proteins. *Neuroscience* **89**: 1189–1208. doi:10.1016/S0306-4522(98)00367-4

Smith Y, Bolam JP. 1990. The neural network of the basal ganglia as revealed by the study of synaptic connections of identified neurones. *Trends Neurosci* **13**: 259–265. doi:10.1016/0166-2236(90)90106-K

Smith Y, Parent A. 1986. Differential connections of caudate nucleus and putamen in the squirrel monkey (*Saimiri sciureus*). *Neuroscience* **18**: 347–371. doi:10.1016/0306-4522(86)90159-4

Smith Y, Raju DV, Pare JF, Sidibé M. 2004. The thalamo-striatal system: a highly specific network of the basal ganglia circuitry. *Trends Neurosci* **27**: 520–527. doi:10.1016/j.tins.2004.07.004

Smith Y, Surmeier DJ, Redgrave P, Kimura M. 2011. Thalamic contributions to basal ganglia-related behavioral switching and reinforcement. *J Neurosci* **31**: 16102–16106. doi:10.1523/JNEUROSCI.4634-11.2011

Spillantini MG, Schmidt ML, Lee VM, Trojanowski JQ, Jakers R, Goedert M. 1997. α-Synuclein in Lewy bodies. *Nature* **388**: 839–840. doi:10.1038/42166

Sugimoto T, Hattori T. 1983. Confirmation of thalamosubthalamic projections by electron microscope autoradiography. *Brain Res* **267**: 335–339. doi:10.1016/0006-8993(83)90885-5

Sugimoto T, Hattori T, Mizuno N, Itoh K, Sato M. 1983. Direct projections from the centre median-parafascicular complex to the subthalamic nucleus in the cat and rat. *J Comp Neurol* **214**: 209–216. doi:10.1002/cne.902140208

Szabo J. 1980. Organization of the ascending striatal afferents in monkeys. *J Comp Neurol* **189**: 307–321. doi:10.1002/cne.901890207

Tepper MJ, Koós T, Wilson CJ. 2004. GABAergic microcircuits in the neostriatum. *Trends Neurosci* **27**: 662–669. doi:10.1016/j.tins.2004.08.007

Thomas TM, Smith Y, Levey AI, Hersch SM. 2000. Cortical inputs to m2-immunoreactive striatal interneurons in rat and monkey. *Synapse* **37**: 252–261. doi:10.1002/1098-2396(20000915)37:4<252::AID-SYN2>3.0.CO;2-A

Van der Kooy D, Hattori T. 1980. Single subthalamic nucleus neurons project to both the globus pallidus and substantia nigra in rat. *J Comp Neurol* **192**: 751–768. doi:10.1002/cne.901920409

Van der Werf YD, Witter MP, Groenewegen HJ. 2002. The intralaminar and midline nuclei of the thalamus. Anatomical and functional evidence for participation in processes of arousal and awareness. *Brain Res Rev* **39**: 107–140. doi:10.1016/S0165-0173(02)00181-9

Vogt C, Vogt O. 1941. Thalamusstudien I–III. I. Sur Einfürung, II. Homogenität und Grenzgestaldung der Grisea des Thalamus, III. Das Griseum Centrale (centrum medianum Luys). *J Physiol Neurol* **50**: 31–154.

Wilson SAK. 1925. Disorders of motility and tone. *Lancet Neurol* **1**: 1–103.

* Wichmann T. 2024. Pathophysiology of motor control abnormalities in Parkinson's disease. *Cold Spring Harb Perspect Biol* doi: 10.1101/cshperspect.a041616

Worbe Y, Baup N, Grabli D, Chaigneau M, Mounayar S, McCairn K, Féger J, Tremblay L. 2009. Behavioral and movement disorders induced by local inhibitory dysfunction in primate striatum. *Cereb Cortex* **19**: 1844–1856. doi:10.1093/cercor/bhn214

Wu Y, Richard S, Parent A. 2000. The organization of the striatal output system: a single-cell juxtacellular labeling study in the rat. *Neurosci Res* **38**: 49–62. doi:10.1016/S0168-0102(00)00140-1

Yashukawa T, Kita T, Xue Y, Kita H. 2004. Rat intralaminar thalamic nuclei projections to the globus pallidus: a biotinylated dextran amine anterograde tracing study. *J Comp Neurol* **471**: 153–167.

Yin HH, Knowlton B. 2006. The role of the basal ganglia in habit formation. *Nat Neurosci Rev* **7**: 464–476. doi:10.1038/nrn1919

Pathophysiology of Motor Control Abnormalities in Parkinson's Disease

Thomas Wichmann

Emory National Primate Research Center and School of Medicine, Emory University, Atlanta, Georgia 30329, USA

Correspondence: twichma@emory.edu

Research in the last few decades has brought us closer to an understanding of the brain circuit abnormalities that underlie parkinsonian motor signs. This article summarizes the current knowledge in this rapidly emerging field. Traditional observations of activity changes of basal ganglia neurons that accompany akinesia and bradykinesia have been supplemented with new knowledge regarding specific pathophysiologic changes that are associated with other parkinsonian signs, such as tremor and gait impairments. New research also emphasizes the role of non-basal ganglia structures in parkinsonism, including the pedunculopontine nucleus, the cerebellum, and the cerebral cortex, and the role of structural and functional neuroplasticity. A more detailed understanding of the brain network abnormalities that result from Parkinson's disease is necessary to arrive at more effective and specific treatments for these symptoms in parkinsonian patients through circuit interventions reaching from deep brain stimulation to genetic and chemogenetic treatments.

The defining manifestations of Parkinson's disease (PD) are abnormalities of movement, including slowness, difficulties with gait and balance, and tremor. The disease also features many nonmotor features (such as cognitive impairments, sleep disturbances, or autonomic problems), which will not be discussed here. After a brief introduction to the motor signs of PD, an overview of the relevant brain circuitry, and a summary of the available pathophysiologic knowledge will be presented. While this paper and Lanciego and Obeso (2024) share some common information about the functional neuroanatomy of the basal ganglia, their pri-

mary focus differs. This work centers on the pathophysiology of PD, using anatomical details primarily as background. In contrast, Lanciego and Obeso (2024) focus on anatomy, with only brief mentions of PD and dyskinesia.

MOTOR FEATURES OF PARKINSON'S DISEASE

Although the overall definition of PD is currently under debate (Siderowf et al. 2023; Cardoso et al. 2024), motor features of manifest PD have long been accepted to include slowness of movement (bradykinesia), tremor at rest, muscle

Copyright © 2025 Cold Spring Harbor Laboratory Press; all rights reserved
Cite this article as *Cold Spring Harb Perspect Med* doi: 10.1101/cshperspect.a041616

rigidity, and gait and balance impairments. The clinical diagnosis of PD is supported by an asymmetric appearance of these motor symptoms, the presence of tremor at rest, and the demonstration of the therapeutic effectiveness of dopamine replacement therapy.

Bradykinesia (i.e., slowing of movements in the absence of weakness) is a fundamental aspect of PD. Bradykinesia is associated with a slow buildup of muscle activation in agonist muscles. The term bradykinesia is also often used to include an overall paucity of movement and difficulty with movement initiation (akinesia). From a patient's perspective, bradykinesia and akinesia are frequently described as a failure of motor automaticity, and a sense of loss of agency.

PD is typically associated with an asymmetric slow tremor at rest, that diminishes with voluntary movement. As discussed below, the pathophysiology of parkinsonian tremor is likely different from that of akinesia and bradykinesia. Another cardinal sign of PD, muscle stiffness (rigidity), is detected during physical examination. Brief interruptions of muscle resistance during passive movement may give rise to a sensation of "cog wheeling."

Postural, balance, and gait abnormalities, ranging from mildly flexed posture to severely impaired postural reflexes and frequent falls, are among the most disabling components of advanced PD. A specifically severe gait disorder, frequently occurring in advanced PD, is termed "freezing of gait" (FOG) and consists of momentary failures of gait control in which the patient shows a hesitation at the start of, or during, walking, often triggered by specific sensory contexts or situations, such as when stepping across thresholds or when walking on specific floor patterns. FOG is often associated with a long disease duration (Giladi et al. 1992, 2001; Perez-Lloret et al. 2014; Amboni et al. 2015) and coincident with the presence of cognitive impairments (Factor et al. 2011, 2014; Sunwoo et al. 2013; Peterson et al. 2015; Picillo et al. 2015; Myers et al. 2017). Similar to tremor, the pathophysiologic basis of FOG may differ from that of the other parkinsonian signs and symptoms.

Basal Ganglia-Thalamocortical Circuitry

Basic Circuit Anatomy

The motor aspects of PD result mostly from pathology in the basal ganglia. These structures participate in multiple spatially segregated parallel networks that link them to the thalamus and frontal cortex with each subcircuit involved in specific functional domains. The overall anatomy of these circuits and their interactions with the cerebellum is shown in Figure 1.

With regard to the pathophysiology of PD, one of these circuits, the "motor circuit," is most relevant. This circuit originates from corticostriatal projecting neurons in the primary motor cortex (M1), supplementary motor area (SMA), premotor cortex, cingulate motor area, and postcentral sensory cortical areas. The corticostriatal projections reach somatotopically organized portions of the postcommissural putamen which is the major input station of the basal ganglia. In primates (including humans), most of the movement-related output of the basal ganglia arises from the internal segment of the globus pallidus (GPi), with lesser contributions from the substantia nigra pars reticulata (SNr). GPi/SNr project to the ventral anterior and anterior ventrolateral nuclei of the thalamus (VA, VL), with collaterals to the movement-related portion of the caudal intralaminar nuclei (especially the centromedian nucleus [CM]), and the pedunculopontine nucleus (PPN). VA/VL, in turn, reciprocally interact with the SMA and M1, thus, at least partially, closing the "motor circuit."

The putamen is linked to GPi/SNr by two anatomically distinct pathways (i.e., the monosynaptic GABAergic "direct" projection of striatal spiny projection neurons to GPi/SNr) and the polysynaptic GABAergic "indirect" projection that links striatal medium spiny neurons to GABAergic neurons in the external pallidal segment (GPe), which in turn project to glutamatergic projection neurons in the subthalamic nucleus (STN) that then innervate GPi/SNr (Fig. 1). Traditionally, these pathways have been seen to oppose one another's functions, with the direct pathway acting to reduce the GABAergic basal ganglia output, and activation of the indirect pathway raising it. As the basal ganglia out-

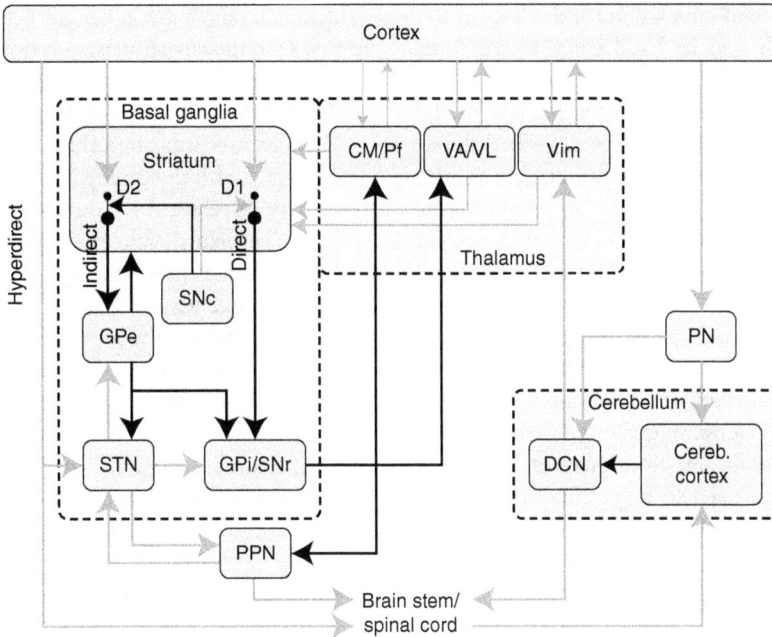

Figure 1. Cortico–subcortical circuits that have been implicated in the pathophysiology of parkinsonism. Dark arrows indicate inhibitory connections; gray arrows indicate excitatory connections. (CM/Pf) Centromedian and parafascicular nuclei of the thalamus, (Cereb. cortex) cerebellar cortex, (DCN) deep cerebellar nuclei, (D1) D1-like DA receptor subtype, (D2) D2-like DA receptor subtype, (GPe) external segment of the globus pallidus, (GPi) internal segment of the globus pallidus, (PN) pontine nuclei, (PPN) pedunculopontine nucleus, (SNc) substantia nigra pars compacta, (SNr) substantia nigra pars reticulata, (STN) subthalamic nucleus, and (VA/VL) ventral anterior and ventral lateral nuclei of the thalamus. (Reprinted, with permission, from Wichmann 2018.)

put nuclei send inhibitory projections to the thalamocortical neurons in the VL/VA, this would translate eventually into a role of the indirect pathway in reducing cortical activity (and movement), and a role of the direct pathway in activating cortical activity (and movement). This view is supported by some recent optogenetic and chemogenetic animal studies (Kravitz et al. 2010; Chu et al. 2017; Chen et al. 2023b). However, recent rodent studies have suggested that direct and indirect pathways may, in fact, be jointly engaged in many routine activities (Cui et al. 2013; Vicente et al. 2016; Yttri and Dudman 2016, 2018). According to these results, the interplay between the direct and indirect pathways may either allow focusing of basal ganglia output to permit only intended movements or regulate the overall amount and "vigor" of movements (Desmurget and Turner 2008; Turner and Desmurget 2010; Dudman and Kra-

kauer 2016; Thura and Cisek 2017; Jurado-Parras et al. 2020; Park et al. 2020; Hanakawa 2021; Lee et al. 2023).

Role of Dopamine

The activity of putamenal spiny projection neurons that give rise to the direct and indirect pathways is differentially regulated by dopaminergic inputs, emerging from the substantia nigra pars compacta (SNc) in the midbrain. The actions of the widely collateralizing nigrostriatal projection (Matsuda et al. 2009) on direct pathway neurons are mediated via (excitatory) dopamine D1 receptors, while those onto indirect pathway neurons use (inhibitory) D2 receptors. At least in first approximation, dopamine release therefore activates the direct pathway and inactivates the indirect pathway neurons, although these actions are more than likely further modulated by axon collateral interactions at the level of the striatum itself and in basal ganglia structures downstream from it.

Dopamine also acts at sites outside of the striatum. While earlier studies described dopamine release at sites in the pallidum, SNr, the STN, and VA/VL, they did not provide clear evidence for a functional role of dopaminergic transmission at these sites (Rommelfanger and Wichmann 2010). Recent work in a mouse model of parkinsonism has, however, suggested that dopamine release at extrastriatal sites is behaviorally important, and that loss of dopamine at these sites is a prerequisite for parkinsonism to develop (González-Rodríguez et al. 2021). These findings need to be confirmed in other animal models (including primates).

Other Basal Ganglia Connections

A separate cortical input to the STN, also known as the "hyperdirect" pathway (Fig. 1), provides a route between the cerebral cortex and GPi/SNr with shorter latency than the direct and indirect pathways. The hyperdirect pathway may serve to quickly stop ongoing detrimental behaviors (Nambu et al. 2002, 2023; Baladron et al. 2019; Polyakova et al. 2020; Diesburg and Wessel 2021). In rodent species, this pathway is a collateral of corticospinal neurons, while at least portions of this projection may arise from separate cortical neurons in primates.

A variety of internal feedback projections also exist. For example, CM heavily projects to the motor striatum (Smith and Parent 1986; Ragsdale and Graybiel 1991; Sadikot et al. 1992; Sidibé and Smith 1999; Matsumoto et al. 2001; Nanda et al. 2009; Sadikot and Rymar 2009; Yamanaka et al. 2018; Ilyas et al. 2019), and the PPN to most of the other basal ganglia (Mena-Segovia et al. 2004; Martinez-Gonzalez et al. 2011; Mena-Segovia and Bolam 2017). A sizeable proportion of GPe neurons engages in feedback projections to the striatum (Mallet et al. 2012; Abdi et al. 2015; Deffains et al. 2016; Glajch et al. 2016; Baker et al. 2023; Courtney et al. 2023; Guilhemsang and Mallet 2024; Nambu and Chiken 2024). In addition, the striatum and pallidum have a large number of local

axon collaterals that regulate the activity of neurons within the structure in question. The physiologic or pathophysiologic relevance of these feedback interactions is not fully established.

Finally, it is important that the basal ganglia also interact with the cerebellum. These interactions may be relevant for our understanding of tremor (see below), and perhaps of aspects of akinesia. Anatomical tracing studies showed that projections between the STN and the cerebellar cortex engage the pontine nuclei, and that cerebellar output reaches the putamen via thalamostriatal projections from the ventral intermediate nucleus of the thalamus (Bostan et al. 2010; Bostan and Strick 2018; Bhuvanasundaram et al. 2022; Yoshida et al. 2022).

Dopamine Loss and Neuroplastic Changes

Dopamine Loss

Degeneration of the dopaminergic nigrostriatal tract is the quintessential anatomical and biochemical abnormality that characterizes PD (Ehringer and Hornykiewicz 1960; Bernheimer et al. 1973). The emergence of parkinsonism signals a failure of the brain's ability to compensate for the ongoing degeneration of the dopaminergic nigrostriatal tract. Other projections, such as catecholaminergic or cholinergic systems, may also contribute, but can only be mentioned in passing here (Buddhala et al. 2015; Albin et al. 2022; Ray Chaudhuri et al. 2023). The nigrostriatal tract is the dominant dopaminergic projection in the brain, linking dopaminergic neurons in the midbrain to striatal spiny projection neurons. Dopamine loss follows a pattern in which the "motor" putamen is first and most strongly affected, followed by other components of the striatum (caudate nucleus, ventral striatum) (Bernheimer et al. 1973; Nyberg et al. 1983; Hornykiewicz and Kish 1987; Kish et al. 1988). Parkinsonian signs appear when 65%–70% of the putamenal terminals of this projection are lost. Most of the parkinsonian signs are (at least initially) reversible by dopamine replacement therapy with levodopa, suggesting (1) that dopamine loss is indeed the prime driver of motor impairments in early manifest PD, (2) that sufficient dopamine receptors in the striatum (and elsewhere, see below) survive even when dopaminergic terminals are lost, and (3) that mechanisms exist by which synthesis of dopamine from levodopa occurs, even in the face of the loss of dopaminergic neurons. Dopamine may be generated at terminals that are not dopaminergic, for example, extrasynaptically, or within serotoninergic terminals (Brown et al. 1999; Navailles et al. 2010; Mosharov et al. 2015). As mentioned, dopamine (and dopamine loss) also affects the basal ganglia-thalamocortical circuitry outside of the striatum.

Neuroplastic Changes

Studies in animal models of dopamine depletion and in patients with PD demonstrated that the loss of dopamine is associated with secondary anatomical and functional synaptic changes in the basal ganglia and related thalamic and cortical areas. Most of these changes were found to occur at glutamatergic terminals. Thus, a change in the dendritic morphology of striatal spiny projection neurons was first described in tissue samples from patients with PD (McNeill et al. 1988). Similar changes, specifically dendritic spine loss, was later confirmed in rodent and primate models of PD (Ingham et al. 1989; Kreitzer and Malenka 2008; Zhang et al. 2013; Villalba and Smith 2018). The animal studies also demonstrated a loss of corticostriatal synapses, and an enlargement of the remaining spines. Another example of synaptic plasticity is a loss of corticosubthalamic terminals in the STN, described in monkey and rodent models of parkinsonism (Mathai et al. 2015; Chu et al. 2017). The connections between the basal ganglia-receiving portions of the thalamus and the cerebral cortex are also sites of substantial plasticity. The density of terminals of the corticothalamic projections to VA and CM is reduced (while individual remaining terminals are larger) in parkinsonian monkeys (Swain et al. 2020). Pyramidal cells in M1 and SMA (but not in other cortical motor areas) undergo glutamatergic denervation from their thalamic inputs, along with dendritic spine loss in parkinsonian

monkeys (Villalba et al. 2021; Chen et al. 2023a).

Plasticity at GABAergic synapses also occurs and likely contributes to the functional changes seen in parkinsonian animals. Thus far, this has been described to occur in the STN (Fan et al. 2012) and in the thalamus (Swain et al. 2020). It has been suggested that the synaptic adjustment at pallidosubthalamic synapses is coupled to altered activity at corticosubthalamic inputs, constituting "heterosynaptic" interactions between GABAergic and glutamatergic terminals (Fan et al. 2012; Chu and Bevan 2013; Chu et al. 2015).

PATHOPHYSIOLOGY OF PARKINSONIAN MOTOR SIGNS

Before discussing the functional changes known to occur in the parkinsonian brain, it is important to emphasize that the neuronal activity changes mentioned below were mostly identified in animal models of PD. These models usually represent the consequences of dopamine loss, but do not cover the full spectrum of abnormalities that occur in patients with PD. As mentioned below, studies in human patients with PD are also available. These studies reflect the full pathology, but the significance of the identified changes can usually not be proven, as control data are not available, and the coexistence of dopaminergic and (generally unspecified) nondopaminergic pathology, and the use of medications, leads to interpretational uncertainty.

Bradykinesia/Akinesia

Changes in Basal Ganglia Firing Rates and Patterns

Experiments in the 1980s and early 1990s resulted in a highly influential model of the emergence of parkinsonism-relevant activity changes in the basal ganglia-thalamocortical circuitry, based on the idea that dopamine loss upsets the balance of activity between the direct and indirect pathways (see above and Galvan and Wichmann 2008; Wichmann 2018), leading to greater activity of the striatal neurons that give rise to the indirect pathway, and reduced activity in those that give rise to the direct pathway, leading to an overall increased inhibitory GPi output. This model explained bradykinesia/akinesia as the result of excessive inhibition of thalamocortical interactions by the raised GPi output.

This model was supported by early electrophysiologic studies (Miller and DeLong 1987; Bergman et al. 1990, 1994), as well as later optogenetic and chemogenetic studies (Kravitz et al. 2010; Chu et al. 2017; Chen et al. 2023b), and by studies of synaptic metabolism in the basal ganglia and their recipient nuclei (Mitchell et al. 1986, 1989; Palombo et al. 1988, 1990). However, firing abnormalities of basal ganglia neurons other than overall rate changes are now considered to be more relevant for the emergence of bradykinesia and akinesia, including an increased tendency of almost all examined populations of neurons in the basal ganglia-thalamocortical network to discharge in bursts, to exhibit oscillatory firing patterns, and to fire with an increased amount of synchrony with neighboring neurons (Mallet et al. 2006, 2008a,b; Galvan and Wichmann 2008; Kita and Kita 2011; Wichmann 2018).

Parkinsonism-related increases in burst firing have been seen in neurons in GPi, GPe, STN, the GPi-recipient regions of the thalamus, and M1 (Goldberg et al. 2002; Wichmann and Soares 2006; Chan et al. 2011; Pasquereau and Turner 2011; Lobb 2014; Lobb and Jaeger 2015; Kammermeier et al. 2016). Burst discharges of individual neurons occur most often in the parkinsonian state in the context of synchronous oscillatory burst firing patterns (Halje et al. 2019), involving an unusual amount of synchrony between the (otherwise mostly independent) neighboring basal ganglia neurons (Goldberg et al. 2002; Hammond et al. 2007; Bergman et al. 2016). A convenient measure of such synchrony is local field potentials (LFPs), which can be recorded in animals and patients with PD (Brown et al. 2001), and are predominately a measure of synaptic potentials, with the amplitude reflective of the number of simultaneously occurring synaptic events.

LFP and single-cell recording studies have demonstrated that oscillatory activity in the 8–35 Hz range (so-called β-band oscillations; see Fig. 2) is very prominent in parkinsonian individuals, and that this can be suppressed by antiparkinsonian treatments, such as dopamine replacement or by DBS (Bergman et al. 1994; Brown et al. 2001; Weinberger et al. 2009; Wichmann and Dostrovsky 2011; Petersson et al. 2020; Yin et al. 2021). It is important to realize that brief bursts of β-band activity, which may stabilize, or truncate movements are an entirely normal phenomenon (Pfurtscheller 1981; Pfurtscheller et al. 2003; Diesburg et al. 2021; Szul et al. 2023; Zich et al. 2023). Recent studies indicate that the overall increase in β-band activity in the parkinsonian state may reflect an increase in the incidence and duration of such β-band bursts (Tinkhauser et al. 2018; O'Keeffe et al. 2020; Duchet et al. 2021; Neuville et al. 2021; Yu et al. 2021).

γ-Band oscillatory activities (frequencies >35 Hz) are often described as "pro-kinetic," and tend to be reduced in the basal ganglia and cortex of (untreated) parkinsonian patients and in dopamine-depleted animals. They re-emerge to a more normal pattern with levodopa treatment (Brown et al. 2001). Interestingly, parkinsonian patients who experience levodopa-induced involuntary movements show sharp spectral peaks of γ-band activity in the 60–75 Hz range (Brown et al. 2001; Salvadè et al. 2016; Swann et al. 2016; Güttler et al. 2021; Schmidt and Grill 2021; Wiest et al. 2021, 2022; Olaru et al. 2024).

β- and high-frequency γ-band activities are not independent of one another. Excessive coupling of these activities has been described as a characteristic of the parkinsonian state. In cortical and STN recordings from parkinsonian patients and research animals, the amplitude of γ-band activity is often entrained to the phase of ongoing β-band activity (de Hemptinne et al. 2013; Miocinovic et al. 2015; Caiola et al. 2019; Devergnas et al. 2019), potentially representing synchronized neuronal bursting (accounting for the γ-band component) with bursts appearing at β-band frequencies (but see Cole et al. 2017). The pathologic phase-amplitude coupling is re-

 Cite this article as *Cold Spring Harb Perspect Med* doi: 10.1101/cshperspect.a041616

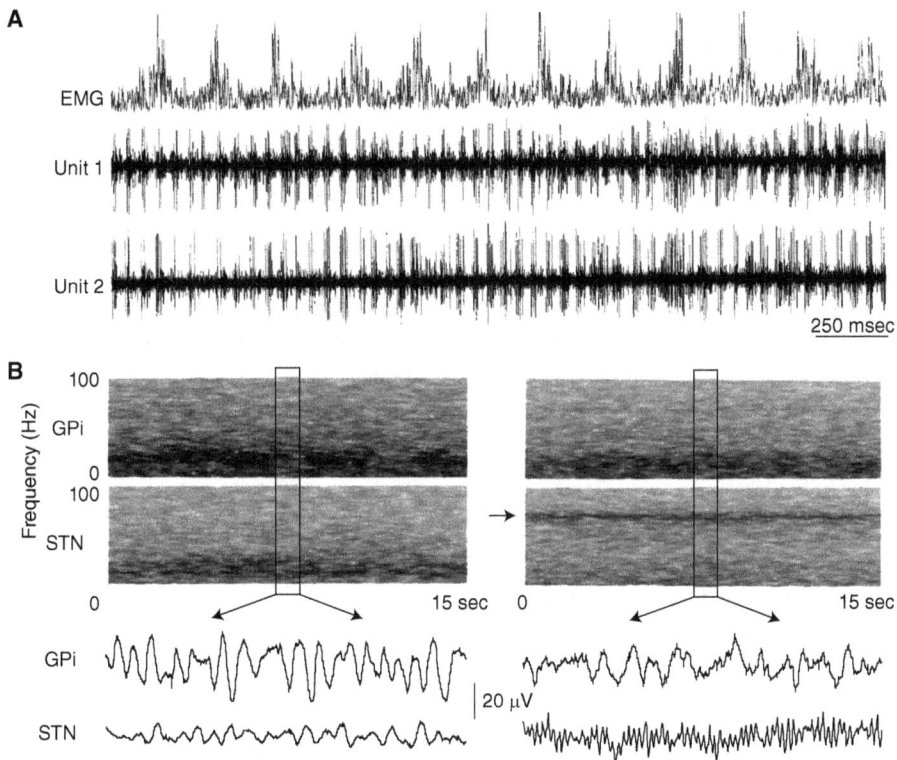

Figure 2. Typical electrophysiologic abnormalities identified in recording studies in patients with Parkinson's disease who were treated with deep brain stimulation. (*A*) Examples of an intraoperative microelectrode recording of two neighboring neurons in the subthalamic nucleus (STN), using a pair of microelectrodes (*bottom*), as well as an electromyography (EMG) recording of a tremulous limb (*top*), demonstrating synchrony between the two cells, as well as lack of synchrony to ongoing tremor. (*B*) Local field potential (LFP) signals recorded in the internal segment of the globus pallidus (GPi) and STN, using deep brain stimulation leads as recording electrodes. The data are shown as traces (*bottom*) and as time-resolved spectrograms (*top*). Without dopaminergic medication (*left*), GPi and STN activity is dominated by β-band oscillations. On levodopa treatment (*right*), the low-frequency activity is reduced in GPi and disappears in the STN. In the STN, a sharply demarcated activity band at ~70 Hz is apparent. (Part *A* was originally published in Levy et al. 2000, © 2000 Society for Neuroscience; part *B* is reprinted, with permission, from Brown et al. 2001, © 2001 Society for Neuroscience.)

sponsive to behaviorally effective antiparkinsonian therapies (Meidahl et al. 2019; O'Keeffe et al. 2020).

Cortical Involvement

With its close connections to the spinal cord, the frontal motor cortical areas are almost certainly involved in the emergence and maintenance of akinesia and bradykinesia. Imaging studies in patients have provided insights into structural and functional cortical activity changes. Morphometric studies demonstrated modest thin-ning of cortical gray matter in (nonmotor) occipital, parietal, and temporal cortices (Laansma et al. 2021), which only in later stages of PD spreads to also involve (motor) frontal cortical regions. While this suggests that the pathophysiology of motor function in PD occurs largely in the absence of gross changes in cortical mass in the relevant brain areas, more subtle biochemical changes certainly do occur earlier. Indeed, changes in cortical catecholamines (dopamine and norepinephrine) have been shown (Ray Chaudhuri et al. 2023), as well as loss of the serotoninergic (Buddhala et al. 2015; Masilamoni et al. 2022), cholinergic (Albin et al. 2022), and glutamatergic innervation (Villalba et al. 2021; Chen et al. 2023a). In general, the evidence for damage of these systems is stronger for cortical areas relevant for cognitive dysfunction rather than motor impairments, however (Blesa et al. 2022).

Studies of glucose metabolism demonstrated elevated metabolism in M1, striatum, and motor thalamus in PD patients off dopaminergic medication (Eidelberg et al. 1994; Feigin et al. 2001; Matthews et al. 2018; Gu et al. 2019), partially reversed with dopaminergic medications (Broussolle et al. 1993). Other studies demonstrated attenuated cerebral blood flow (CBF) in M1 during the performance of motor tasks in unmedicated PD patients (Turner et al. 2003), which is also responsive to treatment with levodopa or surgical treatments (Jenkins et al. 1992; Grafton et al. 1995, 2006; Limousin et al. 1997). Imaging studies also identified changes in functional connectivity between the STN or the striatum and the sensorimotor cortical regions in parkinsonian patients (Ruppert et al. 2020; Steidel et al. 2022; Zang et al. 2022).

Early cortical recording studies in primates suggested that parkinsonian bradykinesia and akinesia is associated with disorganized timing of, and generally reduced, activity of M1 neurons (Doudet et al. 1990; Watts and Mandir 1992). Other studies in primates showed an increased level of synchronous bursting among M1 neurons (Goldberg et al. 2002). A more fine-grain analysis of neuronal activity changes in the primate cortex suggested that corticospinal projecting neurons in M1 are more sensitive to the effects of dopamine depletion than corticostriatal neurons (Pasquereau and Turner 2011). The corticospinal neurons showed a lower rate of firing, increased burst discharges, as well as a more rhythmic β-range firing pattern. The same authors also provided evidence that corticospinal cells in M1 show low-amplitude and abnormally timed movement-related activity (Pasquereau et al. 2016). Recent studies in dopamine-depleted rodents confirmed these results in brain slice recordings (Chen et al. 2021). As in the primate studies, corticostriatal neurons (corresponding to intratelencephalic neurons in

this study) were unaffected by the dopamine depletion, while corticospinal (pyramidal tract) projecting cells did show a significant reduction of their excitability.

Role of Neuroplasticity

It is not clear whether the described activity changes in basal ganglia, thalamus, and cortex emerge because of a lack of striatal or extrastriatal dopamine in an (otherwise healthy) brain, or whether they are a consequence of the secondary plastic changes that are induced by dopamine loss (see above). However, cellular excitability changes that are unlikely to be directly related to the lack of dopamine have been described in brain slice recording studies. In one such study, the excitability of striatal projecting neurons was found to be reduced in direct pathway neurons and increased in indirect pathway neurons in dopamine-depleted animals (Fieblinger et al. 2014). Both changes were only partially reversible by treatment with dopamine replacement drugs. Excitability changes that are not alleviated by (acute) dopamine administration are also known to occur in M1 of dopamine-depleted mice, particularly affecting neurons that project to the spinal cord (Chen et al. 2021). A recent study demonstrated that the cortical neuron responses to their thalamic inputs are reduced in the parkinsonian state (Chen et al. 2023a), in line with the aforementioned anatomically identified reduction of thalamocortical inputs (Villalba et al. 2021).

Are the described changes causal? Promising experimental approaches to provide clarity regarding the link between the observed abnormalities in basal ganglia activity and parkinsonism have relied on an exploration of the temporal relationship between the gradual emergence of parkinsonism in experimental animals and the emergence of firing abnormalities (Gatev et al. 2006; Leblois et al. 2007; Quiroga-Varela et al. 2013; Devergnas et al. 2014). Most of these studies have come to the conclusion that the electrophysiologic changes (especially spectral changes) are linked to the severity of parkinsonian signs (Little et al. 2012; Steiner et al. 2017), although not all studies found a strong correlation between spectral changes and the degree of (early) parkinsonism (Weinberger et al. 2006; Connolly et al. 2015; Devergnas et al. 2019).

Local administration of dopamine receptor antagonists into the striatum in primates (Franco and Turner 2012) demonstrated that bradykinesia and akinesia can be induced by putamenal dopaminergic blockade, in contrast to more recent studies in a progressive genetic model of dopamine loss in which purely striatal dopamine loss was not sufficient to alter behavior (González-Rodríguez et al. 2021). A related approach, the optogenetic or chemogenetic activation of the source neurons of the indirect pathway in the striatum was shown to result in motor slowing, suggesting that massive global activity changes in the striatum can lead to changes in movement (Kravitz et al. 2010; Chu et al. 2017; Chen et al. 2023b). Related to these observations, optogenetic activation of subsets of neurons in the globus pallidus can lead to substantial long-lasting recovery in dopamine-depleted mice (Mastro et al. 2017).

Rigidity

Models of the emergence of rigidity in parkinsonism often focus on alterations in reflex mechanisms. A role of the potentially facilitated stretch reflex system in patients with PD has been discussed (Andrews and Burke 1973; Dietrichson 1973; McLellan 1973; Mortimer and Webster 1979; Cantello et al. 1996; Xia et al. 2009; but see also Noth et al. 1988), as have been changes of the latency and amplitude of long-latency reflexes (Berardelli et al. 1983; Delwaide et al. 1986; Pasquereau and Turner 2013). The severity of rigidity correlates with the degree of striatal dopamine loss (Winogrodzka et al. 2001; Pikstra et al. 2016) and with the power or stability of β-band oscillations in the STN (Kühn et al. 2009; Little et al. 2012), suggesting that altered activity in the basal ganglia and its efferent pathways, either directed at the cerebral cortex (Hirato et al. 1995; Karunanayaka et al. 2016) or at the level of the PPN, are involved in these changes.

Tremor

The pathophysiology of tremor has long been suspected to be different from that of akinesia/bradykinesia and rigidity, as it responds less readily to levodopa treatment, and because parkinsonism with tremor may be associated with a more favorable prognosis than parkinsonism dominated by akinesia and rigidity. It is possible that tremor does not directly relate to nigrostriatal dopamine loss (Hallett 2012, 2014; Helmich and Dirkx 2017; van der Stouwe et al. 2020), but is the result of the degeneration of dopaminergic projections from the retrorubral area to the basal ganglia (specifically the GPi) and VA/VL (Bernheimer et al. 1973; Paulus and Jellinger 1991; Helmich et al. 2011; den Dunnen 2013; Lee et al. 2018). Loss or dysfunction of serotonergic raphe neurons may also play a role (Caretti et al. 2008; Qamhawi et al. 2015; Pasquini et al. 2018; van der Stouwe et al. 2020).

There are no reliable animal models of typical parkinsonian resting tremor. Pathophysiologic considerations are therefore mostly based on studies in human patients, which imposes some limitations on them. A currently popular model (the "dimmer-switch model") is based on neurophysiologic and neuroimaging data (Timmermann et al. 2003; Dirkx et al. 2016, 2017; van der Stouwe et al. 2020). It proposes that tremor is triggered by abnormal interactions between the cerebello-thalamo-cortical and basal ganglia-thalamocortical loops (Helmich et al. 2012; Helmich 2018; Dirkx and Bologna 2022). Transient activation and increased bursting in GPi output may initiate rhythmic bursting in VA/VL (the "switch") (Helmich et al. 2012; Duval et al. 2016; Dirkx and Bologna 2022). Output from VA/VL to M1 and premotor cortices may act upon cortical neurons that also receive input from cerebellar-receiving portions of the thalamus (including the ventral intermedius nucleus [Vim]; see Fig. 1). Engagement of this circuitry may generate tremor-related activity and adjusts the amplitude of tremor (thus acting at the "dimmer") (Helmich et al. 2011; Burciu and Vaillancourt 2018; Underwood and Parr-Brownlie 2021). The central role of M1-thalamic interactions in the production of parkinsonian and other forms of tremor is underscored by the fact that M1 stimulation can serve to reset resting and postural tremor (Ni et al. 2010; Lu et al. 2015), and by the effectiveness of lesions or stimulation of the cerebellar-receiving thalamus in treating tremor. However, it is important to remember that a direct proof of the relevance of these connections remains elusive. It is possible that other mechanisms, such as oscillatory spinal cord circuits (Anastasopoulos 2020) may intersect with the central generation of tremor oscillations. As shown in Figure 2A, a direct link between synchronous oscillatory activity in the basal ganglia and tremor movements has not been established (Levy et al. 2000).

Freezing of Gait

While the general considerations mentioned in the context of akinesia/bradykinesia (above) may also apply to many gait impairments, FOG may have a recognizable pathophysiologic signature that differs from that of other aspects of parkinsonism.

Phenomena similar to parkinsonian FOG have been observed in other conditions, including forms of atypical parkinsonism, hydrocephalus, frontal lobe tumors, vascular disease, or brain irradiation (Meyer and Barron 1960; Petrovici 1968; Sypert et al. 1973; Fisher 1982, 1989; Sudarsky and Simon 1987; Thompson and Marsden 1987; FitzGerald and Jankovic 1989; Imperato et al. 1990; Golbe 1993; Caplan 1995; Giladi et al. 1997; van Zagten et al. 1998; Winikates and Jankovic 1999; Factor et al. 2006; Bompaire et al. 2018), demonstrating that FOG is a relatively common response to a variety of cortical or subcortical pathologies. There is strong evidence that parkinsonian FOG prominently involves cortical functions. Patients typically see worsening of FOG when they simultaneously engage in cognitive activities (Giladi and Hausdorff 2006; Peterson et al. 2015), suggesting competition for cortical resources among these activities. Functional and volumetric MRI studies in parkinsonian patients with FOG have suggested abnormalities in frontal and parietal cortices (Kostić et al. 2012; Tessitore et al. 2012; Shine et al. 2013a,b;

Lewis and Shine 2016; Snijders et al. 2016), and found that cortical amyloid deposition may be a component of the development of FOG in PD (Bohnen et al. 2014; Kim et al. 2019b). Compared to PD patients with FOG, those without FOG show high frontal β-band oscillatory EEG activity and β-band cortico–cortical phase synchronization (Toledo et al. 2014; Scholten et al. 2016). Subcortical components of the basal ganglia-thalamocortical "motor" circuit, as well as cerebellum and PPN may also contribute to FOG in PD (Shine et al. 2013a, Sunwoo et al. 2013; Peterson et al. 2014; Toledo et al. 2014).

While the pathophysiology of FOG is associated with dopamine loss in many patients with PD, as demonstrated by at least partial responsiveness to dopaminergic medications (González-Herrero et al. 2020; Perez Parra et al. 2020; Kwon et al. 2023; McKay et al. 2023), FOG may also involve changes in cholinergic or norepinephrinergic systems (Ono et al. 2016; Bohnen et al. 2019; McKay et al. 2023). The involvement of cholinergic systems is suggested by imaging studies that show loss of cholinergic innervation of cortical regions (Bohnen et al. 2014, 2019) or the striatum (Bohnen et al. 2019) in patients with FOG. Catecholaminergic systems are implicated by many studies that show involvement of the norepinephrinergic locus coeruleus (Ono et al. 2016; Huddleston et al. 2018), and by the (modest) effectiveness of medications that alter catecholamine release in the brain, such as atomoxetine (Jankovic 2009; Revuelta et al. 2015) or L-threo-DOPS (Narabayashi et al. 1982; Katsube et al. 1994; Fukada et al. 2013).

FOG has also been described to occur in animal models of parkinsonism (Revuelta et al. 2012; Gut and Winn 2015; Kucinski et al. 2015; Xiao et al. 2017). In one study of such animals, neurons in the mesencephalic locomotor region neurons (portion of the PPN) had a propensity to discharge in bursts, possibly related to FOG (Goetz et al. 2019). In a study in which a dopamine receptor antagonist was injected into the putamen of monkeys, the animals appeared to develop (arm) freezing (Franco and Turner 2012), further suggesting a role of striatal dopamine in the development of freezing. As in patients (see above), animal studies have also provided evidence for a role of the catecholaminergic locus coeruleus in FOG (Masilamoni et al. 2018), and studies in rodents and monkeys support the notion that severe gait abnormalities (including FOG?) can be elicited by combining lesions of the dopaminergic nigrostriatal system and of cholinergic cells in the pedunculopontine area (Karachi et al. 2010; Grabli et al. 2013; Xiao et al. 2017).

The response of FOG to conventional antiparkinsonian treatments may provide additional clues to its origin. As mentioned, FOG is partially corrected by treatment with dopaminergic medications (Perez-Lloret et al. 2014; Amboni et al. 2015; González-Herrero et al. 2020; Perez Parra et al. 2020; Kwon et al. 2023; McKay et al. 2023). DBS approaches can also be used to address FOG. Different targets have been used for this purpose, including high- or low-frequency stimulation of the STN (Moreau et al. 2008; Vercruysse et al. 2014; Barbe et al. 2018, 2020; Huang et al. 2018; Karachi et al. 2019; Kim et al. 2019a; Di Rauso et al. 2022; Razmkon et al. 2023), either alone or in combination with low-frequency stimulation of the SNr (Valldeoriola et al. 2019), DBS of the PPN area (Niu et al. 2012; Huang et al. 2018; Chang et al. 2021), as well as GPi DBS (Lee et al. 2022). In these studies, the majority of patients appears to benefit from DBS, with the strongest evidence in favor of STN-DBS (although STN-DBS can also worsen freezing [Ferraye et al. 2008]), but long-term efficiency data are lacking (Huang et al. 2018; Di Rauso et al. 2022). Taken together, these data suggest a role of the traditional basal ganglia in freezing, potentially nemphasizing the role of descending pathways that connect them to the PPN.

CONCLUSION

The last few decades have generated a wealth of information regarding the structural and functional brain abnormalities that underlie the emergence and severity of parkinsonian signs. Manifestations of this disease, such as bradykinesia, rigidity, tremor, or FOG, are now understood as expression of different functional or structural physiologic abnormalities that may

be present to varying degrees in individual patients. It is likely (but not proven) that some of the same abnormalities may also affect nonmotor circuits of the brain, contributing to symptoms such as cognition impairments or obsessive–compulsive symptoms.

Alongside the amazing increase of knowledge in this field, gaps in our knowledge are also becoming more clearly defined. For example, within the pathophysiological framework mentioned above, the role of the cerebellum, of the cerebral cortex, or that of the PPN, is underexplored, and the links between abnormalities such as β-band oscillations and movement abnormalities are not understood. Further, nondopaminergic damage to brain and spinal cord circuitry deserves further detailed exploration.

One of the major goals of the pathophysiologic modeling of parkinsonian motor abnormalities is, of course, to identify new symptomatic treatments that may serve to rectify them. With few exceptions, most existing symptomatic antiparkinsonian medications use manipulations of dopaminergic transmission. While this works well, it can also lead to significant side effects, and is associated with an ultimate decline in effectiveness in many patients, as new symptoms develop that do not respond to acute replacement of dopaminergic transmission. Efforts are underway to develop alternative pharmacologic approaches, such as the introduction of T-type calcium channel blockers to reduce neuronal bursting activities (Matthews et al. 2023). DBS approaches, acting on the same circuits as the dopaminergic medications, are also being refined, using insights from the functional studies mentioned above. Exciting brain stimulation approaches that circumvent identified pathophysiologic mechanisms may also become available to treat treatment-resistant motor symptoms. For example, an interesting alternative to therapies targeting the core pathophysiologic changes underlying severe gait impairments, including FOG, was recently published, in which sequenced stimulation of the lumbosacral spinal cord was shown to restore walking and to greatly ameliorate FOG in monkeys as well as in a patient with severe gait abnormalities, utilizing cortical recordings (in nonhuman primates) or joint position measurements (in the human patient) as control signals (Milekovic et al. 2023). It is important to realize that even these approaches require an in-depth understanding of pathophysiologic details of the disease, as they need to separate healthy bio-signals from those degraded by the effects of parkinsonism.

ACKNOWLEDGMENTS

Preparation of this review was funded in part by National Institutes of Health (NIH) research grant P50NS123103 and by an infrastructure grant to the Emory National Primate Research Center (P51OD011132).

REFERENCES

Reference is in this subject collection.

Abdi A, Mallet N, Mohamed FY, Sharott A, Dodson PD, Nakamura KC, Suri S, Avery SV, Larvin JT, Garas FN, et al. 2015. Prototypic and arkypallidal neurons in the dopamine-intact external globus pallidus. *J Neurosci* **35:** 6667–6688. doi:10.1523/JNEUROSCI.4662-14.2015

Albin RL, van der Zee S, van Laar T, Sarter M, Lustig C, Muller M, Bohnen NI. 2022. Cholinergic systems, attentional-motor integration, and cognitive control in Parkinson's disease. *Prog Brain Res* **269:** 345–371. doi:10.1016/bs.pbr.2022.01.011

Amboni M, Stocchi F, Abbruzzese G, Morgante L, Onofrj M, Ruggieri S, Tinazzi M, Zappia M, Attar M, Colombo D, et al. 2015. Prevalence and associated features of self-reported freezing of gait in Parkinson disease: The DEEP FOG study. *Parkinsonism Relat Disord* **21:** 644–649. doi:10.1016/j.parkreldis.2015.03.028

Anastasopoulos D. 2020. Tremor in Parkinson's disease may arise from interactions of central rhythms with spinal reflex loop oscillations. *J Parkinsons Dis* **10:** 383–392. doi:10.3233/JPD-191715

Andrews CJ, Burke D. 1973. Quantitative study of the effect of L-dopa and phenoxybenzamine on the rigidity of Parkinson's disease. *J Neurol Neurosurg Psychiatry* **36:** 321–328. doi:10.1136/jnnp.36.3.321

Baker M, Kang S, Hong SI, Song M, Yang MA, Peyton L, Essa H, Lee SW, Choi DS. 2023. External globus pallidus input to the dorsal striatum regulates habitual seeking behavior in male mice. *Nat Commun* **14:** 4085. doi:10.1038/s41467-023-39545-8

Baladron J, Nambu A, Hamker FH. 2019. The subthalamic nucleus-external globus pallidus loop biases exploratory decisions towards known alternatives: a neuro-computational study. *Eur J Neurosci* **49:** 754–767. doi:10.1111/ejn.13666

Barbe MT, Barthel C, Chen L, Van Dyck N, Brücke T, Seijo F, San Martin ES, Haegelen C, Verin M, Amarell M, et al. 2018. Subthalamic nucleus deep brain stimulation reduces freezing of gait subtypes and patterns in Parkinson's disease. *Brain Stimul* **11:** 1404–1406. doi:10.1016/j.brs.2018.08.016

Barbe MT, Tonder L, Krack P, Debû B, Schüpbach M, Paschen S, Dembek TA, Kühn AA, Fraix V, Brefel-Courbon C, et al. 2020. Deep brain stimulation for freezing of gait in Parkinson's disease with early motor complications. *Mov Disord* **35:** 82–90. doi:10.1002/mds.27892

Berardelli A, Sabra AF, Hallett M. 1983. Physiological mechanisms of rigidity in Parkinson's disease. *J Neurol Neurosurg Psychiatry* **46:** 45–53. doi:10.1136/jnnp.46.1.45

Bergman H, Wichmann T, DeLong MR. 1990. Reversal of experimental parkinsonism by lesions of the subthalamic nucleus. *Science* **249:** 1436–1438. doi:10.1126/science.2402638

Bergman H, Wichmann T, Karmon B, DeLong MR. 1994. The primate subthalamic nucleus. II: Neuronal activity in the MPTP model of parkinsonism. *J Neurophys* **72:** 507–520. doi:10.1152/jn.1994.72.2.507

Bergman H, Zaidel A, Rosin B, Slovik M, Rivlin-Etzion M, Moshel S, Israel Z. 2016. Pathological synchrony of basal ganglia-cortical networks in the systemic MPTP primate model of Parkinson's disease. In *Handbook of basal ganglia structure and function*, 1st ed. (ed. Steiner H, Tseng K), pp. 653–658. Academic, Amsterdam.

Bernheimer H, Birkmayer W, Hornykiewicz O, Jellinger K, Seitelberger F. 1973. Brain dopamine and the syndromes of Parkinson and Huntington. Clinical, morphological and neurochemical correlations. *J Neurol Sci* **20:** 415–455. doi:10.1016/0022-510x(73)90175-5

Bhuvanasundaram R, Krzyspiak J, Khodakhah K. 2022. Subthalamic nucleus modulation of the pontine nuclei and its targeting of the cerebellar cortex. *J Neurosci* **42:** 5538–5551. doi:10.1523/JNEUROSCI.2388-19.2022

Blesa J, Foffani G, Dehay B, Bezard E, Obeso JA. 2022. Motor and non-motor circuit disturbances in early Parkinson disease: which happens first? *Nat Rev Neurosci* **23:** 115–128. doi:10.1038/s41583-021-00542-9

Bohnen NI, Frey KA, Studenski S, Kotagal V, Koeppe RA, Constantine GM, Scott PJ, Albin RL, Müller ML. 2014. Extra-nigral pathological conditions are common in Parkinson's disease with freezing of gait: an in vivo positron emission tomography study. *Mov Disord* **29:** 1118–1124. doi:10.1002/mds.25929

Bohnen NI, Kanel P, Zhou Z, Koeppe RA, Frey KA, Dauer WT, Albin RL, Müller M. 2019. Cholinergic system changes of falls and freezing of gait in Parkinson's disease. *Ann Neurol* **85:** 538–549. doi:10.1002/ana.25430

Bompaire F, Lahutte M, Buffat S, Soussain C, Ardisson AE, Terziev R, Sallansonnet-Froment M, De Greslan T, Edmond S, Saad M, et al. 2018. New insights in radiation-induced leukoencephalopathy: a prospective cross-sectional study. *Support Care Cancer* **26:** 4217–4226. doi:10.1007/s00520-018-4296-9

Bostan AC, Strick PL. 2018. The basal ganglia and the cerebellum: nodes in an integrated network. *Nat Rev Neurosci* **19:** 338–350. doi:10.1038/s41583-018-0002-7

Bostan AC, Dum RP, Strick PL. 2010. The basal ganglia communicate with the cerebellum. *Proc Natl Acad Sci* **107:** 8452–8456. doi:10.1073/pnas.1000496107

Cite this article as *Cold Spring Harb Perspect Med* doi: 10.1101/cshperspect.a041616

Broussolle E, Cinotti L, Pollak P, Landais P, Le Bars D, Galy G, Lavenne F, Khalfallah Y, Chazot G, Mauguière F. 1993. Relief of akinesia by apomorphine and cerebral metabolic changes in Parkinson's disease. *Mov Disord* 8: 459–462. doi:10.1002/mds.870080407

Brown WD, Taylor M, Roberts AD, Oakes TR, Schueller M, Holden JE, Malischke L, DeJesus OT, Nickles RJ. 1999. FluoroDOPA PET shows the nondopaminergic as well as dopaminergic destinations of levodopa. *Neurology* 53: 1212–1212. doi:10.1212/wnl.53.6.1212

Brown P, Oliviero A, Mazzone P, Insola A, Tonali P, Di Lazzaro V. 2001. Dopamine dependency of oscillations between subthalamic nucleus and pallidum in Parkinson's disease. *J Neurosci* 21: 1033–1038. doi:10.1523/JNEUROSCI.21-03-01033.2001

Buddhala C, Loftin SK, Kuley BM, Cairns NJ, Campbell MC, Perlmutter JS, Kotzbauer PT. 2015. Dopaminergic, serotonergic, and noradrenergic deficits in Parkinson disease. *Ann Clin Transl Neurol* 2: 949–959. doi:10.1002/acn3.246

Burciu RG, Vaillancourt DE. 2018. Imaging of motor cortex physiology in Parkinson's disease. *Mov Disord* 33: 1688–1699. doi:10.1002/mds.102

Caiola M, Devergnas A, Holmes MH, Wichmann T. 2019. Empirical analysis of phase-amplitude coupling approaches. *PLoS ONE* 14: e0219264. doi:10.1371/journal.pone.0219264

Cantello R, Gianelli M, Civardi C, Mutani R. 1996. Pathophysiology of Parkinson's disease rigidity. Role of corticospinal motor projections. *Adv Neurol* 69: 129–133.

Caplan LR. 1995. Binswanger's disease—revisited. *Neurology* 45: 626–633. doi:10.1212/wnl.45.4.626

Cardoso F, Goetz CG, Mestre TA, Sampaio C, Adler CH, Berg D, Bloem BR, Burn DJ, Fitts MS, Gasser T, et al. 2024. A statement of the MDS on biological definition, staging, and classification of Parkinson's disease. *Mov Disord* 39: 259–266. doi:10.1002/mds.29683

Caretti V, Stoffers D, Winogrodzka A, Isaias IU, Costantino G, Pezzoli G, Ferrarese C, Antonini A, Wolters EC, Booij J. 2008. Loss of thalamic serotonin transporters in early drug-naive Parkinson's disease patients is associated with tremor: an [(123)I]β-CIT SPECT study. *J Neural Transm* 115: 721–729. doi:10.1007/s00702-007-0015-2

Chan V, Starr PA, Turner RS. 2011. Bursts and oscillations as independent properties of neural activity in the parkinsonian globus pallidus internus. *Neurobiol Dis* 41: 2–10. doi:10.1016/j.nbd.2010.08.012

Chang SJ, Cajigas I, Guest JD, Noga BR, Widerström-Noga E, Haq I, Fisher L, Luca CC, Jagid JR. 2021. Deep brain stimulation of the Cuneiform nucleus for levodopa-resistant freezing of gait in Parkinson's disease: study protocol for a prospective, pilot trial. *Pilot Feasibility Stud* 7: 117. doi:10.1186/s40814-021-00855-7

Chen L, Daniels S, Kim Y, Chu HY. 2021. Cell type-specific decrease of the intrinsic excitability of motor cortical pyramidal neurons in Parkinsonism. *J Neurosci* 41: 5553–5565. doi:10.1523/JNEUROSCI.2694-20.2021

Chen L, Daniels S, Dvorak R, Chu HY. 2023a. Reduced thalamic excitation to motor cortical pyramidal tract neurons in parkinsonism. *Sci Adv* 9: eadg3038. doi:10.1126/sciadv.adg3038

Chen Y, Hong Z, Wang J, Liu K, Liu J, Lin J, Feng S, Zhang T, Shan L, Liu T, et al. 2023b. Circuit-specific gene therapy

reverses core symptoms in a primate Parkinson's disease model. *Cell* 186: 5394–5410.e18. doi:10.1016/j.cell.2023.10.004

Chu HY, Bevan MD. 2013. Long-term potentiation of external globus pallidus-subthalamic nucleus synapses following activation of motor cortical inputs. In the *Society for Neuroscience Annual Meeting Abstract*. Program No. 270.213.

Chu HY, Atherton JF, Wokosin D, Surmeier DJ, Bevan MD. 2015. Heterosynaptic regulation of external globus pallidus inputs to the subthalamic nucleus by the motor cortex. *Neuron* 85: 364–376. doi:10.1016/j.neuron.2014.12.022

Chu HY, McIver EL, Kovaleski RF, Atherton JF, Bevan MD. 2017. Loss of hyperdirect pathway cortico-subthalamic inputs following degeneration of midbrain dopamine neurons. *Neuron* 95: 1306–1318.e5. doi:10.1016/j.neuron.2017.08.038

Cole SR, van der Meij R, Peterson EJ, de Hemptinne C, Starr PA, Voytek B. 2017. Nonsinusoidal β oscillations reflect cortical pathophysiology in Parkinson's disease. *J Neurosci* 37: 4830–4840. doi:10.1523/JNEUROSCI.2208-16.2017

Connolly AT, Jensen AL, Bello EM, Netoff TI, Baker KB, Johnson MD, Vitek JL. 2015. Modulations in oscillatory frequency and coupling in globus pallidus with increasing Parkinsonian severity. *J Neurosci* 35: 6231–6240. doi:10.1523/JNEUROSCI.4137-14.2015

Courtney CD, Pamukcu A, Chan CS. 2023. Cell and circuit complexity of the external globus pallidus. *Nat Neurosci* 26: 1147–1159. doi:10.1038/s41593-023-01368-7

Cui G, Jun SB, Jin X, Pham MD, Vogel SS, Lovinger DM, Costa RM. 2013. Concurrent activation of striatal direct and indirect pathways during action initiation. *Nature* 494: 238–242. doi:10.1038/nature11846

Deffains M, Iskhakova L, Bergman H. 2016. Stop and think about basal ganglia functional organization: the pallidostriatal "stop" route. *Neuron* 89: 237–239. doi:10.1016/j.neuron.2016.01.003

de Hemptinne C, Ryapolova-Webb ES, Air EL, Garcia PA, Miller KJ, Ojemann JG, Ostrem JL, Galifianakis NB, Starr PA. 2013. Exaggerated phase-amplitude coupling in the primary motor cortex in Parkinson disease. *Proc Natl Acad Sci* 110: 4780–4785. doi:10.1073/pnas.1214546110

Delwaide PJ, Sabbatino M, Delwaide C. 1986. Some pathophysiological aspects of the parkinsonian rigidity. *J Neural Transm Suppl* 22: 129–139.

den Dunnen WF. 2013. Neuropathological diagnostic considerations in hyperkinetic movement disorders. *Front Neurol* 4: 7. doi:10.3389/fneur.2013.00007

Desmurget M, Turner RS. 2008. Testing basal ganglia motor functions through reversible inactivations in the posterior internal globus pallidus. *J Neurophys* 99: 1057–1076. doi:10.1152/jn.01010.2007

Devergnas A, Pittard D, Bliwise D, Wichmann T. 2014. Relationship between oscillatory activity in the cortico-basal ganglia network and parkinsonism in MPTP-treated monkeys. *Neurobiol Dis* 68: 156–166. doi:10.1016/j.nbd.2014.04.004

Devergnas A, Caiola M, Pittard D, Wichmann T. 2019. Cortical phase-amplitude coupling in a progressive model of

Parkinsonism in nonhuman primates. *Cereb Cortex* **29:** 167–177. doi:10.1093/cercor/bhx314

Diesburg DA, Wessel JR. 2021. The pause-then-cancel model of human action-stopping: theoretical considerations and empirical evidence. *Neurosci Biobehav Rev* **129:** 17–34. doi:10.1016/j.neubiorev.2021.07.019

Diesburg DA, Greenlee JD, Wessel JR. 2021. Cortico-subcortical β burst dynamics underlying movement cancellation in humans. *eLife* **10:** e70270. doi:10.7554/eLife.70270

Dietrichson P. 1973. The fusimotor system in relation to spasticity and Parkinsonian rigidity. *Scand J Rehabil Med* **5:** 174–178.

Di Rauso G, Cavallieri F, Campanini I, Gessani A, Fioravanti V, Feletti A, Damiano B, Scaltriti S, Bardi E, Corni MG, et al. 2022. Freezing of gait in Parkinson's disease patients treated with bilateral subthalamic nucleus deep brain stimulation: a long-term overview. *Biomedicines* **10:** 2214. doi:10.3390/biomedicines10092214

Dirkx MF, Bologna M. 2022. The pathophysiology of Parkinson's disease tremor. *J Neurol Sci* **435:** 120196. doi:10.1016/j.jns.2022.120196

Dirkx MF, den Ouden H, Aarts E, Timmer M, Bloem BR, Toni I, Helmich RC. 2016. The cerebral network of Parkinson's tremor: an effective connectivity fMRI study. *J Neurosci* **36:** 5362–5372. doi:10.1523/JNEUROSCI.3634-15.2016

Dirkx MF, den Ouden HE, Aarts E, Timmer MH, Bloem BR, Toni I, Helmich RC. 2017. Dopamine controls Parkinson's tremor by inhibiting the cerebellar thalamus. *Brain* **140:** 721–734. doi:10.1093/brain/aww331

Doudet DJ, Gross C, Arluison M, Bioulac B. 1990. Modifications of precentral cortex discharge and EMG activity in monkeys with MPTP-induced lesions of DA nigral neurons. *Exp Brain Res* **80:** 177–188. doi:10.1007/BF00228859

Duchet B, Ghezzi F, Weerasinghe G, Tinkhauser G, Kühn AA, Brown P, Bick C, Bogacz R. 2021. Average β burst duration profiles provide a signature of dynamical changes between the on and off medication states in Parkinson's disease. *PLoS Comput Biol* **17:** e1009116. doi:10.1371/journal.pcbi.1009116

Dudman JT, Krakauer JW. 2016. The basal ganglia: from motor commands to the control of vigor. *Curr Opin Neurobiol* **37:** 158–166. doi:10.1016/j.conb.2016.02.005

Duval C, Daneault JF, Hutchison WD, Sadikot AF. 2016. A brain network model explaining tremor in Parkinson's disease. *Neurobiol Dis* **85:** 49–59. doi:10.1016/j.nbd.2015.10.009

Ehringer H, Hornykiewicz O. 1960. Verteilung von noradrenalin und dopamin (3-Hydroxytyramin) im gehirn des menschen und ihr verhalten bei erkrankungen des extrapyramidalen systems. *Klin Wschr* **38:** 1236–1239. doi:10.1007/BF01485901

Eidelberg D, Moeller JR, Dhawan V, Spetsieris P, Takikawa S, Ishikawa T, Chaly T, Robeson W, Margouleff D, Przedborski S, et al. 1994. The metabolic topography of parkinsonism. *J Cereb Blood Flow Metab* **14:** 783–801. doi:10.1038/jcbfm.1994.99

Factor SA, Higgins DS, Qian J. 2006. Primary progressive freezing gait: a syndrome with many causes. *Neurology* **66:** 411–414. doi:10.1212/01.wnl.0000196469.52995.ab

Factor SA, Steenland NK, Higgins DS, Molho ES, Kay DM, Montimurro J, Rosen AR, Zabetian CP, Payami H. 2011. Postural instability/gait disturbance in Parkinson's disease has distinct subtypes: an exploratory analysis. *JNNP* **82:** 564–568. doi:10.1136/jnnp.2010.222042

Factor SA, Scullin MK, Sollinger AB, Land JO, Wood-Siverio C, Zanders L, Freeman A, Bliwise DL, Goldstein FC. 2014. Freezing of gait subtypes have different cognitive correlates in Parkinson's disease. *Parkinsonism Relat Disord* **20:** 1359–1364. doi:10.1016/j.parkreldis.2014.09.023

Fan KY, Baufreton J, Surmeier DJ, Chan CS, Bevan MD. 2012. Proliferation of external globus pallidus-subthalamic nucleus synapses following degeneration of midbrain dopamine neurons. *J Neurosci* **32:** 13718–13728. doi:10.1523/JNEUROSCI.5750-11.2012

Feigin A, Fukuda M, Dhawan V, Przedborski S, Jackson-Lewis V, Mentis MJ, Moeller JR, Eidelberg D. 2001. Metabolic correlates of levodopa response in Parkinson's disease. *Neurol J Am Hear Assoc* **57:** 2083–2088. doi:10.1212/wnl.57.11.2083

Ferraye MU, Debû B, Pollak P. 2008. Deep brain stimulation effect on freezing of gait. *Mov Disord* **23:** S489–S494. doi:10.1002/mds.21975

Fieblinger T, Graves SM, Sebel LE, Alcacer C, Plotkin JL, Gertler TS, Chan CS, Heiman M, Greengard P, Cenci MA, et al. 2014. Cell type-specific plasticity of striatal projection neurons in parkinsonism and L-DOPA-induced dyskinesia. *Nat Commun* **5:** 5316. doi:10.1038/ncomms6316

Fisher CM. 1982. Hydrocephalus as a cause of disturbances of gait in the elderly. *Neurology* **32:** 1358–1363. doi:10.1212/wnl.32.12.1358

Fisher CM. 1989. Binswanger's encephalopathy: a review. *J Neurol* **236:** 65–79. doi:10.1007/BF00314400

FitzGerald PM, Jankovic J. 1989. Lower body parkinsonism: evidence for vascular etiology. *Mov Disord* **4:** 249–260. doi:10.1002/mds.870040306

Franco V, Turner RS. 2012. Testing the contributions of striatal dopamine loss to the genesis of parkinsonian signs. *Neurobiol Dis* **47:** 114–125. doi:10.1016/j.nbd.2012.03.028

Fukada K, Endo T, Yokoe M, Hamasaki T, Hazama T, Sakoda S. 2013. L-threo-3,4-dihydroxyphenylserine (L-DOPS) co-administered with entacapone improves freezing of gait in Parkinson's disease. *Med Hypotheses* **80:** 209–212. doi:10.1016/j.mehy.2012.11.031

Galvan A, Wichmann T. 2008. Pathophysiology of parkinsonism. *Clin Neurophysiol* **119:** 1459–1474. doi:10.1016/j.clinph.2008.03.017

Gatev P, Darbin O, Wichmann T. 2006. Oscillations in the basal ganglia under normal conditions and in movement disorders. *Mov Disord* **21:** 1566–1577. doi:10.1002/mds.21033

Giladi N, Hausdorff JM. 2006. The role of mental function in the pathogenesis of freezing of gait in Parkinson's disease. *J Neurol Sci* **248:** 173–176. doi:10.1016/j.jns.2006.05.015

Giladi N, McMahon D, Przedborski S, Flaster E, Guillory S, Kostic V, Fahn S. 1992. Motor blocks in Parkinson's disease. *Neurology* **42:** 333–339. doi:10.1212/wnl.42.2.333

Giladi N, Kao R, Fahn S. 1997. Freezing phenomenon in patients with parkinsonian syndromes. *Mov Disord* **12**: 302–305. doi:10.1002/mds.870120307

Giladi N, McDermott MP, Fahn S, Przedborski S, Jankovic J, Stern M, Tanner C; Parkinson Study Group. 2001. Freezing of gait in PD: prospective assessment in the DATA-TOP cohort. *Neurology* **56**: 1712–1721. doi:10.1212/wnl.56.12.1712

Glajch KE, Kelver DA, Hegeman DJ, Cui Q, Xenias HS, Augustine EC, Hernández VM, Verma N, Huang TY, Luo M, et al. 2016. Npas1⁺ pallidal neurons target striatal projection neurons. *J Neurosci* **36**: 5472–5488. doi:10.1523/JNEUROSCI.1720-15.2016

Goetz L, Piallat B, Bhattacharjee M, Mathieu H, David O, Chabardès S. 2019. Spike discharge characteristic of the caudal mesencephalic reticular formation and pedunculopontine nucleus in MPTP-induced primate model of Parkinson disease. *Neurobiol Dis* **128**: 40–48. doi:10.1016/j.nbd.2018.08.002

Golbe LI. 1993. Progressive supranuclear palsy. In *Parkinsonian syndromes* (ed. Stern MB, Koller WB), pp. 227–237. Marcel Dekker, New York.

Goldberg JA, Boraud T, Maraton S, Haber SN, Vaadia E, Bergman H. 2002. Enhanced synchrony among primary motor cortex neurons in the 1-methyl-4-phenyl-1,2,3,6-tetrahydropyridine primate model of Parkinson's disease. *J Neurosci* **22**: 4639–4653. doi:10.1523/JNEUROSCI.22-11-04639.2002

González-Herrero B, Jauma-Classen S, Gómez-Llopico R, Plans G, Calopa M. 2020. Intestinal levodopa/carbidopa infusion as a therapeutic option for unresponsive freezing of gait after deep brain stimulation in Parkinson's disease. *Parkinsons Dis* **2020**: 1627264. doi:10.1155/2020/1627264

González-Rodríguez P, Zampese E, Stout KA, Guzman JN, Ilijic E, Yang B, Tkatch T, Stavarache MA, Wokosin DL, Gao L, et al. 2021. Disruption of mitochondrial complex I induces progressive parkinsonism. *Nature* **599**: 650–656. doi:10.1038/s41586-021-04059-0

Grabli D, Karachi C, Folgoas E, Monfort M, Tande D, Clark S, Civelli O, Hirsch EC, Francois C. 2013. Gait disorders in parkinsonian monkeys with pedunculopontine nucleus lesions: a tale of two systems. *J Neurosci* **33**: 11986–11993. doi:10.1523/JNEUROSCI.1568-13.2013

Grafton ST, Waters C, Sutton J, Lew MF, Couldwell W. 1995. Pallidotomy increases activity of motor association cortex in Parkinson's disease: a positron emission tomographic study. *Ann Neurol* **37**: 776–783. doi:10.1002/ana.410370611

Grafton ST, Turner RS, Desmurget M, Bakay R, Delong M, Vitek J, Crutcher M. 2006. Normalizing motor-related brain activity: subthalamic nucleus stimulation in Parkinson disease. *Neurology* **66**: 1192–1199. doi:10.1212/01.wnl.0000214237.58321.c3

Gu SC, Ye Q, Yuan CX. 2019. Metabolic pattern analysis of ¹⁸F-FDG PET as a marker for Parkinson's disease: a systematic review and meta-analysis. *Rev Neurosci* **30**: 743–756. doi:10.1515/revneuro-2018-0061

Guilhemsang L, Mallet NP. 2024. Arkypallidal neurons in basal ganglia circuits: unveiling novel pallidostriatal loops? *Curr Opin Neurobiol* **84**: 102814. doi:10.1016/j.conb.2023.102814

Gut NK, Winn P. 2015. Deep brain stimulation of different pedunculopontine targets in a novel rodent model of parkinsonism. *J Neurosci* **35**: 4792–4803. doi:10.1523/JNEUROSCI.3646-14.2015

Güttler C, Altschüler J, Tanev K, Böckmann S, Haumesser JK, Nikulin VV, Kühn AA, van Riesen C. 2021. Levodopa-induced dyskinesia are mediated by cortical γ oscillations in experimental Parkinsonism. *Mov Disord* **36**: 927–937. doi:10.1002/mds.28403

Halje P, Brys I, Mariman JJ, da Cunha C, Fuentes R, Petersson P. 2019. Oscillations in cortico-basal ganglia circuits: implications for Parkinson's disease and other neurologic and psychiatric conditions. *J Neurophysiol* **122**: 203–231. doi:10.1152/jn.00590.2018

Hallett M. 2012. Parkinson's disease tremor: pathophysiology. *Parkinsonism Relat Disord* **18**: S85–S86. doi:10.1016/S1353-8020(11)70027-X

Hallett M. 2014. Tremor: pathophysiology. *Parkinsonism Relat Disord* **20**: S118–S122. doi:10.1016/S1353-8020(13)70029-4

Hammond C, Bergman H, Brown P. 2007. Pathological synchronization in Parkinson's disease: networks, models and treatments. *Trends Neurosci* **30**: 357–364. doi:10.1016/j.tins.2007.05.004

Hanakawa T. 2021. Thoughts on vigor in the motor and cognitive domains. *Behav Brain Sci* **44**: e128. doi:10.1017/S0140525X21000133

Helmich RC. 2018. The cerebral basis of Parkinsonian tremor: a network perspective. *Mov Disord* **33**: 219–231. doi:10.1002/mds.27224

Helmich RC, Dirkx MF. 2017. Pathophysiology and management of Parkinsonian tremor. *Semin Neurol* **37**: 127–134. doi:10.1055/s-0037-1601558

Helmich RC, Janssen MJ, Oyen WJ, Bloem BR, Toni I. 2011. Pallidal dysfunction drives a cerebellothalamic circuit into Parkinson tremor. *Ann Neurol* **69**: 269–281. doi:10.1002/ana.22361

Helmich RC, Hallett M, Deuschl G, Toni I, Bloem BR. 2012. Cerebral causes and consequences of parkinsonian resting tremor: a tale of two circuits? *Brain* **135**: 3206–3226. doi:10.1093/brain/aws023

Hirato M, Ishihara J, Horikoshi S, Shibazaki T, Ohye C. 1995. Parkinsonian rigidity, dopa-induced dyskinesia and chorea—dynamic studies on the basal ganglia-thalamocortical motor circuit using PET scan and depth microrecording. *Acta Neurochir Suppl* **64**: 5–8. doi:10.1007/978-3-7091-9419-5_2

Hornykiewicz O, Kish SJ. 1987. Biochemical pathophysiology of Parkinson's disease. *Adv Neurol* **45**: 19–34.

Huang C, Chu H, Zhang Y, Wang X. 2018. Deep brain stimulation to alleviate freezing of gait and cognitive dysfunction in Parkinson's disease: update on current research and future perspectives. *Front Neurosci* **12**: 29. doi:10.3389/fnins.2018.00029

Huddleston D, Goldstein F, Langley J, Tripathi R, Lane M, Crosson B, Hu X, Factor S. 2018. Neuromelanin-sensitive MRI correlates of executive function and verbal memory in Parkinson's disease with freezing of gait. In *4th International Workshop on Freezing of Gait Abstract Volume*, pp. 55–56. Leuven, Belgium.

Ilyas A, Pizarro D, Romeo AK, Riley KO, Pati S. 2019. The centromedian nucleus: anatomy, physiology, and clinical implications. *J Clin Neurosci* **63:** 1–7. doi:10.1016/j.jocn.2019.01.050

Imperato JP, Paleologos NA, Vick NA. 1990. Effects of treatment on long-term survivors with malignant astrocytomas. *Ann Neurol* **28:** 818–822. doi:10.1002/ana.410280614

Ingham CA, Hood SH, Arbuthnott GW. 1989. Spine density on neostriatal neurones changes with 6-hydroxydopamine lesions and with age. *Brain Res* **503:** 334–338. doi:10.1016/0006-8993(89)91686-7

Jankovic J. 2009. Atomoxetine for freezing of gait in Parkinson disease. *J Neurol Sci* **284:** 177–178. doi:10.1016/j.jns.2009.03.022

Jenkins IH, Fernandez W, Playford ED, Lees AJ, Frackowiak RSJ, Passingham RE, Brooks DJ. 1992. Impaired activation of the supplementary motor area in Parkinson's disease is reversed when akinesia is treated with apomorphine. *Ann Neurol* **32:** 749–757. doi:10.1002/ana.410320608

Jurado-Parras MT, Safaie M, Sarno S, Louis J, Karoutchi C, Berret B, Robbe D. 2020. The dorsal striatum energizes motor routines. *Curr Biol* **30:** 4362–4372.e6. doi:10.1016/j.cub.2020.08.049

Kammermeier S, Pittard D, Hamada I, Wichmann T. 2016. Effects of high-frequency stimulation of the internal pallidal segment on neuronal activity in the thalamus in parkinsonian monkeys. *J Neurophysiol* **116:** 2869–2881. doi:10.1152/jn.00104.2016

Karachi C, Grabli D, Bernard FA, Tandé D, Wattiez N, Belaid H, Bardinet E, Prigent A, Nothacker HP, Hunot S, et al. 2010. Cholinergic mesencephalic neurons are involved in gait and postural disorders in Parkinson disease. *J Clin Invest* **120:** 2745–2754. doi:10.1172/JCI42642

Karachi C, Cormier-Dequaire F, Grabli D, Lau B, Belaid H, Navarro S, Vidailhet M, Bardinet E, Fernandez-Vidal S, Welter ML. 2019. Clinical and anatomical predictors for freezing of gait and falls after subthalamic deep brain stimulation in Parkinson's disease patients. *Parkinsonism Relat Disord* **62:** 91–97. doi:10.1016/j.parkreldis.2019.01.021

Karunanayaka PR, Lee EY, Lewis MM, Sen S, Eslinger PJ, Yang QX, Huang X. 2016. Default mode network differences between rigidity- and tremor-predominant Parkinson's disease. *Cortex* **81:** 239–250. doi:10.1016/j.cortex.2016.04.021

Katsube J, Narabayashi H, Hayashi A, Tanaka C, Suzuki T. 1994. Development of l-threo-dops, a norepinephrine precursor amino-acid. *Yakugaku Zasshi* **114:** 823–846. doi:10.1248/yakushi1947.114.11_823

Kim R, Kim HJ, Shin C, Park H, Kim A, Paek SH, Jeon B. 2019a. Long-term effect of subthalamic nucleus deep brain stimulation on freezing of gait in Parkinson's disease. *J Neurosurg* **131:** 1797–1804. doi:10.3171/2018.8.JNS18350

Kim R, Lee J, Kim HJ, Kim A, Jang M, Jeon B, Kang UJ. 2019b. CSF β-amyloid$_{42}$ and risk of freezing of gait in early Parkinson disease. *Neurology* **92:** e40–e47. doi:10.1212/WNL.0000000000006692

Kish SJ, Shannak K, Hornykiewicz O. 1988. Uneven pattern of dopamine loss in the striatum of patients with idiopathic Parkinson's disease. *N Engl J Med* **318:** 876–880. doi:10.1056/NEJM198804073181402

Kita H, Kita T. 2011. Role of striatum in the pause and burst generation in the globus pallidus of 6-OHDA-treated rats. *Front Syst Neurosci* **5:** 42. doi:10.3389/fnsys.2011.00042

Kostić VS, Agosta F, Pievani M, Stefanova E, Ječmenica-Lukić M, Scarale A, Špica V, Filippi M. 2012. Pattern of brain tissue loss associated with freezing of gait in Parkinson disease. *Neurology* **78:** 409–416. doi:10.1212/WNL.0b013e318245d23c

Kravitz AV, Freeze BS, Parker PR, Kay K, Thwin MT, Deisseroth K, Kreitzer AC. 2010. Regulation of parkinsonian motor behaviours by optogenetic control of basal ganglia circuitry. *Nature* **466:** 622–626. doi:10.1038/nature09159

Kreitzer AC, Malenka RC. 2008. Striatal plasticity and basal ganglia circuit function. *Neuron* **60:** 543–554. doi:10.1016/j.neuron.2008.11.005

Kucinski A, Albin RL, Lustig C, Sarter M. 2015. Modeling falls in Parkinson's disease: slow gait, freezing episodes and falls in rats with extensive striatal dopamine loss. *Behav Brain Res* **282:** 155–164. doi:10.1016/j.bbr.2015.01.012

Kühn AA, Tsui A, Aziz T, Ray N, Brücke C, Kupsch A, Schneider GH, Brown P. 2009. Pathological synchronisation in the subthalamic nucleus of patients with Parkinson's disease relates to both bradykinesia and rigidity. *Exp Neurol* **215:** 380–387. doi:10.1016/j.expneurol.2008.11.008

Kwon H, Clifford GD, Genias I, Bernhard D, Esper CD, Factor SA, McKay JL. 2023. An explainable spatial-temporal graphical convolutional network to score freezing of gait in Parkinsonian patients. *Sensors (Basel)* **23:** 1766. doi:10.3390/s23041766

Laansma MA, Bright JK, Al-Bachari S, Anderson TJ, Ard T, Assogna F, Baquero KA, Berendse HW, Blair J, Cendes F, et al. 2021. International multicenter analysis of brain structure across clinical stages of Parkinson's disease. *Mov Disord* **36:** 2583–2594. doi:10.1002/mds.28706

* Laciego JL, Obeso JA. 2024. Functional neuroanatomy of the normal and pathological basal ganglia. *Cold Spring Harb Perspect Med* doi:10.1101/cshperspect.a041617

Leblois A, Meissner W, Bioulac B, Gross CE, Hansel D, Boraud T. 2007. Late emergence of synchronized oscillatory activity in the pallidum during progressive Parkinsonism. *Eur J Neurosci* **26:** 1701–1713. doi:10.1111/j.1460-9568.2007.05777.x

Lee JY, Lao-Kaim NP, Pasquini J, Deuschl G, Pavese N, Piccini P. 2018. Pallidal dopaminergic denervation and rest tremor in early Parkinson's disease: PPMI cohort analysis. *Parkinsonism Relat Disord* **51:** 101–104. doi:10.1016/j.parkreldis.2018.02.039

Lee SH, Lee J, Kim MS, Hwang YS, Jo S, Park KW, Jeon SR, Chung SJ. 2022. Factors correlated with therapeutic effects of globus pallidus deep brain stimulation on freezing of gait in advanced Parkinson's disease: a pilot study. *Parkinsonism Relat Disord* **94:** 111–116. doi:10.1016/j.parkreldis.2021.12.005

Lee D, Liu L, Root CM. 2023. Transformation of valence signaling in a striatopallidal circuit. bioRxiv doi:10.1101/2023.08.01.551547

Levy R, Hutchison WD, Lozano AM, Dostrovsky JO. 2000. High-frequency synchronization of neuronal activity in the subthalamic nucleus of parkinsonian patients with limb tremor. *J Neurosci* **20:** 7766–7775. doi:10.1523/JNEUROSCI.20-20-07766.2000

Levy R, Ashby P, Hutchison WD, Lang AE, Lozano AM, Dostrovsky JO. 2002. Dependence of subthalamic nucleus oscillations on movement and dopamine in Parkinson's disease. *Brain* **125:** 1196–1209. doi:10.1093/brain/awf128

Lewis SJ, Shine JM. 2016. The next step: a common neural mechanism for freezing of gait. *Neuroscientist* **22:** 72–82. doi:10.1177/1073858414559101

Limousin P, Greene J, Pollak P, Rothwell J, Benabid AL, Frackowiak R. 1997. Changes in cerebral activity pattern due to subthalamic nucleus or internal pallidum stimulation in Parkinson's disease. *Ann Neurol* **42:** 283–291. doi:10.1002/ana.410420303

Little S, Pogosyan A, Kuhn AA, Brown P. 2012. β Band stability over time correlates with Parkinsonian rigidity and bradykinesia. *Exp Neurol* **236:** 383–388. doi:10.1016/j.expneurol.2012.04.024

Lobb C. 2014. Abnormal bursting as a pathophysiological mechanism in Parkinson's disease. *Basal Ganglia* **3:** 187–195. doi:10.1016/j.baga.2013.11.002

Lobb CJ, Jaeger D. 2015. Bursting activity of substantia nigra pars reticulata neurons in mouse parkinsonism in awake and anesthetized states. *Neurobiol Dis* **75:** 177–185. doi:10.1016/j.nbd.2014.12.026

Lu MK, Chiou SM, Ziemann U, Huang HC, Yang YW, Tsai CH. 2015. Resetting tremor by single and paired transcranial magnetic stimulation in Parkinson's disease and essential tremor. *Clin Neurophysiol* **126:** 2330–2336. doi:10.1016/j.clinph.2015.02.010

Mallet N, Ballion B, Le Moine C, Gonon F. 2006. Cortical inputs and GABA interneurons imbalance projection neurons in the striatum of parkinsonian rats. *J Neurosci* **26:** 3875–3884. doi:10.1523/JNEUROSCI.4439-05.2006

Mallet N, Pogosyan A, Márton LF, Bolam JP, Brown P, Magill PJ. 2008a. Parkinsonian β oscillations in the external globus pallidus and their relationship with subthalamic nucleus activity. *J Neurosci* **28:** 14245–14258. doi:10.1523/JNEUROSCI.4199-08.2008

Mallet N, Pogosyan A, Sharott A, Csicsvari J, Bolam JP, Brown P, Magill PJ. 2008b. Disrupted dopamine transmission and the emergence of exaggerated β oscillations in subthalamic nucleus and cerebral cortex. *J Neurosci* **28:** 4795–4806. doi:10.1523/JNEUROSCI.0123-08.2008

Mallet N, Micklem BR, Henny P, Brown MT, Williams C, Bolam JP, Nakamura KC, Magill PJ. 2012. Dichotomous organization of the external globus pallidus. *Neuron* **74:** 1075–1086. doi:10.1016/j.neuron.2012.04.027

Martinez-Gonzalez C, Bolam JP, Mena-Segovia J. 2011. Topographical organization of the pedunculopontine nucleus. *Front Neuroanat* **5:** 22. doi:10.3389/fnana.2011.00022

Masilamoni G, Beck G, Factor SA, Nye JA, Huddleston D, Papa SM, Smith Y. 2018. Pathology of brainstem noradrenergic cell groups associated with freezing of gait in MPTP-treated non-human primates. In *4th International Workshop on Freezing of Gait Abstract Volume 11.*

Masilamoni GJ, Weinkle A, Papa SM, Smith Y. 2022. Cortical serotonergic and catecholaminergic denervation in MPTP-treated parkinsonian monkeys. *Cereb Cortex* **32:** 1804–1822. doi:10.1093/cercor/bhab313

Mastro KJ, Zitelli KT, Willard AM, Leblanc KH, Kravitz AV, Gittis AH. 2017. Cell-specific pallidal intervention induces long-lasting motor recovery in dopamine-depleted mice. *Nat Neurosci* **20:** 815–823. doi:10.1038/nn.4559

Mathai A, Ma Y, Paré JF, Villalba RM, Wichmann T, Smith Y. 2015. Reduced cortical innervation of the subthalamic nucleus in MPTP-treated parkinsonian monkeys. *Brain* **138:** 946–962. doi:10.1093/brain/awv018

Matsuda W, Furuta T, Nakamura KC, Hioki H, Fujiyama F, Arai R, Kaneko T. 2009. Single nigrostriatal dopaminergic neurons form widely spread and highly dense axonal arborizations in the neostriatum. *J Neurosci* **29:** 444–453. doi:10.1523/JNEUROSCI.4029-08.2009

Matsumoto N, Minamimoto T, Graybiel AM, Kimura M. 2001. Neurons in the thalamic CM-Pf complex supply striatal neurons with information about behaviorally significant sensory events. *J Neurophysiol* **85:** 960–976. doi:10.1152/jn.2001.85.2.960

Matthews DC, Lerman H, Lukic A, Andrews RD, Mirelman A, Wernick MN, Giladi N, Strother SC, Evans KC, Cedarbaum JM, et al. 2018. FDG PET Parkinson's disease-related pattern as a biomarker for clinical trials in early stage disease. *NeuroImage Clin* **20:** 572–579. doi:10.1016/j.nicl.2018.08.006

Matthews LG, Puryear CB, Correia SS, Srinivasan S, Belfort GM, Pan MK, Kuo SH. 2023. T-type calcium channels as therapeutic targets in essential tremor and Parkinson's disease. *Ann Clin Transl Neurol* **10:** 462–483. doi:10.1002/acn3.51735

McKay JL, Nye J, Goldstein FC, Sommerfeld B, Smith Y, Weinshenker D, Factor SA. 2023. Levodopa responsive freezing of gait is associated with reduced norepinephrine transporter binding in Parkinson's disease. *Neurobiol Dis* **179:** 106048. doi:10.1016/j.nbd.2023.106048

McLellan DL. 1973. Dynamic spindle reflexes and the rigidity of Parkinsonism. *J Neurol Neurosurg Psychiatry* **36:** 342–349. doi:10.1136/jnnp.36.3.342

McNeill TH, Brown SA, Rafols JA, Shoulson I. 1988. Atrophy of medium spiny I striatal dendrites in advanced Parkinson's disease. *Brain Res* **455:** 148–152. doi:10.1016/0006-8993(88)90124-2

Meidahl AC, Moll CKE, van Wijk BCM, Gulberti A, Tinkhauser G, Westphal M, Engel AK, Hamel W, Brown P, Sharott A. 2019. Synchronised spiking activity underlies phase amplitude coupling in the subthalamic nucleus of Parkinson's disease patients. *Neurobiol Dis* **127:** 101–113. doi:10.1016/j.nbd.2019.02.005

Mena-Segovia J, Bolam JP. 2017. Rethinking the pedunculopontine nucleus: from cellular organization to function. *Neuron* **94:** 7–18. doi:10.1016/j.neuron.2017.02.027

Mena-Segovia J, Bolam JP, Magill PJ. 2004. Pedunculopontine nucleus and basal ganglia: distant relatives or part of the same family? *Trends Neurosci* **27:** 585–588. doi:10.1016/j.tins.2004.07.009

Meyer JS, Barron DW. 1960. Apraxia of gait: a clinico-physiological study. *Brain* **83:** 261–284. doi:10.1093/BRAIN/83.2.261

Milekovic T, Moraud EM, Macellari N, Moerman C, Raschellà F, Sun S, Perich MG, Varescon C, Demesmaeker R, Bruel A, et al. 2023. A spinal cord neuroprosthesis for locomotor deficits due to Parkinson's disease. *Nat Med* **29:** 2854–2865. doi:10.1038/s41591-023-02584-1

Miller WC, DeLong MR. 1987. Altered tonic activity of neurons in the globus pallidus and subthalamic nucleus in the primate MPTP model of parkinsonism. In *The basal ganglia II* (ed. Carpenter MB, Jayaraman A), pp. 415–427. Plenum, New York.

Miocinovic S, de Hemptinne C, Qasim S, Ostrem JL, Starr PA. 2015. Patterns of cortical synchronization in isolated dystonia compared with Parkinson disease. *JAMA Neurol* **72:** 1244–1251. doi:10.1001/jamaneurol.2015.2561

Mitchell IJ, Cross AJ, Sambrook MA, Crossman AR. 1986. Neural mechanisms mediating 1-methyl-4-phenyl-1,2,3, 6-tetrahydropyridine-induced parkinsonism in the monkey: relative contributions of the striatopallidal and striatonigral pathways as suggested by 2-deoxyglucose uptake. *Neurosci Lett* **63:** 61–65. doi:10.1016/0304-3940(86)900 13-3

Mitchell IJ, Clarke CE, Boyce S, Robertson RG, Peggs D, Sambrook MA, Crossman AR. 1989. Neural mechanisms underlying parkinsonian symptoms based upon regional uptake of 2-deoxyglucose in monkeys exposed to 1-methyl-4-phenyl-1,2,3,6-tetrahydropyridine. *Neurosci* **32:** 213–226. doi:10.1016/0306-4522(89)90120-6

Moreau C, Defebvre L, Destée A, Bleuse S, Clement F, Blatt JL, Krystkowiak P, Devos D. 2008. STN-DBS frequency effects on freezing of gait in advanced Parkinson disease. *Neurology* **71:** 80–84. doi:10.1212/01.wnl.0000303972 .16279.46

Mortimer JA, Webster DD. 1979. Evidence for a quantitative association between EMG stretch responses and Parkinsonian rigidity. *Brain Res* **162:** 169–173. doi:10.1016/ 0006-8993(79)90768-6

Mosharov EV, Borgkvist A, Sulzer D. 2015. Presynaptic effects of levodopa and their possible role in dyskinesia. *Mov Disord* **30:** 45–53. doi:10.1002/mds.26103

Myers PS, McNeely ME, Koller JM, Earhart GM, Campbell MC. 2017. Cerebellar volume and executive function in Parkinson disease with and without freezing of gait. *J Parkinsons Dis* **7:** 149–157. doi:10.3233/JPD-161029

Nambu A, Chiken S. 2024. External segment of the globus pallidus in health and disease: its interactions with the striatum and subthalamic nucleus. *Neurobiol Dis* **190:** 106362. doi:10.1016/j.nbd.2023.106362

Nambu A, Tokuno H, Takada M. 2002. Functional significance of the cortico-subthalamo-pallidal "hyperdirect" pathway. *Neurosci Res* **43:** 111–117. doi:10.1016/s0168-0102(02)00027-5

Nambu A, Chiken S, Sano H, Hatanaka N, Obeso JA. 2023. Dynamic activity model of movement disorders: the fundamental role of the *hyperdirect* pathway. *Mov Disord* **38:** 2145–2150. doi:10.1002/mds.29646

Nanda B, Galvan A, Smith Y, Wichmann T. 2009. Effects of stimulation of the centromedian nucleus of the thalamus on the activity of striatal cells in awake rhesus monkeys. *Eur J Neurosci* **29:** 588–598. doi:10.1111/j.1460-9568 .2008.06598.x

Narabayashi H, Kondo T, Hayashi A, Suzuki T. 1982. L-threo-3,4-dihydroxyphenylserine treatment for akinesia and freezing of Parkinsonism. *Neurochem Res* **7:** 850–851.

Navailles S, Bioulac B, Gross C, De Deurwaerdère P. 2010. Serotonergic neurons mediate ectopic release of dopamine induced by L-DOPA in a rat model of Parkinson's

disease. *Neurobiol Dis* **38:** 136–143. doi:10.1016/j.nbd .2010.01.012

Neuville RS, Petrucci MN, Wilkins KB, Anderson RW, Hoffman SL, Parker JE, Velisar A, Bronte-Stewart HM. 2021. Differential effects of pathological β burst dynamics between Parkinson's disease phenotypes across different movements. *Front Neurosci* **15:** 733203. doi:10.3389/ fnins.2021.733203

Ni Z, Pinto AD, Lang AE, Chen R. 2010. Involvement of the cerebellothalamocortical pathway in Parkinson disease. *Ann Neurol* **68:** 816–824. doi:10.1002/ana.22221

Niu L, Ji LY, Li JM, Zhao DS, Huang G, Liu WP, Qu Y, Ma LT, Ji XT. 2012. Effect of bilateral deep brain stimulation of the subthalamic nucleus on freezing of gait in Parkinson's disease. *J Int Med Res* **40:** 1108–1113. doi:10.1177/ 147323001204000330

Noth J, Schürmann M, Podoll K, Schwarz M. 1988. Reconsideration of the concept of enhanced static fusimotor drive in rigidity in patients with Parkinson's disease. *Neurosci Lett* **84:** 239–243. doi:10.1016/0304-3940(88) 90415-6

Nyberg P, Nordberg A, Wester P, Winblad B. 1983. Dopaminergic deficiency is more pronounced in putamen than in nucleus caudatus in Parkinson's disease. *Neurochem Pathol* **1:** 193–202. doi:10.1007/BF02834244

O'Keeffe AB, Malekmohammadi M, Sparks H, Pouratian N. 2020. Synchrony drives motor cortex β bursting, waveform dynamics, and phase-amplitude coupling in Parkinson's disease. *J Neurosci* **40:** 5833–5846. doi:10.1523/ JNEUROSCI.1996-19.2020

Olaru M, Cernera S, Hahn A, Wozny TA, Anso J, de Hemptinne C, Little S, Neumann WJ, Abbasi-Asl R, Starr PA. 2024. Motor network γ oscillations in chronic home recordings predict dyskinesia in Parkinson's disease. *Brain* **147:** 2038–2052. doi:10.1093/brain/awae004

Ono SA, Sato T, Muramatsu S. 2016. Freezing of gait in Parkinson's disease is associated with reduced 6-F-18 Fluoro-L-*m*-tyrosine uptake in the locus coeruleus. *Parkinsons Dis* **2016:** 5430920. doi:10.1155/2016/5430920

Palombo E, Porrino LJ, Bankiewicz KS, Crane AM, Kopin IJ, Sokoloff L. 1988. Administration of MPTP acutely increases glucose utilization in the substantia nigra of primates. *Brain Res* **453:** 227–234. doi:10.1016/0006-8993 (88)90162-x

Palombo E, Porrino LJ, Bankiewicz KS, Crane AM, Sokoloff L, Kopin IJ. 1990. Local cerebral glucose utilization in monkeys with hemiparkinsonism induced by intracarotid infusion of the neurotoxin MPTP. *J Neurosci* **10:** 860–869. doi:10.1523/JNEUROSCI.10-03-00860.1990

Park J, Coddington LT, Dudman JT. 2020. Basal ganglia circuits for action specification. *Annu Rev Neurosci* **43:** 485–507. doi:10.1146/annurev-neuro-070918-050452

Pasquereau B, Turner RS. 2011. Primary motor cortex of the parkinsonian monkey: differential effects on the spontaneous activity of pyramidal tract-type neurons. *Cereb Cortex* **21:** 1362–1378. doi:10.1093/cercor/bhq217

Pasquereau B, Turner RS. 2013. Primary motor cortex of the parkinsonian monkey: altered neuronal responses to muscle stretch. *Front Syst Neurosci* **7:** 98. doi:10.3389/ fnsys.2013.00098

Pasquereau B, DeLong MR, Turner RS. 2016. Primary motor cortex of the parkinsonian monkey: altered encoding of active movement. *Brain* **139**: 127–143. doi:10.1093/brain/awv312

Pasquini J, Ceravolo R, Qamhawi Z, Lee JY, Deuschl G, Brooks DJ, Bonuccelli U, Pavese N. 2018. Progression of tremor in early stages of Parkinson's disease: a clinical and neuroimaging study. *Brain* **141**: 811–821. doi:10.1093/brain/awx376

Paulus W, Jellinger K. 1991. The neuropathologic basis of different clinical subgroups of Parkinson's disease. *J Neuropathol Exp Neurol* **50**: 743–755. doi:10.1097/00005072-199111000-00006

Perez-Lloret S, Negre-Pages L, Damier P, Delval A, Derkinderen P, Destée A, Meissner WG, Schelosky L, Tison F, Rascol O. 2014. Prevalence, determinants, and effect on quality of life of freezing of gait in Parkinson disease. *JAMA Neurol* **71**: 884–890. doi:10.1001/jamaneurol.2014.753

Perez Parra S, McKay JL, Factor SA. 2020. Diphasic worsening of freezing of gait in Parkinson's disease. *Mov Disord Clin Pract* **7**: 325–328. doi:10.1002/mdc3.12918

Peterson DS, Pickett KA, Duncan R, Perlmutter J, Earhart GM. 2014. Gait-related brain activity in people with Parkinson disease with freezing of gait. *PLoS ONE* **9**: e90634. doi:10.1371/journal.pone.0090634

Peterson DS, Fling BW, Mancini M, Cohen RG, Nutt JG, Horak FB. 2015. Dual-task interference and brain structural connectivity in people with Parkinson's disease who freeze. *J Neurol Neurosurg Psychiatry* **86**: 786–792. doi:10.1136/jnnp-2014-308840

Petersson P, Kühn AA, Neumann WJ, Fuentes R. 2020. Basal ganglia oscillations as biomarkers for targeting circuit dysfunction in Parkinson's disease. *Prog Brain Res* **252**: 525–557. doi:10.1016/bs.pbr.2020.02.002

Petrovici I. 1968. Apraxia of gait and of trunk movements. *J Neurol Sci* **7**: 229–243. doi:10.1016/0022-510x(68)90145-7

Pfurtscheller G. 1981. Central β rhythm during sensorimotor activities in man. *Electroencephalogr Clin Neurophysiol* **51**: 253–264. doi:10.1016/0013-4694(81)90139-5

Pfurtscheller G, Woertz M, Supp G, Lopes da Silva FH. 2003. Early onset of post-movement β electroencephalogram synchronization in the supplementary motor area during self-paced finger movement in man. *Neurosci Lett* **339**: 111–114. doi:10.1016/s0304-3940(02)01479-9

Picillo M, Dubbioso R, Iodice R, Iavarone A, Pisciotta C, Spina E, Santoro L, Barone P, Amboni M, Manganelli F. 2015. Short-latency afferent inhibition in patients with Parkinson's disease and freezing of gait. *J Neural Transm (Vienna)* **122**: 1533–1540. doi:10.1007/s00702-015-1428-y

Pikstra ARA, van der Hoorn A, Leenders KL, de Jong BM. 2016. Relation of 18-F-Dopa PET with hypokinesia-rigidity, tremor and freezing in Parkinson's disease. *NeuroImage Clinical* **11**: 68–72. doi:10.1016/j.nicl.2016.01.010

Polyakova Z, Chiken S, Hatanaka N, Nambu A. 2020. Cortical control of subthalamic neuronal activity through the hyperdirect and indirect pathways in monkeys. *J Neurosci* **40**: 7451–7463. doi:10.1523/JNEUROSCI.0772-20.2020

Qamhawi Z, Towey D, Shah B, Pagano G, Seibyl J, Marek K, Borghammer P, Brooks DJ, Pavese N. 2015. Clinical correlates of raphe serotonergic dysfunction in early Parkin-son's disease. *Brain* **138**: 2964–2973. doi:10.1093/brain/awv215

Quiroga-Varela A, Walters JR, Brazhnik E, Marin C, Obeso JA. 2013. What basal ganglia changes underlie the parkinsonian state? The significance of neuronal oscillatory activity. *Neurobiol Dis* **58**: 242–248. doi:10.1016/j.nbd.2013.05.010

Ragsdale CW Jr, Graybiel AM. 1991. Compartmental organization of the thalamostriatal connection in the cat. *J Comp Neurol* **311**: 134–167. doi:10.1002/cne.903110110

Ray Chaudhuri K, Leta V, Bannister K, Brooks DJJ, Svenningsson P. 2023. The noradrenergic subtype of Parkinson disease: from animal models to clinical practice. *Nat Rev Neurol* **19**: 333–345. doi:10.1038/s41582-023-00802-2

Razmkon A, Abdollahifard S, Taherifard E, Roshanshad A, Shahrivar K. 2023. Effect of deep brain stimulation on freezing of gait in patients with Parkinson's disease: a systematic review. *Br J Neurosurg* **37**: 3–11. doi:10.1080/02688697.2022.2077308

Revuelta GJ, Uthayathas S, Wahlquist AE, Factor SA, Papa SM. 2012. Non-human primate FOG develops with advanced parkinsonism induced by MPTP treatment. *Exp Neurol* **237**: 464–469. doi:10.1016/j.expneurol.2012.07.021

Revuelta GJ, Embry A, Elm JJ, Gregory C, Delambo A, Kautz S, Hinson VK. 2015. Pilot study of atomoxetine in patients with Parkinson's disease and dopa-unresponsive freezing of gait. *Transl Neurodegener* **4**: 24. doi:10.1186/s40035-015-0047-8

Rommelfanger KS, Wichmann T. 2010. Extrastriatal dopaminergic circuits of the basal ganglia. *Front Neuroanat* **4**: 139. doi:10.3389/fnana.2010.00139

Ruppert MC, Greuel A, Tahmasian M, Schwartz F, Stürmer S, Maier F, Hammes J, Tittgemeyer M, Timmermann L, van Eimeren T, et al. 2020. Network degeneration in Parkinson's disease: multimodal imaging of nigro-striato-cortical dysfunction. *Brain* **143**: 944–959. doi:10.1093/brain/awaa019

Sadikot AF, Rymar VV. 2009. The primate centromedian-parafascicular complex: anatomical organization with a note on neuromodulation. *Brain Res Bull* **78**: 122–130. doi:10.1016/j.brainresbull.2008.09.016

Sadikot AF, Parent A, François C. 1992. Efferent connections of the centromedian and parafascicular thalamic nuclei in the squirrel monkey: a PHA-L study of subcortical projections. *J Comp Neurol* **315**: 137–159. doi:10.1002/cne.903150203

Salvadè A, D'Angelo V, Di Giovanni G, Tinkhauser G, Sancesario G, Städler C, Möller JC, Stefani A, Kaelin-Lang A, Galati S. 2016. Distinct roles of cortical and pallidal β and γ frequencies in hemiparkinsonian and dyskinetic rats. *Exp Neurol* **275**: 199–208. doi:10.1016/j.expneurol.2015.11.005

Schmidt SL, Grill WM. 2021. Levodopa-induced dyskinesia is mediated by cortical γ oscillations in experimental Parkinsonism. *Mov Disord* **36**: 1044–1045. doi:10.1002/mds.28578

Scholten M, Govindan RB, Braun C, Bloem BR, Plewnia C, Krüger R, Gharabaghi A, Weiss D. 2016. Cortical correlates of susceptibility to upper limb freezing in Parkinson's disease. *Clin Neurophysiol* **127**: 2386–2393. doi:10 .1016/j.clinph.2016.01.028

Shine JM, Matar E, Ward PB, Frank MJ, Moustafa AA, Pearson M, Naismith SL, Lewis SJ. 2013a. Freezing of gait in Parkinson's disease is associated with functional decoupling between the cognitive control network and the basal ganglia. *Brain* **136**: 3671–3681. doi:10.1093/brain/awt272

Shine JM, Moustafa AA, Matar E, Frank MJ, Lewis SJ. 2013b. The role of frontostriatal impairment in freezing of gait in Parkinson's disease. *Front Syst Neurosci* **7**: 61. doi:10 .3389/fnsys.2013.00061

Siderowf A, Concha-Marambio L, Lafontant DE, Farris CM, Ma Y, Urenia PA, Nguyen H, Alcalay RN, Chahine LM, Foroud T, et al. 2023. Assessment of heterogeneity among participants in the Parkinson's Progression Markers Initiative cohort using α-synuclein seed amplification: a cross-sectional study. *Lancet Neurol* **22**: 407–417. doi:10 .1016/S1474-4422(23)00109-6

Sidibé M, Smith Y. 1999. Thalamic inputs to striatal interneurons in monkeys: synaptic organization and co-localization of calcium binding proteins. *Neurosci* **89**: 1189–1208. doi:10.1016/s0306-4522(98)00367-4

Smith Y, Parent A. 1986. Differential connections of caudate nucleus and putamen in the squirrel monkey (*Saimiri sciureus*). *Neurosci* **18**: 347–371. doi:10.1016/0306-4522 (86)90159-4

Snijders AH, Takakusaki K, Debu B, Lozano AM, Krishna V, Fasano A, Aziz TZ, Papa SM, Factor SA, Hallett M. 2016. Physiology of freezing of gait. *Ann Neurol* **80**: 644–659. doi:10.1002/ana.24778

Steidel K, Ruppert MC, Greuel A, Tahmasian M, Maier F, Hammes J, van Eimeren T, Timmermann L, Tittgemeyer M, Drzezga A, et al. 2022. Longitudinal trimodal imaging of midbrain-associated network degeneration in Parkinson's disease. *NPJ Park Dis* **8**: 79. doi:10.1038/s41531- 022-00341-8

Steiner LA, Neumann WJ, Staub-Bartelt F, Herz DM, Tan H, Pogosyan A, Kuhn AA, Brown P. 2017. Subthalamic β dynamics mirror Parkinsonian bradykinesia months after neurostimulator implantation. *Mov Disord* **32**: 1183–1190. doi:10.1002/mds.27068

Sudarsky L, Simon S. 1987. Gait disorder in late-life hydrocephalus. *Arch Neurol* **44**: 263–267. doi:10.1001/archneur .1987.00520150019012

Sunwoo MK, Cho KH, Hong JY, Lee JE, Sohn YH, Lee PH. 2013. Thalamic volume and related visual recognition are associated with freezing of gait in non-demented patients with Parkinson's disease. *Parkinsonism Relat Disord* **19**: 1106–1109. doi:10.1016/j.parkreldis.2013.07.023

Swain AJ, Galvan A, Wichmann T, Smith Y. 2020. Structural plasticity of GABAergic and glutamatergic networks in the motor thalamus of parkinsonian monkeys. *J Comp Neurol* **528**: 1436–1456. doi:10.1002/cne.24834

Swann NC, de Hemptinne C, Miocinovic S, Qasim S, Wang SS, Ziman N, Ostrem JL, San Luciano M, Galifianakis NB, Starr PA. 2016. γ Oscillations in the hyperkinetic state detected with chronic human brain recordings in Parkinson's disease. *J Neurosci* **36**: 6445–6458. doi:10.1523/ JNEUROSCI.1128-16.2016

Sypert GW, Leffman H, Ojemann GA. 1973. Occult normal pressure hydrocephalus manifested by parkinsonism-de-

mentia complex. *Neurology* **23**: 234–238. doi:10.1212/wnl .23.3.234

Szul MJ, Papadopoulos S, Alavizadeh S, Daligaut S, Schwartz D, Mattout J, Bonaiuto JJ. 2023. Diverse β burst waveform motifs characterize movement-related cortical dynamics. *Prog Neurobiol* **228**: 102490. doi:10.1016/j.pneurobio .2023.102490

Tessitore A, Amboni M, Cirillo G, Corbo D, Picillo M, Russo A, Vitale C, Santangelo G, Erro R, Cirillo M, et al. 2012. Regional gray matter atrophy in patients with Parkinson disease and freezing of gait. *AJNR Am J Neuroradiol* **33**: 1804–1809. doi:10.3174/ajnr.A3066

Thompson PD, Marsden CD. 1987. Gait disorder of subcortical arteriosclerotic encephalopathy: Binswanger's disease. *Mov Disord* **2**: 1–8. doi:10.1002/mds.870020101

Thura D, Cisek P. 2017. The basal ganglia do not select reach targets but control the urgency of commitment. *Neuron* **95**: 1160–1170.e5. doi:10.1016/j.neuron.2017.07.039

Timmermann L, Gross J, Dirks M, Volkmann J, Freund HJ, Schnitzler A. 2003. The cerebral oscillatory network of parkinsonian resting tremor. *Brain* **126**: 199–212. doi:10.1093/brain/awg022

Tinkhauser G, Torrecillos F, Duclos Y, Tan H, Pogosyan A, Fischer P, Carron R, Welter ML, Karachi C, Vandenberghe W, et al. 2018. β Burst coupling across the motor circuit in Parkinson's disease. *Neurobiol Dis* **117**: 217–225. doi:10.1016/j.nbd.2018.06.007

Toledo JB, López-Azcárate J, Garcia-Garcia D, Guridi J, Valencia M, Artieda J, Obeso J, Alegre M, Rodriguez-Oroz M. 2014. High β activity in the subthalamic nucleus and freezing of gait in Parkinson's disease. *Neurobiol Dis* **64**: 60–65. doi:10.1016/j.nbd.2013.12.005

Turner RS, Desmurget M. 2010. Basal ganglia contributions to motor control: a vigorous tutor. *Curr Opin Neurobiol* **20**: 704–716. doi:10.1016/j.conb.2010.08.022

Turner RS, Grafton ST, McIntosh AR, DeLong MR, Hoffman JM. 2003. The functional anatomy of parkinsonian bradykinesia. *Neuroimage* **19**: 163–179. doi:10.1016/ s1053-8119(03)00059-4

Underwood CF, Parr-Brownlie LC. 2021. Primary motor cortex in Parkinson's disease: functional changes and opportunities for neurostimulation. *Neurobiol Dis* **147**: 105159. doi:10.1016/j.nbd.2020.105159

Valldeoriola F, Munoz E, Rumia J, Roldan P, Camara A, Compta Y, Marti MJ, Tolosa E. 2019. Simultaneous low-frequency deep brain stimulation of the substantia nigra pars reticulata and high-frequency stimulation of the subthalamic nucleus to treat levodopa unresponsive freezing of gait in Parkinson's disease: a pilot study. *Parkinsonism Relat Disord* **60**: 153–157. doi:10.1016/j.park reldis.2018.12.009

van der Stouwe AMM, Nieuwhof F, Helmich RC. 2020. Tremor pathophysiology: lessons from neuroimaging. *Curr Opin Neurol* **33**: 474–481. doi:10.1097/WCO.0000 000000000829

van Zagten M, Lodder J, Kessels F. 1998. Gait disorder and parkinsonian signs in patients with stroke related to small deep infarcts and white matter lesions. *Mov Disord* **13**: 89–95. doi:10.1002/mds.870130119

Vercruysse S, Vandenberghe W, Munks L, Nuttin B, Devos H, Nieuwboer A. 2014. Effects of deep brain stimulation of the subthalamic nucleus on freezing of gait in Parkinson's disease: a prospective controlled study. *J Neurol Neurosurg Psychiatry* **85:** 871–877. doi:10.1136/jnnp-2013-306336

Vicente AM, Galvão-Ferreira P, Tecuapetla F, Costa RM. 2016. Direct and indirect dorsolateral striatum pathways reinforce different action strategies. *Curr Biol* **26:** R267–R269. doi:10.1016/j.cub.2016.02.036

Villalba RM, Smith Y. 2018. Loss and remodeling of striatal dendritic spines in Parkinson's disease: from homeostasis to maladaptive plasticity? *J Neural Transm (Vienna)* **125:** 431–447. doi:10.1007/s00702-017-1735-6

Villalba RM, Behnke JA, Pare JF, Smith Y. 2021. Comparative ultrastructural analysis of thalamocortical innervation of the primary motor cortex and supplementary motor area in control and MPTP-treated parkinsonian monkeys. *Cereb Cortex* **31:** 3408–3425. doi:10.1093/cercor/bhab020

Watts RL, Mandir AS. 1992. The role of motor cortex in the pathophysiology of voluntary movement deficits associated with parkinsonism. *Neurol Clin* **10:** 451–469. doi:10.1016/S0733-8619(18)30221-4

Weinberger M, Mahant N, Hutchison WD, Lozano AM, Moro E, Hodaie M, Lang AE, Dostrovsky JO. 2006. β Oscillatory activity in the subthalamic nucleus and its relation to dopaminergic response in Parkinson's disease. *J Neurophys* **96:** 3248–3256. doi:10.1152/jn.00697.2006

Weinberger M, Hutchison WD, Dostrovsky JO. 2009. Pathological subthalamic nucleus oscillations in PD: can they be the cause of bradykinesia and akinesia? *Exp Neurol* **219:** 58–61. doi:10.1016/j.expneurol.2009.05.014

Wichmann T. 2018. Pathophysiologic basis of movement disorders. *Prog Neurol Surg* **33:** 13–24. doi:10.1159/000480718

Wichmann T, Dostrovsky JO. 2011. Pathological basal ganglia activity in movement disorders. *Neurosci* **198:** 232–244. doi:10.1016/j.neuroscience.2011.06.048

Wichmann T, Soares J. 2006. Neuronal firing before and after burst discharges in the monkey basal ganglia is predictably patterned in the normal state and altered in parkinsonism. *J Neurophys* **95:** 2120–2133. doi:10.1152/jn.01013.2005

Wiest C, Tinkhauser G, Pogosyan A, He S, Baig F, Morgante F, Mostofi A, Pereira EA, Tan H, Brown P, et al. 2021. Subthalamic deep brain stimulation induces finely-tuned γ oscillations in the absence of levodopa. *Neurobiol Dis* **152:** 105287. doi:10.1016/j.nbd.2021.105287

Wiest C, Torrecillos F, Tinkhauser G, Pogosyan A, Morgante F, Pereira EA, Tan H. 2022. Finely-tuned γ oscillations: spectral characteristics and links to dyskinesia. *Exp Neurol* **351:** 113999. doi:10.1016/j.expneurol.2022.113999

Winikates J, Jankovic J. 1999. Clinical correlates of vascular parkinsonism. *Arch Neurol* **56:** 98–102. doi:10.1001/archneur.56.1.98

Winogrodzka A, Wagenaar RC, Bergmans P, Vellinga A, Booij J, van Royen EA, van Emmerik RE, Stoof JC, Wolters EC. 2001. Rigidity decreases resting tremor intensity in Parkinson's disease: a [(123)I]β-CIT SPECT study in early, nonmedicated patients. *Mov Disord* **16:** 1033–1040. doi:10.1002/mds.1205

Xia R, Sun J, Threlkeld AJ. 2009. Analysis of interactive effect of stretch reflex and shortening reaction on rigidity in Parkinson's disease. *Clin Neurophysiol* **120:** 1400–1407. doi:10.1016/j.clinph.2009.05.001

Xiao H, Li M, Cai J, Li N, Zhou M, Wen P, Xie Z, Wang Q, Chang J, Zhang W. 2017. Selective cholinergic depletion of pedunculopontine tegmental nucleus aggravates freezing of gait in parkinsonian rats. *Neurosci Lett* **659:** 92–98. doi:10.1016/j.neulet.2017.08.016

Yamanaka K, Hori Y, Minamimoto T, Yamada H, Matsumoto N, Enomoto K, Aosaki T, Graybiel AM, Kimura M. 2018. Roles of centromedian parafascicular nuclei of thalamus and cholinergic interneurons in the dorsal striatum in associative learning of environmental events. *J Neural Transm (Vienna)* **125:** 501–513. doi:10.1007/s00702-017-1713-z

Yin Z, Zhu G, Zhao B, Bai Y, Jiang Y, Neumann WJ, Kühn AA, Zhang J. 2021. Local field potentials in Parkinson's disease: a frequency-based review. *Neurobiol Dis* **155:** 105372. doi:10.1016/j.nbd.2021.105372

Yoshida J, Oñate M, Khatami L, Vera J, Nadim F, Khodakhah K. 2022. Cerebellar contributions to the basal ganglia influence motor coordination, reward processing, and movement vigor. *J Neurosci* **42:** 8406–8415. doi:10.1523/JNEUROSCI.1535-22.2022

Yttri EA, Dudman JT. 2016. Opponent and bidirectional control of movement velocity in the basal ganglia. *Nature* **533:** 402–406. doi:10.1038/nature17639

Yttri EA, Dudman JT. 2018. A proposed circuit computation in basal ganglia: history-dependent gain. *Mov Disord* **33:** 704–716. doi:10.1002/mds.27321

Yu Y, Escobar Sanabria D, Wang J, Hendrix CM, Zhang J, Nebeck SD, Amundson AM, Busby ZB, Bauer DL, Johnson MD, et al. 2021. Parkinsonism alters β burst dynamics across the basal ganglia-motor cortical network. *J Neurosci* **41:** 2274–2286. doi:10.1523/JNEUROSCI.1591-20.2021

Zang Z, Song T, Li J, Nie B, Mei S, Zhang C, Wu T, Zhang Y, Lu J. 2022. Simultaneous PET/fMRI revealed increased motor area input to subthalamic nucleus in Parkinson's disease. *Cerebr Cortex* **33:** 167–175. doi:10.1093/cercor/bhac059

Zhang Y, Meredith GE, Mendoza-Elias N, Rademacher DJ, Tseng KY, Steece-Collier K. 2013. Aberrant restoration of spines and their synapses in L-DOPA-induced dyskinesia: involvement of corticostriatal but not thalamostriatal synapses. *J Neurosci* **33:** 11655–11667. doi:10.1523/JNEUROSCI.0288-13.2013

Zich C, Quinn AJ, Bonaiuto JJ, O'Neill G, Mardell LC, Ward NS, Bestmann S. 2023. Spatiotemporal organisation of human sensorimotor β burst activity. *eLife* **12:** e80160. doi:10.7554/eLife.80160

Imaging of Disease-Related Networks in Parkinson's Disease

Yoshikazu Nakano,[1] Martin Niethammer,[1,2] and David Eidelberg[1,2,3]

[1]Center for Neurosciences, The Feinstein Institutes for Medical Research, Manhasset, New York 11030, USA

[2]Department of Neurology, Donald and Barbara Zucker School of Medicine at Hofstra/Northwell, Hempstead, New York 11549, USA

[3]Department of Molecular Medicine, Donald and Barbara Zucker School of Medicine at Hofstra/Northwell, Hempstead, New York 11549, USA

Correspondence: deidelberg@northwell.edu

Functional neuroimaging techniques are increasingly being used to advance the diagnosis and management of Parkinson's disease (PD). Methods such as [^{18}F]-fluorodeoxyglucose positron emission tomography (FDG PET), resting-state functional magnetic resonance imaging (rs-fMRI), arterial spin labeling (ASL) MRI, and single-photon emission computed tomography (SPECT) enable the identification of disease-specific patterns like the PD-related pattern (PDRP) and PD cognition-related pattern (PDCP), which correlate with motor and cognitive symptoms. Network analysis using graph theory further elucidates the alterations in brain connectivity associated with PD, providing insights into disease progression and response to treatment. Moreover, these neuroimaging patterns assist in distinguishing PD from atypical parkinsonian syndromes, enhancing diagnostic accuracy. Understanding the impact of genetic variants like *LRRK2* and *GBA1* on functional connectivity highlights the potential for precision medicine in PD. As neuroimaging technologies evolve, their integration into clinical practice will be pivotal in the personalized management of PD, offering improved diagnostic precision and targeted therapeutic interventions.

Parkinson's disease (PD) is a neurodegenerative disorder primarily characterized by motor symptoms such as bradykinesia, rigidity, and tremor, but patients also exhibit a number of nonmotor symptoms including cognitive impairment (Espay et al. 2017). With >10 million people affected globally, PD represents a significant burden on healthcare systems and society (GBD 2019 Diseases and Injuries Collaborators 2020). The development of effective treatments requires not only a detailed understanding of the underlying molecular and anatomical pathways but also accurate clinical diagnosis. While histopathological examination remains the gold standard, in vivo biomarkers, particularly through functional neuroimaging, are increasingly important for early diagnosis and monitoring disease progression (Schindlbeck and Eidelberg 2018; Barbero et al. 2023; Perovnik et al. 2023a).

Functional neuroimaging techniques have become indispensable in the study of neurodegenerative diseases, offering insights into the

Copyright © 2025 Cold Spring Harbor Laboratory Press; all rights reserved
Cite this article as *Cold Spring Harb Perspect Med* doi: 10.1101/cshperspect.a041841

disease's pathophysiology and aiding in the development of targeted therapies (Niethammer and Eidelberg 2012; Barbero et al. 2023; Perovnik et al. 2023a). This review focuses on several key imaging modalities to map neuronal activity in PD such as [^{18}F]-fluorodeoxyglucose positron emission tomography (FDG PET) for cerebral glucose metabolism, and resting-state functional magnetic resonance imaging (rs-fMRI) for blood oxygen-level-dependent (BOLD) signal, as well as cerebral blood flow (CBF) using [^{15}O]-H$_2$O PET and [^{123}I]-IMP and [^{99m}Tc]-ethylcysteinate dimer (ECD) single-photon emission computed tomography (SPECT) techniques, as well as arterial spin labeling (ASL) MRI (Eckert et al. 2007; Ma and Eidelberg 2007; Ma et al. 2007, 2010). These methods enable the visualization and quantification of various aspects of brain function and metabolism, providing critical information that complements clinical and histopathological findings, with particular emphasis on network and functional connectivity analysis. In our previous work (Niethammer et al. 2012), we focus on the use of functional brain imaging and network analysis to evaluate the disease process. This current review also covers new computational methods to characterize abnormal connectivity patterns in metabolic PET and fMRI scans obtained in PD patients. We also review the role of these methods in the evaluation of new therapies for the disorder.

NETWORK ANALYSIS

PD Motor-Related Pattern

The PD-related pattern (PDRP) is a well-established metabolic network identified using FDG PET and SSM–PCA (Spetsieris and Eidelberg 2011; Spetsieris et al. 2015). Initially discovered in a North American population, the PDRP has been validated across numerous cohorts worldwide (Schindlbeck and Eidelberg 2018), demonstrating consistent topographical features: increased glucose metabolism in the putamen, pallidum, thalamus, pons, and cerebellum, and reduced metabolism in the premotor and posterior parietal areas (Fig. 1). These network-level

abnormalities seen in PD do reflect the spatial distribution of the underlying pathogenesis (Ko et al. 2017), as reflected by the accumulation of α-synuclein (Surmeier et al. 2017). This pattern's expression correlates with the severity of motor symptoms such as bradykinesia and rigidity and reflects changes in the striato-pallido-thalamo-cortical motor circuits (Perovnik et al. 2023a).

Importantly, the PDRP has also been identified in drug-naive patients, showing asymmetrical expression levels in both hemispheres regardless of clinical features and dopaminergic dysfunction (Tang et al. 2010a). This suggests that PDRP captures different aspects of the underlying disease process compared to traditional measures of PD pathology. Furthermore, similar network abnormalities have been observed in patients with PD using other imaging modalities such as [^{15}O]-H$_2$O PET, [^{99m}Tc]-ECD SPECT, and ASL MRI (Eckert et al. 2007; Ma et al. 2010; Teune et al. 2014). Recent advancements include the identification of a similar pattern using rs-fMRI (fPDRP) (Vo et al. 2017; Rommal et al. 2021).

The PDRP remains a robust biomarker for PD, consistently correlating with clinical ratings and neurophysiological measures (Eidelberg 2009; Niethammer and Eidelberg 2012; Teune et al. 2013; Tripathi et al. 2013; Meles et al. 2020). To date, it is the only functional network imaging biomarker for PD validated across multiple populations and imaging centers.

PD Cognition-Related Pattern

While the clinical diagnosis of PD involves the motor signs and symptoms (Postuma et al. 2015), cognitive dysfunction can be an important feature in PD. While cognitive impairment typically arises later in the disease and progresses slower, prevalence in PD is high (Aarsland and Kurz 2010), and over time the majority of patients will develop cognitive impairment or dementia (Aarsland and Kurz 2010; Aarsland et al. 2017). Cognitive impairment in PD involves disruptions in normal brain networks, particularly the default mode network (DMN), and increased expression of a PD cognition-related pattern (PDCP) (Huang et al. 2007a, 2008;

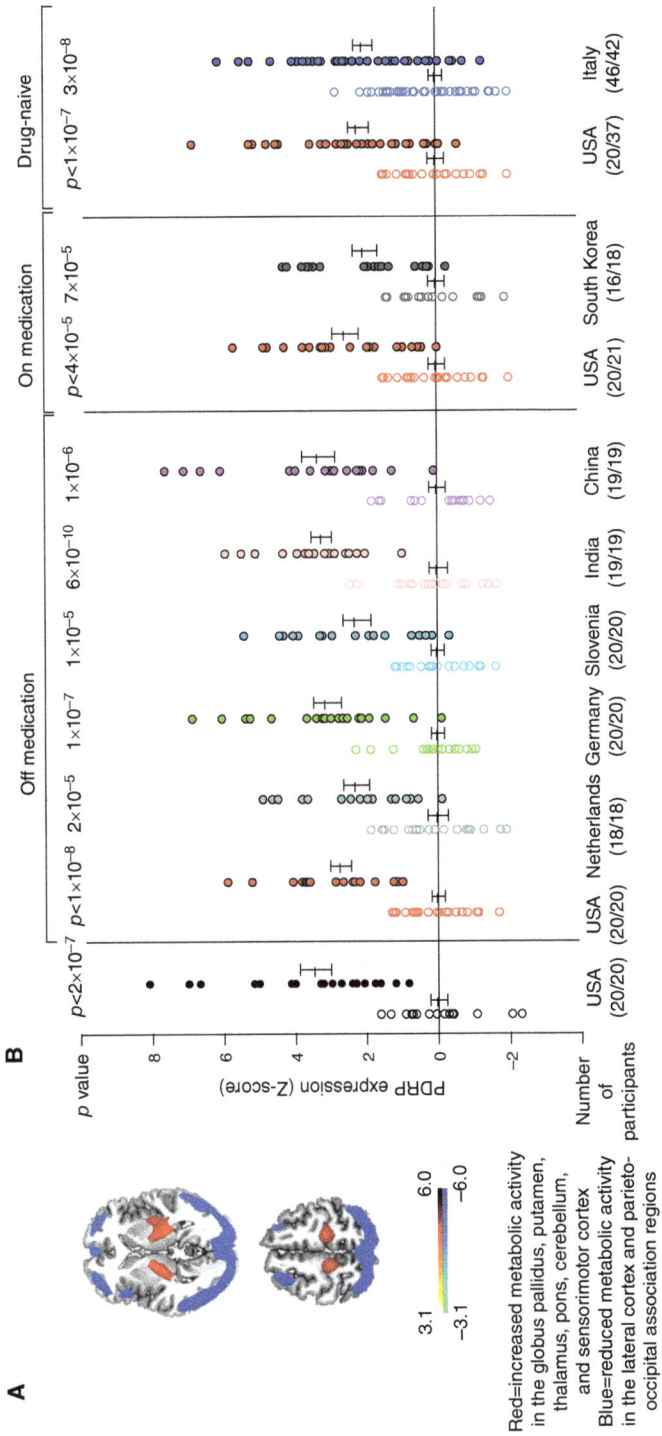

Figure 1. Parkinson's disease–related pattern (PDRP). (*A*) PDRP identified by network analysis from 20 PD patients and 20 age-matched healthy control subjects. (*B*) PDRP expression reliably discriminates PD subjects (closed circles) from healthy subjects (open circles) in multiple independent cohorts. (Reprinted from Schindlbeck and Eidelberg, with permission, from Elsevier, © 2018.)

Mattis et al. 2011). The PDCP is characterized by reduced metabolic activity in the medial frontal and parietal regions and is linked to impaired memory and executive dysfunction. This pattern was initially identified using FDG PET in patients with PD without dementia and has since been validated in multiple independent cohorts (Huang et al. 2007a, 2008). PDCP expression levels correlate consistently with cognitive performance, particularly in executive function and memory, and exhibit high test–retest reliability.

While PDCP expression levels are inversely correlated with the activity of the DMN and PDCP and DMN share some regional variance, they are not interchangeable (Schindlbeck and Eidelberg 2018). Specifically, the PDCP reflects a loss of the ventral DMN component (posterior cingulate region and precuneus), with additional contributions from medial temporal and lateral frontoparietal regions. This network topology suggests that cognitive dysfunction in PD involves multiple mechanisms, including degeneration of dopaminergic and cholinergic pathways and Lewy body pathology in limbic and neocortical areas.

Notably, PDCP and PDRP networks are topographically unrelated, emphasizing the distinct neural underpinnings of motor and cognitive symptoms in PD (Niethammer and Eidelberg 2012; Niethammer et al. 2013; Meles et al. 2015; Mattis et al. 2016; Schindlbeck et al. 2021b). The PDCP has also been identified using rs-fMRI (fPDCP), further supporting its validity (Vo et al. 2017).

The evolution of cognitive deficits in PD, from early executive dysfunction to later posterior cortical syndromes, aligns with the dual syndrome hypothesis, which attributes these changes to the progressive loss of dopaminergic and cholinergic inputs. Unlike Alzheimer's disease, the PDCP does not prominently involve the medial temporal lobes, distinguishing it from Alzheimer's disease–related patterns (Mattis et al. 2016). This specificity makes the PDCP a valuable biomarker for cognitive impairment in PD, offering insights into the distinct neurobiological mechanisms underlying these symptoms.

Graph Theory and Network Metrics

Multivariate approaches such as PCA and ICA provide important information about the specific regional elements (or "nodes") that comprise functional topographies as well as their relative contributions to overall network activity (Spetsieris et al. 2009, 2015). The region-to-region connections (or "edges") that define the internal structure of a network are equally important, particularly in studies of functional adaptations to underlying neurodegenerative pathology. Graph theory, a discrete branch of mathematics, offers a valuable way to describe changes in functional architecture by delineating the specific regions and connections that form the disease network (Niethammer et al. 2018; Schindlbeck et al. 2020; Vo et al. 2023b; Eidelberg et al. 2024). Nodes can be defined in various ways, ranging from individual voxels or clusters to anatomically defined regions-of-interest based on standardized cytoarchitectonic atlases. Connections between nodes are typically defined by distance metrics, among which correlations and partial correlations are most commonly used. Computational algorithms based on graph theory have also been used to isolate small communities of connected regions (termed "modules" or "subgraphs") within the overall network space. Connectivity patterns in different groups of individuals, at different time points and under various treatment conditions can be assessed by rigorously computing relevant graph metrics either for validated disease networks as a whole or for key subgraphs. Connectivity in graphs can be explored on a local or global level.

To assess group differences in connectivity parameters within the relevant network spaces, the following metrics can be computed on weighted graphical links:

1. *Degree centrality:* The number of connections (edges) within the network or subgraph, divided by the total number of nodes in the same space.

2. *Clustering coefficient:* A measure of the likelihood that the nearest neighbors of a node will also be connected.

 Cite this article as *Cold Spring Harb Perspect Med* doi: 10.1101/cshperspect.a041841

3. *Characteristic path length:* The shortest path length between two nodes averaged over all pairs of nodes in a given network. High characteristic path length implies less efficient information transfer through the network (Newman 2010; Rubinov and Sporns 2010).

4. *Small-worldness:* The ratio of clustering coefficient to characteristic path length, normalized to corresponding parameters from an equivalent random graph (Betzel et al. 2017). This measure quantifies the ratio of segregation to integration of information sources in the network space.

5. *Modularity:* The measure that evaluates how the nodes that comprise a given network fall into specific groups (i.e., the degree to which nodes can be assigned to distinct communities [modules] compared to assignments at random). High modularity indicates the presence of dense connections within a community, while connections between communities are sparse (Newman 2010; Schindlbeck et al. 2020).

6. *Assortativity coefficient:* The correlation coefficient between the degrees of all nodes on two opposite ends of a link (Newman 2003; Noldus and Van Mieghem 2015; Barabasi 2016). Assortativity in a group of subjects is considered increased if the assortativity coefficient is significantly elevated compared to values for the same network computed in a reference group. Conversely, assortativity is reduced when the coefficient is lower than in the reference group.

On a local level, the simplest connectivity measure is nodal degree centrality, which is determined by the number of connections an individual node has with other nodes. In contrast, global metrics can be used to describe connectivity patterns for the network as a whole, which in turn influence information flowthrough on the graph. Network properties such as characteristic path length, clustering coefficient, small-worldness, modularity, and assortativity can be used to recognize network-wide connectivity patterns related to disease state and treatment condition. Indeed, at the module level, these measures might help to distinguish between adaptive and pathological connectivity changes within disease networks (Hou et al. 2023; Vo et al. 2023a,b).

Mean PDRP expression increases with disease progression, paralleling the changes in network assortativity seen over time in the same groups of patients. While these measurements relate to a fixed PDRP, it is also possible to identify similar yet topographically more extensive disease patterns at each successive time point (Spetsieris and Eidelberg 2023). In this context, the magnitude of the corresponding dominant principal component eigenvalue rises (Spetsieris and Eidelberg 2011; Spetsieris et al. 2015), with commensurate increases in network assortativity (Noldus and Van Mieghem 2015). Indeed, both of these phenomena have been linked to increasing stress and network instability in physiological systems (Gorban et al. 2010; Scheffer et al. 2012). Advancing neurological symptoms have been associated with monotonic increases in disease network expression in many brain disorders (Niethammer et al. 2023; Perovnik et al. 2023a,b; Rus et al. 2023) and also progressive abnormal elevations in graph assortativity (Niethammer et al. 2023; Perovnik et al. 2023a; Vo et al. 2023b; Eidelberg et al. 2024). Interestingly, however, these phenomena may diverge in response to treatment. Levodopa administration consistently improves motor symptoms in PD patients by repleting nigrostriatal dopamine. Although levodopa lowers PDRP expression levels, network assortativity increased in these individuals during treatment (Vo et al. 2023b). It is conceivable that transient increases in PDRP assortativity occur with each levodopa dose. Under these circumstances, it may be that repeated daily administration of the drug leads over time to the development of a maladaptive connectivity pattern in the network space.

A different set of network changes follows subthalamic adeno-associated virus with the gene for glutamic acid decarboxylase (AAV2-GAD) gene therapy: PDRP expression levels increase over time, consistent with disease progression, yet PDRP assortativity substantially declined in the treatment group (Niethammer et al. 2018). Subthalamic nucleus (STN) AAV2-

GAD gene therapy reduced assortativity in the treatment-induced AAV2-GAD-related pattern (GADRP) space and in PDRP, consistent with extensive remodeling of functional connectivity in both networks (Vo et al. 2023b). These changes contrasted with the absence of GADRP induction and the progression-related increases in PDRP expression seen over time in the parallel sham surgery arm of the study (Niethammer et al. 2018). While PDRP assortativity increased with levodopa and sham surgery and declined with gene therapy, small-worldness declined toward normal after either levodopa or gene therapy. Given that significant clinical improvement occurred with both treatments, it is possible that normalization of PDRP small-worldness is a feature of the symptomatic benefit seen with both interventions (Vo et al. 2023b). The reductions in PDRP assortativity following gene therapy may reflect longer-term adaptive responses that do not occur with acute levodopa treatment. That said, the changes that occur with disease modification are likely to involve a combination of effects: modulation of "generic" disease networks (Oertel et al. 2024), induction of treatment-specific networks (Niethammer et al. 2018; Brakedal et al. 2022; Unadkat et al. 2024), and shifting of network connectivity patterns from pathological to adaptive configurations (Vo et al. 2023b).

In FDG PET studies of PDRP, network metrics were necessarily quantified at the group level. However, to evaluate treatment, these changes should optimally be determined for individual patients. Rs-fMRI provides a useful alternative in that regard. PDRP and PDCP networks closely related to their FDG PET counterparts have already been characterized by this method (Rommal et al. 2021; Schindlbeck et al. 2021b). Unlike FDG PET, however, rs-fMRI time series data can be used to assess network connectivity patterns in individual subjects. While specific graph metrics such as assortativity can be quantified in disease networks using single-subject rs-fMRI, the reliability of these measurements is a topic of ongoing investigation. If successful, the results will support the use of network structure in the assessment of new treatments for PD and related disorders.

APPLICATIONS OF NETWORK ANALYSIS

Genotypic Effect

PD is a clinically heterogeneous disorder with most patients experiencing sporadic disease, while a minority carry gene variants linked to the disorder. These genetic differences can influence disease progression and response to treatments (Davis et al. 2016; Alessi and Sammler 2018; Ben Romdhan et al. 2018), and the specific genetic changes make these patients candidates for clinical trials of neuroprotective agents.

Patients with idiopathic PD might carry mutations such as the G>A change at position 6055 of exon 41 of *LRRK2*, leading to a G2019S substitution, or variants in the gene encoding glucosylceramidase (*GBA1*) (Coukos and Krainc 2024). Imaging studies have highlighted the functional circuit abnormalities associated with these common mutations. For instance, patients with the *LRRK2*-G2019S mutation (PD-LRRK2) generally show slower disease progression and preserved cognitive function compared to those with sporadic PD (Alessi and Sammler 2018; Ben Romdhan et al. 2018). Conversely, patients with *GBA1* variants (PD-GBA) tend to have more aggressive disease progression and more rapid cognitive decline (Alcalay et al. 2015; Brockmann et al. 2015; Ortega et al. 2021). PDRP and PDCP network expression levels are higher in PD-GBA patients than in those with sporadic PD, even when motor symptoms are of similar duration and severity (Schindlbeck et al. 2020). This increased expression suggests a more aggressive pathological process in PD-GBA. On the other hand, PD-LRRK2 patients show lower levels of these patterns, suggesting a less aggressive disease course.

Graph theory also has provided insights into the organization of PDRP relating to these genotypic subtypes. Schindlbeck et al. (2020) showed that the PDRP comprised two distinct subnetworks: a core zone, composed of a set of metabolically active and tightly interconnected nodes located in the putamen, globus pallidus, and thalamus, and a peripheral zone, composed of less metabolically active and relatively loosely connected cortical nodes. In subsequent studies, we substantiated this core–periphery structure of the PDRP using conventional community detection methods (Fig. 2; Schindlbeck et al. 2020; Vo et al. 2023b).

Graph analysis revealed distinct connectivity patterns in these two PDRP modules, along with abnormal information, flow through the PDRP network in patients with PD (Schindlbeck et al. 2020). Genotypic differences are evident in network structure and information flow within PDRP and PDCP subnetworks. In PD-LRRK2 patients, functional connectivity increases within the PDRP core, whereas in PD-GBA patients, changes are seen more in the peripheral regions. These patterns align with the ascending progression of pathological changes. Additionally, graph analysis reveals marked genotypic differences, with PD-LRRK2 patients displaying a more integrated network response, likely reflecting an adaptive mechanism, and PD-GBA patients showing a more fragmented network, indicating a pathological response.

Network assortativity, a measure of the tendency of nodes to connect with similar nodes, also varies with genotype and disease progression. PD-GBA patients exhibit higher assortativity in PDRP/PDCP core zones compared to PD-LRRK2 patients, indicating less efficient information transfer and a more pathological connectivity pattern. In contrast, the reduced assortativity in PD-LRRK2 patients suggests a more efficient and integrated network, potentially contributing to their slower disease progression (Schindlbeck et al. 2020).

The presence of differing network connectivity patterns in two subgroups of patients with genotypes associated with either slowly progressing or rapidly progressing PD suggests that disease networks such as the PDRP exist in different configurations depending on the disease duration and the degree of functional adaptation that has taken place. These findings emphasize the importance of genotype in influencing disease mechanisms and highlight the need for further studies to confirm these observations in larger patient samples. Understanding these genotypic differences can guide the development of targeted therapeutic strategies and improve the prediction of disease progression in PD patients.

Differential Diagnosis

Network methods can significantly improve the differential diagnosis of parkinsonian movement disorders. Parkinsonism can be linked to various neurodegenerative pathologies that are clinically similar in the early stages of the disease. Accurate diagnosis is crucial for confirming idiopathic PD before surgical interventions such as deep brain stimulation (DBS), selecting appropriate participants for clinical trials, and for prognostic purposes.

In the past 12 years, multivariate analytical approaches have been used alongside disease networks and brain imaging scans to improve diagnostic accuracy in both research and clinical settings. For instance, a two-step logistic regression algorithm has been developed based on PDRP expression levels and disease patterns associated with multiple system atrophy (MSA) and progressive supranuclear palsy (PSP), the two most common atypical parkinsonian syndromes (APSs) (Tang et al. 2010b; Poston et al. 2012; Ge et al. 2018; Martí-Andrés et al. 2020; Shen et al. 2020; Tomše et al. 2022). The MSA-related pattern (MSARP) and PSP-related pattern (PSPRP) are characterized by distinct metabolic activity reductions and have been extensively validated in independent populations.

In a study involving 167 patients with uncertain initial clinical diagnoses (96 with PD, 41 with MSA, and 30 with PSP), accurate network-based categorization was achievable 3–4 years before a final diagnosis by an expert clinician masked to the imaging findings (Tang et al. 2010b). These results were replicated in an independent cohort from New Delhi, India, where the positive predictive value (PPV) for discriminating PD from APS was 97%, and 91% for differentiating MSA from PSP (Tripathi et al. 2016). Most of this cohort had a short symptom duration (<2 yr), yet PPV and diagnostic specificity remained high (≥95%) (Fig 3).

Preliminary data suggest a similar logistical categorization approach can distinguish individuals with corticobasal degeneration (CBD) from those with PSP (Niethammer et al. 2014; Nicastro et al. 2019). Since postmortem neuropathology is seldom available to confirm diagnosis, these

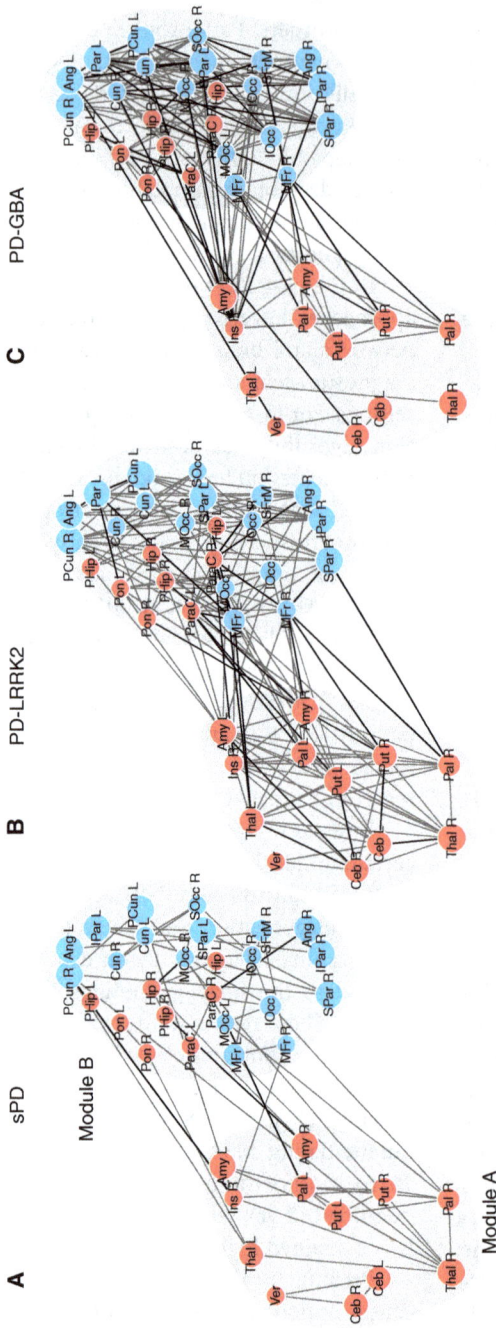

Figure 2. Parkinson's disease–related pattern (PDRP) community structure and connectional gain in sporadic PD (sPD), PD-LRRK2, and PD-GBA. Two major nodal communities were detected within the PDRP space: Module A was composed of 10 interconnected metabolically active nodes (red circles), while module B contained the remaining 10 metabolically active nodes that comprised the network core and the 18 relatively underactive nodes (blue circles) that comprised the network periphery. Significant functional connections (gray lines) linking nodes are mapped for sPD, PD-LRRK2, and PD-GBA (A–C). Bold lines denote connections gained with respect to healthy controls. In PD-LRRK2 (B), approximately half of the connections gained in the PDRP space linked metabolically active nodes within or between modules. In PD-GBA (C), the PDRP connections gained linked underactive module B nodes or linked active module A nodes with underactive module B nodes. (The radius of each circle [node] is proportional to the corresponding regional component of the leading eigenvector of the modularity matrix.) (Reprinted from Schindlbeck et al. 2020, with permission, from Oxford University Press.)

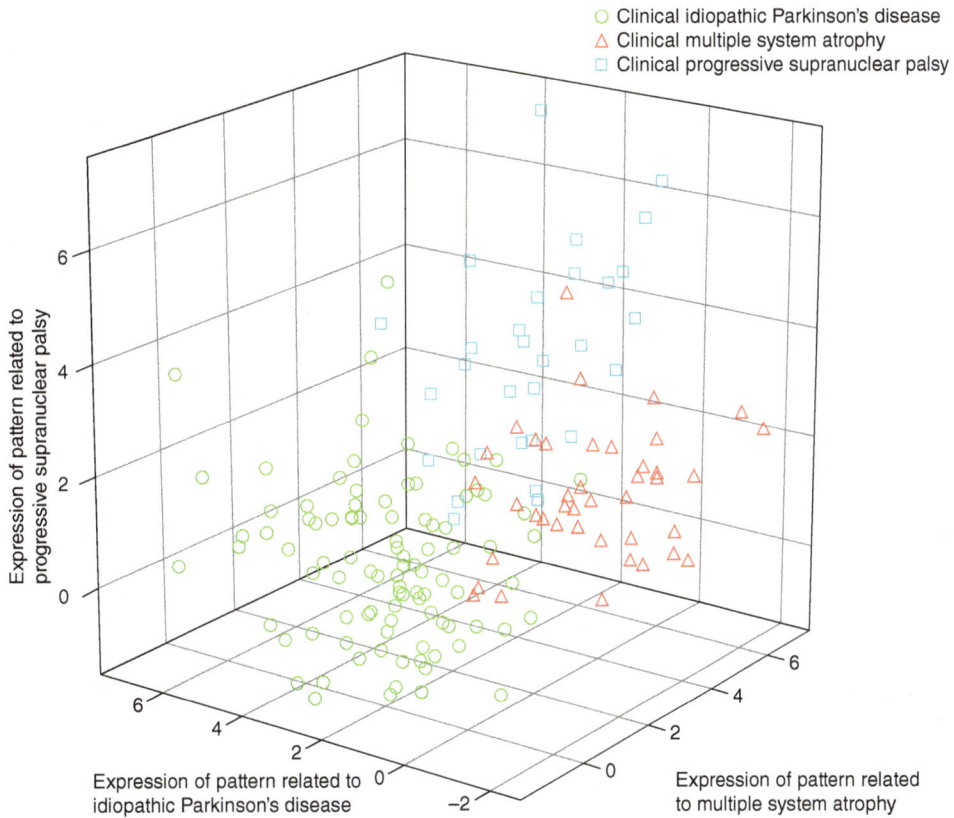

Figure 3. 3D plot of [^{18}F]-fluorodeoxyglucose positron emission tomography (FDG PET) pattern expression. Subject scores of the Parkinson's disease–related pattern (PDRP) (*x*-axis), multiple system atrophy (MSA)-related pattern (MSARP) (*y*-axis), and progressive supranuclear palsy (PSP)-related pattern (PSPRP) (*z*-axis) are shown for 167 patients, including 96 patients with a final clinical diagnosis of IPD (green circles), 41 patients diagnosed as MSA (red triangles), and 30 patients diagnosed as PSP (blue squares). Values were computed from the FDG PET scans of these patients with uncertain clinical diagnosis at the time of referral, which was 2.6 ± 2.4 (mean ± SD) years before final diagnostic ascertainment by a movement disorders specialist. (Reprinted from Tang et al. 2010b, with permission, from Elsevier © 2010.)

studies rely on final clinical diagnosis after at least 2 years of follow-up by a movement disorders specialist, which shows a high concordance with postmortem findings and 99% PPV.

Meta-analyses have found that both observer-dependent and observer-independent methods using metabolic imaging were highly accurate (>90%) in distinguishing PD from APS, provided the observers were highly experienced. Observer-independent approaches, which rely on automated image-based classification, are particularly useful when such expertise is scarce, and are more accurate than initial clinical diag-

nosis, when disease signs symptoms may be mild or nonspecific (Schindlbeck et al. 2021a; Papathoma et al. 2022). Given that clinical diagnosis of PD is inaccurate in as many as 20%–25% of cases, the need for observer-independent approaches is clear, and neuroimaging methods incorporating network algorithms show great promise for improving differential diagnosis in patients with parkinsonism (Tang et al. 2010b; Tripathi et al. 2016).

Although other observer-independent classification algorithms for PET images have been examined, their superiority over the logistic discrimination method is unclear. Several machine learning approaches, such as support vector machines, convolutional neural networks, and decision trees, have been applied to classify patients with parkinsonian syndromes. However, their accuracy has been modest, with specificity ranging from 53% to 76% and sensitivity from 70% to 80%. Improvement is anticipated as larger, well-structured imaging data sets become available for training and validation (Garraux et al. 2013; Mudali et al. 2015). Establishing a large open-source database would be extremely useful for rigorously comparing various classification algorithms.

The classification accuracy of these disease patterns is highly reproducible across populations and clinical phenotypes. A meta-analysis indicated that a two-step algorithm based on PDRP, MSARP, and PSPRP has a pooled sensitivity of 84% and specificity of 96%, supporting its potential use as a diagnostic tool (Papathoma et al. 2022). Both observer-dependent and observer-independent analytical strategies have been employed with metabolic imaging to differentiate PD from APS at the individual patient level. Observer-independent strategies, like automated pattern-based image classification, are especially valuable as they provide a numerical probability for each diagnostic alternative, which is beneficial for evaluating individual patients. It is noteworthy that while PDRP and PDCP are among the best-validated disease networks, less information is available regarding the stability of MSARP and PSPRP across different platforms. Although recent studies point to the replicability of the latter metabolic networks across patient populations and imaging platforms (Perovnik et al. 2023a; Buchert et al. 2024), differentiating PD from APS may also be possible using PDRP and PDCP expression values in combination (Rus et al. 2022). This may be particularly valuable for rs-fMRI, given the absence currently of validated MSARP and PSPRP topographies for this imaging modality.

Despite these advances, the differential diagnosis of parkinsonism remains challenging. While genetic and imaging tests are increasingly used, the diagnosis is still primarily clinical, with nonspecific motor findings leading to low accuracy in the early stages of PD and APS disorders. Dopaminergic imaging typically cannot reliably discriminate between PD and APS or differentiate among APS disorders. However, FDG PET-based network biomarkers can provide critical diagnostic information, showing high accuracy in distinguishing PD from APS when used by experienced observers. Observer-independent methods are particularly useful in regions where such expertise is scarce.

Early Diagnosis

The PDRP is characterized by network-level functional changes involving the brainstem, basal ganglia, and limbic regions, corresponding to Braak stage III–IV pathology (Braak et al. 2003; Braak and Del Tredici 2009). In contrast, the PDCP involves functional network changes in the ventral DMN and other neocortical regions, corresponding to Braak stage V pathology (Fig. 4). Stepwise increases in PDRP and PDCP expression levels occur with disease progression —and in the difference between these values (delta)—with PDRP topography appearing earlier than PDCP topography (Huang et al. 2007a, b; Tang et al. 2010a, 2020; Rus et al. 2022). Typically, in individual idiopathic PD patients, PDRP expression levels are greater than corresponding PDCP values, resulting in a positive delta (PDRP minus PDCP expression levels), which increases over time, consistent with the stereotypical caudorostral sequence of pathological involvement. Mean PDRP and PDCP expression levels, as well as mean delta values, are shown for prodromal (rapid eye movement [REM] sleep behavior disorder [RBD]), early (<4 yr), middle (4–8 yr), and late (>8 yr) stages of PD (Rus et al. 2022). This was true for PDRP and PDCP from PET and rs-fMRI studies. A similar relationship was not observed in APS.

Prodromal PD and Disease Progression

There is considerable interest in identifying markers for prodromal disease stages to enable earlier disease-modifying therapies (Iranzo et al. 2006, 2017). Approximately 50% of individuals with RBD phenoconvert to PD within 5 years.

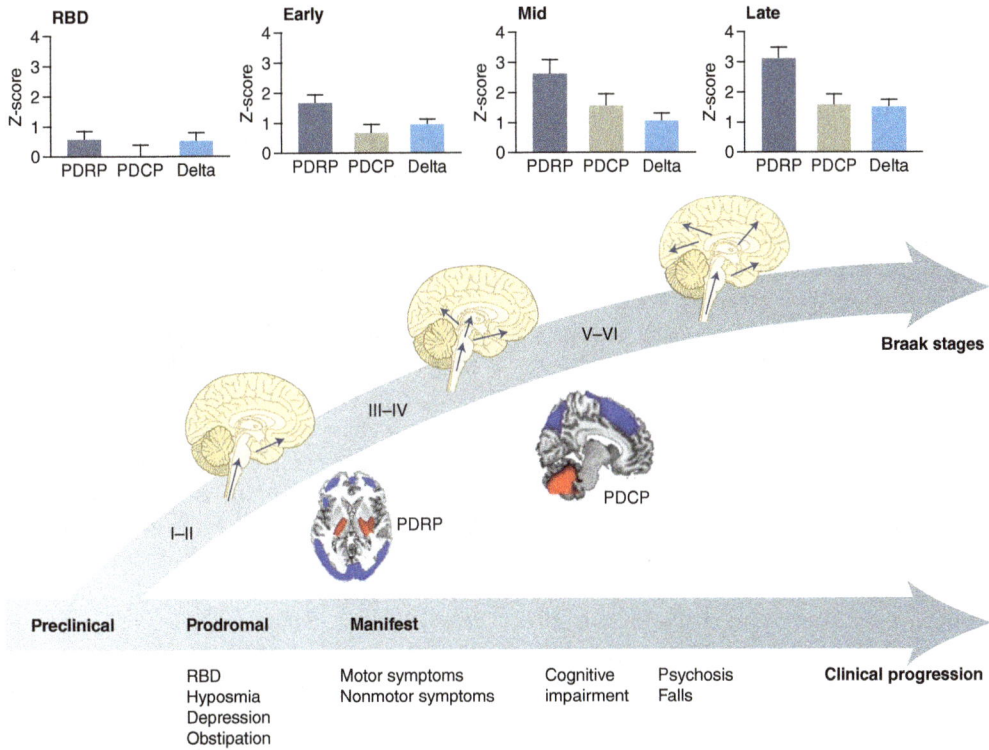

Figure 4. Functional brain network alterations associated with Parkinson's disease (PD). The PD-related pattern (PDRP) is characterized by network-level functional changes involving the brainstem, basal ganglia, and limbic regions, which are the counterpart of Braak stage III–IV pathology. The PD cognition-related pattern (PDCP), by contrast, is characterized by functional network changes involving the ventral default mode network and other neocortical regions associated with Braak stage V pathology. The graphs showing the stepwise increases in PDRP and PDCP expression levels are seen with disease progression, with PDRP topography appearing earlier than PDCP topography. Interestingly, in individual patients with idiopathic PD, PDRP expression levels are typically greater than the corresponding PDCP values, which results in a positive delta (defined as the difference between PDRP and PDCP expression levels). Delta also increases over time, which is consistent with the stereotypical caudorostral sequence of pathological involvement observed in this disorder. Mean PDRP and PDCP expression levels and mean delta values are shown for the prodromal (rapid eye movement [REM] sleep behavior disorder [RBD]), early (<4 yr), middle (4–8 yr), and late (>8 yr) stages of PD in a cross sectional sample of 16 patients with RBD and 96 patients with PD. (Reprinted, with permission, from Perovnik et al. 2023a.)

Indeed, idiopathic RBD is a prodromal syndrome with an increased risk of evolving into clinical α-synucleinopathy-related syndromes, including idiopathic PD, dementia with Lewy bodies (DLB), and MSA (Berg et al. 2015; Fereshtehnejad et al. 2017). Elevated PDRP expression levels in individuals with idiopathic RBD have been observed, falling between those of healthy individuals and patients with early-stage idiopathic PD (Holtbernd et al. 2014; Meles et al. 2017, 2018). Progressive increases in PDRP and

PDCP expression do occur in idiopathic RBD, concurrent with declining caudate and putamen dopamine transporter (DAT) binding (Holtbernd et al. 2014; Huang et al. 2020). PDRP expression is elevated at baseline in idiopathic RBD, while PDCP values reach abnormal levels only at the final imaging time point. Follow-up studies suggest that individuals with idiopathic RBD who subsequently develop PD or DLB have elevated PDRP expression levels at baseline. Limited data suggest individuals with idiopathic RBD and subnormal PDRP expression are more likely to develop MSA than idiopathic PD or DLB, although this requires further confirmation.

A metabolic covariance pattern termed PDRBD-RP, identified in patients with new-onset PD who had premorbid RBD, may better predict phenotypic conversion to PD than the standard PDRP. PDRBD-RP and PDRP do show some overlap in topography but differ mainly in the premotor cortex, which contributes more to PDRBD-RP topography (Meles et al. 2018). Large longitudinal studies are needed to assess the value of these metabolic covariance patterns as predictors of phenotypic conversion. Multiple studies have demonstrated linear progression of PDRP expression with advancing disease. Applying SSM–PCA to FDG PET scans of 30 idiopathic RBD patients, who phenoconverted to an overt α-synucleinopathy (14 PD and 16 DLB), and 46 nonconverters, Mattioli et al. derived a metabolic pattern (iRBDconvRP), which significantly predicted phenoconversion from iRBD to PD or DLB over time (Mattioli et al. 2023; Orso et al. 2023).

In a recent longitudinal study of 13 preclinical iRBD patients who were scanned with FDG PET at baseline, 24 months, and 48 months and followed clinically for an average of 6.5 years after the final imaging time point (Eidelberg et al. 2024), PDRP expression increased linearly over time with abnormally elevated levels at all three time points. PDCP expression reached significantly elevated levels only at the final imaging time point. These changes were accompanied by extensive remodeling of functional connections in both PD networks and by the development of abnormal connections linking their respective nodes. Graph analysis further revealed early reductions in small-worldness and subsequent increases in assortativity for both PDRP and PDCP. These findings suggest that in iRBD, progressive network dysfunction begins at least 5 years before phenoconversion. Moreover, long-term clinical follow-up data supported the hypothesis of an inverse relationship between PDRP expression in iRBD patients at the time from imaging to phenoconversion (Holtbernd et al. 2014; Eidelberg et al. 2024).

Treatment-Induced Networks and the Placebo Response

Symptomatic treatment of PD is associated with reductions in PDRP expression that correlate with clinical outcome. This has been noted in response to acute levodopa administration, STN DBS, and therapeutic subthalamotomy (Hirano et al. 2008; Eidelberg 2009; Pourfar et al. 2009). Given that PDRP expression levels correlated with spontaneous STN firing rates recorded intraoperatively in patients with PD undergoing DBS surgery, a decline in PDRP activity can be expected following this procedure. Levodopa treatment has complex effects on PD network expression, as dopamine affects the activity of basal ganglia–cortical circuitry at multiple levels (Mattis et al. 2011; Mure et al. 2012; Ko et al. 2013; Niethammer et al. 2013; Jourdain et al. 2016). As noted above, antiparkinsonian interventions can induce novel treatment-specific networks that can be used as potential efficacy biomarkers. In addition to demonstrating target engagement and elucidating mechanisms of action for novel treatments, these networks can be used for dose finding and as outcome measures in blinded clinical trials (Barbero et al. 2023). The characterization of treatment-induced networks involves a within-subject supervised PCA algorithm, termed ordinal trends canonical variates analysis (OrT/CVA) (Habeck et al. 2005; Habeck and Stern 2010), that has been described in detail elsewhere (Mure et al. 2011, 2012; Ko et al. 2013, 2014). This approach has proved useful in the study of novel therapies for PD such as subthalamic delivery of AAV2-GAD, which was developed as an alternative to STN DBS or surgical lesioning to normalize network activity and relieve PD symptoms (LeWitt et al. 2011). In a phase 2 clinical trial, the potential efficacy of this intervention was evaluated in a double-blind, sham surgery-controlled multicenter design. The AAV2-GADRP was characterized by covarying treatment-induced reductions involving the caudate nucleus, anterior putamen, globus pallidus, and the ventral anterior and mediodorsal thalamic nuclei, with relative increases in the motor and premotor cortex and the supramarginal gyrus (Niethammer et al. 2017). Expression levels for this pattern increased consistently following gene therapy to a far greater degree than observed following sham surgery or during disease progression. In contrast, PDRP expression increased longitudinally at similar rates following STN AAV2-GAD and sham surgery, suggesting the absence of a significant disease modification effect from the gene therapy. Clinical improvement in the STN AAV2-GAD subjects correlated with increases in GADRP but not with changes in PDRP expression levels. As described above, graph analysis of the longitudinal PET data revealed consistent improvement in GADRP and PDRP network metrics, indicating improved efficiency of information flow through both pathways with gene therapy compared to sham surgery (Vo et al. 2023b). Interestingly, retrospective analysis of earlier phase 1 safety data from PD patients undergoing unilateral STN AAV2-GAD gene therapy (Feigin et al. 2007) revealed stepwise increases in GADRP expression at higher viral vector concentrations (Barbero et al. 2023). This suggests the potential use of network quantification for dose finding. An analogous FDG PET study was performed to identify a treatment network induced by nicotinamide riboside (NR), an oral supplement that augments mitochondrial energetics in PD patients (Brakedal et al. 2022). The resulting NR-related metabolic pattern (NRRP) overlapped topographically with PDRP, and changes in expression levels with 28 days of daily NR correlated with clinical improvement. Significant NRRP changes were not seen with oral placebo, without correlates with clinical outcome under the blind.

Treatment-induced networks have also been identified in response to DBS for PD manifestations. Applying OrT/CVA to FDG PET scan pairs acquired at baseline and during therapeutic ventralis intermediate nucleus (Vim) thalamic DBS disclosed a significant network associated with PD tremor, termed the PD-related tremor pattern (PDTP) (Mure et al. 2011). This topography was characterized by increased metabolic activity in the cerebellar vermis, dentate nucleus, primary motor cortex, and posterior putamen. Subject scores for PDTP, but not PDRP, correlated with accelerometer-measured tremor amplitude recorded in the baseline, off-Vim stimulation condition. STN DBS in an independent group of PD patients with mixed motor symptoms modulated both PDTP and PDRP expression, whereas the therapeutic effect of Vim DBS involved mainly PDTP suppression. Of note, a similar network was identified in response to focused ultrasound of the Vim thalamic nucleus for essential tremor using rs-fMRI and OrT/CVA (Xiong et al. 2022).

In a recent study, OrT/CVA was used to identify and validate a network biomarker of the STN DBS treatment response in PD patients, termed STN StimNet (Unadkat et al. 2024). This network was treatment-specific in that it demonstrated greater modulation with STN DBS compared to other antiparkinsonian interventions. Stimulation-mediated changes in network expression correlated with concurrent motor benefit, and preoperative network expression levels predicted stimulation-mediated motor outcomes. The STN StimNet includes regions modulated by stimulation that are not part of the generic PDRP and PDCP network topographies (see above). Notably, the metabolic network induced by STN stimulation includes the subthalamic target region, its projection to the pontine nuclei, as well as the SMA complex —areas in which local metabolic activity increased with stimulation. Interestingly, the clinical response to STN stimulation correlated with individual differences in STN StimNet expression determined preoperatively using either FDG PET or rs-fMRI. While confirmed at two independent sites, additional research is needed to determine the ultimate role of STN StimNet and analogous networks in selecting patients for DBS surgery.

Another promising application of network analysis in PD is the study of place effects. Placebo effects pose a major challenge in the development of new treatments for PD. Prominent placebo (or sham surgical) responses are frequently encountered in trials of new PD treatments (Goetz et al. 2008; Katsnelson 2011; Quattrone et al. 2018). Expectations of receiving effective treatment lead to striatal dopamine release and changes in fMRI and single cell recordings (Lou 2020). Indeed, an "expensive" placebo produces a stronger response than a "cheap" placebo in patients with PD, deactivating regions such as the left putamen, sensorimotor cortices, and premotor cortex. "Cheap" placebo, by contrast, activates the bilateral anterior and posterior cingulate cortices, left lateral sensorimotor cortex, and right parietal cortex, among other regions (Espay et al. 2015). Ko and colleagues identified and validated a novel metabolic brain network associated with the placebo response in patients with PD who participated in the double-blind sham surgery-controlled phase 2 trial of STN AAV2-GAD mentioned above (LeWitt et al. 2011; Ko et al. 2014). The metabolic topography of the sham surgery-related pattern (SSRP) involves anatomical–functional pathways linking the posterior cerebellar vermis to the limbic cortex via the ventral anterior thalamus, amygdala, and caudate nucleus. Baseline SSRP expression, measured before randomization, correlates with the motor sham response observed under blinded trial conditions. The network changes after sham treatment did not appear with experimental STN gene therapy or levodopa treatment, and were reversed by unblinding. While the clinical impact of the oral placebo employed in the blinded NR study described above (Brakedal et al. 2022) was weaker than for the sham surgery group in the gene therapy trial, baseline SSRP expression predicted the clinical response to placebo under the blind in both randomized trials. Whether SSRP represents a stable predictor of the placebo response in PD clinical trials is a topic for future study.

CONCLUSION

Over the past decade, functional imaging has shown increasing promise to aid in the diagnosis of parkinsonian disorders. Disease-specific networks can be used to assess disease progression and treatment response, especially in the context of clinical trials, where accurate diagnosis and objective outcome measures are critical. Insights into the effects of specific genotypic variants on these networks may aid in the development of more specific treatments as well.

REFERENCES

Aarsland D, Kurz MW. 2010. The epidemiology of dementia associated with Parkinson disease. *J Neurol Sci* **289**: 18–22. doi:10.1016/j.jns.2009.08.034

Aarsland D, Creese B, Politis M, Chaudhuri KR, Ffytche DH, Weintraub D, Ballard C. 2017. Cognitive decline in Parkinson disease. *Nat Rev Neurol* **13**: 217–231. doi:10.1038/nrneurol.2017.27

Alcalay RN, Levy OA, Waters CC, Fahn S, Ford B, Kuo SH, Mazzoni P, Pauciulo MW, Nichols WC, Gan-Or Z, et al. 2015. Glucocerebrosidase activity in Parkinson's disease with and without *GBA* mutations. *Brain* **138**: 2648–2658. doi:10.1093/brain/awv179

Alessi DR, Sammler E. 2018. LRRK2 kinase in Parkinson's disease. *Science* **360**: 36–37. doi:10.1126/science.aar5683

Barabasi AL. 2016. *Network science.* Cambridge University Press, Cambridge.

Barbero JA, Unadkat P, Choi YY, Eidelberg D. 2023. Functional brain networks to evaluate treatment responses in Parkinson's disease. *Neurotherapeutics* **20**: 1653–1668. doi:10.1007/s13311-023-01433-w

Ben Romdhan S, Farhat N, Nasri A, Lesage S, Hdiji O, Ben Djebara M, Landoulsi Z, Stevanin G, Brice A, Damak M, et al. 2018. LRRK2 g2019s Parkinson's disease with more benign phenotype than idiopathic. *Acta Neurol Scand* **138**: 425–431. doi:10.1111/ane.12996

Berg D, Postuma RB, Adler CH, Bloem BR, Chan P, Dubois B, Gasser T, Goetz CG, Halliday G, Joseph L, et al. 2015. MDS research criteria for prodromal Parkinson's disease. *Mov Disord* **30**: 1600–1611. doi:10.1002/mds.26431

Betzel RF, Satterthwaite TD, Gold JI, Bassett DS. 2017. Positive affect, surprise, and fatigue are correlates of network flexibility. *Sci Rep* **7**: 520. doi:10.1038/s41598-017-00425-z

Braak H, Del Tredici K. 2009. *Neuroanatomy and pathology of sporadic Parkinson's disease.* Springer, Heidelberg, Germany.

Braak H, Del Tredici K, Rüb U, De Vos RAI, Jansen Steur ENH, Braak E. 2003. Staging of brain pathology related to sporadic Parkinson's disease. *Neurobiol Aging* **24**: 197–211. doi:10.1016/S0197-4580(02)00065-9

Brakedal B, Dölle C, Riemer F, Ma Y, Nido GS, Skeie GO, Craven AR, Schwarzlmüller T, Brekke N, Diab J, et al. 2022. The NADPARK study: a randomized phase I trial of nicotinamide riboside supplementation in Parkinson's disease. *Cell Metab* **34**: 396–407.e6. doi:10.1016/j.cmet.2022.02.001

Brockmann K, Srulijes K, Pflederer S, Hauser AK, Schulte C, Maetzler W, Gasser T, Berg D. 2015. *GBA*-associated Parkinson's disease: reduced survival and more rapid progression in a prospective longitudinal study. *Mov Disord* **30**: 407–411. doi:10.1002/mds.26071

Buchert R, Huppertz HJ, Wegner F, Berding G, Brendel M, Apostolova I, Buhmann C, Poetter-Nerger M, Dierks A, Katzdobler S, et al. 2024. Added value of FDG-PET for detection of progressive supranuclear palsy. *J Neurol Neurosurg Psychiatry* doi:10.1136/jnnp-2024-333590

Coukos R, Krainc D. 2024. Key genes and convergent pathogenic mechanisms in Parkinson disease. *Nat Rev Neurosci* **25**: 393–413. doi:10.1038/s41583-024-00812-2

Davis MY, Johnson CO, Leverenz JB, Weintraub D, Troja-nowski JQ, Chen-Plotkin A, Van Deerlin VM, Quinn JF, Chung KA, Peterson-Hiller AL, et al. 2016. Association of *GBA* mutations and the E326K polymorphism with motor and cognitive progression in Parkinson disease. *JAMA Neurol* **73:** 1217–1224. doi:10.1001/jamaneurol.2016.2245

Eckert T, Van Laere K, Tang C, Lewis DE, Edwards C, Santens P, Eidelberg D. 2007. Quantification of Parkinson's disease-related network expression with ECD SPECT. *Eur J Nucl Med Mol Imaging* **34:** 496–501. doi:10.1007/s00259-006-0261-9

Eidelberg D. 2009. Metabolic brain networks in neurodegenerative disorders: a functional imaging approach. *Trends Neurosci* **32:** 548–557. doi:10.1016/j.tins.2009.06.003

Eidelberg D, Tang C, Nakano Y, Vo A, Nguyen N, Schindlbeck K, Poston K, Gagnon JF, Postuma R, Niethammer M, et al. 2024. Longitudinal network changes and phenoconversion risk in isolated REM sleep behavior disorder. *Res Sq* doi:10.21203/rs.3.rs-4427198/v1

Espay AJ, Norris MM, Eliassen JC, Dwivedi A, Smith MS, Banks C, Allendorfer JB, Lang AE, Fleck DE, Linke MJ, et al. 2015. Placebo effect of medication cost in Parkinson disease: a randomized double-blind study. *Neurology* **84:** 794–802. doi:10.1212/WNL.0000000000001282

Espay AJ, Brundin P, Lang AE. 2017. Precision medicine for disease modification in Parkinson disease. *Nat Rev Neurol* **13:** 119–126. doi:10.1038/nrneurol.2016.196

Feigin A, Kaplitt MG, Tang C, Lin T, Mattis P, Dhawan V, During MJ, Eidelberg D. 2007. Modulation of metabolic brain networks after subthalamic gene therapy for Parkinson's disease. *Proc Natl Acad Sci* **104:** 19559–19564. doi:10.1073/pnas.0706006104

Fereshtehnejad SM, Montplaisir JY, Pelletier A, Gagnon JF, Berg D, Postuma RB. 2017. Validation of the MDS research criteria for prodromal Parkinson's disease: longitudinal assessment in a REM sleep behavior disorder (RBD) cohort. *Mov Disord* **32:** 865–873. doi:10.1002/mds.26989

Garraux G, Phillips C, Schrouff J, Kreisler A, Lemaire C, Degueldre C, Delcour C, Hustinx R, Luxen A, Destée A, et al. 2013. Multiclass classification of FDG PET scans for the distinction between Parkinson's disease and atypical parkinsonian syndromes. *NeuroImage Clin* **2:** 883–893. doi:10.1016/j.nicl.2013.06.004

GBD 2019 Diseases and Injuries Collaborators. 2020. Global burden of 369 diseases and injuries in 204 countries and territories, 1990–2019: a systematic analysis for the Global Burden of Disease Study 2019. *Lancet* **396:** 1204–1222. doi:10.1016/S0140-6736(20)30925-9

Ge J, Wu J, Peng S, Wu P, Wang J, Zhang H, Guan Y, Eidelberg D, Zuo C, Ma Y. 2018. Reproducible network and regional topographies of abnormal glucose metabolism associated with progressive supranuclear palsy: multivariate and univariate analyses in American and Chinese patient cohorts. *Hum Brain Mapp* **39:** 2842–2858. doi:10.1002/hbm.24044

Goetz CG, Wuu J, McDermott MP, Adler CH, Fahn S, Freed CR, Hauser RA, Olanow WC, Shoulson I, Tandon PK, et al. 2008. Placebo response in Parkinson's disease: comparisons among 11 trials covering medical and surgical interventions. *Mov Disord* **23:** 690–699. doi:10.1002/mds.21894

Gorban AN, Smirnova EV, Tyukina TA. 2010. Correlations, risk and crisis: from physiology to finance. *Phys A Stat Mech Appl* **389:** 3193–3217. doi:10.1016/j.physa.2010.03.035

Habeck C, Stern Y. 2010. Multivariate data analysis for neuroimaging data: overview and application to Alzheimer's disease. *Cell Biochem Biophys* **58:** 53–67. doi:10.1007/s12013-010-9093-0

Habeck C, Krakauer JW, Ghez C, Sackeim HA, Eidelberg D, Stern Y, Moeller JR. 2005. A new approach to spatial covariance modeling of functional brain imaging data: ordinal trend analysis. *Neural Comput* **17:** 1602–1645. doi:10.1162/0899766053723023

Hirano S, Asanuma K, Ma Y, Tang C, Feigin A, Dhawan V, Carbon M, Eidelberg D. 2008. Dissociation of metabolic and neurovascular responses to levodopa in the treatment of Parkinson's disease. *J Neurosci* **28:** 4201–4209. doi:10.1523/JNEUROSCI.0582-08.2008

Holtbernd F, Gagnon JF, Postuma RB, Ma Y, Tang CC, Feigin A, Dhawan V, Vendette M, Soucy JP, Eidelberg D, et al. 2014. Abnormal metabolic network activity in REM sleep behavior disorder. *Neurology* **82:** 620–627. doi:10.1212/WNL.0000000000000130

Hou Y, Feng F, Zhang L, Ou R, Lin J, Gong Q, Shang H. 2023. Disrupted topological organization of resting-state functional brain networks in Parkinson's disease patients with glucocerebrosidase gene mutations. *Neuroradiology* **65:** 361–370. doi:10.1007/s00234-022-03067-9

Huang C, Mattis P, Tang C, Perrine K, Carbon M, Eidelberg D. 2007a. Metabolic brain networks associated with cognitive function in Parkinson's disease. *Neuroimage* **34:** 714–723. doi:10.1016/j.neuroimage.2006.09.003

Huang C, Tang C, Feigin A, Lesser M, Ma Y, Pourfar M, Dhawan V, Eidelberg D. 2007b. Changes in network activity with the progression of Parkinson's disease. *Brain* **130:** 1834–1846. doi:10.1093/brain/awm086

Huang C, Mattis P, Perrine K, Brown N, Dhawan V, Eidelberg D. 2008. Metabolic abnormalities associated with mild cognitive impairment in Parkinson disease. *Neurology* **70:** 1470–1477. doi:10.1212/01.wnl.0000304050.05332.9c

Huang Z, Jiang C, Li L, Xu Q, Ge J, Li M, Guan Y, Wu J, Wang J, Zuo C, et al. 2020. Correlations between dopaminergic dysfunction and abnormal metabolic network activity in REM sleep behavior disorder. *J Cereb Blood Flow Metab* **40:** 552–562. doi:10.1177/0271678X19828916

Iranzo A, Molinuevo JL, Santamaría J, Serradell M, Martí MJ, Valldeoriola F, Tolosa E. 2006. Rapid-eye-movement sleep behaviour disorder as an early marker for a neurodegenerative disorder: a descriptive study. *Lancet Neurol* **5:** 572–577. doi:10.1016/S1474-4422(06)70476-8

Iranzo A, Stefani A, Serradell M, Martí MJ, Lomeña F, Mahlknecht P, Stockner H, Gaig C, Fernández-Arcos A, Poewe W, et al. 2017. Characterization of patients with longstanding idiopathic REM sleep behavior disorder. *Neurology* **89:** 242–248. doi:10.1212/WNL.0000000000004121

Jourdain VA, Tang CC, Holtbernd F, Dresel C, Choi YY, Ma Y, Dhawan V, Eidelberg D. 2016. Flow-metabolism dissociation in the pathogenesis of levodopa-induced dyskinesia. *JCI Insight* **1**: e86615. doi:10.1172/jci.insight.86615

Katsnelson A. 2011. Experimental therapies for Parkinson's disease: why fake it? *Nature* **476**: 142–144. doi:10.1038/476142a

Ko JH, Mure H, Tang CC, Ma Y, Dhawan V, Spetsieris P, Eidelberg D. 2013. Parkinson's disease: increased motor network activity in the absence of movement. *J Neurosci* **33**: 4540–4549. doi:10.1523/JNEUROSCI.5024-12.2013

Ko JH, Feigin A, Mattis PJ, Tang CC, Ma Y, Dhawan V, During MJ, Kaplitt MG, Eidelberg D. 2014. Network modulation following sham surgery in Parkinson's disease. *J Clin Invest* **124**: 3656–3666. doi:10.1172/JCI75073

Ko JH, Lee CS, Eidelberg D. 2017. Metabolic network expression in parkinsonism: clinical and dopaminergic correlations. *J Cereb Blood Flow Metab* **37**: 683–693. doi:10.1177/0271678X16637880

LeWitt PA, Rezai AR, Leehey MA, Ojemann SG, Flaherty AW, Eskandar EN, Kostyk SK, Thomas K, Sarkar A, Siddiqui MS, et al. 2011. AAV2-GAD gene therapy for advanced Parkinson's disease: a double-blind, sham-surgery controlled, randomised trial. *Lancet Neurol* **10**: 309–319. doi:10.1016/S1474-4422(11)70039-4

Lou JS. 2020. Placebo responses in Parkinson's disease. *Int Rev Neurobiol* **153**: 187–211. doi:10.1016/bs.irn.2020.03.031

Ma Y, Eidelberg D. 2007. Functional imaging of cerebral blood flow and glucose metabolism in Parkinson's disease and Huntington's disease. *Mol Imaging Biol* **9**: 223–233. doi:10.1007/s11307-007-0085-4

Ma Y, Tang C, Spetsieris PG, Dhawan V, Eidelberg D. 2007. Abnormal metabolic network activity in Parkinson's disease: test–retest reproducibility. *J Cereb Blood Flow Metab* **27**: 597–605. doi:10.1038/sj.jcbfm.9600358

Ma Y, Huang C, Dyke JP, Pan H, Alsop D, Feigin A, Eidelberg D. 2010. Parkinson's disease spatial covariance pattern: noninvasive quantification with perfusion MRI. *J Cereb Blood Flow Metab* **30**: 505–509. doi:10.1038/jcbfm.2009.256

Martí-Andrés G, van Bommel L, Meles SK, Riverol M, Valentí R, Kogan RV, Renken RJ, Gurvits V, van Laar T, Pagani M, et al. 2020. Multicenter validation of metabolic abnormalities related to PSP according to the MDS-PSP criteria. *Mov Disord* **35**: 2009–2018. doi:10.1002/mds.28217

Mattioli P, Orso B, Liguori C, Famà F, Giorgetti L, Donniaquio A, Massa F, Giberti A, Vállez García D, Meles SK, et al. 2023. Derivation and validation of a phenoconversion-related pattern in idiopathic rapid eye movement behavior disorder. *Mov Disord* **38**: 57–67. doi:10.1002/mds.29236

Mattis PJ, Tang CC, Ma Y, Dhawan V, Eidelberg D. 2011. Network correlates of the cognitive response to levodopa in Parkinson disease. *Neurology* **77**: 858–865. doi:10.1212/WNL.0b013e31822c6224

Mattis PJ, Niethammer M, Sako W, Tang CC, Nazem A, Gordon ML, Brandt V, Dhawan V, Eidelberg D. 2016. Distinct brain networks underlie cognitive dysfunction in Parkinson and Alzheimer diseases. *Neurology* **87**: 1925–1933. doi:10.1212/WNL.0000000000003285

Meles SK, Tang CC, Teune LK, Dierckx RA, Dhawan V, Mattis PJ, Leenders KL, Eidelberg D. 2015. Abnormal metabolic pattern associated with cognitive impairment in Parkinson's disease: a validation study. *J Cereb Blood Flow Metab* **35**: 1478–1484. doi:10.1038/jcbfm.2015.112

Meles SK, Vadasz D, Renken RJ, Sittig-Wiegand E, Mayer G, Depboylu C, Reetz K, Overeem S, Pijpers A, Reesink FE, et al. 2017. FDG PET, dopamine transporter SPECT, and olfaction: combining biomarkers in REM sleep behavior disorder. *Mov Disord* **32**: 1482–1486. doi:10.1002/mds.27094

Meles SK, Renken RJ, Janzen A, Vadasz D, Pagani M, Arnaldi D, Morbelli S, Nobili F, Mayer G, Leenders KL, et al. 2018. The metabolic pattern of idiopathic REM sleep behavior disorder reflects early-stage Parkinson disease. *J Nucl Med* **59**: 1437–1444. doi:10.2967/jnumed.117.202242

Meles SK, Renken RJ, Pagani M, Teune LK, Arnaldi D, Morbelli S, Nobili F, van Laar T, Obeso JA, Rodríguez-Oroz MC, et al. 2020. Abnormal pattern of brain glucose metabolism in Parkinson's disease: replication in three European cohorts. *Eur J Nucl Med Mol Imaging* **47**: 437–450. doi:10.1007/s00259-019-04570-7

Mudali D, Teune LK, Renken RJ, Leenders KL, Roerdink JBTM. 2015. Classification of parkinsonian syndromes from FDG-PET brain data using decision trees with SSM/PCA features. *Comput Math Methods Med* **2015**: 136921. doi:10.1155/2015/136921

Mure H, Hirano S, Tang CC, Isaias IU, Antonini A, Ma Y, Dhawan V, Eidelberg D. 2011. Parkinson's disease tremor-related metabolic network: characterization, progression, and treatment effects. *Neuroimage* **54**: 1244–1253. doi:10.1016/j.neuroimage.2010.09.028

Mure H, Tang CC, Argyelan M, Ghilardi MF, Kaplitt MG, Dhawan V, Eidelberg D. 2012. Improved sequence learning with subthalamic nucleus deep brain stimulation: evidence for treatment-specific network modulation. *J Neurosci* **32**: 2804–2813. doi:10.1523/JNEUROSCI.4331-11.2012

Newman MEJ. 2003. Mixing patterns in networks. *Phys Rev E Top* **67**: 026126. doi:10.1103/PhysRevE.67.026126

Newman MEJ. 2010. *Networks: an introduction.* Oxford University Press, Oxford.

Nicastro N, Wegrzyk J, Preti MG, Fleury V, Van de Ville D, Garibotto V, Burkhard PR. 2019. Classification of degenerative parkinsonism subtypes by support-vector-machine analysis and striatal 123I-FP-CIT indices. *J Neurol* **266**: 1771–1781. doi:10.1007/s00415-019-09330-z

Niethammer M, Eidelberg D. 2012. Metabolic brain networks in translational neurology: concepts and applications. *Ann Neurol* **72**: 635–647. doi:10.1002/ana.23631

Niethammer M, Feigin A, Eidelberg D. 2012. Functional neuroimaging in Parkinson's disease. *Cold Spring Harb Perspect Med* **2**: a009274. doi:10.1101/cshperspect.a009274

Niethammer M, Tang CC, Ma Y, Mattis PJ, Ko JH, Dhawan V, Eidelberg D. 2013. Parkinson's disease cognitive network correlates with caudate dopamine. *Neuroimage* **78**: 204–209. doi:10.1016/j.neuroimage.2013.03.070

Niethammer M, Tang CC, Feigin A, Allen PJ, Heinen L, Hellwig S, Amtage F, Hanspal E, Vonsattel JP, Poston KL, et al. 2014. A disease-specific metabolic brain network associated with corticobasal degeneration. *Brain* **137**: 3036–3046. doi:10.1093/brain/awu256

Niethammer M, Tang CC, LeWitt PA, Rezai AR, Leehey MA, Ojemann SG, Flaherty AW, Eskandar EN, Kostyk SK, Sarkar A, et al. 2017. Long-term follow-up of a randomized AAV2-GAD gene therapy trial for Parkinson's disease. *JCI Insight* **2:** e90133. doi:10.1172/jci.insight.90133

Niethammer M, Tang CC, Vo A, Nguyen N, Spetsieris P, Dhawan V, Ma Y, Small M, Feigin A, During MJ, et al. 2018. Gene therapy reduces Parkinson's disease symptoms by reorganizing functional brain connectivity. *Sci Transl Med* **10:** eaau0713. doi:10.1126/scitranslmed.aau0713

Niethammer M, Tang CC, Jamora RDG, Vo A, Nguyen N, Ma Y, Peng S, Waugh JL, Westenberger A, Eidelberg D. 2023. A network imaging biomarker of X-linked dystonia-parkinsonism. *Ann Neurol* **94:** 684–695. doi:10.1002/ana.26732

Noldus R, Van Mieghem P. 2015. Assortativity in complex networks. *J Complex Netw* **3:** 507–542. doi:10.1093/comnet/cnv005

Oertel WH, Janzen A, Henrich MT, Geibl FF, Sittig E, Meles SK, Carli G, Leenders KL, Booij J, Surmeier DJ, et al. 2024. Acetyl-DL-leucine in two individuals with REM sleep behavior disorder improves symptoms, reverses loss of striatal dopamine-transporter binding and stabilizes pathological metabolic brain pattern—case reports. *Nat Commun* **15:** 7619. doi:10.1038/s41467-024-51502-7

Orso B, Mattioli P, Yoon EJ, Kim YK, Kim H, Shin JH, Kim R, Liguori C, Famà F, Donniaquio A, et al. 2023. Validation of the REM behaviour disorder phenoconversion-related pattern in an independent cohort. *Neurol Sci* **44:** 3161–3168. doi:10.1007/s10072-023-06829-2

Ortega RA, Wang C, Raymond D, Bryant N, Scherzer CR, Thaler A, Alcalay RN, West AB, Mirelman A, Kuras Y, et al. 2021. Association of dual *LRRK2* G2019S and *GBA* variations with Parkinson disease progression. *JAMA Netw Open* **4:** e215845. doi:10.1001/jamanetworkopen.2021.5845

Papathoma PE, Markaki I, Tang C, Lilja Lindström M, Savitcheva I, Eidelberg D, Svenningsson P. 2022. A replication study, systematic review and meta-analysis of automated image-based diagnosis in parkinsonism. *Sci Rep* **12:** 2763. doi:10.1038/s41598-022-06663-0

Perovnik M, Rus T, Schindlbeck KA, Eidelberg D. 2023a. Functional brain networks in the evaluation of patients with neurodegenerative disorders. *Nat Rev Neurol* **19:** 73–90. doi:10.1038/s41582-022-00753-3

Perovnik M, Tang CC, Namías M, Eidelberg D; Alzheimer Disease Neuroimaging Initiative. 2023b. Longitudinal changes in metabolic network activity in early Alzheimer's disease. *Alzheimers Dement* **19:** 4061–4072. doi:10.1002/alz.13137

Poston KL, Tang CC, Eckert T, Dhawan V, Frucht S, Vonsattel JP, Fahn S, Eidelberg D. 2012. Network correlates of disease severity in multiple system atrophy. *Neurology* **78:** 1237–1244. doi:10.1212/WNL.0b013e318250d7fd

Postuma RB, Berg D, Stern M, Poewe W, Olanow CW, Oertel W, Obeso J, Marek K, Litvan I, Lang AE, et al. 2015. MDS clinical diagnostic criteria for Parkinson's disease. *Mov Disord* **30:** 1591–1601. doi:10.1002/mds.26424

Pourfar M, Tang C, Lin T, Dhawan V, Kaplitt MG, Eidelberg DD. 2009. Assessing the microlesion effect of subthalamic deep brain stimulation surgery with FDG PET: clinical article. *J Neurosurg* **110:** 1278–1282. doi:10.3171/2008.12.JNS08991

Quattrone A, Barbagallo G, Cerasa A, Stoessl AJ. 2018. Neurobiology of placebo effect in Parkinson's disease: what we have learned and where we are going. *Mov Disord* **33:** 1213–1227. doi:10.1002/mds.27438

Rommal A, Vo A, Schindlbeck KA, Greuel A, Ruppert MC, Eggers C, Eidelberg D. 2021. Parkinson's disease-related pattern (PDRP) identified using resting-state functional MRI: validation study. *Neuroimage Rep* **1:** 100026. doi:10.1016/j.ynirp.2021.100026

Rubinov M, Sporns O. 2010. Complex network measures of brain connectivity: uses and interpretations. *Neuroimage* **52:** 1059–1069. doi:10.1016/j.neuroimage.2009.10.003

Rus T, Schindlbeck KA, Tang CC, Vo A, Dhawan V, Trošt M, Eidelberg D. 2022. Stereotyped relationship between motor and cognitive metabolic networks in Parkinson's disease. *Mov Disord* **37:** 2247–2256. doi:10.1002/mds.29188

Rus T, Perovnik M, Vo A, Nguyen N, Tang C, Jamšek J, Šurlan Popović K, Grimmer T, Yakushev I, Diehl-Schmid J, et al. 2023. Disease specific and nonspecific metabolic brain networks in behavioral variant of frontotemporal dementia. *Hum Brain Mapp* **44:** 1079–1093. doi:10.1002/hbm.26140

Scheffer M, Carpenter SR, Lenton TM, Bascompte J, Brock W, Dakos V, Van De Koppel J, Van De Leemput IA, Levin SA, Van Nes EH, et al. 2012. Anticipating critical transitions. *Science* **338:** 344–348. doi:10.1126/science.1225244

Schindlbeck KA, Eidelberg D. 2018. Network imaging biomarkers: insights and clinical applications in Parkinson's disease. *Lancet Neurol* **17:** 629–640. doi:10.1016/S1474-4422(18)30169-8

Schindlbeck KA, Vo A, Nguyen N, Tang CC, Niethammer M, Dhawan V, Brandt V, Saunders-Pullman R, Bressman SB, Eidelberg D. 2020. LRRK2 and GBA variants exert distinct influences on Parkinson's disease-specific metabolic networks. *Cereb Cortex* **30:** 2867–2878. doi:10.1093/cercor/bhz280

Schindlbeck KA, Gupta DK, Tang CC, O'Shea SA, Poston KL, Choi YY, Dhawan V, Vonsattel JP, Fahn S, Eidelberg D. 2021a. Neuropathological correlation supports automated image-based differential diagnosis in parkinsonism. *Eur J Nucl Med Mol Imaging* **48:** 3522–3529. doi:10.1007/s00259-021-05302-6

Schindlbeck KA, Vo A, Mattis PJ, Villringer K, Marzinzik F, Fiebach JB, Eidelberg D. 2021b. Cognition-related functional topographies in Parkinson's disease: localized loss of the ventral default mode network. *Cereb Cortex* **31:** 5139–5150. doi:10.1093/cercor/bhab148

Shen B, Wei S, Ge J, Peng S, Liu F, Li L, Guo S, Wu P, Zuo C, Eidelberg D, et al. 2020. Reproducible metabolic topographies associated with multiple system atrophy: network and regional analyses in Chinese and American patient cohorts. *NeuroImage Clin* **28:** 102416. doi:10.1016/j.nicl.2020.102416

Spetsieris PG, Eidelberg D. 2011. Scaled subprofile modeling of resting state imaging data in Parkinson's disease: methodological issues. *Neuroimage* **54:** 2899–2914. doi:10.1016/j.neuroimage.2010.10.025

Spetsieris PG, Eidelberg D. 2023. Parkinson's disease progression: increasing expression of an invariant common core subnetwork. *NeuroImage Clin* **39**: 103488. doi:10 .1016/j.nicl.2023.103488

Spetsieris PG, Ma Y, Dhawan V, Eidelberg D. 2009. Differential diagnosis of parkinsonian syndromes using PCA-based functional imaging features. *Neuroimage* **45**: 1241–1252. doi:10.1016/j.neuroimage.2008.12.063

Spetsieris PG, Ko JH, Tang CC, Nazem A, Sako W, Peng S, Ma Y, Dhawan V, Eidelberg D. 2015. Metabolic resting-state brain networks in health and disease. *Proc Natl Acad Sci* **112**: 2563–2568. doi:10.1073/pnas.1411011112

Surmeier DJ, Obeso JA, Halliday GM. 2017. Selective neuronal vulnerability in Parkinson disease. *Nat Rev Neurosci* **18**: 101–113. doi:10.1038/nrn.2016.178

Tang CC, Poston KL, Dhawan V, Eidelberg D. 2010a. Abnormalities in metabolic network activity precede the onset of motor symptoms in Parkinson's disease. *J Neurosci* **30**: 1049–1056. doi:10.1523/JNEUROSCI.4188-09.2010

Tang CC, Poston KL, Eckert T, Feigin A, Frucht S, Gudesblatt M, Dhawan V, Lesser M, Vonsattel JP, Fahn S, et al. 2010b. Differential diagnosis of parkinsonism: a metabolic imaging study using pattern analysis. *Lancet Neurol* **9**: 149–158. doi:10.1016/S1474-4422(10)70002-8

Tang CC, Holtbernd F, Ma Y, Spetsieris P, Oh A, Fink GR, Timmermann L, Eggers C, Eidelberg D. 2020. Hemispheric network expression in Parkinson's disease: relationship to dopaminergic asymmetries. *J Parkinsons Dis* **10**: 1737–1749. doi:10.3233/JPD-202117

Teune LK, Renken RJ, Mudali D, De Jong BM, Dierckx RA, Roerdink JBTM, Leenders KL. 2013. Validation of parkinsonian disease-related metabolic brain patterns. *Mov Disord* **28**: 547–551. doi:10.1002/mds.25361

Teune LK, Renken RJ, De Jong BM, Willemsen AT, Van Osch MJ, Roerdink JBTM, Dierckx RA, Leenders KL. 2014. Parkinson's disease-related perfusion and glucose metabolic brain patterns identified with PCASL-MRI and FDG-PET imaging. *NeuroImage Clin* **5**: 240–244. doi:10 .1016/j.nicl.2014.06.007

Tomše P, Rebec E, Studen A, Perovnik M, Rus T, Ležaić L, Tang CC, Eidelberg D, Trošt M. 2022. Abnormal metabolic covariance patterns associated with multiple system atrophy and progressive supranuclear palsy. *Phys Med* **98**: 131–138. doi:10.1016/j.ejmp.2022.04.016

Tripathi M, Dhawan V, Peng S, Kushwaha S, Batla A, Jaimini A, D'Souza MM, Sharma R, Saw S, Mondal A. 2013. Differential diagnosis of parkinsonian syndromes using F-18 fluorodeoxyglucose positron emission tomography. *Neuroradiology* **55**: 483–492. doi:10.1007/s00234-012-1132-7

Tripathi M, Tang CC, Feigin A, De Lucia I, Nazem A, Dhawan V, Eidelberg D. 2016. Automated differential diagnosis of early parkinsonism using metabolic brain networks: a validation study. *J Nucl Med* **57**: 60–66. doi:10 .2967/jnumed.115.161992

Unadkat P, Vo A, Ma Y, Peng S, Nguyen N, Niethammer M, Tang CC, Dhawan V, Ramdhani RA, Fenoy A, et al. 2024. Deep brain stimulation of the subthalamic nucleus for Parkinson's disease: a network imaging marker of the treatment response. *Res Sq* doi:10.21203/rs.3.rs-4178 280/v1

Vo A, Sako W, Fujita K, Peng S, Mattis PJ, Skidmore FM, Ma Y, Uluǧ AM, Eidelberg D. 2017. Parkinson's disease-related network topographies characterized with resting state functional MRI. *Hum Brain Mapp* **38**: 617–630. doi:10.1002/hbm.23260

Vo A, Nguyen N, Fujita K, Schindlbeck KA, Rommal A, Bressman SB, Niethammer M, Eidelberg D. 2023a. Disordered network structure and function in dystonia: pathological connectivity vs. adaptive responses. *Cereb Cortex* **33**: 6943–6958. doi:10.1093/cercor/bhad012

Vo A, Schindlbeck KA, Nguyen N, Rommal A, Spetsieris PG, Tang CC, Choi YY, Niethammer M, Dhawan V, Eidelberg D. 2023b. Adaptive and pathological connectivity responses in Parkinson's disease brain networks. *Cereb Cortex* **33**: 917–932. doi:10.1093/cercor/bhac110

Xiong Y, Lin J, Bian X, Lu H, Zhou J, Zhang D, Pan L, Lou X. 2022. Treatment-specific network modulation of MRI-guided focused ultrasound thalamotomy in essential tremor: modulation of ET-related network by MRgFUS thalamotomy. *Neurotherapeutics* **19**: 1920–1931. doi:10 .1007/s13311-022-01294-9

Dopamine Cell-Based Replacement Therapies

Saeed Kayhanian[1,2] and Roger A. Barker[1,3]

[1]Department of Clinical Neurosciences, University of Cambridge, Cambridge CB2 2PY, United Kingdom

[2]Department of Neurosurgery, Cambridge University Hospitals, Cambridge CB2 0QQ, United Kingdom

[3]Department of Neurology, Cambridge University Hospitals, Cambridge CB2 0QQ, United Kingdom

Correspondence: sk776@cam.ac.uk; rab46@cam.ac.uk

Parkinson's disease (PD) is a common disorder that has, as part of its core pathology, the loss of the nigral dopaminergic nerve cells that project to the striatum. Replacing this loss with dopaminergic drugs has been the mainstay of therapy in PD for more than 50 years and while offering significant clinical benefit, especially in early-stage disease, leads to side effects over time. A conceptually more effective way to treat this aspect of the PD pathology would be to replace the missing dopaminergic system with grafts of new dopamine cells. This approach has been investigated for nearly 40 years using a variety of different dopamine cell sources. To date, a proof-of-principle has been shown using human fetal dopamine cells in patients with PD, but the more widespread adoption of this approach has been hampered by logistical reasons around tissue supply, the ethics of the cell source, and, most importantly, by the inconsistent results shown across trials, which in some cases have reported worrying side effects. Reasons for all this have been discussed extensively in the literature and one solution may lie in the development of new human stem cell–derived dopamine cells, which are now just entering first in human clinical trials.

The pathology in Parkinson's disease (PD) involves multiple cell types not only within the central nervous system (CNS) but also including the autonomic and enteric nervous systems. Nevertheless, a core pathology is the loss of ~200,000 dopaminergic neurons in the substantia nigra of the midbrain that project to the ipsilateral striatum. The loss of this dopaminergic nigrostriatal pathway underlies many of the key motor features experienced by people with PD as is evidenced by how such individuals respond to dopaminergic agents, especially early in the disease course.

Oral dopaminergic replacements to treat PD emerged nearly 60 years ago. Over this time, there has been a transformation in both the quality of life and life expectancy for people with Parkinson's, in large part due to the success of pharmacological dopamine replacement (Uitti et al. 1993; Williams-Gray et al. 2013). However, treatment with dopaminergic drugs is not without problems; the treatment responses become more erratic with disease progression as well as causing their own motor side effects such as L-dopa-induced dyskinesias (LIDs) and neuropsychiatric problems such as hallucinations, paranoia, and dopamine dysregulation syndromes. These problems arise from the nonspecific stimulation of relatively intact nonnigrostriatal CNS dopaminergic systems by

Copyright © 2025 Cold Spring Harbor Laboratory Press; all rights reserved
Cite this article as *Cold Spring Harb Perspect Med* doi: 10.1101/cshperspect.a041611

the drugs as well as the non-physiological, pulsatile stimulation of dopaminergic receptors in the striatum, which in turn leads to downstream changes in postsynaptic striatal dopamine receptors. Treatment with dopaminergic agents can also exacerbate other problems seen in PD such as preexisting autonomic dysfunction, which may lead to worsening postural hypotension, for example.

There is, therefore, a significant clinical need for treatments that can better target the diseased dopaminergic pathway in a more physiological manner. One potential way to achieve this would be to replace the population of lost dopaminergic neurons with transplants of new dopamine cells that can synaptically reinnervate the area of dopamine loss in the striatum. There has been significant interest over many decades in this regenerative medicine approach in PD starting with the original preclinical proof-of-principle studies in the early 1980s. In this work, we will explore the scientific rationale underlying this therapeutic approach, the history of cell-replacement strategies for PD, the current status of

this field, and finally a look to its future as a management option in PD (see Fig. 1 for a summary schematic of this work).

HISTORY OF CELL-BASED REPLACEMENT THERAPIES

Scientific Rationale and Preclinical Evidence

The adult human midbrain contains between 400,000 and 500,000 dopaminergic neurons, and contained within it are the A9 nigral cells that project to the ipsilateral striatum. The first motor symptoms of PD emerge when around half of these cells are lost and 70% of the fiber innervation of the striatum (Gibb and Lees 1991; Ma et al. 1997). Replacing these lost cells could restore striatal innervation and dopamine levels back to normal.

Initially, a number of different dopamine cell therapies were considered, including the adrenal medulla—although it became clear that these therapies ultimately failed to produce any significant long-lasting clinical benefits, which matched

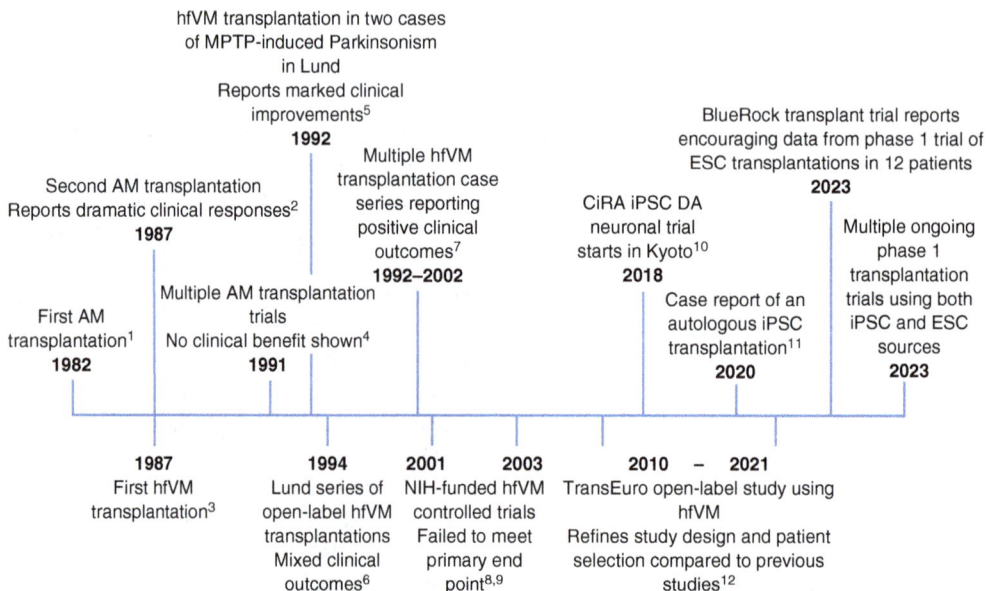

Figure 1. Timeline 1970–2023 of key clinical trials of cell therapies for Parkinson's disease. (AM) Adrenal medullary, (hfVM) human fetal ventral mesencephalon, (iPSC) induced pluripotent stem cell, (ESC) embryonic stem cell. ([1]Lindvall et al. 1987; [2]Madrazo et al. 1987; [3]Lindvall et al. 1989; [4]Goetz et al. 1991; [5]Widner et al. 1992; [6]Lindvall et al. 1994; [7]summarized in Barker et al. 2013; [8]Freed et al. 2001; [9]Olanow et al. 2003; [10]Takahashi 2020; [11]Schweitzer et al. 2020; [12]Barker and TRANSEURO Consortium 2019).

their poor survival at postmortem and the limited functional improvements in the preclinical models (for reviews, see Freed et al. 1990; Barker and Dunnett 1993). This was largely predictable given that adrenal medullary cells produce little dopamine and thus a much more logical cell source would be the dopaminergic cells developing in the fetal ventral mesencephalon (fVM). This tissue was shown to give much more consistent and better results in neurotoxic models of PD beginning with the pioneering studies in Lund in the early 1980s. These studies by Björklund and colleagues during the 1980s established that allografts of fVM tissue in the 6OHDA lesion model of PD showed long-term cell survival with innervation of the host striatum, dopamine release, and recovery of many motor deficits (summarized in Björklund and Lindvall 2017). This was reproduced in many laboratories and as such gave the field confidence to take this forward to clinical trials, albeit with ethical issues, given it would involve using human fetal tissue sourced from the termination of pregnancies.

First Clinical Trials

The first clinical trials in PD around dopamine cell replacement involved adrenal medullary tissue being transplanted into the caudate in two patients in Lund in 1982 (Lindvall et al. 1987). This study failed to demonstrate any clinical improvements but was soon followed by another study taking the same approach, which reported a dramatic clinical improvement within weeks of surgery (Madrazo et al. 1987). This high-profile result led to widespread trials of this approach, particularly in North America, where large numbers of patients were grafted. However, it became clear that the initial results could not be replicated —although this was not finally confirmed until a registry was set up and all the data from such trials analyzed (Goetz et al. 1991). This study and the subsequent postmortem studies revealed that the therapy demonstrated no long-term clinical benefits, poor tissue survival, as well as significant side effects from both the abdominal and intracranial surgery.

Running in parallel to this were clinical trials using human fVM tissue, starting again in Lund

with two patients grafted in 1987 (Lindvall et al. 1989). Over the next decade, a further 15 patients were treated using this approach in an iterative open-label study that sought to improve the method of graft preparation and delivery with trial evolution and data (Lindvall et al. 1994; Wenning et al. 1997). The results of this work were encouraging, demonstrating both safety, in that there were no major serious adverse effects related to the procedure, and proof-of-principle in terms of graft survival, as measured by ^{18}Fluoro-dopa PET imaging and in later postmortem studies, as well as efficacy in some patients (Lindvall et al. 1990; Sawle et al. 1992). However, while clinical outcomes were mixed, in the best cases patients demonstrated significant improvements to the extent that they were able to stop the anti-PD medication completely, with lasting improvements >20 years following transplantation and normalization of their striatal dopamine levels (Kefalopoulou et al. 2014).

The logic and success of this human fVM approach in Sweden led to further similar studies through the 1990s in other sites around Europe and in North America (summarized in Barker et al. 2013). Results in these studies were again mixed, with some patients demonstrating only modest or even no benefits, whereas others showed considerable clinical improvements. However, many of these studies were open label and thus subject to placebo and investigator bias and so in the United States, randomized controlled trials were undertaken to establish unequivocally whether this approach had a true clinical benefit to the patients.

The NIH-Funded Double-Blind Placebo-Controlled Trials

In 1992, the NIH agreed to fund two double-blind, placebo-controlled trials, in which patients with moderately advanced PD would either be grafted with human fVM tissue or receive sham surgery with partial burr holes and no tissue engrafted (Table 1). The two trials, published in 2001 and 2003, both failed to meet their primary end points and reported significant side effects with a number of patients developing graft-induced dyskinesias (GIDs)

Table 1. A summary of the major human fetal ventral mesencephalon (hfVM) trials

	Freed et al. (2001)	Olanow et al. (2003)	TransEuro
Number of patients grafted	20	23	11
Mean age (years) [range]	57 [34–75]	58.5 [30–75]	51.8 [42–67]
Disease duration (mean, years)	14	11	8
Number of hfVM implants per side	2	1 or 4	3
hfVM handling	Implanted as "noodles" stored for up to 28 days before grafting	Implanted as solid pieces stored for up to 2 days before grafting	Implanted as a cell suspension Stored for a maximum of 4 days before grafting
Surgical approach	Trans-frontal, bilateral procedure, two needle tracts per side	Standard approach to putamen, eight needle tracts per side, two unilateral procedures, separated by 1 week	Standard approach to putamen, five needle tracts per side, two unilateral procedures separated by between 1 and 5 months
Immunosuppression	None	Cyclosporin for 5 months	Cyclosporin, azathioprine, and prednisolone for 12 months
Primary end point	Subjective patient questionnaire after 1 year No difference between groups ($P = 0.62$)	UPDRS score in "off" period at 2 years No difference between three groups ($P = 0.24$)	UPDRS score in "off" period at 3 years
Change in "off" UPDRS at primary end point	Transplant at age <60: 59 (pre-graft) to 40 (postgraft) Sham at age <60: 62 (pre-graft) to 60 (postgraft) Transplant at age >60: 59 (pre-graft) to 60 (postgraft) Sham at age >60: 71 (pre-graft) to 70 (postgraft)	One implant: 48–51.5 Four implants: 49–48 Placebo: 51.5–60	Yet to be published
Proportion of patients with graft-induced dyskinesias	15%	56.5%	Yet to be published

(UPDRS) Unified Parkinson's Disease Rating Scale.

(Freed et al. 2001; Olanow et al. 2003). As a result, this approach by 2003 was felt to be without merit, especially given the emerging trial results showing how effective deep brain stimulation (DBS) was for patients entering this phase of their disease course.

However, a closer inspection of these transplant trials did suggest reasons as to why these studies produced the results they did. In the first study by Freed et al. (2001), patients were grafted with human fVM derived from a single fetus (compared with up to seven in the Lund series) delivered using a new *trans-frontal* surgical approach and without immunosuppression postgrafting. The primary end point in this trial was a patient-reported outcome at 1 year postgrafting. This subjective, early outcome was further complicated by the fact that the control, sham arm was lost after 12 months as patients in this arm were all offered transplants after this point, meaning that 33 out of all 40 patients in this trial eventually received transplants.

The second study reported by Olanow et al. (2003) used grafts derived from either one or four fetuses and measured changes in the Unified Parkinson's Disease Rating Scale (UPDRS) score in the defined "off" state 2 years postgrafting. Immunosuppression with cyclosporin A was given but only for 6 months posttransplantation. In this trial, while the primary outcome was not met, patients who had received grafts derived from four fVMs demonstrated a clear trend toward improvement, with ^{18}F-dopa PET imaging showing significant improvements.

In addition to both trials failing to reach their primary end point, a number of patients in each trial developed GIDs (15% and 54%, respectively), which were so severe in some cases that further neurosurgical intervention was needed in the form of DBS. Thus, for many in the PD field, these trials sounded the death knell for dopamine cell-replacement therapies for this condition.

TransEuro

However, in 2006, an international workshop was established to reanalyze the available clinical data on human fVM transplantation and, in particular, to attempt to identify why some patients had significantly benefited from the therapy and others not (see Table 1). In addition, the development of GIDs was a major concern and needed explaining such that if this approach was to be taken on, this risk should be minimized for grafted patients. In this respect, several theories have been put forward to explain the development of GIDs. The first proposed that GIDs resulted from the uneven graft-derived innervation (secondary to dopaminergic "hot spots" developing in the transplanted striatum). The second has suggested that the GIDs (which were originally called runaway dyskinesias) could be due to excessive dopamine produced by the transplants, based on the phenomenology of the dyskinesias, their response to dopamine-blocking drugs and their correlation with clinical response and PET imaging (Greene et al. 2021). The third explanation postulated that GIDs were the result of contaminating serotonin neurons in the grafts—cells that develop adjacent to the midbrain dopamine cells in the human brainstem and that can synthesize dopamine but have no synaptic dopamine transporters (Steece-Collier et al. 2012). As such, all these theories have possible solutions going forward. This can be done by ensuring that not only are 5HT neurons avoided as much as possible in the tissue preparation but that the number and distribution of dopamine cells within the transplant are limited but also delivered evenly across the striatum.

In addition, a limited meta-analysis of all the published trial data suggested several other factors contributed to a positive human fVM graft outcome, including younger age of the recipient; less advanced PD; no significant LIDs pregrafting, preserved dopaminergic innervation on preoperative ^{18}F-dopa PET in the ventral striatum; grafting sufficient tissue to ensure that at least 100,000 dopaminergic nigral neurons per side survived; and that the patients were in receipt of adequate immunosuppression postgrafting (Piccini et al. 2005; Barker et al. 2013).

While this rationalization led to the formulation of a new trial—TransEuro—it was clear (and always had been) that the ethical and practical difficulties in obtaining human fetal brain tissue meant that it would always be challenging to develop this approach into a mainstream therapy or standard of care for people with PD.

The TransEuro study was originally designed to involve an open-label study of 20 transplanted patients, selected from a parallel "natural history" cohort of 150 patients who would then act as a contemporaneous control group. This would be followed by a bigger double-blind placebo-controlled trial powered based on the results of this initial study. In the end, only 11 patients were grafted in a single open-label study mainly due to significant difficulties in sourcing sufficient quantities of human fVM tissue in a timely manner (Barker and TRANSEURO Consortium 2019). The results are still waiting to be published but a preliminary analysis suggests that the trial did not meet its primary end point, which was a significant improvement in the UPDRS Part III at 36 months postgrafting in the defined "off" state. A number of secondary measures were also looked at and, overall, the data suggested that insufficient numbers of cells had survived postgrafting to restore patients and their dopaminergic scans back to normal. Despite this, there was evidence of a significant reduction in the rate of clinical deterioration in some grafted patients and, encouragingly, no patients developed any major, disabling GIDs. The lessons of TransEuro again highlighted the central problem of a reliable cell source but the trial design and the patient cohort selected for inclusion have helped develop the platform for new trials using stem cell–derived dopamine cells.

THE NEW STEM CELL ERA AND CELLULAR REPROGRAMMING

Stem Cell–Derived Dopaminergic Neurons

Despite all the problems with human fVM tissue, the trials of it in PD have demonstrated a proof-of-principle with (in the best cases) restoration to normal of striatal dopamine levels on PET imaging, normalization of dopaminergic innervation in postmortem tissue, and long-term clinical improvements. However, the results have been inconsistent and, in some instances, associated with serious side effects, notably GIDs. A reliable source of midbrain dopamine cells, free from contaminating serotonin neurons, which can be delivered evenly and in appropriate numbers across

the putamen would, in theory, overcome the limitations that have hampered the fVM trials to date.

In 2011/2012, such a reliable cell source became a possibility with the development of "floorplate protocols," which could be used to differentiate pluripotent stem cells, derived either from human embryonic stem cells (hESCs) or induced pluripotent stem cells (iPSCs), into midbrain dopaminergic neurons (Kriks et al. 2011; Kirkeby et al. 2012). This technical advance was built on a revised insight into the developmental origin of dopaminergic neurons, in that it was realized that they developed from a group of cells located in the ventral midline of the neural tube (the "floorplate") and not neuroepithelial progenitors like all other neurons in the brain (Placzek and Briscoe 2005; Ono et al. 2007).

These floorplate protocols have proven a robust, reliable, and highly efficient means to generate dopaminergic neural precursor cells from pluripotent stem cells (Fig. 2). These cells have now undergone extensive preclinical testing in animal models of PD and have been shown to differentiate after transplantation into the relevant midbrain dopamine neuronal subtype (A9), as well as function with equal potency and efficacy to fetal dopaminergic neurons (Grealish et al. 2014; Kirkeby et al. 2017a; summarized in Kirkeby et al. 2017b). Moreover, these cells have now been produced under good manufacturing practice (GMP) conditions, with reliable protocols for cryopreservation and thawing, without significant compromise of the final cell product (Nolbrant et al. 2017). The final safety of the product has also been rigorously tested, demonstrating, in particular, an absence of tumorigenicity and cell migration postgrafting (Piao et al. 2021; Kirkeby et al. 2023).

A number of phase 1 studies have now commenced around the world, using pluripotent stem cell–derived dopaminergic neural precursor cells (Table 2). These studies are all built on the principle of floorplate differentiation protocols, but differ in terms of the pluripotent cell source, the dose of cells engrafted, the method for implanting the cells, the patient cohort, and the immunosuppression regime and so caution is required when drawing conclusions across these early trials. A single case report of an autologous iPSC trans-

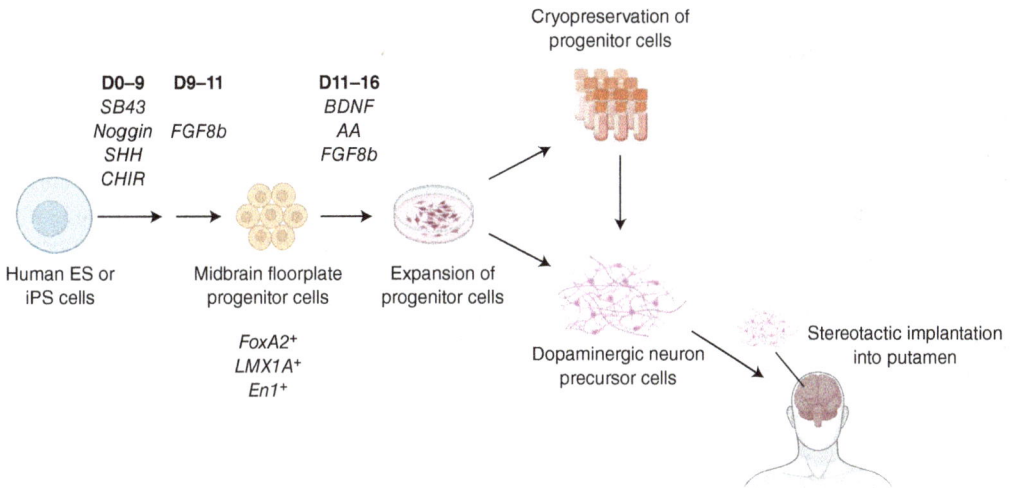

Figure 2. Schematic overview of the floorplate protocol used to generate dopamine neuron progenitor cells in vitro. (ES cell) Embryonic stem cell, (iPS cell) induced pluripotent stem cell.

plant in a patient with PD has been described, although this was undertaken outside of a formal trial and has been the subject of significant debate in the literature (Kordower et al. 2020; Parmar and Björklund 2020; Schweitzer et al. 2020).

These first-in-human trials will provide important evidence on the safety of these cell products before larger trials can explicitly examine efficacy in randomized control trials. Importantly also, the differences in trial protocols should provide some insight into the outstand-

ing technical questions in the field, including issues about the cell product, its delivery, and optimal graft recipient as well as more broadly the relative merits of allogeneic versus autologous iPSC versus ESC-derived dopaminergic precursor cell products.

Direct Reprogramming

A final consideration in this field of cell therapy treatment is the possibility of undertaking "direct

Table 2. A table of planned or ongoing cell therapy clinical trials for Parkinson's disease

Trial	Country	Cell source	Number of patients	Status
BlueRock/Bayer	USA/Canada	ESCs	12	Phase 1 completed
Massachusetts General Hospital case	USA	Autologous iPSCs	1	Single case reported
S.Biomedics	South Korea	ESCs	12	Phase 1 ongoing
STEM-PD	UK/Sweden	ESCs	8	Phase 1 ongoing
Kyoto	Japan	Allogenic iPSCs	7	Phase 1 ongoing
Allife Medical Science and Technology	China	Allogenic iPSCs	10	Phase 1 ongoing
Cyto Therapeutics	Australia	Human parthenogenetic neural stem cells	12	Phase 1 ongoing
Aspen Neuroscience	USA	Autologous iPSCs	To be confirmed	Phase 1 ongoing
Arizona State University	USA	Allogenic iPSCs	To be confirmed	In development

(ESCs) Embryonic stem cells, (iPSCs) induced pluripotent stem cells.

in situ reprogramming" in the PD brain (i.e., using exogenous factors to differentiate cells in the brain into the type of dopaminergic neurons that have been lost through disease). Neurogenesis of nigral dopaminergic cells is thought not to normally occur in the adult mammalian brain so recent approaches have focused on the possibility of transdifferentiating resident glia into dopaminergic neurons. In particular, studies have concentrated on striatal astrocytes, which are present in abundant numbers and are already found to be able to "dedifferentiate" in the adult brain, usually after injury (Magnusson et al. 2014).

The in vitro studies to date have involved the viral delivery of transcription factors or microRNAs that direct the conversion of astrocytes into neurons, either directly or via an intermediate pluripotent cell stage (Torper et al. 2013; Rivetti Di Val Cervo et al. 2017; Wei and Shetty 2021). One recent in vivo study, in a mouse model of PD, reported direct transdifferentiation of midbrain astrocytes into dopaminergic neurons by depletion of an RNA-binding protein PTB (Qian et al. 2020). Subsequent efforts to reproduce these results have not been successful, however, which implied that the lineage-tracing techniques results shown may have been imperfect, with leakage of the cre-recombinase system to surrounding endogenous dopamine neurons (Parmar et al. 2021; Wang et al. 2021).

The potential for direct reprogramming as a therapy for PD, therefore, remains firmly in the preclinical, discovery stage but there is promise in this rapidly developing field and any such potential "in situ" cell therapy approach would present significant advantages over the ex vivo cell manufacturing approaches developed to date.

CONCLUSION

The success of long-term treatment for PD with dopaminergic drugs highlights the clinical utility and rationale for targeting this aspect of the pathology with dopamine cell-replacement therapies. However, the drugs' nonphysiological mode of dopamine replacement and the progressive nature of the disease, both with continued nigral dopaminergic cell loss and the development of extranigral pathology, means that these treatments ultimately fail. Progress with cell-based replacement therapies presents a realistic opportunity to address some of these treatment challenges. The dopaminergic neuronal transplants could allow for a much more physiological release of dopamine in the areas of the brain where it is most required. Proof-of-principle studies using human fVM have confirmed that this approach is logical and can be effective, albeit inconsistently. It is clear also that the successful development of cell-replacement therapies for dopaminergic cell loss is dependent on having a standardized, ethically acceptable, and reliable source of cells. Such a situation now exists with human stem cells and, with this, trials in PD have begun, which will generate interest and present opportunities both for potential new cell-based approaches to the extranigral pathology of PD, as well as for other neurodegenerative and CNS diseases more generally.

Herein we have briefly described the rationale and approaches taken to date with dopamine cell-based therapies for PD and summarized the recent progress in the development of stem cell–based treatments, which have now entered clinical trials. These early trials should of course be interpreted with caution given the differences that exist between them. Nevertheless, the field of dopamine cell replacement in PD is entering an exciting new era and over the next few years, we will know whether such an approach really does have merit in the treatment of this condition.

ACKNOWLEDGMENTS

This work was supported by the NIHR Cambridge Biomedical Research Centre (BRC-1215-20014). S.K. is supported by an NIHR Academic Clinical Fellowship and the Addenbrooke's Charitable Trust.

REFERENCES

Barker R, Dunnett S. 1993. The biology and behaviour of intracerebral adrenal transplants in animals and man. *Rev Neurosci* **4**: 113–146. doi:10.1515/revneuro.1993.4.2.113

Barker RA; TRANSEURO Consortium. 2019. Designing stem-cell-based dopamine cell replacement trials for Parkinson's disease. *Nat Med* **25**: 1045–1053. doi:10.1038/s41591-019-0507-2

Cite this article as *Cold Spring Harb Perspect Med* doi: 10.1101/cshperspect.a041611

Barker RA, Barrett J, Mason SL, Björklund A. 2013. Fetal dopaminergic transplantation trials and the future of neural grafting in Parkinson's disease. *Lancet Neurol* **12:** 84–91. doi:10.1016/S1474-4422(12)70295-8

Björklund A, Lindvall O. 2017. Replacing dopamine neurons in Parkinson's disease: how did it happen? *J Parkinsons Dis* **7:** S21–S31. doi:10.3233/JPD-179002

Freed WJ, Poltorak M, Becker JB. 1990. Intracerebral adrenal medulla grafts: a review. *Exp Neurol* **110:** 139–166. doi:10.1016/0014-4886(90)90026-O

Freed CR, Greene PE, Breeze RE, Tsai WY, DuMouchel W, Kao R, Dillon S, Winfield H, Culver S, Trojanowski JQ, et al. 2001. Transplantation of embryonic dopamine neurons for severe Parkinson's disease. *N Engl J Med* **344:** 710–719. doi:10.1056/NEJM200103083441002

Gibb WRG, Lees AJ. 1991. Anatomy, pigmentation, ventral and dorsal subpopulations of the substantia nigra, and differential cell death in Parkinson's disease. *J Neurol Neurosurg Psychiatry* **54:** 388–396. doi:10.1136/jnnp.54.5.388

Goetz CG, Stebbins GT III, Klawans HL, Koller WC, Grossman RG, Bakay RA, Penn RD. 1991. United Parkinson foundation neurotransplantation registry on adrenal medullary transplants: presurgical, and 1- and 2-year follow-up. *Neurology* **41:** 1719–1722. doi:10.1212/wnl.41.11.1719

Grealish S, Diguet E, Kirkeby A, Mattsson B, Heuer A, Bramoulle Y, Van Camp N, Perrier AL, Hantraye P, Björklund A, et al. 2014. Human ESC-derived dopamine neurons show similar preclinical efficacy and potency to fetal neurons when grafted in a rat model of Parkinson's disease. *Cell Stem Cell* **15:** 653–665. doi:10.1016/j.stem.2014.09.017

Greene PE, Fahn S, Eidelberg D, Bjugstad KB, Breeze RE, Freed CR. 2021. Persistent dyskinesias in patients with fetal tissue transplantation for Parkinson disease. *NPJ Parkinsons Dis* **7:** 38. doi:10.1038/s41531-021-00183-w

Kefalopoulou Z, Politis M, Piccini P, Mencacci N, Bhatia K, Jahanshahi M, Widner H, Rehncrona S, Brundin P, Björklund A, et al. 2014. Long-term clinical outcome of fetal cell transplantation for Parkinson disease: two case reports. *JAMA Neurol* **71:** 83–87. doi:10.1001/jamaneurol.2013.4749

Kirkeby A, Grealish S, Wolf DA, Nelander J, Wood J, Lundblad M, Lindvall O, Parmar M. 2012. Generation of regionally specified neural progenitors and functional neurons from human embryonic stem cells under defined conditions. *Cell Rep* **1:** 703–714. doi:10.1016/j.celrep.2012.04.009

Kirkeby A, Nolbrant S, Tiklova K, Heuer A, Kee N, Cardoso T, Ottosson DR, Lelos MJ, Rifes P, Dunnett SB, et al. 2017a. Predictive markers guide differentiation to improve graft outcome in clinical translation of hESC-based therapy for Parkinson's disease. *Cell Stem Cell* **20:** 135–148. doi:10.1016/j.stem.2016.09.004

Kirkeby A, Parmar M, Barker RA. 2017b. Strategies for bringing stem cell-derived dopamine neurons to the clinic: a European approach (STEM-PD). *Prog Brain Res* **230:** 165–190. doi:10.1016/bs.pbr.2016.11.011

Kirkeby A, Nelander J, Hoban DB, Rogelius N, Bjartmarz H; Novo Nordisk Cell Therapy R&D; Storm P, Fiorenzano A, Adler AF, Vale S, et al. 2023. Preclinical quality, safety, and efficacy of a human embryonic stem cell-derived product for the treatment of Parkinson's disease, STEM-PD. *Cell Stem Cell* **30:** 1299–1314.e9. doi:10.1016/j.stem.2023.08.014

Kordower JH, Okun MS, Jankovic J. 2020. Reply to: "Toward a personalized approach to Parkinson's cell therapy." *Mov Disord* **35:** 2120–2121. doi:10.1002/mds.28329

Kriks S, Shim JW, Piao J, Ganat YM, Wakeman DR, Xie Z, Carrillo-Reid L, Auyeung G, Antonacci C, Buch A, et al. 2011. Dopamine neurons derived from human ES cells efficiently engraft in animal models of Parkinson's disease. *Nature* **480:** 547–551. doi:10.1038/nature10648

Lindvall O, Backlund EO, Farde L, Sedvall G, Freedman R, Hoffer B, Nobin A, Seiger A, Olson L. 1987. Transplantation in Parkinson's disease: two cases of adrenal medullary grafts to the putamen. *Ann Neurol* **22:** 457–468. doi:10.1002/ana.410220403

Lindvall O, Rehncrona S, Brundin P, Gustavii B, Astedt B, Widner H, Lindholm T, Björklund A, Leenders KL, Rothwell JC, et al. 1989. Human fetal dopamine neurons grafted into the striatum in two patients with severe Parkinson's disease. A detailed account of methodology and a 6-month follow-up. *Arch Neurol* **46:** 615–631. doi:10.1001/archneur.1989.00520420033021

Lindvall O, Brundin P, Widner H, Rehncrona S, Gustavii B, Frackowiak R, Leenders KL, Sawle G, Rothwell JC, Marsden CD, et al. 1990. Grafts of fetal dopamine neurons survive and improve motor function in Parkinson's disease. *Science* **247:** 574–577. doi:10.1126/science.2105529

Lindvall O, Sawle G, Widner H, Rothwell JC, Björklund A, Brooks D, Brundin P, Frackowiak R, Marsden CD, Odin P, et al. 1994. Evidence for long-term survival and function of dopaminergic grafts in progressive Parkinson's disease. *Ann Neurol* **35:** 172–180. doi:10.1002/ana.410350208

Ma SY, Röyttä M, Rinne JO, Collan Y, Rinne UK. 1997. Correlation between neuromorphometry in the substantia nigra and clinical features in Parkinson's disease using disector counts. *J Neurol Sci* **151:** 83–87. doi:10.1016/s0022-510x(97)00100-7

Madrazo I, Drucker-Colín R, Díaz V, Martínez-Mata J, Torres C, Becerril JJ. 1987. Open microsurgical autograft of adrenal medulla to the right caudate nucleus in two patients with intractable Parkinson's disease. *N Engl J Med* **316:** 831–834. doi:10.1056/NEJM198704023161402

Magnusson JP, Göritz C, Tatarishvili J, Dias DO, Smith EM, Lindvall O, Kokaia Z, Frisén J. 2014. A latent neurogenic program in astrocytes regulated by Notch signaling in the mouse. *Science* **346:** 237–241. doi:10.1126/science.346.6206.237

Nolbrant S, Heuer A, Parmar M, Kirkeby A. 2017. Generation of high-purity human ventral midbrain dopaminergic progenitors for in vitro maturation and intracerebral transplantation. *Nat Protoc* **12:** 1962–1979. doi:10.1038/nprot.2017.078

Olanow CW, Goetz CG, Kordower JH, Stoessl AJ, Sossi V, Brin MF, Shannon KM, Nauert GM, Perl DP, Godbold J, et al. 2003. A double-blind controlled trial of bilateral fetal nigral transplantation in Parkinson's disease. *Ann Neurol* **54:** 403–414. doi:10.1002/ana.10720

Ono Y, Nakatani T, Sakamoto Y, Mizuhara E, Minaki Y, Kumai M, Hamaguchi A, Nishimura M, Inoue Y, Hayashi

H, et al. 2007. Differences in neurogenic potential in floor plate cells along an anteroposterior location: midbrain dopaminergic neurons originate from mesencephalic floor plate cells. *Development* **134**: 3213–3225. doi:10 .1242/dev.02879

Parmar M, Björklund A. 2020. From skin to brain: a Parkinson's disease patient transplanted with his own cells. *Cell Stem Cell* **27**: 8–10. doi:10.1016/J.STEM.2020.06.008

Parmar M, Björklund A, Björklund T. 2021. In vivo conversion of dopamine neurons in mouse models of Parkinson's disease—a future approach for regenerative therapy? *Curr Opin Genet Dev* **70**: 76–82. doi:10.1016/j.gde .2021.06.002

Piao J, Zabierowski S, Dubose BN, Hill EJ, Navare M, Claros N, Rosen S, Ramnarine K, Horn C, Fredrickson C, et al. 2021. Preclinical efficacy and safety of a human embryonic stem cell-derived midbrain dopamine progenitor product, MSK-DA01. *Cell Stem Cell* **28**: 217–229.e7. doi:10.1016/j.stem.2021.01.004

Piccini P, Pavese N, Hagell P, Reimer J, Björklund A, Oertel WH, Quinn NP, Brooks DJ, Lindvall O. 2005. Factors affecting the clinical outcome after neural transplantation in Parkinson's disease. *Brain* **128**: 2977–2986. doi:10 .1093/brain/awh649

Placzek M, Briscoe J. 2005. The floor plate: multiple cells, multiple signals. *Nat Rev Neurosci* **6**: 230–240. doi:10 .1038/nrn1628

Qian H, Kang X, Hu J, Zhang D, Liang Z, Meng F, Zhang X, Xue Y, Maimon R, Dowdy SF, et al. 2020. Reversing a model of Parkinson's disease with in situ converted nigral neurons. *Nature* **582**: 550–556. doi:10.1038/s41586-020-2388-4

Rivetti di Val Cervo P, Romanov RA, Spigolon G, Masini D, Martín-Montañez E, Toledo EM, La Manno G, Feyder M, Pifl C, Ng YH, et al. Induction of functional dopamine neurons from human astrocytes in vitro and mouse astrocytes in a Parkinson's disease model. *Nat Biotechnol* **35**: 444–452. doi:10.1038/nbt.3835

Sawle GV, Bloomfield PM, Björklund A, Brooks DJ, Brundin P, Leenders KL, Lindvall O, Marsden CD, Rehncrona S, Widner H, et al. 1992. Transplantation of fetal dopamine neurons in Parkinson's disease: PET [^{18}F]6-L-fluorodopa studies in two patients with putaminal implants. *Ann Neurol* **31**: 166–173. doi:10.1002/ana.410310207

Schweitzer JS, Song B, Herrington TM, Park TY, Lee N, Ko S, Jeon J, Cha Y, Kim K, Li Q, et al. 2020. Personalized iPSC-derived dopamine progenitor cells for Parkinson's disease. *N Engl J Med* **382**: 1926–1932. doi:10.1056/NEJ MOA1915872

Steece-Collier K, Rademacher DJ, Soderstrom KE. 2012. Anatomy of graft-induced dyskinesias: circuit remodeling in the Parkinsonian striatum. *Basal Ganglia* **2**: 15–30. doi:10.1016/J.BAGA.2012.01.002

Takahashi J. 2020. iPS cell-based therapy for Parkinson's disease: a Kyoto trial. *Regen Ther* **13**: 18–22. doi:10 .1016/j.reth.2020.06.002

Torper O, Pfisterer U, Wolf DA, Pereira M, Lau S, Jakobsson J, Björklund A, Grealish S, Parmar M. 2013. Generation of induced neurons via direct conversion in vivo. *Proc Natl Acad Sci* **110**: 7038–7043. doi:10.1073/pnas.1303829110

Uitti RJ, Ahlskog JE, Maraganore DM, Muenter MD, Atkinson EJ, Cha RH, O'Brien PC. 1993. Levodopa therapy and survival in idiopathic Parkinson's disease: Olmsted County project. *Neurology* **43**: 1918–1926. doi:10.1212/wnl.43 .10.1918

Wang LL, Serrano C, Zhong X, Ma S, Zou Y, Zhang CL. 2021. Revisiting astrocyte to neuron conversion with lineage tracing in vivo. *Cell* **184**: 5465–5481.e16. doi:10.1016/j .cell.2021.09.005

Wei ZD, Shetty AK. 2021. Treating Parkinson's disease by astrocyte reprogramming: progress and challenges. *Sci Adv* **7**: eabg3198. doi:10.1126/sciadv.abg3198

Wenning GK, Odin P, Morrish P, Rehncrona S, Widner H, Brundin P, Rothwell JC, Brown R, Gustavii B, Hagell P, et al. 1997. Short- and long-term survival and function of unilateral intrastriatal dopaminergic grafts in Parkinson's disease. *Ann Neurol* **42**: 95–107. doi:10.1002/ana.410 420115

Widner H, Tetrud J, Rehncrona S, Snow B, Brundin P, Gustavii B, Björklund A, Lindvall O, Langston JW. 1992. Bilateral fetal mesencephalic grafting in two patients with parkinsonism induced by 1-methyl-4-phenyl-1,2,3,6-tetrahydropyridine (MPTP). *N Engl J Med* **327**: 1556–1563. doi:10.1056/NEJM199211263272203

Williams-Gray CH, Mason SL, Evans JR, Foltynie T, Brayne C, Robbins TW, Barker RA. 2013. The CamPaIGN study of Parkinson's disease: 10-year outlook in an incident population-based cohort. *J Neurol Neurosurg Psychiatry* **84**: 1258–1264. doi:10.1136/jnnp-2013-305277

Addendum to Dopamine Cell-Based Replacement Therapies

Saeed Kayhanian[1,2] and Roger A. Barker[1,3]

[1]Department of Clinical Neurosciences, University of Cambridge, Cambridge CB2 2PY, United Kingdom

[2]Department of Neurosurgery, Cambridge University Hospitals, Cambridge CB2 0QQ, United Kingdom

[3]Department of Neurology, Cambridge University Hospitals, Cambridge CB2 0QQ, United Kingdom

Correspondence: sk776@cam.ac.uk; rab46@cam.ac.uk

Since the publication of our article "Dopamine Cell-Based Replacement Therapies" (Kayhanian and Barker. 2025. *Cold Spring Harb Perspect Med* doi:10.1101/cshperspect .a041611), three significant studies have published results that are important to the progress of the field of dopamine cell-based replacement therapies. In this addendum, we provide an update and short commentary on these results.

In 2025, the results of the 3 year outcomes from the TransEuro study were published (Barker et al. 2025). This open-label study of human fetal ventral mesencephalon tissue grafting in 11 patients with Parkinson's disease (PD) failed to meet its primary end point of significant improvement in UPDRS Part III at 36 months in the defined "off" state. The study again highlighted the biological and logistical difficulties in using fetal tissue as a dopaminergic cell source but has helped to refine the hypothesized optimal cohort for receiving such cell-replacement grafts and demonstrated the feasibility of a nested trial design for such trials (i.e., including a parallel natural history cohort as a contemporaneous control group), which may be useful going forward in the studies of stem cell–derived dopaminergic cell products that are currently in early clinical trials (Table 1).

Also in 2025, two such phase 1 studies of allogeneic stem cell–derived dopaminergic cell products published their first results. Tabar et al. reported 18 month outcomes in a trial of human embryonic stem cell (hESC)-derived dopaminergic precursor cells grafted into 12 patients across North America, and Sawamoto et al. reported 24 month outcomes in seven patients grafted with induced pluripotent stem cell (iPSC)-derived dopaminergic progenitors in Japan (Sawamoto et al. 2025; Tabar et al. 2025). Both trials met their primary end point of safety and tolerability, with some possible signals of efficacy in some patients (as measured by improvements in UPDRS Part III "off" score) in both cohorts. The measures of ^{18}F-DOPA PET uptake in the grafted putamen were also reported to improve in both studies; however, these increases did not appear to approach anything close to the normal levels expected in healthy individuals. Overall, these two trials provide an important first step to validating the safety and feasibility of this stem cell–derived dopaminergic cell replacement approach. The results of several further

Copyright © 2025 Cold Spring Harbor Laboratory Press; all rights reserved
Cite this article as *Cold Spring Harb Perspect Med* doi: 10.1101/cshperspect.a041945

Table 1. Ongoing or recently completed clinical trials of stem cell–derived dopamine cell replacement therapy

Trial	Country	Cell source	Number of patients	Status
BlueRock/Bayer	United States/ Canada	ESCs	12	Phase 1 completed
Kyoto	Japan	Allogenic iPSCs	7	Phase 1 completed
Massachusetts General Hospital case	United States	Autologous iPSCs	1	Single case reported
S.BIOMEDICS	South Korea	ESCs	12	Phase 1 ongoing
STEM-PD	United Kingdom/ Sweden	ESCs	8	Phase 1 ongoing
Royan Institute	Iran	ESCs	4	Phase 1 ongoing
ALLIFE Medical Science and Technology	China	Allogenic iPSCs	10	Phase 1 ongoing
Aspen Neuroscience	United States	Autologous iPSCs	9	Phase 1 ongoing
McLean Hospital	United States	Autologous iPSCs	6	Phase 1 ongoing
iRegene Therapeutics	China	Allogenic iPSCs	40	Phase 1 ongoing
Arizona State University	United States	Allogenic iPSCs	To be confirmed	In development

(ESC) Embryonic stem cell, (iPSC) induced pluripotent stem cell.

phase 1 trials (Table 1) are awaited, which may point toward future refinements in the dose of cells and trial design required going forward.

REFERENCES

Barker RA, Lao-Kaim NP, Guzman NV, Athauda D, Bjartmarz H, Björklund A, Church A, Cutting E, Daft D, Dayal V, et al. 2025. The TransEuro open-label trial of human fetal ventral mesencephali transplantation in patients with moderate Parkinson's disease. *Nat Biotechnol* doi:10.1038/s41587-025-02567-2

Sawamoto N, Doi D, Nakanishi E, Sawamura M, Kikuchi T, Yamakado H, Taruno Y, Shima A, Fushimi Y, Okada T, et al. 2025. Phase I/II trial of iPS-cell-derived dopaminergic cells for Parkinson's disease. *Nature* **641:** 971–977. doi:10.1038/s41586-025-08700-0

Tabar V, Sarva H, Lozano AM, Fasano A, Kalia SK, Yu KKH, Brennan C, Ma Y, Peng S, Eidelberg D, et al. 2025. Phase I trial of hES cell-derived dopaminergic neurons for Parkinson's disease. *Nature* **641:** 978–983. doi:10.1038/s41586-025-08845-y

α-Synuclein Biomarkers for Parkinson's Disease

Alexandra Lodge[1] and Julian Agin-Liebes[2]

[1]University Hospital North Tees, Stockton-on-Tees TS19 8PE, United Kingdom

[2]Department of Neurology, Columbia University Irving Medical School, New York, New York 10032, USA

Correspondence: ja3075@cumc.columbia.edu

α-Synuclein (α-syn) biomarkers show great promise as diagnostic tools for Parkinson's disease (PD). In recent years, a large body of evidence has validated their efficacy as diagnostic tools for PD and other synucleinopathies and has shown potential for use in patients with isolated prodromal symptoms of PD, such as rapid eye movement (REM) sleep behavior disorder and hyposmia, and further illuminates the pathophysiology of both idiopathic and genetic causes. Various detection methods have been deployed, predominantly immunohistochemistry and α-syn seed amplification assays. α-Syn has been shown to be detectable in many different tissues and biofluids in PD patients, each with benefits and limitations for practical use. α-Syn biomarker studies have shown sensitivities for diagnosis of PD and specificity against healthy controls up to 100%. However, lack of standardization of methods of detection currently limits interlaboratory validation of results. Verification of these assays could lead to more widespread inclusion of these modalities to detect α-syn into biological definitions of PD and provide frameworks for developing disease-modifying therapies. In this review, we discuss the current state of α-syn biomarkers and highlight their potential use in clinical practice and research settings, while identifying further work that is needed in this field.

Despite advances in the understanding of the pathophysiology of Parkinson's disease (PD), accurate diagnosis remains a challenge. Other than in homozygous genetic forms of the disease, diagnosis in life is made via clinical diagnostic criteria, with the gold-standard pathological confirmation of disease only possible postmortem. While brain imaging techniques such as magnetic resonance imaging (MRI) and dopamine transporter single photon emission computed tomography (DAT-SPECT) can help to support diagnosis of parkinsonian syndromes, they are not disease specific and therefore cannot be used as standalone diagnostic tools for PD. In recent years, there has been a large volume of research into finding a biochemical biomarker for PD, which could aid diagnosis, provide prognostic information, and be incorporated into a biological definition of PD. Aggregated α-synuclein (α-syn) is the core component of Lewy bodies and Lewy neurites, the pathological hallmarks of PD, and thus emerged as the leading candidate. Accordingly, there has been a growing amount of research into how best to detect pathological strains of the protein in living patients and thus use it as an accurate biomarker, and potentially therapeutic target, in PD and other synucleinopathies.

Copyright © 2025 Cold Spring Harbor Laboratory Press; all rights reserved

Cite this article as *Cold Spring Harb Perspect Med* doi: 10.1101/cshperspect.a041944

This work evaluates the current state of α-syn biomarkers, covering methods of detection in different tissues and biofluids and discussing the benefits and limitations of these methods, with the aim of highlighting challenges of translation to clinical use. The potential for these biomarkers to differentiate PD from other synucleinopathies and common disease mimics such as AD and tauopathies will be discussed, as well as use in prodromal and genetic PD, and current evidence of prognostic value of α-syn biomarkers. The biophysical properties of α-syn protein will be touched on, although this topic will be covered in more detail in Vekrellis et al. (2024). Concluding remarks will cover future research directions and the role of α-syn in emerging biological definitions of PD. For additional information on this subject, please refer to our recently accepted paper (Agin-Liebes et al. 2025).

STRUCTURE AND FUNCTION OF α-SYNUCLEIN

α-Syn is a small acidic neuronal protein, abundantly present in presynaptic terminals, with function in neurotransmitter uptake and release (Maroteaux et al. 1988; Burré et al. 2010, 2018). It is structurally composed of three distinct regions, which are responsible for the different molecular and biological properties of the molecule: the amino terminus, an amphipathic α-helix, the hydrophobic region (also nonamyloid β component [NAC]), and the acidic carboxyl terminus (Burré et al. 2010, 2018; Goedert et al. 2024). The amino terminus is responsible for binding to lipid membranes (George et al. 1995), the hydrophobic region has a preponderance to aggregation (Uéda et al. 1993), and the carboxyl terminus is involved in protein interactions, ion binding, and protection against aggregation (Nielsen et al. 2001; Lautenschläger et al. 2018).

Posttranslational modifications (PTMs) of α-syn alter its function and can lead to pathology. Various α-syn PTMs have been detected in pathological aggregates in postmortem brain tissue of patients with PD, including phosphorylation, acetylation, ubiquitination, nitration, and

carboxyl-terminus and amino-terminus truncation (Zhang et al. 2019). Of these, phosphorylation at Serine 129 (α-syn pS129) is the most common (Altay et al. 2023). PTMs of α-syn lead to a propensity for aggregation into β-pleated sheet-like oligomers (protofibrils), which can further convert into amyloid fibrils and subsequently Lewy bodies, the pathological hallmark of PD and dementia with Lewy bodies (DLBs) (Fig. 1; Calabresi et al. 2023). Pathological α-syn can also be found as glial cytoplasmic inclusions, the pathological hallmark of multiple systems atrophy (MSA) (Calabresi et al. 2023). These diseases are collectively known as "synucleinopathies."

The aggregation of misfolded α-syn species is thought to cause cytotoxicity and disrupts axonal transport and synaptic transmission (Zhang et al. 2019). The clinical manifestations of disease depend on the neurons affected by the interference of pathological α-syn strains. The neurodegeneration in PD is hypothesized to spread from the dorsal motor nucleus of the glossopharyngeal and vagus nerves and anterior olfactory nucleus, spreading through the midbrain to the cortex, with the characteristic clinical symptoms arising due to neuronal death in the substantia nigra (Braak et al. 2003). In contrast, the glial cytoplasmic inclusions of α-syn in oligodendrocytes seen in MSA are hypothesized to propagate in a rostrocaudal fashion, leading to neurodegeneration in the nigrostriatal and olivopontocerebellar regions (Gilman et al. 2008; Koga et al. 2021).

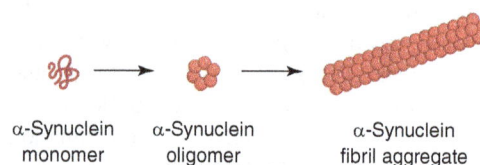

α-Synuclein monomer   α-Synuclein oligomer   α-Synuclein fibril aggregate

Figure 1. Aggregation of α-synuclein (α-syn). Physiological, cytosolic α-syn is predominantly a soluble, unstructured monomer. Various posttranslational changes and intrinsic factors induce aggregation into disordered β-sheet-rich oligomers and further into ordered amyloid fibrils, which form the core of Lewy bodies and glial cytoplasmic inclusions (GCIs). (Figure created in BioRender by Przedborski 2025; https://BioRender.com/0b0v50l.)

    Cite this article as *Cold Spring Harb Perspect Med* doi: 10.1101/cshperspect.a041944

Pathological α-syn can be found in the periphery as well as in the central nervous system (CNS) of patients with synucleinopathies, with abnormal aggregates of α-syn found in the gut, skin, oral mucosa, saliva, serum and gonads, and more (Vacchi et al. 2020). It has been demonstrated that α-syn accumulation in the gut can precede its presence in the CNS in a subset of PD patients, whereas in others, the involvement of the periphery is found later in disease, the so-called "body-first" versus "brain-first" subtypes (Vacchi et al. 2020).

Pathological strains of α-syn propagate in a prion-like manner, templating their conformational form in native α-syn (Guo and Lee 2014). The propagation of the pathological α-syn is dictated both by the structure of the pathological seeds and the structure of the soluble substrate of α-syn they propagate through (Rey et al. 2019; Zhang et al. 2023). Distinct propagation patterns have been shown using different strains of pathological α-syn seeds in mouse models, indicating the clinical heterogeneity of synucleinopathies may be due to different forms of pathological α-syn being implicated in different disease types (Rey et al. 2019).

α-SYN BIOMARKERS

Methods of Detection

As evidence has built for the classification of PD as a synucleinopathy, α-syn was identified as a potential biomarker, which could be used to make a biological diagnosis of disease in life. Research into identifying a suitable α-syn biomarker has been broadly split into studies looking at α-syn in tissues via immunohistochemistry methods, and those exploiting the prion-like propagation of α-syn in seed amplification assays (SAAs), most commonly using samples from cerebrospinal fluid (CSF), but more recently including serum and skin samples.

SAAs were originally developed for diagnosing prion diseases in CSF. They have been adapted for detection of α-syn by exploiting the prion-like properties of misfolded α-syn, in which the protein templates conformational changes in vicinal unfolded protein molecules, which

can be amplified through a cyclical process (Concha-Marambio et al. 2023). SAAs for detecting α-syn aggregates in PD and other synucleinopathies can be divided into real-time quaking-induced conversion (RT-QuIC) assays or protein misfolding cyclic amplification (PMCA) assays. In both cases, seeds of pathological α-syn aggregates are generated via shaking-induced fragmentation (RT-QuIC) or sonification (PMCA) of biofluid or tissue samples. These seeds are added to a recombinant α-syn substrate in which they induce misfolding (Fig. 2). The kinetics of the reaction are monitored via fluorescence, and an assay is deemed to be positive for pathological α-syn when fluorescence reaches a defined intensity threshold (Fig. 2; Concha-Marambio et al. 2023).

Several immune-based techniques have been used to detect pathological strains of α-syn in tissues and extracellular fluids. Immunohistochemistry is most commonly used; however, studies have also looked at enzyme-linked immunosorbent assay (ELISA), western blot, and proximity ligation assay. These methods rely on an expanding toolset of anti-α-syn antibodies, which capture the many confirmations of the protein (Altay et al. 2023).

In the following section, we discuss how these techniques are used in different tissues and biofluids and their potential uses as clinical biomarkers.

Skin

Phosphorylated α-syn (p-α-syn) histopathology has been demonstrated in peripheral as well as central nervous tissue in patients with PD (Beach et al. 2010), which has led to investigations of different organ systems to detect p-α-syn. Skin has been the most promising tissue because of its neural density as well as accessibility making it more suitable for analysis of α-syn pathology compared to other sources such as colonic tissue (Lee et al. 2017). p-α-Syn was first demonstrated in skin biopsies by Ikemura et al. in a 2008 study, which took biopsies from the skin of patients with autopsy confirmed Lewy body disease (Ikemura et al. 2008). The group localized the p-α-syn to the

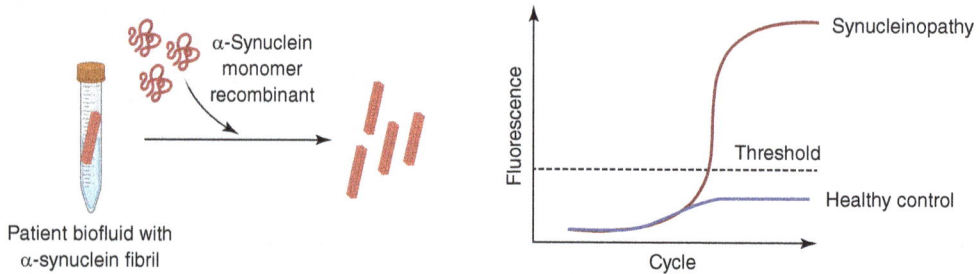

Figure 2. Seed amplification assay. A sample containing aggregated α-syn is added to a substrate of recombinant α-synuclein (α-syn). Aggregated species induce misfolding in neighboring soluble species in the substrate (elongation). Fragmentation via sonification or shaking then occurs, breaking up the newly formed aggregates. This process is repeated cyclically until a critical concentration of aggregated species is formed. The reaction kinetics are monitored via fluorescence; there is an initial lag phase, in which insoluble oligomers begin to form from the native soluble species. As the density of oligomers increases, there is a growth phase as aggregates readily recruit monomers to form fibrils. Finally, there is a plateau as the concentration of monomers is depleted. Assay positivity for synucleinopathy is defined when maximum intensity is over a defined threshold intensity. (Figure created in BioRender by Przedborski 2025; https://BioRender.com/8eq1cs2.)

cutaneous sympathetic nerve fibers. This discovery demonstrated the potential for using immunohistochemistry to detect α-syn in skin as a possible biomarker for PD. Subsequently, skin has been found to have a higher sensitivity and specificity comparing PD to controls than gastrointestinal (GI) and salivary gland samples, which was demonstrated in a meta-analysis by Tsukita et al. (2019).

Since the initial skin biopsy study was published in 2008, many groups have looked at how best to apply this technique in living patients, with a large number of studies published in the subsequent years. The methodology varies between groups but generally involves taking punch biopsies from two or more hairy skin sites, with the most commonly sampled regions being the cervical region, distal thigh, and distal leg (Fig. 3). Lower extremity sites have been traditionally used in skin biopsy for detection of small fiber neuropathy, and this may have influenced the choice of these regions in p-α-syn biopsy studies (Gibbons et al. 2006). Furthermore, it has been shown that combination of both proximal and distal sites is important for accurate differentiation of PD from other synucleinopathies, as studies generally showed a proximodistal gradient of deposition of p-α-syn in PD patients (Doppler et al. 2014; Donadio et al. 2016, 2018a,b; Gibbons et al.

2016), whereas a distal–proximal or more uniform distribution was seen in MSA (Donadio et al. 2018a,b, 2020, 2023; Vacchi et al. 2021; Gibbons et al. 2024). The differences mean that at least two locations are needed to convey sufficient granularity for a useful diagnostic tool.

Most studies have focused on detection of p-α-syn in the cholinergic and adrenergic nerve fibers supplying the sweat glands, small arterioles, arrector pili muscles, and subepidermal plexus, as well as quantifying their corresponding nerve fiber densities as well as intraepidermal nerve fiber density (IENFD) (Vacchi et al. 2021; Giannoccaro et al. 2022; Gibbons et al. 2023; Yuan et al. 2024). However, newer studies are expanding the search to include other localizations such as Schwann cells and macrophages (Oizumi et al. 2022: Donadio et al. 2023).

Skin biopsy immunohistochemistry studies show sensitivity and percentage positivity of p-α-syn detection in patients with PD ranging from 5.3% to 100% with specificity against healthy controls from 80% to 100% (Table 1). Difference in sensitivities and percentage of positivity across laboratories can be attributed in part to differences in patient demographics and disease severity, as well as copathology, which is difficult to exclude without neuropathological confirmation of diagnosis in nongenetic cases. Discrepancies also arise due to the large

Cite this article as *Cold Spring Harb Perspect Med* doi: 10.1101/cshperspect.a041944

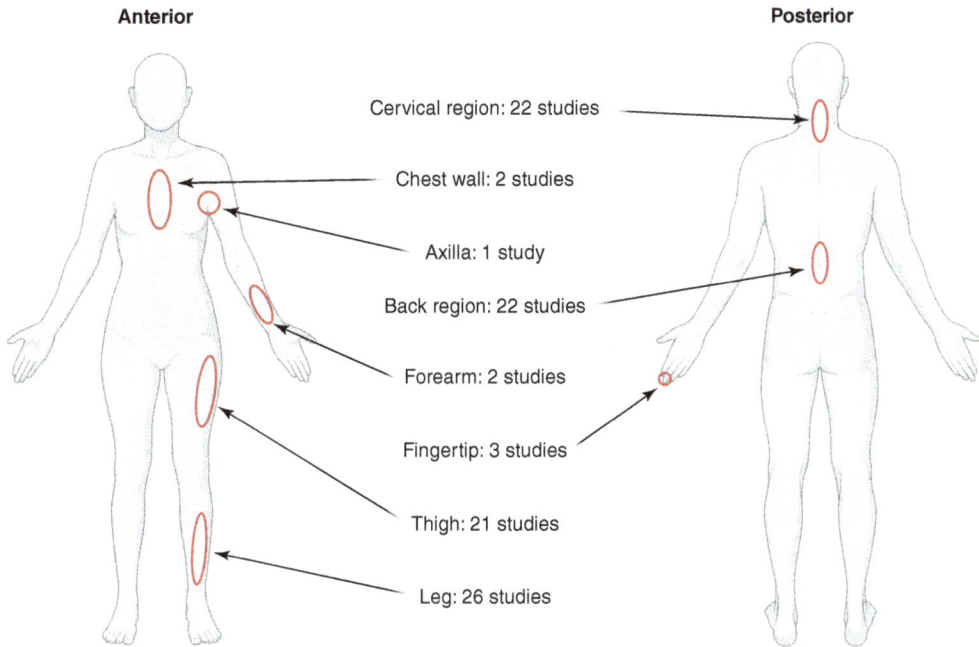

Figure 3. Skin biopsy locations. Locations used in skin biopsy studies for the detection of phosphorylated α-synuclein (p-α-syn). Studies indicated are those in Table 1. Biopsy diameters range from 3 to 6 mm, and samples are then prepared to 3–60 µm thickness. Most studies used a combination of distal and proximal sites. Sites are generally hairy skin sites to allow for sampling of autonomic nerve structures.

variability in methodology, with groups using different biopsy sites, number of samples, samples thicknesses, fixation methods, antibodies and image processing, and acquisition methods. There have been some efforts at standardization, for example, a study compared reliability of the test in different sample thicknesses to determine the optimum sample thickness to use (Wang et al. 2020a); however, more standardization is required to enable regulation of the test and application in clinical practice.

α-Syn deposition in the skin of people with genetic forms of PD has variable results of positivity depending on the genetic variant, although only a few studies published have genetic information available on the participants. A study including participants with several genetic variants including *CHCHD2*, *RAB39B*, *PRKN*, *LRRK2*, and *GBA1* found that detection of p-α-syn was highest in patients with the *LRRK2* G2385R variant (Yuan et al. 2024). This is an interesting finding as it is known that not all patients with *LRRK2*-associated PD demonstrate Lewy body

pathology at autopsy (Kalia et al. 2015), and, thus, the detection of p-α-syn in the skin in these patients could suggest a different disease process in the periphery of these patients or indicate that this particular variant is associated with synuclein pathology. In a study that included *PRKN* linked PD, another variant known to have variable and sometimes absent Lewy body pathology (Schneider and Alcalay 2017), no participants with *PRKN*-linked PD were positive for p-α-syn (Yuan et al. 2024). A study that included *GBA1* heterozygote variants found similar rates of detection and distribution of p-α-syn in these patients as they had in previous studies where genetic information was not available at a frequency of 60% (Doppler et al. 2018). A skin biopsy study looking at patients with *E46K-SNCA* mutations found a sensitivity of 100% (Carmona-Abellan et al. 2019), with higher levels of p-α-syn found in patients with pure autonomic failure (PAF) than DLB and PD. This is concordant with the pathology results published for carriers of *SNCA* mutations or multiplications, which all

Table 1. Summary of studies examining phosphorylated α-synuclein (p-α-syn) in skin samples of live participants with Parkinson's disease (PD)[a]

References	Cases	Controls	Sensitivity or percent positive (in parentheses)	Specificity or percent negative (in parentheses)
Miki et al. 2010	20 PD	N/A	(10%)	N/A
Doppler et al. 2014	31 PD	35 HC	(51.6%)	(100%)
Donadio et al. 2014	21 PD	10 VP 6 AP-tau 4 *parkin* 30 HC	C8 (100%) Thigh (52%) Leg (24%)	(100%)
Doppler et al. 2015	30 PD 12 MSA	15 AP-tau 39 HC	PD 73% MSA 75%	100%
Zange et al. 2015	10 PD 10 MSA	6 ET	PD (100%) MSA (0%)	(100%)
Haga et al. 2015	38 PD	13 MSA	PD (5.3%)	PD vs. MSA 100%
Donadio et al. 2016	16 PD 14 PAF	15 HC	PD, PAF (100%)	(100%)
Doppler et al. 2017	25 PD 18 RBD	20 HC	PD 80% RBD 55.6%	100%
Donadio et al. 2018b	15 PD 12 DLB 5 PAF 12 MSA	10 HC	PD, DLB, PAF (100%) MSA (66.7%)	(100%)
Donadio et al. 2018a	14 PD+OH 14 PD-OH	N/A	PD+OH (90%) PD-OH (38%→52% at follow-up)	N/A
Doppler et al. 2018	10 PD *GBA1*	10 HC	(60%)	(100%)
Melli et al. 2018	19 PD	13 AP (AP-syn: 5 MSA, 2 DLB, AP-tau: 4 PSP, 2 CBD), 17 HC	56%	100%
Kuzkina et al. 2019	27 PD 8 MSA-P	21 HC	PD (82%) MSA (75%)	100%

Continued

Cite this article as *Cold Spring Harb Perspect Med* doi: 10.1101/cshperspect.a041944

Table 1. *Continued*

References	Cases	Controls	Sensitivity or percent positive (in parentheses)	Specificity or percent negative (in parentheses)
Carmona-Abellan et al. 2019	7 *E46K-SNCA* (1 PD, 3 DLB, 2 PAF, 1 AC), 2 *PARK2* PD	2 HC	Glu46Lys-SNCA (100%), PARK2 PD (50%)	(100%)
Donadio et al. 2020	25 PD+OH, 25 MSA-P	N/A	PD (100%), MSA (72%)	N/A
Giannoccaro et al. 2020	21 PD, 13 MSA, 7 DLB, 13 PAF	N/A	PD (95.2%), PAF (100%), DLB 97.5%, MSA (69.2%)	LBD vs. MSA 92.3%
Wang et al. 2020a	29 PD	21 HC	10 μm sample (73%), 20 μm sample (90%), 50 μm sample (100%)	100%
Liu et al. 2020	90 PD	30 HC	(83.3%)	(100%)
Al-Qassabi et al. 2021	20 PD, 28 RBD	10 AP, 21 HC	PD (70%), RBD (82%)	HC (100%), AP (80%)
Vacchi et al. 2021	30 PD, 12 MSA	11 AP-tau, 22 HC	PD 71%, MSA 44.4%	PD vs. HC 80%, PD vs. AP-tau, 72.7%, MSA vs. HC 77.3%
Brumberg et al. 2021	21 PD, 21 MSA	N/A	PD (47.6%), MSA (81%)	PD vs. MSA 81%
Donadio et al. 2021	21 PD, 12 MSA, 6 DLB, 3 PAF	16 AD, 6 VP, 1 DIP, 2 VD, 18 AP-tau, 6 ALS	Synucleinopathies 90%	Synucleinopathies vs. nonsynucleinopathies 100%
Giannoccaro et al. 2022	26 PD	18 PSP, 8 CBD, 26 HC	(100%)	HC (100%), PSP/CBD (92.3%)
Oizumi et al. 2022	10 PD	4 HC	(100%)	(100%)

Continued

Table 1. *Continued*

References	Cases	Controls	Sensitivity or percent positive (in parentheses)	Specificity or percent negative (in parentheses)
Nolano et al. 2022	57 PD 43 MSA-P	100 HC	PD (96%) MSA-P (91%)	(100%)
Gibbons et al. 2023	54 PD 31 MSA	24 HC	MSA vs. PD 96.8%	PD vs. MSA 98.1%
Donadio et al. 2023	34 PD 29 MSA-P 17 MSA-C 16 DLB	50 HC	PD, DLB (100%) MSA (78%)	100%
Gibbons et al. 2024	96 PD 50 DLB 55 MSA 22 PAF	120 HC	PD (92.7%) MSA (98.2%) DLB (96%) PAF (100%)	(96.7%)
Yuan et al. 2024	4 CHCHD2 PD 2 RAB39B PD 16 PRKN PD 14 LRRK2 PD 5 GBA PD 100 PD	20 HC	*PRKN* PD, *RAB39B* PD (0%) *CHCHD2* PD (75%) *LRRK2* PD (78.6%) *GBA1* PD (40%) PD (72%)	(100%)

(MSA) multiple system atrophy ([-P] parkinsonian type, [-C] cerebellar type), (DLB) dementia with Lewy bodies, (PAF) pure autonomic failure, (REM) rapid eye movement, (RBD) REM sleep behavior disorder, (HCs) healthy controls, (AD) Alzheimer's dementia, (VD) vascular dementia, (PSP) progressive supranuclear palsy, (CBD) corticobasal degeneration, (AP) atypical parkinsonism ([-tau] tauopathies, i.e., PSP, CBD, [-syn] synucleinopathies, i.e., MSA, DLB), (VP) vascular parkinsonism, (DIP) drug-induced parkinsonism, (ALS) amyotrophic lateral sclerosis, (ET) essential tremor, (OH) orthostatic hypotension, (AC) asymptomatic carrier, (LBD) Lewy body disease.

[a]Parkinson's disease (PD) (idiopathic unless otherwise specified).

had Lewy bodies on autopsy (Schneider and Alcalay 2017).

Data from studies into the detection of p-α-syn in skin via immunohistochemistry has been used to develop the first commercially available synuclein biomarker detector, Syn-One (Gibbons et al. 2024). This test can be used by clinicians to collect biopsies from three sites, which are then tested for p-α-syn. This is used as a diagnostic test, with the notable limitation being its inability to differentiate between PD and other synucleinopathies. Due to the lack of 100% specificity, a negative result does not guarantee absence of disease, and so test results should only be interpreted in combination with clinical symptoms and other diagnostic tests, such as DAT-SPECT, to ensure patients with false positives are not excluded from consideration of diagnosis and treatment.

More recently, the seeding of pathological α-syn species from skin samples has been demonstrated using α-syn SAAs (Wang et al. 2020b; Donadio et al. 2021). Samples are obtained by punch biopsy from equivalent sites to those used in immunohistochemistry studies, leading to possibility of direct comparison between the methods (Donadio et al. 2021). A recent meta-analysis showed pooled sensitivity of α-syn SAA in skin samples for detecting PD of 89%, which was higher than that of immunology studies included in the same analysis (Zhao et al. 2025). However, there have been mixed results in studies comparing the methods in samples from the same patients, with a recent study showing very high agreement (κ = 0.898) between staining of p-α-syn staining in skin compared to α-syn SAA in a small subset of patients (Kuang et al. 2024), whereas only moderate agreement (κ = 0.6) was seen in an earlier study (Donadio et al. 2021). α-Syn SAA in skin samples showed good interlaboratory reproducibility (Kuang et al. 2024), a potential benefit over immunology studies, which have had mixed reproducibility, as discussed above.

Cerebrospinal Fluid (CSF)

Seeding of α-syn in CSF was first demonstrated in 2016 in a postmortem study using samples from patients with pathologically confirmed PD and DLB (Fairfoul et al. 2016). Since then, numerous studies have been published in which the SAAs have been deployed using samples from live patients, with sensitivities for idiopathic PD ranging from 79% to 97% and specificities against healthy controls from 80% to 100% (Table 2). Variation in methodology is likely to be a contributing factor to the range of results, for example, use of RT-QuIC or PMCA, and difference in the composition and concentration of reaction substrates used. Furthermore, there is no consensus on the cutoff value of fluorescence used to indicate a positive test, as demonstrated in a meta-analysis of α-syn SAA studies (Grossauer et al. 2023). Despite this, a study using randomized CSF samples from the Parkinson's Progression Markers Initiative (PPMI) cohort found good interlaboratory concordance in results across three laboratories using their own α-syn SAA protocols (Russo et al. 2021). Strict characterization of protocol will be required when moving toward application in clinical settings.

Some CSF α-syn SAA studies have looked at genetic causes of synucleinopathies as well as idiopathic forms (Garrido et al. 2019; Rossi et al. 2020; Brockmann et al. 2021; Siderowf et al. 2023). In the 2023 study by Siderowf et al., in addition to a large cohort of idiopathic PD patients, they also looked at symptomatic patients with heterozygous GBA1 and LRRK2 variants. They found that the sensitivity of the test was highest in patients with GBA1 PD (96%) and lowest in those with LRRK2 PD (68%). In all groups, it was found that hyposmia was correlated with increased sensitivity. In the LRRK2 group, the positivity of the test varied by gender, with positivity for female patients more than 20% less than their male counterparts. The youngest female participant in the study was 56, and thus all the female participants are likely to be postmenopausal. It would be interesting to see whether the sensitivity in LRRK2-positive females differs premenopause. The combination of hyposmia and female sex in LRRK2 patients resulted in sensitivity of only 12.5% (Siderowf et al. 2023). The other studies looking at genetic variants of PD also found lower sen-

Table 2. Summary of studies assessing α-synuclein (α-syn) seeding activity in cerebrospinal fluid (CSF) of live participants with Parkinson's disease (PD) using seeding amplification assays using real-time quaking-induced conversion (RT-QuIC) and protein misfolding cyclic amplification (PMCA) methods

References	Cases	Controls	Sensitivity or percent positive (in parentheses)	Specificity or percent negative (in parentheses)
Shahnawaz et al. 2017	76 PD	65 NC 18 NDD 14 AD	88.5%	96.9%
Garrido et al. 2019	10 PD 15 *LRRK2* PD 10 *LRRK2* NMC	10 HC	PD 90% *LRRK2* PD 40% *LRRK2* NMC (18.8%)	80%
van Rumund et al. 2019	53 PD 7 MSA 1 DLB 14 unspecified synucleinopathy	8 PSP 2 other tauopathy 9 VP 14 other/uncertain diagnosis 52 HC	PD 84% MSA 35% DLB 100% Other synucleinopathy 85.7%	HC 98% Nonsynucleinopathy 89%
Rossi et al. 2020	71 PD 34 DLB 31 MSA 28 PAF 18 RBD	43 AD 30 PSP/CBD 62 HC	PD 94.4% DLB 97.1% MSA 6.5% PAF 92.9% RBD 100%	HC 98.4% AD 83.7% PSP/CBD 100%
Shahnawaz et al. 2020	94 PD 75 MSA	56 NC	PD 93.6% MSA 84.6%	100%
Orrù et al. 2021	108 PD	85 HC	97.2%	87.1%
Donadio et al. 2021	21 PD 12 MSA 6 DLB 3 PAF	16 AD 6 VP 1 DIP 2 VD 18 AP-tau 6 ALS	Synucleinopathies 78%	100%

Continued

Table 2. *Continued*

References	Cases	Controls	Sensitivity or percent positive (in parentheses)	Specificity or percent negative (in parentheses)
Mammana et al. 2021	13 PD 15 DLB	12 PN 9 AD 5 VD 4 ALS 4 FTD 1 CBD 6 other	PD, DLB 89.3%	95.1%
Russo et al. 2021	30 PD	30 HC 20 SWEDD	PD 86%–96%	HC 93%–100% SWEDD 80%–85%
Brockmann et al. 2021	108 PD 99 PD *GBA1* 9 PD *LRRK2* 20 PD *parkin/PINK1/DJ1* 33 *DLB* 16 DLB *GBA1*	14 NMC 26 HC	PD (91%) PD *GBA1* (87%) PD *LRRK2* (78%) PD parkin/ *PINK1/DJ*-heterozygous (59%) PD *parkin/PINK1/DJ*-biallelic (0%) DLB (79%) DLB *GBA1* (100%)	HC (92%) NMC (86%)
Poggiolini et al. 2022	74 PD 24 MSA 45 RBD	55 HC	PD 89.2% MSA 75% RBD 64.4%	96.4%
Compta et al. 2022	20 PD 37 MSA	23 PSP 13 CBD 19 HC	PD 81% MSA 9%	HC 100% PSP/CBD 91%
Siderowf et al. 2023	373 PD 123 PD *LRRK2* 49 PD *GBA1* 51 prodromal (33 RBD, 18 hyposmia)	163 HC 54 SWEDD 310 NMC (159 *LRRK2*, 151 *GBA1*)	PD 87.7% PD *LRRK2* 67.5% PD *GBA1* 95.9% Prodromal (86%)	HC 96.3% SWEDD 90.7% NMC 92%
Chahine et al. 2023	59 PD	21 HC	92.6%	90.5%

Continued

Table 2. *Continued*

References	Cases	Controls	Sensitivity or percent positive (in parentheses)	Specificity or percent negative (in parentheses)
Bräuer et al. 2023	28 PD 47 DLB	N/A	PD (78.6%) DLB (85.1%)	N/A
Ma et al. 2024	109 PD 111 MSA 10 DLB 21 RBD	63 NC 83 HC	PD 91% MSA 87% DLB 70% RBD 95%	NC 84% HC 94%
Orrú et al. 2025[a]	928 PD 173 *LRRK2* PD 159 *GBA1* PD 14 *SNCA* PD 18 *PRKN* PD	52 PSP 248 HC	PD (92.7%)[a] *LRRK2* PD (66%) *GBA1* PD (92%) *SNCA* PD (100%) *PRKN* PD (61%)	PSP (85%) HC (91.9%)[a]

(SAA) Seed amplification assay, (RT-QuIC) real-time quaking-induced conversion, protein cyclic misfolding amplification, (NCs) neurological controls, (NDD) neurodegenerative disease, (FTD) frontotemporal dementia, (SWEDD) scans without evidence of dopaminergic deficit (referring to DAT-SPECT), (NMCs) nonmanifesting carriers, (PN) peripheral neuropathy, (SDS) sodium dodecyl sulfate, (ThT) thioflavin T, (Rec.) recombinant, others as defined in Table 1.
[a]Multiple cohorts are used in this study. Results are a combination of UK and PPMI cohorts, and others are PPMI cohort only.

sitivity of the test in *LRRK2* PD compared to *GBA1* and idiopathic PD (Garrido et al. 2019; Rossi et al. 2020; Brockmann et al. 2021; Orrú et al. 2025). This is in contrast to the findings that p-α-syn was more sensitive for *LRRK2* PD than other genetic variants, as previously discussed (Yuan et al. 2024), the pathophysiology underpinning these differences requires further investigation, provided these findings can be validated. Differences in kinetics have been observed between genetic variants of PD, with *GBA1* PD shown to have faster seeding than idiopathic PD, which in turn has faster seeding than *LRRK2* PD (Orrú et al. 2025). This could indicate difference in structure of α-syn fibrils in different generic forms of the disease. This difference in structure could account for clinical heterogeneity between genetic forms of PD and indeed within sporadic PD, considering distinct kinetic profiles have recently been demonstrated in a study using recursive α-syn SAA in idiopathic PD patients (Bräuer et al. 2025).

Although CSF studies have a large body of evidence supporting their use, the fact that CSF must be obtained by lumbar puncture limits its practicality as a biomarker source. This is due to patient aversion and reliance on trained medical professionals to perform the procedure. Compared to other biofluids and tissues used for biomarker detection, this is the greatest limitation of CSF when considering translation to clinical practice.

Blood

Blood is an easily obtained sample and thus is an attractive biofluid for biomarker detection. Blood-based α-syn SAAs have been developed; however, their implementation is complicated by the effect human serum albumin (HSA) and other proteins found in serum have on the aggregation of α-syn (Bellomo et al. 2019). Different groups have looked at how best to mitigate these effects, via immunoprecipitation (Okuzumi et al. 2023), dilution–centrifugation (Kuang et al. 2025), and isolation of neuron-derived extracellular vesicles (EVs) from blood samples (Kluge et al. 2022; Schaeffer et al. 2024). All three methods have proved successful, with

sensitivities to detect PD ranging from 80.4% to 98.8% (Schaeffer et al. 2024; Kuang et al. 2025) and specificities >90% against healthy controls (Okuzumi et al. 2023; Kluge et al. 2024a; Kuang et al. 2025). Of interest, blood-derived EV α-syn SAA was positive in eight of 13 patients with biallelic *PRKN* mutations (Kluge et al. 2024a,b,c), in contrast to an immunoprecipitation study in which 17 patients with biallelic *PRKN* mutations had negative assay results (Okuzumi et al. 2023). In the EV study, the fibrils formed in the reaction of one of the *PRKN* PD patients were comparable in structure to those formed by the patients with idiopathic PD, suggesting that pathophysiology of *PRKN*-related PD may be similar to idiopathic PD in some patients (Kluge et al. 2024a,b,c), despite studies showing an absence of Lewy body pathology in the majority of autopsies with *PRKN* variants (Schneider and Alcalay 2017). Despite these findings, there is current debate on the methodology of EV α-syn SAAs (Bernhardt et al. 2024; Kluge et al. 2024a,b,c) and further refinement is needed to standardize the procedure to validate these findings.

Other Tissues

GI tissue samples have been a target of α-syn biomarker studies due to the knowledge that α-syn accumulates in the enteric nervous system (ENS) in PD (Lebouvier et al. 2008). There have been numerous studies on GI tissue samples using a range of techniques. Studies looking at deposition of total α-syn (t-α-syn) and p-α-syn have shown low specificity, with p-α-syn found to be present in samples from all healthy controls in some studies (Antunes et al. 2016; Corbillé et al. 2017). Higher specificity has been seen in recent α-syn SAA studies using GI tissue samples (Fenyi et al. 2019; Vascellari et al. 2023); if further validated in larger cohorts, α-syn SAA in GI tissue may emerge as a suitable PD biomarker; however, the invasiveness of sampling, requiring an endoscopy or colonoscopy, is a limitation.

α-Syn SAA in olfactory epithelium, and saliva samples are limited and have had varied results. Studies into olfactory mucosal samples have had sensitivities for diagnosing PD in the range of 46%–69% (De Luca et al. 2019; Bargar et al. 2021; Stefani et al. 2021; Kuzkina et al. 2023). It has been proposed that the presence of SARS-CoV-2 virus within the nasal cavity could be a confounding factor in olfactory studies, and thus more postpandemic studies are needed to reduce this effect. Salivary studies have yielded higher sensitivities (Luan et al. 2022, 2024; Vivacqua et al. 2023); however, healthy controls had false-positive results in up to 22% of cases (Vivacqua et al. 2023). In a recent study, combining the salivary α-syn SAA with measurements of salivary microRNA-29a-3p increased the sensitivity for both PD and MSA, and notably increased the specificity against both healthy controls and controls with essential tremor, an important disease mimic (Luan et al. 2024). This demonstrates how evaluating multiple biomarkers from a single sample type can improve diagnostic value. α-syn SAA has also recently been applied in urine using samples from DLB patients with confirmed CSF α-syn SAA positivity; however, sensitivity was low (Bsoul et al. 2025).

Investigating α-syn SAA using samples from different sources could give information on the origin of pathology in PD patients, separating them into "body-first" and "brain-first" subtypes. A 2023 study found differing rates of positivity for skin and neuro-olfactory epithelial α-syn SAAs in PD patients and postulated that the patients who were positive for the neuro-olfactory assay were of the "brain-first" subtype and those with skin positivity were of the "body-first" subtype (Kuzkina et al. 2023). Further work comparing other samples associated with both CNS and PNS pathology could help classify patients into these two subtypes, which could be incorporated into the biological definition of the disease and aid grouping of patients for clinical trial design.

SUCCESSES AND PITFALLS

Differentiating from Disease Mimics

Differentiating between PD and other synucleinopathies can present difficulties in clinical practice as they can present similarly in early stages of disease. Differentiating between them is important as the diseases have different treatments and prognoses and could help guide management. Furthermore, in clinical trial recruitment, it is important to be able to correctly categorize patients by disease type to ensure homogeneity of study participants. Therefore, a useful synuclein biomarker must be able to differentiate between PD and other synucleinopathies, as well as other disease mimics.

p-α-Syn skin biomarkers can differentiate between PD and MSA with up to 100% accuracy (Doppler et al. 2015; Haga et al. 2015; Donadio et al. 2020). Differentiation is achieved by studying differences between the distribution and location of p-α-syn in the biopsy samples. A proximodistal gradient of deposition is reported in PD (Doppler et al. 2014; Donadio et al. 2016, 2018b; Melli et al. 2018; Giannoccaro et al. 2022; Gibbons et al. 2023, 2024). Some studies demonstrate a distal–proximal gradient in MSA (Donadio et al. 2018b, 2020, 2023); however, the largest study to date shows a more uniform distribution (Gibbons et al. 2024). The differences mean that at least two locations are needed for a useful diagnostic tool.

Notable differences have been found between PD and other synucleinopathies in the distribution of pathological α-syn throughout cutaneous structures. In PD and DLB, depositions are most commonly found in autonomic fibers innervating sweat glands, arrector pili muscles, and small arterioles (Doppler et al. 2015; Donadio et al. 2020, 2023; Giannoccaro et al. 2020; Liu et al. 2020; Wang et al. 2020a; Gibbons et al. 2023, 2024). In MSA, depositions have been shown predominantly in the somatic fibers of the subepidermal plexus (Doppler et al. 2015; Donadio et al. 2018b, 2020, 2023; Gibbons et al. 2023, 2024), and a recent study also showed depositions in Schwann cells (Donadio et al. 2023). Length-dependent small fiber neuropathy is found to differ between disease types and is found to be more common in DLB and PD than MSA (Haga et al. 2015; Brumberg et al. 2021; Gibbons et al. 2024). This means that by combining the results of IENFD with distribution of p-α-syn allowed for increased certainty when differentiating between diseases.

Differentiating between disease types when using SAAs is achieved by considering kinetic parameters of the assays. Higher maximum intensity and longer time to reach maximum intensity was seen in PD compared to MSA in CSF studies (Shahnawaz et al. 2020; Singer et al. 2021; Poggiolini et al. 2022). A multicenter cohort CSF α-syn SAA study expanded on this by categorizing the intensities into high fluorescence "type 1" seeds, produced by samples containing Lewy bodies, and intermediate fluorescence "type 2" seeds, produced by samples containing glial cytoplasmic inclusions (Ma et al. 2024). In this study, sensitivity of type 1 seeds for PD and DLB was 97% and 86%, respectively, and type 2 seeds had sensitivity of 77% for MSA. The assay had 94% specificity against healthy controls (Ma et al. 2024). The differences in kinetics are likely to be due to the differences in structure of α-syn fibrils produced from the substrate, as it has been shown that fibrils produced from samples from patients with PD and DLB have distinct morphology to the two polymorphs produced from MSA patient samples (Okuzumi et al. 2023). This corresponds with the knowledge that the core of pathological α-syn filaments in PD and DLB are composed of the Lewy fold, whereas in MSA there are two cores, formed from two protofilaments each (Schweighauser et al. 2020).

Separate from differentiating PD from other synucleinopathies, being able to exclude tauopathies, especially progressive supranuclear palsy (PSP) and corticobasal degeneration (CBD), is an essential function of a biomarker for PD, as overlap in symptoms with these diseases contributes to diagnostic uncertainty. As these diseases have a worse prognosis and different treatment approaches compared with PD, it is essential to prevent misdiagnoses, as it can have a devastating effect on the well-being of patients. Differentiation from tauopathies can be achieved by using tau protein as a biomarker in combination with α-syn. In such a study, quantification of tau amount at two biopsy sites gave sensitivity of 90% for detecting tauopathy and specificity of 69% for differentiating tauopathies from PD (Vacchi et al. 2022). The same biopsy sites were used in this study and the pre-vious α-syn biomarker study by the same group (Vacchi et al. 2021); this gives the possibility in the future of testing biopsy samples for both tau and α-syn, which could help to give greater confidence in diagnosis.

Normal pressure hydrocephalus (NPH) is another condition that often poses a diagnostic challenge due to gait abnormalities reminiscent of those seen in synucleinopathies. In a study of 50 patients with a diagnosis of idiopathic NPH, 14% of the patients ($n = 7$) were positive on α-syn SAA. These patients were more responsive to levodopa and had a wider-based gait than those with negative α-syn SAA results but otherwise did not differ clinically (Fasano et al. 2022). Another study with 143 NPH patients showed a higher α-syn SAA positivity rate of 30.1% (Weber et al. 2025). The patients in this study with positive assay results were more likely to have clinical signs and symptoms associated with synucleinopathies, such as poor olfaction, hallucinations, and upper limb rigidity (Weber et al. 2025). Another study, with 33 patients with NPH, had a similar α-syn SAA positivity rate of 33.3%, and both the synuclein positive and negative groups showed improvement in the timed up and go test and number of steps, but the results for the positive group was not statistically significant (Yu et al. 2025). These findings indicate that NPH patients may have a comorbid synucleinopathy, or rather they may be misdiagnosed as NPH. This highlights the challenge in accurately diagnosing NPH and differentiating it from PD and other synucleinopathies.

The potential confounder of copathology is also demonstrated when considering lack of specificity of α-syn against AD. A study looked at α-syn SAA positivity in 240 AD patients and found positivity in 30% of these patients, compared to 87% of controls with PD/DLB and 9% of healthy controls (Bellomo et al. 2024). Positivity in AD patients was correlated with cognitive decline, visuospatial impairment, and behavioral disturbances (Bellomo et al. 2024). α-Syn copathology may have contributed to the lower specificity of this assay for diagnosing synucleinopathy, as it has been shown that 33% of AD patients have concomitant Lewy body disease (DeTure and Dickson 2019). This result demonstrates the importance of correlating biomarker results with clinical presentation, as in these patients, using the positive biomarker alone would have given an incorrect diagnosis of Lewy body disease, when the primary diagnosis was AD. Testing for α-syn in AD patients may be useful in clinical trials as disease-modifying therapies for AD targeting tau or amyloid may exclude patients with concomitant Lewy body disease, and thus there may be a need for the development of alternative treatments.

Specificity of α-syn biomarkers against other disease mimics, such as essential tremor, vascular parkinsonism, drug-induced parkinsonism, and other neurodegenerative diseases, ranges from 67% to 100% (Donadio et al. 2014, 2017, 2021; Zange et al. 2015; Shahnawaz et al. 2017; van Rumund et al. 2019; Rossi et al. 2020; Perra et al. 2021), with the lower limit seen in patients with vascular induced parkinsonism. In the absence of pathological confirmation of diagnosis, it is hard to quantify the true specificity of α-syn biomarkers against these diseases and rule out copathology. Further longitudinal studies in which biomarker results can be verified postmortem are needed to help give further information on the rate of copathology in patients with negative biomarker results. This validation could lead to refinement of assays, negating the impact of copathology on test results.

Monitoring Disease Progression

There are no biomarkers for disease progression in PD, and α-syn-based assays have been studied for their efficacy in prognostication. Correlations between α-syn deposition and disease severity have been reported in some skin biopsy studies. An early study found correlation between deposition of soluble α-syn and autonomic dysfunction and disease severity (Wang et al. 2013). These findings have been replicated in studies that have looked at p-α-syn deposition, with positive correlation found between total p-α-syn deposition and autonomic dysfunction (Gibbons et al. 2016, 2024; Donadio et al. 2018a) and disease severity scores, such as Hoehn and Yahr and MDS-Unified Parkinson's Disease Rating Scale (MDS-UPDRS) (Doppler et al. 2017; Gibbons

et al. 2023, 2024). However, other studies looking at these measures did not find significant correlations (Zange et al. 2015; Donadio et al. 2016; Kuzkina et al. 2019).

There have been similarly mixed results when correlating test positivity and disease severity scores in α-syn SAA studies. Some studies across different mediums have shown correlation between time to reach maximum intensity and disease rating scales such as Hoehn and Yahr score and MDS-UPDRS Part III (UPDRS-III) (Concha-Marambio et al. 2023; Okuzumi et al. 2023) and disease duration (Okuzumi et al. 2023). In contrast, a recent blood-derived EV α-syn SAA study showed inverse correlation between maximum fluorescence of the reaction and disease duration and UPDRS-III score but no correlation between time to reach this intensity and duration or UPDRS-III score (Schaeffer et al. 2024). Furthermore, a large CSF α-syn SAA study showed no significant correlation between severity scales in idiopathic PD patients and assay parameters (Siderowf et al. 2023).

Correlation between kinetic assay parameters with cognitive performance of patients has been reported (Bräuer et al. 2023, 2025; Orrú et al. 2025). For example, Orrú et al. (2025) demonstrated that faster seeding of α-syn SAA predicted cognitive decline in patients independently to copathology with AD, in accordance with another study using recursive α-syn SAA (Bräuer et al. 2025).

Overall, the conflicting results across studies indicate that no α-syn biomarker currently has strong enough evidence for use as a prognostic marker, meaning we may have to look elsewhere for this. Interestingly, in skin biopsy studies looking at nerve density, IENFD was found to correlate to disease severity and duration in some cases (Donadio et al. 2016; Brumberg et al. 2021; Vacchi et al. 2021). More specific information about how IENFD can relate to the nature of progression of disease was shown in a study following a cohort of PD patients over 24 months (Vacchi et al. 2021). They found that a reduction of IENFD in the cervical region was associated with increased risk of developing cognitive decline, and reduction in IENFD in the distal leg was associated with increased risk of developing motor impairment (Vacchi et al. 2021). This is an exciting discovery, which, if validated in larger studies, would make IENFD an important prognostic marker in PD.

Use in Prodromal Disease

There are overlapping prodromal disorders for the synucleinopathies, such as hyposmia and idiopathic rapid eye movement (REM) sleep behavior disorder (RBD). There is currently no objective way to predict the likelihood of conversion to manifest disease in a patient presenting with RBD. Having more certainty about risk of conversion could allow for patients and clinicians to better plan follow-up over the prodromal phase. Furthermore, prodromal patients would be ideal candidates for potential disease-modifying therapies for PD, giving further value to the study of biomarker validity in these patients.

α-Syn biomarkers have been studied in patients with RBD, with detection of p-α-syn demonstrated in patients with RBD in skin (Antelmi et al. 2017, 2019; Doppler et al. 2017; Al-Qassabi et al. 2021; Liguori et al. 2023), colonic mucosa (Sprenger et al. 2015), and parotid gland samples (Fernández-Arcos et al. 2018), and seeding of skin (Liguori et al. 2023; Kuang et al. 2024), CSF (Rossi et al. 2020, 2021; Iranzo et al. 2021, 2023; Siderowf et al. 2023), serum (Okuzumi et al. 2023), and olfactory mucosal samples (Stefani et al. 2021) demonstrated in α-syn SAAs. Skin and CSF samples have the largest bodies of evidence and sensitivity has been reported as high as 100% using CSF α-syn SAA (Rossi et al. 2020), although a recent comparative study of skin and CSF found detection of p-α-syn in skin to have higher diagnostic accuracy than α-syn SAA in skin and CSF samples (Liguori et al. 2023). More interesting than biomarker positivity in patients with RBD is the risk of phenoconversion to PD or other synucleinopathies. In longitudinal studies, detection of p-α-syn in skin (Doppler et al. 2017; Al-Qassabi et al. 2021) and α-syn SAA positivity in CSF (Iranzo et al. 2021, 2023) and serum samples (Okuzumi et al. 2023) has been shown to correlate with phenoconversion to PD. In one study of α-syn SAA using blood-derived EVs, assay

positivity was found up to 10 years before diagnosis of PD (Kluge et al. 2024c). Presence of p-α-syn in skin (Doppler et al. 2017; Antelmi et al. 2019; Al-Qassabi et al. 2021) and α-syn SAA positivity in CSF (Siderowf et al. 2023) and olfactory mucosa (Stefani et al. 2021) is correlated with hyposmia, and inversely correlated with dopamine transporter binding, with α-syn detection preceding abnormal DAT-SPECT results (Antelmi et al. 2019; Siderowf et al. 2023). Linear regression modeling combining hyposmia scores, DAT-SPECT results, and α-syn detection may be used in future to predict phenoconversion of RBD to manifest PD; however, more longitudinal studies are needed to validate the role of α-syn biomarkers in such a tool.

α-Syn and the Biologic Definition of Parkinson's Disease

In recent years, there has been a movement to establish a biological definition of PD to allow for more precise categorization of the disease, removing the subjectivity of the current clinical definition in life, and negating the need for pathological confirmation postmortem. The aim of such a system is to aid categorization of patients in clinical trial design, allowing for greater homogeneity of participants, as has been done in Alzheimer's disease with the NIA-AA framework (Jack et al. 2018). Two proposed classifications have been published at the time of writing: the "neuronal synuclein disease international staging system (NSD-ISS)," proposed by Simuni et al. (2024), and the "SynNeurGe" classification, proposed by Höglinger et al. (2024). Both systems center on defining disease based on detection of pathological α-syn, genetic factors, and evidence of neurodegeneration, with the SynNeurGe classification also including clinical features. The presence of α-syn is intrinsic to the definition of disease in the NSD-ISS, and so variants of what we currently consider to be PD that do not have Lewy body disease, such as *LRRK2* and *PRKN*-linked PD, are classified as separate disease entities. Despite their clinical features fitting into the current clinical definitions of PD, the detection of α-syn in patients with these variants using assays described in the text could challenge this declaration. Thus, accurate detection of α-syn using assays as described in this work is essential to the validity of such classification systems.

CONCLUSIONS

The development of tools to detect α-syn in vivo have validated its potential for use as a biomarker to accurately diagnose PD, prognosticate disease trajectories and differentiate it from other diseases. There is hope that these tools will be able to be incorporated into biological definitions of PD and allow for categorization of disease, which could improve participant selection for clinical trials and lead to increased hope of developing disease-modifying treatments for the disease. At present, α-syn assays have been applied in a wide range of different biofluids and tissue samples. The challenge of choosing the most suitable biomarker for widespread use in clinical and research settings is to balance factors that impact patient tolerance of the procedure such as invasiveness, with accuracy of the assay, particularly by limiting false positives. Limitations of current assays include lack of standardization of methodology, small sample sizes with lack of diversity in participants, paucity of longitudinal studies, and reliance on clinical criteria rather than validation via postmortem examination, which is the current gold standard for diagnosis of PD.

Considering that on autopsy neurodegenerative disorders, including PD, typically have a diversity of pathologies, the use of one biomarker may not be sufficient for diagnosis during life. We foresee the need for a panel of biomarkers including α-syn, tau, β-amyloid, and other disease-related proteins, possibly from different sample sources, to fully capture the spectrum of pathology seen in these disorders. Further study is needed to investigate which biomarkers should be used to aid in the diagnosis and treatment of these complex disorders.

REFERENCES

*Reference is also in this subject collection.

Agin-Liebes J, Lodge A, Reddy H, Vacchi E, Usseglio J, Honig LS, Melli G, Noble JM, Przedborski S. 2025.

α-Synuclein biomarker assays: bridging research and patient care. *Lancet Neurol* **24:** 681–697. doi:10.1016/S1474-4422(25)00194-2

Al-Qassabi A, Tsao TS, Racolta A, Kremer T, Cañamero M, Belousov A, Santana MA, Beck RC, Zhang H, Meridew J, et al. 2021. Immunohistochemical detection of synuclein pathology in skin in idiopathic rapid eye movement sleep behavior disorder and parkinsonism. *Mov Disord* **36:** 895–904. doi:10.1002/mds.28399

Altay MF, Kumar ST, Burtscher J, Jagannath S, Strand C, Miki Y, Parkkinen L, Holton JL, Lashuel HA. 2023. Development and validation of an expanded antibody toolset that captures alpha-synuclein pathological diversity in Lewy body diseases. *NPJ Parkinsons Dis* **9:** 161. doi:10.1038/s41531-023-00604-y

Antelmi E, Donadio V, Incensi A, Plazzi G, Liguori R. 2017. Skin nerve phosphorylated α-synuclein deposits in idiopathic REM sleep behavior disorder. *Neurology* **88:** 2128–2131. doi:10.1212/WNL.0000000000003989

Antelmi E, Pizza F, Donadio V, Filardi M, Sosero YL, Incensi A, Vandi S, Moresco M, Ferri R, Marelli S, et al. 2019. Biomarkers for REM sleep behavior disorder in idiopathic and narcoleptic patients. *Ann Clin Transl Neurol* **6:** 1872–1876. doi:10.1002/acn3.50833

Antunes L, Frasquilho S, Ostaszewski M, Weber J, Longhino L, Antony P, Baumuratov A, Buttini M, Shannon KM, Balling R, et al. 2016. Similar α-synuclein staining in the colon mucosa in patients with Parkinson's disease and controls. *Mov Disord* **31:** 1567–1570. doi:10.1002/mds.26702

Bargar C, De Luca CMG, Devigili G, Elia AE, Cilia R, Portaleone SM, Wang W, Tramacere I, Bistaffa E, Cazzaniga FA, et al. 2021. Discrimination of MSA-P and MSA-C by RT-QuIC analysis of olfactory mucosa: the first assessment of assay reproducibility between two specialized laboratories. *Mol Neurodegener* **16:** 82. doi:10.1186/s13024-021-00491-y

Beach TG, Adler CH, Sue LI, Vedders L, Lue L, White CL III, Akiyama H, Caviness JN, Shill HA, Sabbagh MN, et al. 2010. Multi-organ distribution of phosphorylated alpha-synuclein histopathology in subjects with Lewy body disorders. *Acta Neuropathol* **119:** 689–702. doi:10.1007/s00401-010-0664-3

Bellomo G, Bologna S, Cerofolini L, Paciotti S, Gatticchi L, Ravera E, Parnetti L, Fragai M, Luchinat C. 2019. Dissecting the interactions between human serum albumin and α-synuclein: new insights on the factors influencing α-synuclein aggregation in biological fluids. *J Phys Chem B* **123:** 4380–4386. doi:10.1021/acs.jpcb.9b02381

Bellomo G, Toja A, Paolini Paoletti F, Ma Y, Farris CM, Gaetani L, Salvadori N, Chiasserini D, Wojdała AL, Concha-Marambio L, et al. 2024. Investigating alpha-synuclein co-pathology in Alzheimer's disease by means of cerebrospinal fluid alpha-synuclein seed amplification assay. *Alzheimers Dement* **20:** 2444–2452. doi:10.1002/alz.13658

Bernhardt AM, Nemati M, Boros FA, Hopfner F, Levin J, Mollenhauer B, Winkler J, Zerr I, Zunke F, Höglinger G. 2024. α-Synuclein seed amplification assays from blood-based extracellular vesicles in Parkinson's disease: an evaluation of the evidence. *Mov Disord* **39:** 1269–1271. doi:10.1002/mds.29923

Braak H, Del Tredici K, Rüb U, de Vos RA, Jansen Steur EN, Braak E. 2003. Staging of brain pathology related to sporadic Parkinson's disease. *Neurobiol Aging* **24:** 197–211. doi:10.1016/S0197-4580(02)00065-9

Bräuer S, Rossi M, Sajapin J, Henle T, Gasser T, Parchi P, Brockmann K, Falkenburger BH. 2023. Kinetic parameters of alpha-synuclein seed amplification assay correlate with cognitive impairment in patients with Lewy body disorders. *Acta Neuropathol Commun* **11:** 162. doi:10.1186/s40478-023-01653-3

Bräuer S, Schniewind I, Dinter E, Falkenburger BH. 2025. Recursive seed amplification detects distinct α-synuclein strains in cerebrospinal fluid of patients with Parkinson's disease. *Acta Neuropathol Commun* **13:** 13. doi:10.1186/s40478-024-01923-8

Brockmann K, Quadalti C, Lerche S, Rossi M, Wurster I, Baiardi S, Roeben B, Mammana A, Zimmermann M, Hauser AK, et al. 2021. Association between CSF alpha-synuclein seeding activity and genetic status in Parkinson's disease and dementia with Lewy bodies. *Acta Neuropathol Commun* **9:** 175. doi:10.1186/s40478-021-01276-6

Brumberg J, Kuzkina A, Lapa C, Mammadova S, Buck A, Volkmann J, Sommer C, Isaias IU, Doppler K. 2021. Dermal and cardiac autonomic fiber involvement in Parkinson's disease and multiple system atrophy. *Neurobiol Dis* **153:** 105332. doi:10.1016/j.nbd.2021.105332

Bsoul R, McWilliam OH, Waldemar G, Hasselbalch SG, Simonsen AH, von Buchwald C, Bech M, Pinborg CH, Pedersen CK, Baungaard SO, et al. 2025. Accurate detection of pathologic α-synuclein in CSF, skin, olfactory mucosa, and urine with a uniform seeding amplification assay. *Acta Neuropathol Commun* **13:** 113. doi:10.1186/s40478-025-02034-8

Burré J, Sharma M, Tsetsenis T, Buchman V, Etherton MR, Südhof TC. 2010. Alpha-synuclein promotes SNARE-complex assembly in vivo and in vitro. *Science* **329:** 1663–1667. doi:10.1126/science.1195227

Burré J, Sharma M, Südhof TC. 2018. Cell biology and pathophysiology of α-synuclein. *Cold Spring Harb Perspect Med* **8:** a024091. doi:10.1101/cshperspect.a024091

Calabresi P, Mechelli A, Natale G, Volpicelli-Daley L, Di Lazzaro G, Ghiglieri V. 2023. Alpha-synuclein in Parkinson's disease and other synucleinopathies: from overt neurodegeneration back to early synaptic dysfunction. *Cell Death Dis* **14:** 176. doi:10.1038/s41419-023-05672-9

Carmona-Abellan M, Gabilondo I, Murueta-Goyena A, Khurana V, Tijero B, Luquin MR, Acera M, Del Pino R, Gardeazabal J, Martínez-Valbuena I, et al. 2019. Small fiber neuropathy and phosphorylated alpha-synuclein in the skin of E46K-SNCA mutation carriers. *Parkinsonism Relat Disord* **65:** 139–145. doi:10.1016/j.parkreldis.2019.05.038

Chahine LM, Beach TG, Adler CH, Hepker M, Kanthasamy A, Appel S, Pritzkow S, Pinho M, Mosovsky S, Serrano GE, et al. 2023. Central and peripheral α-synuclein in Parkinson disease detected by seed amplification assay. *Ann Clin Transl Neurol* **10:** 696–705. doi:10.1002/acn3.51753

Compta Y, Painous C, Soto M, Pulido-Salgado M, Fernández M, Camara A, Sánchez V, Bargalló N, Caballol N, Pont-Sunyer C, et al. 2022. Combined CSF α-SYN RT-QuIC,

CSF NFL and midbrain-pons planimetry in degenerative parkinsonisms: from bedside to bench, and back again. *Parkinsonism Relat Disord* **99:** 33–41. doi:10.1016/j.parkreldis.2022.05.006

Concha-Marambio L, Pritzkow S, Shahnawaz M, Farris CM, Soto C. 2023. Seed amplification assay for the detection of pathologic alpha-synuclein aggregates in cerebrospinal fluid. *Nat Protoc* **18:** 1179–1196. doi:10.1038/s41596-022-00787-3

Corbillé AG, Preterre C, Rolli-Derkinderen M, Coron E, Neunlist M, Lebouvier T, Derkinderen P. 2017. Biochemical analysis of α-synuclein extracted from control and Parkinson's disease colonic biopsies. *Neurosci Lett* **641:** 81–86. doi:10.1016/j.neulet.2017.01.050

De Luca CMG, Elia AE, Portaleone SM, Cazzaniga FA, Rossi M, Bistaffa E, De Cecco E, Narkiewicz J, Salzano G, Carletta O, et al. 2019. Efficient RT-QuIC seeding activity for α-synuclein in olfactory mucosa samples of patients with Parkinson's disease and multiple system atrophy. *Transl Neurodegener* **8:** 24. doi:10.1186/s40035-019-0164-x

DeTure MA, Dickson DW. 2019. The neuropathological diagnosis of Alzheimer's disease. *Mol Neurodegener* **14:** 32. doi:10.1186/s13024-019-0333-5

Donadio V, Incensi A, Leta V, Giannoccaro MP, Scaglione C, Martinelli P, Capellari S, Avoni P, Baruzzi A, Liguori R. 2014. Skin nerve α-synuclein deposits: a biomarker for idiopathic Parkinson disease. *Neurology* **82:** 1362–1369. doi:10.1212/WNL.0000000000000316

Donadio V, Incensi A, Piccinini C, Cortelli P, Giannoccaro MP, Baruzzi A, Liguori R. 2016. Skin nerve misfolded α-synuclein in pure autonomic failure and Parkinson disease. *Ann Neurol* **79:** 306–316. doi:10.1002/ana.24567

Donadio V, Incensi A, Rizzo G, Capellari S, Pantieri R, Stanzani Maserati M, Devigili G, Eleopra R, Defazio G, Montini F, et al. 2017. A new potential biomarker for dementia with Lewy bodies: skin nerve α-synuclein deposits. *Neurology* **89:** 318–326. doi:10.1212/WNL.0000000000004146

Donadio V, Incensi A, Del Sorbo F, Rizzo G, Infante R, Scaglione C, Modugno N, Fileccia E, Elia AE, Cencini F, et al. 2018a. Skin nerve phosphorylated α-synuclein deposits in Parkinson disease with orthostatic hypotension. *J Neuropathol Exp Neurol* **77:** 942–949. doi:10.1093/jnen/nly074

Donadio V, Incensi A, El-Agnaf O, Rizzo G, Vaikath N, Del Sorbo F, Scaglione C, Capellari S, Elia A, Stanzani Maserati M, et al. 2018b. Skin α-synuclein deposits differ in clinical variants of synucleinopathy: an in vivo study. *Sci Rep* **8:** 14246. doi:10.1038/s41598-018-32588-8

Donadio V, Incensi A, Rizzo G, De Micco R, Tessitore A, Devigili G, Del Sorbo F, Bonvegna S, Infante R, Magnani M, et al. 2020. Skin biopsy may help to distinguish multiple system atrophy-parkinsonism from Parkinson's disease with orthostatic hypotension. *Mov Disord* **35:** 1649–1657. doi:10.1002/mds.28126

Donadio V, Wang Z, Incensi A, Rizzo G, Fileccia E, Vacchiano V, Capellari S, Magnani M, Scaglione C, Stanzani Maserati M, et al. 2021. In vivo diagnosis of synucleinopathies: a comparative study of skin biopsy and RT-QuIC. *Neurology* **96:** e2513–e2524. doi:10.1212/WNL.0000000000011935

Donadio V, Incensi A, Rizzo G, Westermark GT, Devigili G, De Micco R, Tessitore A, Nyholm D, Parisini S, Nyman D, et al. 2023. Phosphorylated α-synuclein in skin Schwann cells: a new biomarker for multiple system atrophy. *Brain* **146:** 1065–1074. doi:10.1093/brain/awac124

Doppler K, Ebert S, Üçeyler N, Trenkwalder C, Ebentheuer J, Volkmann J, Sommer C. 2014. Cutaneous neuropathy in Parkinson's disease: a window into brain pathology. *Acta Neuropathol* **128:** 99–109. doi:10.1007/s00401-014-1284-0

Doppler K, Weis J, Karl K, Ebert S, Ebentheuer J, Trenkwalder C, Klebe S, Volkmann J, Sommer C. 2015. Distinctive distribution of phospho-alpha-synuclein in dermal nerves in multiple system atrophy. *Mov Disord* **30:** 1688–1692. doi:10.1002/mds.26293

Doppler K, Jentschke HM, Schulmeyer L, Vadasz D, Janzen A, Luster M, Höffken H, Mayer G, Brumberg J, Booij J, et al. 2017. Dermal phospho-alpha-synuclein deposits confirm REM sleep behaviour disorder as prodromal Parkinson's disease. *Acta Neuropathol* **133:** 535–545. doi:10.1007/s00401-017-1684-z

Doppler K, Brockmann K, Sedghi A, Wurster I, Volkmann J, Oertel WH, Sommer C. 2018. Dermal phospho-alpha-synuclein deposition in patients with Parkinson's disease and mutation of the glucocerebrosidase gene. *Front Neurol* **9:** 1094. doi:10.3389/fneur.2018.01094

Fairfoul G, McGuire LI, Pal S, Ironside JW, Neumann J, Christie S, Joachim C, Esiri M, Evetts SG, Rolinski M, et al. 2016. Alpha-synuclein RT-QuIC in the CSF of patients with alpha-synucleinopathies. *Ann Clin Transl Neurol* **3:** 812–818. doi:10.1002/acn3.338

Fasano A, Martinez-Valbuena I, Azevedo P, da Silva CC, Algarni M, Vasilevskaya A, Anastassiadis C, Taghdiri F, Kongkham P, Radovanovic I, et al. 2022. Alpha-synuclein RT-QuIC in idiopathic normal pressure hydrocephalus. *Ann Neurol* **92:** 985–991. doi:10.1002/ana.26505

Fenyi A, Leclair-Visonneau L, Clairembault T, Coron E, Neunlist M, Melki R, Derkinderen P, Bousset L. 2019. Detection of alpha-synuclein aggregates in gastrointestinal biopsies by protein misfolding cyclic amplification. *Neurobiol Dis* **129:** 38–43. doi:10.1016/j.nbd.2019.05.002

Fernández-Arcos A, Vilaseca I, Aldecoa I, Serradell M, Tolosa E, Santamaría J, Gelpi E, Iranzo A. 2018. Alpha-synuclein aggregates in the parotid gland of idiopathic REM sleep behavior disorder. *Sleep Med* **52:** 14–17. doi:10.1016/j.sleep.2018.08.003

Garrido A, Fairfoul G, Tolosa ES, Martí MJ, Green A, Baracelona LRRK2 Study Group. 2019. Alpha-synuclein RT-QuIC in cerebrospinal fluid of LRRK2-linked Parkinson's disease. *Ann Clin Transl Neurol* **6:** 1024–1032. doi:10.1002/acn3.772

George JM, Jin H, Woods WS, Clayton DF. 1995. Characterization of a novel protein regulated during the critical period for song learning in the zebra finch. *Neuron* **15:** 361–372. doi:10.1016/0896-6273(95)90040-3

Giannoccaro MP, Donadio V, Giannini G, Devigili G, Rizzo G, Incensi A, Cason E, Calandra-Buonaura G, Eleopra R, Cortelli P, et al. 2020. Comparison of 123I-MIBG scintigraphy and phosphorylated α-synuclein skin deposits in synucleinopathies. *Parkinsonism Relat Disord* **81:** 48–53. doi:10.1016/j.parkreldis.2020.10.016

Giannoccaro MP, Avoni P, Rizzo G, Incensi A, Infante R, Donadio V, Liguori R. 2022. Presence of skin α-synuclein deposits discriminates Parkinson's disease from progressive supranuclear palsy and corticobasal syndrome. *J Parkinsons Dis* **12:** 585–591. doi:10.3233/JPD-212904

Gibbons CH, Griffin JW, Polydefkis M, Bonyhay I, Brown A, Hauer PE, McArthur JC. 2006. The utility of skin biopsy for prediction of progression in suspected small fiber neuropathy. *Neurology* **66:** 256–258. doi:10.1212/01.wnl.0000194314.86486.a2

Gibbons CH, Garcia J, Wang N, Shih LC, Freeman R. 2016. The diagnostic discrimination of cutaneous α-synuclein deposition in Parkinson disease. *Neurology* **87:** 505–512. doi:10.1212/WNL.0000000000002919

Gibbons C, Wang N, Rajan S, Kern D, Palma JA, Kaufmann H, Freeman R. 2023. Cutaneous α-synuclein signatures in patients with multiple system atrophy and Parkinson disease. *Neurology* **100:** e1529–e1539. doi:10.1212/WNL.0000000000206772

Gibbons CH, Levine T, Adler C, Bellaire B, Wang N, Stohl J, Agarwal P, Aldridge GM, Barboi A, Evidente VGH, et al. 2024. Skin biopsy detection of phosphorylated α-synuclein in patients with synucleinopathies. *JAMA* **331:** 1298–1306. doi:10.1001/jama.2024.0792

Gilman S, Wenning GK, Low PA, Brooks DJ, Mathias CJ, Trojanowski JQ, Wood NW, Colosimo C, Dürr A, Fowler CJ, et al. 2008. Second consensus statement on the diagnosis of multiple system atrophy. *Neurology* **71:** 670–676. doi:10.1212/01.wnl.0000324625.00404.15

Goedert M, Griesinger C, Outeiro TF, Riek R, Schröder GF, Spillantini MG. 2024. Abandon the NAC in α-synuclein. *Lancet Neurol* **23:** 669. doi:10.1016/S1474-4422(24)00176-5

Grossauer A, Hemicker G, Krismer F, Peball M, Djamshidian A, Poewe W, Seppi K, Heim B. 2023. α-Synuclein seed amplification assays in the diagnosis of synucleinopathies using cerebrospinal fluid—a systematic review and meta-analysis. *Mov Disord Clin Pract* **10:** 737–747. doi:10.1002/mdc3.13710

Guo JL, Lee VM. 2014. Cell-to-cell transmission of pathogenic proteins in neurodegenerative diseases. *Nat Med* **20:** 130–138. doi:10.1038/nm.3457

Haga R, Sugimoto K, Nishijima H, Miki Y, Suzuki C, Wakabayashi K, Baba M, Yagihashi S, Tomiyama M. 2015. Clinical utility of skin biopsy in differentiating between Parkinson's disease and multiple system atrophy. *Parkinsons Dis* **2015:** 167038.

Höglinger GU, Adler CH, Berg D, Klein C, Outeiro TF, Poewe W, Postuma R, Stoessl AJ, Lang AE. 2024. A biological classification of Parkinson's disease: the SynNeurGe research diagnostic criteria. *Lancet Neurol* **23:** 191–204. doi:10.1016/S1474-4422(23)00404-0

Ikemura M, Saito Y, Sengoku R, Sakiyama Y, Hatsuta H, Kanemaru K, Sawabe M, Arai T, Ito G, Iwatsubo T, et al. 2008. Lewy body pathology involves cutaneous nerves. *J Neuropathol Exp Neurol* **67:** 945–953. doi:10.1097/NEN.0b013e318186de48

Iranzo A, Fairfoul G, Ayudhaya ACN, Serradell M, Gelpi E, Vilaseca I, Sanchez-Valle R, Gaig C, Santamaria J, Tolosa E, et al. 2021. Detection of α-synuclein in CSF by RT-QuIC in patients with isolated rapid-eye-movement sleep behaviour disorder: a longitudinal observational study.

Cite this article as *Cold Spring Harb Perspect Med* doi: 10.1101/cshperspect.a041944

Lancet Neurol **20**: 203–212. doi:10.1016/S1474-4422(20)30449-X

Iranzo A, Mammana A, Muñoz-Lopetegi A, Dellavalle S, Mayà G, Rossi M, Serradell M, Baiardi S, Arqueros A, Quadalti C, et al. 2023. Misfolded α-synuclein assessment in the skin and CSF by RT-QuIC in isolated REM sleep behavior disorder. *Neurology* **100**: e1944–e1954. doi:10.1212/WNL.0000000000207147

Jack CR, Bennett DA, Blennow K, Carrillo MC, Dunn B, Haeberlein SB, Holtzman DM, Jagust W, Jessen F, Karlawish J, et al. 2018. NIA-AA research framework: toward a biological definition of Alzheimer's disease. *Alzheimers Dement* **14**: 535–562. doi:10.1016/j.jalz.2018.02.018

Kalia LV, Lang AE, Hazrati LN, Fujioka S, Wszolek ZK, Dickson DW, Ross OA, Van Deerlin VM, Trojanowski JQ, Hurtig HI, et al. 2015. Clinical correlations with Lewy body pathology in *LRRK2*-related Parkinson disease. *JAMA Neurol* **72**: 100–105. doi:10.1001/jamaneurol.2014.2704

Kluge A, Bunk J, Schaeffer E, Drobny A, Xiang W, Knacke H, Bub S, Lückstädt W, Arnold P, Lucius R, et al. 2022. Detection of neuron-derived pathological α-synuclein in blood. *Brain* **145**: 3058–3071. doi:10.1093/brain/awac115

Kluge A, Borsche M, Streubel-Gallasch L, Gül T, Schaake S, Balck A, Prasuhn J, Campbell P, Morris HR, Schapira AH, et al. 2024a. α-Synuclein pathology in *PRKN*-linked Parkinson's disease: new insights from a blood-based seed amplification assay. *Ann Neurol* **95**: 1173–1177. doi:10.1002/ana.26917

Kluge A, Schaeffer E, Berg D. 2024b. Response to viewpoint by Bernhardt et al. *Mov Disord* **39**: 1272–1275. doi:10.1002/mds.29919

Kluge A, Schaeffer E, Bunk J, Sommerauer M, Röttgen S, Schulte C, Roeben B, von Thaler AK, Welzel J, Lucius R, et al. 2024c. Detecting misfolded α-synuclein in blood years before the diagnosis of Parkinson's disease. *Mov Disord* **39**: 1289–1299. doi:10.1002/mds.29766

Koga S, Sekiya H, Kondru N, Ross OA, Dickson DW. 2021. Neuropathology and molecular diagnosis of synucleinopathies. *Mol Neurodegener* **16**: 83. doi:10.1186/s13024-021-00501-z

Kuang Y, Mao H, Gan T, Guo W, Dai W, Huang W, Wu Z, Li H, Huang X, Yang X, et al. 2024. A skin-specific α-synuclein seeding amplification assay for diagnosing Parkinson's disease. *NPJ Parkinsons Dis* **10**: 129. doi:10.1038/s41531-024-00738-7

Kuang Y, Mao H, Dai W, Gan T, Lin H, Li J, Yang X, Xu P. 2025. Enhanced serum-based seed amplification assay for detecting propagative α-synuclein seeds in Parkinson's disease. *Transl Neurodegener* **14**: 24. doi:10.1186/s40035-025-00488-3

Kuzkina A, Schulmeyer L, Monoranu CM, Volkmann J, Sommer C, Doppler K. 2019. The aggregation state of α-synuclein deposits in dermal nerve fibers of patients with Parkinson's disease resembles that in the brain. *Parkinsonism Relat Disord* **64**: 66–72. doi:10.1016/j.parkreldis.2019.03.003

Kuzkina A, Rößle J, Seger A, Panzer C, Kohl A, Maltese V, Musacchio T, Blaschke SJ, Tamgüney G, Kaulitz S, et al. 2023. Combining skin and olfactory α-synuclein seed amplification assays (SAA)—towards biomarker-driven

phenotyping in synucleinopathies. *NPJ Parkinsons Dis* **9**: 79. doi:10.1038/s41531-023-00519-8

Lautenschläger J, Stephens AD, Fusco G, Ströhl F, Curry N, Zacharopoulou M, Michel CH, Laine R, Nespovitaya N, Fantham M, et al. 2018. C-terminal calcium binding of α-synuclein modulates synaptic vesicle interaction. *Nat Commun* **9**: 712. doi:10.1038/s41467-018-03111-4

Lebouvier T, Chaumette T, Damier P, Coron E, Touchefeu Y, Vrignaud S, Naveilhan P, Galmiche JP, Bruley des Varannes S, Derkinderen P, et al. 2008. Pathological lesions in colonic biopsies during Parkinson's disease. *Gut* **57**: 1741–1743. doi:10.1136/gut.2008.162503

Lee JM, Derkinderen P, Kordower JH, Freeman R, Munoz DG, Kremer T, Zago W, Hutten SJ, Adler CH, Serrano GE, et al. 2017. The search for a peripheral biopsy indicator of α-synuclein pathology for Parkinson's disease. *J Neuropathol Exp Neurol* **76**: 2–15.

Liguori R, Donadio V, Wang Z, Incensi A, Rizzo G, Antelmi E, Biscarini F, Pizza F, Zou W, Plazzi G. 2023. A comparative blind study between skin biopsy and seed amplification assay to disclose pathological α-synuclein in RBD. *NPJ Parkinsons Dis* **9**: 34. doi:10.1038/s41531-023-00473-5

Liu X, Yang J, Yuan Y, He Q, Gao Y, Jiang C, Li L, Xu Y. 2020. Optimization of the detection method for phosphorylated α-synuclein in Parkinson disease by skin biopsy. *Front Neurol* **11**: 569446. doi:10.3389/fneur.2020.569446

Luan M, Sun Y, Chen J, Jiang Y, Li F, Wei L, Sun W, Ma J, Song L, Liu J, et al. 2022. Diagnostic value of salivary real-time quaking-induced conversion in Parkinson's disease and multiple system atrophy. *Mov Disord* **37**: 1059–1063. doi:10.1002/mds.28976

Luan M, Wei L, Sun Y, Chen J, Jiang Y, Wu W, Li F, Sun W, Zhu L, Wang Z, et al. 2024. Combining salivary α-synuclein seeding activity and miRNA-29a to distinguish Parkinson's disease and multiple system atrophy. *Parkinsonism Relat Disord* **127**: 107088. doi:10.1016/j.parkreldis.2024.107088

Ma Y, Farris CM, Weber S, Schade S, Nguyen H, Pérez-Soriano A, Giraldo DM, Fernández M, Soto M, Cámara A, et al. 2024. Sensitivity and specificity of a seed amplification assay for diagnosis of multiple system atrophy: a multicentre cohort study. *Lancet Neurol* **23**: 1225–1237. doi:10.1016/S1474-4422(24)00395-8

Mammana A, Baiardi S, Quadalti C, Rossi M, Donadio V, Capellari S, Liguori R, Parchi P. 2021. RT-QuIC detection of pathological α-synuclein in skin punches of patients with Lewy body disease. *Mov Disord* **36**: 2173–2177. doi:10.1002/mds.28651

Maroteaux L, Campanelli JT, Scheller RH. 1988. Synuclein: a neuron-specific protein localized to the nucleus and presynaptic nerve terminal. *J Neurosci* **8**: 2804–2815. doi:10.1523/JNEUROSCI.08-08-02804.1988

Melli G, Vacchi E, Biemmi V, Galati S, Staedler C, Ambrosini R, Kaelin-Lang A. 2018. Cervical skin denervation associates with alpha-synuclein aggregates in Parkinson disease. *Ann Clin Transl Neurol* **5**: 1394–1407. doi:10.1002/acn3.669

Miki Y, Tomiyama M, Ueno T, Haga R, Nishijima H, Suzuki C, Mori F, Kaimori M, Baba M, Wakabayashi K. 2010. Clinical availability of skin biopsy in the diagnosis of

Parkinson's disease. *Neurosci Lett* **469**: 357–359. doi:10
.1016/j.neulet.2009.12.027

Nielsen MS, Vorum H, Lindersson E, Jensen PH. 2001. Ca^{2+}
binding to alpha-synuclein regulates ligand binding and
oligomerization. *J Biol Chem* **276**: 22680–22684. doi:10
.1074/jbc.M101181200

Nolano M, Caporaso G, Manganelli F, Stancanelli A, Borreca
I, Mozzillo S, Tozza S, Dubbioso R, Iodice R, Vitale F, et al.
2022. Phosphorylated α-synuclein deposits in cutaneous
nerves of early parkinsonism. *J Parkinsons Dis* **12**: 2453–
2468. doi:10.3233/JPD-223421

Oizumi H, Yamasaki K, Suzuki H, Ohshiro S, Saito Y, Murayama S, Sugimura Y, Hasegawa T, Fukunaga K, Takeda
A. 2022. Phosphorylated alpha-synuclein in Iba1-positive
macrophages in the skin of patients with Parkinson's disease. *Ann Clin Transl Neurol* **9**: 1136–1146. doi:10.1002/
acn3.51610

Okuzumi A, Hatano T, Matsumoto G, Nojiri S, Ueno SI,
Imamichi-Tatano Y, Kimura H, Kakuta S, Kondo A, Fukuhara T, et al. 2023. Propagative α-synuclein seeds as
serum biomarkers for synucleinopathies. *Nat Med* **29**:
1448–1455. doi:10.1038/s41591-023-02358-9

Orrù CD, Ma TC, Hughson AG, Groveman BR, Srivastava A,
Galasko D, Angers R, Downey P, Crawford K, Hutten SJ,
et al. 2021. A rapid α-synuclein seed assay of Parkinson's
disease CSF panel shows high diagnostic accuracy. *Ann
Clin Transl Neurol* **8**: 374–384. doi:10.1002/acn3.51280

Orrú CD, Vaughan DP, Vijiaratnam N, Real R, Martinez-
Carrasco A, Fumi R, Jensen MT, Hodgson M, Girges C,
Gil-Martinez AL, et al. 2025. Diagnostic and prognostic
value of α-synuclein seed amplification assay kinetic measures in Parkinson's disease: a longitudinal cohort study.
Lancet Neurol **24**: 580–590. doi:10.1016/S1474-4422(25)
00157-7

Perra D, Bongianni M, Novi G, Janes F, Bessi V, Capaldi S,
Sacchetto L, Tagliapietra M, Schenone G, Morbelli S, et al.
2021. Alpha-synuclein seeds in olfactory mucosa and cerebrospinal fluid of patients with dementia with Lewy
bodies. *Brain Commun* **3**: fcab045. doi:10.1093/brain
comms/fcab045

Poggiolini I, Gupta V, Lawton M, Lee S, El-Turabi A, Que-
rejeta-Coma A, Trenkwalder C, Sixel-Döring F, Foubert-
Samier A, Pavy-Le Traon A, et al. 2022. Diagnostic value
of cerebrospinal fluid alpha-synuclein seed quantification
in synucleinopathies. *Brain* **145**: 584–595. doi:10.1093/
brain/awab431

Rey NL, Bousset L, George S, Madaj Z, Meyerdirk L, Schulz
E, Steiner JA, Melki R, Brundin P. 2019. α-Synuclein
conformational strains spread, seed and target neuronal
cells differentially after injection into the olfactory bulb.
Acta Neuropathol Commun **7**: 221. doi:10.1186/s40478-
019-0859-3

Rossi M, Candelise N, Baiardi S, Capellari S, Giannini G,
Orrù CD, Antelmi E, Mammana A, Hughson AG, Calandra-Buonaura G, et al. 2020. Ultrasensitive RT-QuIC assay with high sensitivity and specificity for Lewy body-
associated synucleinopathies. *Acta Neuropathol* **140**: 49–
62. doi:10.1007/s00401-020-02160-8

Rossi M, Baiardi S, Teunissen CE, Quadalti C, van de Beek
M, Mammana A, Stanzani Maserati M, Van der Flier
WM, Sambati L, Zenesini C, et al. 2021. Diagnostic value
of the CSF α-synuclein real-time quaking-induced con-

version assay at the prodromal MCI stage of dementia
with Lewy bodies. *Neurology* **97**: e930–e940. doi:10
.1212/WNL.0000000000012438

Russo MJ, Orru CD, Concha-Marambio L, Giaisi S, Groveman BR, Farris CM, Holguin B, Hughson AG, LaFontant
DE, Caspell-Garcia C, et al. 2021. High diagnostic performance of independent alpha-synuclein seed amplification assays for detection of early Parkinson's disease.
Acta Neuropathol Commun **9**: 179. doi:10.1186/s40478-
021-01282-8

Schaeffer E, Kluge A, Schulte C, Deuschle C, Bunk J, Welzel
J, Maetzler W, Berg D. 2024. Association of misfolded α-
synuclein derived from neuronal exosomes in blood with
Parkinson's disease diagnosis and duration. *J Parkinsons
Dis* **14**: 667–679. doi:10.3233/JPD-230390

Schneider SA, Alcalay RN. 2017. Neuropathology of genetic
synucleinopathies with parkinsonism: review of the literature. *Mov Disord* **32**: 1504–1523. doi:10.1002/mds
.27193

Schweighauser M, Shi Y, Tarutani A, Kametani F, Murzin
AG, Ghetti B, Matsubara T, Tomita T, Ando T, Hasegawa
K, et al. 2020. Structures of α-synuclein filaments from
multiple system atrophy. *Nature* **585**: 464–469. doi:10
.1038/s41586-020-2317-6

Shahnawaz M, Tokuda T, Waragai M, Mendez N, Ishii R,
Trenkwalder C, Mollenhauer B, Soto C. 2017. Development of a biochemical diagnosis of Parkinson disease by
detection of α-synuclein misfolded aggregates in cerebrospinal fluid. *JAMA Neurol* **74**: 163–172. doi:10.1001/jama
neurol.2016.4547

Shahnawaz M, Mukherjee A, Pritzkow S, Mendez N, Raba-
dia P, Liu X, Hu B, Schmeichel A, Singer W, Wu G, et al.
2020. Discriminating α-synuclein strains in Parkinson's
disease and multiple system atrophy. *Nature* **578**: 273–
277. doi:10.1038/s41586-020-1984-7

Siderowf A, Concha-Marambio L, Lafontant DE, Farris CM,
Ma Y, Urenia PA, Nguyen H, Alcalay RN, Chahine LM,
Foroud T, et al. 2023. Assessment of heterogeneity among
participants in the Parkinson's progression markers initiative cohort using α-synuclein seed amplification: a
cross-sectional study. *Lancet Neurol* **22**: 407–417. doi:10
.1016/S1474-4422(23)00109-6

Simuni T, Chahine LM, Poston K, Brumm M, Buracchio T,
Campbell M, Chowdhury S, Coffey C, Concha-Marambio
L, Dam T, et al. 2024. A biological definition of neuronal
α-synuclein disease: towards an integrated staging system
for research. *Lancet Neurol* **23**: 178–190. doi:10.1016/
S1474-4422(23)00405-2

Singer W, Schmeichel AM, Shahnawaz M, Schmelzer JD,
Sletten DM, Gehrking TL, Gehrking JA, Olson AD, Sua-
rez MD, Misra PP, et al. 2021. Alpha-synuclein oligomers
and neurofilament light chain predict phenoconversion
of pure autonomic failure. *Ann Neurol* **89**: 1212–1220.
doi:10.1002/ana.26089

Sprenger FS, Stefanova N, Gelpi E, Seppi K, Navarro-Otano
J, Offner F, Vilas D, Valldeoriola F, Pont-Sunyer C, Aldecoa I, et al. 2015. Enteric nervous system α-synuclein
immunoreactivity in idiopathic REM sleep behavior disorder. *Neurology* **85**: 1761–1768. doi:10.1212/WNL.00
00000000002126

Stefani A, Iranzo A, Holzknecht E, Perra D, Bongianni M,
Gaig C, Heim B, Serradell M, Sacchetto L, Garrido A, et al.

Cite this article as *Cold Spring Harb Perspect Med* doi: 10.1101/cshperspect.a041944

2021. Alpha-synuclein seeds in olfactory mucosa of patients with isolated REM sleep behaviour disorder. *Brain* **144:** 1118–1126. doi:10.1093/brain/awab005

Tsukita K, Sakamaki-Tsukita H, Tanaka K, Suenaga T, Takahashi R. 2019. Value of in vivo α-synuclein deposits in Parkinson's disease: a systematic review and meta-analysis. *Mov Disord* **34:** 1452–1463. doi:10.1002/mds.27794

Uéda K, Fukushima H, Masliah E, Xia Y, Iwai A, Yoshimoto M, Otero DA, Kondo J, Ihara Y, Saitoh T. 1993. Molecular cloning of cDNA encoding an unrecognized component of amyloid in Alzheimer disease. *Proc Natl Acad Sci* **90:** 11282–11286. doi:10.1073/pnas.90.23.11282

Vacchi E, Kaelin-Lang A, Melli G. 2020. Tau and alpha synuclein synergistic effect in neurodegenerative diseases: when the periphery is the core. *Int J Mol Sci* **21:** 5030. doi:10.3390/ijms21145030

Vacchi E, Senese C, Chiaro G, Disanto G, Pinton S, Morandi S, Bertaina I, Bianco G, Staedler C, Galati S, et al. 2021. Alpha-synuclein oligomers and small nerve fiber pathology in skin are potential biomarkers of Parkinson's disease. *NPJ Parkinsons Dis* **7:** 119. doi:10.1038/s41531-021-00262-y

Vacchi E, Lazzarini E, Pinton S, Chiaro G, Disanto G, Marchi F, Robert T, Staedler C, Galati S, Gobbi C, et al. 2022. Tau protein quantification in skin biopsies differentiates tauopathies from alpha-synucleinopathies. *Brain* **145:** 2755–2768. doi:10.1093/brain/awac161

van Rumund A, Green AJE, Fairfoul G, Esselink RAJ, Bloem BR, Verbeek MM. 2019. α-Synuclein real-time quaking-induced conversion in the cerebrospinal fluid of uncertain cases of parkinsonism. *Ann Neurol* **85:** 777–781. doi:10.1002/ana.25447

Vascellari S, Orrù CD, Groveman BR, Parveen S, Fenu G, Pisano G, Piga G, Serra G, Oppo V, Murgia D, et al. 2023. α-Synuclein seeding activity in duodenum biopsies from Parkinson's disease patients. *PLoS Pathog* **19:** e1011456. doi:10.1371/journal.ppat.1011456

* Vekrellis K, Emmanouilidou E, Xilouri M, Stefanis L. 2024. α-synuclein in Parkinson's disease: 12 years later. *Cold Spring Harb Perspect Med* **14:** a041645. doi:10.1101/cshperspect.a041645

Vivacqua G, Mason M, De Bartolo MI, Węgrzynowicz M, Calò L, Belvisi D, Suppa A, Fabbrini G, Berardelli A, Spillantini M. 2023. Salivary α-synuclein RT-QuIC correlates with disease severity in de novo Parkinson's disease. *Mov Disord* **38:** 153–155. doi:10.1002/mds.29246

Wang N, Gibbons CH, Lafo J, Freeman R. 2013. α-Synuclein in cutaneous autonomic nerves. *Neurology* **81:** 1604–1610. doi:10.1212/WNL.0b013e3182a9f449

Wang N, Garcia J, Freeman R, Gibbons CH. 2020a. Phosphorylated alpha-synuclein within cutaneous autonomic nerves of patients with Parkinson's disease: the implications of sample thickness on results. *J Histochem Cytochem* **68:** 669–678. doi:10.1369/0022155420960250

Wang Z, Becker K, Donadio V, Siedlak S, Yuan J, Rezaee M, Incensi A, Kuzkina A, Orrú CD, Tatsuoka C, et al. 2020b. Skin α-synuclein aggregation seeding activity as a novel biomarker for Parkinson disease. *JAMA Neurol* **78:** 30–40. doi:10.1001/jamaneurol.2020.3311

Weber S, Farris CM, Ma Y, Dakna M, Starke M, Schade S, Bartl M, Trenkwalder C, Concha-Marambio L, Mollenhauer B. 2025. Anosmia and upper limb rigidity—a potential phenotype of idiopathic normal pressure hydrocephalus with cerebrospinal fluid α-synuclein seeds. *Mov Disord* **40:** 1206–1213. doi:10.1002/mds.30184

Yu JK, Choi HJ, Kim D, Ngoc PH, Kwak IH, Nguyen HD, Thanh TN, Song SJ, Lee J, Ma HI, et al. 2025. Prevalence and clinical impact of alpha-synuclein pathology in idiopathic normal pressure hydrocephalus: insights from RT-QuIC assay. *Parkinsonism Relat Disord* **136:** 107857. doi:10.1016/j.parkreldis.2025.107857

Yuan Y, Wang Y, Liu M, Luo H, Liu X, Li L, Mao C, Yang T, Li S, Zhang X, et al. 2024. Peripheral cutaneous synucleinopathy characteristics in genetic Parkinson's disease. *Front Neurol* **15:** 1404492. doi:10.3389/fneur.2024.1404492

Zange L, Noack C, Hahn K, Stenzel W, Lipp A. 2015. Phosphorylated α-synuclein in skin nerve fibres differentiates Parkinson's disease from multiple system atrophy. *Brain* **138:** 2310–2321. doi:10.1093/brain/awv138

Zhang J, Li X, Li JD. 2019. The roles of post-translational modifications on α-synuclein in the pathogenesis of Parkinson's diseases. *Front Neurosci* **13:** 381. doi:10.3389/fnins.2019.00381

Zhang S, Zhu R, Pan B, Xu H, Olufemi MF, Gathagan RJ, Li Y, Zhang L, Zhang J, Xiang W, et al. 2023. Author correction: Post-translational modifications of soluble α-synuclein regulate the amplification of pathological α-synuclein. *Nat Neurosci* **26:** 2250. doi:10.1038/s41593-023-01474-6

Zhao Y, Luan M, Liu J, Wang Q, Deng J, Wang Z, Sun Y, Li K. 2025. Skin α-synuclein assays in diagnosing Parkinson's disease: a systematic review and meta-analysis. *J Neurol* **272:** 326. doi:10.1007/s00415-025-12978-5

Modeling Parkinson's Disease in Primates

Erwan Bezard,[1,2,3] Margaux Teil,[1] Marie-Laure Arotcarena,[1] Gregory Porras,[2] Qin Li,[3] and Benjamin Dehay[1]

[1]Université de Bordeaux, Centre National de la Recherche Scientifique (CNRS), IMN, UMR 5293, F-33000 Bordeaux, France

[2]Motac Neuroscience, F-33270 Floirac, France

[3]Motac Beijing Services, PRC-100050 Beijing, China

Correspondence: erwan.bezard@u-bordeaux.fr

Decades of research have identified the pathological and pathophysiological hallmarks of Parkinson's disease (PD): profound deficit in brain dopamine and other monoamines, pathological α-synuclein aggregation, synaptic and neuronal network dysfunction, aberrant proteostasis, altered energy homeostasis, inflammation, and neuronal cell death. The purpose of this contribution is to present the phenocopy aspect, pathogenic, and etiologic nonhuman primate (NHP) models of PD to readers with limited prior knowledge of PD so that they are ready to start working on PD. How NHPs, the closest species to man on which we can model diseases, contribute to the knowledge progress and how these models represent an invaluable translational step in therapeutic development are highlighted.

PHENOMENOLOGY AND PATHOLOGY OF PARKINSON'S DISEASE

The first issue is to recognize what we aim for in modeling Parkinson's disease (PD).[4] PD is characterized, in part, by a progressive loss of dopamine neurons in the substantia nigra pars compacta (SNc) (Poewe et al. 2017). The progressive dorsoventral dopamine depletion of the striatum causes the classical motor signs of bradykinesia, rigidity, postural instability, and resting tremor (Poewe et al. 2017; Blesa et al. 2022). These symptoms are improved by current dopamine-replacement strategies, including levodopa (L-dopa, the precursor of dopamine). After several years of disease progression, gait impairments, balance problems, and freezing-of-gait episodes affect up to 90% of PD patients. Symptomatic relief is also complicated over time by the onset of motor fluctuations and L-dopa-induced dyskinesia (LID), leading to alternating periods of reduced mobility and abnormal involuntary movements (Blesa et al. 2022). Some patients receiving dopaminergic drugs also develop abnormal behaviors, including impulse control disorders or dopamine dysregulation syndrome (Voon et al. 2017).

Besides SNc dopamine neurons, the neurodegenerative process involves other neurotransmitters such as noradrenergic, serotoninergic,

---

[4]This is an update to a previous article published in *Cold Spring Harbor Perspectives in Medicine* [Porras et al. (2012). *Cold Spring Harb Perpect Med* 2: a009308. doi:10.1101/cshperspect.a009308].

Copyright © 2025 Cold Spring Harbor Laboratory Press; all rights reserved
Cite this article as *Cold Spring Harb Perspect Med* doi: 10.1101/cshperspect.a041612

and cholinergic systems (Poewe et al. 2017; Giguère et al. 2018). Nonmotor signs such as depression, neuropsychological deficits (notably with executive functions culminating into dementia in late stages), sleep abnormalities, speech and swallowing difficulties, hyposmia (anosmia), and autonomic failure (which manifests as symptomatic orthostatic hypotension, urinary incontinence, and constipation) are the likely consequences of degeneration of both dopaminergic (SNc and ventral tegmental area) and nondopaminergic systems (Giguère et al. 2018): cholinergic neurons in the pedunculopontine nucleus (PPN), noradrenergic neurons of the locus coeruleus (LC), cholinergic neurons of the nucleus basalis of Meynert (NBM) and of the dorsal motor nucleus of the vagus (DMV), and serotonergic neurons of the raphe nuclei (RN).

In most of the clinically typical PD cases, besides neurodegeneration, the hallmark pathological finding is the presence of Lewy pathology (LP) in the SNc (Braak et al. 2003; Halliday et al. 2011). LP means the presence of α-synuclein-positive intracytoplasmic inclusions present in the neurons, called Lewy bodies (LBs), as well as in axons and dendrites, called Lewy neurites (LNs), and is found across the central, peripheral, and enteric nervous system (CNS, PNS, and ENS) (Poewe et al. 2017). Both cellular inclusions, LB and LN, are predominantly formed of aggregated α-synuclein but also include many different molecules, proteins, and organelles, such as ubiquitin, tubulin, neurofilaments, lipids, and mitochondria (Shahmoradian et al. 2019).

Nonhuman primate (NHP) models of PD, much like other model categories, can be categorized into three main types: phenocopic, pathogenic, and etiologic models (Bezard and Przedborski 2011). Phenocopic models aim at replicating the spectrum of motor and nonmotor symptoms of PD. On the other hand, pathogenic models attempt to replicate the observed pathology in PD—that is, multisystemic neurodegeneration, whole-body LP and central/peripheral neuroinflammation, and etiologic models focus on studying the consequences of genetic manipulation of PARK loci.

MPTP MODELS: PHENOCOPIC MODELS

The literature often refers to "the" 1-methyl-4-phenyl-1,2,3,6-tetrahydropyridine (MPTP) monkey model of PD, but we will learn that we should instead speak about MPTP monkey models of PD. First, a few species have been used over the years. The Old World primate species currently used in PD research feature Chlorocebus sabaeus (African green monkey, also called the vervet monkey), Macaca mulatta (rhesus), and Macaca fascicularis (cynomolgus), whereas the representatives of the New World monkeys are Saimiri sciureus (squirrel monkey) and Callithrix jacchus (marmoset). They have been used to varying degrees and for specific purposes that will be covered in this review. Second, to further add to the complexity, many intoxication regimens or administration methods have been used over the years. Unilateral models have first been developed after intracarotidian administration of MPTP to macaques (Benazzouz et al. 1993; Kordower et al. 1995). This method was soon followed by various regimens of systemic administration, through intramuscular, subcutaneous, or intravenous routes, over a few days, weeks, or months (Langston et al. 1984; Rose et al. 1993; Bezard et al. 1997, 2001; Morissette et al. 1999; Iravani et al. 2003; Jan et al. 2003). Infusion of MPTP with an osmotic minipump can also be applied over a few months, followed by final titration with systemic intramuscular shots (Bélanger et al. 2003).

MPTP-Induced Anatomopathology

The primary effect of MPTP exposure will be to deplete dopamine (DA) levels through the induction of nigrostriatal degeneration. Doubts were raised in the early days (and still persist despite the evidence) about whether the patterns of nigrostriatal fiber loss were similar after MPTP exposure and in PD. Low-dose MPTP exposure produces a pattern of nigrostriatal degeneration characteristic of that seen in PD, in which there is a greater depletion of dopaminergic markers in the putamen than in the caudate nucleus (Moratalla et al. 1992; Jan et al. 2003; Iravani et al. 2005), especially at the posterior rostrocaudal level. Moreover, within the regions of diminished dopaminergic markers, there is substantial preservation in striosomes relative to the surrounding matrix (Moratalla et al. 1992). Higher doses or stronger regimens of intoxication, although generating more comprehensive behavioral phenotypes, abolish these differences with a more homogenous pattern of denervation in putamen and caudate, whereas the ventral striatum remains less affected (Bezard et al. 2001; Guigoni et al. 2005; Fernagut et al. 2010).

Interestingly, the MPTP neurotoxin is often presented as being "selective of DA neurons." Hence, few studies have investigated either the DA loss outside the nigrostriatal system or the loss of other neurotransmitters. Pifl et al. (1991) executed an insightful study in which they measured the tissue concentration of the monoamines DA, noradrenaline, and serotonin in 45 brain regions in macaque monkeys chronically intoxicated with systemic MPTP over a few weeks. None of the three major brain monoamine neuron systems was completely resistant to the neurotoxin. In addition, each brain monoamine had a characteristic regional pattern of MPTP-induced changes. The most significant alterations were found within the nigrostriatal dopamine system (i.e., profound DA loss in the caudate nucleus, putamen, and SNc). However, many extrastriatal regions of the subcortex and brainstem also suffered significant loss of dopamine, with the noradrenaline loss in the regionally subdivided brainstem being less widespread and the serotonin levels least affected. Thus, in the subcortex/brainstem, the ranking order of sensitivity to MPTP was DA > noradrenaline > serotonin. All three monoamine neuron systems suffered widespread statistically significant losses in the cerebral (neo- and limbic) cortex. This study showed that in the macaque monkey, MPTP mimicked some of the extrastriatal DA, noradrenaline, and serotonin changes often seen in the brain of patients with idiopathic PD, in addition to the profound striatal dopamine loss (Pifl et al. 1991). Since these early contributions, we confirmed and extended the findings using MALDI imaging, demonstrating unequivocally that multiple neurotransmitter levels were impacted throughout the brain in a similar pattern

to what has been reported in PD (Shariatgorji et al. 2019; Fridjonsdottir et al. 2021). Finally, MPTP toxicity outside the brain should also be considered. The impact of MPTP on the DA system in the spinal cord (Barraud et al. 2010) and even the ENS has been analyzed in that respect (Chaumette et al. 2009).

A classic opposition to the model is the lack of LB, the pathological landmark of PD, in the MPTP monkey models. Although no LB has so far been observed in these models (Halliday et al. 2009), a few reports have investigated the expression, regulation, and/or pattern of an LB major constituent, α-synuclein, after MPTP exposure. Purisai et al. (2005) were instrumental in this topic, delivering pivotal findings (also, McCormack et al. 2008), later confirmed in other NHP subspecies and in macaques (Kowall et al. 2000; Huang et al. 2018; Deffains et al. 2020). They studied the relationship between toxic injury and α-synuclein expression in the SNc of squirrel monkeys treated with a single injection of MPTP, which were sacrificed 1 week or 1 month later. At 1 week, when stereological cell counting revealed only a small decrease (10%) in the number of dopaminergic neurons, α-synuclein mRNA and protein were markedly enhanced. Increased α-synuclein immunoreactivity was evident at the level of neuronal fibers, whereas nigral cell bodies were devoid of detectable protein. At 1 month post-MPTP, neuronal loss rose to 40%. Both α-synuclein mRNA and protein remained elevated, but noticeably, a robust α-synuclein immunoreactivity characterized many cell bodies. More importantly, they found that the vast majority of α-synuclein-immunoreactive neurons contained neuromelanin granules (Purisai et al. 2005), the hallmark of dopaminergic neurons in primates and man, and ~80% of the dopaminergic cell bodies that survived MPTP toxicity stained positive for α-synuclein.

Following these seminal data, we investigated the issue throughout the brain of MPTP-intoxicated macaques, looking at α-synuclein, tau, and gliosis (Deffains et al. 2020). Our data revealed a widespread immunoreactivity for both α-synuclein and tau in several brain regions of the MPTP-lesioned macaques. These included the SN, pons, medulla oblongata, and cerebellum (dentate nucleus). Glial fibrillary acidic protein (GFAP) staining was marked in the SN and pallidum. Interestingly, although the levodopa treatment did not correct the tauopathy, the immunostaining for α-synuclein was reduced in the pallidum, pons, medulla oblongata, and cerebellum but not in the SN (Deffains et al. 2020). This study extends the validity of the MPTP-lesioned macaque model of PD to pathological findings. In addition, it suggests that although α-synuclein increased expression is levodopa-insensitive in the SN, it is sensitive to dopamine-replacement therapy in other brain regions. These data, therefore, offer a window for studying dopamine–α-synuclein interactions and their consequences on widespread neurodegeneration.

MPTP-Induced Phenocopy

Motor Triad

MPTP-induced parkinsonism in monkeys produces an amazing phenocopy of PD. In the early days of MPTP, most researchers focused on the bradykinesia/akinesia and rigidity analogy to PD with a clear-cut demonstration of the similarities. Quoting the first mechanographic study (Doudet et al. 1985, 1986), the researchers describe that "… after MPTP treatment, alterations in movement parameters and EMG [electromyography] activity were observed. Mean reaction time and movement duration increased by 20%–25% and 25%–30%, respectively. The movements were slower and were associated with a generalized depression in the shape and the amplitude of EMG activity in the agonist muscle." The Unified Parkinson's Disease Rating Scale (UPDRS)-like clinical rating also assesses such behavioral features (Imbert et al. 2000). Mechanographic studies have been overwhelmed by such clinical ratings, despite their soundness (Benazzouz et al. 1993), mostly for the sake of the time and expertise they require. More recent video-based kinematic studies, coupled or not with telemetric EMG activity recording, have rejuvenated the field (Yin et al. 2014; Baufreton et al. 2018; Darricau et al. 2023; Milekovic et al. 2023), providing unique end points fully translatable from NHP to PD patients in early-phase clinical trials.

The third main symptom of PD, tremor, is the only symptom seldom reproduced in monkeys. It seems, however, that it is more a species issue. Indeed, although macaque monkeys rarely display the typical PD rest tremor (but clearly show action and postural tremor) (Bezard et al. 1998), the MPTP-intoxicated *C. sabaeus* (African green monkey, also called the vervet monkey) does display a validated rest tremor (Raz et al. 2000; Guehl et al. 2003).

Cognitive Impairment

After the first years, researchers have turned their interest toward other motor and nonmotor symptoms. Among these, the cognitive disturbances reported in PD patients have retained their interest. To specifically address those cognitive impairments, Schneider and colleagues developed an intoxication paradigm in the macaque monkey, leading to cognitive impairment before the motor symptom occurs (Schneider 1990; Schneider and Kovelowski 1990; Schneider and Roeltgen 1993). MPTP is administered at low doses several times per week for several months. MPTP administration continues until cognitive deficits appear in the absence of Parkinsonian motor impairment. Animals are considered "cognitively impaired" if they show at least a 15% decrease in cognitive task performance (Decamp and Schneider 2004; Decamp et al. 2004). Measurement of cognitive performance is done by a computer-controlled touch-screen battery of tests validated in this model. The so-called "chronic low-dose" MPTP-treated macaque monkeys develop impairments in the performance of spatial delayed responses, delayed matching to sample, delayed alternation, object retrieval, and discrimination reversal tasks, as well as a variety of specific impairments of attentional and executive functions (Schneider 1990; Schneider and Kovelowski 1990; Schneider and Roeltgen 1993; Decamp and Schneider 2004; Decamp et al. 2004). This model has relevance to the evaluation and development of cognition-enhancing drugs to treat the neuropsychological aspects of PD, in which frontostriatal dysfunction has been implicated.

Sleep/Chronobiological Disturbances

Sleep disturbances, excessive daytime sleepiness (EDS), and rapid eye movement (REM) sleep dysregulation are among the most frequent and disabling nonmotor manifestations of PD (Poewe et al. 2017). They may precede the cardinal motor features of the disease by years and may serve as early biomarkers of the premotor phase of PD. To replicate the sleep–wake disorders of PD and to understand the temporal relationship between these sleep disturbances and the occurrence of parkinsonism, Ghorayeb and collaborators performed long-term continuous electroencephalographic monitoring of vigilance states in unrestrained rhesus monkeys using an implanted miniaturized telemetry device and tested the effect of MPTP intoxication on their sleep–wake organization (Barraud et al. 2009). MPTP injection yielded a dramatic disruption of sleep–wake architecture with reduced sleep efficacy that persisted years after MPTP administration. Primary deregulation of REM sleep and increased daytime sleepiness occurring before the emergence of motor symptoms were striking features of the MPTP effect. This was concomitant with a breakdown of dopaminergic homeostasis (Barraud et al. 2009). In the long term, partial reemergence of REM sleep paralleled the partial adaptation to parkinsonism, the latter being known to result from compensatory mechanisms within the dopaminergic system (Barraud et al. 2010; Hyacinthe et al. 2014). Altogether, these findings highlight the suitability of the MPTP macaque model of PD as a tool to model the sleep–wake disturbances of human disease. Ultimately, this may help decipher the specific role of neurotransmitter depletion in these symptoms.

Gait Impairment, Freezing of Gait

About 40% of people with PD experience locomotor deficits, including gait impairments, balance problems, and freezing-of-gait episodes (Ge et al. 2020). For patients with late-stage PD, the prevalence of locomotor deficits rises to >90% (Stolze et al. 2005). Currently, available therapies improve certain features of gait patterns but have limited impact on dopa-resistant components such as gait initiation, balance, postural instability, and freezing of gait (Fasano et al. 2015). Indeed, the frequency and severity of freezing-of-gait episodes can even increase when deep brain stimulation is turned on (Fasano et al. 2015). Consequently, identifying complementary therapies to alleviate locomotor deficits is a priority for patients with late-stage PD (Fasano et al. 2015); hence, the need for a model. However, until recently, the locomotor deficits of MPTP-treated NHPs had not been characterized by high-precision kinematic analyses. Therefore, it remained unclear whether the deficits observed during quadrupedal locomotion in MPTP-treated NHPs resemble those observed during the bipedal gait of people with PD. To enable this comparison, we established comparable gait recording platforms for NHPs and humans that allowed high-resolution recordings of whole-body kinematics during natural locomotion in both species (Yin et al. 2014). We showed that MPTP-treated NHPs classified with late-stage parkinsonism, based on neuronal loss and overall behavioral deficits, displayed gait impairments and balance problems that share many features commonly observed in PD patients (Milekovic et al. 2023). We thus concluded that the MPTP-treated NHP is an appropriate preclinical model for the development and preliminary validation of therapies to alleviate gait impairments and remedy balance problems due to PD (Milekovic et al. 2023).

Dyskinesia

Besides PD symptoms, PD patients experience side effects of their long-term L-dopa therapy. Such therapy is confounded by the development of adverse events related to fluctuations in motor response. Motor fluctuations include on–off fluctuations, sudden, unpredictable changes in mobility, and the wearing-off phenomenon, a decrease in the duration of action of L-dopa. However, the most debilitating class of motor fluctuation is involuntary movements known as LID. Four species have been regularly used (Bastide et al. 2015): namely, the macaque monkeys (*M. mulatta* and *M. fascicularis*), the marmoset (*C. jacchus*), and the squirrel monkeys

(*S. sciureus*), although the most human-like ones are displayed by the macaques (Bastide et al. 2015). These macaques with LID show various combinations of choreic-athetoid (i.e., characterized by constant writhing and jerking motions), dystonic, and even ballistic movements (i.e., large-amplitude flinging, flailing movements), although less frequently for the latter. The repertoire and severity of dyskinesia are not distinguishable from LID occurring in PD patients (Fox et al. 2012; Bastide et al. 2015) and made the MPTP macaque the gold-standard model of LID research.

Behavioral Assessment

The various symptoms must be precisely quantified to demonstrate the efficacy of an experimental therapeutic strategy. Most studies currently rely on UPDRS-like clinical assessments (Imbert et al. 2000; Bezard and Przedborski 2011). These measurements must be performed at least in double-blind and watching post-hoc video recordings of animals behaving in vast environments, allowing proper motor behavior to develop (Fox et al. 2012). Not only does the procedure ensure reproducibility and environmental control, but it also allows further checks by independent examiners or trained neurologists. Such simple and obvious criteria are unfortunately seldom fulfilled.

Clinical assessments must be accompanied by automated man-free measurement of locomotor activity such as infrared monitoring and video-based measurement of traveled distance (Bézard et al. 2003; Ahmed et al. 2010; Porras et al. 2012; Urs et al. 2015), mechanography analysis (Doudet et al. 1985, 1986; Benazzouz et al. 1993), or, even better although still challenging, whole-body kinematics analysis (Yin et al. 2014; Baufreton et al. 2018; Milekovic et al. 2023). To increase the translation ability of the MPTP monkey model, animals could also be submitted to various behavioral tasks that would assess both the motor and nonmotor functions. Such tasks range from a simple food-retrieval task (Bezard et al. 1997) to a comprehensive neuropsychological assessment (i.e., CANTAB-like battery of tests) (Schneider and Roeltgen 1993; Decamp

and Schneider 2004; Decamp et al. 2004), passing by object retrieval tasks with various cognitive loads (Charvin et al. 2018; Rosenblad et al. 2019; Darricau et al. 2023).

Experimental Use

Altogether, these data show the amazing potential of the MPTP monkey model as a (almost perfect) phenocopy of PD. Such a phenocopy is made possible because MPTP is all but specific to nigrostriatal DA neurons and is more obvious in animals rendered parkinsonian with chronic regimens of intoxication.

Selecting the best species and regimen remains, however, a challenge. A few key criteria should be considered: (1) What stage of the disease do you want to study? (2) What symptoms are key to your experiment? (3) What pathophysiological features do you want to obtain? (4) How reproducible should your model be? Ideally, a single model should address all possible questions. However, we have seen that specific symptoms or features could be expressed by a given species (i.e., the African green monkey only displays the PD rest tremor).

As one might need to deal with different stages of the disease (i.e., the prodromal/presymptomatic, the symptomatic, or the dyskinetic phases), a chronic, slowly progressing model should be favored. PD is a bilateral multisystemic disease in nature, and systemic exposure will lead to a variety of lesions, as reviewed above. Although the daily management of animals is complicated by the potential threat to animal health status (e.g., adipsia, aphagia, loss of body weight), systemic exposure to the toxin is required. Trained staff must follow the animal's health status and body weight daily. Gavage should be introduced quite early in the intoxication process as a safety measure with proven efficacy in increasing the survival rate. In a chronic regimen, intoxication should often be stopped before obtaining a full-blown syndrome, as symptoms tend to worsen once they have appeared. It is, therefore, recommended to rely on the expertise of a trained team before starting any new experiment. The more chronic the regimen is, the better it is to introduce dopamine transporter imaging to follow ongoing neurodegenerative processes and stop intoxication at the desired extent of denervation (Prunier et al. 2003; Fernagut et al. 2010). To rely on a model in which the kinetics of nigrostriatal denervation is known, either with in vivo imaging (Fernagut et al. 2010) or with postmortem end points (Bezard et al. 2001; Fridjonsdottir et al. 2021), is an advantage. However, this is seldom the case, as a single model/regimen so far has fulfilled this requirement.

In the same way, primate models of PD have often been criticized for their supposed lack of reproducibility. This is the consequence of the heterogeneity in the primate population used by most investigators. Most laboratories have difficulties accessing NHPs and use animals of both genders, of varying body ranges and ages, and most likely from different genetic backgrounds (e.g., there are differences between *cynomolgus* macaques from Mauritius, China, and the Philippines). In our laboratory, we use exclusively F_2-bred macaques. Using our classic intoxication paradigm replicating PD progression over a month in the female *cynomolgus* macaque (Bezard et al. 2001), symptoms reach the clinical criterion for stopping MPTP at 15.6 ± 1.01 days. Such an amazing result derives from $n = 298$ female *cynomolgus* of 3 years of age and 3.3 ± 0.2 kg. These numbers highlight the need for homogeneity of the monkey population, just like work done using rodents.

MODELS OF SYNUCLEINOPATHIES: PATHOGENIC MODELS

The seminal work of Braak et al. (2003) suggesting that LB pathology follows a predictable pattern of progression within the PD brain, as well as the "host-to-graft" observation (Kordower et al. 2008; Li et al. 2008; Mendez et al. 2008), led to the development of experimental models based on injection with α-synuclein (the most represented protein component of LB) assemblies (Spillantini et al. 1997; Dehay et al. 2016; Recasens et al. 2018). These experimental models suggest that α-synuclein, in pathological conformations such as the one found in LBs, initiates a cascade of events leading to dopaminergic neuron degeneration and cell-to-cell propagation of α-synuclein

pathology through a self-templating mechanism (i.e., in a prion-like manner).

Several studies have suggested that prefibrillar oligomers may represent one of the major neurotoxic entities in PD (Winner et al. 2011; Bourdenx et al. 2017). This notion has been derived primarily from studies using large doses of recombinant α-synuclein applied to cell cultures or injected into adult mice, overexpressing either mutant or wild-type (WT) α-synuclein (Bengoa-Vergniory et al. 2017). Similarly, intracerebral injection of low doses of α-synuclein-containing LB extracts, purified from the SNc of postmortem PD brains, promotes α-synuclein pathology and dopaminergic neurodegeneration in WT mice (Recasens et al. 2014). Importantly, this neuropathological effect was directly linked to the presence of α-synuclein in LB extracts because immunodepletion of α-synuclein from the LB fractions prevented the development of pathology following injection into WT mice.

Developing such pathogenic prion-like models has also reached NHPs with several efforts using both patient-derived extracts and recombinant-α-synuclein-preformed fibrils (PFFs).

Patient-Derived Brain Extract Models

We and other groups have used patient brain–derived material to get closer to the human pathology regarding the nature of inoculated material and the amount of injected α-synuclein. In 2014, Recasens and collaborators used LB-enriched fractions purified from three PD patients' mesencephalon, which contained amyloid-like structures and insoluble aggregated α-synuclein, to inject nanograms of pathological α-synuclein into mice and NHPs. They showed that injection of LB-enriched fractions into mice induced significant dopaminergic neurodegeneration, accompanied by accumulation of proteinase K (PK)-resistant and S129-phosphorylated forms of endogenous α-synuclein in the SNc, the striatum, and the neocortical areas. These data were the first proof-of-concept regarding the induction of the pathogenicity from the human pathogenic material toward the endogenous murine α-synuclein protein. Along with the demonstrated aggregation and propagation of α-synuclein into interconnected brain areas, this study brought strong evidence about the "prion-like" hypothesis of α-synuclein in synucleinopathies (Recasens et al. 2018). More interestingly, LB-derived fractions into the SN or the striatum were injected into four rhesus monkeys. One monkey was previously chronically treated with MPTP 3 years before. Using positron emission tomography (PET) imaging, they demonstrated that striatal and nigral LB-inoculated monkeys exhibited a striatal dopaminergic lesion that appeared at 9 months and lasted up to 12 months. Fourteen months after the LB inoculation, they obtained a more pronounced dopaminergic cell loss in the SNc of striatal-injected monkeys compared to the nigral-inoculated animals. They also showed that, in striatal LB-inoculated primates, PD-derived LB extracts induced a widespread increase of α-synuclein levels in interconnected regions, suggesting a long-distance propagation of α-synuclein pathology, which appeared to be more local for the nigral-injected groups. Interestingly, striatal inoculation of LB fractions in MPTP-treated monkeys did not lead to α-synuclein pathology in the SNc but instead to an aggravated increase of α-synuclein into striatal and efferent areas, suggesting a retrograde transmission of α-synuclein from the striatum to the SNc. This work was the first proof-of-concept showing that LB extracts purified from patients' brains induced a pathological response in NHP, including neurodegeneration and a "prion-like" synucleinopathy. To confirm this result, a follow-up study investigated the consequences of PD-derived LB inoculation on a larger cohort of NHP. Our group and collaborators injected PD patient–derived LB fractions containing large and insoluble α-synuclein aggregates in the striatum of olive baboons (*Papio papio*) (Bourdenx et al. 2020). Two years after the injection, LB-inoculated NHP presented nigrostriatal neurodegeneration associated with α-synuclein pathology localized in different brain regions. More surprisingly, and in contrast with mice, when PD patient–derived no-LB fractions, containing small aggregates and mainly soluble α-synuclein, were injected into NHP, they observed dopaminergic neurodegeneration to the extent of LB-inoculated monkeys, also associated with α-synuclein pa-

thology localized in many brain regions. Taking advantage of a machine-learning approach, we sorted out the 20 variables that constituted the best predictors of neurodegeneration among a data set of 180 measured variables for the two injection groups. Interestingly, unique pathological signatures of induced pathology between LB and no-LB groups were obtained, leading to the same dopaminergic lesion level. This study showed that distinct pathological α-synuclein species led to the same dopaminergic lesion level through different underlying mechanisms, modeling synucleinopathies' multifactorial nature and complexity. Using the same type of patient brain–derived extracts, our group injected LB fractions in the stomach and ventral duodenum wall of five baboon monkeys. The underlying idea was to compare the pathology induced with the one obtained for the striatal-LB-fraction-injected group and to challenge the hypothesis of a caudal–rostral propagation of α-synuclein pathology presumed by Braak and colleagues (Arotcarena et al. 2020). Interestingly, we observed that enteric inoculation of LB fractions in NHP led to central dopaminergic neurodegeneration in the SNc and the striatum, at the same level as for the striatal-LB-injected group, associated with the development of α-synuclein pathology in the CNS. These data suggested that α-synuclein pathology propagated in a caudal–rostral fashion from the ENS toward the CNS. More surprisingly, we found that not only did the enteric LB-injected animals induce a local α-synuclein accumulation in enteric neurons, but so did the striatal-LB-injected animals. Moreover, we observed a significant negative correlation between the number of nigral dopaminergic neurons in the CNS and the amount of α-synuclein in the ENS neurons, suggesting that the enteric α-synuclein pathology extent may reflect the severity of the central dopaminergic lesion. These data demonstrated that α-synuclein pathology also propagated rostrocaudally from the CNS toward the ENS, very much in line with the most recent brain first/body first clinical hypothesis (Borghammer and Van Den Berge 2019; Horsager et al. 2020; Borghammer 2023). Regarding the bidirectional routes of propagation of the synucleinopathy, the vagus nerve was put aside in this experimental model, and biological fluids have been considered as possible alternative routes of α-synuclein pathology spreading.

Striatal injection of LB-enriched fractions in NHPs recapitulates two important neuropathological criteria to model synucleinopathies: neurodegeneration and the presence and spreading of α-synuclein pathology. Hence, LB-enriched inoculation in NHP may provide a relevant model to understand better the mechanisms underlying the pathology in synucleinopathy. However, only subtle behavioral changes assessed by validated ethological evaluation have been observed 2 years after the injection of LB fraction due to a dopaminergic cell loss that does not reach the threshold of the appearance of the motor symptoms. This progressive model can thus be used to mimic the early stages of synucleinopathies to decipher the underlying mechanisms and better understand the "prion-like" properties of α-synuclein. Assessing α-synuclein pathology and neurodegeneration at early time points after LB injection must be performed to evaluate the pathological signature dynamics of the pathology at very early stages.

Similarly, increasing LB postinjection duration could be considered to assess a more advanced picture of neurodegeneration and α-synuclein pathology. One additional aspect to consider is the injection site of Lewy body–enriched fractions, as this can potentially lead to distinct pathologies. Teil et al. (2023) focused on understanding the progression of α-synuclein pathology after prefrontal cortex administration. This study revealed that 1 year after intracortical administration, mesencephalic-derived LB injections resulted in widespread prefrontal neuronal loss, prefrontal inflammation, and S129-phosphorylated α-synuclein immunoreactivity. This prefrontal cortex pathology was accompanied by S129-phosphorylated α-synuclein accumulation in the caudate nucleus but not in the putamen, indicating a unique pattern of pathology and striatal dopaminergic function alteration. This study also highlighted the propagation of α-synuclein pathology from the cortex to the striatum in NHPs, mirroring aspects of dementia with Lewy bodies (DLBs) (Teil et al. 2023). Of note, the fact that CNS administration of glial cytoplasmic inclusion–containing fractions derived from the brain of multiple system

atrophy patients or tau seeds extracted from progressive supranuclear palsy patients in NHPs led to the development of multiple system atrophy-like (Teil et al. 2022) and progressive supranuclear palsy-like pathologies (Darricau et al. 2023), respectively, suggesting that one may envision extending the procedure to all proteinopathies. However, the limited access to human material remains the drawback of using NHP models based on human brain–derived extracts. Considerable collective efforts must be made to find innovative solutions to bypass the need for fresh human material, such as seeding amplification assays.

Recombinant α-Synuclein Preformed Fibrils Models

The injection of α-synuclein PFFs in vitro and in mice has also shown their efficacy in inducing both loss of dopaminergic neurons of the SNc and the formation of aggregates of insoluble α-synuclein (Luk et al. 2009, 2012a,b). Following the demonstration of this in rodents, Shimozawa et al. (2017) endeavored to determine whether this was also the case in marmoset monkeys. For this purpose, they injected marmoset monkeys with mouse recombinant α-synuclein fibrils in the caudate nucleus and putamen. Three months after injection, they observed abundant S129-phosphorylated α-synuclein structures throughout various brain regions, including the striatum, SNc, cortex, amygdala, and thalamus. The inclusions were also positive for human α-synuclein (i.e., endogenous primate synuclein), ubiquitin, and p62 staining. Formation of round S129-phosphorylated α-synuclein-positive inclusions was detected in dopaminergic neurons 3 months postinjection. Neurodegeneration was also observed with the decrease of tyrosine hydroxylase (TH)-positive staining in the SNc. Colocalization of the human-specific α-synuclein antibody (e.g., LB509) and the microglial marker Iba1 suggested that the inclusions were phagocytosed by microglial cells. Altogether, this study showed for the first time that using PFFs could induce the formation of inclusions and dopaminergic neurodegeneration in the SNc of marmoset monkeys. More recently, Chu and colleagues used injections of PFFs in the striatum of macaque monkeys to determine the effects of human recombinant α-synuclein fibrils (Chu et al. 2019; Dehay and Bezard 2019). Cynomolgus monkeys received intrastriatal injections of PFFs, and four control monkeys received sham surgery. After 12–15 months, they observed a loss of TH immunostaining and increased striatal DAT immunostaining. In addition, they observed S129-phosphorylated α-synuclein inclusions in the SNc that presented two different aspects: a more granular aspect or whole-cell staining with absent cytoplasm, suggesting the beginning of LB formation. These nigral neurons containing α-synuclein inclusions had lost their TH and Nurr1 staining, reminiscent of PD pathology. These aggregates indicated a progressive α-synuclein accumulation in both cases, forming larger LB-like inclusions. These studies have shown the possibility of inducing a PD-like pathology in marmoset and macaque monkeys. Despite using mouse or human α-synuclein PFFs, these studies demonstrated the ability of α-synuclein propagation and aggregate formation in NHPs.

Nonetheless, it is important to note that large quantities of α-synuclein fibrils had to be injected to induce these PD-like pathologies. These are still not sufficient to induce the appearance of clinical symptoms. Other PFF delivery methods have recently been tested in cynomolgus monkeys using intranasal injections (Guo et al. 2021). After 1, 4, or 17 months, Guo and colleagues observed an iron accumulation, accompanied by a sparse appearance of S129-phosphorylated α-synuclein, which did not colocalize with iron deposits. Nonetheless, this PFF species did not induce dopaminergic neurodegeneration.

Finally, in a pilot study, Fayard et al. (2023) compared the capacity of two recombinant α-synuclein strains and of patient-derived LB extracts to model synucleinopathies after intraputaminal injection in NHPs. Functional alterations triggered by these injections were evaluated in vivo using glucose positron emission tomography imaging. Postmortem immunohistochemical and biochemical analyses were used to detect neuropathological alterations in the dopaminergic system and α-synuclein pathology propagation. In vivo results revealed a decrease in glucose metabolism more pronounced in

α-synuclein strain–injected animals. Histology showed fewer dopaminergic cells in the SN to different extents according to the inoculum used. Biochemical analysis revealed that α-synuclein-induced aggregation, phosphorylation at S129, and propagation in different brain regions are strain-specific. In conclusion, this study showed that distinct α-synuclein strains can induce specific patterns of synuclein pathology in NHP, changes in the nigrostriatal pathway, and functional alterations that resemble early-stage PD (Fayard et al. 2023).

Altogether, these studies showed the growing use of PFFs in NHP models and their potential, depending on their administration, to cause a PD-like pathology, calling for systematic appraisal with a sound characterization of the PFF isomorph or the nature of LB content. Although the pathology is strikingly similar (and informative) to the human PD pathology, none of these models presents overt parkinsonism like the MPTP macaque model. Although the pathology load varies from one prion-like model to another (e.g., PFF vs. LB), they share a nigrostriatal lesion that remains below the threshold for overt symptomatic manifestations (i.e., 50% of the dopaminergic terminals into the striatum). Consequently, none of these models has shown a reversal of symptoms by levodopa administration, as none displays the classic symptomatology. Such observation is strikingly not commented on at all in the literature. One can consider this as an epiphenomenon due to the slowness of the pathology progression. However, some studies surveyed their prion-like exposed NHPs for 2 years without progression of the nigrostriatal deficit despite the progression of synuclein pathology (Bourdenx et al. 2020). The relative failure in producing a large lesion extent certainly tells us that the prion-like mechanisms are not everything and that this modality must be accompanied by a second hit to lead to overt parkinsonism.

MODELS OF FAMILIAL PD: ETIOLOGIC MODELS

The year 1997 marks a very important milestone in PD research. Spillantini et al. (1997) elegantly demonstrated that LBs and LNs found in the brainstem and cortices of both PD and DLB cases were firmly immunopositive for α-synuclein using various anti-α-synuclein antibodies. Polymeropoulos et al. (1997), concomitantly, reported the genetic association of α-synuclein mutation and familial parkinsonism, launching the race to identify genetic causes for PD. Besides a few monogenic (Mendelian) forms of PD (*SNCA, LRRK2, GBA1*, etc.) (Puschmann et al. 2017; Outeiro et al. 2023), a large variety of genetic risk factors identified in genome-wide association studies now exist (up to 90) (Puschmann 2017; Nalls et al. 2019). Soon after identifying mutated or duplicated/triplicated α-synuclein as causative of PD (Singleton et al. 2003; Chartier-Harlin et al. 2004), truly etiologic models were developed in various species, including NHPs.

Viral Vector–Based Models

Intracerebral injections of adeno-associated viruses (AAVs) were first validated in rodents, particularly in rats, before being tested in NHPs (Kirik et al. 2002; Koprich et al. 2010). Building on their study in rats, Kirik et al. (2003) injected AAV1/2 to overexpress GFP, WT-α-synuclein, or A53T-mutant α-synuclein unilaterally in the right SNc of common marmosets. Three weeks postinjection, they verified their AAV's correct expression in control (GFP) monkeys and found that their virus was specifically expressed in neurons. Following this, they waited 4 months postinjection to assess their various AAV effects on neurodegeneration and α-synuclein. In A53T-α-synuclein marmosets, they observed a loss of TH immunostaining in the SNc with a decrease in VMAT-2. They also observed the presence of TH-positive fragmented neurites and α-synuclein-positive inclusions. This was accompanied by behavioral changes with a bias in the head position test on the ipsilateral side, starting at 6 weeks postinjection. This study was followed by using AAV2/5, also under the same neuronal promoter, targeting GFP, WT-α-synuclein, and A53T-mutant α-synuclein in a larger group of common marmosets (Eslamboli et al. 2007). The long-term effects of these unilateral injections in the SNc were assessed 1 year after the AAVs' injection. During the 1-year live phase,

they observed contralateral motor deficits in A53T-α-synuclein injected monkeys, with worsening general motor coordination over time. Postmortem analysis showed that these monkeys displayed a loss of TH immunostaining in the SNc of WT and A53T-α-synuclein injected monkeys. They also observed total and S129-phosphorylated α-synuclein-positive inclusions, with ubiquitin added to these inclusions in the A53T-injected group. These studies showed that both WT and A53T-mutant α-synuclein had deleterious effects when overexpressed in marmosets, but the A53T-mutant was more potent in inducing a PD-like pathology. These two first studies in marmoset monkeys were essential in demonstrating the ability to overexpress α-synuclein via viral vectors in NHPs, leading to decreased dopaminergic neurons and the formation of α-synuclein inclusions, associated with motor behavior impairment.

Other approaches have included injecting lentiviral vectors to drive the expression of A53T-mutant α-synuclein in the SNc of rhesus monkeys (Yang et al. 2015). Yang and colleagues wanted to assess the effect of injecting lentiviral vectors containing mutant α-synuclein and whether this effect depended on the monkeys' age. After first verifying their lentivirus' correct expression in control and A53T-injected monkeys, they observed the initial formation of small α-synuclein aggregates in long neuronal processes, like LNs, in the A53T-α-synuclein injected monkey. They next injected monkeys of different ages (2, 8, and >15 years old) with either PBS or the A53T-α-synuclein lentivirus. After 8 weeks, A53T-α-synuclein injected monkeys demonstrated the formation of LNs and astroglial activation, which were more abundant in older monkeys than in younger monkeys. These monkeys also presented axonal degeneration and TH immunostaining loss in the SNc, specifically in A53T-α-synuclein injected monkeys. This study suggested that monkeys' age plays a role in neuropathology when combined with lentiviral overexpression of A53T-α-synuclein.

Similarly, Bourdenx et al. (2015) wanted to assess whether age was a factor in developing PD neuropathology. To this end, they injected an AAV2/9 to overexpress the A53T-α-synuclein mutant using a neuronal promoter in the SNc of young and old marmoset monkeys. Eleven weeks postinjection, they observed decreased TH immunostaining in sham-operated old animals in the SNc and striatum. They observed a decrease in TH immunostaining in the SNc and the striatum in both young and aged monkeys in the injected side. Concerning α-synuclein pathology, both young and old monkeys demonstrated increased total and S129-phosphorylated α-synuclein. Surprisingly, old monkeys presented less α-synuclein phosphorylation than young animals. Here, they showed that overexpression of α-synuclein induced a decrease in dopaminergic neurons and fibers accompanied by α-synuclein accumulation. Still, the age of monkeys did not impact this pathological progression. These last two studies aiming at determining the part of age in α-synuclein accumulation have shown quite diverging results. On the one hand, Yang and colleagues demonstrated that A53T overexpression impacts older monkeys more severely, whereas Bourdenx and colleagues did not observe this same effect of age. This could be due to the difference in species used, with one study using rhesus monkeys and the other using common marmosets or lentivirus compared to AAV.

More recently, Koprich et al. (2016) injected AAV1/2 A53T-α-synuclein in the SN of cynomolgus macaques. This study used different parameters to determine the conditions in which sustained expression of α-synuclein would induce neurodegeneration. In the first experiment, no dopaminergic neuron loss was observed when injecting the virus at four sites of the SNc. They used higher titers or larger volumes in their second experiment to optimize their AAV effects. Both high titers and larger volumes induced the loss of dopaminergic neurons in the SNc.

Nonetheless, the injection of a higher titer of the virus had a more substantial effect on the decrease of DAT and dopamine than the injection of larger volumes. On the contrary, injecting larger volumes of virus induced higher levels of putaminal α-synuclein. Altogether, this study showed the impact of both the titer and volume of injection in monkeys, demonstrating once more that diverging results between experiments could not only be due to the species used but also to the injection itself.

Given that aging represents the primary risk factor for synucleinopathies and considering the progressive nature of these pathologies, it is crucial to consider the extended life span of NHP and the age-dependent buildup of the neuronal pigment neuromelanin as critical factors for consideration. Neuromelanin is primate-specific. Although a key feature of human dopamine neurons and likely involved in oxidative stress, it has only been recently scrutinized. This is an innovative aspect investigated by Vila et al. (2019) in rodents (Carballo-Carbajal et al. 2019; Gonzalez-Sepulveda et al. 2023) and very recently in NHPs (Chocarro et al. 2023) where they used an AAV 2/1 encoding the human tyrosinase (hTyr) gene. Following stereotactic nigral injection of AAV-hTyr, and after 4–8 months, the study revealed a time-dependent accumulation of neuromelanin. This accumulation was found to initiate the pathological misfolding of endogenous α-synuclein in pigmented midbrain dopaminergic cells, with the potential to propagate anterogradely toward the cerebral cortex (Chocarro et al. 2023).

Transgenic Models

The discovery of familial cases of PD was concomitant with the boom of mouse transgenesis. With a few exceptions, most mouse lines (of whichever PARK loci) have failed to show neurodegeneration, noticeable LBs, or the accumulation of pS129-α-synuclein observed in PD patients. Given that the first transgenic NHP modeling Huntington's disease displayed very early on typical human-like symptoms with associated pathology (Yang et al. 2008), NHPs transgenic for PD genes soon represented the fascinating possibility of accessing a true model of PD and not simply ultralocalized impairments as those afforded by viral vector delivery of genetic material.

Given the ethical, economic, and technical constraints, very few studies have attempted to create transgenic models of synucleinopathy. From the same team that created the Huntington NHPs, Niu et al. (2015) endeavored to generate a transgenic monkey model by lentiviral vector injection in fertilized monkey eggs. After expressing A53T-mutant α-synuclein in oocytes, 75 eggs were transferred in competent female rhesus monkeys. These transfers resulted in 11 pregnancies and led to the birth of five monkeys. Transgene expression was confirmed, and immunostaining also showed the presence of increased α-synuclein in the SNc, striatum, and cortex, but not of S129-phosphorylated α-synuclein in these A53T-α-synuclein transgenic monkeys. The authors also noted that older monkeys developed cognitive defects and anxiety starting at 2.5 years of age. These defects implicated object recognition, dexterity, and stereotypical circling behavior of these animals, reminiscent of prodromal defects seen in PD patients. Nonetheless, these animals showed no motor abnormalities and no neurodegeneration using magnetic resonance imaging (MRI). Compared to other models, these transgenic monkeys could be more reliable in observing phenotypes and pathological modifications and finding biomarkers for PD. Given that this study only followed the A53T-α-synuclein transgenic monkeys for 2.5 years, it is difficult to say whether they will develop other aspects of PD in the future. Still, this study remained highly interested in observing the age-dependent factors of PD onset.

More recently, with the discovery of CRISPR–Cas9 technology, certain studies have taken advantage of this system to develop transgenic synucleinopathy models. Yang et al. (2019) injected CRISPR–Cas9 directed against the *PINK1* gene in 1-cell stage embryos from rhesus monkeys. After the transfer of the embryos to surrogate rhesus monkeys, 11 fetuses developed (eight *PINK1* mutants and three WT) and were born naturally. Four mutants and one WT monkey died in the first week after birth. Of the four remaining mutant monkeys, one lived for 1.5 years, and the others were terminated 3 years after birth. Certain monkeys with *PINK1* mutations displayed decreased gray matter density in the cortex, and others had decreased movement after 1.5 years. After 1.5 years, decreased neuronal immunostaining and increased astrogliosis compared to WT monkeys were observed. Electron microscopy in one mutant monkey demonstrated degeneration of neurons in the cortex, SNc, and striatum. Despite not demonstrating any modifications in α-synuclein or its distribution, this team showed the possible use of CRISPR-Cas9 for PD in NHPs. Regardless

of the uncommon use of transgenic NHP models in PD, it is important to note their potential importance for future studies. Given the constraints observed, it is unsurprising that few studies currently exist that have created transgenic NHPs. Still, we suspect that the number of studies will grow in the next years to resolve the need to understand intractable diseases such as PD better.

CONCLUDING REMARKS

Although of paramount importance and despite the heuristic value of almost every NHP-relating scientific manuscript, the number of investigators using NHP models is limited. This is the main limiting factor to developing consensual models exploited with proper power and enabling replication by other investigators. The complexity of the models, their cost owing to the animal cost (especially in these post-COVID times), and the time required for the pathology's establishment call for a consortium effort toward the generation and exploitation of consensually defined methodologies (including, but not limited to, most appropriate species, age, procedure, modality, etc.). Nonetheless, in the upcoming years, we will have the opportunity to assess the potential benefits of introducing genetic modifications into marmosets and other NHPs. This approach could facilitate the development of PD models and enable the exploration of genetic factors that contribute to the disease.

ACKNOWLEDGMENTS

L'Institut National de la Santé et de la Recherche Médicale (INSERM), Centre National de la Recherche Scientifique (CNRS), and the University of Bordeaux provided financial and infrastructural support. This review received financial support from the French government in the framework of the University of Bordeaux's IdEx "Investments for the Future" program/GPR BRAIN_2030.

REFERENCES

Ahmed MR, Berthet A, Bychkov E, Porras G, Li Q, Bioulac BH, Carl YT, Bloch B, Kook S, Aubert I, et al. 2010. Lentiviral overexpression of GRK6 alleviates L-dopa-induced dyskinesia in experimental Parkinson's disease. *Sci Transl Med* **2**: 28ra28. doi:10.1126/scitranslmed.3000664

Arotcarena ML, Dovero S, Prigent A, Bourdenx M, Camus S, Porras G, Thiolat ML, Tasselli M, Aubert P, Kruse N, et al. 2020. Bidirectional gut-to-brain and brain-to-gut propagation of synucleinopathy in non-human primates. *Brain* **143**: 1462–1475. doi:10.1093/brain/awaa096

Barraud Q, Lambrecq V, Forni C, McGuire S, Hill M, Bioulac B, Balzamo E, Bezard E, Tison F, Ghorayeb I. 2009. Sleep disorders in Parkinson's disease: the contribution of the MPTP non-human primate model. *Exp Neurol* **219**: 574–582. doi:10.1016/j.expneurol.2009.07.019

Barraud Q, Obeid I, Aubert I, Barrière G, Contamin H, McGuire S, Ravenscroft P, Porras G, Tison F, Bezard E, et al. 2010. Neuroanatomical study of the A11 diencephalospinal pathway in the non-human primate. *PLoS ONE* **5**: e13306. doi:10.1371/journal.pone.0013306

Bastide MF, Meissner WG, Picconi B, Fasano S, Fernagut PO, Feyder M, Francardo V, Alcacer C, Ding Y, Brambilla R, et al. 2015. Pathophysiology of L-dopa-induced motor and non-motor complications in Parkinson's disease. *Prog Neurobiol* **132**: 96–168. doi:10.1016/j.pneurobiol.2015.07.002

Baufreton J, Milekovic T, Li Q, McGuire S, Moraud EM, Porras G, Sun S, Ko WKD, Chazalon M, Morin S, et al. 2018. Inhaling xenon ameliorates L-dopa-induced dyskinesia in experimental parkinsonism. *Mov Disord* **33**: 1632–1642. doi:10.1002/mds.27404

Bélanger N, Grégoire L, Hadj Tahar A, Bédard PJ. 2003. Chronic treatment with small doses of cabergoline prevents dopa-induced dyskinesias in parkinsonian monkeys. *Mov Disord* **18**: 1436–1441. doi:10.1002/mds.10589

Benazzouz A, Gross C, Féger J, Boraud T, Bioulac B. 1993. Reversal of rigidity and improvement in motor performance by subthalamic high-frequency stimulation in MPTP-treated monkeys. *Eur J Neurosci* **5**: 382–389. doi:10.1111/j.1460-9568.1993.tb00505.x

Bengoa-Vergniory N, Roberts RF, Wade-Martins R, Alegre-Abarrategui J. 2017. α-Synuclein oligomers: a new hope. *Acta Neuropathol* **134**: 819–838. doi:10.1007/s00401-017-1755-1

Bezard E, Przedborski S. 2011. A tale on animal models of Parkinson's disease. *Mov Disord* **26**: 993–1002. doi:10.1002/mds.23696

Bezard E, Imbert C, Deloire X, Bioulac B, Gross CE. 1997. A chronic MPTP model reproducing the slow evolution of Parkinson's disease: evolution of motor symptoms in the monkey. *Brain Res* **766**: 107–112. doi:10.1016/S0006-8993(97)00531-3

Bezard E, Imbert C, Gross CE. 1998. Experimental models of Parkinson's disease: from the static to the dynamic. *Rev Neurosci* **9**: 71–90. doi:10.1515/revneuro.1998.9.2.71

Bezard E, Dovero S, Prunier C, Ravenscroft P, Chalon S, Guilloteau D, Crossman AR, Bioulac B, Brotchie JM, Gross CE. 2001. Relationship between the appearance of symptoms and the level of nigrostriatal degeneration in a progressive 1-methyl-4-phenyl-1,2,3,6-tetrahydropyridine-lesioned macaque model of Parkinson's disease. *J Neurosci* **21**: 6853–6861. doi:10.1523/JNEUROSCI.21-17-06853.2001

Bézard E, Ferry S, Mach U, Stark H, Leriche L, Boraud T, Gross C, Sokoloff P. 2003. Attenuation of levodopa-induced dyskinesia by normalizing dopamine D3 receptor function. *Nat Med* **9**: 762–767. doi:10.1038/nm875

Blesa J, Foffani G, Dehay B, Bezard E, Obeso JA. 2022. Motor and non-motor circuit disturbances in early Parkinson disease: which happens first? *Nat Rev Neurosci* **23**: 115–128. doi:10.1038/s41583-021-00542-9

Borghammer P. 2023. The brain-first vs. body-first model of Parkinson's disease with comparison to alternative models. *J Neural Transm (Vienna)* **130**: 737–753. doi:10.1007/s00702-023-02633-6

Borghammer P, Van Den Berge N. 2019. Brain-first versus gut-first Parkinson's disease: a hypothesis. *J Parkinsons Dis* **9**: S281–S295. doi:10.3233/JPD-191721

Bourdenx M, Dovero S, Engeln M, Bido S, Bastide MF, Dutheil N, Vollenweider I, Baud L, Piron C, Grouthier V, et al. 2015. Lack of additive role of ageing in nigrostriatal neurodegeneration triggered by α-synuclein overexpression. *Acta Neuropathol Commun* **3**: 46. doi:10.1186/s40478-015-0222-2

Bourdenx M, Koulakiotis NS, Sanoudou D, Bezard E, Dehay B, Tsarbopoulos A. 2017. Protein aggregation and neurodegeneration in prototypical neurodegenerative diseases: examples of amyloidopathies, tauopathies and synucleinopathies. *Prog Neurobiol* **155**: 171–193. doi:10.1016/j.pneurobio.2015.07.003

Bourdenx M, Nioche A, Dovero S, Arotcarena ML, Camus S, Porras G, Thiolat ML, Rougier NP, Prigent A, Aubert P, et al. 2020. Identification of distinct pathological signatures induced by patient-derived α-synuclein structures in non-human primates. *Sci Adv* **6**: eaaz9165. doi:10.1126/sciadv.aaz9165

Braak H, Del Tredici K, Rüb U, de Vos RA, Jansen Steur EN, Braak E. 2003. Staging of brain pathology related to sporadic Parkinson's disease. *Neurobiol Aging* **24**: 197–211. doi:10.1016/s0197-4580(02)00065-9

Carballo-Carbajal I, Laguna A, Romero-Giménez J, Cuadros T, Bové J, Martinez-Vicente M, Parent A, Gonzalez-Sepulveda M, Peñuelas N, Torra A, et al. 2019. Brain tyrosinase overexpression implicates age-dependent neuromelanin production in Parkinson's disease pathogenesis. *Nat Commun* **10**: 973. doi:10.1038/s41467-019-08858-y

Chartier-Harlin MC, Kachergus J, Roumier C, Mouroux V, Douay X, Lincoln S, Levecque C, Larvor L, Andrieux J, Hulihan M, et al. 2004. α-Synuclein locus duplication as a cause of familial Parkinson's disease. *Lancet* **364**: 1167–1169. doi:10.1016/S0140-6736(04)17103-1

Charvin D, Di Paolo T, Bezard E, Gregoire L, Takano A, Duvey G, Pioli E, Halldin C, Medori R, Conquet F. 2018. An mGlu4-positive allosteric modulator alleviates parkinsonism in primates. *Mov Disord* **33**: 1619–1631. doi:10.1002/mds.27462

Chaumette T, Lebouvier T, Aubert P, Lardeux B, Qin C, Li Q, Accary D, Bézard E, Bruley des Varannes S, Derkinderen P, et al. 2009. Neurochemical plasticity in the enteric nervous system of a primate animal model of experimental Parkinsonism. *Neurogastroenterol Motil* **21**: 215–222. doi:10.1111/j.1365-2982.2008.01226.x

Chocarro J, Rico AJ, Ariznabarreta G, Roda E, Honrubia A, Collantes M, Peñuelas I, Vázquez A, Rodríguez-Pérez AI, Labandeira-García JL, et al. 2023. Neuromelanin accumulation drives endogenous synucleinopathy in non-hu-

Cite this article as *Cold Spring Harb Perspect Med* doi: 10.1101/cshperspect.a041612

man primates. *Brain* **146:** 5000–5014. doi:10.1093/brain/awad331

Chu Y, Muller S, Tavares A, Barret O, Alagille D, Seibyl J, Tamagnan G, Marek K, Luk KC, Trojanowski JQ, et al. 2019. Intrastriatal α-synuclein fibrils in monkeys: spreading, imaging and neuropathological changes. *Brain* **142:** 3565–3579. doi:10.1093/brain/awz296

Darricau M, Katsinelos T, Raschella F, Milekovic T, Crochemore L, Li Q, Courtine G, McEwan WA, Dehay B, Bezard E, et al. 2023. Tau seeds from patients induce progressive supranuclear palsy pathology and symptoms in primates. *Brain* **146:** 2524–2534. doi:10.1093/brain/awac428

Decamp E, Schneider JS. 2004. Attention and executive function deficits in chronic low-dose MPTP-treated non-human primates. *Eur J Neurosci* **20:** 1371–1378. doi:10.1111/j.1460-9568.2004.03586.x

Decamp E, Tinker JP, Schneider JS. 2004. Attentional cueing reverses deficits in spatial working memory task performance in chronic low dose MPTP-treated monkeys. *Behav Brain Res* **152:** 259–262. doi:10.1016/j.bbr.2003.10.007

Deffains M, Canron MH, Teil M, Li Q, Dehay B, Bezard E, Fernagut PO. 2020. L-Dopa regulates α-synuclein accumulation in experimental parkinsonism. *Neuropathol Appl Neurobiol* **47:** 532–543. doi:10.1111/nan.12678

Dehay B, Bezard E. 2019. Intrastriatal injection of α-synuclein fibrils induces Parkinson-like pathology in macaques. *Brain* **142:** 3321–3322. doi:10.1093/brain/awz329

Dehay B, Vila M, Bezard E, Brundin P, Kordower JH. 2016. α-Synuclein propagation: new insights from animal models. *Mov Disord* **31:** 161–168. doi:10.1002/mds.26370

Doudet D, Gross C, Lebrun-Grandie P, Bioulac B. 1985. MPTP primate model of Parkinson's disease: a mechanographic and electromyographic study. *Brain Res* **335:** 194–199. doi:10.1016/0006-8993(85)90294-x

Doudet D, Gross C, Lebrun-Grandie P, Bioulac B. 1986. Effect of increasing regimens of levodopa on chronic MPTP-induced parkinsonism in monkey; mechanographic and electromyographic data. *Electromyogr Clin Neurophysiol* **26:** 711–727.

Eslamboli A, Romero-Ramos M, Burger C, Bjorklund T, Muzyczka N, Mandel RJ, Baker H, Ridley RM, Kirik D. 2007. Long-term consequences of human α-synuclein overexpression in the primate ventral midbrain. *Brain* **130:** 799–815. doi:10.1093/brain/awl382

Fasano A, Aquino CC, Krauss JK, Honey CR, Bloem BR. 2015. Axial disability and deep brain stimulation in patients with Parkinson disease. *Nat Rev Neurol* **11:** 98–110. doi:10.1038/nrneurol.2014.252

Fayard A, Fenyi A, Lavisse S, Dovero S, Bousset L, Bellande T, Lecourtois S, Jouy C, Guillermier M, Jan C, et al. 2023. Functional and neuropathological changes induced by injection of distinct α-synuclein strains: a pilot study in non-human primates. *Neurobiol Dis* **180:** 106086. doi:10.1016/j.nbd.2023.106086

Fernagut PO, Li Q, Dovero S, Chan P, Wu T, Ravenscroft P, Hill M, Chen Z, Bezard E. 2010. Dopamine transporter binding is unaffected by L-dopa administration in normal and MPTP-treated monkeys. *PLoS ONE* **5:** e14053. doi:10.1371/journal.pone.0014053

Fox SH, Johnston TH, Li Q, Brotchie J, Bezard E. 2012. A critique of available scales and presentation of the non-human primate dyskinesia rating scale. *Mov Disord* **27:** 1373–1378. doi:10.1002/mds.25133

Fridjonsdottir E, Shariatgorji R, Nilsson A, Vallianatou T, Odell LR, Schembri LS, Svenningsson P, Fernagut PO, Crossman AR, Bezard E, et al. 2021. Mass spectrometry imaging identifies abnormally elevated brain L-DOPA levels and extrastriatal monoaminergic dysregulation in L-dopa-induced dyskinesia. *Sci Adv* **7:** eabe5948. doi:10.1126/sciadv.abe5948

Ge HL, Chen XY, Lin YX, Ge TJ, Yu LH, Lin ZY, Wu XY, Kang DZ, Ding CY. 2020. The prevalence of freezing of gait in Parkinson's disease and in patients with different disease durations and severities. *Chin Neurosurg J* **6:** 17. doi:10.1186/s41016-020-00197-y

Giguère N, Burke Nanni S, Trudeau LE. 2018. On cell loss and selective vulnerability of neuronal populations in Parkinson's disease. *Front Neurol* **9:** 455. doi:10.3389/fneur.2018.00455

Gonzalez-Sepulveda M, Compte J, Cuadros T, Nicolau A, Guillard-Sirieix C, Peñuelas N, Lorente-Picon M, Parent A, Romero-Giménez J, Cladera-Sastre JM, et al. 2023. In vivo reduction of age-dependent neuromelanin accumulation mitigates features of Parkinson's disease. *Brain* **146:** 1040–1052. doi:10.1093/brain/awac445

Guehl D, Pessiglione M, François C, Yelnik J, Hirsch EC, Féger J, Tremblay L. 2003. Tremor-related activity of neurons in the 'motor' thalamus: changes in firing rate and pattern in the MPTP vervet model of parkinsonism. *Eur J Neurosci* **17:** 2388–2400. doi:10.1046/j.1460-9568.2003.02685.x

Guigoni C, Dovero S, Aubert I, Li Q, Bioulac BH, Bloch B, Gurevich EV, Gross CE, Bezard E. 2005. Levodopa-induced dyskinesia in MPTP-treated macaques is not dependent on the extent and pattern of nigrostrial lesioning. *Eur J Neurosci* **22:** 283–287. doi:10.1111/j.1460-9568.2005.04196.x

Guo JJ, Yue F, Song DY, Bousset L, Liang X, Tang J, Yuan L, Li W, Melki R, Tang Y, et al. 2021. Intranasal administration of α-synuclein preformed fibrils triggers microglial iron deposition in the substantia nigra of *Macaca fascicularis*. *Cell Death Dis* **12:** 81. doi:10.1038/s41419-020-03369-x

Halliday G, Herrero MT, Murphy K, McCann H, Ros-Bernal F, Barcia C, Mori H, Blesa FJ, Obeso JA. 2009. No Lewy pathology in monkeys with over 10 years of severe MPTP Parkinsonism. *Mov Disord* **24:** 1519–1523. doi:10.1002/mds.22481

Halliday GM, Holton JL, Revesz T, Dickson DW. 2011. Neuropathology underlying clinical variability in patients with synucleinopathies. *Acta Neuropathol* **122:** 187–204. doi:10.1007/s00401-011-0852-9

Horsager J, Andersen KB, Knudsen K, Skjærbæk C, Fedorova TD, Okkels N, Schaeffer E, Bonkat SK, Geday J, Otto M, et al. 2020. Brain-first versus body-first Parkinson's disease: a multimodal imaging case-control study. *Brain* **143:** 3077–3088. doi:10.1093/brain/awaa238

Huang B, Wu S, Wang Z, Ge L, Rizak JD, Wu J, Li J, Xu L, Lv L, Yin Y, et al. 2018. Phosphorylated α-synuclein accumulations and Lewy body-like pathology distributed in Parkinson's disease-related brain areas of aged rhesus monkeys treated with MPTP. *Neuroscience* **379:** 302–315. doi:10.1016/j.neuroscience.2018.03.026

Hyacinthe C, Barraud Q, Tison F, Bezard E, Ghorayeb I. 2014. D1 receptor agonist improves sleep-wake parameters in experimental parkinsonism. *Neurobiol Dis* **63:** 20–24. doi:10.1016/j.nbd.2013.10.029

Imbert C, Bezard E, Guitraud S, Boraud T, Gross CE. 2000. Comparison of eight clinical rating scales used for the assessment of MPTP-induced parkinsonism in the Macaque monkey. *J Neurosci Methods* **96:** 71–76. doi:10.1016/s0165-0270(99)00184-3

Iravani MM, Jackson MJ, Kuoppamäki M, Smith LA, Jenner P. 2003. 3,4-Methylenedioxymethamphetamine (ecstasy) inhibits dyskinesia expression and normalizes motor activity in 1-methyl-4-phenyl-1,2,3,6-tetrahydropyridine-treated primates. *J Neurosci* **23:** 9107–9115. doi:10.1523/JNEUROSCI.23-27-09107.2003

Iravani MM, Syed E, Jackson MJ, Johnston LC, Smith LA, Jenner P. 2005. A modified MPTP treatment regime produces reproducible partial nigrostriatal lesions in common marmosets. *Eur J Neurosci* **21:** 841–854. doi:10.1111/j.1460-9568.2005.03915.x

Jan C, Pessiglione M, Tremblay L, Tandé D, Hirsch EC, François C. 2003. Quantitative analysis of dopaminergic loss in relation to functional territories in MPTP-treated monkeys. *Eur J Neurosci* **18:** 2082–2086. doi:10.1046/j.1460-9568.2003.02946.x

Kirik D, Rosenblad C, Burger C, Lundberg C, Johansen TE, Muzyczka N, Mandel RJ, Björklund A. 2002. Parkinson-like neurodegeneration induced by targeted overexpression of α-synuclein in the nigrostriatal system. *J Neurosci* **22:** 2780–2791. doi:10.1523/JNEUROSCI.22-07-02780.2002

Kirik D, Annett LE, Burger C, Muzyczka N, Mandel RJ, Björklund A. 2003. Nigrostriatal α-synucleinopathy induced by viral vector-mediated overexpression of human α-synuclein: a new primate model of Parkinson's disease. *Proc Natl Acad Sci* **100:** 2884–2889. doi:10.1073/pnas.0536383100

Koprich JB, Johnston TH, Reyes MG, Sun X, Brotchie JM. 2010. Expression of human A53T α-synuclein in the rat substantia nigra using a novel AAV1/2 vector produces a rapidly evolving pathology with protein aggregation, dystrophic neurite architecture and nigrostriatal degeneration with potential to model the pathology of Parkinson's disease. *Mol Neurodegener* **5:** 43. doi:10.1186/1750-1326-5-43

Koprich JB, Johnston TH, Reyes G, Omana V, Brotchie JM. 2016. Towards a non-human primate model of α-synucleinopathy for development of therapeutics for Parkinson's disease: optimization of AAV1/2 delivery parameters to drive sustained expression of α synuclein and dopaminergic degeneration in macaque. *PLoS ONE* **11:** e0167235. doi:10.1371/journal.pone.0167235

Kordower JH, Liu YT, Winn S, Emerich DF. 1995. Encapsulated PC 12 cell transplants into hemiparkinsonian monkeys: a behavioral, neuroanatomical, and neurochemical analysis. *Cell Transplant* **4:** 155–171. doi:10.1177/096368979500400203

Kordower JH, Chu Y, Hauser RA, Freeman TB, Olanow CW. 2008. Lewy body-like pathology in long-term embryonic nigral transplants in Parkinson's disease. *Nat Med* **14:** 504–506. doi:10.1038/nm1747

Kowall NW, Hantraye P, Brouillet E, Beal MF, McKee AC, Ferrante RJ. 2000. MPTP induces α-synuclein aggregation in the substantia nigra of baboons. *Neuroreport* **11:** 211–213. doi:10.1097/00001756-200001170-00041

Langston JW, Forno LS, Rebert CS, Irwin I. 1984. Selective nigral toxicity after systemic administration of 1-methyl-4-phenyl-1,2,5,6-tetrahydropyrine (MPTP) in the squirrel monkey. *Brain Res* **292:** 390–394. doi:10.1016/0006-8993(84)90777-7

Li JY, Englund E, Holton JL, Soulet D, Hagell P, Lees AJ, Lashley T, Quinn NP, Rehncrona S, Björklund A, et al. 2008. Lewy bodies in grafted neurons in subjects with Parkinson's disease suggest host-to-graft disease propagation. *Nat Med* **14:** 501–503. doi:10.1038/nm1746

Luk KC, Song C, O'Brien P, Stieber A, Branch JR, Brunden KR, Trojanowski JQ, Lee VM. 2009. Exogenous α-synuclein fibrils seed the formation of Lewy body-like intracellular inclusions in cultured cells. *Proc Natl Acad Sci* **106:** 20051–20056. doi:10.1073/pnas.0908005106

Luk KC, Kehm VM, Zhang B, O'Brien P, Trojanowski JQ, Lee VM. 2012a. Intracerebral inoculation of pathological α-synuclein initiates a rapidly progressive neurodegenerative α-synucleinopathy in mice. *J Exp Med* **209:** 975–986. doi:10.1084/jem.20112457

Luk KC, Kehm V, Carroll J, Zhang B, O'Brien P, Trojanowski JQ, Lee VM. 2012b. Pathological α-synuclein transmission initiates Parkinson-like neurodegeneration in nontransgenic mice. *Science* **338:** 949–953. doi:10.1126/science.1227157

McCormack AL, Mak SK, Shenasa M, Langston WJ, Forno LS, Di Monte DA. 2008. Pathologic modifications of α-synuclein in 1-methyl-4-phenyl-1,2,3,6-tetrahydropyridine (MPTP)-treated squirrel monkeys. *J Neuropathol Exp Neurol* **67:** 793–802. doi:10.1097/NEN.0b013e318180f0bd

Mendez I, Viñuela A, Astradsson A, Mukhida K, Hallett P, Robertson H, Tierney T, Holness R, Dagher A, Trojanowski JQ, et al. 2008. Dopamine neurons implanted into people with Parkinson's disease survive without pathology for 14 years. *Nat Med* **14:** 507–509. doi:10.1038/nm1752

Milekovic T, Moraud EM, Macellari N, Moerman C, Raschellà F, Sun S, Perich MG, Varescon C, Demesmaeker R, Bruel A, et al. 2023. A spinal cord neuroprosthesis for locomotor deficits due to Parkinson's disease. *Nat Med* **29:** 2854–2865. doi:10.1038/s41591-023-02584-1

Moratalla R, Quinn B, DeLanney LE, Irwin I, Langston JW, Graybiel AM. 1992. Differential vulnerability of primate caudate-putamen and striosome-matrix dopamine systems to the neurotoxic effects of 1-methyl-4-phenyl-1,2,3,6-tetrahydropyridine. *Proc Natl Acad Sci* **89:** 3859–3863. doi:10.1073/pnas.89.9.3859

Morissette M, Grondin R, Goulet M, Bédard PJ, Di Paolo T. 1999. Differential regulation of striatal preproenkephalin and preprotachykinin mRNA levels in MPTP-lesioned monkeys chronically treated with dopamine D1 or D2 receptor agonists. *J Neurochem* **72:** 682–692. doi:10.1046/j.1471-4159.1999.0720682.x

Nalls MA, Blauwendraat C, Vallerga CL, Heilbron K, Bandres-Ciga S, Chang D, Tan M, Kia DA, Noyce AJ, Xue A, et al. 2019. Identification of novel risk loci, causal insights, and heritable risk for Parkinson's disease: a meta-analysis of genome-wide association studies. *Lancet Neurol* **18:** 1091–1102. doi:10.1016/S1474-4422(19)30320-5

Niu Y, Guo X, Chen Y, Wang CE, Gao J, Yang W, Kang Y, Si W, Wang H, Yang SH, et al. 2015. Early Parkinson's disease symptoms in α-synuclein transgenic monkeys. *Hum Mol Genet* **24:** 2308–2317. doi:10.1093/hmg/ddu748

Outeiro TF, Alcalay RN, Antonini A, Attems J, Bonifati V, Cardoso F, Chesselet MF, Hardy J, Madeo G, McKeith I, et al. 2023. Defining the riddle in order to solve it: there is more than one "Parkinson's disease." *Mov Disord* **38:** 1127–1142. doi:10.1002/mds.29419

Pifl C, Schingnitz G, Hornykiewicz O. 1991. Effect of 1-methyl-4-phenyl-1,2,3,6-tetrahydropyridine on the regional distribution of brain monoamines in the rhesus monkey. *Neuroscience* **44:** 591–605. doi:10.1016/0306-4522(91)90080-8

Poewe W, Seppi K, Tanner CM, Halliday GM, Brundin P, Volkmann J, Schrag AE, Lang AE. 2017. Parkinson disease. *Nat Rev Dis Primers* **3:** 17013. doi:10.1038/nrdp.2017.13

Polymeropoulos MH, Lavedan C, Leroy E, Ide SE, Dehejia A, Dutra A, Pike B, Root H, Rubenstein J, Boyer R, et al. 1997. Mutation in the α-synuclein gene identified in families with Parkinson's disease. *Science* **276:** 2045–2047. doi:10.1126/science.276.5321.2045

Porras G, Berthet A, Dehay B, Li Q, Ladepeche L, Normand E, Dovero S, Martinez A, Doudnikoff E, Martin-Négrier ML, et al. 2012. PSD-95 expression controls L-dopa dyskinesia through dopamine D1 receptor trafficking. *J Clin Invest* **122:** 3977–3989. doi:10.1172/JCI59426

Prunier C, Payoux P, Guilloteau D, Chalon S, Giraudeau B, Majorel C, Tafani M, Bezard E, Esquerré JP, Baulieu JL. 2003. Quantification of dopamine transporter by 123I-PE2I SPECT and the noninvasive Logan graphical method in Parkinson's disease. *J Nucl Med* **44:** 663–670.

Purisai MG, McCormack AL, Langston WJ, Johnston LC, Di Monte DA. 2005. α-Synuclein expression in the substantia nigra of MPTP-lesioned non-human primates. *Neurobiol Dis* **20:** 898–906. doi:10.1016/j.nbd.2005.05.028

Puschmann A. 2017. New genes causing hereditary Parkinson's disease or parkinsonism. *Curr Neurol Neurosci Rep* **17:** 66. doi:10.1007/s11910-017-0780-8

Raz A, Vaadia E, Bergman H. 2000. Firing patterns and correlations of spontaneous discharge of pallidal neurons in the normal and the tremulous 1-methyl-4-phenyl-1,2,3,6-tetrahydropyridine vervet model of parkinsonism. *J Neurosci* **20:** 8559–8571. doi:10.1523/JNEUROSCI.20-22-08559.2000

Recasens A, Dehay B, Bové J, Carballo-Carbajal I, Dovero S, Pérez-Villalba A, Fernagut PO, Blesa J, Parent A, Perier C, et al. 2014. Lewy body extracts from Parkinson disease brains trigger α-synuclein pathology and neurodegeneration in mice and monkeys. *Ann Neurol* **75:** 351–362. doi:10.1002/ana.24066

Recasens A, Ulusoy A, Kahle PJ, Di Monte DA, Dehay B. 2018. In vivo models of α-synuclein transmission and propagation. *Cell Tissue Res* **373:** 183–193. doi:10.1007/s00441-017-2730-9

Rose S, Nomoto M, Jackson EA, Gibb WR, Jaehnig P, Jenner P, Marsden CD. 1993. Age-related effects of 1-methyl-4-phenyl-1,2,3,6-tetrahydropyridine treatment of common marmosets. *Eur J Pharmacol* **230:** 177–185. doi:10.1016/0014-2999(93)90800-w

Rosenblad C, Li Q, Pioli EY, Dovero S, Antunes AS, Agúndez L, Bardelli M, Linden RM, Henckaerts E, Björklund A, et al. 2019. Vector-mediated l-3,4-dihydroxyphenylalanine delivery reverses motor impairments in a primate model of Parkinson's disease. *Brain* **142:** 2402–2416. doi:10.1093/brain/awz176

Schneider JS. 1990. Chronic exposure to low doses of MPTP. II: Neurochemical and pathological consequences in cognitively impaired, motor asymptomatic monkeys. *Brain Res* **534:** 25–36. doi:10.1016/0006-8993(90)90108-N

Schneider JS, Kovelowski CJ II. 1990. Chronic exposure to low doses of MPTP. I: Cognitive deficits in motor asymptomatic monkeys. *Brain Res* **519:** 122–128. doi:10.1016/0006-8993(90)90069-N

Schneider JS, Roeltgen DP. 1993. Delayed matching-to-sample, object retrieval, and discrimination reversal deficits in chronic low dose MPTP-treated monkeys. *Brain Res* **615:** 351–354. doi:10.1016/0006-8993(93)90049-s

Shahmoradian SH, Lewis AJ, Genoud C, Hench J, Moors TE, Navarro PP, Castaño-Díez D, Schweighauser G, Graff-Meyer A, Goldie KN, et al. 2019. Lewy pathology in Parkinson's disease consists of crowded organelles and lipid membranes. *Nat Neurosci* **22:** 1099–1109. doi:10.1038/s41593-019-0423-2

Shariatgorji M, Nilsson A, Fridjonsdottir E, Vallianatou T, Källback P, Katan L, Sävmarker J, Mantas I, Zhang X, Bezard E, et al. 2019. Comprehensive mapping of neurotransmitter networks by MALDI-MS imaging. *Nat Methods* **16:** 1021–1028. doi:10.1038/s41592-019-0551-3

Shimozawa A, Ono M, Takahara D, Tarutani A, Imura S, Masuda-Suzukake M, Higuchi M, Yanai K, Hisanaga SI, Hasegawa M. 2017. Propagation of pathological α-synuclein in marmoset brain. *Acta Neuropathol Commun* **5:** 12. doi:10.1186/s40478-017-0413-0

Singleton AB, Farrer M, Johnson J, Singleton A, Hague S, Kachergus J, Hulihan M, Peuralinna T, Dutra A, Nussbaum R, et al. 2003. α-Synuclein locus triplication causes Parkinson's disease. *Science* **302:** 841. doi:10.1126/science.1090278

Spillantini MG, Schmidt ML, Lee VM, Trojanowski JQ, Jakes R, Goedert M. 1997. α-Synuclein in Lewy bodies. *Nature* **388:** 839–840. doi:10.1038/42166

Stolze H, Klebe S, Baecker C, Zechlin C, Friege L, Pohle S, Deuschl G. 2005. Prevalence of gait disorders in hospitalized neurological patients. *Mov Disord* **20:** 89–94. doi:10.1002/mds.20266

Teil M, Dovero S, Bourdenx M, Arotcarena ML, Camus S, Porras G, Thiolat ML, Trigo-Damas I, Perier C, Estrada C, et al. 2022. Brain injections of glial cytoplasmic inclusions induce a multiple system atrophy-like pathology. *Brain* **145:** 1001–1017. doi:10.1093/brain/awab374

Teil M, Dovero S, Bourdenx M, Arotcarena ML, Darricau M, Porras G, Thiolat ML, Trigo-Damas I, Perier C, Estrada C, et al. 2023. Cortical Lewy body injections induce long-distance pathogenic alterations in the non-human primate brain. *NPJ Parkinsons Dis* **9:** 135. doi:10.1038/s41531-023-00579-w

Urs NM, Bido S, Peterson SM, Daigle TL, Bass CE, Gainetdinov RR, Bezard E, Caron MG. 2015. Targeting β-arrestin2 in the treatment of L-DOPA-induced dyskinesia in Parkinson's disease. *Proc Natl Acad Sci* **112:** E2517–E2526. doi:10.1073/pnas.1502740112

Vila M, Laguna A, Carballo-Carbajal I. 2019. Intracellular crowding by age-dependent neuromelanin accumulation disrupts neuronal proteostasis and triggers Parkinson disease pathology. *Autophagy* **15:** 2028–2030. doi:10.1080/15548627.2019.1659621

Voon V, Napier TC, Frank MJ, Sgambato-Faure V, Grace AA, Rodriguez-Oroz M, Obeso J, Bezard E, Fernagut PO. 2017. Impulse control disorders and levodopa-induced dyskinesias in Parkinson's disease: an update. *Lancet Neurol* **16:** 238–250. doi:10.1016/S1474-4422(17)30004-2

Winner B, Jappelli R, Maji SK, Desplats PA, Boyer L, Aigner S, Hetzer C, Loher T, Vilar M, Campioni S, et al. 2011. In vivo demonstration that α-synuclein oligomers are toxic. *Proc Natl Acad Sci* **108:** 4194–4199. doi:10.1073/pnas.1100976108

Yang SH, Cheng PH, Banta H, Piotrowska-Nitsche K, Yang JJ, Cheng EC, Snyder B, Larkin K, Liu J, Orkin J, et al. 2008. Towards a transgenic model of Huntington's disease in a non-human primate. *Nature* **453:** 921–924. doi:10.1038/nature06975

Yang W, Wang G, Wang CE, Guo X, Yin P, Gao J, Tu Z, Wang Z, Wu J, Hu X, et al. 2015. Mutant α-synuclein causes age-dependent neuropathology in monkey brain. *J Neurosci* **35:** 8345–8358. doi:10.1523/JNEUROSCI.0772-15.2015

Yang W, Liu Y, Tu Z, Xiao C, Yan S, Ma X, Guo X, Chen X, Yin P, Yang Z, et al. 2019. CRISPR/Cas9-mediated PINK1 deletion leads to neurodegeneration in rhesus monkeys. *Cell Res* **29:** 334–336. doi:10.1038/s41422-019-0142-y

Yin M, Borton DA, Komar J, Agha N, Lu Y, Li H, Laurens J, Lang Y, Li Q, Bull C, et al. 2014. Wireless neurosensor for full-spectrum electrophysiology recordings during free behavior. *Neuron* **84:** 1170–1182. doi:10.1016/j.neuron.2014.11.010

Cite this article as *Cold Spring Harb Perspect Med* doi: 10.1101/cshperspect.a041612

Toxin-Induced Animal Models of Parkinson's Disease

Kim Tieu,[1,2] Said S. Salehe,[1] and Harry J. Brown[1,2]

[1]Department of Environmental Health Sciences; [2]Biomolecular Sciences Institute, Florida International University, Miami, Florida 33199, USA

Correspondence: ktieu@fiu.edu

The debilitating motor symptoms of Parkinson's disease (PD) result primarily from the degenerative nigrostriatal dopaminergic pathway. To elucidate pathogenic mechanisms and evaluate therapeutic strategies for PD, numerous animal models have been developed. Understanding the strengths and limitations of these models can significantly impact the choice of model, experimental design, and data interpretation. Herein, we systematically review the literature over the past decade. Some models no longer serve the purpose of PD models. The primary objectives of this review are: First, to assist new investigators in navigating through available animal models and making appropriate selections based on the objective of the study. Emphasis will be placed on common toxin-induced murine models. And second, to provide an overview of basic technical requirements for assessing the nigrostriatal pathway's pathology, structure, and function.

Parkinson's disease (PD) is the second most common neurodegenerative disorder, after Alzheimer's disease.[3] Currently, there is no cure for PD, and disease-modifying therapies for this devastating disease are urgently needed. To achieve this goal, it is critical to understand its etiology and the underlying mechanisms of neuronal dysfunction and degeneration. However, the cause(s) of most PD cases remains unknown. Less than 10% of PD cases can be directly linked to monogenic mutations. Environmental factors or a combination of both environment and genetic susceptibility have been proposed to play a role in sporadic PD. Accordingly, experimental models using exposure to exogenous neurotoxicants, mutations in genes linked to PD, or a combination of both have been created to study PD and evaluate therapeutic strategies. The success rate of translating this basic research into clinical relevance for PD relies heavily on the extent to which these experimental models accurately recapitulate the pathology, symptoms, and pathogenic mechanisms as seen in PD patients.

Pathologically, the hallmarks of PD include the loss of dopaminergic neurons in the substantia nigra pars compacta (SNpc) and the presence of cytoplasmic protein aggregates known as Lewy bodies in remaining dopaminergic cells (Dauer and Przedborski 2003; Goedert et al. 2017). When degeneration in these neurons results in

---

[3]This is an update to a previous article published in *Cold Spring Harbor Perspectives in Medicine* [Tieu (2011). *Cold Spring Harb Perspect Med* 1: a009316. doi:10.1101/cshperspect.a009316].

Copyright © 2025 Cold Spring Harbor Laboratory Press; all rights reserved
Cite this article as *Cold Spring Harb Perspect Med* doi: 10.1101/cshperspect.a041643

a threshold reduction of ~80% dopamine content in the striatum (Dauer and Przedborski 2003), motor symptoms of PD emerge. Mechanistically, the death of dopaminergic neurons has been linked to mitochondrial dysfunction, oxidative stress, neuroinflammation, and insufficient autophagic or proteasomal protein degradation (Michel et al. 2016; Tansey et al. 2022). In addition to the loss of nigrostriatal dopaminergic structures and function, PD affects many other areas of the central nervous system such as the dorsal motor nucleus of the vagus, the nucleus basalis of Meynert, the locus coeruleus, and the hypothalamus (Hornykiewicz and Kish 1987; Braak et al. 2004). Furthermore, the pathology of PD extends well beyond the central nervous system as evidenced by the presence of α-synuclein aggregates in the peripheral system (Blesa et al. 2022; Metta et al. 2022). Together, these extranigrostriatal regions may account for the observed nonmotor symptoms such as sleep disturbances, depression, cognitive impairment, anosmia, constipation, incontinence, and autonomic dysfunctions (Jain 2011; Pfeiffer 2016; Titova and Chaudhuri 2018).

An ideal model of PD should consist of pathological and clinical features of PD involving both dopaminergic and nondopaminergic systems, the central and peripheral nervous systems, plus motor and nonmotor symptoms. Additionally, the age-dependent onset and progressive nature of PD should be reflected. Unfortunately, none of the existing models currently exhibit all the features associated with PD. Consequently, thus far, our ability to experimentally phenocopy PD remains limited, and we can only emulate certain crucial aspects of the condition. Despite these limitations, animal models have contributed significantly to our current understanding of the disease processes and potential therapeutic targets in PD.

Current animal models of PD can be broadly divided into two categories: genetic and neurotoxic models. Each group has strengths and weaknesses. One strength of the genetic models is that they are created primarily based on identified targets associated with potential mechanisms known to cause PD in humans (Meredith et al. 2008; Bezard and Przedborski 2011). How-

ever, currently available models do not display appreciable neurodegeneration and behavioral phenotypes. This limitation of genetic models, however, can be complemented by the neurotoxic models in which different neurotoxic molecules are used to damage the nigrostriatal pathway. Faced with a wide variety of PD models, a new investigator may find selecting the appropriate one to be a daunting task. Another challenging step is knowing what basic techniques and equipment are required for assessing neurodegeneration and dysfunction in these models. To address these issues, this work will provide both an overview of the strengths and limitations of commonly used animal models as well as a general guide to assessing nigrostriatal damage. Because genetic models of PD and other vertebrate species such as nonhuman primates are discussed by Dawson and Dawson (2024) and Bezard et al. (2024), this work will focus on neurotoxic rodent models, which have been updated (Tieu 2011).

6-Hydroxydopamine

6-Hydroxydopamine (6-OHDA), a hydroxylated analog of dopamine (Fig. 1; see also Table 1), was first identified more than 60 years ago (Senoh and Witkop 1959; Senoh et al. 1959). Initially, 6-OHDA was reported to cause depletion of noradrenaline in the mouse heart (Porter et al. 1963, 1965). The subsequent discovery that 6-OHDA could induce selective degeneration in sympathetic adrenergic nerve terminals (Tranzer and Thoenen 1968, 1973) led to the novel concept of chemical denervation in neurobiology, where a neurotoxic molecule is used to target a specific cell population (Jonsson 1980). 6-OHDA has served as a useful tool to chemically lesion the nigrostriatal dopaminergic system as a model of PD. The detection of this molecule endogenously in the human brain (Curtius et al. 1974) and urine samples (Andrew et al. 1993) lends additional credence to this model. Multiple potential pathways by which 6-OHDA can be formed in vivo have been proposed (Varešlija et al. 2020).

In the brain, 6-OHDA is capable of inducing degeneration of both dopaminergic and noradrenergic neurons (Ungerstedt 1968). These types

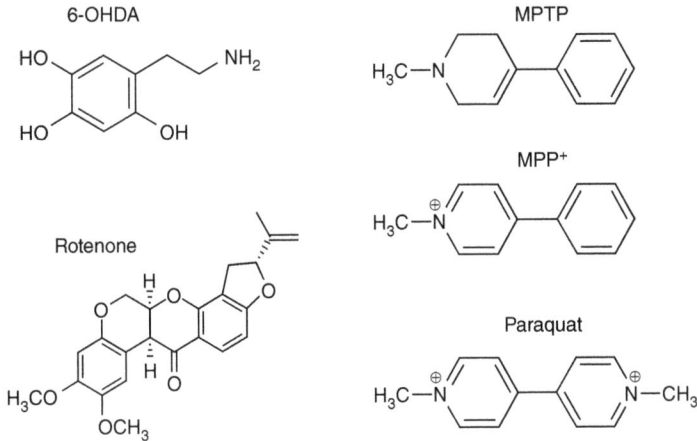

Figure 1. Structures of neurotoxic molecules used to induce nigrostriatal damage in some common animal models of Parkinson's disease (PD).

of neurons are particularly vulnerable to 6-OHDA because their plasma membrane transporters, the dopamine transporter (DAT) and noradrenergic transporter, respectively, have high affinity for this molecule (Luthman et al. 1989). Once taken up into neurons, 6-OHDA accumulates in the cytosol where it is readily oxidized leading to the generation of reactive oxygen species and ultimately oxidative stress-related cytotoxicity (Saner and Thoenen 1971; Graham 1978; Jonsson 1983; Cohen and Werner 1994; Blum et al. 2001). Accordingly, the locus coeruleus and the nigrostriatal region are highly sensitive to this neurotoxin (Jonsson 1980). To target specific neurons and to bypass the blood–brain barrier, 6-OHDA is typically injected stereotaxically into the brain region of interest. Due to the difficulty in targeting small brain structures such as the substantia nigra or medial forebrain bundle (MFB), 6-OHDA is more commonly used in rats than in mice to model PD (Jonsson 1983). Although uncommon, 6-OHDA has also been used in cats, guinea-pigs, dogs, and monkeys (Bezard et al. 1998; Bezard and Przedborski 2011).

The magnitude and characteristics of neurodegeneration induced by 6-OHDA are significantly affected by the site of injection (Agid et al. 1973; Przedborski et al. 1995; Przedborski and Tieu 2006). 6-OHDA is commonly injected unilaterally to the substantia nigra, MFB, or stri-

atum. Although the nerve terminals are more sensitive to 6-OHDA toxicity than the axon and cell body (Malmfors and Sachs 1968; Jonsson 1983), when injected into the nigra or the MFB, 6-OHDA produces a complete and rapid lesion in the nigrostriatal pathway. When injected into the nigra, degeneration of dopaminergic neurons takes place within 12 hours preceding a significant loss of striatal terminals, which occurs 2–3 days later (Faull and Laverty 1969; Jeon et al. 1995). When injected into the MFB, however, 6-OHDA induces degeneration in striatal terminals before dopaminergic cell death occurs (Sarre et al. 2004). In contrast to the nigra and MFB, when delivered to the striatum, 6-OHDA induces slower, progressive, and partial damage to the nigrostriatal structure in a retrograde fashion over a period of up to 3 weeks (Sauer and Oertel 1994; Przedborski et al. 1995). In rats, however, to produce extensive nigrostriatal damage, a four-site injection regimen is required (Kirik et al. 1998). The striatal injection offers four advantages: First, the progressive and less extensive lesion is more relevant to PD. Second, this regimen has been shown to produce nonmotor symptoms of PD including cognitive, psychiatric, and gastrointestinal dysfunction (Branchi et al. 2008; Tadaiesky et al. 2008; Cannon and Greenamyre 2010). Third, the ease of stereotaxically injecting a large structure such as the striatum enhances the likelihood of success in mice.

Table 1. Key features of common neurotoxic models of Parkinson's disease (PD)

Models	Pathology					Behavioral phenotypes	
	Nigrostriatal damage						
	SN cell body	Str. terminals	Str. DA	Extranigral pathology	α-Synuclein pathology	Motor (L-DOPA or apomorphine responsive)	Nonmotor
6-OHDA							
Rat: Stereotaxic injection to SN, MFB, striatum	Yes	Yes	Yes	No	No	Yes	Cognitive, psychiatric, and GI disorders
MPTP							
Nonhuman primates: i.p., i.m., intracarotid infusion	Yes	Yes	Yes	LC	Yes	Yes	Numerous (see Fox and Brotchie 2010; Bezard et al. 2024)
Mouse: Acute, subacute (i.p.)	Yes	Yes	Yes	No	No	Yes	Transient ↑colon motility
Chronic (osmotic minipumps and i.p.)	Yes	Yes	Yes	LC	Conflicting data	Yes	ND
PQ							
Mouse: i.p.	Yes	Conflicting	Yes	LC	Yes	ND	ND
Rat: (osmotic minipumps)	Yes	ND	Yes	ND	Yes	ND	ND
PQ/Maneb							
Mouse: i.p.	Yes	ND	Yes	ND	Yes	ND	ND
Rotenone							
Rat: Infusion via osmotic minipumps	Yes	Yes	Yes	ND	Yes	Yes	↓GI motility
i.p. injection	Yes	Yes	Yes	GI	Yes	Yes	↓GI motility

(SN) Substantia nigra, (Str.) striatal, (LC) locus coeruleus, (GI) gastrointestinal, (ND) not determined, (MFB) medial forebrain bundle, (PQ) paraquat, (6-OHDA) 6-hydroxydopamine, (MPTP) 1-methyl-4-phenyl-1,2,3,6-tetrahydropyridine.

Fourth, this route of injection also induces less dopaminergic neurodegeneration in the ventral tegmental area, consistent with its effects of retrograde cell death in the nigrostriatal pathway (Kirik et al. 1998).

The major advantage of using the 6-OHDA model is its rather unique effect on quantifiable circling motor abnormality in animals (Unger-stedt and Arbuthnott 1970). Typically, unilateral injection into one hemisphere (hemiparkinsonian model) is performed, leaving the unlesioned side as an internal control. Subsequent to this unilateral lesioning, systemic injection of dopamine receptor agonists (such as apomorphine), L-3,4-dihydroxyphenylalanine (L-DOPA, a dopamine precursor) or dopamine-releasing compounds (such as amphetamine) induces asymmetrical rotation (Ungerstedt and Arbuthnott 1970; Hefti et al. 1980). The magnitude of the nigrostriatal lesion correlates with the circling motor behavior (Ungerstedt 1968; Ungerstedt and Arbuthnott 1970; Hefti et al. 1980; Przedborski et al. 1995). The unilateral 6-OHDA rat model has been used as a preclinical model to assess the antiparkinsonian effects and neuroprotection of new pharmacological therapies (Jiang et al. 1993; Chan et al. 2010; Ilijic et al. 2011), as well as the clinical improvement of cell transplantation (Björklund et al. 2002; Kirik et al. 2002; Roy et al. 2006). However, like many other neurotoxic models of PD, the acute neurodegenerative property of the 6-OHDA model lacks the progressive, age-dependent effects of PD. Furthermore, neuroinflammation generated in this model lacks the widespread pattern seen across different brain regions in PD. Lastly, α-synuclein pathology is not present in this model.

1-Methyl-4-phenyl-1,2,3, 6-tetrahydropyridine

The 1-methyl-4-phenyl-1,2,3,6-tetrahydropyridine (MPTP) model originates from discoveries in the early 1980s when four Californian intravenous drug users were admitted to the Stanford University Hospital exhibiting severe symptoms similar to PD (Langston et al. 1983). Further investigations uncovered that these patients had self-administered synthetic meperidine contaminated with MPTP (Langston et al. 1983). The movement abnormalities in these patients were successfully treated with L-DOPA, a cornerstone treatment of PD, suggesting similar underlying neuropathological and biochemical features to those seen in PD patients. Soon after that landmark paper, it was demonstrated in the rhesus monkeys that MPTP caused nigrostriatal dopaminergic cell loss and L-DOPA-responsive locomotor impairment reminiscent of the MPTP-injected patients (Burns et al. 1983). Subsequent postmortem studies confirmed the loss of nigrostriatal structures in these patients (Langston et al. 1999), further strengthening the observation made in the nonhuman primate model. One of the surviving patients showed a significant clinical improvement when treated with deep brain stimulation (Christine et al. 2009), further affirming the damage induced by MPTP in the basal ganglia resembles that in human PD. Since its discovery, MPTP has drastically altered the course of PD research by providing insights into potential pathogenesis and mechanisms for cell death in PD. Studies using this model have led to the concepts such as environmental toxicity as a potential culprit in sporadic PD and mitochondrial dysfunction as a potential pathogenic mechanism. Additionally, work with this model has enabled the development of some of the current treatments for PD (Fox and Brotchie 2010; Langston 2017).

The mechanism of MPTP toxicity has been extensively studied and characterized (Dauer and Przedborski 2003; Rappold and Tieu 2010). Because MPTP is lipophilic, it can rapidly cross the blood–brain barrier. In astrocytes, MPTP is metabolized by monoamine oxidase-B and subsequently converted to the active toxic cation 1-methyl-4-phenylpyridinium (MPP^+) (Fig. 1). MPP^+ is released from the nigral and striatal astrocytes through the organic cation transporter 3 into the extracellular space (Cui et al. 2009; Rappold and Tieu 2010) where it is taken up by the neighboring dopaminergic neurons and terminals through the DAT. Once accumulated in dopaminergic neurons, MPP^+ induces neurotoxicity primarily by inhibiting complex I of the mitochondrial electron transport chain, resulting in ATP depletion and increased oxidative stress (Nicklas et al. 1985; Mizuno et al. 1987).

Various mammalian species, including sheep, dogs, guinea-pigs, cats, mice, rats, and monkeys have been treated with MPTP to model PD (Bezard et al. 1998; Przedborski et al. 2001). Rats are less sensitive to MPTP toxicity than mice (Giovanni et al. 1994). To date, the MPTP-monkey model remains an important preclinical model for testing therapeutic strategies for PD (for a detailed review of the MPTP-monkey model, refer to Bezard et al. 2024). For many researchers, however, the mouse remains a popular choice due to ethical regulations, lack of resources, and trained personnel for the nonhuman primate model. Additionally, available genetic mouse models allow investigators to assess the roles of certain genetic mutations in response to MPTP neurotoxicity.

In both monkeys and mice, MPTP primarily causes damage to the nigrostriatal dopaminergic pathway (Forno et al. 1993; Dauer and Przedborski 2003; Fox and Brotchie 2010). This specific and reproducible neurotoxic effect to the nigrostriatal system is a strength of this model. Although intraneuronal inclusions reminiscent of Lewy bodies have been described in MPTP-injected monkeys (Forno et al. 1986; Kowall et al. 2000), in general, this pathological feature is absent in mice. However, it has been reported that when mice are infused with a chronic low dose of MPTP over a period of 30 days using osmotic minipumps, inclusions immunoreactive for both ubiquitin and α-synuclein can be detected (Fornai et al. 2005). Additionally, with this regimen, Lewy bodies and degeneration of noradrenergic neurons are detected in the locus coeruleus. Using another chronic MPTP regimen, cytoplasmic accumulation of α-synuclein has also been detected (Korecka et al. 2013). Therefore, the ability of MPTP to produce α-synuclein inclusions seems to be related to dosage regimens. However, this observation is controversial as some investigators did not detect α-synuclein pathology using similar chronic MPTP paradigms (Shimoji et al. 2005; Alvarez-Fischer et al. 2008). Behaviorally, motor deficits induced by MPTP have been extensively characterized in both monkeys and mice. These abnormal phenotypes are reversible by L-DOPA or a dopamine agonist, confirming a connection between these symptoms and damage in the nigrostriatal system (Ogawa et al. 1985; Fredriksson and Archer 1994; Rozas et al. 1998; Fernagut et al. 2002). In general, parkinsonian symptoms are better reproduced in monkeys than in mice and a more profound loss of striatal dopamine is required in mice to produce some behavioral deficits. In addition to the nigrostriatal pathway, MPTP has been reported to induce loss of tyrosine hydroxylase (TH)-positive neurons in the enteric nervous system and alter colon motility in mice (Anderson et al. 2007). However, a recent study did not detect any significant dysfunction in gastric emptying rate or intestinal permeability in MPTP-treated monkeys (Delamarre et al. 2020).

In summary, with the caveat of its acute toxic property as seen with other neurotoxic PD models and the lack of consistent detection of α-synuclein aggregation, MPTP will continue to play a major role in PD research based on its ability to produce PD-like symptoms in nonhuman primates, its reproducible L-DOPA-responsive lesion on the nigrostriatal system, and its ease of administration with the typical intraperitoneal injection.

Paraquat

Paraquat (N,N'-dimethyl-4-4′-bipiridinium) is one of the most widely used herbicides globally. It is registered and sold in more than 90 countries, including the United States. However, due to the concerns of its toxicity, this herbicide has been banned in Europe and some other countries (Bastías-Candia et al. 2019). Acute exposure to paraquat (PQ) results in lung fibrosis and liver and kidney damage (Bastías-Candia et al. 2019). In addition to peripheral organ damage, significant brain injury and neuroinflammation were reported in patients with PQ poisoning (Grant et al. 1980; Hughes 1988). The potential link between PQ exposure and PD was highlighted in a search for environmental contaminants with a similar molecular structure to MPP^+ (Fig. 1; Snyder and D'Amato 1985) after the discovery of MPTP (Langston et al. 1983) and the subsequent characterization of MPP^+ as an active metabolite of MPTP (Nicklas et al. 1985). Since then, this divalent cation has been used to model PD in mice.

Despite a similar structure to MPP^+, PQ exhibits different transport properties and mechanism(s) of toxicity. First, although both molecules are cations, in contrast to MPP^+, PQ has been reported to have the ability to penetrate the blood–brain barrier. This property is proposed to be mediated by the neutral amino acid transporter (Shimizu et al. 2001; McCormack and Di Monte 2003). However, the extent to which PQ accumulates in the brain is age-dependent, with the highest levels detected in 2-week-old or >12-month-old rats—suggesting the blood–brain barrier does play a role (Corasaniti et al. 1991) as young and old animals have higher

Cite this article as *Cold Spring Harb Perspect Med* doi: 10.1101/cshperspect.a041643

blood–brain barrier permeability. Second, MPP$^+$ is an excellent substrate for the DAT, but PQ, in its divalent form, is not transported by DAT (Richardson et al. 2005). To be taken up by dopaminergic neurons through DAT, divalent PQ must be reduced to monovalent PQ (Rappold et al. 2011). Third, toxicity induced by PQ is primarily mediated by redox cycling with cellular diaphorases such as NADPH oxidase and nitric oxide synthase (Day et al. 1999), leading to the generation of superoxide. Fourth, inside mitochondria, PQ is not a complex I inhibitor (Richardson et al. 2005), although this is the site where it is reduced to form superoxide (Cochemé and Murphy 2008).

In PQ mouse models, different routes of administration (oral, nasal, injection-intraperitoneal, or stereotaxic delivery into the brain) have been used. However, intraperitoneal injection is the most used and best characterized. When intraperitoneally injected into mice, PQ is accumulated in the ventral midbrain linearly between one and five doses using the regimen of 10 mg/kg, one dose on Monday, Wednesday, and Friday, with an apparent half-life of $\sim$28 days (Prasad et al. 2007). PQ was reported to induce motor deficits and loss of nigral dopaminergic neurons in a dose- (Brooks et al. 1999; McCormack et al. 2002) and age- (McCormack et al. 2002; Thiruchelvam et al. 2003) dependent manner. The effects of PQ are specific to nigral dopaminergic neurons as γ-aminobutyric acid (GABA) neurons in the nigral and striatal regions, glutamate neurons in the hippocampus, and dopaminergic neurons in the ventral tegmental area are not affected (Thiruchelvam et al. 2000b; McCormack et al. 2002). However, the damage induced by PQ in dopaminergic cell bodies and terminals has not been consistently observed (Thiruchelvam et al. 2000b; Cicchetti et al. 2005). Furthermore, even in studies where a loss in nigral dopaminergic neurons is detected, PQ does not have an effect on striatal dopamine levels (Thiruchelvam et al. 2000b; McCormack et al. 2002; Rappold et al. 2011). This lack of dopamine reduction might be related to the compensatory up-regulation of TH activity in the striatum after PQ injection (Thiruchelvam et al. 2000b; McCormack et al. 2002; Os-

sowska et al. 2005). When combined with manganese ethylenebisdithiocarbamate (Maneb), a more significant loss in dopaminergic neurons and a trend reduction ($\sim$20%, not statistically significant) of striatal dopamine are produced (Thiruchelvam et al. 2000a,b). Even with this revised PQ/Maneb protocol, however, it remains unclear how the motor deficits reported in this model can be attributed to such a modest loss of striatal dopamine. A similar argument can be made in a rat model in which PQ is injected weekly for 24 weeks but only $\sim$30% reduction in dopamine is detected (Ossowska et al. 2005). Given that PQ is known to cause peripheral toxicity (Cicchetti et al. 2005; Saint-Pierre et al. 2006; Bastías-Candia et al. 2019), which could affect motor performance, it would be necessary to assess whether L-DOPA can alleviate motor deficits induced by PQ. As discussed, this strategy has been used in the MPTP and 6-OHDA animal models to confirm that the observed motor deficit is linked to the damage of the nigrostriatal pathway. Using an "upgraded" PQ exposure paradigm, male Wistar rats were exposed to a low chronic dose of PQ via subcutaneous osmotic minipumps for 8 weeks (Cristóvão et al. 2020). This model produces a loss of 41% nigral dopaminergic neurons, 40% striatal dopamine content, and a reduction in motor function. Furthermore, this model increases the levels of α-synuclein, pS129-α-synuclein, and protein aggregation.

The strength of the PQ or PQ/Maneb model is its potential relevance to environmental toxicants as a risk factor for developing PD. These chemicals have been used in overlapping geographical areas. Although PQ has been banned in Europe, it is still used in the United States and other parts of the world (Donley 2019). Epidemiological studies have suggested an increased risk for PD after PQ exposure (Hertzman et al. 1990; Liou et al. 1997; Kamel et al. 2007; Ritz et al. 2009; Tanner et al. 2011; Tangamornsuksan et al. 2019). Additionally, PQ treatment increases α-synuclein aggregates reminiscent of Lewy bodies in PD (Manning-Bog et al. 2002; Fernagut et al. 2007; Mak et al. 2010; Cristóvão et al. 2020). PQ is also reported to reduce noradrenergic neurons in the locus coeruleus (Fernagut et al. 2007; Hou et al. 2017), a pathological feature present in PD. Overall, PQ is an environmentally relevant model of PD, but the lack of reproducible effects on the nigrostriatal pathway, especially striatal dopamine depletion, in the commonly used regimens may limit the use of this model to assess neuroprotective therapies for PD. The recently developed rat model (Cristóvão et al. 2020) is promising but its reproducibility needs to be confirmed by other laboratories.

Rotenone

The isoflavone rotenone is a naturally occurring chemical found in plants belonging to the *Leguminosae* family. Traditionally, the indigenous used rotenone to catch fish hereafter being translated to its use to kill pest fish stock in bodies of water around the globe. Because this is a natural product, it was used in organic farming as an insecticide and herbicide; however, rotenone is no longer permitted in the United States and Europe, although it is still being used to study PD.

Due to its high lipophilicity, rotenone can readily cross the blood–brain barrier and enter all cells without being dependent on a specific transporter. The mechanism of toxicity of rotenone is primarily mediated by potent inhibition of the mitochondrial complex I, which in turn leads to the generation of superoxide. Based on the observations that MPP^+ is a complex I inhibitor and that reduced function of this mitochondrial subunit has been reported in PD patients (Parker et al. 1989; Schapira et al. 1989), rotenone exposure has been adopted by researchers as a model to study PD. Initially, when rotenone was first used as a method to model PD, widespread lesions beyond the nigrostriatal system were reported (Heikkila et al. 1985; Ferrante et al. 1997; Rojas et al. 2009). The rotenone model thus received little attention until Greenamyre and colleagues developed a chronic low-dose regimen (Betarbet et al. 2000). When infused continuously via a jugular vein cannula attached to a subcutaneous osmotic minipump, rotenone produces selective nigrostriatal neurodegeneration and α-synuclein-positive cytoplasmic inclusions. This study generated significant interest in

the field because, first, despite widespread complex I inhibition in the brain, selective neurodegeneration occurs in the nigrostriatal pathway, strengthening the notion that nigral dopaminergic neurons are inherently more vulnerable. Second, the relationship between complex I inhibition and the pathogenic mechanism of cell death in PD is reinforced. Third, it suggests that a chronic low-dose neurotoxic regimen may be required to produce Lewy bodies. This rationale was subsequently applied in the chronic model of MPTP, in which continuous delivery of MPTP for 30 days using osmotic minipumps does produce α-synuclein aggregates (Fornai et al. 2005). Fourth, the rotenone model reinforces the theory that environmental toxins may play a role in the pathogenesis of sporadic PD.

Despite its positive features, the rotenone infusion model has not been widely adopted. The primary concern is related to the high variability in animal sensitivity to rotenone and the inability of other investigators to consistently reproduce the parkinsonian neuropathology and phenotype of this model (Höglinger et al. 2003; Fleming et al. 2004b; Lapointe et al. 2004; Zhu et al. 2004). To address these concerns, a revised rotenone model has been developed (Cannon et al. 2009). With this protocol, using the medium-chain triglyceride Miglyol 812N, as a specialized delivery vehicle, rats in three age groups (3, 7, and 12–14 months) received daily intraperitoneal injection of rotenone (2.75 or 3.0 mg/kg) until they developed debilitating phenotypes, which were more pronounced in the 7- and 12- to 14-month-old groups. This regimen is reported to produce a more consistent lesion in the nigrostriatal pathway, accompanied by the presence of α-synuclein and ubiquitin-positive inclusions and motor deficits that are reversed by apomorphine (Cannon et al. 2009). One potential issue of this model is the selected endpoint criterion, which is the onset of debilitating motor function, for when the animal is euthanized. Since the onset of debilitating symptoms varies greatly from one animal to another within the same treatment group (Cannon et al. 2009), perhaps due to peripheral toxicity induced by rotenone, animals in the same study do not receive the same amount of rotenone. Using a lower and more chronic paradigm (2.0 mg/

kg, i.p., for 6 weeks), these investigators also reported a loss of neurons and the appearance of α-synuclein aggregation in the myenteric plexus as well as a decrease in gastrointestinal motility 6 months after treatment (Drolet et al. 2009). Recently, the rotenone rat model has been further refined. Middle-aged rats (7–9 months old) were transiently exposed to rotenone (2.8 mg/kg, i.p.) once daily for 5 consecutive days, followed by motor function and neuropathology evaluation over the next 9 months (Van Laar et al. 2023). The loss of nigral dopaminergic neurons is detectable at 1 month and stable up to 9 months postrotenone injection. Interestingly, although the loss of neurons is not progressive, motor abnormalities emerge at 3 months and progressively worsen thereafter. In neurons of the substantia nigra and frontal cortex, there is a delay in accumulation of total α-synuclein and pS129-α-synuclein, which is maximal at 9 months postrotenone exposure. As compared to other acute and subacute rotenone models, the delay onset of motor abnormalities and α-synuclein in this model provides a window to investigate pathogenic mechanisms and therapeutic interventions. Although promising, as with other newly developed models, the results of this modified rotenone model need to be replicated by other laboratories.

Lipopolysaccharide

Neuroinflammation has been proposed to be involved in PD (Przedborski 2007; Hirsch et al. 2012; Ransohoff 2016; Tansey et al. 2022). Although all the neurotoxic molecules described above can elicit an inflammatory response, it is challenging to delineate whether neuroinflammation is the cause or consequence of injured dopaminergic neurons. The consensus appears to be that neuroinflammation is not the primary cause of cell death in these models, but rather it serves as a secondary event that perpetuates the vicious cycle of neurotoxicity. To study the role of neuroinflammation in causing cell death, the lipopolysaccharide (LPS) model is probably more appropriate.

LPS is a gram-negative bacterial endotoxin that activates microglia through the Toll-like receptor-4 receptor, leading to the production of inflammatory cytokines and chemokines (Skrzypczak-Wiercioch and Sałat 2022). LPS has been administered stereotaxically, intraperitoneally, or intranasally to induce nigrostriatal damage (Deng and Bobrovskaya 2022; Skrzypczak-Wiercioch and Sałat 2022). Stereotaxically, LPS can be delivered into the substantia nigra, MFB, or striatum to produce acute damage (Herrera et al. 2000; García-Revilla et al. 2022). When injected unilaterally into the rat nigra, microglia activation and stable loss of nigral dopamine neurons (~70%–80%) occurs within 24 hours (Iravani et al. 2005) along with a stable reduction (~60%) in striatal dopamine in the ipsilateral side up to 1 year after injection (Herrera et al. 2000). GABAergic and serotonergic neurons do not appear to be affected by LPS exposure, suggesting the neurotoxic effect is specific to dopaminergic neurons (Herrera et al. 2000). To induce a more chronic and progressive cell loss, LPS is infused into the rat supra-nigra for 2 weeks using osmotic minipumps (Gao et al. 2002). Microglia activation occurs within 3 days and reaches a sustained plateau from 2 to 8 weeks after LPS infusion. Significant loss of dopamine neurons (~39%) is detectable at 6 weeks and progressed to 69% loss at 10 weeks after the initiation of LPS infusion. To avoid the technical challenges of stereotaxic surgery, a single systemic intraperitoneal injection of LPS has been described in mice (Qin et al. 2007, 2013). Although microglial activation occurs as early as 3 hours after this peripheral injection, it takes 7 months to observe a modest (23%) and 10 months to have significant (47%) nigral dopaminergic cell loss. Although repeated intraperitoneal injection of LPS mouse models has been developed, the detection of nigrostriatal damage is inconsistent (Deng and Bobrovskaya 2022). The intranasal LPS mouse model has also recently been developed and they appear to be promising. This new model is well described in a recent review by Deng and Bobrovskaya (2022).

Overall, the LPS model has not been widely used. Several reasons may contribute to this lack of popularity: First, stereotaxic injection is a technical deterrent for many laboratories. Second, it takes too long to detect nigrostriatal damage with intraperitoneal injection. Although this progressive cell loss can be an attractive feature, it is not feasible for most neuroprotective studies. Third, protein aggregation is not detectable. Fourth, despite it being a useful tool to establish that neuroinflammation can cause nigral cell loss, its relevance to PD remains uncertain as it is controversial that neuroinflammation is the primary cause of cell death in this disorder. Lastly, the reproducibility of this model may be an issue since the bacterial origin of the LPS used could greatly affect immunogenicity and neurotoxicity. Care must be taken when selecting this endotoxin from a variety of bacteria species, strains, and serotypes (Deng and Bobrovskaya 2022; Skrzypczak-Wiercioch and Sałat 2022).

BASIC ASSESSMENTS OF THE NIGROSTRIATAL DOPAMINERGIC STRUCTURE AND FUNCTION

Depending on the nature of the study and the selected PD model, the methods required for the analysis of neuropathology and function may vary. However, whether planning to study neurodegeneration or neuroprotection using neurotoxic models, it is common to assess the integrity and function of the nigrostriatal pathway. Over the years, there have been some well-developed and accepted techniques that are commonly used for these purposes. This section will highlight such basic methods and equipment with an emphasis on murine models. Detailed step-by-step procedures are beyond the scope of this article and readers are encouraged to consult the recommended references.

Quantification of Dopaminergic Neurons in the Substantia Nigra Par Compacta

In the mesencephalon, there are three groups of dopaminergic neurons classified historically as A8, A9, and A10 (Dahlström and Fuxe 1964; Hokfelt et al. 1984; Smith and Kieval 2000; Zaborszky and Vadasz 2001). The A8 group resides in the retrorubral area, whereas the A9 and A10 group belong to the substantia nigra and ventral tegmental area, respectively (Fig. 2). The estimated average number of dopaminergic neurons in an adult mouse nigra is commonly reported to be ~8000–14,000. The number may vary slightly depending on the strain (Zaborszky and Vadasz 2001). The distribution of dopaminergic neurons in the nigra is not homogenous. As shown in Figure 2, when the nigra is sectioned coronally, there is a significant difference in the density of dopaminergic neurons between the caudal and rostral regions. It is necessary, therefore, to sample the population of dopaminergic neurons at different levels throughout the entire nigra. The past practice of comparing the number of dopaminergic neurons from only one or certain nigral sections between animals should be avoided. One way to sample the entire nigra efficiently is to count dopaminergic neurons systematically at a regular section interval as illustrated in Figure 2. To count these neurons, the gold standard is to use an unbiased stereological cell counting with an optical dissector system (West et al. 1991, 1996; West 1993; Tieu et al. 2003). The major components required for this method are a computerized stereology software and a microscope with a motorized stage.

Quantification of Dopaminergic Terminals in the Striatum

The projection of terminals to the striatum from the nigral dopaminergic neurons has been well established (Joel and Weiner 2000). Although the total density of striatal dopaminergic terminals frequently correlates with the number of their cell bodies in the nigra, it is not uncommon to observe differential damage or protection between these two structures. For example, as discussed, in the PQ model, dopaminergic cell bodies are more vulnerable than their terminals. On the other hand, striatal terminals are more sensitive to MPTP or 6-OHDA toxicity. Quantifying striatal dopaminergic terminals is, therefore, also informative.

Similar to the uneven distribution of dopaminergic neurons throughout the nigra, the density of dopaminergic terminals is also not homogenous in the striatum. As shown in Figure 2 and previous studies (Tassin et al. 1976; Scally et al. 1978; Widmann and Sperk 1986), dopaminergic innervation increases from the caudal to the rostral part. Accordingly, when quantifying

 Cite this article as *Cold Spring Harb Perspect Med* doi: 10.1101/cshperspect.a041643

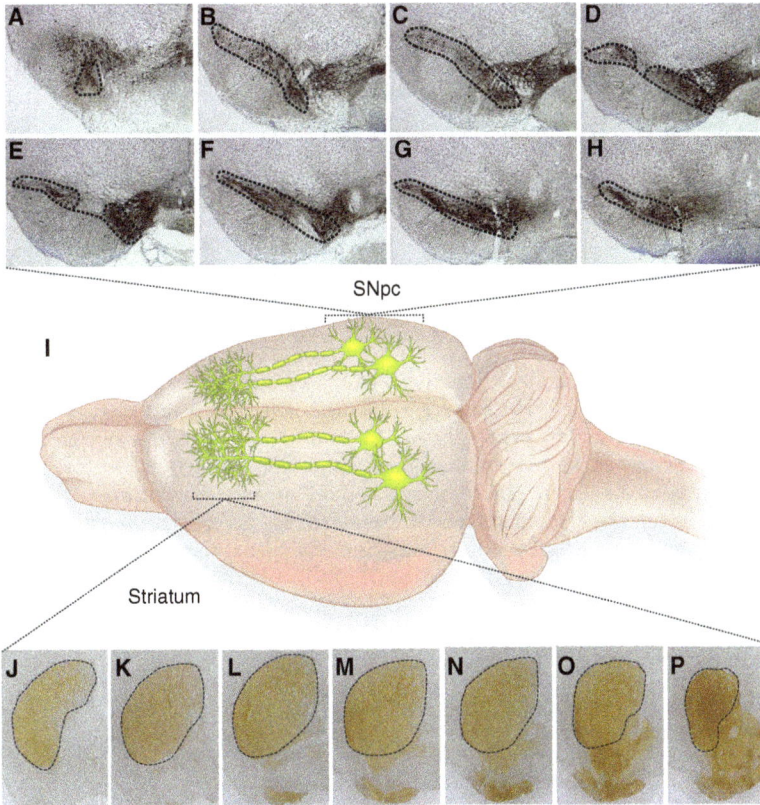

Figure 2. Topographical distribution of dopaminergic neurons in the nigral and striatal regions. The cell bodies of dopaminergic neurons that reside in the substantia nigra pars compacta (SNpc) project their terminals to the dorsal striatum where dopamine is released (*I*). To show the heterogeneous distribution of dopaminergic neurons in the entire SNpc (outlined) spanning from the caudal (*A*) to rostral regions (*H*), coronal sections (30 μm) are presented at every fourth section interval (*A–H*). Similarly, the distribution of dopaminergic terminals is not homogenous in the striatum (*J–P*). The density is relatively low in the caudal (*J*) as compared to the rostral region (*P*). Images are captured at every eighth section interval of the striatum. Dopaminergic neurons and terminals are immunostained with an antibody against tyrosine hydroxylase (TH) and visualized using 3,3′-diaminobenzidine.

dopaminergic terminals, representative regions of the striatum should be assessed (Fig. 2). The density of striatal dopaminergic terminals can be digitized and then quantified based on optical density or fiber density of TH immunoreactivity (Tieu et al. 2003; Fernagut et al. 2007). As an alternative to immunohistochemistry, fresh striatal tissues can be isolated for immunoblotting using TH or DAT as markers to assess for the levels of striatal dopaminergic terminals. Although more objective, some potential disadvantages of this later approach are the limited quantity of striatal tissue and the lack of resolution of the size and pattern of the lesion.

Quantification of Dopamine Content in the Striatum

In addition to structural analysis (cell body and terminal), it is also critical to assess the function of the nigrostriatal pathway. Measuring the amount of dopamine produced in the striatum and assessing motor movement (see below) are some such functional studies. The best way to measure dopamine is to use reverse-phase high-performance liquid chromatography (HPLC) coupled with an electrochemical detector. With this method, samples prepared from striatal tissues are injected into HPLC (Fig. 3). Figure 3

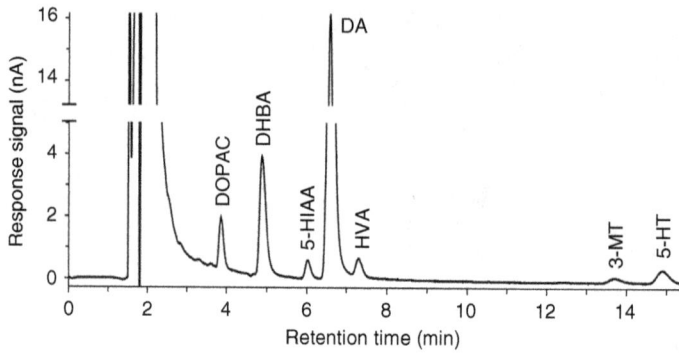

Figure 3. High-performance liquid chromatography (HPLC) chromatogram of catecholamines in striatal tissue. The striatum was freshly dissected and processed for HPLC measurement as described (Cui et al. 2009). Samples were eluted on a narrowbore (ID:2 mm) reverse-phase C18 column (MD-150, ESA) using a 12-channel CoulArray 5600A (ESA). A highly sensitive amperometric microbore cell (model 5041, ESA) was used to analyze the content of dopamine with a potential set at +220 mV. The striatum is a major "input" structure of the basal ganglia. In addition to receiving extensive dopaminergic terminals from the substantia nigra, the striatum also contains serotonergic terminals from the dorsal raphe nuclei. Dopamine and serotonin as well as their metabolites, therefore, are often detectable in the same striatal samples. (DA) Dopamine, (DOPAC) 3,4-dihydroxyphenylacetic acid, a metabolite of DA, (HVA) homovanillic acid, a metabolite of DA, (3-MT) 3-methoxytyramine, a metabolite of DA, (5-HT) serotonin, (5-HIAA) 5-hydroxyindoleacetic acid, a metabolite of 5-HT, (DHBA) 3,4-dihydroxybenzylamine, internal control for the measurement of catecholamines.

provides an example of what to expect when a striatal sample is analyzed using HPLC. To maximize efficiency and consistency as well as to minimize the number of required animals, the brain can be processed in such a way that data for striatal dopamine content, striatal terminal density, and nigral cell counts can be obtained within the same animal. For this approach, freshly removed brains are coronally sectioned at the level of the optic chiasm. The caudal half containing the substantia nigra is used for stereological cell counts, while the rostral half is divided mid-sagittally with one hemisphere used for striatal density and the other hemisphere used for HPLC measurement of dopamine.

Detection of α-Synuclein Pathology

α-Synuclein, a central component of the Lewy bodies, is encoded by *SNCA*, which is the most investigated PD-linked gene to date. Although mutations in this gene are rare, the *SNCA* locus is a significant variant factor for PD development (Lill et al. 2012). Missense mutations as well as gene duplications and triplications of *SNCA* have

been identified in familial PD (Polymeropoulos et al. 1997; Krüger et al. 1998; Singleton et al. 2003; Zarranz et al. 2004; Appel-Cresswell et al. 2013; Lesage et al. 2013). The fact that increasing the gene dosage of *SNCA* by twofold to threefold can cause PD (Singleton et al. 2003) indicates that elevated wild-type α-synuclein alone is sufficient to cause the disease. Indeed, the accumulation of α-synuclein aggregates is a key pathological feature of idiopathic PD. Recognizing the significance of this protein, numerous methods have been developed to detect the levels, different forms, and structures of pathogenic α-synuclein. An in-depth discussion of these techniques is beyond the scope of this paper, but they have been well-described in this recent review (Estaun-Panzano et al. 2023). Briefly, the following are some examples of the techniques that have been used to monitor α-synuclein pathology. Seed amplification assays, also known as real-time quaking-induced conversion, have been used successfully in a recent breakthrough to detect α-synuclein as a biomarker for PD (Siderowf et al. 2023). This method was initially developed to detect self-propagation of the prion protein

(Saborio et al. 2001). Taking advantage of the self-propagating property of misfolded α-synuclein, this method has been adopted to monitor aggregated α-synuclein with a β-sheet-rich conformation (seed) to template the conformational change of native α-synuclein to acquire the same pathological structure as the seed (Concha-Marambio et al. 2023). Other commonly used methods such as thioflavin S or T can be used to identify the presence of fibrillary α-synuclein. Antibodies against different epitopes, conformations, and posttranslational modifications of α-synuclein are great tools for routine techniques such as immunohistochemistry, immunoblotting, and enzyme-linked immunosorbent assay. Because misfolded or fibrillary α-synuclein is relatively resistant to proteolysis, one simple way to detect the presence of α-synuclein aggregates is to treat brain sections with proteinase K, followed by immunohistochemistry using an antibody against α-synuclein. All these techniques and more are discussed in detail in the review by Estaun-Panzano et al. (2023).

Assessment of Motor Function

The functional relationship between the depletion of striatal dopamine and the motor deficits in PD was discovered more than 60 years ago (Carlsson et al. 1957; Carlsson 1959). Since this discovery, subsequent PD animal models have been generated with the objectives of inducing a loss in nigrostriatal dopaminergic structure and dopamine content to accurately reproduce PD pathology and L-DOPA-responsive motor deficits as seen in PD. Many behavioral tests are, therefore, designed to assess motor phenotypes linked to the nigrostriatal function.

Behavioral tests routinely used in animal models of PD include quantification of, for example, locomotor activity, rotation, rotarod, stride length of the paws, and pole tests. For a more in-depth discussion of these methods, including their strengths and weaknesses, readers are encouraged to consult other reviews (Sedelis et al. 2001; Brooks and Dunnett 2009; Taylor et al. 2010). Briefly, one simple and basic way to assess locomotor activity is to use automated open-field chambers. Typically, mice are placed in transparent chambers that are equipped with horizontal and vertical infrared photobeams. The movement of each animal is tracked and quantified based on photobeam breaks, which are registered in a computer and processed for parameters such as jumps, traveled distance, and vertical, ambulatory, or stereotypical movements in different bin sizes over different periods of time. Most PD models have reduced locomotor activity (Taylor et al. 2010). The rotation test has been used for decades to assess the motor asymmetry in the unilateral 6-OHDA lesion rat and more recently in mice (Iancu et al. 2005). As discussed previously, subsequent to the 6-OHDA injection, the magnitude of the nigrostriatal lesion correlates with the circling motor behavior (Ungerstedt 1968; Ungerstedt and Arbuthnott 1970; Hefti et al. 1980; Przedborski et al. 1995). The direction of rotation is ipsilateral to the lesion if dopamine released from the intact terminals is induced by compounds such as amphetamine; however, an agonist such as apomorphine stimulates the sensitized dopaminergic receptors on the lesioned side and will lead to contralateral rotation. This rotation can be quantified using a rotameter test bowl (Brooks and Dunnett 2009). To measure abnormal movement that is analogous to the shuffling gait in PD patients, the "footprint" test can be used to measure the stride length of the paws. With this test, the fore and hind limbs of the animal are inked with different colors, and the stride length is quantified after a walk down the narrow corridor (Fernagut et al. 2002; Brooks and Dunnett 2009). To measure body coordination and balance, either the rotarod or pole test can be used. In the rotarod (Dunham and Miya 1957; Rozas et al. 1998; Brooks and Dunnett 2009), the most commonly used behavioral test, a mouse is placed on a rod that can rotate at a fixed or an accelerating speed. The length of time an animal can maintain balance and stay on the rotating rod is recorded. In the pole test (Ogawa et al. 1985; Fleming et al. 2004a), the animal is placed facing upward on top of a vertical wooden or wire-mesh pole. The time that it takes the animal to orient downward and the total time that it takes to descend to the base of the pole are recorded. With all motor function tests, caution must be taken with data interpretation when the abnormal movements are attributed to the dopaminergic nigrostriatal pathway in models whose peripheral system is also affected (such as intraperitoneal injection of a toxin), especially when striatal dopamine content is not significantly reduced.

CONCLUDING REMARKS

Since the initial discovery by Carlsson (1959) and Carlsson and colleagues (1957) in the 1950s that L-DOPA restored motor deficits induced by reserpine, many animal models of PD have been developed with the primary objective of improving PD-like pathology and phenotypes in existing ones for bettering the utility of these models to elucidate pathogenic mechanisms and to evaluate therapeutics. Together, these models have contributed to the development of some of the current PD treatments such as L-DOPA, dopamine agonists, and monoamine oxidase-B inhibitors. Our current understanding of the basal ganglia circuitry, pathogenic mechanisms, and potential therapeutic targets for PD also greatly benefited from these models. However, despite these accomplishments, current models still have significant shortcomings. Some such limitations are highlighted in the models discussed in this paper.

Confronted with the strengths and weaknesses of numerous PD models, it may be challenging to make a selection. Because there is no perfect model, the decision should be carefully balanced by considerations such as the following: (1) What is the nature and the goal of the study? If the loss of the nigrostriatal pathway is required to address the question, then a neurotoxic model is necessary. However, if a target known to cause PD is the primary interest, then perhaps a genetic model is relevant with the caveats that these animals do not display appreciable neurodegeneration and require time to develop motor deficits. (2) How technically involved is the model? Intraperitoneal injection certainly is more convenient and consistent than stereotaxic delivery into the brain or the implantation of osmotic minipumps. (3) How reproducible is the model? Once

an investigator has navigated through such questions and arrived at a decision, the technical aspects will follow that model. With an overview of the methods commonly used to assess the nigrostriatal structure and function, this work also provides additional technical guidance to a new investigator.

ACKNOWLEDGMENTS

This work was supported in part by the National Institute of Environmental Health Sciences of the National Institutes of Health under Award Number R35ES030523. The content is solely the responsibility of the authors and does not necessarily represent the official views of the National Institutes of Health.

REFERENCES

*Reference is also in this subject collection.

Agid Y, Javoy F, Glowinski J, Bouvet D, Sotelo C. 1973. Injection of 6-hydroxydopamine into the substantia nigra of the rat. II: Diffusion and specificity. *Brain Res* **58:** 291–301. doi:10.1016/0006-8993(73)90002-4

Alvarez-Fischer D, Guerreiro S, Hunot S, Saurini F, Marien M, Sokoloff P, Hirsch EC, Hartmann A, Michel PP. 2008. Modelling Parkinson-like neurodegeneration via osmotic minipump delivery of MPTP and probenecid. *J Neurochem* **107:** 701–711. doi:10.1111/j.1471-4159.2008.05651.x

Anderson G, Noorian AR, Taylor G, Anitha M, Bernhard D, Srinivasan S, Greene JG. 2007. Loss of enteric dopaminergic neurons and associated changes in colon motility in an MPTP mouse model of Parkinson's disease. *Exp Neurol* **207:** 4–12. doi:10.1016/j.expneurol.2007.05.010

Andrew R, Watson DG, Best SA, Midgley JM, Wenlong H, Petty RK. 1993. The determination of hydroxydopamines and other trace amines in the urine of parkinsonian patients and normal controls. *Neurochem Res* **18:** 1175–1177. doi:10.1007/BF00978370

Appel-Cresswell S, Vilarino-Guell C, Encarnacion M, Sherman H, Yu I, Shah B, Weir D, Thompson C, Szu-Tu C, Trinh J, et al. 2013. α-Synuclein p.H50Q, a novel pathogenic mutation for Parkinson's disease. *Mov Disord* **28:** 811–813. doi:10.1002/mds.25421

Bastías-Candia S, Zolezzi JM, Inestrosa NC. 2019. Revisiting the paraquat-induced sporadic Parkinson's disease-like model. *Mol Neurobiol* **56:** 1044–1055. doi:10.1007/s12035-018-1148-z

Betarbet R, Sherer TB, MacKenzie G, Garcia-Osuna M, Panov AV, Greenamyre JT. 2000. Chronic systemic pesticide exposure reproduces features of Parkinson's disease. *Nat Neurosci* **3:** 1301–1306. doi:10.1038/81834

Bezard E, Przedborski S. 2011. A tale on animal models of Parkinson's disease. *Mov Disord* **26:** 993–1002. doi:10.1002/mds.23696

Bezard E, Imbert C, Gross CE. 1998. Experimental models of Parkinson's disease: from the static to the dynamic. *Rev Neurosci* **9:** 71–90. doi:10.1515/revneuro.1998.9.2.71

* Bezard E, Teil M, Arotcarena ML, Porras G, Li Q, Dehay B. 2024. Modeling Parkinson's disease in primates. *Cold Spring Harb Perspect Med.* doi:10.1101/cshperspect.a041612

Björklund LM, Sánchez-Pernaute R, Chung S, Andersson T, Chen IY, McNaught KS, Brownell AL, Jenkins BG, Wahlestedt C, Kim KS, et al. 2002. Embryonic stem cells develop into functional dopaminergic neurons after transplantation in a Parkinson rat model. *Proc Natl Acad Sci* **99:** 2344–2349. doi:10.1073/pnas.022438099

Blesa J, Foffani G, Dehay B, Bezard E, Obeso JA. 2022. Motor and non-motor circuit disturbances in early Parkinson disease: which happens first? *Nat Rev Neurosci* **23:** 115–128. doi:10.1038/s41583-021-00542-9

Blum D, Torch S, Lambeng N, Nissou M, Benabid AL, Sadoul R, Verna JM. 2001. Molecular pathways involved in the neurotoxicity of 6-OHDA, dopamine and MPTP: contribution to the apoptotic theory in Parkinson's disease. *Prog Neurobiol* **65:** 135–172. doi:10.1016/s0301-0082(01)00003-x

Braak H, Ghebremedhin E, Rüb U, Bratzke H, Del TK. 2004. Stages in the development of Parkinson's disease-related pathology. *Cell Tissue Res* **318:** 121–134. doi:10.1007/s00441-004-0956-9

Branchi I, D'Andrea I, Armida M, Cassano T, Pèzzola A, Potenza RL, Morgese MG, Popoli P, Alleva E. 2008. Non-motor symptoms in Parkinson's disease: investigating early-phase onset of behavioral dysfunction in the 6-hydroxydopamine-lesioned rat model. *J Neurosci Res* **86:** 2050–2061. doi:10.1002/jnr.21642

Brooks SP, Dunnett SB. 2009. Tests to assess motor phenotype in mice: a user's guide. *Nat Rev Neurosci* **10:** 519–529. doi:10.1038/nrn2652

Brooks AI, Chadwick CA, Gelbard HA, Cory-Slechta DA, Federoff HJ. 1999. Paraquat elicited neurobehavioral syndrome caused by dopaminergic neuron loss. *Brain Res* **823:** 1–10. doi:10.1016/s0006-8993(98)01192-5

Burns RS, Chiueh CC, Markey SP, Ebert MH, Jacobowitz DM, Kopin IJ. 1983. A primate model of parkinsonism: selective destruction of dopaminergic neurons in the pars compacta of the substantia nigra by N-methyl-4-phenyl-1,2,3,6-tetrahydropyridine. *Proc Natl Acad Sci* **80:** 4546–4550. doi:10.1073/pnas.80.14.4546

Cannon JR, Greenamyre JT. 2010. Neurotoxic in vivo models of Parkinson's disease recent advances. *Prog Brain Res* **184:** 17–33. doi:10.1016/S0079-6123(10)84002-6

Cannon JR, Tapias V, Na HM, Honick AS, Drolet RE, Greenamyre JT. 2009. A highly reproducible rotenone model of Parkinson's disease. *Neurobiol Dis* **34:** 279–290. doi:10.1016/j.nbd.2009.01.016

Carlsson A. 1959. The occurrence, distribution and physiological role of catecholamines in the nervous system. *Pharmacol Rev* **11:** 490–493.

Carlsson A, Lindqvist M, Magnusson T. 1957. 3,4-Dihydroxyphenylalanine and 5-hydroxytryptophan as reserpine antagonists. *Nature* **180:** 1200. doi:10.1038/1801200a0

Chan H, Paur H, Vernon AC, Zabarsky V, Datla KP, Croucher MJ, Dexter DT. 2010. Neuroprotection and functional recovery associated with decreased microglial activation following selective activation of mGluR2/3 receptors in a rodent model of Parkinson's disease. *Parkinsons Dis* **2010:** 190450. doi:10.4061/2010/190450

Christine CW, Langston JW, Turner RS, Starr PA. 2009. The neurophysiology and effect of deep brain stimulation in a patient with 1-methyl-4-phenyl-1,2,3,6-tetrahydropyridine–induced parkinsonism. *J Neurosurg* **110:** 234–238. doi:10.3171/2008.8.JNS08882

Cicchetti F, Lapointe N, Roberge-Tremblay A, Saint-Pierre M, Jimenez L, Ficke BW, Gross RE. 2005. Systemic exposure to paraquat and maneb models early Parkinson's disease in young adult rats. *Neurobiol Dis* **20:** 360–371. doi:10.1016/j.nbd.2005.03.018

Cochemé HM, Murphy MP. 2008. Complex I is the major site of mitochondrial superoxide production by paraquat. *J Biol Chem* **283:** 1786–1798. doi:10.1074/jbc.M708597200

Cohen G, Werner P. 1994. Free radicals, oxidative stress, and neurodegeneration. In *Neurodegenerative diseases* (ed. Calne DB), pp. 139–161. W.B. Saunders, Philadelphia.

Concha-Marambio L, Pritzkow S, Shahnawaz M, Farris CM, Soto C. 2023. Seed amplification assay for the detection of pathologic α-synuclein aggregates in cerebrospinal fluid. *Nat Protoc* **18:** 1179–1196. doi:10.1038/s41596-022-00787-3

Corasaniti MT, Defilippo R, Rodino P, Nappi G, Nistico G. 1991. Evidence that paraquat is able to cross the blood-brain barrier to a different extent in rats of various age. *Funct Neurol* **6:** 385–391.

Cristóvão AC, Campos FL, Je G, Esteves M, Guhathakurta S, Yang L, Beal MF, Fonseca BM, Salgado AJ, Queiroz J, et al. 2020. Characterization of a Parkinson's disease rat model using an upgraded paraquat exposure paradigm. *Eur J Neurosci* **52:** 3242–3255. doi:10.1111/ejn.14683

Cui M, Aras R, Christian WV, Rappold PM, Hatwar M, Panza J, Jackson-Lewis V, Javitch JA, Ballatori N, Przedborski S, et al. 2009. The organic cation transporter-3 is a pivotal modulator of neurodegeneration in the nigrostriatal dopaminergic pathway. *Proc Natl Acad Sci* **106:** 8043–8048. doi:10.1073/pnas.0900358106

Curtius HC, Wolfensberger M, Steinmann B, Redweik U, Siegfried J. 1974. Mass fragmentography of dopamine and 6-hydroxydopamine. Application to the determination of dopamine in human brain biopsies from the caudate nucleus. *J Chromatogr* **99:** 529–540. doi:10.1016/s0021-9673(00)90882-3

Dahlström A, Fuxe K. 1964. Localization of monoamines in the lower brain stem. *Experientia* **20:** 398–399. doi:10.1007/BF02147990

Dauer W, Przedborski S. 2003. Parkinson's disease: mechanisms and models. *Neuron* **39:** 889–909. doi:10.1016/s0896-6273(03)00568-3

* Dawson T, Dawson V. 2024. Animal models of Parkinson's disease. *Cold Spring Harb Perspect Med.* doi:10.1101/cshperspect.a041644

Day BJ, Patel M, Calavetta L, Chang LY, Stamler JS. 1999. A mechanism of paraquat toxicity involving nitric oxide synthase. *Proc Natl Acad Sci* **96:** 12760–12765. doi:10.1073/pnas.96.22.12760

Delamarre A, MacSweeney C, Suzuki R, Brown AJ, Li Q, Pioli EY, Bezard E. 2020. Gastrointestinal and metabolic function in the MPTP-treated macaque model of Parkinson's disease. *Heliyon* **6:** e05771. doi:10.1016/j.heliyon.2020.e05771

Deng I, Bobrovskaya L. 2022. Lipopolysaccharide mouse models for Parkinson's disease research: a critical appraisal. *Neural Regen Res* **17:** 2413–2417. doi:10.4103/1673-5374.331866

Donley N. 2019. The USA lags behind other agricultural nations in banning harmful pesticides. *Environ Health* **18:** 44. doi:10.1186/s12940-019-0488-0

Drolet RE, Cannon JR, Montero L, Greenamyre JT. 2009. Chronic rotenone exposure reproduces Parkinson's disease gastrointestinal neuropathology. *Neurobiol Dis* **36:** 96–102. doi:10.1016/j.nbd.2009.06.017

Dunham NW, Miya TS. 1957. A note on a simple apparatus for detecting neurological deficit in rats and mice. *J Am Pharm Assoc Am Pharm Assoc* **46:** 208–209. doi:10.1002/jps.3030460322

Estaun-Panzano J, Arotcarena ML, Bezard E. 2023. Monitoring α-synuclein aggregation. *Neurobiol Dis* **176:** 105966. doi:10.1016/j.nbd.2022.105966

Faull RL, Laverty R. 1969. Changes in dopamine levels in the corpus striatum following lesions in the substantia nigra. *Exp Neurol* **23:** 332–340. doi:10.1016/0014-4886(69)90081-8

Fernagut PO, Diguet E, Labattu B, Tison F. 2002. A simple method to measure stride length as an index of nigrostriatal dysfunction in mice. *J Neurosci Methods* **113:** 123–130. doi:10.1016/s0165-0270(01)00485-x

Fernagut PO, Hutson CB, Fleming SM, Tetreaut NA, Salcedo J, Masliah E, Chesselet MF. 2007. Behavioral and histopathological consequences of paraquat intoxication in mice: effects of α-synuclein over-expression. *Synapse* **61:** 991–1001. doi:10.1002/syn.20456

Ferrante RJ, Schulz JB, Kowall NW, Beal MF. 1997. Systemic administration of rotenone produces selective damage in the striatum and globus pallidus, but not in the substantia nigra. *Brain Res* **753:** 157–162. doi:10.1016/s0006-8993(97)00008-5

Fleming SM, Salcedo J, Fernagut PO, Rockenstein E, Masliah E, Levine MS, Chesselet MF. 2004a. Early and progressive sensorimotor anomalies in mice overexpressing wild-type human α-synuclein. *J Neurosci* **24:** 9434–9440. doi:10.1523/JNEUROSCI.3080-04.2004

Fleming SM, Zhu C, Fernagut PO, Mehta A, DiCarlo CD, Seaman RL, Chesselet MF. 2004b. Behavioral and immunohistochemical effects of chronic intravenous and subcutaneous infusions of varying doses of rotenone. *Exp Neurol* **187:** 418–429. doi:10.1016/j.expneurol.2004.01.023

Fornai F, Schlüter OM, Lenzi P, Gesi M, Ruffoli R, Ferrucci M, Lazzeri G, Busceti CL, Pontarelli F, Battaglia G, et al. 2005. Parkinson-like syndrome induced by continuous MPTP infusion: convergent roles of the ubiquitin-proteasome system and α-synuclein. *Proc Natl Acad Sci* **102:** 3413–3418. doi:10.1073/pnas.0409713102

Forno LS, Langston JW, DeLanney LE, Irwin I, Ricaurte GA. 1986. Locus ceruleus lesions and eosinophilic inclusions in MPTP-treated monkeys. *Ann Neurol* **20:** 449–455. doi:10.1002/ana.410200403

Forno LS, DeLanney LE, Irwin I, Langston JW. 1993. Similarities and differences between MPTP-induced parkinsonism and Parkinson's disease: neuropathologic considerations. *Adv Neurol* **60:** 600–608.

Fox SH, Brotchie JM. 2010. The MPTP-lesioned non-human primate models of Parkinson's disease. Past, present, and future. *Prog Brain Res* **184:** 133–157. doi:10.1016/S0079-6123(10)84007-5

Fredriksson A, Archer T. 1994. MPTP-induced behavioural and biochemical deficits: a parametric analysis. *J Neural Transm Park Dis Dement Sect* **7:** 123–132. doi:10.1007/BF02260967

Gao HM, Jiang J, Wilson B, Zhang W, Hong JS, Liu B. 2002. Microglial activation-mediated delayed and progressive degeneration of rat nigral dopaminergic neurons: relevance to Parkinson's disease. *J Neurochem* **81:** 1285–1297. doi:10.1046/j.1471-4159.2002.00928.x

García-Revilla J, Herrera AJ, de Pablos RM, Venero JL. 2022. Inflammatory animal models of Parkinson's disease. *J Parkinsons Dis* **12:** S165–S182. doi:10.3233/JPD-213138

Giovanni A, Sieber BA, Heikkila RE, Sonsalla PK. 1994. Studies on species sensitivity to the dopaminergic neurotoxin 1-methyl-4-phenyl-1,2,3,6-tetrahydropyridine. Part 1: systemic administration. *J Pharmacol Exp Ther* **270:** 1000–1007.

Goedert M, Jakes R, Spillantini MG. 2017. The synucleinopathies: twenty years on. *J Parkinsons Dis* **7:** S51–S69. doi:10.3233/JPD-179005

Graham DG. 1978. Oxidative pathways for catecholamines in the genesis of neuromelanin and cytotoxic quinones. *Mol Pharmacol* **14:** 633–643.

Grant H, Lantos PL, Parkinson C. 1980. Cerebral damage in paraquat poisoning. *Histopathology* **4:** 185–195. doi:10.1111/j.1365-2559.1980.tb02911.x

Hefti F, Melamed E, Wurtman RJ. 1980. Partial lesions of the dopaminergic nigrostriatal system in rat brain: biochemical characterization. *Brain Res* **195:** 123–137. doi:10.1016/0006-8993(80)90871-9

Heikkila RE, Nicklas WJ, Vays I, Duvoisin RC. 1985. Dopaminergic toxicity of rotenone and the 1-methyl-4-phenylpyridinium ion after their stereotaxic administration to rats: implication for the mechanism of 1-methyl-4-phenyl-1,2,3,6-tetrahydropyridine toxicity. *Neurosci Lett* **62:** 389–394. doi:10.1016/0304-3940(85)90580-4

Herrera AJ, Castaño A, Venero JL, Cano J, Machado A. 2000. The single intranigral injection of LPS as a new model for studying the selective effects of inflammatory reactions on dopaminergic system. *Neurobiol Dis* **7:** 429–447. doi:10.1006/nbdi.2000.0289

Hertzman C, Wiens M, Bowering D, Snow B, Calne D. 1990. Parkinson's disease: a case-control study of occupational and environmental risk factors. *Am J Ind Med* **17:** 349–355. doi:10.1002/ajim.4700170307

Hirsch EC, Vyas S, Hunot S. 2012. Neuroinflammation in Parkinson's disease. *Parkinsonism Relat Disord* **18** (Suppl 1): S210–S212. doi:10.1016/S1353-8020(11)70065-7

Höglinger GU, Féger J, Prigent A, Michel PP, Parain K, Champy P, Ruberg M, Oertel WH, Hirsch EC. 2003. Chronic systemic complex I inhibition induces a hypokinetic multisystem degeneration in rats. *J Neurochem* **84:** 491–502. doi:10.1046/j.1471-4159.2003.01533.x

Hokfelt T, Martensson R, Bjorklund A, Kleinau S, Goldstein M. 1984. Distributional maps of tyrosine-hydroxylase-immunoreactive neurons in the rat brain. In *Handbook of chemical neuroanatomy classical transmitters in the CNS. Part I* (ed. Bjorklund A, Hokfelt T), pp. 277–379. Elsevier, Amsterdam.

Hornykiewicz O, Kish SJ. 1987. Biochemical pathophysiology of Parkinson's disease. In *Parkinson's disease* (ed. Yahr M, Bergmann KJ), pp. 19–34. Raven, New York.

Hou L, Zhang C, Wang K, Liu X, Wang H, Che Y, Sun F, Zhou X, Zhao X, Wang Q. 2017. Paraquat and maneb co-exposure induces noradrenergic locus coeruleus neurodegeneration through NADPH oxidase-mediated microglial activation. *Toxicology* **380:** 1–10. doi:10.1016/j.tox.2017.02.009

Hughes JT. 1988. Brain damage due to paraquat poisoning: a fatal case with neuropathological examination of the brain. *Neurotoxicology* **9:** 243–248.

Iancu R, Mohapel P, Brundin P, Paul G. 2005. Behavioral characterization of a unilateral 6-OHDA-lesion model of Parkinson's disease in mice. *Behav Brain Res* **162:** 1–10. doi:10.1016/j.bbr.2005.02.023

Ilijic E, Guzman JN, Surmeier DJ. 2011. The L-type channel antagonist isradipine is neuroprotective in a mouse model of Parkinson's disease. *Neurobiol Dis* **43:** 364–371. doi:10.1016/j.nbd.2011.04.007

Iravani MM, Leung CC, Sadeghian M, Haddon CO, Rose S, Jenner P. 2005. The acute and the long-term effects of nigral lipopolysaccharide administration on dopaminergic dysfunction and glial cell activation. *Eur J Neurosci* **22:** 317–330. doi:10.1111/j.1460-9568.2005.04220.x

Jain S. 2011. Multi-organ autonomic dysfunction in Parkinson disease. *Parkinsonism Relat Disord* **17:** 77–83. doi:10.1016/j.parkreldis.2010.08.022

Jeon BS, Jackson-Lewis V, Burke RE. 1995. 6-Hydroxydopamine lesion of the rat substantia nigra: time course and morphology of cell death. *Neurodegeneration* **4:** 131–137. doi:10.1006/neur.1995.0016

Jiang H, Jackson-Lewis V, Muthane U, Dollison A, Ferreira M, Espinosa A, Parsons B, Przedborski S. 1993. Adenosine receptor antagonists potentiate dopamine receptor agonist-induced rotational behavior in 6-hydroxydopamine-lesioned rats. *Brain Res* **613:** 347–351. doi:10.1016/0006-8993(93)90925-d

Joel D, Weiner I. 2000. The connections of the dopaminergic system with the striatum in rats and primates: an analysis with respect to the functional and compartmental organization of the striatum. *Neuroscience* **96:** 451–474. doi:10.1016/s0306-4522(99)00575-8

Jonsson G. 1980. Chemical neurotoxins as denervation tools in neurobiology. *Annu Rev Neurosci* **3:** 169–187. doi:10.1146/annurev.ne.03.030180.001125

Jonsson G. 1983. Chemical lesioning techniques: monoamine neurotoxins. In *Handbook of chemical neuroanatomy, Vol. 1, Methods in chemical neuroanatomy* (ed. Björklund A, Hökfelt T), pp. 463–507. Elsevier, Amsterdam.

Kamel F, Tanner C, Umbach D, Hoppin J, Alavanja M, Blair A, Comyns K, Goldman S, Korell M, Langston J, et al. 2007. Pesticide exposure and self-reported Parkinson's disease in the agricultural health study. *Am J Epidemiol* **165:** 364–374. doi:10.1093/aje/kwk024

Kirik D, Rosenblad C, Björklund A. 1998. Characterization of behavioral and neurodegenerative changes following partial lesions of the nigrostriatal dopamine system induced by intrastriatal 6-hydroxydopamine in the rat. *Exp Neurol* **152:** 259–277. doi:10.1006/exnr.1998.6848

Kirik D, Georgievska B, Burger C, Winkler C, Muzyczka N, Mandel RJ, Björklund A. 2002. Reversal of motor impairments in parkinsonian rats by continuous intrastriatal delivery of L-dopa using rAAV-mediated gene transfer. *Proc Natl Acad Sci* **99:** 4708–4713. doi:10.1073/pnas.062047599

Korecka JA, Eggers R, Swaab DF, Bossers K, Verhaagen J. 2013. Modeling early Parkinson's disease pathology with chronic low dose MPTP treatment. *Restor Neurol Neurosci* **31:** 155–167. doi:10.3233/RNN-110222

Kowall NW, Hantraye P, Brouillet E, Beal MF, McKee AC, Ferrante RJ. 2000. MPTP induces α-synuclein aggregation in the substantia nigra of baboons. *Neuroreport* **11:** 211–213. doi:10.1097/00001756-200001170-00041

Krüger R, Kuhn W, Müller T, Woitalla D, Graeber M, Kösel S, Przuntek H, Epplen JT, Schols L, Riess O. 1998. Ala30Pro mutation in the gene encoding α-synuclein in Parkinson's disease. *Nat Genet* **18:** 106–108. doi:10.1038/ng0298-106

Langston JW. 2017. The MPTP story. *J Parkinsons Dis* **7:** S11–S19. doi:10.3233/JPD-179006

Langston JW, Ballard P, Tetrud JW, Irwin I. 1983. Chronic parkinsonism in humans due to a product of meperidine-analog synthesis. *Science* **219:** 979–980. doi:10.1126/science.6823561

Langston JW, Forno LS, Tetrud J, Reeves AG, Kaplan JA, Karluk D. 1999. Evidence of active nerve cell degeneration in the substantia nigra of humans years after 1-methyl-4-phenyl-1,2,3,6-tetrahydropyridine exposure. *Ann Neurol* **46:** 598–605. doi:10.1002/1531-8249(199910)46:4<598::aid-ana7>3.0.co;2-f

Lapointe N, St-Hilaire M, Martinoli MG, Blanchet J, Gould P, Rouillard C, Cicchetti F. 2004. Rotenone induces non-specific central nervous system and systemic toxicity. *FASEB J* **18:** 717–719. doi:10.1096/fj.03-0677fje

Lesage S, Anheim M, Letournel F, Bousset L, Honoré A, Rozas N, Pieri L, Madiona K, Dürr A, Melki R, et al. 2013. G51d α-synuclein mutation causes a novel parkinsonian–pyramidal syndrome. *Ann Neurol* **73:** 459–471. doi:10.1002/ana.23894

Lill CM, Roehr JT, McQueen MB, Kavvoura FK, Bagade S, Schjeide BM, Schjeide LM, Meissner E, Zauft U, Allen NC, et al. 2012. Comprehensive research synopsis and systematic meta-analyses in Parkinson's disease genetics: the PDGene database. *PLoS Genet* **8:** e1002548. doi:10.1002510.1001371/journal.pgen.1002548

Liou HH, Tsai MC, Chen CJ, Jeng JS, Chang YC, Chen SY, Chen RC. 1997. Environmental risk factors and Parkinson's disease: a case-control study in Taiwan. *Neurology* **48:** 1583–1588. doi:10.1212/wnl.48.6.1583

Luthman J, Fredriksson A, Sundström E, Jonsson G, Archer T. 1989. Selective lesion of central dopamine or noradrenaline neuron systems in the neonatal rat: motor behavior and monoamine alterations at adult stage. *Behav Brain Res* **33:** 267–277. doi:10.1016/s0166-4328(89)80121-4

Mak SK, McCormack AL, Manning-Boğ AB, Cuervo AM, Di Monte DA. 2010. Lysosomal degradation of α-synuclein in vivo. *J Biol Chem* **285:** 13621–13629. doi:10.1074/jbc.M109.074617

Malmfors T, Sachs C. 1968. Degeneration of adrenergic nerves produced by 6-hydroxydopamine. *Eur J Pharmacol* **3:** 89–92. doi:10.1016/0014-2999(68)90056-3

Manning-Bog AB, McCormack AL, Li J, Uversky VN, Fink AL, Di Monte DA. 2002. The herbicide paraquat causes up-regulation and aggregation of α-synuclein in mice. *J Biol Chem* **277:** 1641–1644. doi:10.1074/jbc.C100560200

McCormack AL, Di Monte DA. 2003. Effects of L-dopa and other amino acids against paraquat-induced nigrostriatal degeneration. *J Neurochem* **85:** 82–86. doi:10.1046/j.1471-4159.2003.01621.x

McCormack AL, Thiruchelvam M, Manning-Bog AB, Thiffault C, Langston JW, Cory-Slechta DA, Di Monte DA. 2002. Environmental risk factors and Parkinson's disease: selective degeneration of nigral dopaminergic neurons caused by the herbicide paraquat. *Neurobiol Dis* **10:** 119–127. doi:10.1006/nbdi.2002.0507

Meredith GE, Sonsalla PK, Chesselet MF. 2008. Animal models of Parkinson's disease progression. *Acta Neuropathol* **115:** 385–398. doi:10.1007/s00401-008-0350-x

Metta V, Leta V, Mrudula KR, Prashanth LK, Goyal V, Borgohain R, Chung-Faye G, Chaudhuri KR. 2022. Gastrointestinal dysfunction in Parkinson's disease: molecular pathology and implications of gut microbiome, probiotics, and fecal microbiota transplantation. *J Neurol* **269:** 1154–1163. doi:10.1007/s00415-021-10567-w

Michel PP, Hirsch EC, Hunot S. 2016. Understanding dopaminergic cell death pathways in Parkinson disease. *Neuron* **90:** 675–691. doi:10.1016/j.neuron.2016.03.038

Mizuno Y, Sone N, Saitoh T. 1987. Effects of 1-methyl-4-phenyl-1,2,3,6-tetrahydropyridine and 1-methyl-4-phenylpyridinium ion on activities of the enzymes in the electron transport system in mouse brain. *J Neurochem* **48:** 1787–1793. doi:10.1111/j.1471-4159.1987.tb05737.x

Nicklas WJ, Vyas I, Heikkila RE. 1985. Inhibition of NADH-linked oxidation in brain mitochondria by 1-methyl-4-phenyl-pyridine, a metabolite of the neurotoxin, 1-methyl-4-phenyl-1,2,5,6-tetrahydropyridine. *Life Sci* **36:** 2503–2508. doi:10.1016/0024-3205(85)90146-8

Ogawa N, Hirose Y, Ohara S, Ono T, Watanabe Y. 1985. A simple quantitative bradykinesia test in MPTP-treated mice. *Res Commun Chem Pathol Pharmacol* **50:** 435–441.

Ossowska K, Wardas J, Śmiałowska M, Kuter K, Lenda T, Wierońska JM, Zięba B, Nowak P, Dąbrowska J, Bortel A, et al. 2005. A slowly developing dysfunction of dopaminergic nigrostriatal neurons induced by long-term paraquat administration in rats: an animal model of preclinical stages of Parkinson's disease? *Eur J Neurosci* **22:** 1294–1304. doi:10.1111/j.1460-9568.2005.04301.x

Parker WD Jr, Boyson SJ, Parks JK. 1989. Abnormalities of the electron transport chain in idiopathic Parkinson's disease. *Ann Neurol* **26:** 719–723. doi:10.1002/ana.410260606

Pfeiffer RF. 2016. Non-motor symptoms in Parkinson's disease. *Parkinsonism Relat Disord* **22** (Suppl 1): S119–S122. doi:10.1016/j.parkreldis.2015.09.004

Polymeropoulos MH, Lavedan C, Leroy E, Ide SE, Dehejia A, Dutra A, Pike B, Root H, Rubenstein J, Boyer R, et al. 1997. Mutation in the α-synuclein gene identified in families with Parkinson's disease. *Science* **276:** 2045–2047. doi:10.1126/science.276.5321.2045

Porter CC, Totaro JA, Stone CA. 1963. Effect of 6-hydroxydopamine and some other compounds on the concentration of norepinephrine in the hearts of mice. *J Pharmacol Exp Ther* **140:** 308–316.

Porter CC, Totaro JA, Burcin A. 1965. The relationship between radioactivity and norepinephrine concentrations in the brains and hearts of mice following administration of labeled methyldopa or 6-hydroxydopamine. *J Pharmacol Exp Ther* **150:** 17–22.

Prasad K, Winnik B, Thiruchelvam MJ, Buckley B, Mirochnitchenko O, Richfield EK. 2007. Prolonged toxicokinetics and toxicodynamics of paraquat in mouse brain. *Environ Health Perspect* **115:** 1448–1453. doi:10.1289/ehp.9932

Przedborski S. 2007. Neuroinflammation and Parkinson's disease. *Handb Clin Neurol* **83:** 535–551. doi:10.1016/S0072-9752(07)83026-0

Przedborski S, Tieu K. 2006. Toxic animal models. In *Neurodegenerative diseases* (ed. Beal MF, Lang AE, Ludolph AC), pp. 1196–1221. Cambridge University Press, Cambridge.

Przedborski S, Levivier M, Jiang H, Ferreira M, Jackson-Lewis V, Donaldson D, Togasaki DM. 1995. Dose-dependent lesions of the dopaminergic nigrostriatal pathway induced by instrastriatal injection of 6-hydroxydopamine. *Neuroscience* **67:** 631–647. doi:10.1016/0306-4522(95)00066-r

Przedborski S, Jackson-Lewis V, Naini A, Jakowec M, Petzinger G, Miller R, Akram M. 2001. The parkinsonian toxin 1-methyl-4-phenyl-1,2,3,6-tetrahydropyridine (MPTP): a technical review of its utility and safety. *J Neurochem* **76:** 1265–1274. doi:10.1046/j.1471-4159.2001.00183.x

Qin L, Wu X, Block ML, Liu Y, Breese GR, Hong JS, Knapp DJ, Crews FT. 2007. Systemic LPS causes chronic neuroinflammation and progressive neurodegeneration. *Glia* **55:** 453–462. doi:10.1002/glia.20467

Qin L, Liu Y, Hong JS, Crews FT. 2013. NADPH oxidase and aging drive microglial activation, oxidative stress, and dopaminergic neurodegeneration following systemic LPS administration. *Glia* **61:** 855–868. doi:10.1002/glia.22479

Ransohoff RM. 2016. How neuroinflammation contributes to neurodegeneration. *Science* **353:** 777–783. doi:10.1126/science.aag2590

Rappold PM, Tieu K. 2010. Astrocytes and therapeutics for Parkinson's disease. *Neurotherapeutics* **7:** 413–423. doi:10.1093/toxsci/kfi304

Rappold PM, Cui M, Chesser AS, Tibbett J, Grima JC, Duan L, Sen N, Javitch JA, Tieu K. 2011. Paraquat neurotoxicity is mediated by the dopamine transporter and organic cation transporter-3. *Proc Natl Acad Sci* **108:** 20766–20771. doi:10.1073/pnas.1115141108

Richardson JR, Quan Y, Sherer TB, Greenamyre JT, Miller GW. 2005. Paraquat neurotoxicity is distinct from that of MPTP and rotenone. *Toxicol Sci* **88:** 193–201. doi:10.1093/toxsci/kfi304

Ritz BR, Manthripragada AD, Costello S, Lincoln SJ, Farrer MJ, Cockburn M, Bronstein J. 2009. Dopamine transporter genetic variants and pesticides in Parkinson's disease. *Environ Health Perspect* **117:** 964–969. doi:10.1289/ehp.0800277

Rojas JC, Simola N, Kermath BA, Kane JR, Schallert T, Gonzalez-Lima F. 2009. Striatal neuroprotection with methylene blue. *Neuroscience* **163:** 877–889. doi:10.1016/j.neuroscience.2009.07.012

Roy NS, Cleren C, Singh SK, Yang L, Beal MF, Goldman SA. 2006. Functional engraftment of human ES cell-derived dopaminergic neurons enriched by coculture with telomerase-immortalized midbrain astrocytes. *Nat Med* **12:** 1259–1268. doi:10.1038/nm1495

Rozas G, López-Martín E, Guerra MJ, Labandeira-García JL. 1998. The overall rod performance test in the MPTP-treated-mouse model of parkinsonism. *J Neurosci Methods* **83:** 165–175. doi:10.1016/s0165-0270(98)00078-8

Saborio GP, Permanne B, Soto C. 2001. Sensitive detection of pathological prion protein by cyclic amplification of protein misfolding. *Nature* **411:** 810–813. doi:10.1038/350 81095

Saint-Pierre M, Tremblay ME, Sik A, Gross RE, Cicchetti F. 2006. Temporal effects of paraquat/maneb on microglial activation and dopamine neuronal loss in older rats. *J Neurochem* **98:** 760–772. doi:10.1111/j.1471-4159.2006.03923.x

Saner A, Thoenen H. 1971. Model experiments on the molecular mechanism of action of 6-hydroxydopamine. *Mol Pharmacol* **7:** 147–154.

Sarre S, Yuan H, Jonkers N, Van HA, Ebinger G, Michotte Y. 2004. in vivo characterization of somatodendritic dopamine release in the substantia nigra of 6-hydroxydopamine-lesioned rats. *J Neurochem* **90:** 29–39. doi:10.1111/j.1471-4159.2004.02471.x

Sauer H, Oertel WH. 1994. Progressive degeneration of nigrostriatal dopamine neurons following intrastriatal terminal lesions with 6-hydroxydopamine: a combined retrograde tracing and immunocytochemical study in the rat. *Neuroscience* **59:** 401–415. doi:10.1016/0306-4522(94)90605-x

Scally MC, Ulus IH, Wurtman RJ, Pettibone DJ. 1978. Regional distribution of neurotransmitter-synthesizing enzymes and substance P within the rat corpus striatum. *Brain Res* **143:** 556–560. doi:10.1016/0006-8993(78)90367-0

Schapira AH, Cooper JM, Dexter D, Jenner P, Clark JB, Marsden CD. 1989. Mitochondrial complex I deficiency in Parkinson's disease. *Lancet* **1:** 1269. doi:10.1016/s0140-6736(89)92366-0

Sedelis M, Schwarting RK, Huston JP. 2001. Behavioral phenotyping of the MPTP mouse model of Parkinson's disease. *Behav Brain Res* **125:** 109–125. doi:10.1016/s0166-4328(01)00309-6

Senoh S, Witkop B. 1959. Nonenzymatic conversions of dopamine to norepinephrine and trihydroxyphenethylamine. *J Am Chem Soc* **81:** 6222–6231. doi:10.1021/ja01532a028

Senoh S, Creveling CR, Udenfriend S, Witkop B. 1959. Chemical, enzymatic and metabolic studies on the mechanism of oxidation of dopamine. *J Am Chem Soc* **81:** 6236–6240. doi:10.1021/ja01532a030

Shimizu K, Ohtaki K, Matsubara K, Aoyama K, Uezono T, Saito O, Suno M, Ogawa K, Hayase N, Kimura K, et al. 2001. Carrier-mediated processes in blood–brain barrier penetration and neural uptake of paraquat. *Brain Res* **906:** 135–142. doi:10.1016/s0006-8993(01)02577-x

Shimoji M, Zhang L, Mandir AS, Dawson VL, Dawson TM. 2005. Absence of inclusion body formation in the MPTP mouse model of Parkinson's disease. *Brain Res Mol Brain Res* **134:** 103–108. doi:10.1016/j.molbrainres.2005.01.012

Siderowf A, Concha-Marambio L, Lafontant DE, Farris CM, Ma Y, Urenia PA, Nguyen H, Alcalay RN, Chahine LM, Foroud T, et al. 2023. Assessment of heterogeneity among participants in the Parkinson's progression markers initiative cohort using α-synuclein seed amplification: a cross-sectional study. *Lancet Neurol* **22:** 407–417. doi:10.1016/S1474-4422(23)00109-6

Singleton AB, Farrer M, Johnson J, Singleton A, Hague S, Kachergus J, Hulihan M, Peuralinna T, Dutra A, Nussbaum R, et al. 2003. α-Synuclein locus triplication causes Parkinson's disease. *Science* **302:** 841. doi:10.1126/science.1090278

Skrzypczak-Wiercioch A, Sałat K. 2022. Lipopolysaccharide-induced model of neuroinflammation: mechanisms of action, research application and future directions for its use. *Molecules* **27:** 5481. doi:10.3390/molecules27175481

Smith Y, Kieval JZ. 2000. Anatomy of the dopamine system in the basal ganglia. *Trends Neurosci* **23:** S28–S33. doi:10.1016/s1471-1931(00)00023-9

Snyder SH, D'Amato RJ. 1985. Predicting Parkinson's disease. *Nature* **317:** 198–199. doi:10.1038/317198a0

Tadaiesky MT, Dombrowski PA, Figueiredo CP, Cargnin-Ferreira E, Da CC, Takahashi RN. 2008. Emotional, cognitive and neurochemical alterations in a premotor stage model of Parkinson's disease. *Neuroscience* **156:** 830–840. doi:10.1016/j.neuroscience.2008.08.035

Tangamornsuksan W, Lohitnavy O, Sruamsiri R, Chaiyakunapruk N, Norman Scholfield C, Reisfeld B, Lohitnavy M. 2019. Paraquat exposure and Parkinson's disease: a systematic review and meta-analysis. *Arch Environ Occup Health* **74:** 225–238. doi:10.1080/19338244.2018.1492894

Tanner CM, Kamel F, Ross GW, Hoppin JA, Goldman SM, Korell M, Marras C, Bhudhikanok GS, Kasten M, Chade A, et al. 2011. Rotenone, paraquat, and Parkinson's disease. *Environ Health Perspect* **119:** 866–872. doi:10.1289/ehp.1002839

Tansey MG, Wallings RL, Houser MC, Herrick MK, Keating CE, Joers V. 2022. Inflammation and immune dysfunction in Parkinson disease. *Nat Rev Immunol* **22:** 657–673. doi:10.1038/s41577-022-00684-6

Tassin JP, Cheramy A, Blanc G, Thierry AM, Glowinski J. 1976. Topographical distribution of dopaminergic innervation and of dopaminergic receptors in the rat striatum. I. Microestimation of [³H] dopamine uptake and dopamine content in microdiscs. *Brain Res* **107:** 291–301. doi:10.1016/0006-8993(76)90227-4

Taylor TN, Greene JG, Miller GW. 2010. Behavioral phenotyping of mouse models of Parkinson's disease. *Behav Brain Res* **211:** 1–10. doi:10.1016/j.bbr.2010.03.004

Thiruchelvam M, Brockel BJ, Richfield EK, Baggs RB, Cory-Slechta DA. 2000a. Potentiated and preferential effects of combined paraquat and maneb on nigrostriatal dopamine systems: environmental risk factors for Parkinson's disease? *Brain Res* **873:** 225–234. doi:10.1016/s0006-8993(00)02496-3

Thiruchelvam M, Richfield EK, Baggs RB, Tank AW, Cory-Slechta DA. 2000b. The nigrostriatal dopaminergic system as a preferential target of repeated exposures to combined paraquat and maneb: implications for Parkinson's disease. *J Neurosci* **20:** 9207–9214. doi:10.1523/JNEUROSCI.20-24-09207.2000

Thiruchelvam M, McCormack A, Richfield EK, Baggs RB, Tank AW, Di Monte DA, Cory-Slechta DA. 2003. Age-related irreversible progressive nigrostriatal dopaminergic neurotoxicity in the paraquat and maneb model of the Parkinson's disease phenotype. *Eur J Neurosci* **18:** 589–600. doi:10.1046/j.1460-9568.2003.02781.x

Tieu K. 2011. A guide to neurotoxic animal models of Parkinson's disease. *Cold Spring Harb Perspect Med* **1:** a009316. doi:10.1101/cshperspect.a009316

Tieu K, Perier C, Caspersen C, Teismann P, Wu DC, Yan SD, Naini A, Vila M, Jackson-Lewis V, Ramasamy R, et al. 2003. D-β-hydroxybutyrate rescues mitochondrial respiration and mitigates features of Parkinson's disease. *J Clin Invest* **112:** 892–901. doi:10.1172/JCI18797

Titova N, Chaudhuri KR. 2018. Non-motor Parkinson disease: new concepts and personalised management. *Med J Aust* **208:** 404–409. doi:10.5694/mja17.00993

Tranzer JP, Thoenen H. 1968. An electron microscopic study of selective, acute degeneration of sympathetic nerve terminals after administration of 6-hydroxydopamine. *Experientia* **24:** 155–156. doi:10.1007/BF02146956

Tranzer JP, Thoenen H. 1973. Selective destruction of adrenergic nerve terminals by chemical analogues of 6-hydroxydopamine. *Experientia* **29:** 314–315. doi:10.1007/BF01926498

Ungerstedt U. 1968. 6-Hydroxydopamine induced degeneration of central monoamine neurons. *Eur J Pharmacol* **5:** 107–110. doi:10.1016/0014-2999(68)90164-7

Ungerstedt U, Arbuthnott G. 1970. Quantitative recording of rotational behaviour in rats after 6-hydroxydopamine lesions of the nigrostriatal dopamine system. *Brain Res* **24:** 485–493. doi:10.1016/s0165-0270(01)00359-4

Van Laar AD, Webb KR, Keeney MT, Van Laar VS, Zharikov A, Burton EA, Hastings TG, Glajch KE, Hirst WD, Greenamyre JT, et al. 2023. Transient exposure to rotenone causes degeneration and progressive parkinsonian motor deficits, neuroinflammation, and synucleinopathy. *NPJ Parkinsons Dis* **9:** 121. doi:10.1038/s41531-023-00561-6

Varešlija D, Tipton KF, Davey GP, McDonald AG. 2020. 6-Hydroxydopamine: a far from simple neurotoxin. *J Neural Transm (Vienna)* **127:** 213–230. doi:10.1007/s00702-019-02133-6

West MJ. 1993. New stereological methods for counting neurons. *Neurobiol Aging* **14:** 275–285. doi:10.1016/0197-4580(93)90112-o

West MJ, Slomianka L, Gundersen HJ. 1991. Unbiased stereological estimation of the total number of neurons in the subdivisions of the rat hippocampus using the optical fractionator. *Anat Rec* **231:** 482–497. doi:10.1002/ar.1092310411

West MJ, Ostergaard K, Andreassen OA, Finsen B. 1996. Estimation of the number of somatostatin neurons in the striatum: an in situ hybridization study using the optical fractionator method. *J Comp Neurol* **370:** 11–22. doi:10.1002/(SICI)1096-9861(19960617)370:1<11::AID-CNE2>3.0.CO;2-O

Widmann R, Sperk G. 1986. Topographical distribution of amines and major amine metabolites in the rat striatum. *Brain Res* **367:** 244–249. doi:10.1016/0006-8993(86)91598-2

Zaborszky L, Vadasz C. 2001. The midbrain dopaminergic system: anatomy and genetic variation in dopamine neuron number of inbred mouse strains. *Behav Genet* **31:** 47–59. doi:10.1023/a:1010257808945

Zarranz JJ, Alegre J, Gómez-Esteban JC, Lezcano E, Ros R, Ampuero I, Vidal L, Hoenicka J, Rodriguez O, Atarés B, et al. 2004. The new mutation, E46K, of α-synuclein causes Parkinson and Lewy body dementia. *Ann Neurol* **55:** 164–173. doi:10.1002/ana.10795

Zhu C, Vourc'h P, Fernagut PO, Fleming SM, Lacan S, DiCarlo CD, Seaman RL, Chesselet MF. 2004. Variable effects of chronic subcutaneous administration of rotenone on striatal histology. *J Comp Neurol* **478:** 418–426. doi:10.1002/cne.20305

Animal Models of Parkinson's Disease

Valina L. Dawson and Ted M. Dawson

NeuroRegeneration and Stem Cell Programs, Institute for Cell Engineering; Department of Physiology, Pharmacology and Therapeutics; Department of Neurology; Solomon H. Snyder Department of Neuroscience, Johns Hopkins University School of Medicine, Baltimore, Maryland 21205, USA

Correspondence: tdawson@jhmi.edu

Parkinson's disease (PD) is a complex genetic disorder that is associated with environmental risk factors and aging. Vertebrate genetic models, especially in mice, has aided the study of autosomal-dominant and autosomal-recessive PD. Mice are capable of exhibiting a broad range of phenotypes and coupled with their conserved genetic and anatomical structures provides unparalleled molecular and pathological tool to model human disease. These models used in combination with aging and PD-associated toxins have expanded our understanding of PD pathogenesis. Attempts to refine PD animal models using conditional approaches have yielded in vivo nigrostriatal degeneration that is instructive in ordering pathogenic signaling and in developing therapeutic strategies to cure or halt the disease. α-Synuclein preformed fibril (PFF) injections, which induce the aggregation of endogenous α-synuclein, remarkably recapitulate pathological processes observed in human PD. Here, we provide an overview of the generation and characterization of transgenic and knockout mice and the α-synuclein PFF models used to study PD followed by molecular insights that have been gleamed these PD mouse models.

arkinson's disease (PD) is the most common neurodegenerative movement disorder with a characteristic degeneration of dopamine (DA) producing neurons in the substantia nigra (SN) (Pirooznia et al. 2021; Park et al. 2025).[1] Surviving DA neurons of the SN and neurons in other affected areas carry α-synuclein (α-syn) containing Lewy bodies, implying a pathological role for α-syn in PD progression (Spillantini et al. 1997). Despite the relative selective loss of DA neurons in the SN that account for the underlying motor deficit in PD, patients commonly also develop nonmotor symptoms with substantial neurodegeneration and Lewy body formation in broad brain regions (Churchyard and Lees 1997; Chaudhuri et al. 2007; Pirooznia et al. 2021). DA supplementation alleviates the major motor symptoms in PD (Savitt et al. 2006), but its effectiveness declines over time. Unfortunately, no effective treatments are available to stop or prevent the disease progression due to a poor understanding of the molecular mechanisms of selective and progressive neurodegeneration in PD.

---

[1]This is an update to a previous article published in *Cold Spring Harbor Perspectives in Medicine* [Lee et al. (2012). *Cold Spring Harb Perspect Med* **2:** a009324. doi:10.1101/cshperspect.a009324].

Copyright © 2025 Cold Spring Harbor Laboratory Press; all rights reserved
Cite this article as *Cold Spring Harb Perspect Med* doi: 10.1101/cshperspect.a041644

In rare inherited PD, mutations in several genes cause PD pathologies that are often indistinguishable from sporadic PD (Gasser 2004; Martin et al. 2011; Pirooznia et al. 2021). The identification of PD-linked genes has prompted focused studies of molecular signals that cause PD. Moreover, PD genes provide a rational basis to model the disease in cells or animals by genetic manipulations.

Animal models that faithfully recapitulate the characteristic neurodegeneration and pathological hallmarks of PD are necessary to validate pathogenic molecular pathways in vivo and also test therapeutic strategies in more controlled physiologic systems. Many genetic PD models have been informative in understanding molecular pathways and pathological changes that may be PD-relevant (Dawson et al. 2010; Cramb et al. 2023). For additional information, please refer to our previously published paper (Lee et al. 2012).

MODELING PD IN MICE: OVERVIEW ON THEIR GENERATION AND CHARACTERIZATION

Researchers generally prefer mouse models to simulate human genetic disorders because mice possess similar neuronal networks and disease-associated gene homologs (Waterston et al. 2002). Although the ability to study human DA neurons using inducible pluripotent stem cells (IPSCs) has changed the scientific landscape to model and study PD in relevant human model systems (Hu et al. 2020), the evolutionarily conserved neuronal networks and PD gene homologs indicate that findings in PD mouse models are likely to mirror the pathological events occurring in human PD. The ease and advances in genetic manipulation has enabled the generation and refinement of genetic mouse models (Figs. 1 and 2). These include conventional transgenic approaches in which a gene is overexpressed under the control of a promoter that drives expression in the brain. Promoters can be selected that provide tissue-specific or cell-type-specific expression. Conditional temporal expression of a transgene in which a gene's expression is controlled by a drug-based switch allows

temporal fine-tuning of the expression (Landel et al. 1990). One can also knockout (KO) gene and knockin (KI) discrete mutations by homologous recombination (Table 1; Capecchi 1989). The availability of inbred strains of mice with similar genetic backgrounds facilitates the comparison of control and genetically manipulated mice in the same genetic background. Moreover, the relatively long life span (2 years) compared to simpler organisms offers the opportunity to assess pathological changes across the aging process, which is the single most prominent risk factor for PD in humans. Finally, the broad spectrum of phenotypic readout in mice better represents the complex characteristics of human PD (Rosenthal and Brown 2007).

GENETIC MOUSE MODEL TECHNIQUES

Transgenic Mouse Models

Even though "transgenic animal" defines a rather broad spectrum (Rülicke et al. 2007), it is commonly used to describe animals that overexpress a foreign protein. Transgenic technology has successfully modeled many human diseases in mice by introducing human disease genes or mutants (Rockenstein et al. 2007). For example, many neurodegenerative disease such as Alzheimer's disease (AD), Huntington's disease (HD), and PD share a common etiology of aberrant protein accumulation and aggregation, and mice expressing disease proteins (i.e., amyloid-β in AD, expanded glutamine repeats in huntingtin in HD, α-syn in PD) in brain mirror human disease conditions and serve as instructive genetic models (Rockenstein et al. 2007).

The transgenic construct comprises structural elements and promoter sequences. Usually dominant mutants (i.e., A53T, A30P, and E45K for α-syn; G2019S and R1441C/G mutants for LRRK2) are preferred because the mode of inheritance supports a gain of toxicity or exaggeration of its endogenous function (Fig. 2). For the induction of the transgene, the promoter directly governs expression levels, pattern, and timing. Considering the progressive feature of neurodegenerative diseases, the promoter for modeling the disorder should be constitutively active

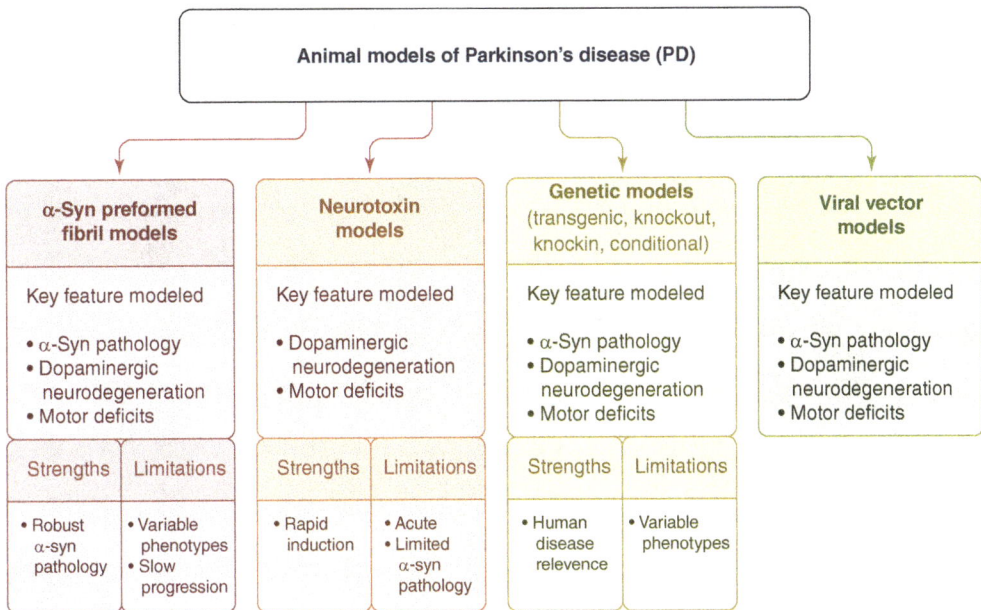

Figure 1. Animal models of Parkinson's disease (PD). This diagram provides an overview of the principal animal models used to study PD pathogenesis and therapy development. The models are grouped into three major categories. (1) α-Synuclein preformed fibril (PFF) models: Injection of synthetic or human-derived α-synuclein fibrils into the brain induces endogenous α-synuclein aggregation, prion-like spread of pathology, and progressive loss of dopaminergic neurons, closely mimicking human PD progression. (2) Genetic models: Transgenic over-expression—mice engineered to overexpress human PD-related genes such as α-synuclein (wild-type or mutant forms), which model protein aggregation and some motor deficits, or LRRK2 (mutant forms), which model dopaminergic deficits. Knockout (KO) models: Mice lacking genes implicated in familial PD (e.g., Parkin, PINK1, DJ-1), used to study loss-of-function effects and mitochondrial dysfunction. Knockin (KI) models: Mice with precise human PD mutations introduced into the endogenous gene, allowing study of disease mechanisms at physiological expression levels. Conditional/inducible models: Use of Cre-LoxP, tetracycline (Tet)-off, or Cre-ER systems to achieve tissue-specific or adult-onset gene manipulation, overcoming developmental compensation. (3) Viral vector models: Use of recombinant adenoassociated virus (rAAV), herpes simplex virus (HSV), or lentivirus to deliver PD-related genes or induce gene KO in specific brain regions, enabling rapid and spatially controlled modeling of PD pathology. Arrows in the diagram indicate the relationships between model types and the key pathological features they recapitulate, such as dopaminergic neuron degeneration, Lewy body formation, and motor deficits. The diagram highlights that while each model captures specific aspects of PD, no single model fully recapitulates the human disease, underscoring the importance of using complementary approaches in PD research.

throughout the lifetime of mice and rather strong to induce sufficient levels of the transgene.

Tet Off Conditional Transgenic Models

A more advanced technique is the tetracycline (Tet)-regulated transgenic switch that uses two separate components acting in *trans* (i.e., responder component, tet-promoter [tetP] + transgene; activator or driver component, tissue-specific promoter + tTA) (Gossen and Bu-

jard 1992). Tight regulation of tetP by Tet-regulated transcription activator (tTA) drives transgene expression. Therefore, expression of the transgene should follow the activity pattern of the promoter in the driver construct. The ability of tTA to change its conformation and affinity for tetP by doxycyline allows temporal on/off control of transgene induction (Gossen and Bujard 1992; Kistner et al. 1996; Sprengel and Hasan 2007). In addition to temporal and spatial regulation, this system offers amplifica-

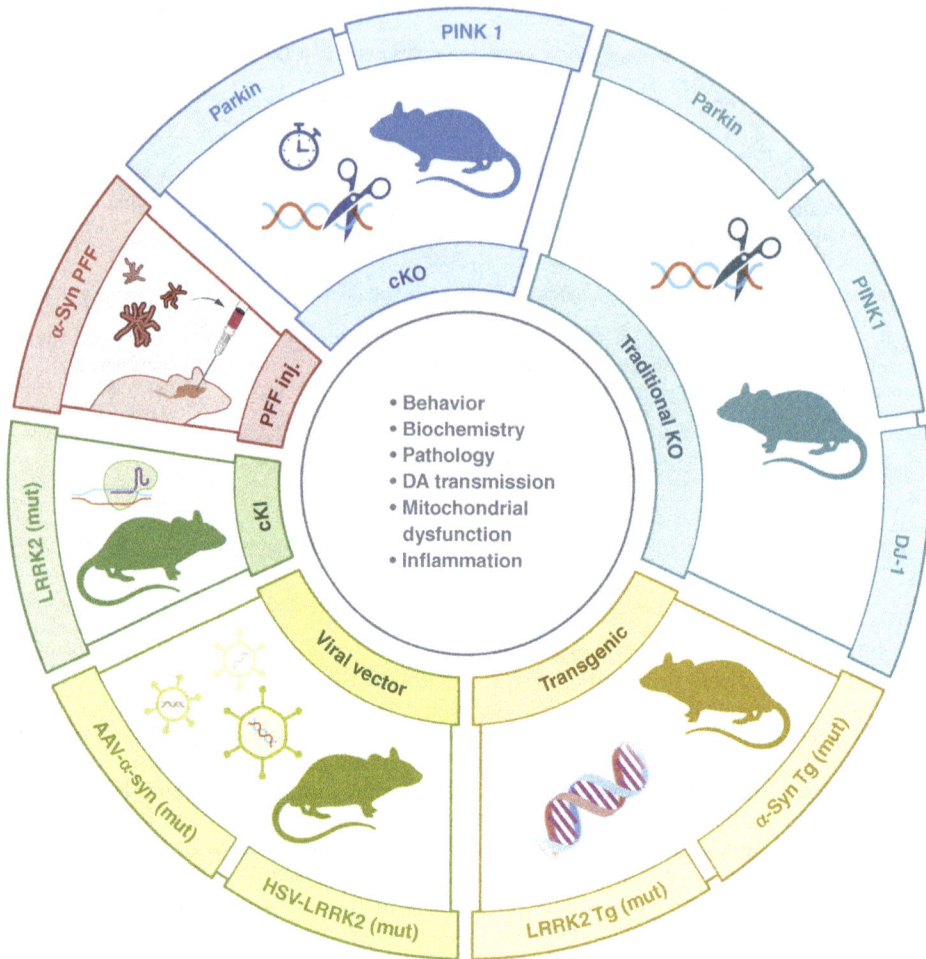

Figure 2. Overview of mouse models used in Parkinson's disease (PD) research. The diagram shows the major categories of mouse models used to study PD, highlighting genetic, viral, and protein-based approaches. The outer ring segments represent different model types: conditional knockout (cKO) models for PINK1 and Parkin (blue, *upper left*), traditional KO models for Parkin, PINK1, and DJ-1 (blue, *upper right*), transgenic models expressing mutant LRRK2 or α-synuclein (yellow, *lower right*), viral vector-mediated models (yellow-green, *lower center*), conditional knockin (cKI) LRRK2 mutant models (green, *lower left*), and α-synuclein preformed fibril (PFF) injection models (red, *upper left*). Icons within each segment depict the method or target gene/protein modification. The central text summarizes key research end points assessed in these models: behavior, biochemistry, pathology, dopamine (DA) transmission, mitochondrial dysfunction, and inflammation. This figure emphasizes the diversity of PD mouse models and their utility in dissecting disease mechanisms and therapeutic targets.

tion of tissue-specific promoter strength by the relaying action of tTA onto tetP, thus resulting in more robust induction of the transgene compared to that of conventional transgenesis where the transgene expression is driven by the upstream tissue-specific promoter alone (Bond et al. 2000). The potential refinement of previ-

ous unsatisfactory transgenic models by use of the Tet off genetic switch provides the possibility of creating mice that exhibit appropriate pathologies. For instance, the development of Tet-off conditional PD mouse models expressing wildtype (WT) and mutant forms of α-syn may have a strong tendency for significant loss of DA neu-

Table 1. Key features and strengths and limitations of Parkinson's disease (PD) animal models

Model type	Genetic target/approach	Key features modeled	Strengths	Limitations
Transgenic (overexpression)	α-Synuclein, LRRK2, (mutants), and other	Protein aggregation, partial dopamine (DA) dysfunction	Model-dominant PD, aggregation pathology	Rarely full DA neuron loss
Knockout (KO)	Parkin, PINK1, DJ-1	Loss-of-function, mild DA signaling defects	Model-recessive PD, subtle mitochondrial role	Minimal degeneration in germline KO
Knockin (KI)	LRRK2, Parkin (precise mutation)	Physiological expression, subtle phenotypes	Mimics human mutation context	Often mild, slow phenotypes
Conditional KO/ expression	Cre-LoxP, Tet-off, Cre-ER	Adult-onset, tissue-specific effects	Overcomes developmental compensation	Requires complex breeding/ induction
Viral vector models	Recombinant adenoassociated virus (rAAV), herpes simplex virus (HSV), lentivirus	Acute gene delivery, region-specific effects	Rapid, spatial/ temporal control	Packaging limits, variable expression
α-Synuclein preformed fibril (PFF) models	Synthetic or human PFF injection	Prion-like spread, robust DA neuron loss	Closely mimics human PD progression	Technical, not genetic etiology

rons, particularly when the expression is turned on in adulthood. Moreover, conditional models allow the opportunity to determine whether behavioral and pathologic changes are reversible when the offending protein expression is reduced or turned off (Dawson et al. 2010).

KO Models

Recessive genetic mutations seem to lead to PD pathogenesis via loss of function. The association of heterozygous mutations with increased susceptibility to develop PD supports the notion that autosomal-recessive PD genes exert protective roles in PD-related pathology. As such, targeted deletion of functionally important exons of the autosomal-recessive gene or introduction of premature termination should be able to simulate early-onset PD in genetic mouse models much in the same way homozygous autosomal-recessive PD gene mutations cause PD in humans (Fig. 2). For the induction of the transgene, the promoter directly governs expression levels, pattern, and timing. However, germline deletion of parkin, PINK1, and DJ-1 has yielded

mice with minimal phenotypes (Dawson et al. 2010). Even knocking out all three genes in the same mouse provided no substantial phenotype (Kitada et al. 2009). Most mouse models of recessive PD using a conventional KO approach exhibit mainly defects in dopaminergic signaling and do not exhibit a degenerative phenotype (Cramb et al. 2023). Thus, compensatory mechanisms preventing neurodegeneration make it necessary to use more sophisticated conditional models. The Cre-LoxP-mediated conditional KO approach is widely used when embryonic lethality prevents studying deletion of a gene in adult animals (Fig. 2). For the induction of the transgene, the promoter directly governs expression levels, pattern, and timing. By using well-characterized tissue-specific Cre-expressing lines, exons or genes flanked by loxP sites in conditional mice can be deleted in desired tissues. Here, the promoter upstream of Cre governs the timing as well as the region(s) of gene deletion. Recent advances have coupled the Cre-LoxP system to a tamoxifen-sensitive Cre (Cre-ER). Tamoxifen binding can activate an otherwise dormant Cre-ER providing temporal regulation of gene deletion. Alternatively, viruses carrying a Cre expression cassette can provide the temporal and spatial resolution to delete genes from select brain regions at various ages. This approach was used successfully to create the first genetic model of PD that exhibits progressive degeneration of DA neurons by conditionally knocking out parkin in adult mice (Shin et al. 2011). Adult knockdown of PINK1 also leads to degeneration of DA neurons (Lee et al. 2017).

Virus-Induced Models

Stereotaxic viral injection is another tool that enables acute and robust induction of the transgene in desired regions of brain (Van der Perren et al. 2015). Different viral systems (e.g., lentivirus, recombinant adenoassociated virus [rAAV], and herpes simplex virus [HSV]) are available and can be chosen according to the gene structure and the purpose of experiments (Fig. 2). For the induction of the transgene, the promoter directly governs expression levels, pattern, and timing. For instance, the rAAV system has the advantage to transduce broad regions of tissue. A single stereotaxic injection can transduce most DA neurons in the SN. Different rAAV serotypes can be used to transduce different cell types preferentially (e.g., AAV2 is more specific for neurons vs. glia; AAV1 and 5 are highly diffusible but not selective). A limitation of rAAV is that it has packaging capacity of ~4.7 kb, so large genes cannot be efficiently packaged by rAAV. HSV, on the other hand, can package large genes such as LRRK2 and transduce neurons very efficiently. HSV also has the advantage of retrograde transport, in which HSV virus can be stereotaxically injected into the striatum leading to efficient transduction of DA neurons in the SN (Lee et al. 2010). The ability of HSV to package large genes such as LRRK2 made possible the successful modeling of LRRK2-induced DAergic neuronal toxicity and further testing of potential kinase inhibitors in vivo (Lee et al. 2010). Similar success was demonstrated with adenoviral-mediated expression of G201S LRRK2 (Tsika et al. 2015). Notably, the successful expression of the transgene and PD modeling by viral approaches heavily depends on the quality and titer of virus particle. High titer

and infectivity ensures success. Viruses can also be used to cluster regularly interspaced short palindromic repeats (CRISPR)-CRISPR-associated protein 9 (Cas9) to accomplish efficient gene knockdown in vivo (Hana et al. 2021). Moreover, using the Cre-LoxP system, Cre expressing viruses can acutely silence genes when injected into conditional KO mice that have LoxP flanked sites.

Characterization of PD Mouse Models

Our understanding of PD has been advanced by an integrated approach of in vitro, cellular, and animal studies in connection with findings from clinical studies in PD patients (Moore et al. 2005). Moreover, validation of PD mouse models has relied on how faithfully they can recapitulate the major characteristic clinical and pathological features found in human PD. Animal models are reliable and have a predictive value only when they reproduce the abnormal phenotypic changes of human PD (Dawson et al. 2010; Cramb et al. 2023). A number of features characterize PD (Savitt et al. 2006; Pirooznia et al. 2021) and various models exhibit some but not all the features of human PD (Figs. 1 and 2).

Diverse approaches and tools used in human studies can similarly be applied to characterize animal models in detail. For example, application of new methods in expression profiling, proteomics, and metabolomics can be used for detailed characterization of PD mouse models, thus providing a global network assessment of pathogenic pathways in a controlled physiologic environment with the same genetic background (Simunovic et al. 2009; Cloutier and Wellstead 2010).

WHAT WE HAVE LEARNED FROM PD MOUSE MODELS

α-Synuclein Transgenic PD Models

Familial mutations and elevated levels of α-syn due to multiplication increases α-syn's tendency to form aggregates that correlates with the severity of pathology in PD (Polymeropoulos et al. 1997; Singleton et al. 2003). Haplosufficiency and a dominant mode of inheritance also support a toxic gain-of-function model for α-syn

 Cite this article as *Cold Spring Harb Perspect Med* doi: 10.1101/cshperspect.a041644

pathogenesis theory. Interestingly, α-syn aggregation characterizes many disorders designated as α-synucleinopathies (Dev et al. 2003). Thus, Lewy bodies or oligomer, protofibril, and fibril intermediates may contribute to progression of many neurological disorders through damage in different regions and types of cells depending on where α-syn aggregates. In this regard, mice expressing α-syn can be used to model α-synucleinopathy-induced neuronal degeneration (Figs. 1 and 2; Table 1).

Many transgenic lines have used panneuronal or DA-specific promoters to express α-syn WT or mutants. Notably, the severity and age of onset of disease depends heavily on the promoter and levels of transgene expression. Many of the models exhibit neurodegeneration and are instructive for studying the molecular mechanism by which α-syn aggregation and neurodegeneration occurs in vivo. However, most of these models fail to exhibit DA neuron loss, the key pathological feature of PD, although there are subtle functional abnormalities in the nigrostriatal system and neurodegeneration in other anatomical circuits (Fernagut and Chesselet 2004).

Among the many α-syn transgenic mice, the mouse prion promoter (mPrP) A53T transgenic mice recapitulate most of the pathologies linked to α-synucleinopathies including phosphorylation, ubiquitination, and aggregation of α-syn leading to progressive neurodegeneration (Giasson et al. 2002; Lee et al. 2002). Even though most of the neurodegeneration occurs in spinal motor neurons not in SN DA neurons, the study of α-syn-induced cell death pathways in these other neuronal systems may provide insight into mechanisms of α-syn-induced degeneration that has relevance for degeneration of DA neurons (Forno 1987; Churchyard and Lees 1997).

Because there is no obvious DA neuronal loss in most of α-syn transgenic mice, mechanistic studies on DA neuronal death by α-syn have been aided through the study of the impact of α-syn expression following 1-methyl-4-phenyl-1,2,3,6-tetrahydropyridine (MPTP) intoxication (Song et al. 2004; Nieto et al. 2006). Mitochondrial dysfunction appears to be sufficient to induce α-syn aggregation and downstream toxicity for DA neuronal loss in MPTP mouse models. As expected, DA neurons of α-syn transgenic show more sensitivity for mitochondrial toxins (Nieto et al. 2006). Interestingly, DA neurons of α-syn KOs are resistant to MPTP intoxication (Dauer et al. 2002), which is complemented by WT and mutant human α-syn, demonstrating the requirement of α-syn aggregation for DA neuron loss downstream from mitochondrial dysfunction (Thomas et al. 2011). Conversely, α-syn transgenic mice display mitochondrial abnormalities in degenerating neurons (Martin et al. 2006).

The mechanism of α-syn-induced mitochondrial dysfunction might involve sequestration of mitochondrial proteins or yet unidentified pathways (Olzscha et al. 2011), including interactions of α-syn with the mitochondria-associated membranes (Guardia-Laguarta et al. 2014; Barbuti et al. 2025). The dysfunction of mitochondria with α-syn aggregation may enhance the toxic aggregate formation in a feedforward signal amplification manner. In this context, inhibition of either mitochondria damage or α-syn aggregation may halt or delay degenerative processes set in motion in PD animal models if restorative protective signaling pathways remain intact. The amelioration of progression of motor deficits in Tet-off α-syn transgenics by doxycycline-induced cessation of α-syn expression supports this notion (Nuber et al. 2008).

In addition to mitochondrial dysfunction in α-syn transgenic mice, one downstream consequence of α-syn aggregation is likely to be a DNA damage response, leading to cell death. MPTP toxicity for DA neurons is protected not only by α-syn deletion but also by PARP1 deletion (Mandir et al. 1999). Since PARP1 overactivation upon DNA damage is sufficient to induce cell death (Koh et al. 2005), α-syn aggregation may induce toxicity by trapping and sequestering protein complexes that are important for the regulation of PARP1. Consistent with this notion is the observation that mice lacking PARP1 are dramatically resistant to MPTP intoxication (Mandir et al. 1999).

α-Synuclein PFF Models

Direct injection of recombinant α-syn preformed (Luk et al. 2012; Kim et al. 2019) or pathological human α-syn from PD patients (Masuda-Suzukake et al. 2013) into WT mouse brains induced widespread α-syn inclusion formation, neurodegeneration, and parkinsonian deficits (Figs. 1 and 2; Table 1). Supporting the prion-like mechanism hypothesis, endogenous α-syn was shown to be essential for transmission: α-Syn-deficient mice injected with preformed fibril (PFF) developed neither inclusions nor neurobehavioral abnormalities, and neurodegeneration was absent (Luk et al. 2012; Kim et al. 2019). While anatomical connectivity predicts α-syn pathology spread (Henderson et al. 2019; Pandya et al. 2019; Mezias et al. 2020), certain neuronal populations avoid toxicity and instead function as conduits for neural network transmission (Henrich et al. 2020). This model closely mimics human PD progression and is now widely used for mechanistic and therapeutic studies.

LRRK2 Transgenic PD Models

Dominant mutations in LRRK2 are the most common cause of familial PD (Goldwurm et al. 2005). Structurally, LRRK2 is a very interesting protein, which has guanosine triphosphatase (GTPase) and kinase domains in addition to leucine-rich repeat domains, forming a large protein kinase. The aberrant kinase activity of LRRK2 by common mutations such as G2019S, R1441C, and R1441G is thought to be the culprit leading to DA neuron degeneration in PD (West et al. 2005, 2007). In this context, many groups have generated LRRK2-related PD mouse models expressing LRRK2 WT or PD-associated mutant LRRK2 to simulate aberrant kinase activity (Figs. 1 and 2; Table 1; Dawson et al. 2010). In spite of several transgenic techniques for LRRK2-related PD modeling in mice (i.e., conventional, BAC transgenic, mutant LRRK2 KI, and Tet inducible transgenic), most LRRK2 transgenic animals manifest deficits in DA transmission and DA responsive behavior, and one of the LRRK2 models reproduce age-dependent nigral DA neuron death (Ramonet et al. 2011). Mice engineered to carry endogenous LRRK2 mutations (G2019S and R1441C/G) replicate the genetic context of human disease. These KI models allow for the study of LRRK2 function and kinase activity under physiological regulation, providing insights into subtle synaptic and cellular changes. LRRK2 KO mice are used to study loss-of-function effects and compensatory mechanisms, often revealing roles in immune regulation and lysosomal function (Seegobin et al. 2020).

rAAV and HSV vectors enable targeted, high-level expression of LRRK2 (including large genes) in specific brain regions. HSV-mediated expression of G2019S LRRK2 in the SN induces DA neuron degeneration in a kinase-dependent manner, offering a robust platform for testing kinase inhibitors and dissecting pathogenic mechanisms (Xiong et al. 2017).

Recent advances have significantly refined LRRK2 animal models, particularly through the use of transgenic and conditional expression systems. Notably, Xiong et al. (2018) has developed and characterized transgenic mouse models expressing human LRRK2 G2019S under the control of the tyrosine hydroxylase (TH) promoter, using a Tet-sensitive (Tet-off) system. These models demonstrate that LRRK2 G2019S expression leads to age- and kinase-dependent, cell-autonomous degeneration of both DA and norepinephrine (NE) neurons (Figs. 1 and 2; Table 1). Importantly, these mice exhibit DA-dependent behavioral deficits and α-syn pathology, closely mirroring human PD features. Transmission electron microscopy revealed a significant reduction in synaptic vesicle number and increased clathrin-coated vesicles in DA neurons, all dependent on LRRK2 kinase activity. The use of kinase-dead LRRK2 G2019S controls confirmed the necessity of kinase activity for neurodegeneration, establishing these models as a substantial advance for mechanistic and therapeutic studies of LRRK2-associated PD (Xiong et al. 2018).

A major advance in LRRK2 research was the identification of Rab GTPases as direct, physiological substrates of LRRK2 kinase activity (Steger et al. 2016; Jeong et al. 2018; Pfeffer 2018). Rabs

Cite this article as *Cold Spring Harb Perspect Med* doi: 10.1101/cshperspect.a041644

are master regulators of intracellular membrane trafficking, and their phosphorylation by LRRK2 has emerged as a key pathogenic mechanism in PD. LRRK2 phosphorylates a subset of Rab proteins, including Rab3A/B/C/D, Rab8A/B, Rab10, Rab12, Rab29, Rab35, and Rab43. Endogenous phosphorylation has been confirmed for several of these, notably Rab8A, Rab10, and Rab12. LRRK2-mediated phosphorylation typically occurs at a conserved residue in the switch-II domain of Rabs, interfering with their normal function in vesicular trafficking. This disruption in Rab activity impairs endolysosomal and Golgi trafficking, contributing to cellular stress and neurodegeneration. Aberrant Rab phosphorylation serves as a robust biomarker for LRRK2 kinase activity and is a key readout in both animal models and clinical studies.

Martin and colleagues identified ribosomal protein s15 as a critical pathogenic substrate of LRRK2 (Martin et al. 2014). Using *Drosophila* and human neuron PD models, they demonstrated that LRRK2 phosphorylates s15 at threonine 136. This phosphorylation is necessary for the neurotoxic effects of mutant LRRK2 (G2019S), as a phosphodeficient s15 mutant (T136A) rescues DA neuron degeneration and age-related locomotor deficits. Pathogenic LRRK2 stimulates both cap-dependent and cap-independent mRNA translation, resulting in increased bulk protein synthesis—a process that is normalized by phosphodeficient s15. These findings establish a direct mechanistic link between LRRK2 kinase activity, aberrant protein synthesis, and neurodegeneration in PD. Notably, s15 phosphorylation is elevated in both LRRK2 transgenic *Drosophila* and human G2019S carrier brain tissue, and LRRK2 kinase inhibitors block this phosphorylation and its downstream toxicity (Martin et al. 2014).

Parkin Genetic Models

Many groups have generated parkin KO mice to model PD due to parkin loss (Goldberg et al. 2003; Itier et al. 2003; Von Coelln et al. 2004). Initial characterization of parkin null strains failed to see any signs of nigral DA neuronal death although abnormal DA metabolism was noted. Interestingly, catecholaminergic neurons in locus coeruleus (LC) regions of mice with parkin catalytic domain deletion degenerated with an accompanying deficit in the startle response (Von Coelln et al. 2004). Proteomic studies using parkin null mice showed marked reduction of mitochondrial respiratory chain proteins and stress response proteins when compared with littermate controls. Moreover, immunoblot analysis showed several parkin substrates (AIMP2, FBP1, and PARIS) accumulate especially in the ventral midbrain of parkin null mice (Ko et al. 2005, 2006; Shin et al. 2011). These cellular changes may be responsible for subtle deficits in DA metabolism and behavior. Surprisingly, MPTP intoxication in parkin null mice causes a similar level of DA neuronal toxicity compared with WT mice, although parkin overexpression provides protection against MPTP (Perez et al. 2005; Paterna et al. 2007; Thomas et al. 2007). A compensatory remodeling in parkin-deficient DA neurons may have complemented parkin-related protective roles against mitochondrial dysfunction.

A conditional parkin KO model develops DA neurodegeneration. Lentiviral-Cre nigral injection into adult parkin$^{flox/flox}$ mice causes acute parkin deletion and progressive nigral neuron death over 10 months after the gene deletion (Figs. 1 and 2; Table 1). There are pathogenic sequential events due to accumulation of PARIS including PGC1-α downregulation, and ultimately mitochondrial dysfunction (Shin et al. 2011). Considering many reports of parkin's role in the mitochondria (Clark et al. 2006; Wild and Dikic 2010), the pathways revealed in this adult conditional parkin KO model may be quite relevant to PD pathogenesis and may represent a promising model to test new therapies.

Recent KI models, such as the parkin R275W mouse, recapitulate early-onset DA neuron dysfunction, age-dependent SN DA neuron loss, decreased striatal DA, and progressive motor impairment—mirroring autosomal-recessive juvenile parkinsonism (ARJP) (Regoni et al. 2024). Parkin-deficient nonhuman primates generated by CRISPR-Cas9 display robust SN neurodegeneration and α-syn pathology, further validating the role of parkin in DA neuron survival (Han et al. 2024).

PINK1 Genetic Models

Two PINK1-targeted KOs (Kitada et al. 2007; Gispert et al. 2009) and shRNA-mediated knockdown models (Zhou et al. 2007) have been reported. In contrast to robust degenerative phenotypes and mitochondrial defects that have been observed in genetic fly models (Clark et al. 2006), PINK1 KO and knockdown mice failed to replicate these features. Nevertheless, subtle deficits of nigrostriatal DA transmission and accompanying mild mitochondrial abnormalities (i.e., impairment of mitochondrial respiration and electrochemical potential) were observed in PINK1 KO mice (Figs. 1 and 2; Table 1). One PINK1 KO model also exhibits mild, but significantly less DA content in the striatum and reduced spontaneous voluntary activities, which were age dependent (Gispert et al. 2009). The mild deficit in presynaptic function of PINK1 KO mice resembles that of several parkin KO mice.

Although loss of PINK1 kinase activity due to PD-associated mutations causes early-onset PD in human, the lack of nigral neurodegeneration in KO models prevents further investigation of the role of physiologic and pathophysiologic PINK1 phosphosubstrate that may contribute to PD pathogenesis in vivo. Despite the lack of degeneration in the PINK1 KO mouse, there are indeed mild abnormalities in mitochondrial function in one of the PINK1 KO models. Conditional knockdown of PINK1 in adult mouse brains leads to progressive DA neuron loss, which is dependent on PARIS accumulation— linking PINK1 to the parkin-PARIS-PGC-1α axis (Lee et al. 2017). PINK1 KO mice injected with α-syn PFFs exhibit accelerated α-syn aggregation, enhanced glial activation, earlier and more severe DA neuron loss, and worsened motor deficits compared to WT controls (Nguyen et al. 2022).

DJ-1 Genetic Models

Diverse biologic roles for DJ-1 have been proposed from many cell-based studies. The results collectively emphasize DJ-1's role as a redox-sensitive molecular chaperone that provides protection against cellular stresses (Moore et al. 2003). DJ-1 seems to function as an atypical peroxiredoxin-like protease (Andres-Mateos et al. 2007). Despite its diverse involvement in cell-protective processes, DJ-1 null mice failed to display any overt signs of nigral neurodegeneration (Chen et al. 2005; Goldberg et al. 2005). Similarly to parkin and PINK1 KO mice, DJ-1 KO mice also develop mild deficits in DA neurotransmission and mitochondrial dysfunction even in the absence of DA neuron loss (Figs. 1 and 2; Table 1). Still, consistent with its protective role against stress, DA neurons with DJ-1 deletion exhibit increased susceptibility for MPTP (Kim et al. 2005). Moreover, recent electrophysiological studies for DA neurons of DJ-1 KO mice demonstrated elevated mitochondrial oxidant stress due to compromised uncoupling of mitochondria following basal pacemaking potential (Guzman et al. 2010). Although this altered mitochondrial potential and oxidative stress persists without DA neurodegeneration, these changes may predispose DA neurons, so that additional hits such as MPTP or the aging processes surpass the threshold for cell death initiation in these neurons. These studies underscore DJ-1's central role as a redox sensor, mitochondrial modulator, and neuroprotective agent in PD. While DJ-1 KO mice do not show robust nigral neurodegeneration under baseline conditions, they display increased vulnerability to oxidative and mitochondrial stress, subtle dopaminergic deficits, and, as newly described, peripheral sensory neuropathy and pain (Lee et al. 2025). Mechanistic studies highlight the importance of DJ-1 dimerization and its interaction with mitochondrial ATP synthase in maintaining neuronal health. DJ-1 remains a promising therapeutic target and potential biomarker for PD (Abulimiti et al. 2025; Lv et al. 2025).

Recent advances in animal modeling of recessive PD have revealed that developmental compensation limits the phenotypes of germline parkin and PINK1 KO mice. However, conditional and adult-onset gene ablation strategies now yield progressive DA neuron loss, providing robust models for mechanistic and therapeutic studies. Central to these models is the transcriptional repressor PARIS, which accumulates when parkin or PINK1 function is lost, driving mitochondrial dysfunction and DA neuron degeneration through repression of PGC-1α. Targeting PARIS, either genetically or pharmacologically (e.g., with farnesol), represents a promising disease-modifying strategy for PD (Jo et al. 2021).

Refinement or Reinvention of PD Animal Models of DA Neurodegeneration

None of the current genetic mouse PD models recapitulate all the features of PD. Additionally, only a few models develop DA neurodegeneration. The most parsimonious explanation for the lack of DA neurodegeneration in genetic PD mouse models are compensatory mechanisms that may result from adaptive changes during development, thus making it hard to observe the degenerative phenotype over the life span of mice. This kind of compensatory mechanism is prevalent in higher mammalian species, and this may be evolutionarily necessary for survival of a species against rather frequent spontaneous mutations in some essential genes (Gao and Zhang 2003). Presumably, similar compensatory pathway develops in familial human PD patients. The late-onset and strong age association in human PD may support the successful makeover of genetic lesions by these adaptive remodeling mechanisms of DA neuron physiology. The successful modeling of DA neuron loss in lower genetic models such as *Drosophila* and *Caenorhabditis elegans*, in addition to viral-induced acute mouse models as well as adult conditional KO of genes, is shedding light on how the remarkable compensation in genetic mouse models can be overcome, thus creating faithful recapitulation of key PD features in the lifetime of mice.

From various transgenic models expressing autosomal-dominant PD genes, it has become clear that robust expression of the transgene is critical in inducing neurodegeneration. In addition, neuron-specific promoters used in previous models tend to be turned on during embry-

onic neuronal development, thus increasing the chance of adaptation in developing DA neurons against genetic manipulation. Considering these issues together, the best option for improvement of current models is likely a Tet-off conditional transgenic approach that can enable temporal and regional control, together with relatively robust capacity for transgene induction (Sprengel and Hasan 2007). Because several α-syn and LRRK2 Tet responder mice are available, efforts should be focused on developing strong driver lines that can drive constitutively high levels of the transgene in relevant neuronal populations, like had been shown with TH expression of mutant G2019S LRRK2 (Xiong et al. 2018). Another option would be to use Cre lines to initiate transgene expression (Moeller et al. 2005). In this scenario, we would need to generate transgenic lines that have a stuffer sequence that is flanked by loxP and placed between a constitutively strong promoter (e.g., CAG) and the downstream transgene (e.g., synuclein and LRRK2). Again, transgene expression could be controlled either by using Cre-ER lines (Brocard et al. 1997) where Cre activity can be controlled by tamoxifen injection or by using additional elements in the transgene itself such as tTA-spacer-tetP just downstream from 3' loxP sequence. More available lines of Cre and the property of reversible switch by Tet off system favor the latter strategy.

As discussed in the previous section, Lenti-Cre delivery to loxP parkin mice successfully reproduces DA neuronal loss in the lifetime of mice. Likewise, Cre-ER-mediated acute deletion of recessive PD genes in adulthood would be promising in generating recessive PD mouse models with DA neurodegeneration. Alternatively transgenic mouse models expressing relevant substrates of parkin or PINK1 could be used to study PD pathogenesis, such as overexpression of PARIS, which exhibits robust DA neuron degeneration (Jo et al. 2021; Kim et al. 2025). Because signal exaggeration is more achievable using the previously noted transgenic approaches, substrates transgenic mice are likely to produce desired pathologies at earlier time points. Furthermore, the substrate model will narrow down the pathogenic signaling of recessive PD gene mutations that, once proven, may provide better therapeutic options for the treatment of PD. The identification of PARIS as a common substrate of parkin and PINK1 provides unifying transgenic model for these two major recessive PD genes.

CONCLUDING REMARKS

Mice remain the preferred mammalian model due to their genetic tractability, conserved neuronal networks, and broad phenotypic repertoire. Advances in transgenic, KO, and conditional gene manipulation, combined with the ability to model aging, have enabled increasingly sophisticated PD models. Transgenic mice overexpressing human PD genes (e.g., α-syn and LRRK2) under neuronal or DA-specific promoters have been central to modeling dominant PD. Newer models use inducible systems (e.g., Tet-off and Cre-LoxP) for temporal and spatial control, allowing for adult-onset expression and reducing developmental compensation. Tet-regulated and Cre-ER (tamoxifen-inducible) systems now enable reversible or temporally controlled gene expression or deletion, critical for modeling age-dependent neurodegeneration and dissecting early versus late pathogenic events. KO and KI approaches for recessive PD genes (parkin, PINK1, and DJ-1) have provided insight into loss-of-function mechanisms. However, most germline KOs show mild phenotypes due to compensatory mechanisms. Conditional KOs and viral vector-mediated gene deletion in adult mice now yield progressive DA neuron loss, better mirroring human PD. AAV and HSV vectors allow for targeted, robust gene delivery or knockdown in specific brain regions. These approaches have been refined for higher efficiency and are used to model both dominant (e.g., LRRK2 and α-syn) and recessive (e.g., parkin) PD in rodents and nonhuman primates (NHPs). Injection of synthetic α-syn PFFs into rodent brains induces endogenous α-syn aggregation, progressive DA neuron loss, and prion-like spread of Lewy pathology. This model closely mimics human PD progression and is now widely used for mechanistic and therapeutic studies.

The past decade has seen transformative advances in PD animal modeling, with new genetic, viral, and PFF-based approaches offering more faithful recapitulation of human disease. While challenges remain, especially in modeling sporadic and late-onset PD, the integration of aging, multihit strategies, and advanced molecular tools holds promise for uncovering disease mechanisms and developing effective therapies.

ACKNOWLEDGMENTS

T.M.D. is the Leonard and Madlyn Abramson Professor in Neurodegenerative Diseases. We thank Gilber Chen for the illustrations.

REFERENCES

Abulimiti A, Bae H, Ali A, Balakrishnan S, Tsujishita M, Gveric D, Tierney TS, Jonas EA, Smith PJS, Gentleman SM, et al. 2025. Reduced DJ-1-F1Fo ATP synthase association correlates with midbrain dopaminergic neuron vulnerability in idiopathic Parkinson's disease. *Sci Adv* 11: eads3051. doi:10.1126/sciadv.ads3051

Andres-Mateos E, Perier C, Zhang L, Blanchard-Fillion B, Greco TM, Thomas B, Ko HS, Sasaki M, Ischiropoulos H, Przedborski S, et al. 2007. DJ-1 gene deletion reveals that DJ-1 is an atypical peroxiredoxin-like peroxidase. *Proc Natl Acad Sci* 104: 14807–14812. doi:10.1073/pnas.0703219104

Barbuti PA, Guardia-Laguarta C, Yun T, Chatila ZK, Flowers X, Wong C, Santos BFR, Larsen SB, Lotti JS, Hattori N, et al. 2025. The role of alpha-synuclein in synucleinopathy: impact on lipid regulation at mitochondria-ER membranes. *NPJ Parkinsons Dis* 11: 103. doi:10.1038/s41531-025-00960-x

Bond CT, Sprengel R, Bissonnette JM, Kaufmann WA, Pribnow D, Neelands T, Storck T, Baetscher M, Jerecic J, Maylie J, et al. 2000. Respiration and parturition affected by conditional overexpression of the Ca^{2+}-activated K^+ channel subunit, SK3. *Science* 289: 1942–1946. doi:10.1126/science.289.5486.1942

Brocard J, Warot X, Wendling O, Messaddeq N, Vonesch JL, Chambon P, Metzger D. 1997. Spatio-temporally controlled site-specific somatic mutagenesis in the mouse. *Proc Natl Acad Sci* 94: 14559–14563. doi:10.1073/pnas.94.26.14559

Capecchi MR. 1989. The new mouse genetics: altering the genome by gene targeting. *Trends Genet* 5: 70–76. doi:10.1016/0168-9525(89)90029-2

Chaudhuri A, Bowling K, Funderburk C, Lawal H, Inamdar A, Wang Z, O'Donnell JM. 2007. Interaction of genetic and environmental factors in a *Drosophila* parkinsonism model. *J Neurosci* 27: 2457–2467. doi:10.1523/JNEUROSCI.4239-06.2007

Chen L, Cagniard B, Mathews T, Jones S, Koh HC, Ding Y, Carvey PM, Ling Z, Kang UJ, Zhuang X. 2005. Age-de-

Cite this article as *Cold Spring Harb Perspect Med* doi: 10.1101/cshperspect.a041644

pendent motor deficits and dopaminergic dysfunction in DJ-1 null mice. *J Biol Chem* **280**: 21418–21426. doi:10 .1074/jbc.M413955200

Churchyard A, Lees AJ. 1997. The relationship between dementia and direct involvement of the hippocampus and amygdala in Parkinson's disease. *Neurology* **49**: 1570–1576. doi:10.1212/WNL.49.6.1570

Clark IE, Dodson MW, Jiang C, Cao JH, Huh JR, Seol JH, Yoo SJ, Hay BA, Guo M. 2006. *Drosophila* pink1 is required for mitochondrial function and interacts genetically with parkin. *Nature* **441**: 1162–1166. doi:10.1038/nature04779

Cloutier M, Wellstead P. 2010. The control systems structures of energy metabolism. *J R Soc Interface* **7**: 651–665. doi:10.1098/rsif.2009.0371

Cramb KML, Beccano-Kelly D, Cragg SJ, Wade-Martins R. 2023. Impaired dopamine release in Parkinson's disease. *Brain* **146**: 3117–3132. doi:10.1093/brain/awad064

Dauer W, Kholodilov N, Vila M, Trillat AC, Goodchild R, Larsen KE, Staal R, Tieu K, Schmitz Y, Yuan CA, et al. 2002. Resistance of α-synuclein null mice to the parkinsonian neurotoxin MPTP. *Proc Natl Acad Sci* **99**: 14524–14529. doi:10.1073/pnas.172514599

Dawson TM, Ko HS, Dawson VL. 2010. Genetic animal models of Parkinson's disease. *Neuron* **66**: 646–661. doi:10.1016/j.neuron.2010.04.034

Dev KK, Hofele K, Barbieri S, Buchman VL, van der Putten H. 2003. Part II: alpha-synuclein and its molecular pathophysiological role in neurodegenerative disease. *Neuropharmacology* **45**: 14–44. doi:10.1016/S0028-3908(03)00140-0

Fernagut PO, Chesselet MF. 2004. Alpha-synuclein and transgenic mouse models. *Neurobiol Dis* **17**: 123–130. doi:10.1016/j.nbd.2004.07.001

Forno LS. 1987. The Lewy body in Parkinson's disease. *Adv Neurol* **45**: 35–43.

Gao L, Zhang J. 2003. Why are some human disease-associated mutations fixed in mice? *Trends Genet* **19**: 678–681. doi:10.1016/j.tig.2003.10.002

Gasser T. 2004. Genetics of Parkinson's disease. *Dialogues Clin Neurosci* **6**: 295–301. doi:10.31887/DCNS.2004.6.3/tgasser

Giasson BI, Duda JE, Quinn SM, Zhang B, Trojanowski JQ, Lee VM. 2002. Neuronal alpha-synucleinopathy with severe movement disorder in mice expressing A53T human alpha-synuclein. *Neuron* **34**: 521–533. doi:10.1016/S0896-6273(02)00682-7

Gispert S, Ricciardi F, Kurz A, Azizov M, Hoepken HH, Becker D, Voos W, Leuner K, Müller WE, Kudin AP, et al. 2009. Parkinson phenotype in aged PINK1-deficient mice is accompanied by progressive mitochondrial dysfunction in absence of neurodegeneration. *PLoS One* **4**: e5777. doi:10.1371/journal.pone.0005777

Goldberg MS, Fleming SM, Palacino JJ, Cepeda C, Lam HA, Bhatnagar A, Meloni EG, Wu N, Ackerson LC, Klapstein GJ, et al. 2003. Parkin-deficient mice exhibit nigrostriatal deficits but not loss of dopaminergic neurons. *J Biol Chem* **278**: 43628–43635. doi:10.1074/jbc.M308947200

Goldberg MS, Pisani A, Haburcak M, Vortherms TA, Kitada T, Costa C, Tong Y, Martella G, Tscherter A, Martins A, et al. 2005. Nigrostriatal dopaminergic deficits and hypoki-

nesia caused by inactivation of the familial parkinsonism-linked gene DJ-1. *Neuron* **45**: 489–496. doi:10.1016/j .neuron.2005.01.041

Goldwurm S, Di Fonzo A, Simons EJ, Rohe CF, Zini M, Canesi M, Tesei S, Zecchinelli A, Antonini A, Mariani C, et al. 2005. The G6055A (G2019S) mutation in *LRRK2* is frequent in both early and late onset Parkinson's disease and originates from a common ancestor. *J Med Genet* **42**: e65. doi:10.1136/jmg.2005.035568

Gossen M, Bujard H. 1992. Tight control of gene expression in mammalian cells by tetracycline-responsive promoters. *Proc Natl Acad Sci* **89**: 5547–5551. doi:10.1073/pnas .89.12.5547

Guardia-Laguarta C, Area-Gomez E, Rüb C, Liu Y, Magrané J, Becker D, Voos W, Schon EA, Przedborski S. 2014. α-Synuclein is localized to mitochondria-associated ER membranes. *J Neurosci* **34**: 249–259. doi:10.1523/JNEUROSCI.2507-13.2014

Guzman JN, Sanchez-Padilla J, Wokosin D, Kondapalli J, Ilijic E, Schumacker PT, Surmeier DJ. 2010. Oxidant stress evoked by pacemaking in dopaminergic neurons is attenuated by DJ-1. *Nature* **468**: 696–700. doi:10 .1038/nature09536

Han R, Wang Q, Xiong X, Chen X, Tu Z, Li B, Zhang F, Chen C, Pan M, Xu T, et al. 2024. Deficiency of parkin causes neurodegeneration and accumulation of pathological alpha-synuclein in monkey models. *J Clin Invest* **134**: e179633. doi:10.1172/JCI179633

Hana S, Peterson M, McLaughlin H, Marshall E, Fabian AJ, McKissick O, Koszka K, Marsh G, Craft M, Xu S, et al. 2021. Highly efficient neuronal gene knockout in vivo by CRISPR-Cas9 via neonatal intracerebroventricular injection of AAV in mice. *Gene Ther* **28**: 646–658. doi:10.1038/s41434-021-00224-2

Henderson MX, Cornblath EJ, Darwich A, Zhang B, Brown H, Gathagan RJ, Sandler RM, Bassett DS, Trojanowski JQ, Lee VMY. 2019. Spread of α-synuclein pathology through the brain connectome is modulated by selective vulnerability and predicted by network analysis. *Nat Neurosci* **22**: 1248–1257. doi:10.1038/s41593-019-0457-5

Henrich MT, Geibl FF, Lakshminarasimhan H, Stegmann A, Giasson BI, Mao X, Dawson VL, Dawson TM, Oertel WH, Surmeier DJ. 2020. Determinants of seeding and spreading of α-synuclein pathology in the brain. *Sci Adv* **6**: eabc2487. doi:10.1126/sciadv.abc2487

Hu X, Mao C, Fan L, Luo H, Hu Z, Zhang S, Yang Z, Zheng H, Sun H, Fan Y, et al. 2020. Modeling Parkinson's disease using induced pluripotent stem cells. *Stem Cells Int* **2020**: 1061470. doi:10.1155/2020/1061470

Itier JM, Ibanez P, Mena MA, Abbas N, Cohen-Salmon C, Bohme GA, Laville M, Pratt J, Corti O, Pradier L, et al. 2003. Parkin gene inactivation alters behaviour and dopamine neurotransmission in the mouse. *Hum Mol Genet* **12**: 2277–2291. doi:10.1093/hmg/ddg239

Jeong GR, Jang EH, Bae JR, Jun S, Kang HC, Park CH, Shin JH, Yamamoto Y, Tanaka-Yamamoto K, Dawson VL, et al. 2018. Dysregulated phosphorylation of Rab GTPases by LRRK2 induces neurodegeneration. *Mol Neurodegener* **13**: 8. doi:10.1186/s13024-018-0240-1

Jo A, Lee Y, Kam TI, Kang SU, Neifert S, Karuppagounder SS, Khang R, Kang H, Park H, Chou SC, et al. 2021. PARIS farnesylation prevents neurodegeneration in models of

Parkinson's disease. *Sci Transl Med* 13: eaax8891. doi:10 .1126/scitranslmed.aax8891

Kim RH, Smith PD, Aleyasin H, Hayley S, Mount MP, Pownall S, Wakeham A, You-Ten AJ, Kalia SK, Horne P, et al. 2005. Hypersensitivity of DJ-1-deficient mice to 1-methyl-4-phenyl-1,2,3,6-tetrahydropyridine (MPTP) and oxidative stress. *Proc Natl Acad Sci* 102: 5215–5220. doi:10.1073/pnas.0501282102

Kim S, Kwon SH, Kam TI, Panicker N, Karuppagounder SS, Lee S, Lee JH, Kim WR, Kook M, Foss CA, et al. 2019. Transneuronal propagation of pathologic α-synuclein from the gut to the brain models Parkinson's disease. *Neuron* 103: 627–641.e7. doi:10.1016/j.neuron.2019.05 .035

Kim JH, Yang S, Kim H, Vo DK, Maeng HJ, Jo A, Shin JH, Shin JH, Baek HM, Lee GH, et al. 2025. Preclinical studies and transcriptome analysis in a model of Parkinson's disease with dopaminergic ZNF746 expression. *Mol Neurodegener* 20: 24. doi:10.1186/s13024-025-00814-3

Kistner A, Gossen M, Zimmermann F, Jerecic J, Ullmer C, Lübbert H, Bujard H. 1996. Doxycycline-mediated quantitative and tissue-specific control of gene expression in transgenic mice. *Proc Natl Acad Sci* 93: 10933–10938. doi:10.1073/pnas.93.20.10933

Kitada T, Pisani A, Porter DR, Yamaguchi H, Tscherter A, Martella G, Bonsi P, Zhang C, Pothos EN, Shen J. 2007. Impaired dopamine release and synaptic plasticity in the striatum of PINK1-deficient mice. *Proc Natl Acad Sci* 104: 11441–11446. doi:10.1073/pnas.0702717104

Kitada T, Tong Y, Gautier CA, Shen J. 2009. Absence of nigral degeneration in aged Parkin/DJ-1/PINK1 triple knockout mice. *J Neurochem* 111: 696–702. doi:10 .1111/j.1471-4159.2009.06350.x

Ko HS, von Coelln R, Sriram SR, Kim SW, Chung KK, Pletnikova O, Troncoso J, Johnson B, Saffary R, Goh EL, et al. 2005. Accumulation of the authentic parkin substrate aminoacyl-tRNA synthetase cofactor, p38/JTV-1, leads to catecholaminergic cell death. *J Neurosci* 25: 7968–7978. doi:10.1523/JNEUROSCI.2172-05.2005

Ko HS, Kim SW, Sriram SR, Dawson VL, Dawson TM. 2006. Identification of far upstream element-binding protein-1 as an authentic Parkin substrate. *J Biol Chem* 281: 16193–16196. doi:10.1074/jbc.C600041200

Koh DW, Dawson TM, Dawson VL. 2005. Mediation of cell death by poly(ADP-ribose) polymerase-1. *Pharmacol Res* 52: 5–14. doi:10.1016/j.phrs.2005.02.011

Landel CP, Chen SZ, Evans GA. 1990. Reverse genetics using transgenic mice. *Annu Rev Physiol* 52: 841–851. doi:10 .1146/annurev.ph.52.030190.004205

Lee MK, Stirling W, Xu Y, Xu X, Qui D, Mandir AS, Dawson TM, Copeland NG, Jenkins NA, Price DL. 2002. Human alpha-synuclein-harboring familial Parkinson's disease-linked Ala-53→Thr mutation causes neurodegenerative disease with alpha-synuclein aggregation in transgenic mice. *Proc Natl Acad Sci* 99: 8968–8973. doi:10.1073/ pnas.132197599

Lee BD, Shin JH, VanKampen J, Petrucelli L, West AB, Ko HS, Lee YI, Maguire-Zeiss KA, Bowers WJ, Federoff HJ, et al. 2010. Inhibitors of leucine-rich repeat kinase-2 protect against models of Parkinson's disease. *Nat Med* 16: 998–1000. doi:10.1038/nm.2199

Lee Y, Dawson VL, Dawson TM. 2012. Animal models of Parkinson's disease: vertebrate genetics. *Cold Spring Harb Perspect Med* 2: a009324. doi:10.1101/cshperspect.a0 09324

Lee Y, Stevens DA, Kang SU, Jiang H, Lee YI, Ko HS, Scarffe LA, Umanah GE, Kang H, Ham S, et al. 2017. PINK1 primes parkin-mediated ubiquitination of PARIS in dopaminergic neuronal survival. *Cell Rep* 18: 918–932. doi:10 .1016/j.celrep.2016.12.090

Lee SH, Tonello R, Lee K, Roh J, Prudente AS, Kim YH, Park CK, Berta T. 2025. The Parkinson's disease *DJ-1/PARK7* gene controls peripheral neuronal excitability and painful neuropathy. *Brain* 148: 1639–1651. doi:10.1093/brain/ awae341

Luk KC, Kehm V, Carroll J, Zhang B, O'Brien P, Trojanowski JQ, Lee VM. 2012. Pathological α-synuclein transmission initiates Parkinson-like neurodegeneration in nontransgenic mice. *Science* 338: 949–953. doi:10.1126/science .1227157

Lv L, Zhang H, Tan J, Wang C. 2025. Neuroprotective role and mechanistic insights of DJ-1 dimerization in Parkinson's disease. *Cell Commun Signal* 23: 129. doi:10.1186/ s12964-025-02136-9

Mandir AS, Przedborski S, Jackson-Lewis V, Wang ZQ, Simbulan-Rosenthal CM, Smulson ME, Hoffman BE, Guastella DB, Dawson VL, Dawson TM. 1999. Poly(ADP-ribose) polymerase activation mediates 1-methyl-4-phenyl-1, 2,3,6-tetrahydropyridine (MPTP)-induced parkinsonism. *Proc Natl Acad Sci* 96: 5774–5779. doi:10 .1073/pnas.96.10.5774

Martin LJ, Pan Y, Price AC, Sterling W, Copeland NG, Jenkins NA, Price DL, Lee MK. 2006. Parkinson's disease alpha-synuclein transgenic mice develop neuronal mitochondrial degeneration and cell death. *J Neurosci* 26: 41–50. doi:10.1523/JNEUROSCI.4308-05.2006

Martin I, Dawson VL, Dawson TM. 2011. Recent advances in the genetics of Parkinson's disease. *Annu Rev Genomics Hum Genet* 12: 301–325. doi:10.1146/annurev-genom-082410-101440

Martin I, Kim JW, Lee BD, Kang HC, Xu JC, Jia H, Stankowski J, Kim MS, Zhong J, Kumar M, et al. 2014. Ribosomal protein s15 phosphorylation mediates LRRK2 neurodegeneration in Parkinson's disease. *Cell* 157: 472–485. doi:10.1016/j.cell.2014.01.064

Masuda-Suzukake M, Nonaka T, Hosokawa M, Oikawa T, Arai T, Akiyama H, Mann DM, Hasegawa M. 2013. Prion-like spreading of pathological α-synuclein in brain. *Brain* 136: 1128–1138. doi:10.1093/brain/awt037

Mezias C, Rey N, Brundin P, Raj A. 2020. Neural connectivity predicts spreading of alpha-synuclein pathology in fibril-injected mouse models: involvement of retrograde and anterograde axonal propagation. *Neurobiol Dis* 134: 104623. doi:10.1016/j.nbd.2019.104623

Moeller MJ, Soofi A, Sanden S, Floege J, Kriz W, Holzman LB. 2005. An efficient system for tissue-specific overexpression of transgenes in podocytes in vivo. *Am J Physiol Renal Physiol* 289: F481–F488. doi:10.1152/ajprenal .00332.2004

Moore DJ, Dawson VL, Dawson TM. 2003. Genetics of Parkinson's disease: what do mutations in DJ-1 tell us? *Ann Neurol* 54: 281–282. doi:10.1002/ana.10740

Moore DJ, West AB, Dawson VL, Dawson TM. 2005. Molecular pathophysiology of Parkinson's disease. *Annu Rev Neurosci* **28:** 57–87. doi:10.1146/annurev.neuro.28.061604.135718

Nguyen TT, Kim YJ, Lai TT, Nguyen PT, Koh YH, Nguyen LTN, Ma HI, Kim YE. 2022. PTEN-induced putative kinase 1 dysfunction accelerates synucleinopathy. *J Parkinsons Dis* **12:** 1201–1217. doi:10.3233/JPD-213065

Nieto M, Gil-Bea FJ, Dalfó E, Cuadrado M, Cabodevilla F, Sánchez B, Catena S, Sesma T, Ribé E, Ferrer I, et al. 2006. Increased sensitivity to MPTP in human alpha-synuclein A30P transgenic mice. *Neurobiol Aging* **27:** 848–856. doi:10.1016/j.neurobiolaging.2005.04.010

Nuber S, Petrasch-Parwez E, Winner B, Winkler J, von Hörsten S, Schmidt T, Boy J, Kuhn M, Nguyen HP, Teismann P, et al. 2008. Neurodegeneration and motor dysfunction in a conditional model of Parkinson's disease. *J Neurosci* **28:** 2471–2484. doi:10.1523/JNEUROSCI.3040-07.2008

Olzscha H, Schermann SM, Woerner AC, Pinkert S, Hecht MH, Tartaglia GG, Vendruscolo M, Hayer-Hartl M, Hartl FU, Vabulas RM. 2011. Amyloid-like aggregates sequester numerous metastable proteins with essential cellular functions. *Cell* **144:** 67–78. doi:10.1016/j.cell.2010.11.050

Pandya S, Zeighami Y, Freeze B, Dadar M, Collins DL, Dagher A, Raj A. 2019. Predictive model of spread of Parkinson's pathology using network diffusion. *Neuroimage* **192:** 178–194. doi:10.1016/j.neuroimage.2019.03.001

Park H, Kam TI, Dawson VL, Dawson TM. 2025. α-Synuclein pathology as a target in neurodegenerative diseases. *Nat Rev Neurol* **21:** 32–47. doi:10.1038/s41582-024-01043-w

Paterna JC, Leng A, Weber E, Feldon J, Büeler H. 2007. DJ-1 and Parkin modulate dopamine-dependent behavior and inhibit MPTP-induced nigral dopamine neuron loss in mice. *Mol Ther* **15:** 698–704. doi:10.1038/sj.mt.6300067

Perez FA, Curtis WR, Palmiter RD. 2005. Parkin-deficient mice are not more sensitive to 6-hydroxydopamine or methamphetamine neurotoxicity. *BMC Neurosci* **6:** 71. doi:10.1186/1471-2202-6-71

Pfeffer SR. 2018. LRRK2 and Rab GTPases. *Biochem Soc Trans* **46:** 1707–1712. doi:10.1042/BST20180470

Pirooznia SK, Rosenthal LS, Dawson VL, Dawson TM. 2021. Parkinson disease: translating insights from molecular mechanisms to neuroprotection. *Pharmacol Rev* **73:** 33–97. doi:10.1124/pharmrev.120.000189

Polymeropoulos MH, Lavedan C, Leroy E, Ide SE, Dehejia A, Dutra A, Pike B, Root H, Rubenstein J, Boyer R, et al. 1997. Mutation in the alpha-synuclein gene identified in families with Parkinson's disease. *Science* **276:** 2045–2047. doi:10.1126/science.276.5321.2045

Ramonet D, Daher JP, Lin BM, Stafa K, Kim J, Banerjee R, Westerlund M, Pletnikova O, Glauser L, Yang L, et al. 2011. Dopaminergic neuronal loss, reduced neurite complexity and autophagic abnormalities in transgenic mice expressing G2019S mutant LRRK2. *PLoS One* **6:** e18568. doi:10.1371/journal.pone.0018568

Regoni M, Zanetti L, Sevegnani M, Domenicale C, Magnabosco S, Patel JC, Fernandes MK, Feeley RM, Monzani E, Mini C, et al. 2024. Dopamine neuron dysfunction and loss in the *Prkn*R275W mouse model of juvenile parkin-

sonism. *Brain* **147:** 4017–4025. doi:10.1093/brain/awae276

Rockenstein E, Crews L, Masliah E. 2007. Transgenic animal models of neurodegenerative diseases and their application to treatment development. *Adv Drug Deliv Rev* **59:** 1093–1102. doi:10.1016/j.addr.2007.08.013

Rosenthal N, Brown S. 2007. The mouse ascending: perspectives for human-disease models. *Nat Cell Biol* **9:** 993–999. doi:10.1038/ncb437

Rülicke T, Montagutelli X, Pintado B, Thon R, Hedrich HJ. 2007. FELASA guidelines for the production and nomenclature of transgenic rodents. *Lab Anim* **41:** 301–311. doi:10.1258/002367707781282758

Savitt JM, Dawson VL, Dawson TM. 2006. Diagnosis and treatment of Parkinson disease: molecules to medicine. *J Clin Invest* **116:** 1744–1754. doi:10.1172/JCI29178

Seegobin SP, Heaton GR, Liang D, Choi I, Blanca Ramirez M, Tang B, Yue Z. 2020. Progress in LRRK2-associated Parkinson's disease animal models. *Front Neurosci* **14:** 674. doi:10.3389/fnins.2020.00674

Shin JH, Ko HS, Kang H, Lee Y, Lee YI, Pletinkova O, Troconso JC, Dawson VL, Dawson TM. 2011. PARIS (ZNF746) repression of PGC-1α contributes to neurodegeneration in Parkinson's disease. *Cell* **144:** 689–702. doi:10.1016/j.cell.2011.02.010

Simunovic F, Yi M, Wang Y, Macey L, Brown LT, Krichevsky AM, Andersen SL, Stephens RM, Benes FM, Sonntag KC. 2009. Gene expression profiling of substantia nigra dopamine neurons: further insights into Parkinson's disease pathology. *Brain* **132:** 1795–1809. doi:10.1093/brain/awn323

Singleton AB, Farrer M, Johnson J, Singleton A, Hague S, Kachergus J, Hulihan M, Peuralinna T, Dutra A, Nussbaum R, et al. 2003. Alpha-synuclein locus triplication causes Parkinson's disease. *Science* **302:** 841. doi:10.1126/science.1090278

Song DD, Shults CW, Sisk A, Rockenstein E, Masliah E. 2004. Enhanced substantia nigra mitochondrial pathology in human alpha-synuclein transgenic mice after treatment with MPTP. *Exp Neurol* **186:** 158–172. doi:10.1016/S0014-4886(03)00342-X

Spillantini MG, Schmidt ML, Lee VM, Trojanowski JQ, Jakes R, Goedert M. 1997. Alpha-synuclein in Lewy bodies. *Nature* **388:** 839–840. doi:10.1038/42166

Sprengel R, Hasan MT. 2007. Tetracycline-controlled genetic switches. *Handb Exp Pharmacol* **178:** 49–72. doi:10.1007/978-3-540-35109-2_3

Steger M, Tonelli F, Ito G, Davies P, Trost M, Vetter M, Wachter S, Lorentzen E, Duddy G, Wilson S, et al. 2016. Phosphoproteomics reveals that Parkinson's disease kinase LRRK2 regulates a subset of Rab GTPases. *eLife* **5:** e12813. doi:10.7554/eLife.12813

Thomas B, von Coelln R, Mandir AS, Trinkaus DB, Farah MH, Leong Lim K, Calingasan NY, Flint Beal M, Dawson VL, Dawson TM. 2007. MPTP and DSP-4 susceptibility of substantia nigra and locus coeruleus catecholaminergic neurons in mice is independent of parkin activity. *Neurobiol Dis* **26:** 312–322. doi:10.1016/j.nbd.2006.12.021

Thomas B, Mandir AS, West N, Liu Y, Andrabi SA, Stirling W, Dawson VL, Dawson TM, Lee MK. 2011. Resistance to MPTP-neurotoxicity in α-synuclein knockout mice is complemented by human α-synuclein and associated with increased β-synuclein and Akt activation. *PLoS One* **6:** e16706. doi:10.1371/journal.pone.0016706

Tsika E, Nguyen AP, Dusonchet J, Colin P, Schneider BL, Moore DJ. 2015. Adenoviral-mediated expression of G2019S LRRK2 induces striatal pathology in a kinase-dependent manner in a rat model of Parkinson's disease. *Neurobiol Dis* **77:** 49–61. doi:10.1016/j.nbd.2015.02.019

Van der Perren A, Van den Haute C, Baekelandt V. 2015. Viral vector-based models of Parkinson's disease. *Curr Top Behav Neurosci* **22:** 271–301. doi:10.1007/785 4_2014_310

Von Coelln R, Thomas B, Savitt JM, Lim KL, Sasaki M, Hess EJ, Dawson VL, Dawson TM. 2004. Loss of locus coeruleus neurons and reduced startle in parkin null mice. *Proc Natl Acad Sci* **101:** 10744–10749. doi:10.1073/pnas .0401297101

Waterston RH, Lindblad-Toh K, Birney E, Rogers J, Abril JF, Agarwal P, Agarwala R, Ainscough R, Alexandersson M, An P, et al. 2002. Initial sequencing and comparative analysis of the mouse genome. *Nature* **420:** 520–562. doi:10.1038/nature01262

West AB, Moore DJ, Biskup S, Bugayenko A, Smith WW, Ross CA, Dawson VL, Dawson TM. 2005. Parkinson's disease-associated mutations in leucine-rich repeat kinase 2 augment kinase activity. *Proc Natl Acad Sci* **102:** 16842–16847. doi:10.1073/pnas.0507360102

West AB, Moore DJ, Choi C, Andrabi SA, Li X, Dikeman D, Biskup S, Zhang Z, Lim KL, Dawson VL, et al. 2007. Parkinson's disease-associated mutations in LRRK2 link enhanced GTP-binding and kinase activities to neuronal toxicity. *Hum Mol Genet* **16:** 223–232. doi:10.1093/hmg/ ddl471

Wild P, Dikic I. 2010. Mitochondria get a Parkin' ticket. *Nat Cell Biol* **12:** 104–106. doi:10.1038/ncb0210-104

Xiong Y, Dawson TM, Dawson VL. 2017. Models of LRRK2-associated Parkinson's disease. *Adv Neurobiol* **14:** 163–191. doi:10.1007/978-3-319-49969-7_9

Xiong Y, Neifert S, Karuppagounder SS, Liu Q, Stankowski JN, Lee BD, Ko HS, Lee Y, Grima JC, Mao X, et al. 2018. Robust kinase- and age-dependent dopaminergic and norepinephrine neurodegeneration in LRRK2 G2019S transgenic mice. *Proc Natl Acad Sci* **115:** 1635–1640. doi:10.1073/pnas.1712648115

Zhou H, Falkenburger BH, Schulz JB, Tieu K, Xu Z, Xia XG. 2007. Silencing of the *Pink1* gene expression by conditional RNAi does not induce dopaminergic neuron death in mice. *Int J Biol Sci* **3:** 242–250. doi:10.7150/ijbs.3.242

The Mitochondrial Connection in Parkinson's Disease

Eric Schon,[1,2] Diana Matheoud,[3] and Serge Przedborski[1,4,5]

[1]Department of Neurology, Vagelos College of Physicians and Surgeons, Columbia University, New York, New York 10032, USA

[2]Department of Genetics and Development, Vagelos College of Physicians and Surgeons, Columbia University, New York, New York 10032, USA

[3]Département de Neurosciences, Université de Montréal, CRCHUM, Montréal, Québec H2X 0A9, Canada

[4]Department of Pathology and Cell Biology, Vagelos College of Physicians and Surgeons, Columbia University, New York, New York 10032, USA

[5]Department of Neuroscience, Columbia University, New York, New York 10032, USA

Correspondence: sp30@cumc.columbia.edu

Mitochondria are highly dynamic organelles with complex structural features that perform several essential cellular functions, including energy production by oxidative phosphorylation, regulation of calcium and lipid homeostasis, and control of programmed cell death. Given their critical role, alterations in mitochondrial biology can lead to neuronal dysfunction and death. Defects in mitochondrial respiration, especially in oxidative energy production, have long been thought to be implicated in the etiology and pathogenesis of Parkinson's disease. However, given the multifaceted roles of mitochondria in health and diseases, the putative role of mitochondria in Parkinson's disease likely extends well beyond defective respiration. As such, mitochondrial dysfunction represents a promising target for disease-modifying therapies in Parkinson's disease and related conditions.

The mitochondrial hypothesis of Parkinson's disease (PD) is one of the most resilient ideas regarding the cause and mechanisms of this common neurodegenerative disorder since its description by James Parkinson in the nineteenth century (Przedborski 2017).[6] Over the years, however, this hypothesis has evolved, driven in part by a growing understanding of the diverse and essential functions mitochon- dria play in cells, from bioenergetics to, more recently, quality control. This has led to various iterations of the mitochondrial hypothesis of PD, which will be explored in this work. Before embarking on this review, however, it is important to acknowledge that it is an updated version of Perier and Vila (2012), which not only incorporates recent publications but also re- flects our own understanding and interpreta-

---

[6]This is an update to a previous article published in *Cold Spring Harbor Perspectives in Medicine* [Perier and Vila (2012). *Cold Spring Harb Perspect Med* **2:** a009332. doi:10.1101/cshperspect.a009332].

Copyright © 2025 Cold Spring Harbor Laboratory Press; all rights reserved
Cite this article as *Cold Spring Harb Perspect Med* doi: 10.1101/cshperspect.a041891

tion of the role of mitochondria in neurodegeneration.

Mitochondria are intracellular membrane-enclosed organelles present in the hundreds in most eukaryotic cells, which, as we will discuss below, perform a number of crucial functions. From an evolutionary point of view, the origin of mitochondria is explained by two main competing hypotheses: the classical "endosymbiotic theory" and the "hydrogen hypothesis." The endosymbiotic theory, formalized by Ivan Wallin in 1925 (Wallin 1925) and restated in modified form in 1967 by Lynn Margulis Sagan (Sagan 1967), suggests that mitochondria originated from a symbiotic relationship established more than a billion years ago between primordial eukaryotic cells lacking the ability to use oxygen metabolically and primitive aerobic bacteria capable of oxidative phosphorylation (OxPhos). In contrast, the hydrogen hypothesis proposes that the host cell was an anaerobic archaeon that relied on hydrogen as an energy source, while the bacterium that lived inside it was a facultative anaerobe that produced hydrogen as a metabolic byproduct (Martin and Müller 1998). This latter model explains the origin of mitochondria as part of a mutualistic relationship based on hydrogen transfer, rather than on oxygen-based respiration. While both hypotheses acknowledge the prokaryotic ancestry of mitochondria, the hydrogen hypothesis provides an alternative view that integrates early anaerobic metabolic interactions as a key driving force in mitochondrial evolution. In both hypotheses, the symbiotic relationship between the bacterium and the host cell became permanent, leading to the evolution of the bacterium into the mitochondrion. As a result, the host cell gained the ability to metabolize oxygen, a far more efficient method of energy production compared to anaerobic glycolysis. Yet, new data continue to further improve our understanding of mitochondrial origins and evolution. For those interested in this topic, please refer to Gray's paper on mitochondrial evolution (Gray 2012).

Structurally, mitochondria contain four compartments: the outer mitochondrial membrane (OMM), the inner mitochondrial membrane (IMM), the intermembrane space (IMS), and the matrix (i.e., the region inside the inner membrane) (Fig. 1A). The IMM, in which the electron transport chain (ETC, also called the respiratory chain [RC]) is embedded, is highly folded, and protrudes into the matrix by invaginations called cristae, which greatly increase the surface area of the IMM and thus the efficiency of both the ETC and ATP production. Mitochondria are the only organelles of the cell besides the nucleus that contain their own DNA (i.e., mitochondrial DNA [mtDNA]), and their own machinery for synthesizing RNA and proteins. Within the matrix, each mitochondrion contains several copies of the small circular mitochondrial genome (16.6 kb in humans), which encodes for 13 proteins, all of which are components of the OxPhos system. The vast majority of proteins—almost 2000—required to build and maintain functional mitochondria are therefore encoded by nuclear DNA, synthesized in the cytosol, and imported into the mitochondria, where they are targeted to one of the four mitochondrial compartments.

In addition to their function in supplying cellular energy, mitochondria play a vital role in calcium homeostasis and also contain several proteins involved in programmed cell death (PCD). Furthermore, mitochondria are dynamic organelles that actively divide and fuse with one another, and undergo regulated turnover, all of which is important for the maintenance of mitochondrial function and for quality control. In neurons, mitochondria are actively transported throughout axons and dendrites to facilitate their recruitment to critical subcellular compartments distant from the cell body.

From this brief introduction to mitochondria, we can already envision the numerous potential mechanisms that, when disrupted, could lead to neuronal dysfunction and ultimately cell death. This, not surprisingly, explains the field's longstanding fascination with mitochondrial dysfunction, especially since the sequencing of the human mitochondrial genome in 1981 (Anderson et al. 1981), an achievement that served as a catalyst for investigations into mitochondria's role in PD (Przedborski 2017).

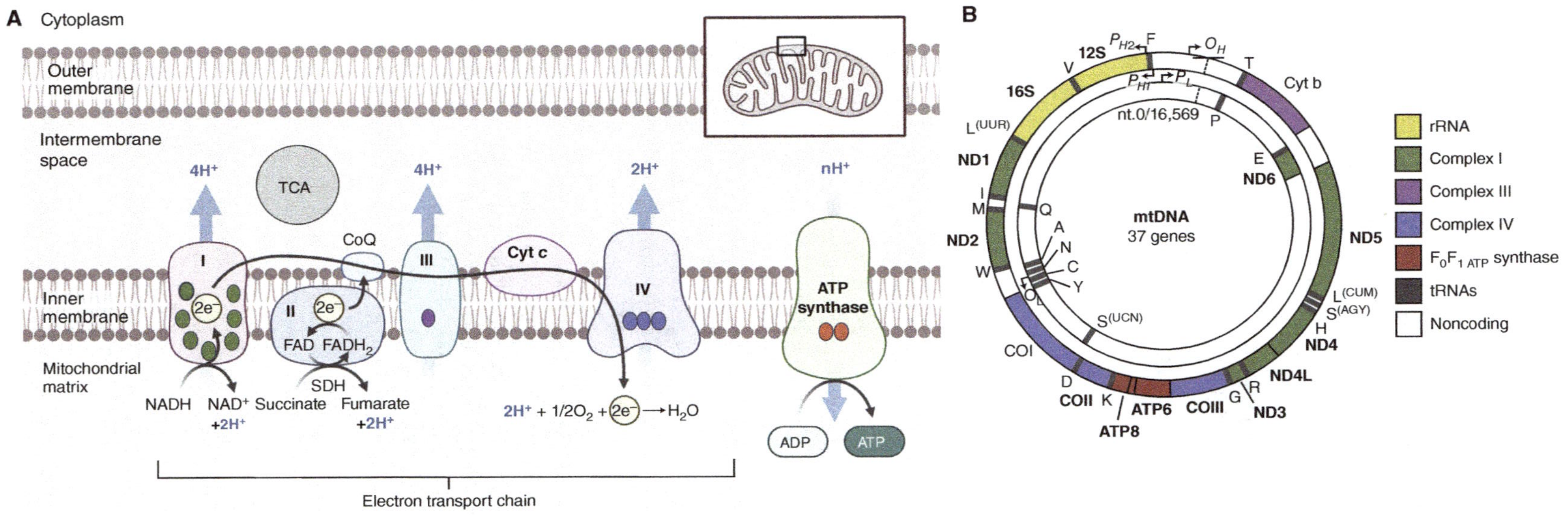

Figure 1. Schematic representation of mitochondrial bioenergetics. (*A*) Mitochondria are divided into four compartments: the outer mitochondrial membrane (OMM), the intermembrane space (IMS), the inner mitochondrial membrane (IMM), and the matrix. The respiratory chain/oxidative phosphorylation (OxPhos) system is located at the IMM, whereas the mitochondrial DNA (mtDNA; see *B*) is located in the matrix. The citric acid cycle (or Krebs cycle or tricarboxylic acid [TCA] cycle) also takes place within the matrix. The respiratory chain, also known as the electron transport chain (ETC), is composed of ~100 proteins, 13 of which are encoded by mtDNA (small colored circles). The remaining components are encoded by the nuclear DNA and imported into the mitochondria. OxPhos consists of five protein complexes (I–V): complex I (nicotinamide adenine dinucleotide [NADH] dehydrogenase [NDI]) and complex II (succinate dehydrogenase [SDH]) receive electrons (2e⁻) from intermediary metabolism, which are transferred first to coenzyme Q (also called ubiquinone), then to cytochrome *b* (Cyt *b*) within complex III (cytochrome *c* reductase), then to complex IV (cytochrome *c* oxidase [COI]), and finally to molecular oxygen, which is reduced to water. Electron transport is coupled to proton pumping across the IMM by complexes I, III, and IV. The resulting proton gradient drives ATP synthesis through complex V (ATP synthase [ATP]). (*B*) Human mtDNA. The genome is a 16.6 double-stranded circle present in multiple copies/cell. It encodes two ribosomal RNAs (rRNAs) (12S and 16S), 22 transfer RNAs (tRNAs) (one-letter code), and 13 polypeptide subunits of complexes I, III, and V (colored genes, corresponding to the colored circles in *A*). (Figure generated with BioRender; https://biorender.com.)

THE MITOCHONDRIAL OXIDATIVE PHOSPHORYLATION SYSTEM

Since the mitochondrial hypothesis of PD originated from the notion of a defect in OxPhos (Przedborski 2017), let us begin by introducing this complex cell machinery. Mitochondria are usually considered the "powerhouses of the cell" because of the production of ATP via the combined efforts of the TCA cycle (also known as Krebs cycle) and the RC/OxPhos system. The OxPhos system, embedded in the IMM, consists of five multimeric protein complexes: reduced nicotinamide adenine dinucleotide (NADH) dehydrogenase–ubiquinone oxidoreductase (complex I, ~46 subunits), succinate dehydrogenase–ubiquinone oxidoreductase (complex II, four subunits), ubiquinone–cytochrome c oxidoreductase (complex III, 11 subunits), cytochrome c oxidase (complex IV, 13 subunits), and ATP synthase (complex V, ~16 subunits) (Fig. 1A). The RC also requires two small electron carriers, ubiquinone (also called coenzyme Q [CoQ]) and cytochrome c. ATP synthesis involves two coordinated processes: electrons derived from energy substrates (such as NADH [processed at complex I] and FADH$_2$ [processed at complex II]) are transported through the mitochondrial respiratory complexes to molecular oxygen, thereby producing water. At the same time, protons are pumped across the mitochondrial inner membrane (i.e., from the matrix to the IMS) by complexes I, III, and IV, generating an electrochemical gradient (termed the mitochondrial membrane potential, $\Delta\psi$). ATP is produced through complex V (ATP synthase) using the influx of these protons back from the IMS into the mitochondrial matrix as the thermodynamic driver (i.e., the proton motive force) converting adenosine diphosphate (ADP) and free phosphate (P$_i$) to ATP. ATP is the main storage form of energy used by the cell, and, once produced in the mitochondrion, it is exported to the cytosol by the adenine nucleotide translocator (ANT) in exchange for cytosolic ADP.

Defective mitochondrial respiration, in particular at the level of complex I, has long been associated with the pathogenesis of PD. Evidence of this involvement first emerged following the observation that accidental exposure of drug abusers to 1-methyl-4-phenyl-1,2,3,4-tetrahydropyridine (MPTP), an inhibitor of mitochondrial complex I, resulted in an acute and irreversible parkinsonian syndrome almost indistinguishable from PD (Langston et al. 1983). It was subsequently shown that when injected into nonhuman primates and mice, MPTP selectively kills dopaminergic neurons of the substantia nigra pars compacta (SNpc), the type of cells that preferentially degenerate in PD (Dauer and Przedborski 2003). Similarly, chronic infusion of the potent complex I inhibitor rotenone in rats produced nigrostriatal dopaminergic neurodegeneration (Betarbet et al. 2000). A link between complex I dysfunction and PD was further established when several groups reported reduced complex I activity in the brain, platelets, and skeletal muscle of patients with sporadic PD (Parker et al. 1989; Schapira et al. 1990). In addition, cell lines engineered to contain mitochondria derived from platelets of PD patients (called cytoplasmic hybrids, or cybrids) were also shown to exhibit reduced complex I activity (Swerdlow et al. 1996). Supporting an instrumental role for complex I dysfunction in PD-related dopaminergic neurodegeneration, the feeding of the mitochondrial ETC directly at complex II by means of the ketone body D-β-hydroxybutyrate was shown to bypass complex I blockade, enhance OxPhos, and attenuate dopaminergic neurodegeneration in MPTP-intoxicated mice (Tieu et al. 2003). Also, virally mediated expression of the single-polypeptide or -component NADH-quinone oxidoreductase derived from the yeast *Saccharomyces cerevisiae*, which is insensitive to complex I inhibitors, into the substantia nigra of rats was shown to protect against rotenone-induced dopaminergic nigrostriatal impairment (Marella et al. 2008). Finally, methylene blue, an alternative electron carrier capable of delivering electrons directly from NADH to cytochrome c, thus bypassing complex I blockade, attenuated mitochondrial dysfunction, behavioral alterations, and dopaminergic neurodegeneration in rotenone-intoxicated rats (Wen et al. 2011). Reinforcing a potential role for complex I defects in PD, most of the pesticides that have been epidemiologically linked to an increased risk of PD cause

Cite this article as *Cold Spring Harb Perspect Med* doi: 10.1101/cshperspect.a041891

complex I dysfunction (Sherer et al. 2002; Schuh et al. 2005, 2009; Richardson et al. 2009).

Consequences of Complex I Blockade

One of the expected consequences of impaired mitochondrial respiration is a reduction in ATP production and subsequent bioenergetic failure. Supporting this view, MPP^+ (MPTP's active metabolite) caused a rapid and profound depletion of cellular ATP levels in isolated hepatocytes (Di Monte et al. 1986), in brain synaptosomal preparations (Scotcher et al. 1990), and in whole mouse brain tissues (Chan et al. 1991). In mice, however, MPTP caused only a mild ($\sim$20%) and transient reduction in striatal and midbrain ATP levels (Chan et al. 1991). These latter results call into question the relevance of the bioenergetic inhibitory function of MPTP in vivo, as it appears that complex I activity must be reduced by more than 50% to cause significant ATP depletion in nonsynaptic brain mitochondria (Davey and Clark 1996). Because complex I activity is only reduced by 25%–30% in tissue preparations from postmortem PD ventral midbrain samples (Parker et al. 1989; Schapira et al. 1990), one may wonder whether such a limited defect in bioenergetics can realistically drive PD-related neurodegeneration. Even more perplexing is the fact that postmortem substantia nigra samples from end-stage PD are profoundly depleted in dopaminergic neurons. Therefore, the reduced complex I activity previously reported likely does not reflect an intrinsic bioenergetic deficit in these neurons. Given that glial cells exhibit lower OxPhos activity than neurons (Bonvento and Bolaños 2021), it is more plausible that the observed reduction in complex I activity reflects a shift in cellular composition, namely, an increased glial-to-neuronal ratio, rather than a primary mitochondrial dysfunction in dopaminergic neurons.

Another consequence of impaired mitochondrial respiration is an increased production of reactive oxygen species (ROS). In the normal situation, small amounts of molecular oxygen in the mitochondria, rather than being converted to water, are reduced to ROS such as superoxide radicals (Zhou et al. 2008). However, thanks to the arsenal of antioxidants present inside the mitochondria, including the mitochondrial isoform of the ROS-scavenging enzyme superoxide dismutase (SOD2), the basal levels of ROS byproducts of mitochondrial respiration are minimal (Zhou et al. 2008). Following complex I blockade, however, the amount of ROS generated by the ETC increases dramatically, likely because of a higher rate of molecular oxygen reduction into superoxide radical in response to the hampered terminal step of electron transfer from the highest potential iron–sulfur cluster of complex I to ubiquinone (Ramsay et al. 1987). In agreement with this, MPP^+ and rotenone increased ROS production in isolated brain mitochondria in proportion to the degree of complex I inhibition (Perier et al. 2005).

Two related aspects of RC biology may help explain why inhibition of complex I can affect ROS production in PD. The first is reverse electron transfer (RET), a situation in which electrons in the RC are transferred "backward" from reduced CoQ to complex I, reducing NAD^+ to NADH and generating ROS in the process (Scialò et al. 2017). The second is the transient switch in the conformation of complex I from an "activated" to a "deactivated" state to counteract the increased amount of ROS generated under RET, because in the deactivated conformation electrons cannot access the ubiquinone binding site (Dröse et al. 2016). While this dynamic behavior is a normal process, prolonged RET can result in reduced metabolic flexibility, oxidative damage, increased superoxide and H_2O_2, and compromised bioenergetics, including reduced complex I activity (Pryde and Hirst 2011; Stepanova et al. 2019). From this point of view, complex I deficiency might be both a result and a cause of elevated ROS in PD.

Increased ROS can oxidatively damage virtually all biological macromolecules, including proteins, lipids, and DNA. For instance, complex I inhibition results in the inactivation of the mitochondrial TCA cycle enzyme aconitase, which is essential to maintain normal metabolic function, by oxidation of the iron–sulfur clusters contained in this enzyme (Liang and Patel 2004). Also, oxidative damage to catalytic subunits of complex I, which correlates with mis-assembly and dysfunction of this complex, has been observed in frontal cortex postmortem samples from PD patients (Keeney et al. 2006). Furthermore, MPTP intoxication in mice leads to the peroxidation of the IMM phospholipid cardiolipin, thereby disrupting the normal binding of cytochrome c to the mitochondrial inner membrane and facilitating the proapoptotic release of cytochrome c to the cytosol (Perier et al. 2005). Mitochondria-derived ROS have also been shown to damage lysosomal membranes in MPTP-intoxicated mice, leading to an impairment of lysosomal function and defective autophagic degradation in these animals (Dehay et al. 2010). In addition to proteins and lipids, MPTP-intoxicated mice also exhibit oxidative damage to nuclear and mtDNA (Hoang et al. 2009). Relevant to PD, oxidative damage to proteins, lipids, and DNA has been observed in postmortem brain samples from PD patients (Dauer and Przedborski 2003). In addition, the PD-linked protein DJ-1, mutations in which cause an autosomal-recessive form of PD (Bonifati et al. 2003), has been identified as a mitochondrial peroxiredoxin-like peroxidase capable of scavenging mitochondrial ROS (Andres-Mateos et al. 2007), and its deficiency in mutant mice results in increased mitochondrial ROS production (Andres-Mateos et al. 2007). Supporting a pathogenic role for mitochondria-derived ROS in the context of PD, transgenic mice overexpressing human catalase (an antioxidant enzyme normally localized in the peroxisome) targeted specifically to the mitochondria exhibited an attenuation of MPTP-induced mitochondrial ROS and reduced dopaminergic cell death (Perier et al. 2010). It is important to note, however, that increased ROS levels in the context of PD may also emanate from sources other than mitochondria, including neighboring glial cells (Zhou et al. 2008).

MITOCHONDRIAL AND UNCOUPLING PROTEINS

Although ATP production is the main energy form produced by mitochondria, extreme conditions such as hypothermia can shunt OxPhos to produce heat (thermogenesis) instead of ATP by dissipating the proton gradient (Okamatsu-Ogura et al. 2020). This process called mitochondria uncoupling is operated by uncoupling proteins (UCPs) that localize in the mitochondrial inner membrane (Demine et al. 2019). Normally, as discussed above, in OxPhos, protons are pumped across the mitochondrial membrane, and their flow back through ATP synthase generates ATP. However, when UCPs are activated, they allow protons to flow back into the mitochondrial matrix without going through ATP synthase. UCP2, 4, and 5 are found in the central nervous system (CNS) and play important roles in oxidative stress regulation and calcium homeostasis (Kumar et al. 2022). Indeed, they prevent overload of calcium in mitochondria by inhibiting Ca^{2+} channels (Koshenov et al. 2020), while UCP2 reduces the levels of ROS production in the SNpc by reducing ETC activity (Andrews et al. 2005). DJ-1 has been suggested to regulate the level of UCP4 and 5 expression (Dolgacheva et al. 2019), and these proteins are important for the regulation of mitochondria membrane potential and oxidative stress against MPTP toxicity (Wu et al. 2003). UCP1 converts energy from fat oxidation directly into heat after short-circuiting the proton gradient, leading to the energy being released as heat instead of being stored as ATP (Ikeda and Yamada 2020). It involves the breakdown of lipid droplets (primarily triglycerides) into fatty acids, which are then oxidized to produce heat. This mechanism is particularly important in brown adipose tissues where UCP1 plays a major role. Lipid droplets are not just passive fat storage sites; they actively participate in protecting cells from oxidative stress, especially in high-stress environments, by storing damaged lipids and limiting lipid peroxidation (Jarc and Petan 2019). As discussed in detail in Area-Gómez et al. (2025), lipid metabolism alterations have been observed in PD patients, as well as abnormal accumulation of lipid droplets in dopamine neurons, microglia, and astrocytes (Girard et al. 2021). Although UCP1 expression localizes in brown adipose tissue mitochondria, some conditions can induce its expression in the brain (Laursen et al. 2015; Claflin et al. 2022). Therefore, it would be interesting to study in more depth the potential role of thermogenesis in PD as a potential defective mechanism, as impaired thermoregulation is observed in PD (Coon and Low 2018).

MITOCHONDRIAL DNA

The human mitochondrial genome (mtDNA) consists of a 16.6 kb multicopy, double-stranded circular molecule containing 37 genes, encoding two ribosomal RNAs and 22 transfer RNAs required for mtDNA translation within the organelle, and 11 messenger RNAs (mRNAs) specifying 13 RC subunits (Fig. 1B; Anderson et al. 1981). Mitochondrial genetics differs from Mendelian genetics in several aspects (DiMauro and Schon 2003). For instance, mammalian mtDNA is transmitted through the maternal line, which implies that a mother carrying an mtDNA mutation will pass it on to all her children, but only her daughters will transmit it to their progeny. Also, each mitochondrion contains several copies of mtDNA, resulting in thousands of mtDNA molecules per cell. In normal subjects, essentially all of the mtDNAs are identical (homoplasmy). In contrast, pathogenic mtDNA mutations are usually present in some, but not all, of these genomes; in this situation, cells and tissues, and even individual mitochondria, can harbor both normal and mutant mtDNAs (heteroplasmy). In the case of heteroplasmy, a minimal number of mutant mtDNAs is required to cause mitochondrial dysfunction and clinical signs, a phenomenon known as the threshold effect (Fig. 2). The threshold for disease is lower in tissues that are highly dependent on oxidative metabolism, such as brain, heart, and skeletal muscle, rendering these tissues especially vulnerable to the effects of pathogenic mtDNA mutations. Finally, during cell division mitochondria pass stochastically from mother to daughter cells. Consequently, in the case of heteroplasmy, the proportion of mutant mtDNAs can drift (both upward and downward), so that the phenotype resulting from exceeding a threshold number of mutant DNAs can shift correspondingly, both in space (e.g., among cells or tissues) and time (e.g., during development). This phenomenon is known as mitotic segregation (Fig. 2).

Transfer of mtDNA from platelets of PD patients into cells depleted of their own mtDNA

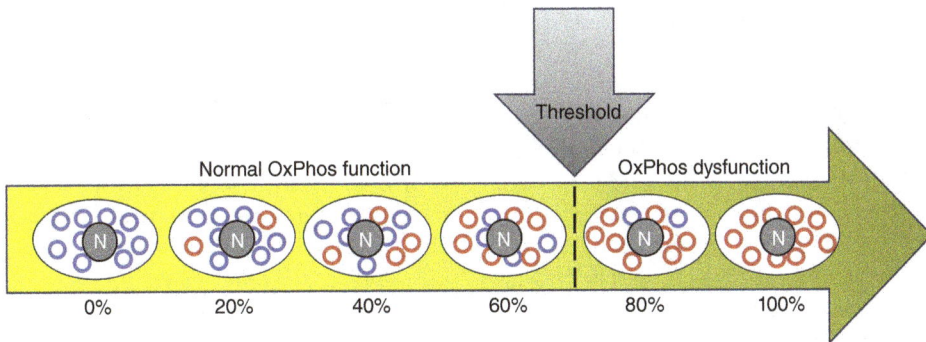

Figure 2. Mitotic segregation and the threshold effect. In the normal situation (e.g., the cell at extreme *left*), all mtDNAs (blue circles) within the cell's mitochondria are identical (homoplasmy) and are phenotypically normal. In a situation known as heteroplasmy, cells harbor a mixture of mitochondria that typically contain two different mtDNA genotypes: wild-type (blue circles) and mutant (red circles). In a pathogenic situation, these mutant mtDNAs have the potential to cause disease, but only if the % mutation (listed below each cell) exceeds a minimal number (the threshold effect). Because mitochondria (and their mtDNAs) are distributed stochastically into the daughter cells following cell division, after multiple divisions the % mutation can drift, a situation known as mitotic segregation. The effects of mitotic segregation and the threshold effect render the diagnosis of mitochondrial disease particularly difficult, as both the mtDNA genotype and the clinical/biochemical phenotype can vary both in space (e.g., among different tissues) and time (e.g., during development). (N) Nucleus. (Figure generated with BioRender; https://biorender.com.)

(i.e., cybrids) recapitulates complex I deficiency and other pathogenic features of PD (Swerdlow et al. 1996; Gu et al. 1998; Trimmer et al. 2004). This observation indicates that PD-derived mtDNA encodes pathogenic information, raising the possibility that mitochondrial alterations in PD may be inherited through the mitochondrial genome or related to somatic mtDNA alterations acquired during aging (Arnheim and Cortopassi 1992). Whereas maternally inherited parkinsonism associated with mtDNA mutations is rare (Thyagarajan et al. 2000), several studies suggest that acquired mtDNA abnormalities may contribute to PD pathogenesis. For example, mutations in the genes encoding mtDNA polymerase-γ, the enzyme responsible for the synthesis (*POLG*) and proofreading (*POLG2*) of mtDNA, are associated with levodopa-responsive parkinsonism, usually as part of a more complex syndrome (Luoma et al. 2004; Davidzon et al. 2006; Van Maldergem et al. 2017). In these families, affected individuals exhibit reduced striatal dopamine uptake on positron emission tomography (PET) analysis and severe loss of pigmented neurons in the substantia nigra at postmortem examination (Luoma et al. 2004), which, as discussed in the articles in this collections on functional imaging and neuropathology in PD, are evidence of damage to the nigrostriatal pathway, as typically seen in this disorder. In all of these patients, *POLG* mutations result in the accumulation of multiple mtDNA deletions in muscle (Luoma et al. 2004; Davidzon et al. 2006; Van Maldergem et al. 2017). Remarkably, mtDNA deletions have been observed in individual dopaminergic neurons microdissected from the substantia nigra of postmortem human brains from aged individuals and from nongenetic PD patients (Bender et al. 2006; Kraytsberg et al. 2006). Of note, different SNpc dopaminergic neurons from the same individual contained unique mtDNA deletions, indicating that these individual deletions were acquired throughout life and then expanded clonally (Bender et al. 2006). Furthermore, high levels of mtDNA deletions in these neurons were associated with decreased histochemical activity of mitochondrial complex IV, suggesting that the accumulation of mtDNA deletions over a certain threshold in SNpc dopaminergic neurons may

cause mitochondrial functional defects (Bender et al. 2006; Kraytsberg et al. 2006). As for etiology, mitochondrial ROS may be responsible for the high level of somatic mtDNA alterations occurring in SNpc dopaminergic neurons. Because of its proximity to the site of ROS production within the mitochondria, mtDNA can be oxidatively damaged during aging or following increased mitochondrial ROS production linked to complex I defects. Consistent with this view, mutant mice overexpressing the antioxidant enzyme catalase specifically in the mitochondria exhibited reduced accumulation of mtDNA mutations during aging (Vermulst et al. 2007). In addition, increased ROS production following MPTP intoxication to mice resulted in oxidative damage to striatal mtDNA (Hoang et al. 2009). Although it remains to be determined whether mtDNA alterations represent a primary or secondary event in PD, mice with a conditional disruption of mitochondrial transcription factor A (TFAM), which regulates mtDNA transcription in dopaminergic neurons, exhibited reduced mtDNA expression, RC defects, and slowly progressive levodopa-responsive motor deficits associated with progressive nigrostriatal denervation (Ekstrand et al. 2007), suggesting that impaired mtDNA expression may contribute to the pathogenesis of PD.

MITOCHONDRIAL DYNAMICS

Mitochondria are not autonomous, rigidly structured organelles, but rather are highly dynamic structures that continually fuse and divide, move on cytoskeletal tracks along the cell, and undergo regulated turnover. All of these dynamic behaviors ensure adequate mitochondrial function at the appropriate time and subcellular location to adapt to changes in cellular requirements. Stressing the importance of these processes, defects in mitochondrial dynamics lead to neurological diseases and may contribute to PD.

Mitochondrial Fusion and Fission

The hundreds of mitochondria within a cell undergo continual cycles of fusion (the joining of two mitochondria into a single organelle) and fission (the separation of long, tubular mitochondria into two or more smaller organelles), resulting in a wide range of mitochondrial morphologies (Fig. 3A; Detmer and Chan 2007; Knott et al. 2008). The proper balance between fusion and fission is crucial for the maintenance of mitochondrial function. For instance, mitochondrial fusion is required for the proper respiratory activity of the mitochondria and has been associated with cell survival (Chen et al. 2007). In addition, the functionality of damaged mitochondria can be restored by exchanging mitochondrial genomes and gene products via fusion with neighboring, intact mitochondria, thereby attenuating the potential deleterious effects of misfolded proteins or mutated mtDNAs (Detmer and Chan 2007; Schon and Przedborski 2011). The proper localization of mitochondria to nerve terminals also depends on the correct balance between mitochondrial fusion and fission, as fragmentation of the mitochondrial network by fission appears to facilitate the recruitment of mitochondria to nerve terminals (Brown et al. 2006).

At the molecular level, mitochondrial fusion and fission are regulated by a series of guanosine triphosphatases (GTPases): mitofusin (MFN)1 and MFN2 for outer membrane fusion, optic atrophy I (OPA1) for inner membrane fusion, and dynamin-related protein I (DRP1; gene *DNM1L*) for mitochondrial fission (Fig. 3A; Chen and Chan 2009). Alterations in this molecular machinery lead to defects in mitochondrial function and can cause cell death and disease. For instance, mutations in MFN2 and OPA1 result in defective mitochondrial fusion and cause inherited neurodegenerative disorders such as Charcot–Marie–Tooth disease type 2A and dominant optic atrophy, respectively (Delettre et al. 2000; Züchner et al. 2004). Also, defective mitochondrial fission linked to mutations in DRP1 results in neurons with elongated mitochondria largely absent from synapses and unable to maintain normal neurotransmission during intense stimulation (Verstreken et al. 2005). On the other hand, excessive DRP1-mediated mitochondrial fission is associated with apoptosis, probably resulting from the enhanced release of mitochondrial proapoptotic molecules, such as cytochrome c, to the cytosol (Frank et al. 2001).

In the context of PD, it has been shown that cultured dopaminergic neurons exposed to neurotoxins 6-hydroxydopamine (6-OHDA), rotenone, and MPP^+ commonly used to generate models of PD are associated with mitochondrial fragmentation (Barsoum et al. 2006; Meuer et al. 2007; Gomez-Lazaro et al. 2008). Supporting a pathogenic role for this observed mitochondrial fragmentation, genetic inhibition of pro-fission DRP1 or overexpression of pro-fusion MFN1 prevents cell death induced by these neurotoxins (Barsoum et al. 2006; Meuer et al. 2007; Gomez-Lazaro et al. 2008). Furthermore, pathogenic mutations in Parkin (an E3 ubiquitin ligase) and PTEN-induced putative kinase-1 (PINK1, a mitochondrially targeted kinase), which cause autosomal-recessive forms of PD (Vila and Przedborski 2004), are associated with an increase in dysfunctional, fragmented mitochondria that can be rescued by pharmacological or genetic inactivation of DRP1 (Lutz et al. 2009; Cui et al. 2010). Also, DJ-1 deficiency in cultured cells results in ROS-dependent mitochondrial fragmentation, which can be prevented by PINK1 and Parkin overexpression (Irrcher et al. 2010). In addition, α-synuclein binding to mitochondrial membranes causes a DRP1-independent mitochondrial fragmentation that can be prevented by PINK1, Parkin, or DJ-1 overexpression (Kamp et al. 2010; Nakamura et al. 2011). However, alterations in mitochondrial fusion/fission balance have not yet been directly shown to occur in PD.

Mitochondrial Motility and Regional Distribution

The basis of the selective vulnerability of certain neurons in PD remains unclear (Brichta and Greengard 2014). However, the dopaminergic neurons in the SNpc, which are particularly susceptible in PD, have long, thin axons with little or no myelination, resulting in high energy demands and a strong reliance on mitochondrial dynamics (Vives-Bauza et al. 2010a). In line with this, SNpc dopaminergic neurons, compared to nondopaminergic neurons, exhibit (1) three times slower

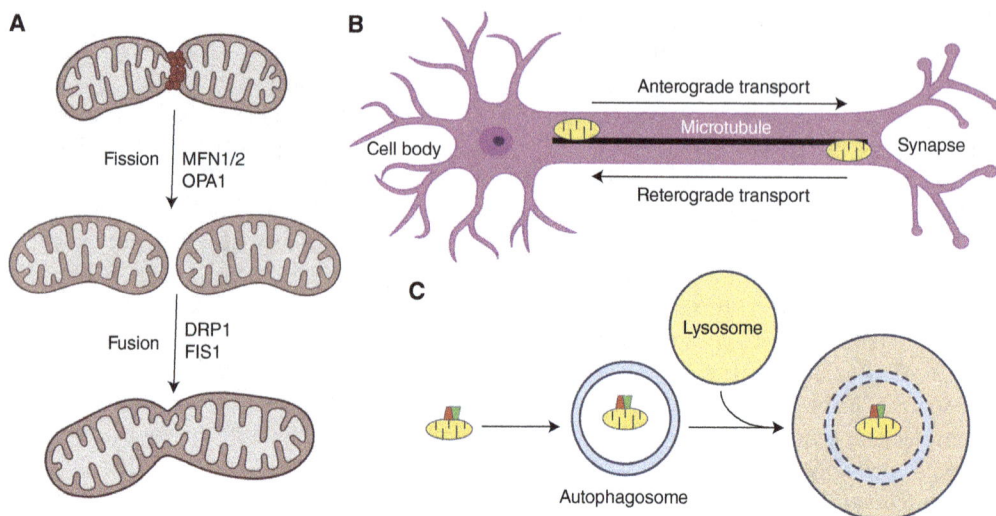

Figure 3. Mitochondrial dynamics. (*A*) Mitochondrial fusion and fission control mitochondrial number and size. Fission is mediated by dynamin-related protein 1 (Drp1) and mitochondrial fission-1 protein (Fis1). Mitofusins (MFN)1 and 2 are involved in the fusion of the outer mitochondrial membrane (OMM), whereas the protein optic atrophy type 1 (OPA1) regulates the fusion of the inner mitochondrial membrane (IMM). (*B*) In neurons, mitochondria are recruited to subcellular compartments distant from the cell body, such as axons and dendrites, by active transport along microtubules and actin filaments. Distinct molecular motors transport the mitochondria in anterograde or retrograde directions. (*C*) Selective autophagic degradation of mitochondria (i.e., mitophagy) involves the recruitment of damaged mitochondria into a pre-autophagosome structure via a PTEN-induced kinase 1 (PINK1) (red)/Parkin (green)-dependent process. Targeted mitochondria are then sequestered into double-membrane-bounded autophagosomes and subsequently delivered to lysosomes for degradation. (Figure generated with BioRender; https://biorender.com.)

mitochondrial axonal transport (Kim-Han et al. 2011), (2) a reduced number of mitochondria in the cell body and in dendrites (Liang et al. 2007), and (3) smaller mitochondrial size (Kim-Han et al. 2011). Whether these factors contribute to their heightened vulnerability in PD is a compelling possibility that warrants formal testing.

Mitochondria move along cytoskeletal tracks, such as microtubules and actin filaments, using distinct molecular motors (Fig. 3B). Short-range mitochondrial transport along actin filaments requires myosin motors, whereas long-range transport on microtubules requires kinesins for anterograde transport and dynein/dynactin for retrograde transport (Hollenbeck 1996; Jung et al. 2004; Schon and Przedborski 2011). Relevant to PD, MPP$^+$ was shown to impair kinesin-mediated anterograde fast axonal transport in isolated squid axoplasm (Morfini et al. 2007), and impaired overall axonal motility of mito-

chondria, but not of other moving particles, in murine mesencephalic cultures (Kim-Han et al. 2011). In addition, proteins linked to familial forms of PD, such as PINK1, Parkin, α-synuclein, and leucine-rich repeat kinase 2 (LRRK2), have been reported to interact with, and potentially impair, microtubule-mediated trafficking (Schon and Przedborski 2011). However, direct evidence of mitochondrial trafficking alterations in PD patients is still lacking, probably because of the technical difficulties in analyzing mitochondrial transport in human-derived material.

To reach nerve terminals, tubular mitochondria are disaggregated by fission into smaller, more motile fragments (Brown et al. 2006). Accordingly, synaptic mitochondria appear mostly punctate, rather than tubular (Brown et al. 2006). However, punctate synaptic mitochondria have limited ability to buffer Ca^{2+} compared to elongated nonsynaptic mitochondria and appear more susceptible to Ca^{2+} overload (Brown et al. 2006). Synaptic mitochondria are also more sensitive to complex I inhibition than their nonsynaptic counterparts: in nonsynaptic mitochondria, complex I can be inhibited by 70% before major changes in mitochondrial respiration and ATP production are detected (Davey and Clark 1996). In contrast, in synaptic mitochondria, this threshold is lowered to 25%, which is within the range of complex I impairment found in PD patients (Davey and Clark 1996). Synaptic mitochondria also exhibit lower levels of cardiolipin than their nonsynaptic counterparts (Kiebish et al. 2008), which may lower the threshold for the proapoptotic release of cytochrome c. These observations may explain, at least in part, the apparent increased susceptibility of striatal dopaminergic terminals, compared to dopaminergic nigral cell bodies, to the degenerative process in PD (Cheng et al. 2010). Based on the aforementioned findings, one might infer that elongated mitochondria are "good" and punctate mitochondria are "bad." In reality, however, the physical shape of mitochondria is determined more by the needs of a particular cell than it is a marker of organelle health. A noteworthy example is in the immunological realm: effector T cells that rely on aerobic glycolysis have fundamentally punctate mitochondria, whereas memory T cells that rely more on fatty acid metabolism have networks of elongated fused mitochondria (Buck et al. 2016).

Mitochondrial Turnover

Selective autophagic degradation of mitochondria, termed mitophagy, is necessary for the steady-state turnover of mitochondria, for the adjustment of mitochondrial numbers to changing metabolic demands, or for the removal of damaged mitochondria (Kim et al. 2007; Youle and Narendra 2011), Using the core autophagic machinery, mitophagy involves the sequestration of targeted mitochondria into double-membrane-bounded structures known as autophagosomes (Fig. 3C). Subsequently, autophagosomes fuse with lysosomes (i.e., cytoplasmic membrane-enclosed organelles that contain a wide variety of hydrolytic enzymes) in which sequestered mitochondria are degraded. Func-

tional mitochondrial alterations, such as loss of $\Delta\psi$ or permeabilization of the OMM, trigger mitophagy, probably in an attempt to limit potential deleterious effects associated with damaged mitochondria, such as excessive ROS production or enhanced release of mitochondrial proapoptotic factors (Tait and Green 2010).

In PD, overall autophagic degradation, including mitophagy, seems to be impaired, Indeed, in experimental PD models and postmortem PD brain samples, abnormal mitochondria readily accumulate in the cytosol of affected neurons (Dehay et al. 2010; Vila et al. 2011), indicating that they cannot be efficiently degraded through mitophagy. Accumulation of dysfunctional mitochondria may contribute to neurodegeneration by, for example, a loss of bioenergetic capacity, an increased production of ROS, and an enhanced release of mitochondrial apoptogenic factors (Vila et al. 2001; Vila and Przedborski 2003; Perier et al. 2005, 2007).

Defective autophagy in PD originates, at least in part (Martinez-Vicente et al. 2008), from a pathogenic reduction in the amount of functional lysosomes (Chu et al. 2009; Dehay et al. 2010). Lysosomal breakdown in PD appears secondary to the abnormal permeabilization of lysosomal membranes by mitochondrially driven oxidative attack (Dehay et al. 2010), leading to a vicious cycle in which increased ROS production from dysfunctional mitochondria contributes to defective autophagy by oxidatively damaging lysosomal membranes, thereby resulting in a further accumulation of altered mitochondria that cannot be degraded through mitophagy. Supporting a pathogenic role for decreased autophagy/mitophagy in PD, pharmacological restoration of lysosomal-mediated degradation by rapamycin is able to reduce the cytosolic accumulation of undegraded autophagosomes (which contain abnormal mitochondria) and to attenuate dopaminergic neurodegeneration in MPTP-intoxicated mice (Dehay et al. 2010; Bové et al. 2011). In addition, besides a general impairment of autophagic degradation, specific defects in mitophagy may also occur in PD. For instance, PD-linked mutations in PINK1 and Parkin have been shown to disrupt the coordinated normal regulatory role of these molecules at promoting autophagic degradation of dysfunctional mitochondria, thereby leading to defective mitophagy (Geisler et al. 2010; Narendra et al. 2010; Vives-Bauza et al. 2010b).

MITOCHONDRIA AND CALCIUM HOMEOSTASIS

Intracellular Ca^{2+} regulates an array of cellular processes and is important for signal transduction. In neurons, Ca^{2+} acts as the main second messenger to transmit depolarization status and synaptic activity to the biochemical machinery of neurons (Gleichmann and Mattson 2011). The concentration of cytosolic-free Ca^{2+} in resting neurons (~100 nM) is 10,000-fold lower than the concentration of Ca^{2+} in the extracellular space (~1.2 mM) (Gleichmann and Mattson 2011). This concentration gradient leads to a significant increase in cytosolic Ca^{2+} after neuron depolarization, rendering Ca^{2+} regulation a critical process in neurons. To maintain Ca^{2+} homeostasis, Ca^{2+} entering neurons is rapidly sequestered in intracellular organelles, such as the mitochondria and the endoplasmic reticulum (ER) or is pumped back across the PM concentration gradient, all of which require high levels of energy in the form of ATP. The ability to accumulate, retain, and release Ca^{2+} is a fundamental property of mitochondria. Accumulation of Ca^{2+} within the mitochondrial matrix depends on both Ca^{2+} uptake into the organelle through an electrogenic uniporter, as well as extrusion of Ca^{2+} from the mitochondria through Na^+/Ca^{2+} and H^+/Ca^{2+} antiporters (Szabadkai et al. 2006; De Stefani et al. 2011). The most significant intracellular storage site for Ca^{2+}, however, is the ER, and there is a significant interplay between mitochondria and ER in relation to Ca^{2+} trafficking. These two organelles are linked, both biochemically and physically (Csordás et al. 2006; de Brito and Scorrano 2008), which facilitates efficient Ca^{2+} transmission from the ER to the mitochondria. The accumulation of Ca^{2+} in the mitochondria leads to the activation of OxPhos and the subsequent increase in ATP production (Gleichmann and Mattson 2011), thereby helping to meet the metabolic demands associated with neuronal electrical activity.

As already mentioned above, in PD, only specific subsets of neurons degenerate (Brichta and Greengard 2014). A particularly notable example of this differential susceptibility is the high vulnerability of dopaminergic neurons in the SNpc compared to the much more resilient ones in the ventral tegmental area (VTA) (Dauer and Przedborski 2003). While the underlying reason for this difference between SNpc and VTA neurons remains uncertain, studies linking Ca^{2+} metabolism and mitochondria may provide a possible explanation. During normal synaptic activity, intracellular Ca^{2+} concentrations increase transiently (on a timescale of seconds to a few minutes) and have no adverse effects on neurons. However, unlike most neurons, ventral midbrain dopaminergic neurons—specifically those in the SNpc and the VTA—are autonomously active, generating action potentials in a clock-like manner even in the absence of synaptic input (Chan et al. 2007). The pacemaking activity of these neurons is driven by voltage-dependent L-type Ca^{2+} channels (Chan et al. 2007). However, it was reported that SNpc dopaminergic neurons have a higher proportion of L-type Ca^{2+} channels containing the Cav1.3 subunit than those in the VTA, resulting in greater Ca^{2+} conductance across the plasma membrane (PM) and elevated cytosolic Ca^{2+} concentrations (Chan et al. 2007). Those authors argued that the substantial Ca^{2+} buffering burden caused by the higher Cav1.3 density in SNpc dopaminergic neurons forces mitochondria to work harder to produce the ATP needed to maintain Ca^{2+} homeostasis which, in turn, generates more ROS, leading to oxidative stress and ultimately neuronal degeneration (Guzman et al. 2010). However, principles of bioenergetics challenge this interpretation, as increased ATP production should normally result in lower, not higher, ROS generation—unless there is a defect in the ETC. Indeed, if the ETC must work harder, the increased flux of electrons should reduce the probability of electron leakage to molecular oxygen, thereby decreasing ROS formation. This is well illustrated by the inverse relationship between mitochondrial coupling and ROS production (Cadenas 2018). Even if the nature of the precise mechanism remains debatable, a pathogenic role for L-type Ca^{2+} channels linked to pacemaking activity has been supported by preclinical studies. Indeed, the L-type Ca^{2+} channel antagonist isradipine has been reported to attenuate rotenone-induced dendritic loss in adult ventral midbrain slices and to reduce SNpc dopaminergic neurodegeneration in MPTP-intoxicated mice (Chan et al. 2007). These preclinical observations suggest that sustained mitochondrial Ca^{2+} overload in adult SNpc dopaminergic neurons may render these cells selectively vulnerable to PD. However, isradipine treatment in clinical trials for early PD patients has generated conflicting results (Parkinson Study Group STEADY-PD III Investigators 2020; Surmeier et al. 2022) attributed, at least in part, to the drug formulation and to the methods of assessment of symptoms. Irrespective of the actual reasons for these inconsistent results, the current idea that targeting L-type Ca^{2+} channels to normalize cytosolic Ca^{2+} could have overt disease-modifying properties remains to be established. In addition to modifying cellular Ca^{2+} entry, it might also be of value to increase the cytosolic Ca^{2+}-buffering capacity. In support of this view, expression of the Ca^{2+}-buffering protein calbindin in selected SNpc dopaminergic populations was negatively correlated with PD-linked cell loss (German et al. 1992; Damier et al. 1999).

Other studies further support a role for alteration in mitochondrial Ca^{2+} homeostasis in PD. For instance, following carbachol-stimulated Ca^{2+} entry, cybrid cells containing mtDNA from PD patients exhibited lower mitochondrial Ca^{2+} sequestration than did control cells (Sheehan et al. 1997). Similarly, the parkinsonian neurotoxins MPP^+ and rotenone caused diminished mitochondrial Ca^{2+} uptake and increased cytosolic-free Ca^{2+} in cultured cells (Frei and Richter 1986; Sousa et al. 2003; Wang and Xu 2005). Also, exogenously applied oligomeric, but not monomeric, α-synuclein to cultured dopaminergic neurons was shown to increase intracellular Ca^{2+} levels through a pore-mediated influx of extracellular Ca^{2+}, leading to increased mitochondrial Ca^{2+}-buffering burden and apoptotic cell death (Danzer et al. 2007). Furthermore, the PD-related protein PINK1 ap-

pears to regulate the physiological release of Ca^{2+} from mitochondria via the mitochondrial Na^+/Ca^{2+} exchanger (Gandhi et al. 2009). Indeed, ablation of PINK1 in dopaminergic neurons leads to impaired Ca^{2+} efflux from mitochondria, accumulation of mitochondrial Ca^{2+}, increased production of mitochondrial ROS, decreased mitochondrial respiration, reduced $\Delta\psi$, and a lowered threshold for Ca^{2+}-dependent opening of the mitochondrial permeability transition pore complex, overall resulting in increased apoptosis (Gandhi et al. 2009). Supporting a role for increased Ca^{2+} load in PD-related cell death in vivo, pharmacological or genetic inhibition of Ca^{2+}-sensitive proteases (i.e., calpains) has been shown to attenuate dopaminergic neurodegeneration in MPTP-intoxicated mice (Crocker et al. 2003).

MITOCHONDRIAL INTERORGANELLE COMMUNICATION

In the last decade, it has become increasingly clear that mitochondria are not self-sufficient organelles present in isolated splendor within the cell. Rather, they communicate, both physically and biochemically, with other cellular organelles at membrane contact sites (MCSs) (Prinz et al. 2020), including the PM (Montes de Oca Balderas 2021), lysosomes (Wong et al. 2018), endolysosomes (Soto-Heredero et al. 2017), peroxisomes (Shai et al. 2018), lipid droplets (Ma et al. 2021), and relevant to our discussion here, the ER (Marchi et al. 2017), where they are denoted as mitochondria-associated ER membranes, or MAMs (Hayashi et al. 2009). While the identification of MAM as a distinct, yet transient, subcellular compartment of the ER was made more than 30 years ago (Rusiñol et al. 1994), it is only relatively recently that we have begun to appreciate its roles in cellular function, in both the normal and pathological situations.

The tethering of ER (i.e., MAM) to the mitochondrial outer membrane is typically accomplished by pairs of proteins located at the MAM–OMM interface. For example, the ER-localized Ca^{2+} channel inositol 1,4,5-trisphosphate receptor (IP3R) interacts with the OMM-localized voltage-dependent anion channel 1 (VDAC1). Another example, with pathological ramifications, is the tethering of ER-localized vesicle-associated membrane protein-associated protein B/C (VAPB) to OMM-localized protein tyrosine phosphatase interacting protein 51 (PTPIP51; gene RMD3) (De Vos et al. 2012). Notably, disruption of this tether is a cause of the common adult-onset paralytic disorder, amyotrophic lateral sclerosis (Stoica et al. 2014; Hartopp et al. 2022), likely as a result of altered OxPhos substrate utilization in patient motor neurons (Larrea et al. 2025).

With respect to PD (Rodríguez-Arribas et al. 2017), it is notable that all of the five known genes associated with familial PD—α-synuclein (gene SNCA) (Calì et al. 2012; Guardia-Laguarta et al. 2014), DJ-1 (gene PARK7) (Ottolini et al. 2013; Liu et al. 2019), LRRK2 (Toyofuku et al. 2020), PINK1 (Gelmetti et al. 2017), and Parkin (gene PARK2) (Gautier et al. 2016)—affect MAM either directly or indirectly (Gómez-Suaga et al. 2018), as they are either localized to MAM or affect MAM structure/function, or both. While the specific mechanisms of action of the five gene products likely differ, the common denominator appears to be that mutations in these proteins all affect MAM behavior.

Finally, one other MAM-related protein deserves special mention: MFN2. Like its homolog MFN1, MFN2 is an OMM protein required for the fusion of mitochondria to form tubular mitochondria (Chen and Chan 2009). However, MFN2 has a second function, namely, to tether mitochondria to ER by virtue of its localization to both the OMM and the ER, either as an MFN2-MFN2 homodimer or as an MFN1-MFN2 heterodimer (Chen and Chan 2009). The "choice" that the cell makes in deciding whether MFN2 connects mitochondria to either other mitochondria or to ER-MAM is dependent upon the particular splice variant(s) generated during MFN2 transcription (Naón et al. 2023). Notably, MFN2 is ubiquitinated by PINK1/Parkin to promote the "untethering" of MAM to mitochondria, thereby steering those organelles toward destruction by mitophagy (McLelland et al. 2018; Barazzuol et al. 2020; Vranas et al. 2022).

Thus, given the intimate connections between mitochondrial function and dynamics described above, it might be useful to view the numerous "hits" found in genome-wide association studies (GWASs) of sporadic PD patients (Nalls et al. 2019; Kim et al. 2024)—especially those hits in protein-coding genes resulting in nonsynonymous amino acid changes—in light of their potential effects on MAM behavior. For example, two sporadic PD risk genes are TMEM175, a lysosomal proton channel (Hu et al. 2022), and glucocerebrosidase (GCAse; gene GBA1), a lysosome-localized enzyme, neither of which has an obvious connection to MAM. It turns out, however, that mutations in TMEM175 affect not only lysosomal function—including reduced GCAse activity (Jinn et al. 2017)—but also cause neuronal deposition of α-synuclein fibrils and impaired mitochondrial function (Jinn et al. 2017), which could conceivably affect, or be affected by, MAM behavior.

MITOCHONDRIA AND INFLAMMATION

The idea that neuroinflammation plays a role in the pathogenesis of PD has a long history (McGeer et al. 1988; Brochard et al. 2009), but the exact mechanism underlying the inflammatory process is still unclear (Hirsch and Standaert 2021). Relevant to the focus of this article, it is important to highlight that mitochondria have been implicated in neuroinflammation in PD, initially through studies emphasizing free radical damage, particularly that caused by neurotoxins (Beal 2003). More recently, it has been found that familial PD-related mutations impair mitochondrial function and appear to provoke an inflammatory response, as evinced by the upregulation of numerous inflammatory markers (Tansey et al. 2022), including cyclooxygenase-2, nuclear factor κB, tumor necrosis factor-α, and various interleukins (Trudler et al. 2015). In addition, activated microglia help drive PD pathology, in part via the promotion of inflammation resulting from increased mitochondrial fission (Sarkar et al. 2017; Lawrence et al. 2022).

Mounting evidence indicates that mitochondria are intimately linked to immune function. Their dynamic behavior influences immu-

Cite this article as *Cold Spring Harb Perspect Med* doi: 10.1101/cshperspect.a041891

nity (Cervantes-Silva et al. 2021), and they act as key platforms for immune signaling (Weinberg et al. 2015). Mitochondria participate in inflammasome assembly (Zhou et al. 2011) and antiviral responses through mitochondrial-associated proteins such as MAVS. Of particular interest is the NLRP3-inflammasome, a key player in inflammation that initiates the inflammatory response, is activated by localization to the MAM (Zhou et al. 2011; Misawa et al. 2017). Specifically, the complex, which is composed of the receptor NOD-like receptor protein, pyrin domain–containing 3 (NLRP3), which is regulated by mitochondrial antiviral signaling protein (MAVS) (Horner et al. 2011; Park et al. 2013) on the ER side and the adaptor ASC (apoptosis-associated speck-like protein containing a caspase recruitment domain [CARD]) on the mitochondrial side, induces caspase-1-dependent maturation of proinflammatory cytokines such as interleukin (IL)-1β and IL-18 (Liu et al. 2018). Notably, several MAM-localized proteins have been associated with inflammation, including glucose-regulated proteins 75 (GRP75; gene *HSPA9*) and 78 (GRP78; gene *HSPA8*) (Voloboueva et al. 2013), further reinforcing the importance of ER–mitochondrial communication in neuroinflammation and PD (Missiroli et al. 2018).

In response to dysfunction, mitochondria activate multiple quality control mechanisms, including the mitochondrial unfolded protein response (Shpilka and Haynes 2018), mitochondrial-derived vesicles (Neuspiel et al. 2008), and mitophagy (Li et al. 2022), depending on the level of cellular stress and damage (Neuspiel et al. 2008). These mitochondrial quality control (MQC) systems play a critical role in modulating inflammation and shaping immune responses. When MQC components are impaired, inflammation is often elevated (Sliter et al. 2018), and mitochondrial antigens can be presented via mitochondrial-derived vesicles (Matheoud et al. 2016).

Immune dysregulation is also a hallmark of PD beyond the CNS. Peripheral immune alterations include elevated levels of proinflammatory cytokines in the serum (Dobbs et al. 1999) and changes in immune cell activation and phenotype (Tansey et al. 2022). The CD4$^+$/CD8$^+$ T-cell ratio is often reduced (Baba et al. 2005), and there is an increase in Th1 and Th17 proinflammatory subsets (Kustrimovic et al. 2018), consistent with heightened cytotoxic and autoimmune responses. Recent studies have identified α-synuclein- and PINK1-specific CD8$^+$ and Th2T cells in PD patients, further suggesting a T-cell-mediated autoimmune component (Sulzer et al. 2017; Lindestam Arlehamn et al. 2020; Williams et al. 2024; Furusawa-Nishii et al. 2025). An increased frequency of Th17 cells—implicated in autoimmunity—has been observed in PD blood samples relative to controls (Sommer et al. 2019; Furusawa-Nishii et al. 2025). Moreover, dopaminergic neurons express higher levels of MHC molecules, making them potential targets for T-cell attack (Cebrián et al. 2014).

Compounding this, altered blood–brain barrier permeability (Lau et al. 2024) facilitates T-cell infiltration into the substantia nigra (Brochard et al. 2009), and inflammatory monocytes capable of promoting neuroinflammation are elevated in PD patients (Grozdanov et al. 2014). Several PD-linked genes, including *LRRK2* (Härtlova et al. 2018), *GBA* (Panicker et al. 2014), *PINK1*, and *Parkin* (Sliter et al. 2018), play important roles in regulating immune responses, particularly in monocytes and macrophages. Immune system changes are also present during the prodromal stages of PD, such as in patients with rapid eye movement (REM) sleep behavior disorder (Mondello et al. 2018), suggesting that immune dysregulation is an early and active contributor to disease development.

Given the intimate connections between mitochondrial function, ER–mitochondrial communication, and immune regulation, further research into the mechanisms linking mitochondrial dysfunction to inflammation in PD is highly warranted.

MITOCHONDRIA AND AGING

Increasing age is the most consistent risk factor for PD, and mitochondria have long been suspected to play an essential role in aging. Indeed, it has been postulated that mitochondrial ROS accumulation in multiple tissues over the years may result in mtDNA alterations, mitochondrial dysfunction, and cell death, leading to the decline in tissue function associated with aging (Wallace 2005). Consistent with this view, genetic ablation of the proapoptotic mitochondrial ROS-producing protein p66Shc extended the life span of mutant mice (Migliaccio et al. 1999; Pinton et al. 2007). In addition, transgenic mice overexpressing the antioxidant enzyme catalase specifically in the mitochondria exhibited reduced accumulation of mtDNA mutations (Vermulst et al. 2007) and increased life span (Schriner et al. 2005). Conversely, mutant mice expressing proofreading-deficient forms of POLG accumulated high levels of mtDNA alterations in all tissues and exhibited decreased mitochondrial respiration, increased apoptosis, accelerated aging, and reduced life span compared to wild-type controls (Trifunovic et al. 2004; Kujoth et al. 2005).

Some of these molecular events may underlie the link between mitochondria, aging, and PD. For instance, as mentioned earlier, mutant mice overexpressing mitochondrial catalase not only exhibited extended life span but were also more resistant to MPTP-induced dopaminergic cell death (Perier et al. 2010). Also, high levels of mtDNA deletions, which are responsible for the premature aging phenotype and shortened life span observed in the POLG mutant mice (Vermulst et al. 2007, 2008), have also been detected in SNpc dopaminergic neurons from postmortem human brains from aged individuals and from PD patients (Bender et al. 2006; Kraytsberg et al. 2006). Furthermore, mutations in POLG are associated with parkinsonism in humans (Luoma et al. 2004; Davidzon et al. 2006).

Other potential links among mitochondria, aging, and PD are provided by members of the sirtuin family of protein deacetylases, which promote longevity in several organisms. Three of the seven mammalian sirtuins (SIRT3, 4, and 5) are targeted to mitochondria, and SIRT1 promotes mitochondrial biogenesis by deacetylating and activating peroxisome proliferator-activated receptor-γ coactivator 1α (PGC-1α), a transcriptional coactivator of nuclear genes encoding mitochondrial proteins (Guarente

2008). Pharmacological activation of SIRT1 protects dopaminergic neurons in midbrain slice cultures against MPP$^+$ intoxication (Okawara et al. 2007), and ablation of PGC-1α in mice increased the sensitivity of these animals to MPTP-induced dopaminergic cell death (St-Pierre et al. 2006). Conversely, overexpression of PGC-1α in cellular disease models resulted in increased expression of nuclear-encoded subunits of the mitochondrial RC and blocked neuronal loss induced by mutant α-synuclein, rotenone, or paraquat (St-Pierre et al. 2006; Zheng et al. 2010). Further linking PGC-1α with PD, a partner of the PD-related protein Parkin, named Parkin-interacting substrate (PARIS), was shown to be a repressor of PGC-1α expression (Shin et al. 2011). Supporting a pathogenic role for PARIS-induced PGC-1α repression, overexpression of PARIS in the substantia nigra of mice led to dopaminergic cell death that was reversed by co-expression of PGC-1α (Shin et al. 2011). Moreover, conditional expression of PARIS in dopaminergic neurons led to selective neurodegeneration, mitochondrial dysfunction, and L-DOPA-responsive motor deficits (Kim et al. 2025). Finally, indicating links among MQC, aging, and PD, treatment with rapamycin, a pharmacological compound able to activate autophagy/mitophagy, has been shown to extend life span in several species, including aged mice (Harrison et al. 2009; Anisimov et al. 2010), and to attenuate dopaminergic neurodegeneration in MPTP-intoxicated mice (Dehay et al. 2010; Bové et al. 2011).

MITOCHONDRIA AND PROGRAMMED CELL DEATH

Lastly, mitochondria may also play a direct role in the mechanisms of neuronal death in pathological conditions. For more than 50 years, PCD, a physiological process in which molecular programs intrinsic to the cell are activated to cause its own destruction has been recognized as a fundamental property of all multicellular organisms and is crucial for their development, organ morphogenesis, tissue homeostasis, and defense against infected or damaged cells (Kerr et al. 1972). However, more recently, the notion that excessive PCD or abnormal reactivation of PCD in adulthood can lead to neurodegeneration has gained tremendous momentum (Vila and Przedborski 2003; Pedrão et al. 2024), especially given the growing recognition of the diversity of PCD forms beyond the most well-known one, apoptosis (Nguyen et al. 2023). Of great relevance to this article is the mounting evidence indicating that mitochondria play a central role in regulating various forms of PCD (Nguyen et al. 2023). However, to date, the majority of investigations on PCD, mitochondria, and PD have focused on apoptosis, which will therefore be the focus of this section.

Regarding apoptosis, it has been known for decades that mitochondria contain several pro-apoptotic factors that, when abnormally released into the cytosol following mitochondrial outer membrane permeabilization (MOMP), activate apoptosis pathways that are either caspase-dependent or caspase-independent (Fig. 4). MOMP represents the point-of-no-return in mitochondria-dependent apoptosis and is highly regulated by a series of proteins of the Bcl-2 family that either prevent (e.g., Bcl-2 and Bcl-xL) or promote (e.g., Bax and Bak) MOMP and subsequent cell death (Vila and Przedborski 2003; Pedrão et al. 2024). Although the exact mechanism by which proapoptotic proteins, such as Bax, induce MOMP is still a matter of debate, it requires the translocation and insertion of these proteins into mitochondrial membranes, whence they can elicit the release of mitochondrial apoptogenic factors, such as cytochrome c, by at least two distinct described mechanisms, one involving the opening of the so-called mitochondrial permeability transition pore complex and the other dependent on the formation of channels directly into mitochondrial membranes (Galluzzi et al. 2009).

In experimental PD models, dopaminergic neurodegeneration appears to occur, at least in part, through activation of mitochondria-dependent apoptosis (Vila and Przedborski 2003; Pedrão et al. 2024). In MPTP-intoxicated mice, there is a time-dependent, region-specific mitochondrial release of cytochrome c followed by activation of caspase-9, caspase-3, and apoptotic nigral cell death (Perier et al. 2005). All of these

MPTP-induced molecular events, including dopaminergic neurodegeneration, are regulated by the proapoptotic protein Bax, as they coincide with the translocation of Bax to mitochondria and are prevented by genetic ablation of Bax (Vila et al. 2001; Perier et al. 2005, 2007). Further supporting the involvement of mitochondria-dependent apoptosis in PD, dopaminergic neurodegeneration caused by MPTP in mice can also be attenuated by targeting other molecules of this pathway, such as caspase-9 or Apaf-1 (Mochizuki et al. 2001; Viswanath et al. 2001) or by overexpressing Bcl-2 (Offen et al. 1998; Yang et al. 1998). Importantly, complex I inhibition by either MPP$^+$, rotenone, or pathogenic complex I mutations does not directly trigger mitochondrial cytochrome c release but instead increases the "releasable" soluble pool of cytochrome c in the mitochondrial IMS that can subsequently be released to the cytosol by activated Bax (Perier et al. 2005). This effect is mediated by peroxidation of the IMM phospholipid cardiolipin, which disrupts the normal binding of cytochrome c to the IMM (Perier et al. 2005). In addition to its detachment from the IMM, cytochrome c release also requires, in other cellular settings, a remodeling of mitochondrial cristae mediated by OPA1 (Scorrano et al. 2002; Frezza et al. 2006).

A role for mitochondria-dependent apoptosis in PD is further reinforced by the finding that many of the mutated nuclear genes associated with familial forms of PD affect mitochondria-dependent apoptosis pathways, either directly or indirectly (Vila and Przedborski 2003; Pedrão et al. 2024). For instance, overexpression of α-synuclein in vivo kills dopaminergic neurons by apoptosis through activation of caspase-9 and caspase-3 (Yamada et al. 2004). Furthermore, aggregated, but not nonaggregated, α-synuclein induces cytochrome c release in isolated rat brain mitochondria (Parihar et al. 2008). In cell lines, PD-linked mutations in LRRK2 lead to mitochondria-dependent apoptosis through the release of cytochrome c (Iaccarino et al. 2007). More recently, the protein gasdermin D has emerged as instrumental in triggering apoptosis and other forms of PCD, not only by forming permeability pores in the PM (Rogers et al. 2019), but more importantly, in the context

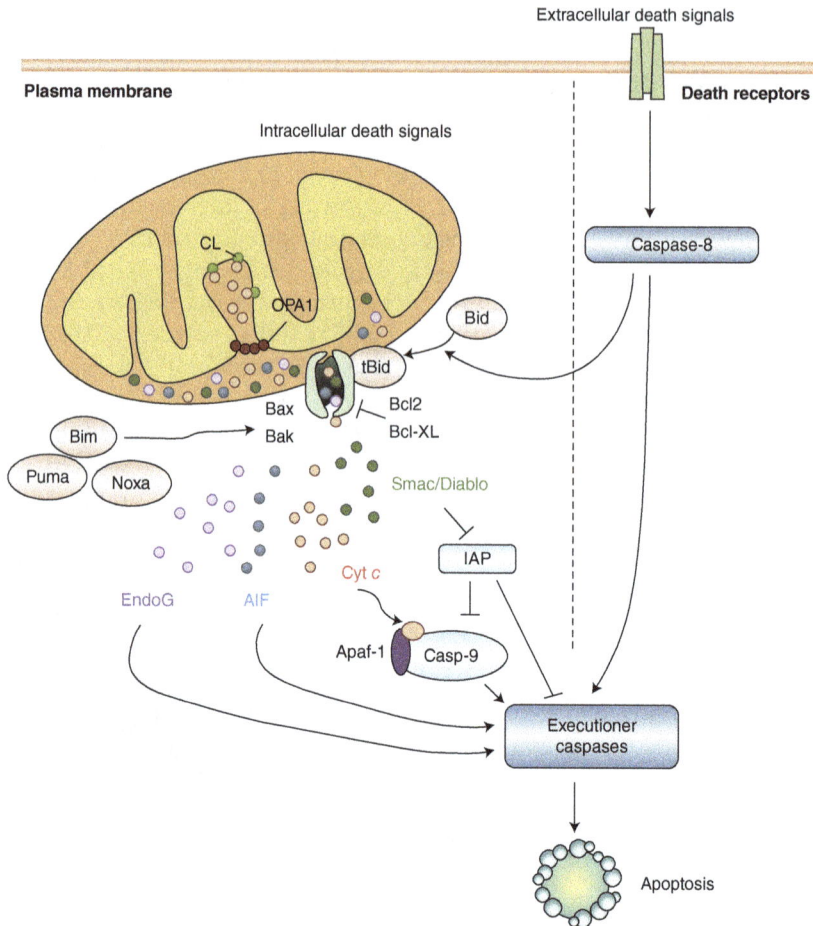

Figure 4. Mitochondrial-dependent apoptosis. Apoptosis can result from the activation of two distinct molecular cascades, known as the extrinsic (or death receptor) and the intrinsic (or mitochondrial) pathways. Both pathways, which can converge at the level of mitochondria, involve the activation of initiator caspases (caspase-8 and -9, respectively) that catalyze the proteolytic maturation of downstream executioner caspases, such as caspase-3, which are the final effectors of cell death. Mitochondrial outer membrane permeabilization (MOMP) represents the point-of-no-return in the mitochondrial apoptotic pathway. Following MOMP, mitochondrial apoptogenic factors, such as cytochrome *c*, Smac/Diablo, endonuclease G, and apoptosis-inducing factor (AIF), are released into the cytosol. Once in the cytosol, these factors can initiate cell death in a caspase-dependent or caspase-independent manner. Released cytochrome *c* interacts with two other cytosolic proteins, Apaf-1 and procaspase-9, to activate caspase-3. Smac/Diablo can interact with several inhibitors of apoptosis (IAPs), thereby relieving the inhibitory effect of IAPs on initiator (e.g., caspase-9) and effector (e.g., caspase-3) caspases. AIF and endonuclease G can translocate to the nucleus and induce caspase-independent DNA fragmentation. MOMP is highly regulated by antiapoptotic (e.g., Bcl-2 and Bcl-xL) and proapoptotic (e.g., Bax and Bak) protein members of the Bcl-2 family. Structurally, all of these proteins share up to four Bcl-2-homology domains (BH1–BH4). In addition to multidomain Bcl-2 family members, there are molecules that share sequence homology only with the BH3 domain (such as Bid, Bim, Puma, and Noxa), which can induce cell death either by activating multidomain proapoptotic proteins or by inactivating antiapoptotic proteins. Bid is activated following its cleavage by caspase-8 (forming truncated Bid [tBid]), thereby linking the extrinsic and intrinsic pathways at the level of the mitochondria. Whereas several components of the mitochondrial apoptotic pathway have been implicated in the pathogenesis of PD, the participation of the extrinsic pathway in PD has not been consistently shown (Perier et al. 2012). (Casp-9) Caspase-9, (CL) cardiolipin, (OPA1) optic atrophy type 1, (Cyt *c*) cytochrome *c*, (EndoG) endonuclease G. (Figure generated with BioRender; https://biorender.com.)

of LRRK2 mutations, by stimulating MOMP (Weindel et al. 2022). Also, overexpression of wild-type PINK1, but not of PD-associated PINK1 mutants, is able to attenuate cytochrome *c* release, caspase activation, and apoptosis induced by parkinsonian neurotoxins or hydrogen peroxide in cultured cells (Petit et al. 2005; Wang et al. 2007). Similarly, Parkin was shown to prevent ceramide-induced cytochrome *c* release, caspase activation, and apoptotic cell death in vitro (Darios et al. 2003), an effect that was abolished by PD-causing Parkin mutations (Darios et al. 2003). Overall, these results suggest that activation of mitochondria-dependent apoptosis pathways contributes to dopaminergic neurodegeneration in PD. We note that research on the potential role of other forms of PCD, such as ferroptosis, necroptosis, and pyroptosis, is beginning to emerge (Pedrão et al. 2024). However, the relationship of these other forms of PCD to mitochondria remains to be established.

CONCLUSIONS

Given the essential role of mitochondria in cell viability, alterations in mitochondria biology can lead to cell dysfunction and cell death. Neurons are particularly vulnerable to mitochondrial impairment because of their dependence for energy on the mitochondrial metabolism of pyruvate produced from glucose by the glycolytic pathway and their need to recruit mitochondria to axons and dendrites. Dopaminergic SNpc neurons are especially susceptible to mitochondrial alterations because of the increased mitochondrial Ca^{2+}-buffering burden created by autonomous pacemaking activity in these cells and their long, poorly myelinated axons. And indeed, defects in mitochondrial respiration have long been implicated in the pathogenesis of PD.

However, the role of mitochondria in PD seems to extend well beyond the sole deficit in energy production, including alterations in mitochondrial dynamics, quality control, and interactions with other subcellular organelles. Although it remains to be shown whether mitochondrial alterations in PD constitute a primary or secondary event, or are just part of a larger multifactorial pathogenic process, the targeting of mitochondrial dysfunction holds promise for the development of therapeutic strategies aimed at altering or slowing down the progression of dopaminergic neurodegeneration in this currently incurable neurodegenerative disorder.

ACKNOWLEDGMENTS

We are grateful to Estela Area-Gomez for her comments on this article. We also thank the NIH (E.S.: NS117538; S.P.: NS117583, MH129381, and AG085144), the U.S. Department of Defense (S.P.: HT9425-1-0958), the Parkinson's Foundation (S.P.), the Chan Zuckerberg Initiative (S.P.), and the Starr Foundation (S.P.) for their support.

REFERENCES

*Reference is also in this subject collection.

Anderson S, Bankier AT, Barrell BG, de Bruijn MH, Coulson AR, Drouin J, Eperon IC, Nierlich DP, Roe BA, Sanger F, et al. 1981. Sequence and organization of the human mitochondrial genome. *Nature* **290:** 457–465. doi:10.1038/290457a0

Andres-Mateos E, Perier C, Zhang L, Blanchard-Fillion B, Greco TM, Thomas B, Ko HS, Sasaki M, Ischiropoulos H, Przedborski S, et al. 2007. DJ-1 gene deletion reveals that DJ-1 is an atypical peroxiredoxin-like peroxidase. *Proc Natl Acad Sci* **104:** 14807–14812. doi:10.1073/pnas.0703219104

Andrews ZB, Diano S, Horvath TL. 2005. Mitochondrial uncoupling proteins in the CNS: in support of function and survival. *Nat Rev Neurosci* **6:** 829–840. doi:10.1038/nrn1767

Anisimov VN, Zabezhinski MA, Popovich IG, Piskunova TS, Semenchenko AV, Tyndyk ML, Yurova MN, Antoch MP, Blagosklonny MV. 2010. Rapamycin extends maximal lifespan in cancer-prone mice. *Am J Pathol* **176:** 2092–2097. doi:10.2353/ajpath.2010.091050

* Area-Gómez E, Fanning S, Dettmer U. 2025. Lipid alterations and pathogenic roles in synucleinopathies. *Cold Spring Harb Perspect Med* doi:10.1101/cshperspect.a041646

Arnheim N, Cortopassi G. 1992. Deleterious mitochondrial DNA mutations accumulate in aging human tissues. *Mutat Res* **275:** 157–167. doi:10.1016/0921-8734(92)90020-P

Baba Y, Kuroiwa A, Uitti RJ, Wszolek ZK, Yamada T. 2005. Alterations of T-lymphocyte populations in Parkinson disease. *Parkinsonism Relat Disord* **11:** 493–498. doi:10.1016/j.parkreldis.2005.07.005

Barazzuol L, Giamogante F, Brini M, Calì T. 2020. PINK1/Parkin mediated mitophagy, Ca^{2+} signalling, and ER-mitochondria contacts in Parkinson's disease. *Int J Mol Sci* **21:** 1772. doi:10.3390/ijms21051772

Barsoum MJ, Yuan H, Gerencser AA, Liot G, Kushnareva Y, Gräber S, Kovacs I, Lee WD, Waggoner J, Cui J, et al. 2006. Nitric oxide-induced mitochondrial fission is regulated by dynamin-related GTPases in neurons. *EMBO J* **25:** 3900–3911. doi:10.1038/sj.emboj.7601253

Beal MF. 2003. Mitochondria, oxidative damage, and inflammation in Parkinson's disease. *Ann NY Acad Sci* **991:** 120–131. doi:10.1111/j.1749-6632.2003.tb07470.x

Bender A, Krishnan KJ, Morris CM, Taylor GA, Reeve AK, Perry RH, Jaros E, Hersheson JS, Betts J, Klopstock T, et al. 2006. High levels of mitochondrial DNA deletions in substantia nigra neurons in aging and Parkinson disease. *Nat Genet* **38:** 515–517. doi:10.1038/ng1769

Betarbet R, Sherer TB, MacKenzie G, Garcia-Osuna M, Panov AV, Greenamyre JT. 2000. Chronic systemic pesticide exposure reproduces features of Parkinson's disease. *Nat Neurosci* **3:** 1301–1306. doi:10.1038/81834

Bonifati V, Rizzu P, van Baren MJ, Schaap O, Breedveld GJ, Krieger E, Dekker MC, Squitieri F, Ibanez P, Joosse M, et al. 2003. Mutations in the *DJ-1* gene associated with autosomal recessive early-onset parkinsonism. *Science* **299:** 256–259. doi:10.1126/science.1077209

Bonvento G, Bolaños JP. 2021. Astrocyte-neuron metabolic cooperation shapes brain activity. *Cell Metab* **33:** 1546–1564. doi:10.1016/j.cmet.2021.07.006

Bové J, Martínez-Vicente M, Vila M. 2011. Fighting neurodegeneration with rapamycin: mechanistic insights. *Nat Rev Neurosci* **12:** 437–452. doi:10.1038/nrn3068

Brichta L, Greengard P. 2014. Molecular determinants of selective dopaminergic vulnerability in Parkinson's disease: an update. *Front Neuroanat* **8:** 152. doi:10.3389/fnana.2014.00152

Brochard V, Combadiére B, Prigent A, Laouar Y, Perrin A, Beray-Berthat V, Bonduelle O, Alvarez-Fischer D, Callebert J, Launay JM, et al. 2009. Infiltration of CD4+ lymphocytes into the brain contributes to neurodegeneration in a mouse model of Parkinson disease. *J Clin Invest* **119:** 182–192. doi:10.1172/JCI36470

Brown MR, Sullivan PG, Geddes JW. 2006. Synaptic mitochondria are more susceptible to Ca^{2+} overload than nonsynaptic mitochondria. *J Biol Chem* **281:** 11658–11668. doi:10.1074/jbc.M510303200

Buck MD, O'Sullivan D, Klein Geltink RI, Curtis JD, Chang CH, Sanin DE, Qiu J, Kretz O, Braas D, van der Windt GJ, et al. 2016. Mitochondrial dynamics controls T cell fate through metabolic programming. *Cell* **166:** 63–76. doi:10.1016/j.cell.2016.05.035

Cadenas S. 2018. Mitochondrial uncoupling, ROS generation and cardioprotection. *Biochim Biophys Acta Bioenerg* **1859:** 940–950. doi:10.1016/j.bbabio.2018.05.019

Calì T, Ottolini D, Negro A, Brini M. 2012. α-Synuclein controls mitochondrial calcium homeostasis by enhancing endoplasmic reticulum–mitochondria interactions. *J Biol Chem* **287:** 17914–17929. doi:10.1074/jbc.M111 .302794

Cebrián C, Zucca FA, Mauri P, Steinbeck JA, Studer L, Scherzer CR, Kanter E, Budhu S, Mandelbaum J, Vonsattel JP, et al. 2014. MHC-I expression renders catecholaminergic neurons susceptible to T-cell-mediated degeneration. *Nat Commun* **5:** 3633. doi:10.1038/ncomms4633

Cervantes-Silva MP, Cox SL, Curtis AM. 2021. Alterations in mitochondrial morphology as a key driver of immunity

and host defence. *EMBO Rep* **22:** e53086. doi:10.15252/embr.202153086

Chan P, DeLanney LE, Irwin I, Langston JW, Di Monte D. 1991. Rapid ATP loss caused by 1-methyl-4-phenyl-1,2,3,6-tetrahydropyridine in mouse brain. *J Neurochem* **57:** 348–351. doi:10.1111/j.1471-4159.1991.tb02134.x

Chan CS, Guzman JN, Ilijic E, Mercer JN, Rick C, Tkatch T, Meredith GE, Surmeier DJ. 2007. "Rejuvenation" protects neurons in mouse models of Parkinson's disease. *Nature* **447:** 1081–1086. doi:10.1038/nature05865

Chen H, Chan DC. 2009. Mitochondrial dynamics—fusion, fission, movement, and mitophagy—in neurodegenerative diseases. *Hum Mol Genet* **18:** R169–R176. doi:10 .1093/hmg/ddp326

Chen H, McCaffery JM, Chan DC. 2007. Mitochondrial fusion protects against neurodegeneration in the cerebellum. *Cell* **130:** 548–562. doi:10.1016/j.cell.2007.06.026

Cheng HC, Ulane CM, Burke RE. 2010. Clinical progression in Parkinson disease and the neurobiology of axons. *Ann Neurol* **67:** 715–725. doi:10.1002/ana.21995

Chu Y, Dodiya H, Aebischer P, Olanow CW, Kordower JH. 2009. Alterations in lysosomal and proteasomal markers in Parkinson's disease: relationship to alpha-synuclein inclusions. *Neurobiol Dis* **35:** 385–398. doi:10.1016/j .nbd.2009.05.023

Claflin KE, Flippo KH, Sullivan AI, Naber MC, Zhou B, Neff TJ, Jensen-Cody SO, Potthoff MJ. 2022. Conditional gene targeting using UCP1-Cre mice directly targets the central nervous system beyond thermogenic adipose tissues. *Mol Metab* **55:** 101405. doi:10.1016/j.molmet.2021.101 405

Coon EA, Low PA. 2018. Thermoregulation in Parkinson disease. *Handb Clin Neurol* **157:** 715–725. doi:10.1016/ B978-0-444-64074-1.00043-4

Crocker SJ, Smith PD, Jackson-Lewis V, Lamba WR, Hayley SP, Grimm E, Callaghan SM, Slack RS, Melloni E, Przedborski S, et al. 2003. Inhibition of calpains prevents neuronal and behavioral deficits in an MPTP mouse model of Parkinson's disease. *J Neurosci* **23:** 4081–4091. doi:10 .1523/JNEUROSCI.23-10-04081.2003

Csordás G, Renken C, Várnai P, Walter L, Weaver D, Buttle KF, Balla T, Mannella CA, Hajnóczky G. 2006. Structural and functional features and significance of the physical linkage between ER and mitochondria. *J Cell Biol* **174:** 915–921. doi:10.1083/jcb.200604016

Cui M, Tang X, Christian WV, Yoon Y, Tieu K. 2010. Perturbations in mitochondrial dynamics induced by human mutant PINK1 can be rescued by the mitochondrial division inhibitor mdivi-1. *J Biol Chem* **285:** 11740–11752. doi:10.1074/jbc.M109.066662

Damier P, Hirsch EC, Agid Y, Graybiel AM. 1999. The substantia nigra of the human brain. II: Patterns of loss of dopamine-containing neurons in Parkinson's disease. *Brain* **122:** 1437–1448. doi:10.1093/brain/122.8.1437

Danzer KM, Haasen D, Karow AR, Moussaud S, Habeck M, Giese A, Kretzschmar H, Hengerer B, Kostka M. 2007. Different species of α-synuclein oligomers induce calcium influx and seeding. *J Neurosci* **27:** 9220–9232. doi:10 .1523/JNEUROSCI.2617-07.2007

Darios F, Corti O, Lucking CB, Hampe C, Muriel MP, Abbas N, Gu WJ, Hirsch EC, Rooney T, Ruberg M, et al. 2003. Parkin prevents mitochondrial swelling and cytochrome *c* release in mitochondria-dependent cell death. *Hum Mol Genet* **12**: 517–526. doi:10.1093/hmg/ddg044

Dauer W, Przedborski S. 2003. Parkinson's disease: mechanisms and models. *Neuron* **39**: 889–909. doi:10.1016/S0896-6273(03)00568-3

Davey GP, Clark JB. 1996. Threshold effects and control of oxidative phosphorylation in nonsynaptic rat brain mitochondria. *J Neurochem* **66**: 1617–1624. doi:10.1046/j.1471-4159.1996.66041617.x

Davidzon G, Greene P, Mancuso M, Klos KJ, Ahlskog JE, Hirano M, DiMauro S. 2006. Early-onset familial parkinsonism due to *POLG* mutations. *Ann Neurol* **59**: 859–862. doi:10.1002/ana.20831

de Brito OM, Scorrano L. 2008. Mitofusin 2 tethers endoplasmic reticulum to mitochondria. *Nature* **456**: 605–610. doi:10.1038/nature07534

Dehay B, Bové J, Rodríguez-Muela N, Perier C, Recasens A, Boya P, Vila M. 2010. Pathogenic lysosomal depletion in Parkinson's disease. *J Neurosci* **30**: 12535–12544. doi:10.1523/JNEUROSCI.1920-10.2010

Delettre C, Lenaers G, Griffoin JM, Gigarel N, Lorenzo C, Belenguer P, Pelloquin L, Grosgeorge J, Turc-Carel C, Perret E, et al. 2000. Nuclear gene *OPA1*, encoding a mitochondrial dynamin-related protein, is mutated in dominant optic atrophy. *Nat Genet* **26**: 207–210. doi:10.1038/79936

Demine S, Renard P, Arnould T. 2019. Mitochondrial uncoupling: a key controller of biological processes in physiology and diseases. *Cells* **8**: 795. doi:10.3390/cells8080795

De Stefani D, Raffaello A, Teardo E, Szabò I, Rizzuto R. 2011. A forty-kilodalton protein of the inner membrane is the mitochondrial calcium uniporter. *Nature* **476**: 336–340. doi:10.1038/nature10230

Detmer SA, Chan DC. 2007. Functions and dysfunctions of mitochondrial dynamics. *Nat Rev Mol Cell Biol* **8**: 870–879. doi:10.1038/nrm2275

De Vos KJ, Mórotz GM, Stoica R, Tudor EL, Lau KF, Ackerley S, Warley A, Shaw CE, Miller CC. 2012. VAPB interacts with the mitochondrial protein PTPIP51 to regulate calcium homeostasis. *Hum Mol Genet* **21**: 1299–1311. doi:10.1093/hmg/ddr559

DiMauro S, Schon EA. 2003. Mitochondrial respiratory-chain diseases. *N Engl J Med* **348**: 2656–2668. doi:10.1056/NEJMra022567

Di Monte D, Jewell SA, Ekström G, Sandy MS, Smith MT. 1986. 1-Methyl-4-phenyl-1,2,3,6-tetrahydropyridine (MPTP) and 1-methyl-4-phenylpyridine (MPP$^+$) cause rapid ATP depletion in isolated hepatocytes. *Biochem Biophys Res Commun* **137**: 310–315. doi:10.1016/0006-291X(86)91211-8

Dobbs RJ, Charlett A, Purkiss AG, Dobbs SM, Weller C, Peterson DW. 1999. Association of circulating TNF-α and IL-6 with ageing and parkinsonism. *Acta Neurol Scand* **100**: 34–41. doi:10.1111/j.1600-0404.1999.tb00721.x

Dolgacheva LP, Berezhnov AV, Fedotova EI, Zinchenko VP, Abramov AY. 2019. Role of DJ-1 in the mechanism of pathogenesis of Parkinson's disease. *J Bioenerg Biomembr* **51**: 175–188. doi:10.1007/s10863-019-09798-4

Dröse S, Stepanova A, Galkin A. 2016. Ischemic A/D transition of mitochondrial complex I and its role in ROS generation. *Biochim Biophys Acta* **1857**: 946–957. doi:10.1016/j.bbabio.2015.12.013

Ekstrand MI, Terzioglu M, Galter D, Zhu S, Hofstetter C, Lindqvist E, Thams S, Bergstrand A, Hansson FS, Trifunovic A, et al. 2007. Progressive parkinsonism in mice with respiratory-chain-deficient dopamine neurons. *Proc Natl Acad Sci* **104**: 1325–1330. doi:10.1073/pnas.0605208103

Frank S, Gaume B, Bergmann-Leitner ES, Leitner WW, Robert EG, Catez F, Smith CL, Youle RJ. 2001. The role of dynamin-related protein 1, a mediator of mitochondrial fission, in apoptosis. *Dev Cell* **1**: 515–525. doi:10.1016/S1534-5807(01)00055-7

Frei B, Richter C. 1986. *N*-methyl-4-phenylpyridine (MMP$^+$) together with 6-hydroxydopamine or dopamine stimulates Ca^{2+} release from mitochondria. *FEBS Lett* **198**: 99–102. doi:10.1016/0014-5793(86)81192-9

Frezza C, Cipolat S, Martins de Brito O, Micaroni M, Beznoussenko GV, Rudka T, Bartoli D, Polishuck RS, Danial NN, De Strooper B, et al. 2006. OPA1 controls apoptotic cristae remodeling independently from mitochondrial fusion. *Cell* **126**: 177–189. doi:10.1016/j.cell.2006.06.025

Furusawa-Nishii E, Solongo B, Rai K, Yoshikawa S, Chiba A, Okuzumi A, Ueno SI, Hoshino Y, Imamichi-Tatano Y, Kimura H, et al. 2025. α-Synuclein orchestrates Th17 responses as antigen and adjuvant in Parkinson's disease. *J Neuroinflammation* **22**: 38. doi:10.1186/s12974-025-03359-w

Galluzzi L, Blomgren K, Kroemer G. 2009. Mitochondrial membrane permeabilization in neuronal injury. *Nat Rev Neurosci* **10**: 481–494. doi:10.1038/nrn2665

Gandhi S, Wood-Kaczmar A, Yao Z, Plun-Favreau H, Deas E, Klupsch K, Downward J, Latchman DS, Tabrizi SJ, Wood NW, et al. 2009. PINK1-associated Parkinson's disease is caused by neuronal vulnerability to calcium-induced cell death. *Mol Cell* **33**: 627–638. doi:10.1016/j.molcel.2009.02.013

Gautier CA, Erpapazoglou Z, Mouton-Liger F, Muriel MP, Cormier F, Bigou S, Duffaure S, Girard M, Foret B, Iannielli A, et al. 2016. The endoplasmic reticulum-mitochondria interface is perturbed in PARK2 knockout mice and patients with PARK2 mutations. *Hum Mol Genet* **25**: 2972–2984. doi:10.1093/hmg/ddw148

Geisler S, Holmström KM, Skujat D, Fiesel FC, Rothfuss OC, Kahle PJ, Springer W. 2010. PINK1/Parkin-mediated mitophagy is dependent on VDAC1 and p62/SQSTM1. *Nat Cell Biol* **12**: 119–131. doi:10.1038/ncb2012

Gelmetti V, De Rosa P, Torosantucci L, Marini ES, Romagnoli A, Di Rienzo M, Arena G, Vignone D, Fimia GM, Valente EM. 2017. PINK1 and BECN1 relocalize at mitochondria-associated membranes during mitophagy and promote ER-mitochondria tethering and autophagosome formation. *Autophagy* **13**: 654–669. doi:10.1080/15548627.2016.1277309

German DC, Manaye KF, Sonsalla PK, Brooks BA. 1992. Midbrain dopaminergic cell loss in Parkinson's disease and MPTP-induced parkinsonism: sparing of calbindin-D28k-containing cells. *Ann NY Acad Sci* **648**: 42–62. doi:10.1111/j.1749-6632.1992.tb24523.x

Girard V, Jollivet F, Knittelfelder O, Celle M, Arsac JN, Chatelain G, Van den Brink DM, Baron T, Shevchenko A, Kühnlein RP, et al. 2021. Abnormal accumulation of lipid droplets in neurons induces the conversion of alpha-synuclein to proteolytic resistant forms in a *Drosophila* model of Parkinson's disease. *PLoS Genet* **17**: e1009921. doi:10.1371/journal.pgen.1009921

Gleichmann M, Mattson MP. 2011. Neuronal calcium homeostasis and dysregulation. *Antioxid Redox Signal* **14**: 1261–1273. doi:10.1089/ars.2010.3386

Gomez-Lazaro M, Bonekamp NA, Galindo MF, Jordán J, Schrader M. 2008. 6-Hydroxydopamine (6-OHDA) induces Drp1-dependent mitochondrial fragmentation in SH-SY5Y cells. *Free Radic Biol Med* **44**: 1960–1969. doi:10.1016/j.freeradbiomed.2008.03.009

Gómez-Suaga P, Bravo-San Pedro JM, González-Polo RA, Fuentes JM, Niso-Santano M. 2018. ER-mitochondria signaling in Parkinson's disease. *Cell Death Dis* **9**: 337. doi:10.1038/s41419-017-0079-3

Gray MW. 2012. Mitochondrial evolution. *Cold Spring Harb Perspect Biol* **4**: a011403. doi:10.1101/cshperspect.a011403

Grozdanov V, Bliederhaeuser C, Ruf WP, Roth V, Fundel-Clemens K, Zondler L, Brenner D, Martin-Villalba A, Hengerer B, Kassubek J, et al. 2014. Inflammatory dysregulation of blood monocytes in Parkinson's disease patients. *Acta Neuropathol* **128**: 651–663. doi:10.1007/s00401-014-1345-4

Gu M, Cooper JM, Taanman JW, Schapira AH. 1998. Mitochondrial DNA transmission of the mitochondrial defect in Parkinson's disease. *Ann Neurol* **44**: 177–186. doi:10.1002/ana.410440207

Guardia-Laguarta C, Area-Gomez E, Rüb C, Liu Y, Magrané J, Becker D, Voos W, Schon EA, Przedborski S. 2014. α-Synuclein is localized to mitochondria-associated ER membranes. *J Neurosci* **34**: 249–259. doi:10.1523/JNEUROSCI.2507-13.2014

Guarente L. 2008. Mitochondria—a nexus for aging, calorie restriction, and sirtuins? *Cell* **132**: 171–176. doi:10.1016/j.cell.2008.01.007

Guzman JN, Sanchez-Padilla J, Wokosin D, Kondapalli J, Ilijic E, Schumacker PT, Surmeier DJ. 2010. Oxidant stress evoked by pacemaking in dopaminergic neurons is attenuated by DJ-1. *Nature* **468**: 696–700. doi:10.1038/nature09536

Harrison DE, Strong R, Sharp ZD, Nelson JF, Astle CM, Flurkey K, Nadon NL, Wilkinson JE, Frenkel K, Carter CS, et al. 2009. Rapamycin fed late in life extends lifespan in genetically heterogeneous mice. *Nature* **460**: 392–395. doi:10.1038/nature08221

Härtlova A, Herbst S, Peltier J, Rodgers A, Bilkei-Gorzo O, Fearns A, Dill BD, Lee H, Flynn R, Cowley SA, et al. 2018. LRRK2 is a negative regulator of *Mycobacterium tuberculosis* phagosome maturation in macrophages. *EMBO J* **37**: e98694. doi:10.15252/embj.201798694

Hartopp N, Lau DHW, Martin-Guerrero SM, Markovinovic A, Mórotz GM, Greig J, Glennon EB, Troakes C, Gomez-Suaga P, Noble W, et al. 2022. Disruption of the VAPB-PTPIP51 ER-mitochondria tethering proteins in postmortem human amyotrophic lateral sclerosis. *Front Cell Dev Biol* **10**: 950767. doi:10.3389/fcell.2022.950767

Hayashi T, Rizzuto R, Hajnoczky G, Su TP. 2009. MAM: more than just a housekeeper. *Trends Cell Biol* **19**: 81–88. doi:10.1016/j.tcb.2008.12.002

Hirsch EC, Standaert DG. 2021. Ten unsolved questions about neuroinflammation in Parkinson's disease. *Mov Disord* **36**: 16–24. doi:10.1002/mds.28075

Hoang T, Choi DK, Nagai M, Wu DC, Nagata T, Prou D, Wilson GL, Vila M, Jackson-Lewis V, Dawson VL, et al. 2009. Neuronal NOS and cyclooxygenase-2 contribute to DNA damage in a mouse model of Parkinson disease. *Free Radic Biol Med* **47**: 1049–1056. doi:10.1016/j.freeradbiomed.2009.07.013

Hollenbeck PJ. 1996. The pattern and mechanism of mitochondrial transport in axons. *Front Biosci* **1**: d91–d102. doi:10.2741/A118

Horner SM, Liu HM, Park HS, Briley J, Gale M Jr. 2011. Mitochondrial-associated endoplasmic reticulum membranes (MAM) form innate immune synapses and are targeted by hepatitis C virus. *Proc Natl Acad Sci* **108**: 14590–14595. doi:10.1073/pnas.1110133108

Hu M, Li P, Wang C, Feng X, Geng Q, Chen W, Marthi M, Zhang W, Gao C, Reid W, et al. 2022. Parkinson's disease-risk protein TMEM175 is a proton-activated proton channel in lysosomes. *Cell* **185**: 2292–2308.e20. doi:10.1016/j.cell.2022.05.021

Iaccarino C, Crosio C, Vitale C, Sanna G, Carrì MT, Barone P. 2007. Apoptotic mechanisms in mutant LRRK2-mediated cell death. *Hum Mol Genet* **16**: 1319–1326. doi:10.1093/hmg/ddm080

Ikeda K, Yamada T. 2020. UCP1 dependent and independent thermogenesis in brown and beige adipocytes. *Front Endocrinol (Lausanne)* **11**: 498. doi:10.3389/fendo.2020.00498

Irrcher I, Aleyasin H, Seifert EL, Hewitt SJ, Chhabra S, Phillips M, Lutz AK, Rousseaux MW, Bevilacqua L, Jahani-Asl A, et al. 2010. Loss of the Parkinson's disease-linked gene DJ-1 perturbs mitochondrial dynamics. *Hum Mol Genet* **19**: 3734–3746. doi:10.1093/hmg/ddq288

Jarc E, Petan T. 2019. Lipid droplets and the management of cellular stress. *Yale J Biol Med* **92**: 435–452.

Jinn S, Drolet RE, Cramer PE, Wong AH, Toolan DM, Gretzula CA, Voleti B, Vassileva G, Disa J, Tadin-Strapps M, et al. 2017. TMEM175 deficiency impairs lysosomal and mitochondrial function and increases α-synuclein aggregation. *Proc Natl Acad Sci* **114**: 2389–2394. doi:10.1073/pnas.1616332114

Jung C, Chylinski TM, Pimenta A, Ortiz D, Shea TB. 2004. Neurofilament transport is dependent on actin and myosin. *J Neurosci* **24**: 9486–9496. doi:10.1523/JNEUROSCI.1665-04.2004

Kamp F, Exner N, Lutz AK, Wender N, Hegermann J, Brunner B, Nuscher B, Bartels T, Giese A, Beyer K, et al. 2010. Inhibition of mitochondrial fusion by α-synuclein is rescued by PINK1, Parkin and DJ-1. *EMBO J* **29**: 3571–3589. doi:10.1038/emboj.2010.223

Keeney PM, Xie J, Capaldi RA, Bennett JP Jr. 2006. Parkinson's disease brain mitochondrial complex I has oxidatively damaged subunits and is functionally impaired and misassembled. *J Neurosci* **26**: 5256–5264. doi:10.1523/JNEUROSCI.0984-06.2006

Kerr JF, Wyllie AH, Currie AR. 1972. Apoptosis: a basic biological phenomenon with wide-ranging implications in tissue kinetics. *Br J Cancer* **26**: 239–257. doi:10.1038/bjc.1972.33

Cite this article as *Cold Spring Harb Perspect Med* doi: 10.1101/cshperspect.a041891

Kiebish MA, Han X, Cheng H, Lunceford A, Clarke CF, Moon H, Chuang JH, Seyfried TN. 2008. Lipidomic analysis and electron transport chain activities in C57BL/6J mouse brain mitochondria. *J Neurochem* **106:** 299–312. doi:10.1111/j.1471-4159.2008.05383.x

Kim I, Rodriguez-Enriquez S, Lemasters JJ. 2007. Selective degradation of mitochondria by mitophagy. *Arch Biochem Biophys* **462:** 245–253. doi:10.1016/j.abb.2007.03.034

Kim JJ, Vitale D, Otani DV, Lian MM, Heilbron K; 23andMe Research Team; Iwaki H, Lake J, Solsberg CW, Leonard H, et al. 2024. Multi-ancestry genome-wide association meta-analysis of Parkinson's disease. *Nat Genet* **56:** 27–36. doi:10.1038/s41588-023-01584-8

Kim JH, Yang S, Kim H, Vo DK, Maeng HJ, Jo A, Shin JH, Shin JH, Baek HM, Lee GH, et al. 2025. Preclinical studies and transcriptome analysis in a model of Parkinson's disease with dopaminergic ZNF746 expression. *Mol Neurodegener* **20:** 24. doi:10.1186/s13024-025-00814-3

Kim-Han JS, Antenor-Dorsey JA, O'Malley KL. 2011. The parkinsonian mimetic, MPP$^+$, specifically impairs mitochondrial transport in dopamine axons. *J Neurosci* **31:** 7212–7221. doi:10.1523/JNEUROSCI.0711-11.2011

Knott AB, Perkins G, Schwarzenbacher R, Bossy-Wetzel E. 2008. Mitochondrial fragmentation in neurodegeneration. *Nat Rev Neurosci* **9:** 505–518. doi:10.1038/nrn2417

Koshenov Z, Oflaz FE, Hirtl M, Bachkoenig OA, Rost R, Osibow K, Gottschalk B, Madreiter-Sokolowski CT, Waldeck-Weiermair M, Malli R, et al. 2020. The contribution of uncoupling protein 2 to mitochondrial Ca^{2+} homeostasis in health and disease—a short revisit. *Mitochondrion* **55:** 164–173. doi:10.1016/j.mito.2020.10.003

Kraytsberg Y, Kudryavtseva E, McKee AC, Geula C, Kowall NW, Khrapko K. 2006. Mitochondrial DNA deletions are abundant and cause functional impairment in aged human substantia nigra neurons. *Nat Genet* **38:** 518–520. doi:10.1038/ng1778

Kujoth GC, Hiona A, Pugh TD, Someya S, Panzer K, Wohlgemuth SE, Hofer T, Seo AY, Sullivan R, Jobling WA, et al. 2005. Mitochondrial DNA mutations, oxidative stress, and apoptosis in mammalian aging. *Science* **309:** 481–484. doi:10.1126/science.1112125

Kumar R, Amruthanjali T, Singothu S, Singh SB, Bhandari V. 2022. Uncoupling proteins as a therapeutic target for the development of new era drugs against neurodegenerative disorder. *Biomed Pharmacother* **147:** 112656. doi:10.1016/j.biopha.2022.112656

Kustrimovic N, Comi C, Magistrelli L, Rasini E, Legnaro M, Bombelli R, Aleksic I, Blandini F, Minafra B, Riboldazzi G, et al. 2018. Parkinson's disease patients have a complex phenotypic and functional Th1 bias: cross-sectional studies of CD4$^+$ Th1/Th2/T17 and Treg in drug-naive and drug-treated patients. *J Neuroinflammation* **15:** 205. doi:10.1186/s12974-018-1248-8

Langston JW, Ballard P, Tetrud JW, Irwin I. 1983. Chronic Parkinsonism in humans due to a product of meperidine-analog synthesis. *Science* **219:** 979–980. doi:10.1126/science.6823561

Larrea D, Tamucci KA, Kabra K, Velasco KR, Yun TD, Pera M, Montesinos J, Agrawal RR, Paradas C, Smerdon JW, et al. 2025. Altered mitochondria-associated ER membrane (MAM) function shifts mitochondrial metabolism in amyotrophic lateral sclerosis (ALS). *Nat Commun* **16:** 379. doi:10.1038/s41467-024-51578-1

Lau K, Kotzur R, Richter F. 2024. Blood-brain barrier alterations and their impact on Parkinson's disease pathogenesis and therapy. *Transl Neurodegener* **13:** 37. doi:10.1186/s40035-024-00430-z

Laursen WJ, Mastrotto M, Pesta D, Funk OH, Goodman JB, Merriman DK, Ingolia N, Shulman GI, Bagriantsev SN, Gracheva EO. 2015. Neuronal UCP1 expression suggests a mechanism for local thermogenesis during hibernation. *Proc Natl Acad Sci* **112:** 1607–1612. doi:10.1073/pnas.1421419112

Lawrence G, Holley CL, Schroder K. 2022. Parkinson's disease: connecting mitochondria to inflammasomes. *Trends Immunol* **43:** 877–885. doi:10.1016/j.it.2022.09.010

Li G, Yin W, Yang Y, Yang H, Chen Y, Liang Y, Zhang W, Xie T. 2022. Bibliometric insights of global research landscape in mitophagy. *Front Mol Biosci* **9:** 851966. doi:10.3389/fmolb.2022.851966

Liang LP, Patel M. 2004. Iron-sulfur enzyme mediated mitochondrial superoxide toxicity in experimental Parkinson's disease. *J Neurochem* **90:** 1076–1084. doi:10.1111/j.1471-4159.2004.02567.x

Liang CL, Wang TT, Luby-Phelps K, German DC. 2007. Mitochondria mass is low in mouse substantia nigra dopamine neurons: implications for Parkinson's disease. *Exp Neurol* **203:** 370–380. doi:10.1016/j.expneurol.2006.08.015

Lindestam Arlehamn CS, Dhanwani R, Pham J, Kuan R, Frazier A, Rezende Dutra J, Phillips E, Mallal S, Roederer M, Marder KS, et al. 2020. α-Synuclein-specific T cell reactivity is associated with preclinical and early Parkinson's disease. *Nat Commun* **11:** 1875. doi:10.1038/s41467-020-15626-w

Liu Q, Zhang D, Hu D, Zhou X, Zhou Y. 2018. The role of mitochondria in NLRP3 inflammasome activation. *Mol Immunol* **103:** 115–124. doi:10.1016/j.molimm.2018.09.010

Liu Y, Ma X, Fujioka H, Liu J, Chen S, Zhu X. 2019. DJ-1 regulates the integrity and function of ER-mitochondria association through interaction with IP3R3-Grp75-VDAC1. *Proc Natl Acad Sci* **116:** 25322–25328. doi:10.1073/pnas.1906565116

Luoma P, Melberg A, Rinne JO, Kaukonen JA, Nupponen NN, Chalmers RM, Oldfors A, Rautakorpi I, Peltonen L, Majamaa K, et al. 2004. Parkinsonism, premature menopause, and mitochondrial DNA polymerase gamma mutations: clinical and molecular genetic study. *Lancet* **364:** 875–882. doi:10.1016/S0140-6736(04)16983-3

Lutz AK, Exner N, Fett ME, Schlehe JS, Kloos K, Lämmermann K, Brunner B, Kurz-Drexler A, Vogel F, Reichert AS, et al. 2009. Loss of parkin or PINK1 function increases Drp1-dependent mitochondrial fragmentation. *J Biol Chem* **284:** 22938–22951. doi:10.1074/jbc.M109.035774

Ma X, Qian H, Chen A, Ni HM, Ding WX. 2021. Perspectives on mitochondria-ER and mitochondria-lipid droplet contact in hepatocytes and hepatic lipid metabolism. *Cells* **10:** 2273. doi:10.3390/cells10092273

Marchi S, Bittremieux M, Missiroli S, Morganti C, Patergnani S, Sbano L, Rimessi A, Kerkhofs M, Parys JB, Bultynck G, et al. 2017. Endoplasmic reticulum-mitochondria communication through Ca^{2+} signaling: the importance of mitochondria-associated membranes (MAMs). *Adv Exp Med Biol* 997: 49–67. doi:10.1007/978-981-10-4567-7_4

Marella M, Seo BB, Nakamaru-Ogiso E, Greenamyre JT, Matsuno-Yagi A, Yagi T. 2008. Protection by the NDI1 gene against neurodegeneration in a rotenone rat model of Parkinson's disease. *PLoS One* 3: e1433. doi:10.1371/journal.pone.0001433

Martin W, Müller M. 1998. The hydrogen hypothesis for the first eukaryote. *Nature* 392: 37–41. doi:10.1038/32096

Martinez-Vicente M, Talloczy Z, Kaushik S, Massey AC, Mazzulli J, Mosharov EV, Hodara R, Fredenburg R, Wu DC, Follenzi A, et al. 2008. Dopamine-modified α-synuclein blocks chaperone-mediated autophagy. *J Clin Invest* 118: 777–788. doi:10.1172/JCI32806

Matheoud D, Sugiura A, Bellemare-Pelletier A, Laplante A, Rondeau C, Chemali M, Fazel A, Bergeron JJ, Trudeau LE, Burelle Y, et al. 2016. Parkinson's disease-related proteins PINK1 and Parkin repress mitochondrial antigen presentation. *Cell* 166: 314–327. doi:10.1016/j.cell.2016.05.039

McGeer PL, Itagaki S, Boyes BE, McGeer EG. 1988. Reactive microglia are positive for HLA-DR in the substantia nigra of Parkinson's and Alzheimer's disease brains. *Neurology* 38: 1285–1291. doi:10.1212/WNL.38.8.1285

McLelland GL, Goiran T, Yi W, Dorval G, Chen CX, Lauinger ND, Krahn AI, Valimehr S, Rakovic A, Rouiller I, et al. 2018. Mfn2 ubiquitination by PINK1/parkin gates the p97-dependent release of ER from mitochondria to drive mitophagy. *eLife* 7: 32866. doi:10.7554/eLife.32866

Meuer K, Suppanz IE, Lingor P, Planchamp V, Göricke B, Fichtner L, Braus GH, Dietz GP, Jakobs S, Bähr M, et al. 2007. Cyclin-dependent kinase 5 is an upstream regulator of mitochondrial fission during neuronal apoptosis. *Cell Death Differ* 14: 651–661. doi:10.1038/sj.cdd.4402087

Migliaccio E, Giorgio M, Mele S, Pelicci G, Reboldi P, Pandolfi PP, Lanfrancone L, Pelicci PG. 1999. The p66shc adaptor protein controls oxidative stress response and life span in mammals. *Nature* 402: 309–313. doi:10.1038/46311

Misawa T, Takahama M, Saitoh T. 2017. Mitochondria-endoplasmic reticulum contact sites mediate innate immune responses. *Adv Exp Med Biol* 997: 187–197. doi:10.1007/978-981-10-4567-7_14

Missiroli S, Patergnani S, Caroccia N, Pedriali G, Perrone M, Previati M, Wieckowski MR, Giorgi C. 2018. Mitochondria-associated membranes (MAMs) and inflammation. *Cell Death Dis* 9: 329. doi:10.1038/s41419-017-0027-2

Mochizuki H, Hayakawa H, Migita M, Shibata M, Tanaka R, Suzuki A, Shimo-Nakanishi Y, Urabe T, Yamada M, Tamayose K, et al. 2001. An AAV-derived Apaf-1 dominant negative inhibitor prevents MPTP toxicity as antiapoptotic gene therapy for Parkinson's disease. *Proc Natl Acad Sci* 98: 10918–10923. doi:10.1073/pnas.191107398

Mondello S, Kobeissy F, Mechref Y, Zhao J, Talih FR, Cosentino F, Antelmi E, Moresco M, Plazzi G, Ferri R. 2018. Novel biomarker signatures for idiopathic REM sleep behavior disorder: a proteomic and system biology approach. *Neurology* 91: e1710–e1715. doi:10.1212/WNL.0000000000006439

Montes de Oca Balderas P. 2021. Mitochondria-plasma membrane interactions and communication. *J Biol Chem* 297: 101164. doi:10.1016/j.jbc.2021.101164

Morfini G, Pigino G, Opalach K, Serulle Y, Moreira JE, Sugimori M, Llinás RR, Brady ST. 2007. 1-Methyl-4-phenylpyridinium affects fast axonal transport by activation of caspase and protein kinase C. *Proc Natl Acad Sci* 104: 2442–2447. doi:10.1073/pnas.0611231104

Nakamura K, Nemani VM, Azarbal F, Skibinski G, Levy JM, Egami K, Munishkina L, Zhang J, Gardner B, Wakabayashi J, et al. 2011. Direct membrane association drives mitochondrial fission by the Parkinson disease-associated protein α-synuclein. *J Biol Chem* 286: 20710–20726. doi:10.1074/jbc.M110.213538

Nalls MA, Blauwendraat C, Vallerga CL, Heilbron K, Bandres-Ciga S, Chang D, Tan M, Kia DA, Noyce AJ, Xue A, et al. 2019. Identification of novel risk loci, causal insights, and heritable risk for Parkinson's disease: a meta-analysis of genome-wide association studies. *Lancet Neurol* 18: 1091–1102. doi:10.1016/S1474-4422(19)30320-5

Naón D, Hernández-Alvarez MI, Shinjo S, Wieczor M, Ivanova S, Martins de Brito O, Quintana A, Hidalgo J, Palacín M, Aparicio P, et al. 2023. Splice variants of mitofusin 2 shape the endoplasmic reticulum and tether it to mitochondria. *Science* 380: eadh9351. doi:10.1126/science.adh9351

Narendra DP, Jin SM, Tanaka A, Suen DF, Gautier CA, Shen J, Cookson MR, Youle RJ. 2010. PINK1 is selectively stabilized on impaired mitochondria to activate Parkin. *PLoS Biol* 8: e1000298. doi:10.1371/journal.pbio.1000298

Neuspiel M, Schauss AC, Braschi E, Zunino R, Rippstein P, Rachubinski RA, Andrade-Navarro MA, McBride HM. 2008. Cargo-selected transport from the mitochondria to peroxisomes is mediated by vesicular carriers. *Curr Biol* 18: 102–108. doi:10.1016/j.cub.2007.12.038

Nguyen TT, Wei S, Nguyen TH, Jo Y, Zhang Y, Park W, Gariani K, Oh CM, Kim HH, Ha KT, et al. 2023. Mitochondria-associated programmed cell death as a therapeutic target for age-related disease. *Exp Mol Med* 55: 1595–1619. doi:10.1038/s12276-023-01046-5

Offen D, Beart PM, Cheung NS, Pascoe CJ, Hochman A, Gorodin S, Melamed E, Bernard R, Bernard O. 1998. Transgenic mice expressing human Bcl-2 in their neurons are resistant to 6-hydroxydopamine and 1-methyl-4-phenyl-1,2,3,6- tetrahydropyridine neurotoxicity. *Proc Natl Acad Sci* 95: 5789–6794. doi:10.1073/pnas.95.10.5789

Okamatsu-Ogura Y, Kuroda M, Tsutsumi R, Tsubota A, Saito M, Kimura K, Sakaue H. 2020. UCP1-dependent and UCP1-independent metabolic changes induced by acute cold exposure in brown adipose tissue of mice. *Metab Clin Exp* 113: 154396. doi:10.1016/j.metabol.2020.154396

Okawara M, Katsuki H, Kurimoto E, Shibata H, Kume T, Akaike A. 2007. Resveratrol protects dopaminergic neurons in midbrain slice culture from multiple insults. *Biochem Pharmacol* 73: 550–560. doi:10.1016/j.bcp.2006.11.003

Ottolini D, Cali T, Negro A, Brini M. 2013. The Parkinson disease-related protein DJ-1 counteracts mitochondrial impairment induced by the tumour suppressor protein p53 by enhancing endoplasmic reticulum-mitochondria tethering. *Hum Mol Genet* **22**: 2152–2168. doi:10.1093/hmg/ddt068

Panicker LM, Miller D, Awad O, Bose V, Lun Y, Park TS, Zambidis ET, Sgambato JA, Feldman RA. 2014. Gaucher iPSC-derived macrophages produce elevated levels of inflammatory mediators and serve as a new platform for therapeutic development. *Stem Cells* **32**: 2338–2349. doi:10.1002/stem.1732

Parihar MS, Parihar A, Fujita M, Hashimoto M, Ghafourifar P. 2008. Mitochondrial association of α-synuclein causes oxidative stress. *Cell Mol Life Sci* **65**: 1272–1284. doi:10.1007/s00018-008-7589-1

Park S, Juliana C, Hong S, Datta P, Hwang I, Fernandes-Alnemri T, Yu JW, Alnemri ES. 2013. The mitochondrial antiviral protein MAVS associates with NLRP3 and regulates its inflammasome activity. *J Immunol* **191**: 4358–4366. doi:10.4049/jimmunol.1301170

Parker WD Jr, Boyson SJ, Parks JK. 1989. Abnormalities of the electron transport chain in idiopathic Parkinson's disease. *Ann Neurol* **26**: 719–723. doi:10.1002/ana.410260606

Parkinson Study Group STEADY-PD III Investigators. 2020. Isradipine versus placebo in early Parkinson disease: a randomized trial. *Ann Intern Med* **172**: 591–598. doi:10.7326/M19-2534

Pedrão L, Medeiros POS, Leandro EC, Falquetto B. 2024. Parkinson's disease models and death signaling: what do we know until now? *Front Neuroanat* **18**: 1419108. doi:10.3389/fnana.2024.1419108

Perier C, Vila M. 2012. Mitochondrial biology and Parkinson's disease. *Cold Spring Harb Perspect Med* **2**: a009332. doi:10.1101/cshperspect.a009332

Perier C, Tieu K, Guégan C, Caspersen C, Jackson-Lewis V, Carelli V, Martinuzzi A, Hirano M, Przedborski S, Vila M. 2005. Complex I deficiency primes Bax-dependent neuronal apoptosis through mitochondrial oxidative damage. *Proc Natl Acad Sci* **102**: 19126–19131. doi:10.1073/pnas.0508215102

Perier C, Bové J, Wu DC, Dehay B, Choi DK, Jackson-Lewis V, Rathke-Hartlieb S, Bouillet P, Strasser A, Schulz JB, et al. 2007. Two molecular pathways initiate mitochondria-dependent dopaminergic neurodegeneration in experimental Parkinson's disease. *Proc Natl Acad Sci* **104**: 8161–8166. doi:10.1073/pnas.0609874104

Perier C, Bové J, Dehay B, Jackson-Lewis V, Rabinovitch PS, Przedborski S, Vila M. 2010. Apoptosis-inducing factor deficiency sensitizes dopaminergic neurons to parkinsonian neurotoxins. *Ann Neurol* **68**: 184–192. doi:10.1002/ana.22034

Perier C, Bové J, Vila M. 2012. Mitochondria and programmed cell death in Parkinson's disease: apoptosis and beyond. *Antioxid Redox Signal* **16**: 883–895. doi:10.1089/ars.2011.4074

Petit A, Kawarai T, Paitel E, Sanjo N, Maj M, Scheid M, Chen F, Gu Y, Hasegawa H, Salehi-Rad S, et al. 2005. Wild-type PINK1 prevents basal and induced neuronal apoptosis, a protective effect abrogated by Parkinson disease-related mutations. *J Biol Chem* **280**: 34025–34032. doi:10.1074/jbc.M505143200

Pinton P, Rimessi A, Marchi S, Orsini F, Migliaccio E, Giorgio M, Contursi C, Minucci S, Mantovani F, Wieckowski MR, et al. 2007. Protein kinase C β and prolyl isomerase 1 regulate mitochondrial effects of the life-span determinant p66[Shc]. *Science* **315**: 659–663. doi:10.1126/science.1135380

Prinz WA, Toulmay A, Balla T. 2020. The functional universe of membrane contact sites. *Nat Rev Mol Cell Biol* **21**: 7–24. doi:10.1038/s41580-019-0180-9

Pryde KR, Hirst J. 2011. Superoxide is produced by the reduced flavin in mitochondrial complex I: a single, unified mechanism that applies during both forward and reverse electron transfer. *J Biol Chem* **286**: 18056–18065. doi:10.1074/jbc.M110.186841

Przedborski S. 2017. The two-century journey of Parkinson disease research. *Nat Rev Neurosci* **18**: 251–259. doi:10.1038/nrn.2017.25

Ramsay RR, Kowal AT, Johnson MK, Salach JI, Singer TP. 1987. The inhibition site of MPP[+], the neurotoxic bioactivation product of 1-methyl-4-phenyl-1,2,3,6-tetrahydropyridine is near the Q-binding site of NADH dehydrogenase. *Arch Biochem Biophys* **259**: 645–649. doi:10.1016/0003-9861(87)90531-5

Richardson JR, Shalat SL, Buckley B, Winnik B, O'Suilleabhain P, Diaz-Arrastia R, Reisch J, German DC. 2009. Elevated serum pesticide levels and risk of Parkinson disease. *Arch Neurol* **66**: 870–875. doi:10.1001/archneurol.2009.89

Rodríguez-Arribas M, Yakhine-Diop SMS, Pedro JMB, Gómez-Suaga P, Gómez-Sánchez R, Martínez-Chacón G, Fuentes JM, González-Polo RA, Niso-Santano M. 2017. Mitochondria-associated membranes (MAMs): overview and its role in Parkinson's disease. *Mol Neurobiol* **54**: 6287–6303. doi:10.1007/s12035-016-0140-8

Rogers C, Erkes DA, Nardone A, Aplin AE, Fernandes-Alnemri T, Alnemri ES. 2019. Gasdermin pores permeabilize mitochondria to augment caspase-3 activation during apoptosis and inflammasome activation. *Nat Commun* **10**: 1689. doi:10.1038/s41467-019-09397-2

Rusiñol AE, Cui Z, Chen MH, Vance JE. 1994. A unique mitochondria-associated membrane fraction from rat liver has a high capacity for lipid synthesis and contains pre-Golgi secretory proteins including nascent lipoproteins. *J Biol Chem* **269**: 27494–27502. doi:10.1016/S0021-9258(18)47012-3

Sagan L. 1967. On the origin of mitosing cells. *J Theor Biol* **14**: 255–274. doi:10.1016/0022-5193(67)90079-3

Sarkar S, Malovic E, Harishchandra DS, Ghaisas S, Panicker N, Charli A, Palanisamy BN, Rokad D, Jin H, Anantharam V, et al. 2017. Mitochondrial impairment in microglia amplifies NLRP3 inflammasome proinflammatory signaling in cell culture and animal models of Parkinson's disease. *NPJ Parkinsons Dis* **3**: 30. doi:10.1038/s41531-017-0032-2

Schapira AH, Cooper JM, Dexter D, Clark JB, Jenner P, Marsden CD. 1990. Mitochondrial complex I deficiency in Parkinson's disease. *J Neurochem* **54**: 823–827. doi:10.1111/j.1471-4159.1990.tb02325.x

Schon EA, Przedborski S. 2011. Mitochondria: the next (neurode)generation. *Neuron* **70**: 1033–1053. doi:10.1016/j.neuron.2011.06.003

Schriner SE, Linford NJ, Martin GM, Treuting P, Ogburn CE, Emond M, Coskun PE, Ladiges W, Wolf N, Van Remmen H, et al. 2005. Extension of murine life span by overexpression of catalase targeted to mitochondria. *Science* **308**: 1909–1911. doi:10.1126/science.1106653

Schuh RA, Kristián T, Gupta RK, Flaws JA, Fiskum G. 2005. Methoxychlor inhibits brain mitochondrial respiration and increases hydrogen peroxide production and CREB phosphorylation. *Toxicol Sci* **88**: 495–504. doi:10.1093/toxsci/kfi334

Schuh RA, Richardson JR, Gupta RK, Flaws JA, Fiskum G. 2009. Effects of the organochlorine pesticide methoxychlor on dopamine metabolites and transporters in the mouse brain. *Neurotoxicology* **30**: 274–280. doi:10.1016/j.neuro.2008.12.015

Scialò F, Fernández-Ayala DJ, Sanz A. 2017. Role of mitochondrial reverse electron transport in ROS signaling: potential roles in health and disease. *Front Physiol* **8**: 428. doi:10.3389/fphys.2017.00428

Scorrano L, Ashiya M, Buttle K, Weiler S, Oakes SA, Mannella CA, Korsmeyer SJ. 2002. A distinct pathway remodels mitochondrial cristae and mobilizes cytochrome *c* during apoptosis. *Dev Cell* **2**: 55–67. doi:10.1016/S1534-5807(01)00116-2

Scotcher KP, Irwin I, DeLanney LE, Langston JW, Di Monte D. 1990. Effects of 1-methyl-4-phenyl-1,2,3,6-tetrahydropyridine and 1-methyl-4-phenylpyridinium ion on ATP levels of mouse brain synaptosomes. *J Neurochem* **54**: 1295–1301. doi:10.1111/j.1471-4159.1990.tb01962.x

Shai N, Yifrach E, van Roermund CWT, Cohen N, Bibi C, IJlst L, Cavellini L, Meurisse J, Schuster R, Zada L, et al. 2018. Systematic mapping of contact sites reveals tethers and a function for the peroxisome-mitochondria contact. *Nat Commun* **9**: 1761. doi:10.1038/s41467-018-03957-8

Sheehan JP, Swerdlow RH, Parker WD, Miller SW, Davis RE, Tuttle JB. 1997. Altered calcium homeostasis in cells transformed by mitochondria from individuals with Parkinson's disease. *J Neurochem* **68**: 1221–1233. doi:10.1046/j.1471-4159.1997.68031221.x

Sherer TB, Betarbet R, Greenamyre JT. 2002. Environment, mitochondria, and Parkinson's disease. *Neuroscientist* **8**: 192–197. doi:10.1177/1073858402008003004

Shin JH, Ko HS, Kang H, Lee Y, Lee YI, Pletinkova O, Troconso JC, Dawson VL, Dawson TM. 2011. PARIS (ZNF746) repression of PGC-1α contributes to neurodegeneration in Parkinson's disease. *Cell* **144**: 689–702. doi:10.1016/j.cell.2011.02.010

Shpilka T, Haynes CM. 2018. The mitochondrial UPR: mechanisms, physiological functions and implications in ageing. *Nat Rev Mol Cell Biol* **19**: 109–120. doi:10.1038/nrm.2017.110

Sliter DA, Martinez J, Hao L, Chen X, Sun N, Fischer TD, Burman JL, Li Y, Zhang Z, Narendra DP, et al. 2018. Parkin and PINK1 mitigate STING-induced inflammation. *Nature* **561**: 258–262. doi:10.1038/s41586-018-0448-9

Sommer A, Marxreiter F, Krach F, Fadler T, Grosch J, Maroni M, Graef D, Eberhardt E, Riemenschneider MJ, Yeo GW, et al. 2019. Th17 lymphocytes induce neuronal cell death in a human iPSC-based model of Parkinson's disease. *Cell Stem Cell* **24**: 1006. doi:10.1016/j.stem.2019.04.019

Soto-Heredero G, Baixauli F, Mittelbrunn M. 2017. Interorganelle communication between mitochondria and the endolysosomal system. *Front Cell Dev Biol* **5**: 95. doi:10.3389/fcell.2017.00095

Sousa SC, Maciel EN, Vercesi AE, Castilho RF. 2003. Ca^{2+}-induced oxidative stress in brain mitochondria treated with the respiratory chain inhibitor rotenone. *FEBS Lett* **543**: 179–183. doi:10.1016/S0014-5793(03)00421-6

Stepanova A, Konrad C, Guerrero-Castillo S, Manfredi G, Vannucci S, Arnold S, Galkin A. 2019. Deactivation of mitochondrial complex I after hypoxia-ischemia in the immature brain. *J Cereb Blood Flow Metab* **39**: 1790–1802. doi:10.1177/0271678X18770331

Stoica R, De Vos KJ, Paillusson S, Mueller S, Sancho RM, Lau KF, Vizcay-Barrena G, Lin WL, Xu YF, Lewis J, et al. 2014. ER-mitochondria associations are regulated by the VAPB-PTPIP51 interaction and are disrupted by ALS/FTD-associated TDP-43. *Nat Commun* **5**: 3996. doi:10.1038/ncomms4996

St-Pierre J, Drori S, Uldry M, Silvaggi JM, Rhee J, Jäger S, Handschin C, Zheng K, Lin J, Yang W, et al. 2006. Suppression of reactive oxygen species and neurodegeneration by the PGC-1 transcriptional coactivators. *Cell* **127**: 397–408. doi:10.1016/j.cell.2006.09.024

Sulzer D, Alcalay RN, Garretti F, Cote L, Kanter E, Agin-Liebes J, Liong C, McMurtrey C, Hildebrand WH, Mao X, et al. 2017. T cells from patients with Parkinson's disease recognize α-synuclein peptides. *Nature* **546**: 656–661. doi:10.1038/nature22815

Surmeier DJ, Nguyen JT, Lancki N, Venuto CS, Oakes D, Simuni T, Wyse RK. 2022. Re-analysis of the STEADY-PD II trial—evidence for slowing the progression of Parkinson's disease. *Mov Disord* **37**: 334–342. doi:10.1002/mds.28850

Swerdlow RH, Parks JK, Miller SW, Tuttle JB, Trimmer PA, Sheehan JP, Bennett JP Jr, Davis RE, Parker WD Jr. 1996. Origin and functional consequences of the complex I defect in Parkinson's disease. *Ann Neurol* **40**: 663–671. doi:10.1002/ana.410400417

Szabadkai G, Simoni AM, Bianchi K, De Stefani D, Leo S, Wieckowski MR, Rizzuto R. 2006. Mitochondrial dynamics and Ca^{2+} signaling. *Biochim Biophys Acta* **1763**: 442–449. doi:10.1016/j.bbamcr.2006.04.002

Tait SW, Green DR. 2010. Mitochondria and cell death: outer membrane permeabilization and beyond. *Nat Rev Mol Cell Biol* **11**: 621–632. doi:10.1038/nrm2952

Tansey MG, Wallings RL, Houser MC, Herrick MK, Keating CE, Joers V. 2022. Inflammation and immune dysfunction in Parkinson disease. *Nat Rev Immunol* **22**: 657–673. doi:10.1038/s41577-022-00684-6

Thyagarajan D, Bressman S, Bruno C, Przedborski S, Shanske S, Lynch T, Fahn S, DiMauro S. 2000. A novel mitochondrial 12SrRNA point mutation in parkinsonism, deafness, and neuropathy. *Ann Neurol* **48**: 730–736. doi:10.1002/1531-8249(200011)48:5<730::AID-ANA6>3.0.CO;2-0

Tieu K, Perier C, Caspersen C, Teismann P, Wu DC, Yan SD, Naini A, Vila M, Jackson-Lewis V, Ramasamy R, et al. 2003. D-β-hydroxybutyrate rescues mitochondrial respiration and mitigates features of Parkinson disease. *J Clin Invest* **112**: 892–901. doi:10.1172/JCI200318797

Toyofuku T, Okamoto Y, Ishikawa T, Sasawatari S, Kumanogoh A. 2020. LRRK2 regulates endoplasmic reticulum-mitochondrial tethering through the PERK-mediated ubiquitination pathway. *EMBO J* **39**: e100875. doi:10.15252/embj.2018100875

Trifunovic A, Wredenberg A, Falkenberg M, Spelbrink JN, Rovio AT, Bruder CE, Bohlooly YM, Gidlöf S, Oldfors A, Wibom R, et al. 2004. Premature ageing in mice expressing defective mitochondrial DNA polymerase. *Nature* **429:** 417–423. doi:10.1038/nature02517

Trimmer PA, Borland MK, Keeney PM, Bennett JP Jr, Parker WD Jr. 2004. Parkinson's disease transgenic mitochondrial cybrids generate Lewy inclusion bodies. *J Neurochem* **88:** 800–812. doi:10.1046/j.1471-4159.2003.02168.x

Trudler D, Nash Y, Frenkel D. 2015. New insights on Parkinson's disease genes: the link between mitochondria impairment and neuroinflammation. *J Neural Transm* **122:** 1409–1419. doi:10.1007/s00702-015-1399-z

Van Maldergem L, Besse A, De Paepe B, Blakely EL, Appadurai V, Humble MM, Piard J, Craig K, He L, Hella P, et al. 2017. POLG2 deficiency causes adult-onset syndromic sensory neuropathy, ataxia and parkinsonism. *Ann Clin Transl Neurol* **4:** 4–14. doi:10.1002/acn3.361

Vermulst M, Bielas JH, Kujoth GC, Ladiges WC, Rabinovitch PS, Prolla TA, Loeb LA. 2007. Mitochondrial point mutations do not limit the natural lifespan of mice. *Nat Genet* **39:** 540–543. doi:10.1038/ng1988

Vermulst M, Wanagat J, Kujoth GC, Bielas JH, Rabinovitch PS, Prolla TA, Loeb LA. 2008. DNA deletions and clonal mutations drive premature aging in mitochondrial mutator mice. *Nat Genet* **40:** 392–394. doi:10.1038/ng.95

Verstreken P, Ly CV, Venken KJ, Koh TW, Zhou Y, Bellen HJ. 2005. Synaptic mitochondria are critical for mobilization of reserve pool vesicles at *Drosophila* neuromuscular junctions. *Neuron* **47:** 365–378. doi:10.1016/j.neuron.2005.06.018

Vila M, Przedborski S. 2003. Targeting programmed cell death in neurodegenerative diseases. *Nat Rev Neurosci* **4:** 365–375. doi:10.1038/nrn1100

Vila M, Przedborski S. 2004. Genetic clues to the pathogenesis of Parkinson's disease. *Nat Med* **10:** S58–S62. doi:10.1038/nm1068

Vila M, Jackson-Lewis V, Vukosavic S, Djaldetti R, Liberatore G, Offen D, Korsmeyer SJ, Przedborski S. 2001. Bax ablation prevents dopaminergic neurodegeneration in the 1-methyl-4-phenyl-1,2,3,6-tetrahydropyridine mouse model of Parkinson's disease. *Proc Natl Acad Sci* **98:** 2837–2842. doi:10.1073/pnas.051633998

Vila M, Bové J, Dehay B, Rodríguez-Muela N, Boya P. 2011. Lysosomal membrane permeabilization in Parkinson disease. *Autophagy* **7:** 98–100. doi:10.4161/auto.7.1.13933

Viswanath V, Wu Y, Boonplueang R, Chen S, Stevenson FF, Yantiri F, Yang L, Beal MF, Andersen JK. 2001. Caspase-9 activation results in downstream caspase-8 activation and bid cleavage in 1-methyl-4-phenyl-1,2,3,6-tetrahydropyridine-induced Parkinson's disease. *J Neurosci* **21:** 9519–9528. doi:10.1523/JNEUROSCI.21-24-09519.2001

Vives-Bauza C, Tocilescu M, Devries RL, Alessi DM, Jackson-Lewis V, Przedborski S. 2010a. Control of mitochondrial integrity in Parkinson's disease. *Prog Brain Res* **183:** 99–113. doi:10.1016/S0079-6123(10)83006-7

Vives-Bauza C, Zhou C, Huang Y, Cui M, de Vries RL, Kim J, May J, Tocilescu MA, Liu W, Ko HS, et al. 2010b. PINK1-dependent recruitment of Parkin to mitochondria in mitophagy. *Proc Natl Acad Sci* **107:** 378–383. doi:10.1073/pnas.0911187107

Volobueva LA, Emery JF, Sun X, Giffard RG. 2013. Inflammatory response of microglial BV-2 cells includes a glycolytic shift and is modulated by mitochondrial glucose-regulated protein 75/mortalin. *FEBS Lett* **587:** 756–762. doi:10.1016/j.febslet.2013.01.067

Vranas M, Lu Y, Rasool S, Croteau N, Krett JD, Sauvé V, Gehring K, Fon EA, Durcan TM, Trempe JF. 2022. Selective localization of Mfn2 near PINK1 enables its preferential ubiquitination by Parkin on mitochondria. *Open Biol* **12:** 210255. doi:10.1098/rsob.210255

Wallace DC. 2005. A mitochondrial paradigm of metabolic and degenerative diseases, aging, and cancer: a dawn for evolutionary medicine. *Annu Rev Genet* **39:** 359–407. doi:10.1146/annurev.genet.39.110304.095751

Wallin I. 1925. On the nature of mitochondria. IX: Demonstration of the bacterial nature of mitochondria. *Am J Anat* **36:** 131–139. doi:10.1002/aja.1000360106

Wang XJ, Xu JX. 2005. Possible involvement of Ca^{2+} signaling in rotenone-induced apoptosis in human neuroblastoma SH-SY5Y cells. *Neurosci Lett* **376:** 127–132. doi:10.1016/j.neulet.2004.11.041

Wang HL, Chou AH, Yeh TH, Li AH, Chen YL, Kuo YL, Tsai SR, Yu ST. 2007. PINK1 mutants associated with recessive Parkinson's disease are defective in inhibiting mitochondrial release of cytochrome c. *Neurobiol Dis* **28:** 216–226. doi:10.1016/j.nbd.2007.07.010

Weinberg SE, Sena LA, Chandel NS. 2015. Mitochondria in the regulation of innate and adaptive immunity. *Immunity* **42:** 406–417. doi:10.1016/j.immuni.2015.02.002

Weindel CG, Martinez EL, Zhao X, Mabry CJ, Bell SL, Vail KJ, Coleman AK, VanPortfliet JJ, Zhao B, Wagner AR, et al. 2022. Mitochondrial ROS promotes susceptibility to infection via gasdermin D-mediated necroptosis. *Cell* **185:** 3214–3231.e23. doi:10.1016/j.cell.2022.06.038

Wen Y, Li W, Poteet EC, Xie L, Tan C, Yan LJ, Ju X, Liu R, Qian H, Marvin MA, et al. 2011. Alternative mitochondrial electron transfer as a novel strategy for neuroprotection. *J Biol Chem* **286:** 16504–16515. doi:10.1074/jbc.M110.208447

Williams GP, Freuchet A, Michaelis T, Frazier A, Tran NK, Rodrigues Lima-Junior J, Phillips EJ, Mallal SA, Litvan I, Goldman JG, et al. 2024. PINK1 is a target of T cell responses in Parkinson's disease. *J Clin Invest* **135:** e180478. doi:10.1172/JCI180478

Wong YC, Ysselstein D, Krainc D. 2018. Mitochondria-lysosome contacts regulate mitochondrial fission via RAB7 GTP hydrolysis. *Nature* **554:** 382–386. doi:10.1038/nature25486

Wu DC, Teismann P, Tieu K, Vila M, Jackson-Lewis V, Ischiropoulos H, Przedborski S. 2003. NADPH oxidase mediates oxidative stress in the 1-methyl-4-phenyl-1,2,3,6-tetrahydropyridine model of Parkinson's disease. *Proc Natl Acad Sci* **100:** 6145–6150. doi:10.1073/pnas.0937239100

Yamada M, Iwatsubo T, Mizuno Y, Mochizuki H. 2004. Overexpression of α-synuclein in rat substantia nigra results in loss of dopaminergic neurons, phosphorylation of alpha-synuclein and activation of caspase-9: resemblance to pathogenetic changes in Parkinson's disease. *J Neurochem* **91:** 451–461. doi:10.1111/j.1471-4159.2004.02728.x

Yang L, Matthews RT, Schulz JB, Klockgether T, Liao AW, Martinou JC, Penney JB Jr, Hyman BT, Beal MF. 1998. 1-Methyl-4-phenyl-1,2,3,6-tetrahydropyride neurotoxicity is attenuated in mice overexpressing Bcl-2. *J Neurosci* **18:** 8145–8152. doi:10.1523/JNEUROSCI.18-20-08145.1998

Youle RJ, Narendra DP. 2011. Mechanisms of mitophagy. *Nat Rev Mol Cell Biol* **12:** 9–14. doi:10.1038/nrm3028

Zheng B, Liao Z, Locascio JJ, Lesniak KA, Roderick SS, Watt ML, Eklund AC, Zhang-James Y, Kim PD, Hauser MA, et al. 2010. *PGC-1α*, a potential therapeutic target for early intervention in Parkinson's disease. *Sci Transl Med* **2:** 52ra73. doi:10.1126/scitranslmed.3001059

Zhou C, Huang Y, Przedborski S. 2008. Oxidative stress in Parkinson's disease: a mechanism of pathogenic and therapeutic significance. *Ann NY Acad Sci* **1147:** 93–104. doi:10.1196/annals.1427.023

Zhou R, Yazdi AS, Menu P, Tschopp J. 2011. A role for mitochondria in NLRP3 inflammasome activation. *Nature* **469:** 221–225. doi:10.1038/nature09663

Züchner S, Mersiyanova IV, Muglia M, Bissar-Tadmouri N, Rochelle J, Dadali EL, Zappia M, Nelis E, Patitucci A, Senderek J, et al. 2004. Mutations in the mitochondrial GTPase mitofusin 2 cause Charcot–Marie–Tooth neuropathy type 2A. *Nat Genet* **36:** 449–451. doi:10.1038/ng1341

Lipid Alterations and Pathogenic Roles in Synucleinopathies

Estela Area-Gómez,[1,2] Saranna Fanning,[3] and Ulf Dettmer[3]

[1]Department of Biomedicine, Margarita Salas Center for Biological Research, CIBERNED, CSIC, Madrid 28040, Spain

[2]Department of Neurology, Columbia University Medical Campus, New York, New York 10032, USA

[3]Ann Romney Center for Neurologic Diseases, Brigham and Women's Hospital and Harvard Medical School, Boston, Massachusetts 02115, USA

Correspondence: udettmer@bwh.harvard.edu

Mounting evidence highlights a role for lipid alterations and defects in lipid signaling in age-related neurodegenerative diseases such as Parkinson's disease (PD) and related conditions (collectively referred to as synucleinopathies). This growing interest is driven by several key findings: (1) lipid membranes are components of Lewy bodies and Lewy neurites, which are prototypical proteinaceous intraneuronal inclusions of PD and other synucleinopathies, primarily composed of α-synuclein (αS); (2) αS shares structural similarities with lipid-binding proteins and has been reported to bind to lipids; (3) glucocerebrosidase, a key enzyme in sphingolipid metabolism, is a major PD risk factor; (4) other enzymes involved in glycolipid and phospholipid regulation, such as diacylglycerol kinase-θ and fatty acid elongase-7, also contribute to PD risk; (5) αS alterations impact lipid homeostasis; (6) αS transiently binds lipid membranes, affecting its conformation. Given these findings, we review what is known about the role of lipids in normal αS biology as well as in the pathogenesis of PD and related conditions. We also highlight areas where further research is warranted.

Growing numbers of studies report dysregulated lipid composition of tissues and biofluids in neurodegenerative diseases, including synucleinopathies such as Parkinson's disease (PD), Parkinson's disease dementia, dementia with Lewy bodies (DLB), and multiple system atrophy (MSA). These conditions are characterized by pathological aggregates of α-synuclein (αS), which are increasingly linked to lipid disturbances that may influence disease pathogenesis. While advancements in technology for lipid detection and quantification have been achieved, understanding the role of lipids in the occurrence and progression of synucleinopathies is still in its infancy.

Emerging evidence reveals statistical associations between fluctuations in lipid species and disease occurrence and progression. However, identifying disease-specific lipids as individual entities, divorced from the cellular context in which they arise, risks misinterpreting their role in disease mechanisms. Membrane lipids

Copyright © 2025 Cold Spring Harbor Laboratory Press; all rights reserved
Cite this article as *Cold Spring Harb Perspect Med* doi: 10.1101/cshperspect.a041646

act in concert within a tight coregulatory network, and changes in individual lipid species, when viewed in isolation, offer limited biological insight. Nonetheless, we recognize that lipid changes as markers of specific metabolic disturbances could help identify new therapeutic targets and metabolic signatures for clinical biomarker development.

Cellular membranes are complex and highly dynamic entities involved in almost all cellular functions. Lipidome fluctuations allow membranes to form specialized domains that compartmentalize cellular functions, including regulation of membrane channels and receptors, reorganization of the cytoskeleton, migration and polarity, and the clustering of specific subsets of lipid-binding proteins (Leonard et al. 2023). The unique native composition of membrane lipids is thus critical for cellular physiology, with disruptions leading to defects in transmembrane protein conformation, membrane fluidity, and intracellular signaling pathways (Shimokawa and Takagi 2024). For instance, the fatty acid (FA) composition of phospholipids modulates membrane fluidity: phospholipids enriched with monounsaturated FAs (MUFAs) or polyunsaturated FAs (PUFAs) increase fluidity, while those with saturated FAs (SFAs) decrease it (Ibarguren et al. 2014). Moreover, lipid domains of the membrane such as phosphatidylserine (PtdSer)-enriched regions also serve in signaling roles mediating recognition and interaction between cell types (e.g., "eat me" signals in inflammation) (Segawa and Nagata 2015). Collectively, these findings underscore the need to understand the contribution of cellular membrane physiology and lipidome maintenance in the context of disease.

All cell types in the central nervous system (CNS), and neurons in particular, maintain a high rate of membrane activity to sustain neurotransmission (Montesinos et al. 2020), making them particularly vulnerable to lipid metabolism dysregulation (Isik and Cizmecioglu 2023).

In this work, we focus on the interaction between αS and lipid metabolism, outlining lipid mechanisms relevant to PD and related conditions. A complex landscape emerges that comprises a bidirectional relationship between both physiological and pathological states of αS and lipid homeostasis (Schepers et al. 2024). We highlight potential therapeutic approaches premised on the αS-lipid interplay and discuss mitochondria-associated endoplasmic reticulum (ER) membranes (MAMs) as an area of lipid biology that could be important for mechanistic understanding of disease.

CHANGES IN LIPID COMPOSITION AFFECT THE INTERACTION OF αS WITH CELLULAR MEMBRANES

The 14 kDa protein αS plays a critical role in PD, DLB, MSA, and certain forms of Alzheimer's disease (AD). Lewy bodies (LBs) and Lewy neurites (LNs), the intracellular hallmark lesions of synucleinopathies, are rich in aggregates of αS (Spillantini et al. 1997). αS is highly expressed in the brain, the second-most lipid-rich organ (Sastry 1985), and has been proposed to be a lipid-binding protein that physiologically interacts with specific membrane domains as well as with free FAs (reviewed by Fanning et al. 2021). Alterations in αS interactions with cellular membranes may trigger its aggregation via primary nucleation (Galvagnion et al. 2015). αS excess at membranes has also been proposed to be toxic in the absence of aggregation (Volles and Lansbury 2007; Dettmer et al. 2017). Additional complexity stems from αS found in LBs potentially being less fibrillar than previously thought and instead associated with large amounts of lipids, membranes, and cellular organelles (Shahmoradian et al. 2019; Moors et al. 2021; Moors and Milovanovic 2024).

In an aqueous solution, unaggregated αS is an unfolded protein (Weinreb et al. 1996). However, its helical fold increases from ~3% to ~80% in the presence of small acidic phospholipid vesicles (Davidson et al. 1998). This switch seems to be driven by the formation of membrane-induced amphipathic helices when αS contacts a lipid membrane, due to an 11-amino acid repeat motif with the core consensus sequence KTKEGV (Fig. 1A). This motif appears imperfectly—six to nine times in the first two-thirds of αS—and resembles lipid-binding

   Cite this article as *Cold Spring Harb Perspect Med* doi: 10.1101/cshperspect.a041646

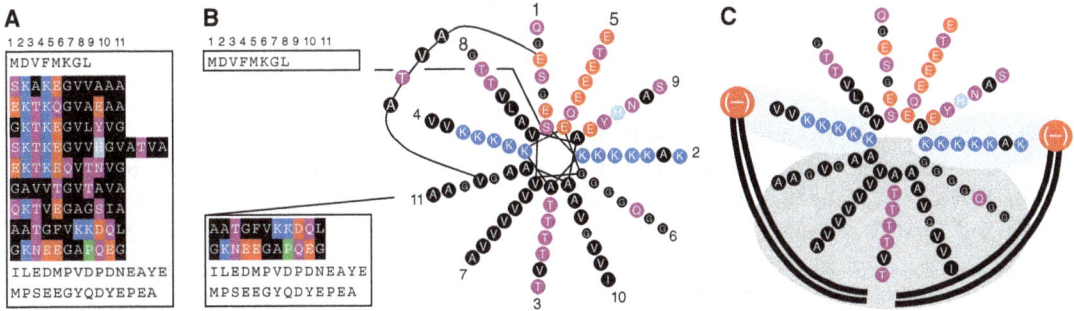

Figure 1. Electrostatic and hydrophobic interactions govern transient α-synuclein (αS) membrane binding. (*A*) Amino acid sequence of human wild-type (WT) αS displayed by aligning the KTKEGV motifs, color-coded residues. (black) Uncharged, (red) negatively charged, (dark blue) positively charged, (light blue) histidine, (purple) uncharged and polar, (green) proline. (*B*) Color-coded schematic of repeats 1–7 (omitting "ATVA" between repeats 4 and 5) in a 3/11 helical wheel. (*C*) Simplified schematic of the forces that attract a membrane-induced αS helix to a membrane: a proposed electrostatic interaction between positive charges of αS lysines and the negatively charged phospholipid headgroups of the membrane (blue area); hydrophobic interactions between the hydrophobic half of the αS membrane-induced amphipathic helix and fatty acyl chains of the membrane lipids (gray area).

domains present in apolipoproteins (George et al. 1995). The membrane conformation of αS is known as a 3–11 helix (3 turns/11 amino acids), where nonpolar amino acids in the hydrophobic half of the amphipathic helix interact with the nonpolar fatty acyl chains of the membrane lipid bilayer (Fig. 1B,C; Jao et al. 2008).

It has been shown that αS preferentially associates with vesicles of smaller diameter (20–25 nm) as opposed to larger vesicles, consistent with a role in vesicle biology at the presynaptic terminal (Nuscher et al. 2004). This preference may arise from lipid-packing defects intrinsic to small vesicles, where the hydrophobic FA tails of membrane lipids become exposed to the cytosol due to curvature effects, enabling interactions with the hydrophobic half of the αS helix (Nuscher et al. 2004). However, αS also possesses polar residues within the hydrophobic half of its helix, notably threonine (see Fig. 1A–C). This imperfection seems to underlie the transient nature of αS-membrane binding, as the replacement of key threonine residues with hydrophobic leucines renders αS fully membrane-associated (Dettmer et al. 2015b).

The interaction of αS with cellular membranes appears to be dynamic as exemplified by two observations: (1) sequential protein extraction from primary neurons determined αS at a ∼1:1 distribution between cytosol and membrane at body temperature while at room temperature most αS is localized at membranes (Ramalingam and Dettmer 2021); and (2) single amino acid exchanges such as the familial PD-linked A30P and G51D are sufficient to render αS almost entirely cytosolic (Ramalingam et al. 2023). Interestingly, it has been suggested that αS does not immediately lose its helical fold upon dissociation from membranes, but may retain it, thereby forming native αS–αS assemblies (Rovere et al. 2018). These observations align with previous descriptions of soluble αS tetramers and related multimers (Bartels et al. 2011; Wang et al. 2011), which may be metastable (Dettmer et al. 2013). Membrane-associated αS multimers have also been reported, potentially mediating SNARE complex assembly (Burré et al. 2014) or vesicle clustering (Wang et al. 2014). This leads to a model of dynamic cellular αS behavior in which the different αS species discussed herein are in an equilibrium—and maintaining the right extent of αS-membrane interaction is at the core of preventing αS perturbation that is associated with PD and related diseases (Burré et al. 2015; Dettmer et al. 2017).

CHANGES IN αS BINDING TO MEMBRANES ARE RELATED TO LB FORMATION AND αS TOXICITY

While the transient membrane interaction of native αS has been widely accepted, the ability of aggregated αS to bind to membranous organelles and lipids has been debated. When αS aggregates were initially isolated from human LBs and subjected to immunogold–electron microscopy (Spillantini et al. 1998), fibrillar structures strongly positive for αS were observed, implying that synucleinopathies are classical protein misfolding diseases (Fig. 2A). This view was challenged by more recent studies (Shahmoradian et al. 2019; reviewed by Moors and Milovanovic 2024), which analyzed PD brain tissue using correlative light and electron microscopy, a technique that allows immunohistological and ultrastructural analysis of the same lesion. The authors found LBs to be medleys of αS and clusters of various membranous structures (Fig. 2B). The core of LBs consisted of various vesicle clusters coated with large amounts of nonfibrillar αS. Only a minority (<20%) of LBs was found to contain apparent amyloid fibrils (>5 nm diameter, >25 nm in length). Raman scattering and infrared spectroscopy identified large amounts of lipids in the LB core; and sphingomyelin and phosphatidylcholine (PC) were abundantly detected by mass spectrometry. Previous LB analyses (e.g., Braak et al. 2003) had largely relied on immunohistochemical staining at the light microscopic level, potentially creating a bias for areas that contain defined fibrillar structures (Bartels 2019). Of note, historical reports also identified vesicle/membrane components in LBs in addition to filamentous aggregates (Roy and Wolman 1969; Forno and Norville 1976), which was corroborated by later studies (Dickson et al. 1989; Hayashida et al. 1993; Nishimura et al. 1994).

The concept of lipid-rich αS pathology has the potential to change how we model PD pathogenesis and design therapeutics and biomarkers, but wider confirmation is needed. Interestingly, when human αS is expressed in *Saccharomyces cerevisiae*, undocked vesicles of

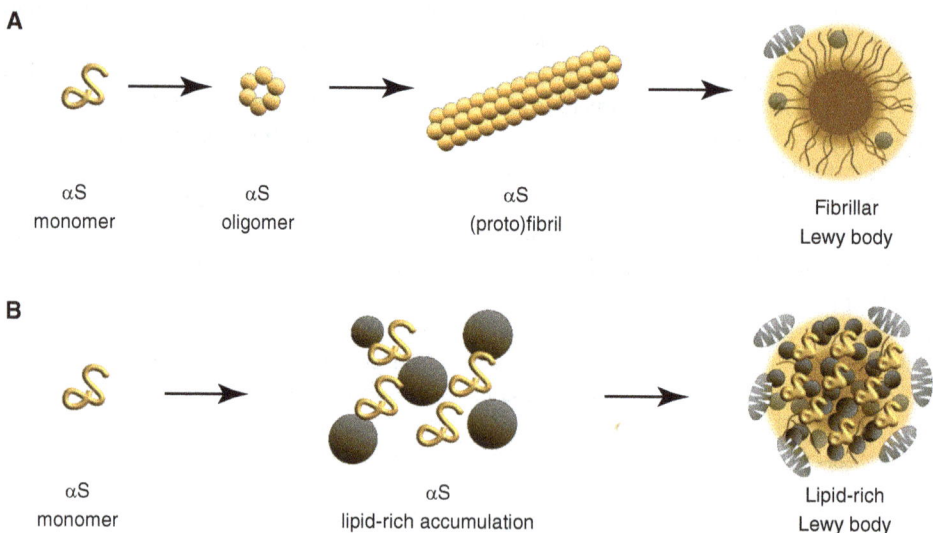

Figure 2. Contrasting α-synuclein (αS) fibrillar and membrane/lipid-rich aggregation forms. (*A*) "Classical" fibrillar Lewy body as the result of proteinaceous αS "amyloid" aggregation. (*B*) Lipid-rich Lewy body as the result of αS accumulation at vesicles, other membranes, lipid droplets, and other organelles. These two scenarios are not mutually exclusive. Lipid-rich aggregates could, for example, be precursors of fibrillar aggregates. Golden shapes represent αS species, dark-gray round shapes represent vesicles or lipid droplets, and gray oval shapes represent mitochondria.

different subcellular origins coalesce into large clusters in an αS dose-dependent manner, with αS fibrils being absent (Gitler et al. 2008; Soper et al. 2008). In cultured human neurons, massive αS/vesicle clustering can be achieved by expressing engineered αS variants such as "3K" (E35K + E46K + E61K), a highly membrane-enriched amplification of the familial PD-causing E46K (Dettmer et al. 2015a, 2017). Because the trafficking and clustering of synaptic vesicles have been suggested to be a normal function of αS (Abeliovich et al. 2000; Chandra et al. 2004; Vargas et al. 2014; Wang et al. 2014), synuclein-opathy-relevant vesicle accumulation may be interpreted as excessive αS function.

The folding state of αS in vesicle-rich inclusions remains unclear. A study of αS point mutants concluded that fibrillization rate in the test tube and toxicity in yeast do not correlate. All noncytotoxic αS sequence variants, however, reduced membrane interactions (Volles and Lansbury 2007). The authors hypothesized that αS cytotoxicity in yeast is caused by the protein binding to membranes at levels sufficient to nonspecifically disrupt membrane homeostasis independent of aggregation (Fig. 3A).

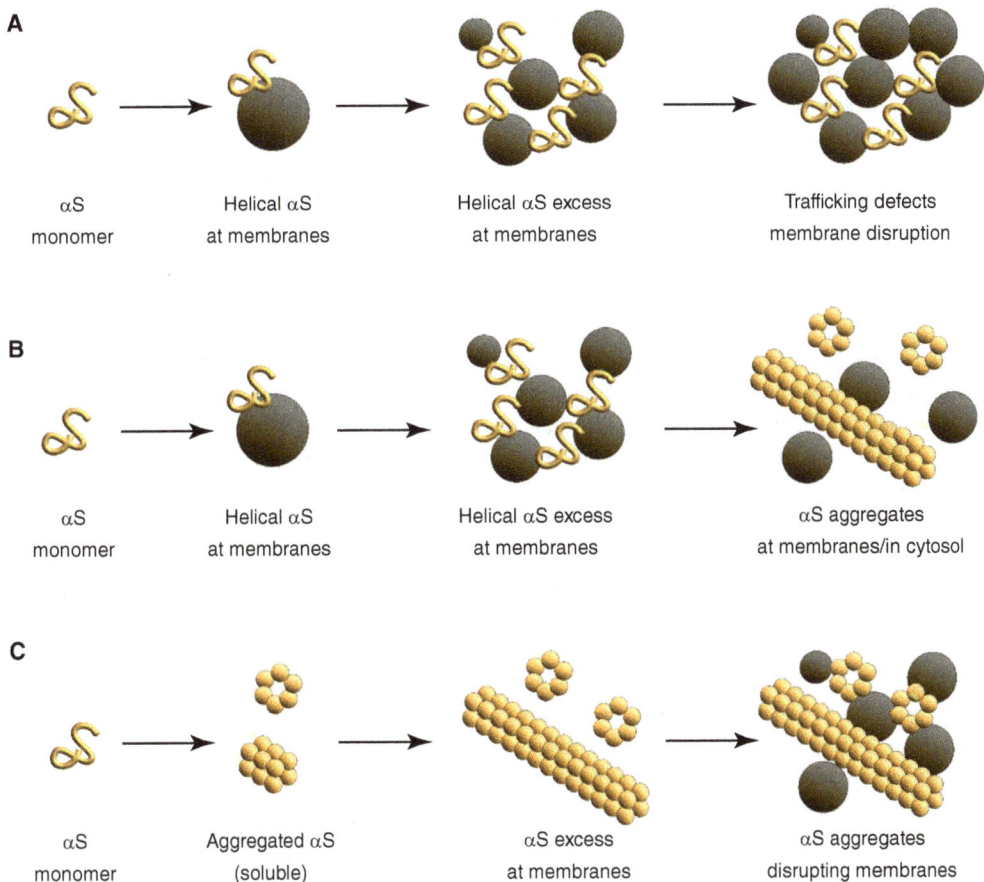

Figure 3. Different scenarios of pathological α-synuclein (αS)-membrane interaction. (A) αS accumulates at membranes and exerts toxicity (membrane disruption, trafficking defects) prior to or independent of aggregation. (B) Excess αS-membrane interactions cause primary nucleation and subsequent aggregation of αS, which then accumulates either at membranes or in the cytosol. (C) αS aggregates in the cytosol and then binds to phospholipids, disrupting membranes. Golden shapes represent αS species and dark-gray round shapes represent vesicles.

Related to that, vesicle trafficking defects were observed in αS A53T and αS triplication iPSC-derived human neurons (Chung et al. 2013). A potential connection between αS-membrane interaction and aggregation was noted in a biophysical study that described how the local enrichment of αS at lipid bilayers can lead to the formation of "classic" β-sheet aggregates via primary nucleation (Galvagnion et al. 2015), which then accumulate either at membranes or in the cytosol (Fig. 3B). It has been proposed that αS aggregates, which may initially form in the cytosol, can disrupt membranes (e.g., via pore formation) (Lashuel et al. 2002). In this scenario, aggregation is likely to be upstream of membrane interaction (Fig. 3C). Thus, the connection between αS-membrane interaction, αS folding state, and cytotoxicity is insufficiently understood and requires further research. A deeper understanding is further complicated by the notion that not only do lipids affect αS biology, but—conversely—αS has been shown to affect lipids.

αS modulates lipid metabolism and the lipid compositions of cellular membranes by a yet unknown mechanism. As proposed in a growing number of studies using models from αS-expressing yeast to patient-derived neurons and mouse models, perturbations in αS result in alterations in the levels of several lipid species, and imbalances in the lipidome of cellular membranes. For example, αS protein levels impact FA synthesis and the saturation of phospholipids (Fanning et al. 2019), which in turn modulates membrane fluidity (e.g., Sharon et al. 2001, 2003; Bodner et al. 2009, 2010; Westphal and Chandra 2013; Runwal and Edwards 2021). Specifically, αS increases MUFA (oleic and palmitoleic acid) synthesis in yeast, primary rat cortical neurons, and human neurons overexpressing αS (Vincent et al. 2018; Fanning et al. 2019). Patient-derived iPSC neurons expressing PD-related triplication of the αS locus or the A53T αS mutation have increased MUFAs (Chung et al. 2013; Fanning et al. 2022). This FA metabolism dysregulation is associated with changes in diglycerides (DGs), triglycerides (TGs), and lipid droplets (LDs) (Fanning et al. 2019, 2022; Nuber et al. 2021, 2022). As mentioned above,

when incorporated into phospholipid membranes, MUFAs alter membrane composition and fluidity as well as protein–membrane interactions.

Missense mutations in the αS gene (SNCA), such as A53T, E46K, and A30P, have also been associated with alterations in phospholipids, such as PC, the most abundant phospholipid in cellular membranes (van Meer et al. 2008; O'Leary et al. 2018). In addition, treating rats with the dopaminergic neurotoxin 6-hydroxydopamine was reported to cause early changes in PC levels in vulnerable brain areas (Farmer et al. 2015). Decreases in phosphatidylinositol (PI) species, particularly those with more SFA, also occured αS dose- and time-dependently in a rat cortical neuron model of αS overexpression (Fanning et al. 2019). In contrast, αS appears to increase specific phosphatidylserine (PS) species, namely, PS with long and unsaturated fatty acyl sidechains (Fabelo et al. 2011; Lou et al. 2017). Other changes in lipids associated with αS and PD have been reviewed in detail (Fanning et al. 2021). Below, we focus specifically on the role of MAM lipids in αS biology and PD.

What emerges from this insight is a scenario in which changes in the cell's lipidome affect the binding of αS to specific membrane domains; in turn, the binding of αS contributes directly or indirectly to the regulation of various lipid pathways. This bidirectional communication between αS and lipid metabolism could help explain the dysregulation of lipid metabolism driven by mutations in SNCA and the aggregation of endogenous αS driven by defects in membrane composition. However, the exact mechanism by which αS contributes to the concomitant regulation of these different lipid routes is unclear.

MITOCHONDRIA-ASSOCIATED ER MEMBRANE AND αS

Lipid homeostasis requires the simultaneous regulation of multiple lipid pathways in response to changes in the environment to maintain the composition and function of cellular membranes. The convergence of many of these lipid pathways occurs in temporary lipid-raft domains formed in the ER, called MAMs (Scorrano et al. 2019). The formation of these domains stimulates the interaction and activation of key regulatory enzymes for the control of metabolic pathways, including lipid homeostasis (Rusiñol et al. 1994). Specifically, the formation of MAMs modulates enzymes involved in the synthesis and turnover of sphingolipids (Serine-Palmitoyl Transferase Long Chain Base Subunit-1 and -2 [SPTLC1/2], Delta 4-Desaturase, Sphingolipid 1 [DEGS1], or neutral Sphingomyelinase 2 [SMPD3]), the synthesis and acylation of FAs and phospholipids (Phosphatidylserine Synthase 1 and 2 [PSS1/2]; Acyl-CoA Synthase 4, ACSL4; Stearoyl-CoA Desaturase [SCD]), the production of TGs (Diacylglycerol O-Acyltransferase 2 [DGAT2]), and the esterification of cholesterol (Sterol O-Acyltransferase 1 [SOAT1]) (Vance 2014). Notably, αS binds to MAM domains and αS overexpression and/or missense mutations in animal models and neuronal cell lines are associated with alterations in MAM domains and their enzy-

matic activities (Fig. 4; Guardia-Laguarta et al. 2014).

It has also been suggested that mutations in αS provoke significant changes in the lipid profile and the biophysical properties of membranes of vulnerable cell populations via perturbation of MAM domains (Guardia-Laguarta et al. 2014). In this scenario, αS contributes to the regulation of lipid homeostasis either by stimulating the formation of the MAM or by direct interaction with lipid enzymes in the ER domain. This is consistent with the preferential binding of αS to lipid-raft domains. Here, αS may impact MAM-resident enzymes and their products (including SCD and MUFA, PSS and PS, DGAT2 and TGs, ACSL4 and esterified PUFAs), thereby affecting lipid saturation and membrane fluidity.

The association between αS and MAMs can also explain the previously reported localization of αS to mitochondria (Guardia-Laguarta et al. 2014). The role of MAM domains in the regulation of metabolic pathways, such as bioenergetics, calcium levels, and autophagy flux, may

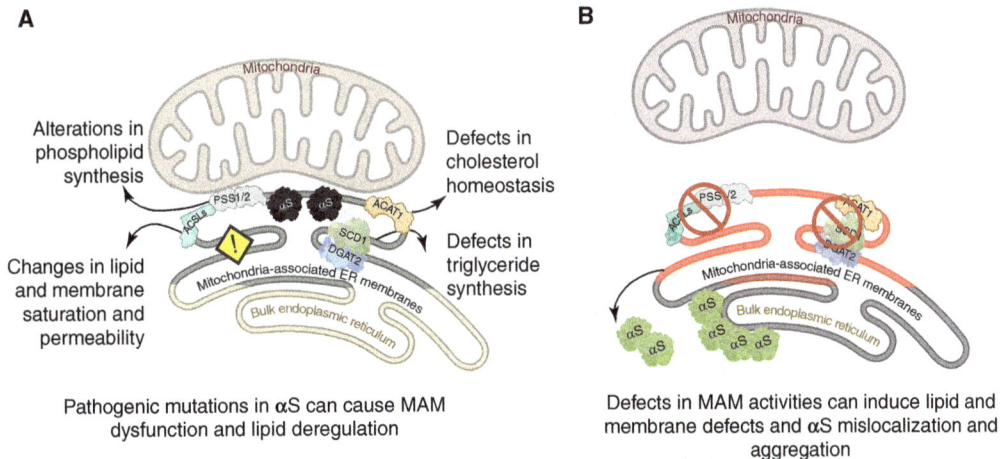

Figure 4. Schematic representation of the dynamic interaction between αS and mitochondria-associated endoplasmic reticulum (ER) membrane (MAM). Both play a role in lipid metabolism and disturbances in either can lead to cellular dysfunction, possibly relevant to neurodegenerative diseases. (A) The MAM is described as a temporary domain in the ER that gathers multiple enzymes crucial for lipid metabolism regulation. αS may influences the regulation of MAM domains. Mutations in αS could potentially disrupt the modulation of lipid-metabolizing enzymes localized within these domains, suggesting a role for αS in maintaining lipid homeostasis at the MAM. (B) Conversely, defects in MAM formation and activity can lead to changes in the lipid composition of cellular membranes. These alterations may result in the mislocalization and aggregation of αS, hinting at a feedback loop where MAM dysfunction affects αS behavior and vice versa.

be the reason why so many of these pathways are altered in the context of synucleinopathies.

We further posit that the activity of specific MAM-resident lipid enzymes will be particularly sensitive to αS pathology and thus the concentration of their enzymatic lipid substrates and products, which may serve as reporters of αS pathology. The identification of those lipid species that optimally stratify with defects in MAM regulation in synucleinopathies may be particularly promising strategies to detect and monitor these neurological disorders. Furthermore, insight into the αS–MAM interplay may also contribute to lipid-related therapeutic targets. One such candidate therapeutic target is outlined below.

OVERCOMING αS TOXICITY BY MODULATING LIPID METABOLISM: STEAROYL-CoA DESATURASE

Insights on the role of αS and lipid metabolism in the context of synucleinopathies has given rise to lipid-centered therapeutic strategies that are being tested in preclinical and clinical studies. A prominent example for a lipid-centered strategy to overcome αS toxicity and perturbation is the inhibition of the MAM-enriched lipid enzyme SCD. Based on the initial observation that αS expression in yeast is associated with MUFA increase, SCD was identified as a candidate therapeutic target that eventually entered PD clinical trials. Dampening expression of the yeast SCD homolog (Ole1) protected against αS-induced toxicity in yeast (Fanning et al. 2019). Genetic reduction or pharmacological inhibition of SCD reversed PD-relevant phenotypes, including αS toxicity in primary rat cortical neurons overexpressing human αS and αS inclusion formation in a 3K αS neuroblastoma model. SCD reduction decreased pSer129 αS in E46K and 3K αS expressing neuroblastoma cells, patient-derived αS triplication neurons, patient neurons with LRRK2 G2019S or R1441C mutations, and cortical neurospheres of patient-derived A53T iPSC neurons (Vincent et al. 2018; Fanning et al. 2019; Nuber et al. 2021, 2022; Fonseca-Ornelas et al. 2022; Raja et al. 2022). SCD inhibition corrected dysregulated αS tetra-

mer:monomer ratios in human cells expressing 3K or E46K αS (Fanning et al. 2019; Imberdis et al. 2019) as well as in patient neurons carrying very different PD-related mutations, namely, LRRK2 G2019S or R1441C (Fonseca-Ornelas et al. 2022). The aberrant increase in the unfolded protein response associated with triplication of the αS locus (Heman-Ackah et al. 2017) was corrected by reducing SCD activity (Fanning et al. 2022). The SCD inhibitor YTX-7739, which entered clinical trials (Tardiff et al. 2022), reduced the desaturation index for C16 and C18 species. This was associated with reduced neuron death from αS toxicity elicited by αS A53T mutant expression in human iPSC neurons (Nuber et al. 2022). Reducing MUFAs in phospholipid membranes by SCD inhibition relocalized αS from a disproportionate membrane-associated/insoluble state to a more physiological cytosolic/soluble state (Imberdis et al. 2019; Nuber et al. 2022). Reducing desaturases fat-5/fat-7 improved wild-type (WT) and A53T αS-associated stress phenotypes in *Caenorhabditis elegans* (Maulik et al. 2019). In the dopamine-responsive αS 3K mouse model of PD (Nuber et al. 2018), the genetic reduction or pharmacological inhibition of SCD ameliorated several disease-associated phenotypes including neuropathology and motor deficits (Nuber et al. 2021, 2022). In keeping with findings in cellular systems, the SCD inhibitor 5b increased cytosolic soluble αS and αS tetramer:monomer ratio, while also decreasing pSer129 αS deposits and reducing proteinase K-resistant lipid-rich aggregates in the brains of WT and 3K αS expressing mice (Nuber et al. 2021). Reestablishing MUFA homeostasis via SCD reduction rescued gait abnormalities including gait asymmetry in WT and 3K αS mice and prevented progressive motor deficits in the 3K αS mice. In another trial, treating 3K αS mice with YTX-7739 reversed PD-relevant pathologies and motor phenotypes (Nuber et al. 2022). The YTX-7739 compound was confirmed as crossing the blood–brain barrier with a brain:plasma ratio of 1:1. The resulting MUFA reduction was associated with improved rotarod and pole climbing performances in 3K αS mice and enhanced motor skill learning in WT and 3K αS mice. Aligned

Cite this article as *Cold Spring Harb Perspect Med* doi: 10.1101/cshperspect.a041646

with findings for 5b treatment, YTX-7739 reestablished the physiological αS tetramer:monomer ratio, increased cytosolic/soluble αS and decreased insoluble αS in 3K αS mice and reduced pSer129 αS and proteinase K–resistant aggregates in WT and 3K αS mice (Nuber et al. 2022).

OVERCOMING αS TOXICITY BY MODULATING LIPID METABOLISM: OTHERS

Other approaches to correcting aberrant phospholipid membrane MUFA composition provide distinct candidate therapeutic avenues. Increased MUFA synthesis can result in LD accumulation (storing excess FAs). Multiple components of the neutral lipids pathway (DG, TG, and LDs) are elevated by excess mutant αS expression (Fanning et al. 2019). The lipid degradation pathway (via lipases) clears accumulated neutral lipids to generate free FA pools. These free FAs can be incorporated into phospholipids and alter αS:membrane dynamics. Hormone-sensitive lipase (HSL/LIPE) functions in a degradative pathway controlling release of FAs stored in neutral lipid storage forms. Pharmacological inhibition or genetic reduction of LIPE reversed several PD-associated αS phenotypes (membrane excess, pS129 elevation, monomer accumulation) in cellular synucleinopathy models including patient-derived αS triplication iPSC neurons (Fanning et al. 2022) and in vivo in the 3K synucleinopathy model specifically in males (Adom et al. 2024). Moreover, a yeast screen revealed that reducing DGAT activity ameliorates the toxic effects of αS, which was confirmed in other systems (Soste et al. 2019). The most significant known risk factor for PD besides aging and genetic variants such as *SNCA* (αS) and *LRRK2* is *GBA1* (GCase). Homozygous *GBA1* mutations block proper sphingomyelin metabolism and cause Gaucher's disease, a developmental disorder characterized by lysosomal dysfunction. Heterozygous mutations (i.e., in Gaucher's carriers) have repeatedly been found to increase PD risk (Clark et al. 2007; Nichols et al. 2009). A lack of GCase and the resultant glucosylceramide buildup promotes increased formation of abnormal oligomers of αS (Mazzulli et al. 2011). In turn, *SNCA* duplication and increased αS dosage result in reduced lysosomal GCase activity, which further stabilizes αS oligomers. There are currently several clinical trials trying to overcome a lack of GCase activity in disease, based on different strategies (reviewed by Smith and Schapira 2022). While the lipid connection is less clear in the case of LRRK2, it has been demonstrated, among other things, that SCD inhibition can overcome LRRK2-related perturbation of αS biology (Fonseca-Ornelas et al. 2022).

CONCLUSIONS

PD has traditionally been approached as a classical "proteinopathy" (i.e., a disease caused by protein misfolding to β-sheet-rich fibrillar aggregates). Affected neurons are considered to have imbalanced protein synthesis, folding, and degradation resulting in neuronal dysfunction and death. Reduced (Burré et al. 2015) or excess (Galvagnion et al. 2015) αS-membrane interactions and specific FA interactions (Sharon et al. 2003) are considered potential triggers for toxic αS oligomer and fibril formation, in keeping with the idea of a lipid-induced proteinopathy. The latest data on LB analysis are also consistent with the opposite (i.e., a synucleinopathy being a protein-induced lipidopathy). This supports cellular lipid dyshomeostasis as the driver of neurotoxicity and αS dyshomeostasis (e.g., excess or reduced αS vesicle binding with αS fibrillar aggregation) as an initiator. Synucleinopathies may be both proteinopathies and lipidopathies with a feedback cycle of dyshomeostasis in lipid metabolism and protein folding, either of which could be triggered by early changes in the subtleties of protein dynamics or the lipidome.

Emerging support for αS accumulation in LBs suggests fibrillar αS may not be the sole contributor to disease. The "PFF" model of αS fibrillization and proteotoxicity (Volpicelli-Daley et al. 2011; Luk et al. 2012) may have to be reconsidered and evaluated in collaboration with models of endogenous αS and intracellular membrane-mediated αS aggregation (Dettmer et al. 2015a, 2017; Lam et al. 2024) and transgenic mice (Nuber et al. 2018).

A shift to approaching synucleinopathies in terms of protein and lipid alterations, could help explore new ideas for therapeutics (e.g., drugs that alter lipid and FA homeostasis may become promising targets) (Vincent et al. 2018; Fanning et al. 2019; Zheng et al. 2019). SCD inhibition, a new candidate therapeutic target, may counteract both changes in lipid metabolism triggered by αS accumulation and prevent negative feedback of these lipid changes on αS conformations (Fanning et al. 2019). The mechanistic link between αS, MAM regulation, and lipid homeostasis establishes a new framework to think about the source and relevance of the lipid alterations frequently observed in the disease and opens new avenues to think about diagnosis/biomarkers.

ACKNOWLEDGMENTS

We thank the members of the Area-Gómez (Margarita Salas-CSIC, Columbia), Fanning, Dettmer, Selkoe, and Nuber laboratories (at BWH/HMS) for advice and scientific discussions. We thank Christiane Lantermann for her contributions to the illustrations. Our work on αS is supported by the National Institutes of Health (grant numbers NS121826, NS099328, NS109209, and NS133979 to U.D. and NS133243 to S.F.); Spanish Ministry of Education, La Caixa Health and Luzon Foundations (PID2021-126818NB-I00; HR23-00124 and OTR11254 to E.A.-G.); Michael J. Fox grant MJFF-021178 (S.F.); the Ellison Foundation of Boston (S.F.). The funders had no role in study design, data collection and analysis, decision to publish, or preparation of the manuscript.

REFERENCES

Abeliovich A, Schmitz Y, Fariñas I, Choi-Lundberg D, Ho WH, Castillo PE, Shinsky N, Verdugo JM, Armanini M, Ryan A, et al. 2000. Mice lacking α-synuclein display functional deficits in the nigrostriatal dopamine system. *Neuron* **25**: 239–252. doi:10.1016/S0896-6273(00)80886-7

Adom MA, Hahn WN, McCaffery TD, Moors TE, Zhang X, Svenningsson P, Selkoe DJ, Fanning S, Nuber S. 2024. Reducing the lipase LIPE in mutant α-synuclein mice improves Parkinson-like deficits and reveals sex differences in fatty acid metabolism. *Neurobiol Dis* **199**: 106 593. doi:10.1016/j.nbd.2024.106593

Bartels T. 2019. A traffic jam leads to Lewy bodies. *Nat Neurosci* **22**: 1043–1045. doi:10.1038/s41593-019-0435-y

Bartels T, Choi JG, Selkoe DJ. 2011. α-Synuclein occurs physiologically as a helically folded tetramer that resists aggregation. *Nature* **477**: 107–110. doi:10.1038/nature 10324

Bodner CR, Dobson CM, Bax A. 2009. Multiple tight phospholipid-binding modes of α-synuclein revealed by solution NMR spectroscopy. *J Mol Biol* **390**: 775–790. doi:10.1016/j.jmb.2009.05.066

Bodner CR, Maltsev AS, Dobson CM, Bax A. 2010. Differential phospholipid binding of α-synuclein variants implicated in Parkinson's disease revealed by solution NMR spectroscopy. *Biochemistry* **49**: 862–871. doi:10.1021/bi901723p

Braak H, Del Tredici K, Rüb U, de Vos RA, Jansen Steur EN, Braak E. 2003. Staging of brain pathology related to sporadic Parkinson's disease. *Neurobiol Aging* **24**: 197–211. doi:10.1016/S0197-4580(02)00065-9

Burré J, Sharma M, Südhof TC. 2014. α-Synuclein assembles into higher-order multimers upon membrane binding to promote SNARE complex formation. *Proc Natl Acad Sci* **111**: E4274–E4283. doi:10.1073/pnas.1416598111

Burré J, Sharma M, Südhof TC. 2015. Definition of a molecular pathway mediating α-synuclein neurotoxicity. *J Neurosci* **35**: 5221–5232. doi:10.1523/JNEUROSCI.4650-14.2015

Chandra S, Fornai F, Kwon HB, Yazdani U, Atasoy D, Liu X, Hammer RE, Battaglia G, German DC, Castillo PE, et al. 2004. Double-knockout mice for α- and β-synucleins: effect on synaptic functions. *Proc Natl Acad Sci* **101**: 14966–14971. doi:10.1073/pnas.0406283101

Chung CY, Khurana V, Auluck PK, Tardiff DF, Mazzulli JR, Soldner F, Baru V, Lou Y, Freyzon Y, Cho S, et al. 2013. Identification and rescue of α-synuclein toxicity in Parkinson patient-derived neurons. *Science* **342**: 983–987. doi:10.1126/science.1245296

Clark LN, Ross BM, Wang Y, Mejia-Santana H, Harris J, Louis ED, Cote LJ, Andrews H, Fahn S, Waters C, et al. 2007. Mutations in the glucocerebrosidase gene are associated with early-onset Parkinson disease. *Neurology* **69**: 1270–1277. doi:10.1212/01.wnl.0000276989.17578.02

Davidson WS, Jonas A, Clayton DF, George JM. 1998. Stabilization of α-synuclein secondary structure upon binding to synthetic membranes. *J Biol Chem* **273**: 9443–9449. doi:10.1074/jbc.273.16.9443

Dettmer U, Newman AJ, Luth ES, Bartels T, Selkoe D. 2013. In vivo cross-linking reveals principally oligomeric forms of α-synuclein and β-synuclein in neurons and non-neural cells. *J Biol Chem* **288**: 6371–6385. doi:10.1074/jbc.M112.403311

Dettmer U, Newman AJ, Soldner F, Luth ES, Kim NC, von Saucken VE, Sanderson JB, Jaenisch R, Bartels T, Selkoe D. 2015a. Parkinson-causing α-synuclein missense mutations shift native tetramers to monomers as a mecha-

nism for disease initiation. *Nat Commun* **6:** 7314. doi:10
.1038/ncomms8314

Dettmer U, Newman AJ, von Saucken VE, Bartels T, Selkoe
D. 2015b. KTKEGV repeat motifs are key mediators
of normal α-synuclein tetramerization: their mutation
causes excess monomers and neurotoxicity. *Proc Natl
Acad Sci* **112:** 9596–9601. doi:10.1073/pnas.1505953112

Dettmer U, Ramalingam N, von Saucken VE, Kim TE, Newman AJ, Terry-Kantor E, Nuber S, Ericsson M, Fanning S,
Bartels T, et al. 2017. Loss of native α-synuclein multimerization by strategically mutating its amphipathic helix
causes abnormal vesicle interactions in neuronal cells.
Hum Mol Genet **26:** 3466–3481. doi:10.1093/hmg/ddx
227

Dickson DW, Crystal H, Mattiace LA, Kress Y, Schwagerl A,
Ksiezak-Reding H, Davies P, Yen SH. 1989. Diffuse Lewy
body disease: light and electron microscopic immunocy-
tochemistry of senile plaques. *Acta Neuropathol* **78:** 572–
584. doi:10.1007/BF00691284

Fabelo N, Martín V, Santpere G, Marín R, Torrent L, Ferrer I,
Díaz M. 2011. Severe alterations in lipid composition of
frontal cortex lipid rafts from Parkinson's disease and
incidental Parkinson's disease. *Mol Med* **17:** 1107–1118.
doi:10.2119/molmed.2011.00119

Fanning S, Haque A, Imberdis T, Baru V, Barrasa MI, Nuber
S, Termine D, Ramalingam N, Ho GPH, Noble T, et al.
2019. Lipidomic analysis of α-synuclein neurotoxicity
identifies stearoyl CoA desaturase as a target for Parkinson treatment. *Mol Cell* **73:** 1001–1014.e8. doi:10.1016/j
.molcel.2018.11.028

Fanning S, Selkoe D, Dettmer U. 2021. Vesicle trafficking
and lipid metabolism in synucleinopathy. *Acta Neuropathol* **141:** 491–510. doi:10.1007/s00401-020-02177-z

Fanning S, Cirka H, Thies JL, Jeong J, Niemi SM, Yoon J, Ho
GPH, Pacheco JA, Dettmer U, Liu L, et al. 2022. Lipase
regulation of cellular fatty acid homeostasis as a Parkinson's disease therapeutic strategy. *NPJ Parkinsons Dis* **8:**
74. doi:10.1038/s41531-022-00335-6

Farmer K, Smith CA, Hayley S, Smith J. 2015. Major alterations of phosphatidylcholine and lysophosphotidylcho-
line lipids in the substantia nigra using an early stage
model of Parkinson's disease. *Int J Mol Sci* **16:** 18865–
18877. doi:10.3390/ijms160818865

Fonseca-Ornelas L, Stricker JMS, Soriano-Cruz S, Weykopf
B, Dettmer U, Muratore CR, Scherzer CR, Selkoe DJ.
2022. Parkinson-causing mutations in LRRK2 impair
the physiological tetramerization of endogenous α-synuclein in human neurons. *NPJ Parkinsons Dis* **8:** 118.
doi:10.1038/s41531-022-00380-1

Forno LS, Norville RL. 1976. Ultrastructure of Lewy bodies
in the stellate ganglion. *Acta Neuropathol* **34:** 183–197.
doi:10.1007/BF00688674

Galvagnion C, Buell AK, Meisl G, Michaels TCT, Vendruscolo M, Knowles TPJ, Dobson CM. 2015. Lipid vesicles
trigger α-synuclein aggregation by stimulating primary
nucleation. *Nat Chem Biol* **11:** 229–234. doi:10.1038/
nchembio.1750

George JM, Jin H, Woods WS, Clayton DF. 1995. Characterization of a novel protein regulated during the critical
period for song learning in the zebra finch. *Neuron* **15:**
361–372. doi:10.1016/0896-6273(95)90040-3

Gitler AD, Bevis BJ, Shorter J, Strathearn KE, Hamamichi S,
Su LJ, Caldwell KA, Caldwell GA, Rochet JC, McCaffery
JM, et al. 2008. The Parkinson's disease protein α-synu-
clein disrupts cellular Rab homeostasis. *Proc Natl Acad
Sci* **105:** 145–150. doi:10.1073/pnas.0710685105

Guardia-Laguarta C, Area-Gomez E, Rüb C, Liu Y, Magrané
J, Becker D, Voos W, Schon EA, Przedborski S. 2014.
α-Synuclein is localized to mitochondria-associated ER
membranes. *J Neurosci* **34:** 249–259. doi:10.1523/JNEUR
OSCI.2507-13.2014

Hayashida K, Oyanagi S, Mizutani Y, Yokochi M. 1993. An
early cytoplasmic change before Lewy body maturation:
an ultrastructural study of the substantia nigra from an
autopsy case of juvenile parkinsonism. *Acta Neuropathol*
85: 445–448. doi:10.1007/BF00334457

Heman-Ackah SM, Manzano R, Hoozemans JJM, Scheper
W, Flynn R, Haerty W, Cowley SA, Bassett AR, Wood
MJA. 2017. α-Synuclein induces the unfolded protein
response in Parkinson's disease SNCA triplication
iPSC-derived neurons. *Hum Mol Genet* **26:** 4441–4450.
doi:10.1093/hmg/ddx331

Ibarguren M, López DJ, Escribá PV. 2014. The effect of
natural and synthetic fatty acids on membrane structure,
microdomain organization, cellular functions and human
health. *Biochim Biophys Acta* **1838:** 1518–1528. doi:10
.1016/j.bbamem.2013.12.021

Imberdis T, Negri J, Ramalingam N, Terry-Kantor E, Ho
GPH, Fanning S, Stirtz G, Kim TE, Levy OA, Young-
Pearse TL, et al. 2019. Cell models of lipid-rich α-synuclein aggregation validate known modifiers of α-synuclein biology and identify stearoyl-CoA desaturase. *Proc
Natl Acad Sci* **116:** 20760–20769. doi:10.1073/pnas.1903
216116

Isik OA, Cizmecioglu O. 2023. Rafting on the plasma membrane: lipid rafts in signaling and disease. *Adv Exp Med
Biol* **1436:** 87–108. doi:10.1007/5584_2022_759

Jao CC, Hegde BG, Chen J, Haworth IS, Langen R. 2008.
Structure of membrane-bound α-synuclein from site-di-
rected spin labeling and computational refinement. *Proc
Natl Acad Sci* **105:** 19666–19671. doi:10.1073/pnas.0807
826105

Lam I, Ndayisaba A, Lewis AJ, Fu Y, Sagredo GT, Kuzkina A,
Zaccagnini L, Celikag M, Sandoe J, Sanz RL, et al. 2024.
Rapid iPSC inclusionopathy models shed light on formation, consequence, and molecular subtype of α-synuclein
inclusions. *Neuron* **112:** 2886–2909.e16. doi:10.1016/j
.neuron.2024.06.002

Lashuel HA, Hartley D, Petre BM, Walz T, Lansbury PT.
2002. Neurodegenerative disease: amyloid pores from
pathogenic mutations. *Nature* **418:** 291. doi:10.1038/
418291a

Leonard TA, Loose M, Martens S. 2023. The membrane
surface as a platform that organizes cellular and biochemical processes. *Dev Cell* **58:** 1315–1332. doi:10.1016/j
.devcel.2023.06.001

Lou X, Kim J, Hawk BJ, Shin YK. 2017. α-Synuclein may
cross-bridge v-SNARE and acidic phospholipids to facilitate SNARE-dependent vesicle docking. *Biochem J* **474:**
2039–2049. doi:10.1042/BCJ20170200

Luk KC, Kehm VM, Zhang B, O'Brien P, Trojanowski JQ,
Lee VMY. 2012. Intracerebral inoculation of pathological
α-synuclein initiates a rapidly progressive neurodegener-

ative α-synucleinopathy in mice. *J Exp Med* **209**: 975–986. doi:10.1084/jem.20112457

Maulik M, Mitra S, Basmayor AM, Lu B, Taylor BE, Bult-Ito A. 2019. Genetic silencing of fatty acid desaturases modulates α-synuclein toxicity and neuronal loss in Parkinson-like models of *C. elegans*. *Front Aging Neurosci* **11**: 207. doi:10.3389/fnagi.2019.00207

Mazzulli JR, Xu YH, Sun Y, Knight AL, McLean PJ, Caldwell GA, Sidransky E, Grabowski GA, Krainc D. 2011. Gaucher disease glucocerebrosidase and α-synuclein form a bidirectional pathogenic loop in synucleinopathies. *Cell* **146**: 37–52. doi:10.1016/j.cell.2011.06.001

Montesinos J, Guardia-Laguarta C, Area-Gomez E. 2020. The fat brain. *Curr Opin Clin Nutr Metab Care* **23**: 68–75. doi:10.1097/MCO.0000000000000634

Moors TE, Milovanovic D. 2024. Defining a Lewy body: running up the hill of shifting definitions and evolving concepts. *J Parkinsons Dis* **14**: 17–33. doi:10.3233/JPD-230183

Moors TE, Maat CA, Niedieker D, Mona D, Petersen D, Timmermans-Huisman E, Kole J, El-Mashtoly SF, Spycher L, Zago W, et al. 2021. The subcellular arrangement of α-synuclein proteoforms in the Parkinson's disease brain as revealed by multicolor STED microscopy. *Acta Neuropathol* **142**: 423–448. doi:10.1007/s00401-021-02329-9

Nichols WC, Pankratz N, Marek DK, Pauciulo MW, Elsaesser VE, Halter CA, Rudolph A, Wojcieszek J, Pfeiffer RF, Foroud T, et al. 2009. Mutations in GBA are associated with familial Parkinson disease susceptibility and age at onset. *Neurology* **72**: 310–316. doi:10.1212/01.wnl.0000327823.81237.d1

Nishimura M, Tomimoto H, Suenaga T, Nakamura S, Namba Y, Ikeda K, Akiguchi I, Kimura J. 1994. Synaptophysin and chromogranin A immunoreactivities of Lewy bodies in Parkinson's disease brains. *Brain Res* **634**: 339–344. doi:10.1016/0006-8993(94)91940-2

Nuber S, Rajsombath M, Minakaki G, Winkler J, Müller CP, Ericsson M, Caldarone B, Dettmer U, Selkoe DJ. 2018. Abrogating native α-synuclein tetramers in mice causes a L-DOPA-responsive motor syndrome closely resembling Parkinson's disease. *Neuron* **100**: 75–90.e5. doi:10.1016/j.neuron.2018.09.014

Nuber S, Nam AY, Rajsombath MM, Cirka H, Hronowski X, Wang J, Hodgetts K, Kalinichenko LS, Müller CP, Lambrecht V, et al. 2021. A stearoyl-coenzyme A desaturase inhibitor prevents multiple Parkinson disease phenotypes in α-synuclein mice. *Ann Neurol* **89**: 74–90. doi:10.1002/ana.25920

Nuber S, Chung CY, Tardiff DF, Bechade PA, McCaffery TD, Shimanaka K, Choi J, Chang B, Raja W, Neves E, et al. 2022. A brain-penetrant stearoyl-CoA desaturase inhibitor reverses α-synuclein toxicity. *Neurotherapeutics* **19**: 1018–1036. doi:10.1007/s13311-022-01199-7

Nuscher B, Kamp F, Mehnert T, Odoy S, Haass C, Kahle PJ, Beyer K. 2004. α-Synuclein has a high affinity for packing defects in a bilayer membrane: a thermodynamics study. *J Biol Chem* **279**: 21966–21975. doi:10.1074/jbc.M401076200

O'Leary EI, Jiang Z, Strub MP, Lee JC. 2018. Effects of phosphatidylcholine membrane fluidity on the conformation and aggregation of N-terminally acetylated α-synuclein. *J*

Biol Chem **293**: 11195–11205. doi:10.1074/jbc.RA118.002780

Raja WK, Neves E, Burke C, Jiang X, Xu P, Rhodes KJ, Khurana V, Scannevin RH, Chung CY. 2022. Patient-derived three-dimensional cortical neurospheres to model Parkinson's disease. *PLoS One* **17**: e0277532. doi:10.1371/journal.pone.0277532

Ramalingam N, Dettmer U. 2021. Temperature is a key determinant of α- and β-synuclein membrane interactions in neurons. *J Biol Chem* **296**: 100271. doi:10.1016/j.jbc.2021.100271

Ramalingam N, Brontesi L, Jin SX, Selkoe DJ, Dettmer U. 2023. Dynamic reversibility of α-synuclein serine-129 phosphorylation is impaired in synucleinopathy models. *EMBO Rep* **24**: e57145. doi:10.15252/embr.202357145

Rovere M, Sanderson JB, Fonseca-Ornelas L, Patel DS, Bartels T. 2018. Refolding of helical soluble α-synuclein through transient interaction with lipid interfaces. *FEBS Lett* **592**: 1464–1472. doi:10.1002/1873-3468.13047

Roy S, Wolman L. 1969. Ultrastructural observations in Parkinsonism. *J Pathol* **99**: 39–44. doi:10.1002/path.1710990106

Runwal G, Edwards RH. 2021. The membrane interactions of synuclein: physiology and pathology. *Annu Rev Pathol* **16**: 465–485. doi:10.1146/annurev-pathol-031920-092547

Rusiñol AE, Cui Z, Chen MH, Vance JE. 1994. A unique mitochondria-associated membrane fraction from rat liver has a high capacity for lipid synthesis and contains pre-Golgi secretory proteins including nascent lipoproteins. *J Biol Chem* **269**: 27494–27502. doi:10.1016/S0021-9258(18)47012-3

Sastry PS. 1985. Lipids of nervous tissue: composition and metabolism. *Prog Lipid Res* **24**: 69–176. doi:10.1016/0163-7827(85)90011-6

Schepers J, Löser T, Behl C. 2024. Lipids and α-synuclein: adding further variables to the equation. *Front Mol Biosci* **11**: 1455817. doi:10.3389/fmolb.2024.1455817

Scorrano L, De Matteis MA, Emr S, Giordano F, Hajnóczky G, Kornmann B, Lackner LL, Levine TP, Pellegrini L, Reinisch K, et al. 2019. Coming together to define membrane contact sites. *Nat Commun* **10**: 1287. doi:10.1038/s41467-019-09253-3

Segawa K, Nagata S. 2015. An apoptotic "eat me" signal: phosphatidylserine exposure. *Trends Cell Biol* **25**: 639–650. doi:10.1016/j.tcb.2015.08.003

Shahmoradian SH, Lewis AJ, Genoud C, Hench J, Moors TE, Navarro PP, Castaño-Díez D, Schweighauser G, Graff-Meyer A, Goldie KN, et al. 2019. Lewy pathology in Parkinson's disease consists of crowded organelles and lipid membranes. *Nat Neurosci* **22**: 1099–1109. doi:10.1038/s41593-019-0423-2

Sharon R, Goldberg MS, Bar-Josef I, Betensky RA, Shen J, Selkoe DJ. 2001. α-Synuclein occurs in lipid-rich high molecular weight complexes, binds fatty acids, and shows homology to the fatty acid-binding proteins. *Proc Natl Acad Sci* **98**: 9110–9115. doi:10.1073/pnas.171300598

Sharon R, Bar-Joseph I, Frosch MP, Walsh DM, Hamilton JA, Selkoe DJ. 2003. The formation of highly soluble oligomers of α-synuclein is regulated by fatty acids and enhanced in Parkinson's disease. *Neuron* **37**: 583–595. doi:10.1016/S0896-6273(03)00024-2

Shimokawa N, Takagi M. 2024. Biomimetic lipid raft: domain stability and interaction with physiologically active molecules. *Adv Exp Med Biol* **1461:** 15–32. doi:10.1007/978-981-97-4584-5_2

Smith L, Schapira AHV. 2022. GBA variants and Parkinson disease: mechanisms and treatments. *Cells* **11:** 1261. doi:10.3390/cells11081261

Soper JH, Roy S, Stieber A, Lee E, Wilson RB, Trojanowski JQ, Burd CG, Lee VMY. 2008. α-Synuclein-induced aggregation of cytoplasmic vesicles in *Saccharomyces cerevisiae*. *Mol Biol Cell* **19:** 1093–1103. doi:10.1091/mbc.e07-08-0827

Soste M, Charmpi K, Lampert F, Gerez JA, van Oostrum M, Malinovska L, Boersema PJ, Prymaczok NC, Riek R, Peter M, et al. 2019. Proteomics-based monitoring of pathway activity reveals that blocking diacylglycerol biosynthesis rescues from α-synuclein toxicity. *Cell Syst* **9:** 309–320.e8. doi:10.1016/j.cels.2019.07.010

Spillantini MG, Schmidt ML, Lee VM, Trojanowski JQ, Jakes R, Goedert M. 1997. α-Synuclein in Lewy bodies. *Nature* **388:** 839–840. doi:10.1038/42166

Spillantini MG, Crowther RA, Jakes R, Hasegawa M, Goedert M. 1998. α-Synuclein in filamentous inclusions of Lewy bodies from Parkinson's disease and dementia with Lewy bodies. *Proc Natl Acad Sci* **95:** 6469–6473. doi:10.1073/pnas.95.11.6469

Tardiff DF, Lucas M, Wrona I, Chang B, Chung CY, Le Bourdonnec B, Rhodes KJ, Scannevin RH. 2022. Nonclinical pharmacology of YTX-7739: a clinical stage stearoyl-CoA desaturase inhibitor being developed for Parkinson's disease. *Mol Neurobiol* **59:** 2171–2189. doi:10.1007/s12035-021-02695-1

Vance JE. 2014. MAM (mitochondria-associated membranes) in mammalian cells: lipids and beyond. *Biochim Biophys Acta* **1841:** 595–609. doi:10.1016/j.bbalip.2013.11.014

van Meer G, Voelker DR, Feigenson GW. 2008. Membrane lipids: where they are and how they behave. *Nat Rev Mol Cell Biol* **9:** 112–124. doi:10.1038/nrm2330

Vargas KJ, Makani S, Davis T, Westphal CH, Castillo PE, Chandra SS. 2014. Synucleins regulate the kinetics of synaptic vesicle endocytosis. *J Neurosci* **34:** 9364–9376. doi:10.1523/JNEUROSCI.4787-13.2014

Vincent BM, Tardiff DF, Piotrowski JS, Aron R, Lucas MC, Chung CY, Bacherman H, Chen Y, Pires M, Subramaniam R, et al. 2018. Inhibiting stearoyl-CoA desaturase ameliorates α-synuclein cytotoxicity. *Cell Rep* **25:** 2742–2754.e31. doi:10.1016/j.celrep.2018.11.028

Volles MJ, Lansbury PT. 2007. Relationships between the sequence of α-synuclein and its membrane affinity, fibrillization propensity, and yeast toxicity. *J Mol Biol* **366:** 1510–1522. doi:10.1016/j.jmb.2006.12.044

Volpicelli-Daley LA, Luk KC, Patel TP, Tanik SA, Riddle DM, Stieber A, Meaney DF, Trojanowski JQ, Lee VM-Y. 2011. Exogenous α-synuclein fibrils induce Lewy body pathology leading to synaptic dysfunction and neuron death. *Neuron* **72:** 57–71. doi:10.1016/j.neuron.2011.08.033

Wang W, Perovic I, Chittuluru J, Kaganovich A, Nguyen LTT, Liao J, Auclair JR, Johnson D, Landeru A, Simorellis AK, et al. 2011. A soluble α-synuclein construct forms a dynamic tetramer. *Proc Natl Acad Sci* **108:** 17797–17802. doi:10.1073/pnas.1113260108

Wang L, Das U, Scott DA, Tang Y, McLean PJ, Roy S. 2014. α-Synuclein multimers cluster synaptic vesicles and attenuate recycling. *Curr Biol* **24:** 2319–2326. doi:10.1016/j.cub.2014.08.027

Weinreb PH, Zhen W, Poon AW, Conway KA, Lansbury PT. 1996. NACP, a protein implicated in Alzheimer's disease and learning, is natively unfolded. *Biochemistry* **35:** 13709–13715. doi:10.1021/bi961799n

Westphal CH, Chandra SS. 2013. Monomeric synucleins generate membrane curvature. *J Biol Chem* **288:** 1829–1840. doi:10.1074/jbc.M112.418871

Zheng J, Jeon S, Jiang W, Burbulla LF, Ysselstein D, Oevel K, Krainc D, Silverman RB. 2019. Conversion of quinazoline modulators from inhibitors to activators of β-glucocerebrosidase. *J Med Chem* **62:** 1218–1230. doi:10.1021/acs.jmedchem.8b01294

Autophagy and Protein Quality Control in Parkinson's Disease

Marta Martinez-Vicente[1] and Miquel Vila[1,2,3,4]

[1]Neurodegenerative Diseases Research Group, Vall d'Hebron Research Institute (VHIR) Network Center for Biomedical Research in Neurodegenerative Diseases (CIBERNED), 08035 Barcelona, Spain

[2]Catalan Institution for Research and Advanced Studies (ICREA), 08010 Barcelona, Spain

[3]Neuroscience Institute-Autonomous University of Barcelona (INc-UAB), 08193 Barcelona, Spain

[4]Aligning Science Across Parkinson's Collaborative Research Network (ASAP-CRN), Chevy Chase, Maryland 20815, USA

Correspondence: marta.martinez@vhir.org

Autophagy is a vital cellular process responsible for the degradation of proteins, organelles, and other cellular components within lysosomes. In neurons, basal autophagy is indispensable for maintaining cellular homeostasis and protein quality control. Accordingly, lysosomal dysfunction has been proposed to be associated with neurodegeneration, and with Parkinson's disease (PD) in particular. Aging, dopamine metabolism, and PD-linked genetic mutations are thought to impair the autophagic–lysosomal pathway, disrupt cellular proteostasis, and contribute to PD pathogenesis. These alterations represent an opportunity to identify potential new therapeutic targets and disease biomarkers, thus laying the groundwork for the development of novel disease-modifying strategies for PD that are aimed at restoring cellular proteostasis and quality control systems.

Neuronal homeostasis depends on the proper functioning of quality control systems. All intracellular components, including proteins and organelles, are subject to continuous turnover through coordinated synthesis, degradation, and recycling of their constituent elements. Cells have developed complex mechanisms to degrade intracellular material and preserve this constant process of maintenance. The two major degradative systems in eukaryotic cells are autophagy and the ubiquitin-proteasome system (UPS). Autophagy is the catabolic mechanism by which intracellular cytosolic components, comprising proteins, organelles, aggregates, and any other intracellular materials, are delivered to lysosomes for degradation. This process collectively forms the autophagic-lysosomal pathway (ALP) (Fig. 1).

In neurons, basal autophagy is essential to maintain homeostasis and avoid neurodegeneration. Neurons are postmitotic cells that lack the ability to dilute unwanted cellular components through division. They also possess an elaborate morphology and need to achieve degradation at

Copyright © 2025 Cold Spring Harbor Laboratory Press; all rights reserved
Cite this article as *Cold Spring Harb Perspect Med* doi: 10.1101/cshperspect.a041619

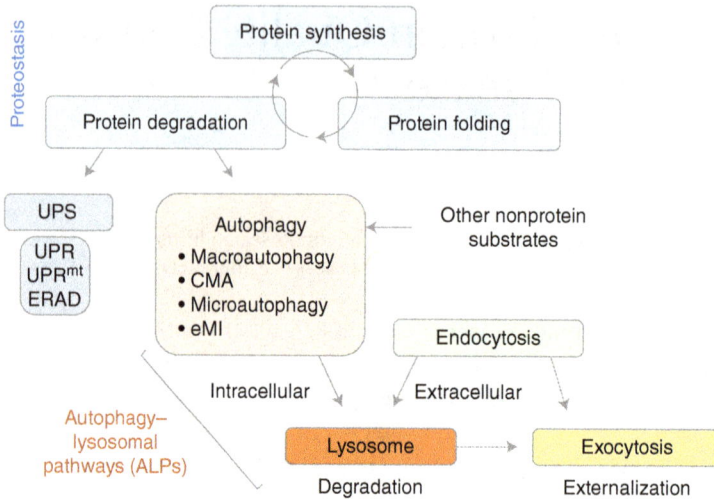

Figure 1. Schematic representation of the overview of the autophagy–lysosomal pathways (ALPs), proteostasis, and other related and interconnected pathways.

distant sites. Neurons therefore rely heavily on autophagy to manage their unique energetic, metabolic, and morphological challenges and avoid the accumulation of nondegraded materials (Scrivo et al. 2018). Dysfunction in autophagy has been implicated in various neurodegenerative diseases, including Parkinson's disease (PD). Autophagy dysfunction contributes to the accumulation of misfolded proteins and damaged organelles, emphasizing the importance of understanding its potential pathogenic role. Additionally, aging reduces the effectiveness of autophagy so that as we age, the nervous system becomes more vulnerable and increasingly dependent on autophagic pathways (Kaushik et al. 2021).

AUTOPHAGIC–LYSOSOMAL PATHWAY MECHANISMS

Lysosomes are specialized organelles containing acid hydrolases, which are capable of degrading all types of macromolecules, such as proteins, polysaccharides, lipids, carbohydrates, and nucleic acids. The main function of lysosomes is the degradation of cellular components: These are broken down into their primary constituents

to be recycled and used to build new cellular components (Xu and Ren 2015; Bonam et al. 2019). It is important to note that, in addition to their pivotal role as degradative and recycling organelles, lysosomes also have an important role in cell metabolism as nutrient-sensing and metabolic signal transduction hubs (Ballabio and Bonifacino 2020).

Lysosomal dysfunction, caused by different factors like aging, disease-associated mutant proteins, or some PD-associated gene variants, can prevent normal degradation of substrates. The subsequent accumulation of undigested cellular components can contribute to neurotoxicity and neurodegeneration. In mammalian cells, three different and coexisting autophagy mechanisms have been described: macroautophagy, chaperone-mediated autophagy (CMA), and microautophagy. These three mechanisms differ in their regulation, selectivity, and recognition of substrates, and mechanism to deliver these substrates to the lysosome where they are finally degraded (Fig. 1; Mizushima et al. 2008).

Macroautophagy is commonly referred to as autophagy. It is an intricate process that involves the formation of double-membraned vesicles, known as autophagosomes, that encapsulate in-

Cite this article as *Cold Spring Harb Perspect Med* doi: 10.1101/cshperspect.a041619

tracellular components destined for degradation within lysosomes. Macroautophagy involves various stages, including initiation, autophagosome formation, cargo recognition, trafficking, and fusion with lysosomes (Fig. 2; Mizushima and Levine 2020). Macroautophagy exhibits versatility, existing in two functional forms: "bulk" and "selective." Bulk macroautophagy engulfs large and random portions of the cytoplasm, while selective macroautophagy targets specific cellular components. In both cases, substrates are finally degraded inside the lysosome after fusion of the loaded autophagosomes and lysosome (Mizushima and Levine 2020).

Selective macroautophagy can degrade various types of organelles, aggregated components, or pathogens. Different forms of selective macroautophagy are described according to the substrate involved: mitophagy (mitochondria), aggrephagy (protein aggregates), ER-phagy (endoplasmic reticulum [ER]), ferritinophagy (ferritin), lysophagy (lysosomes), glycophagy (glycogen), lipophagy (lipid droplets), pexophagy (peroxisomes), ribophagy (ribosomes), RN/DNautophagy (acid nucleics), or xenophagy (external pathogens) (Fig. 2; Levine and Kroemer 2019; Lamark and Johansen 2021). These different forms of selective macroautophagy rely on receptors to recognize their target substrate and facilitate its sequestration within autophagosomes. Failure of selective degradation by macroautophagy of mitochondria (mitophagy) and of protein aggregates (aggrephagy) has important pathophysiological consequences in neurodegenerative diseases. Notably, some PD-associated proteins, such as PINK1 and Parkin, play an important role in mitophagy machinery (see below).

Macroautophagy is induced in response to diverse cellular stressors, including nutrient deprivation, hypoxia, and genotoxic stress. Mammalian target of rapamycin (mTOR) kinase, as the principal component mTORC1 complex, is the main negative regulator of macroautophagy: It inhibits autophagosome initiation under basal conditions by inactivating the ULK1/ATG13/FIP200 complex. Under different forms of stress (e.g., amino acid deprivation [starvation], insulin, growth factors, low ATP levels, or hypoxia), mTOR is inactivated, and autophagy can be induced (Jung et al. 2010). Additionally, AMPK signaling can activate macroautophagy under conditions of energy deficiency, through the phosphorylation of different synchronized complexes involved in macroautophagy initiation, such as the Beclin1–VPS34 complex (Fig. 2). Transcription factor EB (TFEB) emerges as a key transcriptional regulator of macroautophagy. TFEB enhances the degradative process through the coordinated expression of genes involved in both lysosomal biogenesis and in the initiation and formation of new autophagosomes (Settembre et al. 2011; Abokyi et al. 2023).

CMA is characterized by the specificity of its substrates, which consist exclusively of cytosolic proteins carrying a specific CMA motif. Unlike macroautophagy—which requires the formation of autophagosomes to sequester substrates—CMA does not involve any double-membrane vesicle formation. Instead, substrate proteins are directly translocated into lysosomes for degradation through a lysosomal–membrane complex (Kaushik and Cuervo 2018).

The specific CMA motif is a pentapeptide related to the amino acid sequence KFERQ. This sequence in the substrate protein is recognized by the HSC70 chaperone, which binds the protein to create an HSC70–substrate complex. Then, with the help of other cochaperones, the complex is targeted to the lysosomal membrane. The CMA motif is quite flexible, admitting certain alternative amino acid and posttranslational modifications. Approximately 40% of mammalian proteins contain a recognized CMA motif and are potential candidates for CMA degradation (Kaushik and Cuervo 2018).

The HSC70–substrate complex interacts directly with the transmembrane protein LAMP-2A (lysosome-associated membrane protein type 2A) on the lysosomal membrane (Chiang and Dice 1979; Cuervo and Dice 1979). LAMP-2A plays a pivotal role in CMA as the key translocation protein, acting as the limiting factor for CMA activity. The interaction of LAMP-2A with HSC70 triggers the assembly and stabilization of the translocation complex. During this process, LAMP-2A transitions from an inactive

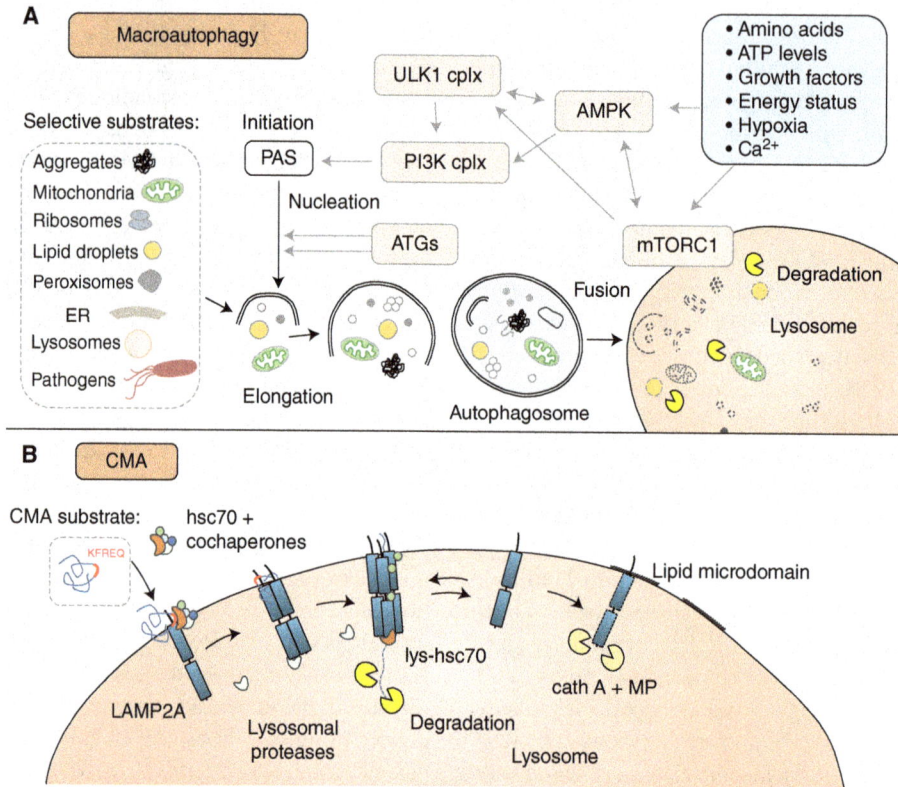

Figure 2. Autophagy pathways. (*A*) Macroautophagy sequesters cytosolic cargo (in bulk and selective cargo) by a delimiting membrane that forms through a complex multistep process involving Atgs proteins and lipids. The membrane then seals into an autophagosome that is fused with lysosomes to degrade the trapped cargo. Macroautophagy is induced in response to diverse cellular stressors (blue). The mTORC1 complex is the main negative regulator of macroautophagy, modulating the pathway through phosphorylation of different synchronized complexes (orange). (*B*) Chaperone-mediated autophagy (CMA): Cytosolic proteins containing the KFERQ-like motif in their sequence are recognized by Hsc70 and brought to the lysosomal membrane for translocation across the lysosome-associated membrane protein type 2A (LAMP-2A) multimeric complex. Lysosomal Hsc70 aids in the translocation of the substrate protein, which is rapidly degraded once inside the lysosomal lumen. LAMP-2A levels and CMA activity are modulated by the localization of the LAMP-2A protein; outside the lipid-enriched microdomains LAMP-2A can form the oligomeric active translocation complex while inside the enriched-lipid microdomain LAMP-2A is found in an inactive monomeric form that can be degraded.

monomeric conformation located in lipid-enriched microdomain regions to an oligomeric conformation outside the lipid-enriched microdomains. This conformational change creates a channel in the lysosomal membrane through which the substrate protein is unfolded and translocated into the lysosome for degradation (Fig. 2).

CMA activity hinges on LAMP-2A abundance on the lysosomal membrane. Different conditions, such as oxidative stress, genotoxic damage, hypoxia, or changes in lipid metabolism, can modulate LAMP-2A levels through different mechanisms. These include the promotion of LAMP-2A expression, LAMP-2A trafficking to lysosomes, and LAMP-2A degradation as a monomer in the membrane (Kaushik et al. 2006; Rodriguez-Navarro et al. 2012; Anguiano et al. 2013; Zhang et al. 2017; Gong et al. 2018; Bourdenx et al. 2021; Chen et al. 2021; Gomez-Sintes et al. 2022; Navarro-Romero et al. 2022; Sohn et al. 2023).

Cite this article as *Cold Spring Harb Perspect Med* doi: 10.1101/cshperspect.a041619

Among neurodegenerative diseases, PD was the first to be associated with dysfunctional CMA. This association is due mainly to the strong reliance of α-synuclein turnover on CMA activity (Cuervo et al. 2004); consequently, any alterations in CMA directly impact α-synuclein levels. Dysfunctional CMA activity, linked to α-synuclein accumulation, has been reported in numerous in vitro and in vivo PD models, as well as in samples derived from PD patients (Chu et al. 2009; Alvarez-Erviti et al. 2010; Dehay et al. 2010). The observed failure of CMA activity in PD may have multiple causes, including age-dependent decline of CMA and deleterious effects on the CMA machinery caused by pathogenic proteins, such as mutant α-synuclein, UCHL1, LRRK2, or GBA (see below).

Microautophagy involves the direct engulfment of cytoplasmic components by lysosomes or late endosomes (LEs) without the formation of autophagosomes. Among the various types of autophagy, microautophagy has received the least research attention and information about it is limited. In fact, most microautophagy research to date has been conducted in yeast, and results cannot be translated into mammalian cells because most molecular components in yeast do not preserve the same function in mammals (Wang et al. 2023). Nevertheless, mammalian cells are known to exhibit a subtype of microautophagy named endosomal microautophagy (eMI), where cytosolic components are internalized directly in LE by microautophagy (Sahu et al. 2011). eMi is a highly selective mechanism. Its substrates might contain the KFERQ-like motif, the same as in CMA, because HSC70 chaperone is the receptor that targets eMi substrates to LE (Tekirdag and Cuervo 2018). Once the substrate is in the LE membrane, and with the help of ESCRT machinery, the LE membrane is deformed and the substrate is incorporated into intraluminal endosomes. Upon the fusion of LE with lysosomes, the substrates are degraded (Sahu et al. 2011).

In summary, most studies on autophagy in neurodegenerative diseases, including PD, have focused on macroautophagy and CMA. While currently there is less direct evidence on eMI in the context of neurodegeneration, it is important to note that the autophagic process as a whole is intricately connected. Different forms of autophagy may compensate for one another, and their dysregulation collectively contributes to cellular dysfunction observed in neurodegenerative diseases.

AUTOPHAGIC–LYSOSOMAL PATHWAY ALTERATIONS IN PARKINSON'S DISEASE

Proteostasis and the Autophagic–Lysosomal Pathway in PD

Proteostasis is a broad concept that encompasses the maintenance of protein homeostasis in cells, from synthesis to degradation. Proteostasis involves the interaction of a network of fundamental cellular processes that ensure balance between protein synthesis, folding, and degradation, thereby maintaining a healthy and functional proteome. The ALP and proteostasis are interconnected but represent distinct cellular networks (Fig. 1).

The ALP is responsible for the degradation of intracellular macromolecules: proteins, organelles, and external pathogens. ALP proteolytic activity therefore plays an essential role in the proteostasis network, and it has a particularly significant role in preventing protein misfolding and aggregation. Other proteolytic pathways are also part of the proteostasis network. For example, the UPS degrades mostly short half-life cytosolic proteins; it coexists with CMA and macroautophagy to manage the turnover of cytosolic proteins.

Under normal physiological conditions, the active proteostasis network can control and prevent the intrinsic ability of proteins to convert from their native conformation to fibrillar aggregates. When this balance is lost, first proteostasis defenses—chaperones, UPS, and CMA—are overcome and abnormal and potentially toxic proteins are accumulated (Dobson 2003; Chiti and Dobson 2017; Hartl 2017). At this point, macroautophagy is the only proteolytic pathway able to accomplish selective elimination of aggregates, through aggrephagy (Fig. 1; Lamark and Johansen 2012).

Recently, CMA has been shown to be instrumental in avoiding neuronal proteotoxicity and guaranteeing proteostasis maintenance (Bourdenx et al. 2021). CMA failure, due to aging or to the presence of pathogenic proteins associated with neurodegenerative diseases such as PD (see below), has a strong impact on the neuronal proteome, promoting protein insolubility and aggregate formation. UPS dysfunction has also been associated with PD, mostly because pharmacological inhibition of the UPS can reproduce some PD phenotypes, generating different in vitro and in vivo PD models (Le 2014; Behl et al. 2022; Liang et al. 2023). Although UPS activity contributes to maintain neuronal proteostasis, the exact role of UPS in PD pathology is controversial and is still not fully understood.

In addition to cytosolic systems formed by chaperones, UPS, and autophagy pathways, the proteostasis network also includes additional organelle-specific systems to handle proteotoxicity inside the organelles, such as the ER and mitochondria. Upon accumulation of unfolded or aberrant proteins in the ER, the unfolded protein response (UPR) is initiated to assist in the identification and degradation of these abnormal unfolded proteins and prevent protein aggregation within the ER (Hetz and Glimcher 2009; Hetz et al. 2020; Phillips et al. 2020). If the UPR response is insufficient to avoid accumulation of unfolded proteins, the endoplasmic reticulum–associated protein degradation (ERAD) system retrotranslocates the unfolded proteins to the cytosol where they can be degraded by UPS (Wu and Rapoport 2018; Needham et al. 2019).

When proteotoxicity in the ER is not handled properly, chronic ER stress rises. This important cellular stressor is associated with many age-related disorders, including PD (Hetz and Saxena 2017; Lehtonen et al. 2019). Indeed, ER stress is a common feature that contributes to PD pathology: Abnormal levels of ER stress are associated with neuronal degeneration in PD models and in postmortem brain tissue from PD patients (Colla et al. 2012; Lehtonen et al. 2019).

The mitochondrial UPR system (UPRmt) is analogous to the ER UPR and is activated in response to mitochondrial proteotoxic stress. Specifically, the UPRmt is a mechanism involving mitochondrial-to-nucleus communication. In response to proteotoxic stress within the mitochondria, the UPRmt activates the expression of a transcriptional program for mitochondrial homeostasis, enhancing mitochondrial chaperones and reducing protein translation (Martinus et al. 1996; Zhao et al. 2002). Dysregulation of the UPRmt has been described in PD models, converging with alterations in other mitochondrial quality control systems, such as mitophagy impairment and excessive ROS production (Martinez et al. 2017).

Interestingly, all systems in the proteolytic network are coordinated. In addition to the redundancy of some pathways under certain conditions and tissues, the continuous cross talk between different proteolytic pathways can compensate for impairment in one system (Massey et al. 2006; Pandey et al. 2007; Kaushik et al. 2008; Matus et al. 2008; Senft and Ronai 2015). However, in the presence of other aggravating factors (aging, pathogenic proteins, or oxidative stress), this compensatory effect between proteostasis systems might be diminished (Schneider et al. 2014; Gavilán et al. 2015) and the resulting failure of proteolytic systems may lead to proteotoxicity, especially in neurons (Bourdenx et al. 2021).

Endocytosis and the Autophagic–Lysosomal Pathway in PD

Endocytosis is a fundamental cellular process crucial for the internalization of extracellular materials. It plays a pivotal role in nutrient uptake, receptor recycling, and cellular signaling. This intricate mechanism involves the engulfment of substances from the cell's external environment, followed by their transport into the cell through vesicle formation. After internalization through different routes, vesicular cargo is delivered in early endosomes that act as a sorting station, targeting cargo toward different destinations. Vesicle contents may be recycled back to the plasma membrane or sorted by retrograde trafficking to the *trans*-Golgi network. Alternatively, early endosomes may mature to

Cite this article as *Cold Spring Harb Perspect Med* doi: 10.1101/cshperspect.a041619

become LE, which fuse to lysosomes to degrade their cargo (Polo and Di Fiore 2006).

Endocytosis and autophagy are cellular pathways that converge at the lysosome. Although their cargo and regulatory mechanisms differ, both pathways use lysosomes as the final destination for cargo processing. Notably, not all cargo internalized through endocytosis undergoes degradation; some are recycled back to the cell surface or directed to various intracellular compartments, emphasizing the versatility of this cellular process. Thus, endocytosis extends beyond its degradative role: It regulates the surface expression of receptors, modulates cellular signaling cascades, and participates in the maintenance of plasma membrane composition.

Endocytosis has a special relevance in the brain. In addition to its role in degrading, recycling, and sorting proteins and lipids between the plasma membrane and intracellular compartments, there is a subtype of endocytosis in neurons called synaptic vesicle endocytosis (SVE). SVE involves the regeneration of synaptic vesicles from the plasma membrane following neurotransmission. This process is essential as a strategic mechanism in neuronal signaling and communication, maintenance of synaptic homeostasis, and synaptic plasticity and adaptation (Dittman and Ryan 2009; Cosker and Segal 2014; Zou et al. 2021).

Endocytosis is regulated by an intricate myriad of mechanisms that orchestrate the delicate balance between membrane internalization and recycling. Some of the key players in endocytosis regulation include membrane trafficking regulators, such as Rab GTPases, the endosomal sorting complexes required for transport (ESCRT), and the retromer complex. Several different components and regulators of the endosome trafficking machinery are identified as proteins codified by PD-linked genes. These include *LRRK2*, *VPS35*, *GAK15*, *RAB29*, *Rab7L1*, *DNAJC6*, *DNAJC13*, *SYNJ1*, *ATP6AP2*, and *SH3GL2* (Edvardson et al. 2012; Gan-Or et al. 2012; Korvatska et al. 2013; MacLeod et al. 2013; Nalls et al. 2014; Vilariño-Güell et al. 2014; Wang et al. 2014; Giovedì et al. 2020). Pathogenic mutations in these genes affect different steps of the intricate endocytic mechanisms and lead

to dysfunctional endocytosis. Some of these PD-associated proteins are also involved in autophagy pathways, as their functions include the regulation of common mechanisms of dynamics and fusion of intracellular vesicles (Abeliovich and Gitler 2016; Schreij et al. 2016). Among these proteins, the role of LRRK2 stands out for being one of the main proteins involved in familial PD, and for its multiple roles in autophagy and endocytosis regulation (see below).

Exocytosis and the Autophagic-Lysosomal Pathway in PD

While autophagy is involved in degradation, and endocytosis is crucial for internalization, exocytosis is required for external release. Exocytosis encompasses the fusion of intracellular vesicles with the plasma membrane, leading to the discharge of vesicular cargo into the extracellular space. This dynamic mechanism is fundamental to numerous physiological functions, and, importantly, in neurons exocytosis takes center stage in synaptic transmission.

Lysosomal exocytosis is a subtype of exocytosis: While conventional exocytosis involves the release of vesicle contents from secretory vesicles, lysosomal exocytosis specifically refers to the fusion of lysosomes with the cell membrane, leading to the extracellular release of lysosomal contents. In the PD context, lysosomal exocytosis is described as the mechanism able to release pathogenic α-synuclein species from neurons. This pathway seems to be a response mechanism to clear toxic α-synuclein species when ALP mechanisms are not sufficient to handle α-synuclein proteotoxicity. However, this release by exocytosis may promote cell-to-cell transmission of toxic α-synuclein species and spread α-synuclein pathology in α-synucleinopathies, including PD (Xie et al. 2022). Thus, alterations in the ALP may lead not only to pathological intracellular accumulation of nondegraded material and overloaded vesicles, but also to an increase in the number of released exosomes (Alvarez-Erviti et al. 2011; Poehler et al. 2014; Fussi et al. 2018). In PD models, lysosomal dysfunction, such as that caused by mutant GBA, is associated with increased release of α-synuclein by exocytosis (Fernandes et al. 2016; Thomas et al. 2018; Gegg et al. 2020). Furthermore, pharmacological inhibition of the ALP has been shown to increase the level of extracellular α-synuclein (Sagini et al. 2021; Xie et al. 2022).

In summary, the intricate interplay between autophagy and lysosomal function, and also between endocytosis and exocytosis, orchestrates a finely tuned cellular equilibrium, regulating the dynamic processes of material turnover, cellular clearance, and extracellular communication. Several genes/proteins linked to PD have a direct role in the regulation of these mechanisms, and/or their pathogenic variants are known to impair the proper function of this intricate network, contributing to the neurodegeneration process.

EVIDENCE LINKING AUTOPHAGIC–LYSOSOMAL DYSFUNCTION TO PD

A critical association between disturbances in the ALP and PD pathogenesis has been well established in past decades. Both familial and sporadic PD cases exhibit alterations in these pathways. Additionally, mutations in genes related to Mendelian inheritance (Lesage and Brice 2009; Bandres-Ciga et al. 2020), or identified as risk factors in PD (Simón-Sánchez et al. 2009; Nalls et al. 2014, 2019; Chang et al. 2017), code for proteins directly or indirectly involved in the ALP machinery, thus underscoring the significance of the ALP in disease pathogenesis (Fig. 3).

α-Synuclein

Various lines of evidence, including observations in human postmortem material, transgenic mice, and cellular/animal models of PD, consistently point to a correlation between α-synuclein accumulation and disturbances in the ALP (Xilouri et al. 2016).

Alterations in any of the mechanisms regulating α-synuclein turnover lead to an increase in cytoplasmic concentration of α-synuclein and promote the formation of oligomeric and fibrillary α-synuclein species. These changes result in an accumulation of potentially toxic α-

synuclein species that may contribute to PD pathology (Lashuel et al. 2013; Klein and Mazzulli 2018). Although α-synuclein can be degraded by the UPS, the ALP has been proposed as the main player in α-synuclein turnover. CMA, in particular, is the key pathway involved in soluble α-synuclein basal turnover (Webb et al. 2003; Cuervo et al. 2004; Vogiatzi et al. 2008; Gan-Or et al. 2015). Oligomeric, fibrillary, or aggregated α-synuclein cannot be degraded by CMA or UPS, since only cytosolic soluble proteins can be unfolded and digested by these machineries. In this context, macroautophagy is the only proteolytic pathway able to sequester and degrade toxic α-synuclein species (Martinez-Vicente 2012). Accordingly, alterations in the ALP system due to aging, external factors, or mutant PD-associated genes, affect the normal turnover of α-synuclein and promote pathological accumulation and aggregation of α-synuclein. These aberrant species, in turn, hinder the normal functioning of the ALP, creating a detrimental cycle that exacerbates α-synuclein accumulation and helps to amplify the impact of α-synuclein on the ALP. The resulting cascade of events ultimately contributes to neurotoxicity and neurodegeneration.

α-Synuclein possesses the CMA motif, that is recognized by the chaperone HSC70. This motif initiates the recruitment of the protein–chaperone complex to the lysosomal membrane, where interaction with the CMA receptor LAMP-2A occurs. Subsequently, α-synuclein undergoes translocation into the lysosomal lumen for degradation. However, mutations or posttranslational modifications in α-synuclein can disrupt its turnover via CMA (Cuervo et al. 2004; Vogiatzi et al. 2008; Mak et al. 2010). Certain mutant forms linked to familial PD cases, such as p.A30P and p.A53T, exhibit inefficient degradation through CMA. Despite their high-affinity binding to LAMP-2A on the lysosomal surface, these mutants fail to internalize into lysosomes, impeding their degradation and blocking CMA-dependent degradation of other substrates (Cuervo et al. 2004; Vogiatzi et al. 2008; Xilouri et al. 2009). Various posttranslational modifications found in PD patients, including phosphorylated, nitrated, oxi-

Cite this article as *Cold Spring Harb Perspect Med* doi: 10.1101/cshperspect.a041619

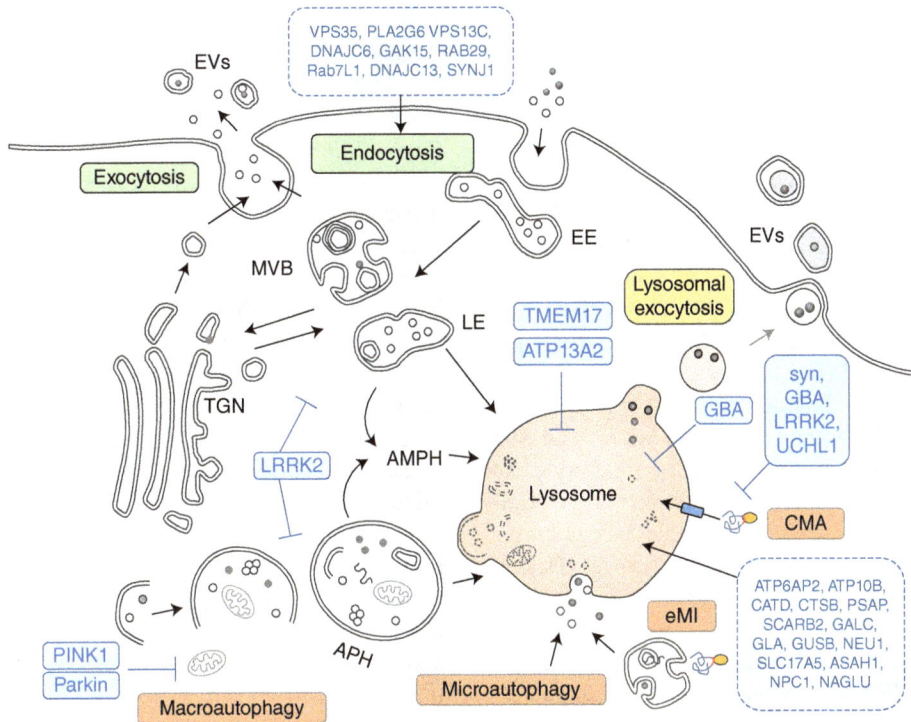

Figure 3. Lysosomes are the terminal compartment where the intracellular macromolecules delivered by macroautophagy, microautophagy, chaperone-mediated autophagy (CMA), and endosomal microautophagy (eMI) are degraded. Extracellular macromolecules can be delivered to lysosomes for degradation through endocytosis. Parkinson's disease (PD)-related proteins are listed alongside the pathway or organelle in which they are involved.

dized, oligomeric, and dopamine-modified α-synuclein forms, also alter CMA-degradation rates (Martinez-Vicente et al. 2008). Reduced efficiency in translocating and eliminating α-synuclein species through the CMA pathway enhances the accumulation of soluble forms in the cytosol, facilitating the development of oligomeric protofibril intermediates that typically advance into insoluble α-synuclein fibrils.

Both in vitro and in vivo investigations have substantiated the significant involvement of CMA in regulating α-synuclein turnover. In addition, diminished levels of CMA markers have been identified in postmortem nigral samples from individuals with PD (Vogiatzi et al. 2008; Xilouri et al. 2008, 2009; Alvarez-Erviti et al. 2010; Mak et al. 2010; Murphy et al. 2015). However, CMA defects that lead to α-synuclein accumulation can also be caused by other PD-re-

lated proteins, such as UCH-L1, LRRK2, GBA, and VPS35 (see below).

GBA

Biallelic mutations in the *GBA* gene cause the autosomal recessive Gaucher's disease (GD), the most common lysosomal storage disorder (LSD) (Smith and Schapira 2022). Conversely, heterozygous *GBA* variants represent the most prevalent genetic risk factor for PD and Lewy body dementia (LBD) (Neumann et al. 2009; Sidransky et al. 2009; Blandini et al. 2019). Situated at 1q21, the *GBA* gene encodes glucocerebrosidase (GCase), a lysosomal hydrolase crucial for sphingolipid degradation (Brady et al. 1965). Synthesized in the ER, GBA, upon translation, interacts with its specific transporter LIMP-2 (Reczek et al. 2007; Malini et al. 2015) and is transported to lysosomes (Blanz et al. 2015), where the GBA-LIMP-2 complex dissociates. GBA can then associate with its coactivator, saposin C (Aerts et al. 1990; Wilkening et al. 1998; Atrian et al. 2008; Abdul-Hammed et al. 2017).

Approximately 300 *GBA* mutations have been identified. These are stratified as risk variants (e.g., p.E326K and p.T369M), mild variants (e.g., p. N370S and p. R496H), and severe variants (e.g., p.L444P, p.D409H, p.V394L, and RecTL) (Lerche et al. 2021, 2019; Höglinger et al. 2022; Parlar et al. 2023). Reduced GBA activity leads to the accumulation of glucosylceramide (GlcCer) and to the generation of GlcSph (also known as lyso-Gb1), a subproduct of GlcCer. GlcSph is the sphingolipid that predominately accumulates in cells and tissues and is the specific biomarker for GD diagnosis (Dekker et al. 2011; Rolfs et al. 2013; Murugesan et al. 2016).

A well-established connection exists between GBA alterations and α-synuclein pathology (Kinghorn 2011; Stojkovska et al. 2018). While the precise mechanisms linking decreased GBA activity to PD pathogenesis remain incompletely understood, a reduction in GBA activity correlates with elevated α-synuclein levels. Likewise, high α-synuclein levels impede GBA activity, as observed both in in vivo and in vitro models and in GBA-PD patient samples (Xu et al. 2011; Schöndorf et al. 2014; Rocha et al. 2015b; Rockenstein et al. 2016; Pradas and Martinez-Vicente 2023).

In recent years, several hypotheses have proposed alternative and complementary pathological mechanisms to explain why lysosomal enzyme mutations lead to α-synuclein accumulation and become an important risk factor in PD etiology. Classic loss of GBA activity has been linked to a dysfunctional ALP and a subsequent decrease in autophagy-dependent α-synuclein turnover. However, several other pathological mechanisms underlying GBA-associated parkinsonism have been suggested.

Different research groups have proposed new mechanisms to explain links between loss of GBA function, CMA failure, and elevated α-synuclein levels in a GBA-PD context. Some

authors found that loss of GBA activity generated a broad lysosomal dysfunction and promoted the enrichment of sphingolipids and cholesterol in the lysosomal membrane, favoring the formation of more lipid microdomains (Navarro-Romero et al. 2022). As LAMP-2A levels and assembly (from monomer to oligomeric) depend on its localization either inside or outside lipid microdomains, higher levels of sphingolipids and cholesterol avoid the stability of the CMA-translocation complex and promote LAMP-2A degradation by cathepsin A (Kaushik et al. 2006). Lower levels of LAMP-2A at the lysosomal membrane lead to a decrease in CMA activity and contribute to abnormal accumulation of α-synuclein (Navarro-Romero et al. 2022). An alternative mechanism of CMA inhibition based on the gain of function of mutant GBA protein has also been proposed (Kuo et al. 2022). In this model, the GBA protein is retrotranslocated from the ER to the cytosol and then degraded via CMA. However, in contrast to wild-type GBA, mutant GBA binds to CMA machinery but is not translocated into the lysosome, thereby blocking CMA machinery. This mechanism is analogous to previously observed mechanisms with mutant α-synuclein, mutant LRRK2, and UCHL-1 (Cuervo et al. 2004; Kabuta et al. 2008; Orenstein et al. 2013).

While GBA is not directly implicated in macroautophagy, conflicting results surround the association between GBA deficiency and macroautophagy dysfunction (Pradas and Martinez-Vicente 2023). Studies with total GBA activity depletion often report macroautophagy failure (Osellame et al. 2013; Du et al. 2015; Rocha et al. 2015a; Kinghorn et al. 2016; Magalhaes et al. 2016), whereas models or specimens with partial GBA loss in GBA-PD patients tend to find preserved macroautophagic flux (Sardi et al. 2011; McNeill et al. 2014; Fernandes et al. 2016; Li et al. 2019; Johnson et al. 2021; Polinski et al. 2021; Kuo et al. 2022; Navarro-Romero et al. 2022).

Apart from a direct effect of pathogenic GBA protein on the ALP, mutant GBA can also affect the levels, conformation, and localization of α-synuclein. This is because sphingolipids, accumulated as a consequence of GBA

Cite this article as *Cold Spring Harb Perspect Med* doi: 10.1101/cshperspect.a041619

deficiency, can impact α-synuclein metabolism. α-Synuclein interacts with lipid membranes and presents a marked preference for lipid microdomains (Kamp and Beyer 2006; Lee et al. 2011; Leftin et al. 2013). This interaction can affect membrane stability, since α-synuclein-lipid-binding sites can act as seeding points for α-synuclein oligomerization and fibrillation (Martinez et al. 2007; Gaspar et al. 2018; O'Leary et al. 2018; Perissinotto et al. 2019). Thus, accumulation of GlcCer and GlcSph, as a consequence of GBA loss of function, can induce α-synuclein aggregation both in vitro and in vivo (Taguchi et al. 2017; Kim et al. 2018; Paul et al. 2021).

In addition to these effects, mutant GBAs contribution to PD etiology extends to other systems. These include endocytosis, exocytosis, and lysosomal-independent mechanisms such as ER stress, mitochondrial dysfunction, oxidative stress, and dysregulated calcium metabolism.

LRRK2

LRRK2 variants are implicated in both familial and sporadic PD (Sosero and Gan-Or 2023). LRRK2 is a large multidomain enzyme, including a catalytic kinase and GTPase domains and multiple protein–protein interaction domains. LRRK2 has been implicated in numerous cellular pathways, reflecting its capacity to interact with multiple proteins and exert influence over various cellular processes, including the ALP, endocytosis, inflammation, apoptosis, ciliogenesis, and others (Cookson 2015; Manzoni 2017; Sosero and Gan-Or 2023). Consequently, mutations in this gene have been implicated in multiple pathogenic mechanisms in PD, including dysfunction of CMA and macroautophagy, among others.

Regarding the role of LRRK2 in the modulation of α-synuclein degradation by CMA, LRKK2 protein contains the CMA motif and is susceptible to degradation through CMA. However, when the p.G2019S LRRK2-mutated protein is transported to the lysosomal membrane, it interacts with LAMP-2A and—in a mechanism equivalent to mutant-synuclein—remains stacked on the surface of the lysosomes. This impedes its own degradation and also hinders the breakdown of other CMA substrates, including α-synuclein; the latter tends to form toxic oligomeric species (Orenstein et al. 2013).

LRRK2 has also been associated with various stages of macroautophagy (Manzoni 2017), but there are conflicting reports of its roles meaning that these challenging and controversial findings have created a complex and unclear scenario (Roosen and Cookson 2016). In some cell models, the common G2019S mutation is associated with reduced macroautophagy flux and disrupted endosome trafficking, while LRRK2 deletion increases both macroautophagy and lysosomal degradation (Plowey et al. 2008). However, in other in vitro models, LRRK2 overexpression enhances macroautophagy, and its silencing diminishes degradation (Henry et al. 2015; Wallings et al. 2015; Boecker et al. 2021). In conclusion, LRRK2 mutations exhibit a complex impact on macroautophagy, presenting a possible dual role in its regulation.

On the contrary, evidence of LRRK2's role in endocytosis is more robust. LRRK2's kinase activity can phosphorylate different proteins involved in endocytosis machinery, such as several Rab proteins (Beilina et al. 2014; Liu et al. 2018; Madero-Pérez et al. 2018; Bae and Lee 2020; Kuwahara and Iwatsubo 2020), VPS35 (Williams et al. 2017), or VPS52 (Beilina et al. 2020). Consequently, mutations in LRRK2 have been reported to result in a defective endosomal system (Boecker 2023).

PINK1 and Parkin

Among monogenic forms of PD, two autosomal recessive genes, PINK1 and PRKN (PARK2), are associated with early-onset PD (Shimura et al. 2000; Valente et al. 2004). Their products, PINK1 and Parkin, work in a coordinated manner to act as cellular stress sensors in different pathways, including mitophagy (Martinez-Vicente 2015).

Mitophagy is the degradation and elimination through macroautophagy of unwanted (dysfunctional or unnecessary) mitochondria. As part of mitochondrial quality control, mitophagy can be selectively activated under different stress conditions, such as mitochondrial depolarization, oxidative stress, hypoxia, and nitrogen starvation (Sun et al. 2016). Numerous types of mitophagy mechanisms to eliminate unwanted mitochondria by macroautophagy have been described in mammalian cells (Martinez-Vicente 2017; Palikaras et al. 2018). All of them involve the selective recognition of targeted mitochondria by a mitophagy receptor and subsequent binding to nascent autophagosomes, which elongate until they totally engulf the mitochondria. However, there are differences in the molecular machinery for labeling the mitochondria (UB-dependent or independent), and in the mitophagy receptor that connects the mitochondria with the autophagosomes (Martinez-Vicente 2017; Palikaras et al. 2018; Li et al. 2023).

PINK1 is a protein kinase that localizes to the mitochondrial membrane. Under normal conditions, PINK1 is newly synthesized in the cytosol and continuously imported into healthy mitochondria, translocated across the outer mitochondrial membrane (OMM), and inserted in the inner mitochondrial membrane (IMM), where it is subsequently cleaved by proteases and degraded by the proteasome in the cytosol (Jin et al. 2010; Liu et al. 2017). When a mitochondrion becomes damaged or depolarized due to stress or other factors, the importation of PINK1 is impaired, preventing its cleavage and degradation and allowing it to become an active kinase on the mitochondria outer membrane. PINK1, stabilized on the damaged mitochondria, phosphorylates ubiquitin (UB) and other proteins on the OMM, including Parkin. Parkin is an E3 UB ligase that requires PINK1-mediated phosphorylation for activation (Vives-Bauza et al. 2009; Matsuda et al. 2010; Narendra and Youle 2011). Then, active Parkin can ubiquitinate a variety of substrates present on the OMM.

Parkin-mediated ubiquitination and PINK1-mediated phosphorylation of UB and Parkin together feed a positive feedback loop that amplifies phosphorylated poly(UB) chains on the mitochondrial surface. The mitochondrial surface chains in turn serve as signals for the recruitment of autophagy receptors, such as

p62/SQSTM1, OPTN, NBR1, NDP52, and TAX1BP1 (Lazarou et al. 2015). These mitophagy receptors bind and connect ubiquitinated mitochondrial to double-membraned nascent autophagosomes, which engulf the damaged mitochondria. The autophagosome then fuses with a lysosome and the contents of the autophagosome, including the damaged mitochondria, are degraded by lysosomal enzymes (Kane et al. 2014; Koyano et al. 2014; Kazlauskaite et al. 2015; Lazarou et al. 2015; Palikaras et al. 2018).

The presence of mutations in *PINK1* and *PRKN* genes highlights the consequences of ALP dysfunction and mitochondrial dysfunction in the pathogenesis of PD. However, it is important to note that cells have different mitophagy mechanisms, and several stimuli can promote mitophagy through multiple mechanisms (Matsuda et al. 2010; Narendra and Youle 2011).

Additionally, PINK1 and Parkin work in a coordinated manner to sense mitochondrial damage as part of the mitochondrial quality control system (Eldeeb et al. 2022); labeling mitochondrial proteins with phosphorylated-UB chains also has functions beyond mitophagy, and so mutations in PINK1 or Parkin might contribute to mitochondrial dysfunction through mitophagy-independent pathways.

In fact, in mouse models, knockout of PINK1 and Parkin generates mitochondrial dysfunction but fails to reproduce PD pathology (Blesa and Przedborski 2014), possibly because other processes compensate for loss of PINK1/Parkin. In contrast, in *Drosophila*, mutations in PINK1 and Parkin homologs are enough to cause mitochondrial dysfunction and loss of dopaminergic neurons (Clark et al. 2006; Yang et al. 2006). In humans, both *PINK1* and *PRKN* are autosomal recessive genes; pathogenic mutation in just one is sufficient to cause early-onset PD.

ATP13A2

The *ATP13A2* gene (PARK9) encodes for a lysosomal transmembrane ATPase pump that is involved in cation (H^+ and K^+) and polyamine transport (van Veen et al. 2020; Fujii et al. 2023). Mutations in this gene cause Kufor–Rakeb syn-

Cite this article as *Cold Spring Harb Perspect Med* doi: 10.1101/cshperspect.a041619

drome, an autosomal recessive atypical form of early-onset PD (Ramirez et al. 2006; Fujii et al. 2023). Some studies have linked heterozygous variants in the *ATP13A2* gene to early-onset PD (Di Fonzo et al. 2007; Lin et al. 2008; Djarmati et al. 2009; Chen et al. 2011) and some rare variants have also been identified as risk factors for PD (Hopfner et al. 2020).

ATP13A2 functions as a polyamine exporter in lysosomes (van Veen et al. 2020) and also as an H^+,K^+-ATPase in lysosomes regulating lysosomal pH (Ramonet et al. 2012). PD-associated mutations or deletion of the *ATP13A2* gene cause lysosomal alkalinization and general lysosomal dysfunction (Dehay et al. 2012; Ramonet et al. 2012; Usenovic et al. 2012; Lopes da Fonseca et al. 2016; Fujii et al. 2023), which are associated with α-synuclein accumulation and aggregation. In this context of ATP13A2 loss of function, abnormal levels of α-synuclein can be promoted not only by lysosomal dysfunction, but also by the abnormal cytoplasmic polyamine level (Antony et al. 2003) and externalization of α-synuclein (Si et al. 2021).

TMEM175

The *TMEM175* gene has been defined as a genetic risk factor for PD (Chang et al. 2017; Nalls et al. 2019; Hopfner et al. 2020). This gene encodes a lysosomal transmembrane protein, described as a lysosomal proton release channel, that regulates lysosome pH within the range of 4.5–5.0 required for optimal hydrolytic activity of lysosomal enzymes (Hu et al. 2022; Yang et al. 2023). TMEM175-mediated proton efflux is needed to balance out the proton influx produced by V-ATPases (e.g., ATP13A2 and others). Under TMEM175-deficient conditions, lysosomes are overacidified due to an unrestricted proton influx produced by V-ATPases. This results in impaired lysosomal hydrolysis, promoting α-synuclein aggregation in different cellular models (Jinn et al. 2017).

Other Autophagic-Lysosomal-Related Proteins

Other genes encoding lysosomal proteins have also been proposed as PD genetic risk factors or candidates for PD susceptibility (Navarro-Romero et al. 2020). Interestingly, most of these genes are identified as lysosomal storage disease (LSD) genes that are responsible for inherited metabolic disorders caused by lysosomal enzyme deficiencies leading to the accumulation of undegraded substrate. LSD are mostly autosomal recessive disorders where biallelic mutations cause each particular LSD. However, some heterozygous variants can become PD risk factors. Among them, *GBA* is the main example (see above), but several other genes, such as *CTSD, CTSB, PSAP, SCARB2, GALC, GLA, GUSB, NEU1, SLC17A5, ASAH1, NPC1*, and *NAGLU* have been proposed as risk factors or candidates that convey higher susceptibility for PD (extensively reviewed in Navarro-Romero et al. 2020).

Different studies have confirmed the high number of pathogenic variants in LSD genes present in sporadic PD patients (over 50%) (Robak et al. 2017). A bidirectional association between PD and LSD is suggested by the presence of neurodegeneration and α-synuclein inclusions in the brain regions of animal models and in patients with different LSD (Saito et al. 2003, 2004; Wong et al. 2004; Suzuki et al. 2007; Smith et al. 2014; Bae et al. 2015; Del Tredici et al. 2020).

Recently, the list of ALP-related proteins associated with PD and atypical parkinsonian syndromes was expanded and now includes the lysosomal ATPase complex proteins, ATP10B (Martin et al. 2020) and ATP6AP2 (Ramser et al. 2005; Korvatska et al. 2013). In addition, several proteins involved in endolysosomal vesicular trafficking, such as VPS35, PLA2G6 (Paisan-Ruiz et al. 2009; Yoshino et al. 2010; Kauther et al. 2011; Dehnavi et al. 2023), VPS13C (Darvish et al. 2018; Schormair et al. 2018), DNAJC6 (Edvardson et al. 2012; Köroğlu et al. 2013; Olgiati et al. 2016), and SYNJ1 (Quadri et al. 2013; Fasano et al. 2018; Lunati et al. 2018) have been newly associated with early-onset PD and different parkinsonian syndromes.

Although for many of these genes, the exact mechanistic association between pathogenic variants, α-synuclein accumulation, and neurodegeneration remains to be established, these findings reinforce the role of lysosomal system dysfunction in PD (Smolders and Van Broeckhoven 2020).

ALP-BASED BIOMARKERS AND THERAPIES IN PD

The changes observed in the ALP as PD develops and advances are evident in a range of clinical biomarkers linked to the disease. Simultaneously, these alterations present an opportunity to identify potential therapeutic targets, laying the groundwork for the development of innovative and more effective treatment strategies for this currently incurable disease.

ALP Dysfunction in Biomarker Development

A fundamental aspect of PD is its remarkable heterogeneity. This manifests as variable symptoms, disease progression, and treatment responses, which are also reflected in the assessment of biomarkers. PD patients carrying mutations in ALP-related genes may exhibit a specific clinical phenotype, as observed in PD-GBA patients (Gan-Or et al. 2008; Anheim et al. 2012; Cilia et al. 2016; Liu et al. 2016; Thaler et al. 2018). These characteristic features can be translated into distinctive outcomes during the assessment of different PD biomarkers, when comparing PD-GBA and idiopathic PD (iPD).

When PD patients are stratified and analyzed according to GBA status, significant differences can be observed between iPD and PD-GBA subgroups. For example, a decrease in total α-synuclein levels in cerebrospinal fluid (CSF) samples from PD patients is expected compared to healthy controls (Kwon et al. 2022; Xiang et al. 2022). Interestingly, the lowest levels of total α-synuclein correspond to the most severe GBA variants compared to the mild variants or iPD (Lerche et al. 2020). Recently, α-synuclein seed amplification assays (SAAs) using CSF have emerged as the most specific method to distinguish PD patients from healthy controls —by detecting the presence of α-synuclein spe-

cies with seeding capacity to amplify α-synuclein aggregation in vitro (Russo et al. 2021). CSF samples from GBA-PD subjects have the highest proportion of positive α-synuclein SAA results (Brockmann et al. 2021; Siderowf et al. 2023). This may be because GBA mutations can generate a more intense pathology, with higher levels of α-synuclein that translate into higher seeding capacity.

Currently, efforts are underway to identify new biomarkers of ALP dysfunction in PD, with the aim to identify and stratify subjects according to the severity of ALP dysfunction. Classical ALP markers, such as protein levels of LC3, p62, or LAMP-2 in peripheral biosamples, have led to inconsistent study findings (Xicoy et al. 2019). Present alternatives, based primarily on enzymatic lysosomal activities (Xicoy et al. 2019; Drobny et al. 2022) and alterations in lipid metabolism in peripheral biofluids (plasma and CSF) (Esfandiary et al. 2022), aim to identify or stratify alterations in the ALP.

Understanding the biochemical changes associated with ALP dysfunction not only enhances our comprehension of PD pathogenesis but also holds promise for the development of reliable biomarkers. Diverse alterations in the ALP are not only indicative of the disease's heterogeneity but also emphasize the importance of considering genetic factors in assessing biomarkers. Future developments can contribute significantly to the diagnosis and stratification of PD, paving the way for more targeted and personalized therapeutic approaches.

Enhancement of the Autophagic–Lysosomal Pathway as a Therapeutic Approach in PD

At a preclinical level, it is well established that overexpression of various ALP-related genes, such as *LAMP-2A* (Vogiatzi et al. 2008; Xilouri et al. 2013; Navarro-Romero et al. 2022), *beclin 1* (Decressac et al. 2013), or *GBA* (Sardi et al. 2011, 2013; Yun et al. 2018; Glajch et al. 2021; Sucunza et al. 2021) can contribute to the elimination of toxic α-synuclein species and alleviate several PD-like symptoms (reviewed by Ejlerskov et al. 2019). Notably, gene therapy overexpressing

Cite this article as *Cold Spring Harb Perspect Med* doi: 10.1101/cshperspect.a041619

TFEB stands out as a promising avenue in treating neurodegenerative diseases, harnessing its capacity to activate the ALP system through lysosome biogenesis and augmented production of autophagosomes (Sardiello et al. 2009; Settembre et al. 2011). A neuroprotective effect of TFEB has been observed in various animal models of PD—MPTP model (Torra et al. 2018), overexpression of α-synuclein model (Decressac et al. 2013), neuromelanin-producing model (Carballo-Carbajal et al. 2019)—as well as in other proteinopathy models relating to Alzheimer's disease and Huntington's disease (Napolitano and Ballabio 2016; Jiao et al. 2023).

Pharmacological macroautophagy activators are predominantly targeting the AMPK/mTOR-signaling pathway, beclin-1, or TFEB, and have undergone preclinical validation over the last decade. Noteworthy drugs include rapamycin and its derivatives, metformin, and speridin, among many others (reviewed in Bonam et al. 2021). Although these compounds lack high specificity as ALP activators, a substantial body of evidence has accumulated through in vitro and in vivo models of PD to support the therapeutic benefits of many of these compounds.

Recent efforts have focused on developing gene-specific therapies to address the heterogeneity of PD, primarily targeting GBA and LRRK2 proteins. Different small-molecule LRRK2 inhibitors and antisense oligonucleotides have been developed to treat PD-LRRK2 patients specifically, with clinical trials of LRRK2-targeted therapies already underway (Tolosa et al. 2020). Similarly, the reciprocal relationship between GBA levels/activity and α-synuclein levels suggests that increasing GBA activity can be beneficial in reducing neurotoxic α-synuclein accumulation, presenting a potential therapeutic option for PD-GBA patients. Indeed, genetic overexpression of GBA in various in vivo models results in the reversal of some PD-like symptoms (Sardi et al. 2011, 2013; Yun et al. 2018; Glajch et al. 2021; Sucunza et al. 2021).

Alternatively, GBA activity can also be targeted through enzyme replacement therapy (ERT), substrate inhibition therapy (SRT), and GBA-specific chaperones (Smith and Schapira 2022). ERT and SRT are effective therapies in GD to treat nonneurological symptoms. ERT involves long-term treatment via injection of GBA recombinant enzyme, while SRT involves oral administration of glucosylceramide synthetase inhibitors to reduce the production of glucocerebroside and avoid its accumulation (Bennett and Mohan 2013). The use of these therapies for treating neurological symptoms in GD and in PD is under study, and several research efforts aim to adapt these approved therapies for use in PD patients (Smith and Schapira 2022). Finally, small molecular chaperones that specifically bind the GBA enzyme can facilitate GBA folding, stability, and transport to the lysosome, and these currently represent a significant focus in PD therapeutic strategies. Some of these GBA chaperones (e.g., ambroxol) are undergoing clinical evaluation for use in PD (Bonam et al. 2021).

Shifting focus to the pharmacological activation of CMA, recent developments have shown that different inhibitors and antagonists of RARα compounds, such as AR7, GR2, QX77, CA77.1, and CA 39, exhibit promising potential as CMA activators. These compounds have undergone testing in in vitro PD models (Navarro-Romero et al. 2022) and in in vivo models of neurodegenerative diseases, such as Alzheimer's disease and retinal neurodegeneration (Bourdenx et al. 2021; Gomez-Sintes et al. 2022), and offer an attractive future strategy for PD treatment.

Finally, new targeted protein degradation (TPD) tools based on the degradation of targeted proteins by the ALP are currently in development. Innovative techniques like LYTACs, AbTAC, AUTAC, AUTOTAC, and ATTEC platforms (Takahashi et al. 2019; Hoon Ji et al. 2022; Zhang et al. 2022) enable selective ALP-dependent protein degradation using various chimeric systems. The first autophagy-targeting chimera has already demonstrated its ability to target and degrade α-synuclein aggregates in cellular and rodent PD models (Lee et al. 2023). These TPD platforms represent a novel and disruptive strategy, and their application in neurodegenerative diseases holds significant promise.

CONCLUDING REMARKS

Multiple lines of evidence point to alterations in the ALP and cellular proteostasis as major pathogenic events in PD. A number of autosomal dominant and recessive genes associated with PD, as well as several genetic risk factors, encode for lysosomal and autophagic proteins. Mutations in these PD-linked genes can cause lysosomal dysfunction and impair α-synuclein turnover, contributing to its pathological accumulation and aggregation. In addition, age-related decline of the ALP can also contribute to impaired proteostasis and trigger PD pathology. Recent studies have also highlighted the bidirectional link between PD and LSDs. Overall, strategies to enhance or restore the ALP and cellular proteostasis may provide unprecedented therapeutic opportunities to prevent, halt, or delay neuronal dysfunction and degeneration linked to PD.

ACKNOWLEDGMENTS

Aligning Science Across Parkinson's through The Michael J. Fox Foundation for Parkinson's Research, USA (ASAP-020505 to M.V.); The Michael J. Fox Foundation for Parkinson's Research, USA (MJFF-007184 and MJFF-001059 to M.V., MJFF-008096 to M.M.-V.); Ministry of Science and Innovation (MICINN), Spain (PID2020-116339RB-I00 to M.V.); EU Joint Programme Neurodegenerative Disease Research (JPND), Instituto de Salud Carlos III, EU/Spain (PI20/00728 to M.M.-V., PI24/00062 to M.M.-V., AC20/00121 to M.V.); Centres of Excellence in Neurodegeneration (CoEN4016 to M.V.); Fundación BBVA (Nano-ERT to M.M.-V.); and La Caixa Bank Foundation, Spain (Health Research Grant HR22-00602 to M.M.-V., CaixaImpulse CI24-20222 to M.M.-V., Health Research grant, ID 100010434 under the agreement LCF/PR/HR17/52150003 to M.V.).

REFERENCES

Abdul-Hammed M, Breiden B, Schwarzmann G, Sandhoff K. 2017. Lipids regulate the hydrolysis of membrane bound glucosylceramide by lysosomal β-glucocerebrosidase. *J Lipid Res* **58**: 563–577. doi:10.1194/jlr.M073510

Abeliovich A, Gitler AD. 2016. Defects in trafficking bridge Parkinson's disease pathology and genetics. *Nature* **539**: 207–216. doi:10.1038/nature20414

Abokyi S, Ghartey-Kwansah G, Tse DYY. 2023. TFEB is a central regulator of the aging process and age-related diseases. *Ageing Res Rev* **89**: 101985. doi:10.1016/j.arr.2023.101985

Aerts JMFG, Sa Miranda MC, Brouwer-Kelder EM, Van Weely S, Barranger JA, Tager JM. 1990. Conditions affecting the activity of glucocerebrosidase purified from spleens of control subjects and patients with type 1 Gaucher disease. *Biochim Biophys Acta* **1041**: 55–63. doi:10.1016/0167-4838(90)90122-V

Alvarez-Erviti L, Rodriguez-Oroz MC, Cooper JM, Caballero C, Ferrer I, Obeso JA, Schapira AHV. 2010. Chaperone-mediated autophagy markers in Parkinson disease brains. *Arch Neurol* **67**: 1464–1472. doi:10.1001/archneurol.2010.198

Alvarez-Erviti L, Seow Y, Schapira AH, Gardiner C, Sargent IL, Wood MJA, Cooper JM. 2011. Lysosomal dysfunction increases exosome-mediated α-synuclein release and transmission. *Neurobiol Dis* **42**: 360–367. doi:10.1016/j.nbd.2011.01.029

Anguiano J, Garner TP, Mahalingam M, Das BC, Gavathiotis E, Cuervo AM. 2013. Chemical modulation of chaperone-mediated autophagy by retinoic acid derivatives. *Nat Chem Biol* **9**: 374–382. doi:10.1038/nchembio.1230

Anheim M, Elbaz A, Lesage S, Durr A, Condroyer C, Viallet F, Pollak P, Bonaïti B, Bonaïti-Pellié C, Brice A. 2012. Penetrance of Parkinson disease in glucocerebrosidase gene mutation carriers. *Neurology* **78**: 417–420. doi:10.1212/WNL.0b013e318245f476

Antony T, Hoyer W, Cherny D, Heim G, Jovin TM, Subramaniam V. 2003. Cellular polyamines promote the aggregation of α-synuclein. *J Biol Chem* **278**: 3235–3240. doi:10.1074/jbc.M208249200

Atrian S, López-Viñas E, Gómez-Puertas P, Chabás A, Vilageliu L, Grinberg D. 2008. An evolutionary and structure-based docking model for glucocerebrosidase–saposin C and glucocerebrosidase–substrate interactions—relevance for Gaucher disease. *Proteins Struct Funct Bioinforma* **70**: 882–891. doi:10.1002/prot.21554

Bae EJ, Lee SJ. 2020. The LRRK2-RAB axis in regulation of vesicle trafficking and α-synuclein propagation. *Biochim Biophys Acta* **1866**: 165632. doi:10.1016/j.bbadis.2019.165632

Bae EJ, Yang NY, Lee C, Kim S, Lee HJ, Lee SJ. 2015. Haploinsufficiency of cathepsin D leads to lysosomal dysfunction and promotes cell-to-cell transmission of α-synuclein aggregates. *Cell Death Dis* **6**: e1901–e1901. doi:10.1038/cddis.2015.283

Ballabio A, Bonifacino JS. 2020. Lysosomes as dynamic regulators of cell and organismal homeostasis. *Nat Rev Mol Cell Biol* **21**: 101–118. doi:10.1038/s41580-019-0185-4

Bandres-Ciga S, Diez-Fairen M, Kim JJ, Singleton AB. 2020. Genetics of Parkinson's disease: an introspection of its journey towards precision medicine. *Neurobiol Dis* **137**: 104782. doi:10.1016/j.nbd.2020.104782

Behl T, Kumar S, Althafar ZM, Sehgal A, Singh S, Sharma N, Badavath VN, Yadav S, Bhatia S, Al-Harrasi A, et al. 2022.

Exploring the role of ubiquitin–proteasome system in Parkinson's disease. *Mol Neurobiol* **59**: 4257–4273. doi:10.1007/s12035-022-02851-1

Beilina A, Rudenko IN, Kaganovich A, Civiero L, Chau H, Kalia SK, Kalia LV, Lobbestael E, Chia R, Ndukwe K, et al. 2014. Unbiased screen for interactors of leucine-rich repeat kinase 2 supports a common pathway for sporadic and familial Parkinson disease. *Proc Natl Acad Sci* **111**: 2626–2631. doi:10.1073/pnas.1318306111

Beilina A, Bonet-Ponce L, Kumaran R, Kordich JJ, Ishida M, Mamais A, Kaganovich A, Saez-Atienzar S, Gershlick DC, Roosen DA, et al. 2020. The Parkinson's disease protein LRRK2 interacts with the GARP complex to promote retrograde transport to the trans-Golgi network. *Cell Rep* **31**: 107614. doi:10.1016/j.celrep.2020.107614

Bennett LL, Mohan D. 2013. Gaucher disease and its treatment options. *Ann Pharmacother* **47**: 1182–1193. doi:10.1177/1060028013500469

Blandini F, Cilia R, Cerri S, Pezzoli G, Schapira AHV, Mullin S, Lanciego JL. 2019. Glucocerebrosidase mutations and synucleinopathies: toward a model of precision medicine. *Mov Disord* **34**: 9–21. doi:10.1002/mds.27583

Blanz J, Zunke F, Markmann S, Damme M, Braulke T, Saftig P, Schwake M. 2015. Mannose 6-phosphate-independent lysosomal sorting of LIMP-2. *Traffic* **16**: 1127–1136. doi:10.1111/tra.12313

Blesa J, Przedborski S. 2014. Parkinson's disease: animal models and dopaminergic cell vulnerability. *Front Neuroanat* **8**: 155. doi:10.3389/fnana.2014.00155

Boecker CA. 2023. The role of LRRK2 in intracellular organelle dynamics. *J Mol Biol* **435**: 167998. doi:10.1016/j.jmb.2023.167998

Boecker CA, Goldsmith J, Dou D, Cajka GG, Holzbaur ELF. 2021. Increased LRRK2 kinase activity alters neuronal autophagy by disrupting the axonal transport of autophagosomes. *Curr Biol* **31**: 2140–2154.e6. doi:10.1016/j.cub.2021.02.061

Bonam SR, Wang F, Muller S. 2019. Lysosomes as a therapeutic target. *Nat Rev Drug Discov* **18**: 923–948. doi:10.1038/s41573-019-0036-1

Bonam SR, Tranchant C, Muller S. 2021. Autophagy–lysosomal pathway as potential therapeutic target in Parkinson's disease. *Cells* **10**: 3547. doi:10.3390/cells10123547

Bourdenx M, Martín-Segura A, Scrivo A, Rodriguez-Navarro JA, Kaushik S, Tasset I, Diaz A, Storm NJ, Xin Q, Juste YR, et al. 2021. Chaperone-mediated autophagy prevents collapse of the neuronal metastable proteome. *Cell* **184**: 2696–2714.e25. doi:10.1016/j.cell.2021.03.048

Brady RO, Kanfer J, Shapiro D. 1965. The metabolism of glucocerebrosides. I: Purification and properties of a glucocerebroside-cleaving enzyme from spleen tissue. *J Biol Chem* **240**: 39–43. doi:10.1016/S0021-9258(18)97611-8

Brockmann K, Quadalti C, Lerche S, Rossi M, Wurster I, Baiardi S, Roeben B, Mammana A, Zimmermann M, Hauser AK, et al. 2021. Association between CSF α-synuclein seeding activity and genetic status in Parkinson's disease and dementia with Lewy bodies. *Acta Neuropathol Commun* **9**: 175. doi:10.1186/s40478-021-01276-6

Carballo-Carbajal I, Laguna A, Romero-Giménez J, Cuadros T, Bové J, Martinez-Vicente M, Parent A, Gonzalez-Sepulveda M, Peñuelas N, Torra A, et al. 2019. Brain tyrosinase overexpression implicates age-dependent neurome-lanin production in Parkinson's disease pathogenesis. *Nat Commun* **10**: 973. doi:10.1038/s41467-019-08858-y

Chang D, Nalls MA, Hallgrímsdóttir IB, Hunkapiller J, van der Brug M, Cai F, Kerchner GA, Ayalon G, Bingol B, Sheng M, et al. 2017. A meta-analysis of genome-wide association studies identifies 17 new Parkinson's disease risk loci. *Nat Genet* **49**: 1511–1516. doi:10.1038/ng.3955

Chen CM, Lin CH, Juan HF, Hu FJ, Hsiao YC, Chang HY, Chao CY, Chen IC, Lee LC, Wang TW, et al. 2011. ATP13A2 variability in Taiwanese Parkinson's disease. *Am J Med Genet Part B Neuropsychiatr Genet* **156**: 720–729. doi:10.1002/ajmg.b.31214

Chen J, Mao K, Yu H, Wen Y, She H, Zhang H, Liu L, Li M, Li W, Zou F. 2021. p38-TFEB pathways promote microglia activation through inhibiting CMA-mediated NLRP3 degradation in Parkinson's disease. *J Neuroinflammation* **18**: 295. doi:10.1186/s12974-021-02349-y

Chiang H, Dice JF. 1979. A role for a 70 KDa heat shock protein in lysosomal degradation of intracellular protein. *Science* **246**: 382–385. doi:10.1126/science.2799391

Chiti F, Dobson CM. 2017. Protein misfolding, amyloid formation, and human disease: a summary of progress over the last decade. *Annu Rev Biochem* **86**: 27–68. doi:10.1146/annurev-biochem-061516-045115

Chu Y, Dodiya H, Aebischer P, Olanow CW, Kordower JH. 2009. Alterations in lysosomal and proteasomal markers in Parkinson's disease: relationship to α-synuclein inclusions. *Neurobiol Dis* **35**: 385–398. doi:10.1016/j.nbd.2009.05.023

Cilia R, Tunesi S, Marotta G, Cereda E, Siri C, Tesei S, Zecchinelli AL, Canesi M, Mariani CB, Meucci N, et al. 2016. Survival and dementia in GBA-associated Parkinson's disease: the mutation matters. *Ann Neurol* **80**: 662–673. doi:10.1002/ana.24777

Clark IE, Dodson MW, Jiang C, Cao JH, Huh JR, Seol JH, Yoo SJ, Hay BA, Guo M. 2006. *Drosophila* pink1 is required for mitochondrial function and interacts genetically with parkin. *Nature* **441**: 1162–1166. doi:10.1038/nature04779

Colla E, Jensen PH, Pletnikova O, Troncoso JC, Glabe C, Lee MK. 2012. Accumulation of toxic α-synuclein oligomer within endoplasmic reticulum occurs in α-synucleinopathy in vivo. *J Neurosci* **32**: 3301–3305. doi:10.1523/JNEUROSCI.5368-11.2012

Cookson MR. 2015. LRRK2 pathways leading to neurodegeneration. *Curr Neurol Neurosci Rep* **15**: 42. doi:10.1007/s11910-015-0564-y

Cosker KE, Segal RA. 2014. Neuronal signaling through endocytosis. *Cold Spring Harb Perspect Biol* **6**: a020669. doi:10.1101/cshperspect.a020669

Cuervo AM, Dice JF. 1979. A receptor for the selective uptake and degradation of proteins by lysosomes. *Science* **273**: 501–503. doi:10.1126/science.273.5274.501

Cuervo AM, Stafanis L, Fredenburg R, Lansbury PT, Sulzer D. 2004. Impaired degradation of mutant α-synuclein by chaperone-mediated autophagy. *Science* **305**: 1292–1295. doi:10.1126/science.1101738

Darvish H, Bravo P, Tafakhori A, Azcona LJ, Ranji-Burachaloo S, Johari AH, Paisán-Ruiz C. 2018. Identification of a large homozygous VPS13C deletion in a patient with early-onset parkinsonism. *Mov Disord* **33**: 1968–1970. doi:10.1002/mds.27516

Decressac M, Mattsson B, Weikop P, Lundblad M, Jakobsson J, Björklund A. 2013. TFEB-mediated autophagy rescues midbrain dopamine neurons from α-synuclein toxicity. *Proc Natl Acad Sci* 110: E1817–E1826. doi:10.1073/pnas.1305623110

Dehay B, Bové J, Rodríguez-Muela N, Perier C, Recasens A, Boya P, Vila M. 2010. Pathogenic lysosomal depletion in Parkinson's disease. *J Neurosci* 30: 12535–12544. doi:10.1523/JNEUROSCI.1920-10.2010

Dehay B, Ramirez A, Martinez-Vicente M, Perier C, Canron MHMH, Doudnikoff E, Vital A, Vila M, Klein C, Bezard E. 2012. Loss of P-type ATPase ATP13A2/PARK9 function induces general lysosomal deficiency and leads to Parkinson disease neurodegeneration. *Proc Natl Acad Sci* 109: 9611–9616. doi:10.1073/pnas.1112368109

Dehnavi AZ, Bemanalizadeh M, Kahani SM, Ashrafi MR, Rohani M, Toosi MB, Heidari M, Hosseinpour S, Amini B, Zokaei S, et al. 2023. Phenotype and genotype heterogeneity of PLA2G6-associated neurodegeneration in a cohort of pediatric and adult patients. *Orphanet J Rare Dis* 18: 177. doi:10.1186/s13023-023-02780-9

Dekker N, van Dussen L, Hollak CEM, Overkleeft H, Scheij S, Ghauharali K, van Breemen MJ, Ferraz MJ, Groener JEM, Maas M, et al. 2011. Elevated plasma glucosylsphingosine in Gaucher disease: relation to phenotype, storage cell markers, and therapeutic response. *Blood* 118: e118–e127. doi:10.1182/blood-2011-05-352971

Del Tredici K, Ludolph AC, Feldengut S, Jacob C, Reichmann H, Bohl JR, Braak H. 2020. Fabry disease with concomitant Lewy body disease. *J Neuropathol Exp Neurol* 79: 378–392. doi:10.1093/jnen/nlz139

Di Fonzo A, Chien HF, Socal M, Giraudo S, Tassorelli C, Iliceto G, Fabbrini G, Marconi R, Fincati E, Abbruzzese G, et al. 2007. ATP13A2 missense mutations in juvenile parkinsonism and young onset Parkinson disease. *Neurology* 68: 1557–1562. doi:10.1212/01.wnl.0000260963.08711.08

Dittman J, Ryan TA. 2009. Molecular circuitry of endocytosis at nerve terminals. *Annu Rev Cell Dev Biol* 25: 133–160. doi:10.1146/annurev.cellbio.042308.113302

Djarmati A, Hagenah J, Reetz K, Winkler S, Behrens MI, Pawlack H, Lohmann K, Ramirez A, Tadić V, Brüggemann N, et al. 2009. ATP13A2 variants in early-onset Parkinson's disease patients and controls. *Mov Disord* 24: 2104–2111. doi:10.1002/mds.22728

Dobson CM. 2003. Protein folding and misfolding. *Nature* 426: 884–890. doi:10.1038/nature02261

Drobny A, Prieto Huarcaya S, Dobert J, Kluge A, Bunk J, Schlothauer T, Zunke F. 2022. The role of lysosomal cathepsins in neurodegeneration: mechanistic insights, diagnostic potential and therapeutic approaches. *Biochim Biophys Acta* 1869: 119243. doi:10.1016/j.bbamcr.2022.119243

Du TT, Wang L, Duan CL, Lu LL, Zhang JL, Gao G, Qiu XB, Wang XM, Yang H. 2015. GBA deficiency promotes SNCA/α-synuclein accumulation through autophagic inhibition by inactivated PPP2A. *Autophagy* 11: 1–43.

Edvardson S, Cinnamon Y, Ta-Shma A, Shaag A, Yim YI, Zenvirt S, Jalas C, Lesage S, Brice A, Taraboulos A, et al. 2012. A deleterious mutation in DNAJC6 encoding the neuronal-specific clathrin-uncoating co-chaperone aux-

ilin, is associated with juvenile parkinsonism. *PLoS ONE* 7: e36458. doi:10.1371/journal.pone.0036458

Ejlerskov P, Ashkenazi A, Rubinsztein DC. 2019. Genetic enhancement of macroautophagy in vertebrate models of neurodegenerative diseases. *Neurobiol Dis* 122: 3–8. doi:10.1016/j.nbd.2018.04.001

Eldeeb MA, Thomas RA, Ragheb MA, Fallahi A, Fon EA. 2022. Mitochondrial quality control in health and in Parkinson's disease. *Physiol Rev* 102: 1721–1755. doi:10.1152/physrev.00041.2021

Esfandiary A, Finkelstein DI, Voelcker NH, Rudd D. 2022. Clinical sphingolipids pathway in Parkinson's disease: from GCase to integrated-biomarker discovery. *Cells* 11: 1353. doi:10.3390/cells11081353

Fasano D, Parisi S, Pierantoni GM, De Rosa A, Picillo M, Amodio G, Pellecchia MT, Barone P, Moltedo O, Bonifati V, et al. 2018. Alteration of endosomal trafficking is associated with early-onset parkinsonism caused by SYNJ1 mutations. *Cell Death Dis* 9: 385. doi:10.1038/s41419-018-0410-7

Fernandes HJRR, Hartfield EM, Christian HC, Emmanouilidou E, Zheng Y, Booth H, Bogetofte H, Lang C, Ryan BJ, Sardi SP, et al. 2016. ER stress and autophagic perturbations lead to elevated extracellular α-synuclein in GBA-N370S Parkinson's iPSC-derived dopamine neurons. *Stem Cell Reports* 6: 342–356. doi:10.1016/j.stemcr.2015.11.013

Fujii T, Nagamori S, Wiriyasermkul P, Zheng S, Yago A, Shimizu T, Tabuchi Y, Okumura T, Fujii T, Takeshima H, et al. 2023. Parkinson's disease-associated ATP13A2/PARK9 functions as a lysosomal H⁺,K⁺-ATPase. *Nat Commun* 14: 2174. doi:10.1038/s41467-023-37815-z

Fussi N, Höllerhage M, Chakroun T, Nykänen NPP, Rösler TW, Koeglsperger T, Wurst W, Behrends C, Höglinger GU. 2018. Exosomal secretion of α-synuclein as protective mechanism after upstream blockage of macroautophagy. *Cell Death Dis* 9: 757. doi:10.1038/s41419-018-0816-2

Gan-Or Z, Giladi N, Rozovski U, Shifrin C, Rosner S, Gurevich T, Bar-Shira A, Orr-Urtreger A. 2008. Genotype-phenotype correlations between GBA mutations and Parkinson disease risk and onset. *Neurology* 70: 2277–2283. doi:10.1212/01.wnl.0000304039.11891.29

Gan-Or Z, Bar-Shira A, Dahary D, Mirelman A, Kedmi M, Gurevich T, Giladi N, Orr-Urtreger A. 2012. Association of sequence alterations in the putative promoter of RAB7L1 with a reduced Parkinson disease risk. *Arch Neurol* 69: 105. doi:10.1001/archneurol.2011.924

Gan-Or Z, Dion A, Rouleau GA. 2015. Genetic perspective on the role of the autophagy-lysosome pathway in Parkinson disease. *Autophagy* 11: 1443–1457. doi:10.1080/15548627.2015.1067364

Gaspar R, Pallbo J, Weininger U, Linse S, Sparr E. 2018. Ganglioside lipids accelerate α-synuclein amyloid formation. *Biochim Biophys Acta* 1866: 1062–1072. doi:10.1016/j.bbapap.2018.07.004

Gavilán E, Pintado C, Gavilan MP, Daza P, Sánchez-Aguayo I, Castaño A, Ruano D. 2015. Age-related dysfunctions of the autophagy lysosomal pathway in hippocampal pyramidal neurons under proteasome stress. *Neurobiol Aging* 36: 1953–1963. doi:10.1016/j.neurobiolaging.2015.02.025

Cite this article as *Cold Spring Harb Perspect Med* doi: 10.1101/cshperspect.a041619

Gegg ME, Verona G, Schapira AHV. 2020. Glucocerebrosidase deficiency promotes release of α-synuclein fibrils from cultured neurons. *Hum Mol Genet* **29:** 1716–1728. doi:10.1093/hmg/ddaa085

Giovedì S, Ravanelli MM, Parisi B, Bettegazzi B, Guarnieri FC. 2020. Dysfunctional autophagy and endolysosomal system in neurodegenerative diseases: relevance and therapeutic options. *Front Cell Neurosci* **14:** 602116. doi:10.3389/fncel.2020.602116

Glajch KE, Moors TE, Chen Y, Bechade PA, Nam AY, Rajsombath MM, McCaffery TD, Dettmer U, Weihofen A, Hirst WD, et al. 2021. Wild-type GBA1 increases the α-synuclein tetramer-monomer ratio, reduces lipid-rich aggregates, and attenuates motor and cognitive deficits in mice. *Proc Natl Acad Sci* **118:** 1–10. doi:10.1073/pnas.2103425118

Gomez-Sintes R, Xin Q, Ignacio Jimenez-Loygorri J, McCabe M, Diaz A, Garner TP, Cotto-Rios XM, Wu Y, Dong S, Reynolds CA, et al. 2022. Targeting retinoic acid receptor α-corepressor interaction activates chaperone-mediated autophagy and protects against retinal degeneration. *Nat Commun* **13:** 4220. doi:10.1038/s41467-022-31869-1

Gong Z, Tasset I, Diaz A, Anguiano J, Tas E, Cui L, Kuliawat R, Liu H, Kühn B, Cuervo AM, et al. 2018. Humanin is an endogenous activator of chaperonemediated autophagy. *J Cell Biol* **217:** 635–647. doi:10.1083/jcb.201606095

Hartl FU. 2017. Protein misfolding diseases. *Annu Rev Biochem* **86:** 21–26. doi:10.1146/annurev-biochem-061516-044518

Henry AG, Aghamohammadzadeh S, Samaroo H, Chen Y, Mou K, Needle E, Hirst WD. 2015. Pathogenic LRRK2 mutations, through increased kinase activity, produce enlarged lysosomes with reduced degradative capacity and increase ATP13A2 expression. *Hum Mol Genet* **24:** 6013–6028. doi:10.1093/hmg/ddv314

Hetz C, Glimcher LH. 2009. Fine-tuning of the unfolded protein response: assembling the IRE1α interactome. *Mol Cell* **35:** 551–561. doi:10.1016/j.molcel.2009.08.021

Hetz C, Saxena S. 2017. ER stress and the unfolded protein response in neurodegeneration. *Nat Rev Neurol* **13:** 477–491. doi:10.1038/nrneurol.2017.99

Hetz C, Zhang K, Kaufman RJ. 2020. Mechanisms, regulation and functions of the unfolded protein response. *Nat Rev Mol Cell Biol* **21:** 421–438. doi:10.1038/s41580-020-0250-z

Höglinger G, Schulte C, Jost WH, Storch A, Woitalla D, Krüger R, Falkenburger B, Brockmann K. 2022. GBA-associated PD: chances and obstacles for targeted treatment strategies. *J Neural Transm* **129:** 1219–1233. doi:10.1007/s00702-022-02511-7

Hoon Ji C, Yeon Kim H, Ju Lee M, Jung Heo A, Youngjae Park D, Lim S, Shin S, Seung Yang W, An Jung C, Young Kim K, et al. 2022. The AUTOTAC chemical biology platform for targeted protein degradation via the autophagy-lysosome system. *Nat Commun* **13:** 904. doi:10.1038/s41467-022-28520-4

Hopfner F, Mueller SH, Szymczak S, Junge O, Tittmann L, May S, Lohmann K, Grallert H, Lieb W, Strauch K, et al. 2020. Rare variants in specific lysosomal genes are associated with Parkinson's disease. *Mov Disord* **35:** 1245–1248. doi:10.1002/mds.28037

Hu M, Li P, Wang C, Feng X, Geng Q, Chen W, Marthi M, Zhang W, Gao C, Reid W, et al. 2022. Parkinson's disease-risk protein TMEM175 is a proton-activated proton channel in lysosomes. *Cell* **185:** 2292–2308.e20. doi:10.1016/j.cell.2022.05.021

Jiao F, Zhou B, Meng L. 2023. The regulatory mechanism and therapeutic potential of transcription factor EB in neurodegenerative diseases. *CNS Neurosci Ther* **29:** 37–59. doi:10.1111/cns.13985

Jin SM, Lazarou M, Wang C, Kane LA, Narendra DP, Youle RJ. 2010. Mitochondrial membrane potential regulates PINK1 import and proteolytic destabilization by PARL. *J Cell Biol* **191:** 933–942. doi:10.1083/jcb.201008084

Jinn S, Drolet RE, Cramer PE, Wong AHK, Toolan DM, Gretzula CA, Voleti B, Vassileva G, Disa J, Tadin-Strapps M, et al. 2017. TMEM175 deficiency impairs lysosomal and mitochondrial function and increases α-synuclein aggregation. *Proc Natl Acad Sci* **114:** 2389–2394. doi:10.1073/pnas.1616332114

Johnson ME, Bergkvist L, Stetzik L, Steiner JA, Meyerdirk L, Schulz E, Wolfrum E, Luk KC, Wesson DW, Krainc D, et al. 2021. Heterozygous GBA D409V and ATP13a2 mutations do not exacerbate pathological α-synuclein spread in the prodromal preformed fibrils model in young mice. *Neurobiol Dis* **159:** 105513. doi:10.1016/j.nbd.2021.105513

Jung CH, Ro SH, Cao J, Otto NM, Kim DH. 2010. mTOR regulation of autophagy. *FEBS Lett* **584:** 1287–1295. doi:10.1016/j.febslet.2010.01.017

Kabuta T, Setsuie R, Mitsui T, Kinugawa A, Sakurai M, Aoki S, Uchida K, Wada K. 2008. Aberrant molecular properties shared by familial Parkinson's disease-associated mutant UCH-L1 and carbonyl-modified UCH-L1. *Hum Mol Genet* **17:** 1482–1496. doi:10.1093/hmg/ddn037

Kamp F, Beyer K. 2006. Binding of α-synuclein affects the lipid packing in bilayers of small vesicles. *J Biol Chem* **281:** 9251–9259. doi:10.1074/jbc.M512292200

Kane LA, Lazarou M, Fogel AI, Li Y, Yamano K, Sarraf SA, Banerjee S, Youle RJ. 2014. PINK1 phosphorylates ubiquitin to activate parkin E3 ubiquitin ligase activity. *J Cell Biol* **205:** 143–153. doi:10.1083/jcb.201402104

Kaushik S, Cuervo AM. 2018. The coming of age of chaperone-mediated autophagy. *Nat Rev Mol Cell Biol* **19:** 365–381. doi:10.1038/s41580-018-0001-6

Kaushik S, Massey AC, Cuervo AM. 2006. Lysosome membrane lipid microdomains: novel regulators of chaperone-mediated autophagy. *EMBO J* **25:** 3921–3933. doi:10.1038/sj.emboj.7601283

Kaushik S, Massey AC, Mizushima N, Cuervo AM. 2008. Constitutive activation of chaperone-mediated autophagy in cells with impaired macroautophagy. *Mol Biol Cell* **19:** 2179–2192. doi:10.1091/mbc.e07-11-1155

Kaushik S, Tasset I, Arias E, Pampliega O, Wong E, Martinez-Vicente M, Cuervo AM. 2021. Autophagy and the hallmarks of aging. *Ageing Res Rev* **72:** 101468. doi:10.1016/j.arr.2021.101468

Kauther KM, Höft C, Rissling I, Oertel WH, Möller JC. 2011. The PLA2G6 gene in early-onset Parkinson's disease. *Mov Disord* **26:** 2415–2417. doi:10.1002/mds.23851

Kazlauskaite A, Martínez-torres RJ, Wilkie S, Kumar A, Peltier J, Johnson C, Zhang J, Hope AG, Peggie M, Trost M, et al. 2015. Binding to serine 65-phosphorylated ubiquitin primes parkin for optimal PINK1-dependent phosphorylation and activation. *EMBO Rep* **16**: 939–954. doi:10.15252/embr.201540352

Kim MJ, Jeon S, Burbulla LF, Krainc D. 2018. Acid ceramidase inhibition ameliorates α-synuclein accumulation upon loss of GBA1 function. *Hum Mol Genet* **27**: 1972–1988. doi:10.1093/hmg/ddy105

Kinghorn KJ. 2011. Pathological looping in the synucleinopathies: investigating the link between Parkinson's disease and Gaucher disease. *DMM Dis Model Mech* **4**: 713–715. doi:10.1242/dmm.008615

Kinghorn KJ, Grönke S, Castillo-Quan JI, Woodling NS, Li L, Sirka E, Gegg M, Mills K, Hardy J, Bjedov I, et al. 2016. A *Drosophila* model of neuronopathic Gaucher disease demonstrates lysosomal-autophagic defects and altered mTOR signalling and is functionally rescued by rapamycin. *J Neurosci* **36**: 11654–11670. doi:10.1523/JNEUROSCI.4527-15.2016

Klein AD, Mazzulli JR. 2018. Is Parkinson's disease a lysosomal disorder? *Brain* **141**: 2255–2262. doi:10.1093/brain/awy147

Köroğlu Ç, Baysal L, Cetinkaya M, Karasoy H, Tolun A. 2013. DNAJC6 is responsible for juvenile parkinsonism with phenotypic variability. *Parkinsonism Relat Disord* **19**: 320–324. doi:10.1016/j.parkreldis.2012.11.006

Korvatska O, Strand NS, Berndt JD, Strovas T, Chen DHH, Leverenz JB, Kiianitsa K, Mata IF, Karakoc E, Greenup JL, et al. 2013. Altered splicing of ATP6AP2 causes X-linked parkinsonism with spasticity (XPDS). *Hum Mol Genet* **22**: 3259–3268. doi:10.1093/hmg/ddt180

Koyano F, Okatsu K, Kosako H, Tamura Y, Go E, Kimura M, Kimura Y, Tsuchiya H, Yoshihara H, Hirokawa T, et al. 2014. Ubiquitin is phosphorylated by PINK1 to activate parkin. *Nature* **510**: 162–166. doi:10.1038/nature13392

Kuo SH, Tasset I, Cheng MM, Diaz A, Pan MK, Lieberman OJ, Hutten SJ, Alcalay RN, Kim S, Ximénez-Embún P, et al. 2022. Mutant glucocerebrosidase impairs α-synuclein degradation by blockade of chaperone-mediated autophagy. *Sci Adv* **8**: eabm6393. doi:10.1126/sciadv.abm6393

Kuwahara T, Iwatsubo T. 2020. The emerging functions of LRRK2 and Rab GTPases in the endolysosomal system. *Front Neurosci* **14**: 227. doi:10.3389/fnins.2020.00227

Kwon EH, Tennagels S, Gold R, Gerwert K, Beyer L, Tönges L. 2022. Update on CSF biomarkers in Parkinson's disease. *Biomolecules* **12**: 329. doi:10.3390/biom12020329

Lamark T, Johansen T. 2012. Aggrephagy: selective disposal of protein aggregates by macroautophagy. *Int J Cell Biol* **2012**: 1–21. doi:10.1155/2012/736905

Lamark T, Johansen T. 2021. Mechanisms of selective autophagy. *Annu Rev Cell Dev Biol* **37**: 143–169. doi:10.1146/annurev-cellbio-120219-035530

Lashuel HA, Overk CR, Oueslati A, Masliah E. 2013. The many faces of α-synuclein: from structure and toxicity to therapeutic target. *Nat Rev Neurosci* **14**: 38–48. doi:10.1038/nrn3406

Lazarou M, Sliter DA, Kane LA, Sarraf SA, Wang C, Burman JL, Sideris DP, Fogel AI, Youle RJ. 2015. The ubiquitin kinase PINK1 recruits autophagy receptors to induce mitophagy. *Nature* **524**: 309–314. doi:10.1038/nature14893

Le W. 2014. Role of iron in UPS impairment model of Parkinson's disease. *Park Relat Disord* **20**(Suppl 1): S158–S161. doi:10.1016/S1353-8020(13)70038-5

Lee YJ, Wang S, Slone SR, Yacoubian TA, Witt SN. 2011. Defects in very long chain fatty acid synthesis enhance α-synuclein toxicity in a yeast model of Parkinson's disease. *PLoS ONE* **6**: e15946. doi:10.1371/journal.pone.0015946

Lee J, Sung KW, Bae EJ, Yoon D, Kim D, Lee JS, Park DH, Park DY, Mun SR, Kwon SC, et al. 2023. Targeted degradation of α-synuclein aggregates in Parkinson's disease using the AUTOTAC technology. *Mol Neurodegener* **18**: 1–21. doi:10.1186/s13024-023-00630-7

Leftin A, Job C, Beyer K, Brown MF. 2013. Solid-state 13C NMR reveals annealing of raft-like membranes containing cholesterol by the intrinsically disordered protein α-synuclein. *J Mol Biol* **425**: 2973–2987. doi:10.1016/j.jmb.2013.04.002

Lehtonen Š, Sonninen TM, Wojciechowski S, Goldsteins G, Koistinaho J. 2019. Dysfunction of cellular proteostasis in Parkinson's disease. *Front Neurosci* **13**: 1–19. doi:10.3389/fnins.2019.00457

Lerche S, Wurster I, Roeben B, Zimmermann M, Riebenbauer B, Deuschle C, Hauser A, Schulte C, Berg D, Maetzler W. 2019. Parkinson's disease: glucocerebrosidase 1 mutation severity is associated with CSF α-synuclein profiles. *Mov Disord* **35**: 495–499. doi:10.1002/mds.27884

Lerche S, Wurster I, Roeben B, Zimmermann M, Riebenbauer B, Deuschle C, Hauser A, Schulte C, Berg D, Maetzler W, et al. 2020. Parkinson's disease: glucocerebrosidase 1 mutation severity is associated with CSF α-synuclein profiles. *Mov Disord* **35**: 495–499. doi:10.1002/mds.27884

Lerche S, Schulte C, Wurster I, Machetanz G, Roeben B, Zimmermann M, Deuschle C, Hauser A-KK, Böhringer J, Krägeloh-Mann I, et al. 2021. The mutation matters: CSF profiles of GCase, sphingolipids, α-synuclein in PDGBA. *Mov Disord* **36**: 1216–1228. doi:10.1002/mds.28472

Lesage S, Brice A. 2009. Parkinson's disease: from monogenic forms to genetic susceptibility factors. *Hum Mol Genet* **18**: R48–R59. doi:10.1093/hmg/ddp012

Levine B, Kroemer G. 2019. Biological functions of autophagy genes: a disease perspective. *Cell* **176**: 11–42. doi:10.1016/j.cell.2018.09.048

Li H, Ham A, Ma C, Kuo SHH, Kanter E, Kim D, Ko HS, Quan Y, Sardi SP, Li A, et al. 2019. Mitochondrial dysfunction and mitophagy defect triggered by heterozygous GBA mutations. *Autophagy* **15**: 113–130. doi:10.1080/15548627.2018.1509818

Li J, Yang D, Li Z, Zhao M, Wang D, Sun Z, Wen P, Dai Y, Gou F, Ji Y, et al. 2023. PINK1/Parkin-mediated mitophagy in neurodegenerative diseases. *Ageing Res Rev* **84**: 101817. doi:10.1016/j.arr.2022.101817

Liang Y, Zhong G, Ren M, Sun T, Li Y, Ye M, Ma C, Guo Y, Liu C. 2023. The role of ubiquitin–proteasome system and mitophagy in the pathogenesis of Parkinson's disease. *Neuro Molecular Med* **25**: 471–488. doi:10.1007/s12017-023-08755-0

Lin CH, Tan EK, Chen ML, Tan LC, Lim HQ, Chen GS, Wu RM. 2008. Novel ATP13A2 variant associated with Parkinson disease in Taiwan and Singapore. *Neurology* **71**: 1727–1732. doi:10.1212/01.wnl.0000335167.72412.68

Liu G, Boot B, Locascio JJ, Jansen IE, Winder-Rhodes S, Eberly S, Elbaz A, Brice A, Ravina B, van Hilten JJ, et al. 2016. Specifically neuropathic Gaucher's mutations accelerate cognitive decline in Parkinson's. *Ann Neurol* **80**: 674–685. doi:10.1002/ana.24781

Liu Y, Guardia-Laguarta C, Yin J, Erdjument-Bromage H, Martin B, James M, Jiang X, Przedborski S. 2017. The ubiquitination of PINK1 is restricted to its mature 52-kDa form. *Cell Rep* **20**: 30–39. doi:10.1016/j.celrep.2017.06.022

Liu Z, Bryant N, Kumaran R, Beilina A, Abeliovich A, Cookson MR, West AB. 2018. LRRK2 phosphorylates membrane-bound Rabs and is activated by GTP-bound Rab7L1 to promote recruitment to the *trans*-Golgi network. *Hum Mol Genet* **27**: 385–395. doi:10.1093/hmg/ddx410

Lopes da Fonseca T, Pinho R, Outeiro TF. A familial ATP13A2 mutation enhances α-synuclein aggregation and promotes cell death. *Hum Mol Genet* **25**: 2959–2971. doi:10.1093/hmg/ddw147

Lunati A, Lesage S, Brice A. 2018. The genetic landscape of Parkinson's disease. *Rev Neurol (Paris)* **174**: 628–643. doi:10.1016/j.neurol.2018.08.004

MacLeod DA, Rhinn H, Kuwahara T, Zolin A, Di Paolo G, McCabe BD, Marder KS, Honig LS, Clark LN, Small SA, et al. 2013. RAB7L1 interacts with LRRK2 to modify intraneuronal protein sorting and Parkinson's disease risk. *Neuron* **77**: 425–439. doi:10.1016/j.neuron.2012.11.033

Madero-Pérez J, Fdez E, Fernández B, Lara Ordóñez AJ, Blanca Ramírez M, Gómez-Suaga P, Waschbüsch D, Lobbestael E, Baekelandt V, Nairn AC, et al. 2018. Parkinson disease-associated mutations in LRRK2 cause centrosomal defects via Rab8a phosphorylation. *Mol Neurodegener* **13**: 3. doi:10.1186/s13024-018-0235-y

Magalhaes J, Gegg ME, Migdalska-Richards A, Doherty MK, Whitfield PD, Schapira AHV. 2016. Autophagic lysosome reformation dysfunction in glucocerebrosidase deficient cells: relevance to Parkinson disease. *Hum Mol Genet* **25**: 3432–3445. doi:10.1093/hmg/ddw185

Mak SK, McCormack AL, Manning-Bog AB, Cuervo AM, Di Monte DA. 2010. Lysosomal degradation of α-synuclein in vivo. *J Biol Chem* **285**: 13621–13629. doi:10.1074/jbc.M109.074617

Malini E, Zampieri S, Deganuto M, Romanello M, Sechi A, Bembi B, Dardis A. 2015. Role of LIMP-2 in the intracellular trafficking of β-glucosidase in different human cellular models. *FASEB J* **29**: 3839–3852. doi:10.1096/fj.15-271148

Manzoni C. 2017. The LRRK2-macroautophagy axis and its relevance to Parkinson's disease. *Biochem Soc Trans* **45**: 155–162. doi:10.1042/BST20160265

Martin S, Smolders S, Van den Haute C, Heeman B, van Veen S, Crosiers D, Beletchi I, Verstraeten A, Gossye H, Gelders G, et al. 2020. Mutated ATP10B increases Parkinson's disease risk by compromising lysosomal glucosylceramide export. *Acta Neuropathol* **139**: 1001–1024. doi:10.1007/s00401-020-02145-7

Martinez Z, Zhu M, Han S, Fink AL. 2007. GM1 specifically interacts with α-synuclein and inhibits fibrillation. *Biochemistry* **46**: 1868–1877. doi:10.1021/bi061749a

Martinez BA, Petersen DA, Gaeta AL, Stanley SP, Caldwell GA, Caldwell KA. 2017. Dysregulation of the mitochondrial unfolded protein response induces non-apoptotic dopaminergic neurodegeneration in *C. elegans* models of Parkinson's disease. *J Neurosci* **37**: 11085–11100. doi:10.1523/JNEUROSCI.1294-17.2017

Martinez-Vicente M. 2012. Multiple ways for α-synuclein degradation. *Mov Disord* **27**: 345–345. doi:10.1002/mds.24932

Martinez-Vicente M. 2015. Autophagy in neurodegenerative diseases: from pathogenic dysfunction to therapeutic modulation. *Semin Cell Dev Biol* **40**: 115–126. doi:10.1016/j.semcdb.2015.03.005

Martinez-Vicente M. 2017. Neuronal mitophagy in neurodegenerative diseases. *Front Mol Neurosci* **10**: 1–13. doi:10.3389/fnmol.2017.00064

Martinez-Vicente M, Talloczy Z, Kaushik S, Massey AC, Mazzulli J, Mosharov EV, Hodara R, Fredenburg R, Wu DC, Follenzi A, et al. 2008. Dopamine-modified α-synuclein blocks chaperone-mediated autophagy. *J Clin Invest* **118**: 777–778. doi:10.1172/JCI32806

Martinus RD, Garth GP, Webster TL, Cartwright P, Naylor DJ, Høj PB, Hoogenraad NJ. 1996. Selective induction of mitochondrial chaperones in response to loss of the mitochondrial genome. *Eur J Biochem* **240**: 98–103. doi:10.1111/j.1432-1033.1996.0098h.x

Massey AC, Kaushik S, Sovak G, Kiffin R, Cuervo AM. 2006. Consequences of the selective blockage of chaperone-mediated autophagy. *Proc Natl Acad Sci* **103**: 5805–5810. doi:10.1073/pnas.0507436103

Matsuda N, Sato S, Shiba K, Okatsu K, Saisho K, Gautier CA, Sou YS, Saiki S, Kawajiri S, Sato F, et al. 2010. PINK1 stabilized by mitochondrial depolarization recruits parkin to damaged mitochondria and activates latent parkin for mitophagy. *J Cell Biol* **189**: 211–221. doi:10.1083/jcb.200910140

Matus S, Lisbona F, Torres M, Leon C, Thielen P, Hetz C. 2008. The stress rheostat: an interplay between the unfolded protein response (UPR) and autophagy in neurodegeneration. *Curr Mol Med* **8**: 157–172. doi:10.2174/156652408784221324

McNeill A, Magalhaes J, Shen C, Chau KYY, Hughes D, Mehta A, Foltynie T, Cooper JM, Abramov AY, Gegg M, et al. 2014. Ambroxol improves lysosomal biochemistry in glucocerebrosidase mutation-linked Parkinson disease cells. *Brain* **137**: 1481–1495. doi:10.1093/brain/awu020

Mizushima N, Levine B. 2020. Autophagy in human diseases. *N Engl J Med* **383**: 1564–1576. doi:10.1056/NEJMra2022774

Mizushima N, Levine B, Cuervo AM, Klionsky DJ. 2008. Autophagy fights disease through cellular self-digestion. *Nature* **451**: 1069–1075. doi:10.1038/nature06639

Murphy KE, Gysbers AM, Abbott SK, Spiro AS, Furuta A, Cooper A, Garner B, Kabuta T, Halliday GM. 2015. Lysosomal-associated membrane protein 2 isoforms are differentially affected in early Parkinson's disease. *Mov Disord* **30**: 1639–1647. doi:10.1002/mds.26141

Murugesan V, Chuang WL, Liu J, Lischuk A, Kacena K, Lin H, Pastores GM, Yang R, Keutzer J, Zhang K, et al. 2016. Glucosylsphingosine is a key biomarker of Gaucher disease. *Am J Hematol* **91:** 1082–1089. doi:10.1002/ajh .24491

Nalls MA, Pankratz N, Lill CM, Do CB, Hernandez DG, Saad M, DeStefano AL, Kara E, Bras J, Sharma M, et al. 2014. Large-scale meta-analysis of genome-wide association data identifies six new risk loci for Parkinson's disease. *Nat Genet* **46:** 989–993. doi:10.1038/ng.3043

Nalls MA, Blauwendraat C, Vallerga CL, Heilbron K, Bandres-Ciga S, Chang D, Tan M, Kia DA, Noyce AJ, Xue A, et al. 2019. Identification of novel risk loci, causal insights, and heritable risk for Parkinson's disease: a meta-analysis of genome-wide association studies. *Lancet Neurol* **18:** 1091–1102. doi:10.1016/S1474-4422(19)30320-5

Napolitano G, Ballabio A. 2016. TFEB at a glance. *J Cell Sci* **129:** 2475–2481. doi:10.1242/jcs.146365

Narendra DP, Youle RJ. 2011. Targeting mitochondrial dysfunction: role for PINK1 and parkin in mitochondrial quality control. *Antioxid Redox Signal* **14:** 1929–1938. doi:10.1089/ars.2010.3799

Navarro-Romero A, Montpeyó M, Martinez-Vicente M. 2020. The emerging role of the lysosome in Parkinson's disease. *Cells* **9:** 2399. doi:10.3390/cells9112399

Navarro-Romero A, Fernandez-Gonzalez I, Riera J, Montpeyo M, Albert-Bayo M, Lopez-Royo T, Castillo-Sanchez P, Carnicer-Caceres C, Arranz-Amo JA, Castillo-Ribelles L, et al. 2022. Lysosomal lipid alterations caused by glucocerebrosidase deficiency promote lysosomal dysfunction, chaperone-mediated-autophagy deficiency, and α-synuclein pathology. *NPJ Park Dis* **8:** 126. doi:10.1038/s41531-022-00397-6

Needham PG, Guerriero CJ, Brodsky JL. 2019. Chaperoning endoplasmic reticulum–associated degradation (ERAD) and protein conformational diseases. *Cold Spring Harb Perspect Biol* **11:** a033928. doi: 10.1101/cshperspect.a033928

Neumann J, Bras J, Deas E, O'sullivan SS, Parkkinen L, Lachmann RH, Li A, Holton J, Guerreiro R, Paudel R, et al. 2009. Glucocerebrosidase mutations in clinical and pathologically proven Parkinson's disease. *Brain* **132:** 1783–1794. doi:10.1093/brain/awp044

O'Leary EI, Jiang Z, Strub MP, Lee JC. 2018. Effects of phosphatidylcholine membrane fluidity on the conformation and aggregation of N-terminally acetylated α-synuclein. *J Biol Chem* **293:** 11195–11205. doi:10.1074/jbc.RA118.002780

Olgiati S, Quadri M, Fang M, Rood JPMA, Saute JA, Chien HF, Bouwkamp CG, Graafland J, Minneboo M, Breedveld GJ, et al. 2016. DNAJC6 mutations associated with early-onset Parkinson's disease. *Ann Neurol* **79:** 244–256. doi:10.1002/ana.24553

Orenstein SJ, Kuo SH, Tasset I, Arias E, Koga H, Fernandez-Carasa I, Cortes E, Honig LS, Dauer W, Consiglio A, et al. 2013. Interplay of LRRK2 with chaperone-mediated autophagy. *Nat Neurosci* **16:** 394–406. doi:10.1038/nn.3350

Osellame LD, Rahim AA, Hargreaves IP, Gegg ME, Richard-Londt A, Brandner S, Waddington SN, Schapira AHV, Duchen MR. 2013. Mitochondria and quality control defects in a mouse model of Gaucher disease—links to Parkinson's disease. *Cell Metab* **17:** 941–953. doi:10.1016/j.cmet.2013.04.014

Paisan-Ruiz C, Bhatia KP, Li A, Hernandez D, Davis M, Wood NW, Hardy J, Houlden H, Singleton A, Schneider SA. 2009. Characterization of PLA2G6 as a locus for dystonia-parkinsonism. *Ann Neurol* **65:** 19–23. doi:10.1002/ana.21415

Palikaras K, Lionaki E, Tavernarakis N. 2018. Mechanisms of mitophagy in cellular homeostasis, physiology and pathology. *Nat Cell Biol* **20:** 1013–1022. doi:10.1038/s41556-018-0176-2

Pandey UB, Batlevi Y, Baehrecke EH, Taylor JP. 2007. HDAC6 at the intersection of autophagy, the ubiquitin-proteasome system and neurodegeneration. *Autophagy* **3:** 643–645. doi:10.4161/auto.5050

Parlar SC, Grenn FP, Kim JJ, Baluwendraat C, Gan-Or Z. 2023. Classification of GBA1 variants in Parkinson's disease: the GBA1-PD browser. *Mov Disord* **38:** 489–495. doi:10.1002/mds.29314

Paul A, Jacoby G, Laor Bar-Yosef D, Beck R, Gazit E, Segal D. 2021. Glucosylceramide associated with Gaucher disease forms amyloid-like twisted ribbon fibrils that induce α-synuclein aggregation. *ACS Nano* **15:** 11854–11868. doi:10.1021/acsnano.1c02957

Perissinotto F, Rondelli V, Parisse P, Tormena N, Zunino A, Almásy L, Merkel DG, Bottyán L, Sajti S, Casalis L. 2019. GM1 ganglioside role in the interaction of α-synuclein with lipid membranes: morphology and structure. *Biophys Chem* **255:** 106272. doi:10.1016/j.bpc.2019.106272

Phillips BP, Gomez-Navarro N, Miller EA. 2020. Protein quality control in the endoplasmic reticulum. *Curr Opin Cell Biol* **65:** 96–102. doi:10.1016/j.ceb.2020.04.002

Plowey ED, Cherra SJ III, Liu YJ, Chu CT. 2008. Role of autophagy in G2019S-LRRK2-associated neurite shortening in differentiated SH-SY5Y cells. *J Neurochem* **105:** 1048–1056. doi:10.1111/j.1471-4159.2008.05217.x

Poehler AM, Xiang W, Spitzer P, May VEL, Meixner H, Rockenstein E, Chutna O, Outeiro TF, Winkler J, Masliah E, et al. 2014. Autophagy modulates SNCA/α-synuclein release, thereby generating a hostile microenvironment. *Autophagy* **10:** 2171–2192. doi:10.4161/auto.36436

Polinski NK, Martinez TN, Gorodinsky A, Gareus R, Sasner M, Herberth M, Switzer R, Ahmad SO, Cosden M, Kandebo M, et al. 2021. Decreased glucocerebrosidase activity and substrate accumulation of glycosphingolipids in a novel GBA1 D409V knock-in mouse model. *PLoS ONE* **16:** e0252325. doi:10.1371/journal.pone.0252325

Polo S, Di Fiore PP. 2006. Endocytosis conducts the cell signaling orchestra. *Cell* **124:** 897–900. doi:10.1016/j.cell.2006.02.025

Pradas E, Martinez-Vicente M. 2023. The consequences of GBA deficiency in the autophagy–lysosome system in Parkinson's disease associated with GBA. *Cells* **12:** 191. doi:10.3390/cells12010191

Quadri M, Fang M, Picillo M, Olgiati S, Breedveld GJ, Graafland J, Wu B, Xu F, Erro R, Amboni M, et al. 2013. Mutation in the SYNJ1 gene associated with autosomal recessive, early-onset parkinsonism. *Hum Mutat* **34:** 1208–1215. doi:10.1002/humu.22373

Ramirez A, Heimbach A, Gründemann J, Stiller B, Hampshire D, Cid LP, Goebel I, Mubaidin AF, Wriekat AL, Roeper J, et al. 2006. Hereditary parkinsonism with dementia is caused by mutations in ATP13A2, encoding a lysosomal type 5 P-type ATPase. *Nat Genet* **38:** 1184–1191. doi:10.1038/ng1884

Cite this article as *Cold Spring Harb Perspect Med* doi: 10.1101/cshperspect.a041619

Ramonet D, Podhajska A, Stafa K, Sonnay S, Trancikova A, Tsika E, Pletnikova O, Troncoso JC, Glauser L, Moore DJ. 2012. PARK9-associated ATP13A2 localizes to intracellular acidic vesicles and regulates cation homeostasis and neuronal integrity. *Hum Mol Genet* **21**: 1725–1743. doi:10.1093/hmg/ddr606

Ramser J, Abidi FE, Burckle CA, Lenski C, Toriello H, Wen G, Lubs HA, Engert S, Stevenson RE, Meindl A, et al. 2005. A unique exonic splice enhancer mutation in a family with X-linked mental retardation and epilepsy points to a novel role of the renin receptor. *Hum Mol Genet* **14**: 1019–1027. doi:10.1093/hmg/ddi094

Reczek D, Schwake M, Schröder J, Hughes H, Blanz J, Jin X, Brondyk W, Van Patten S, Edmunds T, Saftig P. 2007. LIMP-2 is a receptor for lysosomal mannose-6-phosphate-independent targeting of β-glucocerebrosidase. *Cell* **131**: 770–783. doi:10.1016/j.cell.2007.10.018

Robak LA, Jansen IE, van Rooij J, Uitterlinden AG, Kraaij R, Jankovic J, Heutink P, Shulman JM, Nalls MA, Plagnol V, et al. 2017. Excessive burden of lysosomal storage disorder gene variants in Parkinson's disease. *Brain* **140**: 3191–3203. doi:10.1093/brain/awx285

Rocha EM, Smith GA, Park E, Cao H, Brown E, Hayes MA, Beagan J, McLean JR, Izen SC, Perez-Torres E, et al. 2015a. Glucocerebrosidase gene therapy prevents α-synucleinopathy of midbrain dopamine neurons. *Neurobiol Dis* **82**: 495–503. doi:10.1016/j.nbd.2015.09.009

Rocha EM, Smith GA, Park E, Cao H, Graham A-R, Brown E, McLean JR, Hayes MA, Beagan J, Izen SC, et al. 2015b. Sustained systemic glucocerebrosidase inhibition induces brain α-synuclein aggregation, microglia and complement C1q activation in mice. *Antioxid Redox Signal* **23**: 550–564. doi:10.1089/ars.2015.6307

Rockenstein E, Clarke J, Viel C, Panarello N, Treleaven CM, Kim C, Spencer B, Adame A, Park H, Dodge JC, et al. 2016. Glucocerebrosidase modulates cognitive and motor activities in murine models of Parkinson's disease. *Hum Mol Genet* **25**: 2645–2660.

Rodriguez-Navarro JA, Kaushik S, Koga H, Dall'Armi C, Shui G, Wenk MR, Di Paolo G, Cuervo AM. 2012. Inhibitory effect of dietary lipids on chaperone-mediated autophagy. *Proc Natl Acad Sci* **109**: E705–E714. doi:10.1073/pnas.1113036109

Rolfs A, Giese AK, Grittner U, Mascher D, Elstein D, Zimran A, Böttcher T, Lukas J, Hübner R, Gölnitz U, et al. 2013. Glucosylsphingosine is a highly sensitive and specific biomarker for primary diagnostic and follow-up monitoring in Gaucher disease in a non-Jewish, caucasian cohort of Gaucher disease patients. *PLoS ONE* **8**: e79732. doi:10.1371/journal.pone.0079732

Roosen DA, Cookson MR. 2016. LRRK2 at the interface of autophagosomes, endosomes and lysosomes. *Mol Neurodegener* **11**: 73. doi:10.1186/s13024-016-0140-1

Russo MJ, Orru CD, Concha-Marambio L, Giaisi S, Groveman BR, Farris CM, Holguin B, Hughson AG, LaFontant DE, Caspell-Garcia C, et al. 2021. High diagnostic performance of independent α-synuclein seed amplification assays for detection of early Parkinson's disease. *Acta Neuropathol Commun* **9**: 179. doi:10.1186/s40478-021-01282-8

Sagini K, Buratta S, Delo F, Pellegrino RM, Giovagnoli S, Urbanelli L, Emiliani C. 2021. Drug-induced lysosomal impairment is associated with the release of extracellular vesicles carrying autophagy markers. *Int J Mol Sci* **22**: 12922. doi:10.3390/ijms222312922

Sahu R, Kaushik S, Clement CC, Cannizzo ES, Scharf B, Follenzi A, Potolicchio I, Nieves E, Cuervo AM, Santambrogio L. 2011. Microautophagy of cytosolic proteins by late endosomes. *Dev Cell* **20**: 131–139. doi:10.1016/j.devcel.2010.12.003

Saito Y, Kawashima A, Ruberu NN, Fujiwara H, Koyama S, Sawabe M, Arai T, Nagura H, Yamanouchi H, Hasegawa M, et al. 2003. Accumulation of phosphorylated α-synuclein in aging human brain. *J Neuropathol Exp Neurol* **62**: 644–654. doi:10.1093/jnen/62.6.644

Saito Y, Suzuki K, Hulette CM, Murayama S. 2004. Aberrant phosphorylation of α-synuclein in human Niemann-Pick type C1 disease. *J Neuropathol Exp Neurol* **63**: 323–328. doi:10.1093/jnen/63.4.323

Sardi SP, Clarke J, Kinnecom C, Tamsett TJ, Li L, Stanek LM, Passini MA, Grabowski GA, Schlossmacher MG, Sidman RL, et al. 2011. CNS expression of glucocerebrosidase corrects α-synuclein pathology and memory in a mouse model of Gaucher-related synucleinopathy. *Proc Natl Acad Sci* **108**: 12101–12106. doi:10.1073/pnas.1108197108

Sardi SP, Clarke J, Viel C, Chan M, Tamsett TJ, Treleaven CM, Bu J, Sweet L, Passini MA, Dodge JC, et al. 2013. Augmenting CNS glucocerebrosidase activity as a therapeutic strategy for parkinsonism and other Gaucher-related synucleinopathies. *Proc Natl Acad Sci* **110**: 3537–3542. doi:10.1073/pnas.1220464110

Sardiello M, Palmieri M, di Ronza A, Medina DL, Valenza M, Gennarino VA, Di Malta C, Donaudy F, Embrione V, Polishchuk RS, et al. 2009. A gene network regulating lysosomal biogenesis and function. *Science* **325**: 473–477. doi:10.1126/science.1174447

Schneider JL, Suh Y, Cuervo AM. 2014. Deficient chaperone-mediated autophagy in liver leads to metabolic dysregulation. *Cell Metab* **20**: 417–432. doi:10.1016/j.cmet.2014.06.009

Schöndorf DC, Aureli M, McAllister FE, Hindley CJ, Mayer F, Schmid B, Sardi SP, Valsecchi M, Hoffmann S, Schwarz LK, et al. 2014. IPSC-derived neurons from GBA1-associated Parkinson's disease patients show autophagic defects and impaired calcium homeostasis. *Nat Commun* **5**: 4028. doi:10.1038/ncomms5028

Schormair B, Kemlink D, Mollenhauer B, Fiala O, Machetanz G, Roth J, Berutti R, Strom TM, Haslinger B, Trenkwalder C, et al. 2018. Diagnostic exome sequencing in early-onset Parkinson's disease confirms VPS13C as a rare cause of autosomal-recessive Parkinson's disease. *Clin Genet* **93**: 603–612. doi:10.1111/cge.13124

Schreij AMAA, Fon EA, McPherson PS. 2016. Endocytic membrane trafficking and neurodegenerative disease. *Cell Mol Life Sci* **73**: 1529–1545. doi:10.1007/s00018-015-2105-x

Scrivo A, Bourdenx M, Pampliega O, Cuervo AM. 2018. Selective autophagy as a potential therapeutic target for neurodegenerative disorders. *Lancet Neurol* **17**: 802–815. doi:10.1016/S1474-4422(18)30238-2

Senft D, Ronai ZA. 2015. UPR, autophagy, and mitochondria crosstalk underlies the ER stress response. *Trends Biochem Sci* **40**: 141–148. doi:10.1016/j.tibs.2015.01.002

Settembre C, Di Malta C, Polito VA, Garcia Arencibia M, Vetrini F, Erdin S, Erdin SU, Huynh T, Medina D, Colella P, et al. 2011. TFEB links autophagy to lysosomal biogenesis. *Science* **332:** 1429–1433. doi:10.1126/science .1204592

Shimura H, Hattori N, Kubo S, Mizuno Y, Asakawa S, Minoshima S, Shimizu N, Iwai K, Chiba T, Tanaka K, et al. 2000. Familial Parkinson disease gene product, parkin, is a ubiquitin-protein ligase. *Nat Genet* **25:** 302–305. doi:10 .1038/77060

Si J, Van den Haute C, Lobbestael E, Martin S, van Veen S, Vangheluwe P, Baekelandt V. 2021. ATP13A2 regulates cellular α-synuclein multimerization, membrane association, and externalization. *Int J Mol Sci* **22:** 2689. doi:10 .3390/ijms22052689

Siderowf A, Concha-Marambio L, Lafontant DE, Farris CM, Ma Y, Urenia PA, Nguyen H, Alcalay RN, Chahine LM, Foroud T, et al. 2023. Assessment of heterogeneity among participants in the Parkinson's progression markers initiative cohort using α-synuclein seed amplification: a cross-sectional study. *Lancet Neurol* **22:** 407–417. doi:10 .1016/S1474-4422(23)00109-6

Sidransky E, Samaddar T, Tayebi N. 2009. MUTATIONS in GBA are associated with familial Parkinson disease susceptibility and age at onset. *Neurology* **73:** 1424–1426. doi:10.1212/WNL.0b013e3181b28601

Simón-Sánchez J, Schulte C, Bras JM, Sharma M, Gibbs JR, Berg D, Paisan-Ruiz C, Lichtner P, Scholz SW, Hernandez DG, et al. 2009. Genome-wide association study reveals genetic risk underlying Parkinson's disease. *Nat Genet* **41:** 1308–1312. doi:10.1038/ng.487

Smith L, Schapira AHVV. 2022. GBA variants and Parkinson disease: mechanisms and treatments. *Cells* **11:** 1261. doi:10.3390/cells11081261

Smith BR, Santos MB, Marshall MS, Cantuti-Castelvetri L, Lopez-Rosas A, Li G, Van Breemen R, Claycomb KI, Gallea JI, Celej MS, et al. 2014. Neuronal inclusions of α-synuclein contribute to the pathogenesis of Krabbe disease. *J Pathol* **232:** 509–521. doi:10.1002/path.4328

Smolders S, Van Broeckhoven C. 2020. Genetic perspective on the synergistic connection between vesicular transport, lysosomal and mitochondrial pathways associated with Parkinson's disease pathogenesis. *Acta Neuropathol Commun* **8:** 1–28. doi:10.1186/s40478-020-00935-4

Sohn EJ, Kim JH, Oh SO, Kim JY. 2023. Regulation of self-renewal in ovarian cancer stem cells by fructose via chaperone-mediated autophagy. *Biochim Biophys Acta* **1869:** 166723. doi:10.1016/j.bbadis.2023.166723

Sosero YL, Gan-Or Z. 2023. LRRK2 and Parkinson's disease: from genetics to targeted therapy. *Ann Clin Transl Neurol* **10:** 850–864. doi:10.1002/acn3.51776

Stojkovska I, Krainc D, Mazzulli JR. 2018. Molecular mechanisms of α-synuclein and GBA1 in Parkinson's disease. *Cell Tissue Res* **373:** 51–60. doi:10.1007/s00441-017-2704-y

Sucunza D, Rico AJ, Roda E, Collantes M, González-Aseguinolaza G, Rodríguez-Pérez AI, Peñuelas I, Vázquez A, Labandeira-García JL, Broccoli V, et al. 2021. Glucocerebrosidase gene therapy induces α-synuclein clearance and neuroprotection of midbrain dopaminergic neurons in mice and macaques. *Int J Mol Sci* **22:** 4825. doi:10.3390/ijms22094825

Sun N, Youle RJ, Finkel T. 2016. The mitochondrial basis of aging. *Mol Cell* **61:** 654–666. doi:10.1016/j.molcel.2016.01 .028

Suzuki K, Iseki E, Togo T, Yamaguchi A, Katsuse O, Katsuyama K, Kanzaki S, Shiozaki K, Kawanishi C, Yamashita S, et al. 2007. Neuronal and glial accumulation of α- and β-synucleins in human lipidoses. *Acta Neuropathol* **114:** 481–489. doi:10.1007/s00401-007-0264-z

Taguchi YV, Liu J, Ruan J, Pacheco J, Zhang X, Abbasi J, Keutzer J, Mistry PK, Chandra SS. 2017. Glucosylsphingosine promotes α-synuclein pathology in mutant GBA-associated Parkinson's disease. *J Neurosci* **37:** 9617–9631. doi:10.1523/JNEUROSCI.1525-17.2017

Takahashi D, Moriyama J, Nakamura T, Miki E, Takahashi E, Sato A, Akaike T, Itto-Nakama K, Arimoto H. 2019. AUTACs: cargo-specific degraders using selective autophagy. *Mol Cell* **76:** 797–810.e10. doi:10.1016/j.molcel .2019.09.009

Tekirdag K, Cuervo AM. 2018. Chaperone-mediated autophagy and endosomal microautophagy: joint by a chaperone. *J Biol Chem* **293:** 5414–5424. doi:10.1074/jbc.R117 .818237

Thaler A, Bregman N, Gurevich T, Shiner T, Dror Y, Zmira O, Gan-Or Z, Bar-Shira A, Gana-Weisz M, Orr-Urtreger A, et al. 2018. Parkinson's disease phenotype is influenced by the severity of the mutations in the GBA gene. *Parkinsonism Relat Disord* **55:** 45–49. doi:10.1016/j.parkreldis .2018.05.009

Thomas RE, Vincow ES, Merrihew GE, MacCoss MJ, Davis MY, Pallanck LJ. 2018. Glucocerebrosidase deficiency promotes protein aggregation through dysregulation of extracellular vesicles. *PLoS Genet* **14:** e1007694. doi:10 .1371/journal.pgen.1007694

Tolosa E, Vila M, Klein C, Rascol O. 2020. LRRK2 in Parkinson disease: challenges of clinical trials. *Nat Rev Neurol* **16:** 97–107. doi:10.1038/s41582-019-0301-2

Torra A, Parent A, Cuadros T, Rodríguez-Galván B, Ruiz-Bronchal E, Ballabio A, Bortolozzi A, Vila M, Bové J. 2018. Overexpression of TFEB drives a pleiotropic neurotrophic effect and prevents Parkinson's disease-related neurodegeneration. *Mol Ther* **26:** 1552–1567. doi:10 .1016/j.ymthe.2018.02.022

Usenovic M, Tresse E, Mazzulli JR, Taylor JP, Krainc D. 2012. Deficiency of ATP13A2 leads to lysosomal dysfunction, α-synuclein accumulation, and neurotoxicity. *J Neurosci* **32:** 4240–6. doi:10.1523/JNEUROSCI.5575-11.2012

Valente EM, Abou-Sleiman PM, Caputo V, Muqit MM, Harvey K, Gispert S, Ali Z, Del Turco D, Bentivoglio AR, Healy DG, et al. 2004. Hereditary early-onset Parkinson's disease caused by mutations in PINK1. *Science* **304:** 1158–1160. doi:10.1126/science.1096284

van Veen S, Martin S, Van den Haute C, Benoy V, Lyons J, Vanhoutte R, Kahler JP, Decuypere JP, Gelders G, Lambie E, et al. 2020. ATP13A2 deficiency disrupts lysosomal polyamine export. *Nature* **578:** 419–424. doi:10.1038/ s41586-020-1968-7

Vilariño-Güell C, Rajput A, Milnerwood AJ, Shah B, Szu-Tu C, Trinh J, Yu I, Encarnacion M, Munsie LN, Tapia L, et al. 2014. DNAJC13 mutations in Parkinson disease. *Hum Mol Genet* **23:** 1794–1801. doi:10.1093/hmg/ddt570

Vives-Bauza C, Zhou C, Huang Y, Cui M, de Vries RLA, Kim J, May J, Tocilescu MA, Liu W, Ko HS, et al. 2010. PINK1-

Cite this article as *Cold Spring Harb Perspect Med* doi: 10.1101/cshperspect.a041619

dependent recruitment of parkin to mitochondria in mitophagy. *Proc Natl Acad Sci* **107**: 378–383. doi:10.1073/pnas.0911187107

Vogiatzi T, Xilouri M, Vekrellis K, Stefanis L. 2008. Wild type α-synuclein is degraded by chaperone-mediated autophagy and macroautophagy in neuronal cells. *J Biol Chem* **283**: 23542–23556. doi:10.1074/jbc.M801992200

Wallings R, Manzoni C, Bandopadhyay R. 2015. Cellular processes associated with LRRK2 function and dysfunction. *FEBS J* **282**: 2806–2826. doi:10.1111/febs.13305

Wang S, Ma Z, Xu X, Wang Z, Sun L, Zhou Y, Lin X, Hong W, Wang T. 2014. A role of Rab29 in the integrity of the *trans*-Golgi network and retrograde trafficking of mannose-6-phosphate receptor. *PLoS ONE* **9**: e96242. doi:10.1371/journal.pone.0096242

Wang L, Klionsky DJ, Shen H-M. 2023. The emerging mechanisms and functions of microautophagy. *Nat Rev Mol Cell Biol* **24**: 186–203. doi:10.1038/s41580-022-00529-z

Webb JL, Ravikumar B, Atkins J, Skepper JN, Rubinsztein DC. 2003. α-Synuclein is degraded by both autophagy and the proteasome. *J Biol Chem* **278**: 25009–25013. doi:10.1074/jbc.M300227200

Wilkening G, Linke T, Sandhoff K. 1998. Lysosomal degradation on vesicular membrane surfaces. *J Biol Chem* **273**: 30271–30278. doi:10.1074/jbc.273.46.30271

Williams ET, Chen X, Moore DJ. 2017. VPS35, the retromer complex and Parkinson's disease. *J Parkinsons Dis* **7**: 219–233. doi:10.3233/JPD-161020

Wong K, Sidransky E, Verma A, Mixon T, Sandberg GD, Wakefield LK, Morrison A, Lwin A, Colegial C, Allman JM, et al. 2004. Neuropathology provides clues to the pathophysiology of Gaucher disease. *Mol Genet Metab* **82**: 192–207. doi:10.1016/j.ymgme.2004.04.011

Wu X, Rapoport TA. 2018. Mechanistic insights into ER-associated protein degradation. *Curr Opin Cell Biol* **53**: 22–28. doi:10.1016/j.ceb.2018.04.004

Xiang C, Cong S, Tan X, Ma S, Liu Y, Wang H, Cong S. 2022. A meta-analysis of the diagnostic utility of biomarkers in cerebrospinal fluid in Parkinson's disease. *NPJ Park Dis* **8**: 165. doi:10.1038/s41531-022-00431-7

Xicoy H, Peñuelas N, Vila M, Laguna A. 2019. Autophagic- and lysosomal-related biomarkers for Parkinson's disease: lights and shadows. *Cells* **8**: 1–20.

Xie YX, Naseri NN, Fels J, Kharel P, Na Y, Lane D, Burré J, Sharma M. 2022. Lysosomal exocytosis releases pathogenic α-synuclein species from neurons in synucleinopathy models. *Nat Commun* **13**: 4918. doi:10.1038/s41467-022-32625-1

Xilouri M, Vogiatzi T, Vekrellis K, Stefanis L. 2008. α-Synuclein degradation by autophagic pathways: a potential key to Parkinson's disease pathogenesis. *Autophagy* **4**: 917–919. doi:10.4161/auto.6685

Xilouri M, Vogiatzi T, Vekrellis K, Park D, Stefanis L. 2009. Abberant α-synuclein confers toxicity to neurons in part

through inhibition of chaperone-mediated autophagy. *PLoS ONE* **4**: e5515. doi:10.1371/journal.pone.0005515

Xilouri M, Brekk OR, Kirik D, Stefanis L. 2013. LAMP2A as a therapeutic target in Parkinson disease. *Autophagy* **9**: 2166–2168. doi:10.4161/auto.26451

Xilouri M, Brekk OR, Stefanis L. 2016. Autophagy and α-synuclein: relevance to Parkinson's disease and related synucleopathies. *Mov Disord* **31**: 178–192. doi:10.1002/mds.26477

Xu H, Ren D. 2015. Lysosomal physiology. *Annu Rev Physiol* **77**: 57–80. doi:10.1146/annurev-physiol-021014-071649

Xu YH, Sun Y, Ran H, Quinn B, Witte D, Grabowski GA. 2011. Accumulation and distribution of α-synuclein and ubiquitin in the CNS of Gaucher disease mouse models. *Mol Genet Metab* **102**: 436–447. doi:10.1016/j.ymgme.2010.12.014

Yang Y, Gehrke S, Imai Y, Huang Z, Ouyang Y, Wang JW, Yang L, Beal MF, Vogel H, Lu B. 2006. Mitochondrial pathology and muscle and dopaminergic neuron degeneration caused by inactivation of *Drosophila* Pink1 is rescued by parkin. *Proc Natl Acad Sci* **103**: 10793–10798. doi:10.1073/pnas.0602493103

Yang C, Tian F, Hu M, Kang C, Ping M, Liu Y, Hu M, Xu H, Yu Y, Gao Z, et al. 2023. Characterization of the role of TMEM175 in an in vitro lysosomal H+ fluxes model. *FEBS J* **290**: 4641–4659. doi:10.1111/febs.16814

Yoshino H, Tomiyama H, Tachibana N, Ogaki K, Li Y, Funayama M, Hashimoto T, Takashima S, Hattori N. 2010. Phenotypic spectrum of patients with PLA2G6 mutation and PARK14-linked parkinsonism. *Neurology* **75**: 1356–1361. doi:10.1212/WNL.0b013e3181f73649

Yun SP, Kim D, Kim S, Kim S, Karuppagounder SS, Kwon SH, Lee S, Kam TI, Lee S, Ham S, et al. 2018. α-Synuclein accumulation and GBA deficiency due to L444P GBA mutation contributes to MPTP-induced parkinsonism. *Mol Neurodegener* **13**: 1–19. doi:10.1186/s13024-017-0233-5

Zhang J, Johnson JL, He J, Napolitano G, Ramadass M, Rocca C, Kiosses WB, Bucci C, Xin Q, Gavathiotis E, et al. 2017. Cystinosin, the small GTPase Rab11, and the Rab7 effector RILP regulate intracellular trafficking of the chaperone-mediated autophagy receptor LAMP2A. *J Biol Chem* **292**: 10328–10346. doi:10.1074/jbc.M116.764076

Zhang T, Liu C, Li W, Kuang J, Qiu XY, Min L, Zhu L. 2022. Targeted protein degradation in mammalian cells: a promising avenue toward future. *Comput Struct Biotechnol J* **20**: 5477–5489. doi:10.1016/j.csbj.2022.09.038

Zhao Q, Wang J, Levichkin IV, Stasinopoulos S, Ryan MT, Hoogenraad NJ. 2002. A mitochondrial specific stress response in mammalian cells. *EMBO J* **21**: 4411–4419. doi:10.1093/emboj/cdf445

Zou L, Tian Y, Zhang Z. 2021. Dysfunction of synaptic vesicle endocytosis in Parkinson's disease. *Front Integr Neurosci* **15**: 619160. doi: 10.3389/fnint.2021.619160

The Gut–Brain Axis in Parkinson's Disease

Virginia Gao,[1,2,3] Carl V. Crawford,[4] and Jacqueline Burré[1]

[1]Appel Institute for Alzheimer's Disease Research and Brain and Mind Research Institute, Weill Cornell Medicine, New York, New York 10021, USA

[2]Parkinson's Disease and Movement Disorders Institute, Department of Neurology, Weill Cornell Medicine, New York, New York 10065, USA

[3]Division of Movement Disorders, The Neurological Institute of New York, Columbia University Irving Medical Center, New York, New York 10033, USA

[4]Division of Gastroenterology and Hepatology, Weill Cornell Medicine, New York, New York 10065, USA

Correspondence: vig9070@med.cornell.edu

Parkinson's disease (PD) involves both the central nervous system (CNS) and enteric nervous system (ENS), and their interaction is important for understanding both the clinical manifestations of the disease and the underlying disease pathophysiology. Although the neuroanatomical distribution of pathology strongly suggests that the ENS is involved in disease pathophysiology, there are significant gaps in knowledge about the underlying mechanisms. In this article, we review the clinical presentation and management of gastrointestinal dysfunction in PD. In addition, we discuss the current understanding of disease pathophysiology in the gut, including controversies about early involvement of the gut in disease pathogenesis. We also review current knowledge about gut α-synuclein and the microbiome, discuss experimental models of PD-linked gastrointestinal pathophysiology, and highlight areas for further research. Finally, we discuss opportunities to use the gut–brain axis for the development of biomarkers and disease-modifying treatments.

Parkinson's disease (PD) is the fastest-growing neurological disorder worldwide (Dorsey et al. 2018), and this growth is driven both by an increase in population age and longevity and by environmental factors. Although the disease is classically defined by its motor manifestations (bradykinesia, rigidity, rest tremor, and postural instability) due to degeneration of substantia nigra dopamine neurons, PD is now understood to involve both the central nervous system (CNS) and peripheral nervous system, including the enteric nervous system (ENS) (Fasano et al. 2015).

The ENS controls the function of the gastrointestinal tract and is a main division of the peripheral nervous system. It is likely involved early in the disease course, before motor symptoms manifest, as gastrointestinal dysfunction can precede motor dysfunction in PD by decades (Kalia and Lang 2015). The gut provides a route for environmental exposure and can serve as a conduit for pathogenesis to the CNS, given the lack of a blood–ganglia barrier. This idea is further supported by the epidemiological link between PD and environmental factors, especially

Copyright © 2025 Cold Spring Harbor Laboratory Press; all rights reserved
Cite this article as *Cold Spring Harb Perspect Med* doi: 10.1101/cshperspect.a041618

those related to industrialization, including pesticides, industrial solvents, and heavy metals (Goldman 2014).

James Parkinson's first formal description of the disease in 1817 occurred during the Industrial Revolution. He highlighted early gastrointestinal symptoms—"The bowels, which had been all along torpid, now, in most cases, demand stimulating medicines of very considerable power"—and postulated that pathology in the bowels could induce pathology in the CNS—"Although unable to trace the connection by which a disordered state of the stomach and bowels may induce a morbid action in a part of the medulla spinalis … little hesitation need be employed before we determine on the probability of such occurrence" (Parkinson 2002). Despite this early recognition of the involvement of the ENS in PD, significant gaps remain in our understanding about the underlying mechanisms.

Here, we discuss clinical manifestations, evaluation, and treatment of gastrointestinal dysfunction in PD, and review our understanding to date of the gastrointestinal pathophysiology underlying the disease, highlighting current gaps in knowledge and areas for further investigation.

GASTROINTESTINAL DYSFUNCTION IN PARKINSON'S DISEASE

Gastrointestinal symptoms are among the earliest manifestations of PD and precede the onset of motor symptoms (Fig. 1). Gastrointestinal mo-

tor impairments and dysautonomia are intrinsic to the underlying disease process in PD (Fasano et al. 2015; Warnecke et al. 2022). This pathophysiology also underlies treatment-related motor fluctuations (wearing off and dyskinesia) due to altered absorption of medication. We begin by reviewing the neuroanatomy of the ENS (Fig. 2) to lay the foundation for understanding its interaction with the CNS and how it may be involved in disease pathogenesis. We then discuss the symptomatic manifestations of PD throughout the gastrointestinal tract and the evaluation and management of these symptoms.

NEUROANATOMY OF THE ENS

The gastrointestinal tract is innervated from the esophagus to the anus by the ENS, the largest part of the peripheral nervous system containing hundreds of millions of neurons and even more glia (Furness 2007). Its functions include controlling movement in the gastrointestinal tract, regulating the shift of fluid across the epithelial lining, maintaining epithelial barrier integrity, regulating local blood flow and nutrient handling, and interacting with the immune and endocrine systems. The gastrointestinal tract is a unique peripheral organ in that its extensive intrinsic nervous system can function autonomously without extrinsic input (Bayliss and Starling 1899), although in practice, neuronal control of gastrointestinal function is modulated by both intrinsic and extrinsic fibers.

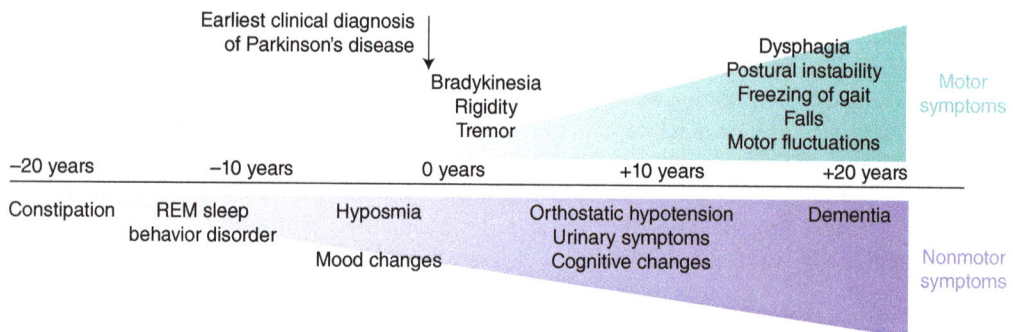

Figure 1. Systemic manifestations of prodromal, early, and established Parkinson's disease (PD). Time line and severity of nonmotor and motor manifestations of PD, and typical timing of diagnosis during the course of the disease are shown.

Cite this article as *Cold Spring Harb Perspect Med* doi: 10.1101/cshperspect.a041618

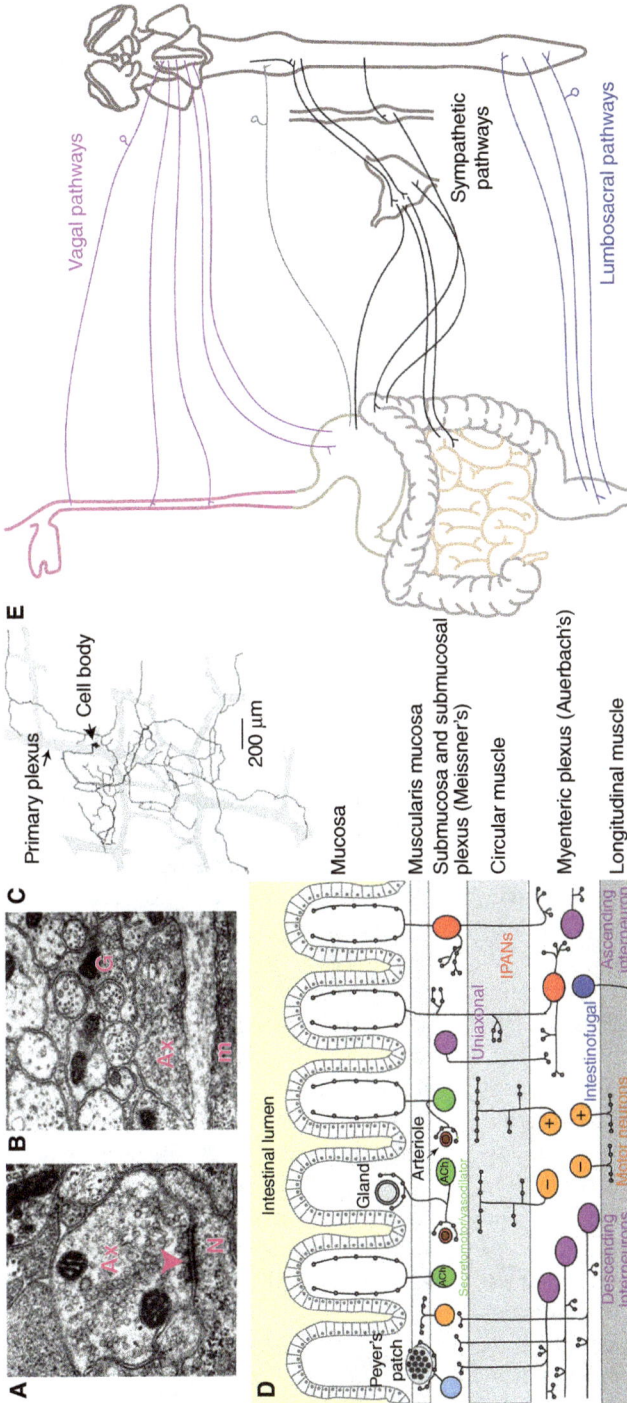

Figure 2. Enteric nervous system (ENS) neuroanatomy. (A) Electron micrograph of an enteric axon (Ax) varicosity containing transmitter vesicles on a myenteric nerve (N) cell with pre- and postsynaptic densities at the synapse (arrows). (B) Electron micrograph of an enteric Ax at the surface of a nerve fiber bundle facing muscle (m), juxtaposed to the processes of glial (G) cells. (C) The processes arising from a single motor neuron axon innervating the longitudinal muscle of guinea pig ileum. The neuron was filled with marker dye through an intracellular electrode placed on the cell soma. The ganglia and connected strands of the myenteric plexus are shown in gray. (D) Types of neurons in the ENS defined by their function, key transmitters, and projections to targets. Excitatory motor neurons are marked + and inhibitory motor neurons are marked –. (ACh) Acetylcholine, (IPANs) intrinsic primary afferent neurons. (E) Extrinsic fibers connecting the central nervous system (CNS) and ENS. (A–D adapted from Furness 2007; E is based on data from Furness 2012.)

Extrinsic or CNS fibers include the vagus nerve, sympathetic fibers from the prevertebral ganglia of the thoracic spinal cord, and the lumbosacral pathways (Fig. 2E; Furness 2012). The vagus nerve contains 50,000 fibers, the majority of which are afferent fibers. The efferent vagal fibers from the dorsal motor nucleus of the vagus (DMV) and the nucleus ambiguus provide parasympathetic motor stimuli that regulate motility. Extrinsic modulation of gastrointestinal motility is more prominent in the esophagus and stomach (Travagli and Anselmi 2016), whereas intrinsic circuits are dominant in the small and large intestines and consist of autonomous reflex circuits (Furness et al. 2014). The intrinsic enteric neurons and glia cluster into interconnected ganglia (Fig. 2C) in two autonomically distinct regions, the myenteric (Auerbach's) plexus, which resides between the outer longitudinal and circular muscle layers, and the submucosal (Meissner's) plexus (Fig. 2D). The enteric ganglia are tightly packed, and ultrastructural imaging has shown that neurons within the myenteric and submucosal plexuses receive synaptic inputs showing typical presynaptic vesicles and postsynaptic densities (Fig. 2A,B; Cook and Burnstock 1976).

Enteric neurons have been classified by their morphology and physiological properties, their innervation, the transmitters they use, and, more recently, by transcriptomic profiling. Types of neurons include (1) motor neurons, including secretomotor and vasomotor neurons, (2) intrinsic primary afferent neurons (IPANs), also referred to as intrinsic sensory neurons, (3) interneurons, and (4) intestinofugal neurons (Fig. 3D; Furness 2007). The primary transmitters for excitatory motor neurons, which innervate the muscular layers, are acetylcholine and tachykinins, whereas inhibitory motor neurons release multiple transmitters including nitric oxide (NO), vasoactive intestinal polypeptide (VIP), ATP, and pituitary adenylate-cyclase-activating polypeptide (PACAP). Secretomotor and vasomotor neurons control fluid exchange. Types of secretomotor and vasomotor neurons include VIP-ergic neurons, calretinin-containing cholinergic neurons, and neuropeptide Y (NPY)-containing cholinergic neurons (Furness et al. 2003). There are also motor neurons that innervate enteric endo-crine cells and immune tissue, including lymphoid tissue, such as Peyer's patches, lymphocytes, and mast cells (Furness 2007). The IPANs respond to distension and mechanical stimulation from luminal contents to control movement, blood flow, and secretion. They contain the neurotransmitters acetylcholine, calcitonin gene-related peptide (CGRP), and tachykinin (Schneider et al. 2022). Interneurons include cholinergic interneurons containing NO synthase (NOS) and serotonin (5-HT) that mediate local motility reflexes, and cholinergic interneurons containing somatostatin that mediate migrating motor complexes, recurring cyclical patterns of motility that occur during fasting, a key feature of intrinsic ENS neural activity. There are also small populations of VIP and NOS-containing interneurons (Furness 2007). Finally, there is a small population of intestinofugal afferent fibers to the sympathetic ganglia that contain acetylcholine and VIP.

In summary, the ENS contains hundreds of millions of neurons comprised mostly of a unique intrinsic nervous system capable of autonomous function, in addition to a smaller number of extrinsic fibers that connect it to the CNS. Classes of enteric neurons are classically defined by their morphology, location, and neurotransmitters, and include motor neurons, IPANs, interneurons, and intestinofugal neurons. Although ultrastructural imaging has suggested that enteric neurons form synapses similar to those seen in the brain, synaptic morphology and function in the ENS have not been systematically investigated. More recent technical advances have suggested a greater neuronal diversity than currently described. Transcriptomic profiling of enteric neurons at the single-cell level has revealed dozens of types of neurons, with numbers differing between regions (Drokhlyansky et al. 2020), allowing for further future functional dissection of ENS physiology. A better understanding of ENS physiology is critical for dissecting the pathophysiology of PD given its early role in the disease process.

Dysphagia

Swallowing involves the neuromuscular coordination of oral, pharyngeal, and esophageal phases. The causes of dysphagia, or difficulty swallowing,

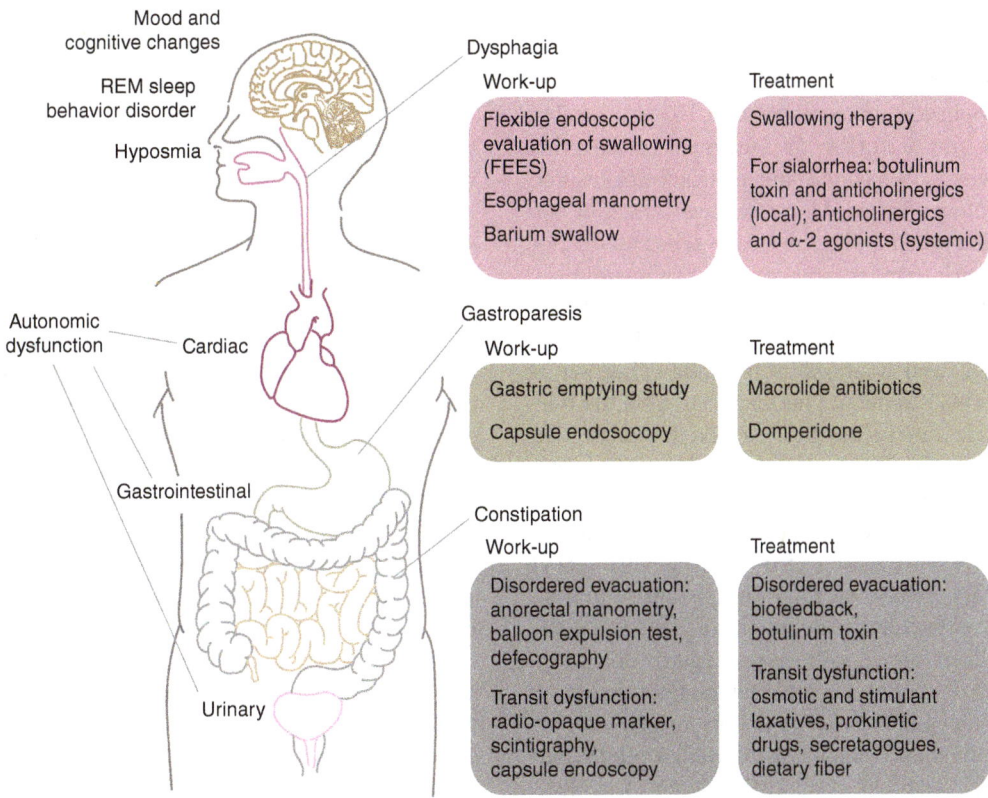

Figure 3. Nonmotor symptoms and management of gastrointestinal symptoms. Depiction of organ systems involved in nonmotor manifestations and current management strategy for gastrointestinal symptoms.

are multifactorial in PD. The common complaint of sialorrhea is due to both swallowing difficulties and difficulties maintaining saliva in the oral cavity. The excess saliva and associated drooling are not due to increased secretion, but the inability to handle normal secretory volumes in the mouth (Proulx et al. 2005). This is affected by characteristic slowed facial movements and reduction of facial expression, or hypomimia, unintentional mouth opening, and disease-related changes in head and neck posture. Sialorrhea is associated with an increased risk of aspiration (Nóbrega et al. 2008b) and respiratory infections (Nóbrega et al. 2008a), which are further associated with significantly increased mortality (Won et al. 2021).

Objective testing of swallowing demonstrates dysphagia in >80% of patients (Kalf et al. 2012), occurring early in the course of the disease (Pflug et al. 2018). The sensory innervation of the pharynx includes sensory branches of the glossopha-

ryngeal nerve (cranial nerve IX) and vagus nerve (cranial nerve X). These sensory branches are crucial for triggering swallowing and upper airway protective reflexes, and they demonstrate Lewy pathology postmortem (Mu et al. 2013). Many patients may be unaware of symptoms, possibly because of local somatosensory deficits, as there are higher sensory detection thresholds and increased oropharyngeal residue on endoscopic evaluation without a difference in self-reported swallow function in PD compared to control (Hammer et al. 2013). Peripheral mechanisms involving the pharyngeal nerves (Mu et al. 2012), including reduction of the neuropeptide substance P (Shröder et al. 2019), may also contribute. As the disease progresses, swallowing is also affected by higher cortical areas through the involvement of the supplementary motor areas and anterior cingulate cortex (Kikuchi et al. 2013).

The gold standard for workup of dysphagia is videofluoroscopy or flexible endoscopic evaluation of swallowing (FEES) (Langmore 2003). Esophageal manometry and barium swallow are also used (Fig. 3). Treatment of sialorrhea includes behavioral and pharmacologic interventions. Targeted local treatments include botulinum toxin, which decreases saliva production when injected into the parotid or submandibular glands (Jost et al. 2019), and local anticholinergics, such as sublingual atropine (Hyson et al. 2002) and ipratropium bromide (Thomsen et al. 2007). Systemic pharmacologic treatments include anticholinergics like glycopyrrolate (Arbouw et al. 2010) and the α-2 adrenergic agonist clonidine (Serrano-Duenas 2003). Levodopa and other dopaminergic medications also improve some aspects of swallowing (Chang et al. 2021). Behavioral interventions are commonly recommended, including head positioning and deliberate swallowing by use of gum or hard candies. Evidence-based behavioral management of swallowing includes expiratory muscle strength training and video-assisted swallowing therapy (Fig. 3; Troche et al. 2010; Manor et al. 2013; Athukorala et al. 2014; Claus et al. 2021).

Gastroparesis

The majority of patients with PD have impaired gastric emptying, which can be present at any point in the disease, although not all affected patients present with symptoms (Marrinan et al. 2014). Gastroparesis also affects treatment: It delays levodopa absorption in the small intestine, and, further, levodopa itself may delay gastric emptying (Hardoff et al. 2001; Pfeiffer et al. 2020). How levodopa plasma levels and efficacy are affected has not been studied in detail.

Associated symptoms include nausea, vomiting, early satiety, and abdominal distention, and further workup should be considered in symptomatic patients and to rule out obstruction or gall bladder disease. Gastroparesis can contribute to malnutrition and weight loss seen frequently in PD. Diagnosis of gastroparesis is made using a quantitative technetium-99 labeled gastric emptying study (Camilleri and Shin 2013) or capsule endoscopy (Fig. 3). Gastroparesis is classified ac-

cording to gastric retention at 4 h, and includes mild (10%–15%), moderate (15%–35%), and severe (>35%), with severity guiding treatment. The first-line therapy for gastroparesis, metoclopramide, a dopamine antagonist, is contraindicated in PD because of its extrapyramidal side effects. However, imaging is useful to guide the use of another dopamine antagonist, domperidone, which does not cross the blood–brain barrier. Although domperidone is not readily available in the United States, it can be prescribed to patients through a Food and Drug Administration (FDA) Investigational New Drug application. Domperidone can also increase the risk of cardiac arrhythmias. Other treatment choices include macrolide antibiotics (Fig. 3), including erythromycin, and motilin agonists, although their use is limited to 4 weeks at a time because of tachyphylaxis, reduced responsiveness with repeated use. Serotonin 5-HT4 receptor agonists can be used chronically, stimulate motility, and accelerate gastric emptying, although there are currently only investigational formulations available in the United States. Ghrelin agonists have shown promise in diabetic gastroparesis (Shin and Wo 2015). Botulinum toxin injection into the pyloric sphincter (Gil et al. 2011) and subthalamic nucleus deep brain stimulation (Arai et al. 2012) have also been proposed to provide benefits for gastroparesis. Other therapies for gastroparesis include surgical pyloroplasty, gastric peroral endoscopic myotomy (G-POEM), gastric drainage and jejunal feeding tube placement, or gastric electrical stimulation or gastric bypass or resection, but these have not been well-studied in PD.

Constipation

Constipation is the most common gastrointestinal symptom in PD. It occurs in up to 80% of patients (Knudsen et al. 2017) and is part of the disease prodrome, emerging before the onset of motor symptoms. It is also one of the earliest autonomic features of the disease, developing up to 15 years before the onset of motor symptoms, and the risk of developing PD increases with constipation severity (Lin et al. 2014). Neurodegeneration may be involved, as reduced myenteric neuron numbers can be present in

Cite this article as *Cold Spring Harb Perspect Med* doi: 10.1101/cshperspect.a041618

patients with severe, chronic constipation (Wedel et al. 2002), although neuron loss is not necessarily observed in PD (Annerino et al. 2012). Constipation can be further worsened by drugs used to treat motor symptoms, including anticholinergics and dopamine agonists.

Workup should distinguish between disordered evacuation, or dyssynergic dysfunction, and a colonic transit abnormality. Although not validated in patients with PD specifically, this includes anorectal manometry, balloon expulsion test, and barium or magnetic resonance imaging (MRI) defecography for outlet dysfunction, and a radio-opaque markers study, scintigraphy, or capsule endoscopy for transit dysfunction (Fig. 3; Fasano et al. 2015). Treatment for outlet dysfunction includes biofeedback and botulinum neurotoxin injection to the anal sphincter. Treatment for transit dysfunction includes an increase in dietary fiber, osmotic and stimulant laxatives, prokinetic drugs, and secretagogues (Fig. 3; Fasano et al. 2015).

Systemic Manifestations and Treatment Considerations

Malnutrition and weight loss occur in PD due to dysphagia, levodopa-induced nausea, and disruption of regular meals to limit interference with levodopa absorption. Endocrine mechanisms also contribute. Ghrelin, which is secreted in physiological conditions when the stomach is empty to increase motility in preparation for a meal, is reduced in PD (Song et al. 2017). Other nonmotor symptoms, including depression, apathy, and olfactory loss, can also contribute. On the other hand, weight gain can also occur and is associated with treatment with dopamine agonists (Nirenberg and Waters 2006) and deep brain stimulation in the subthalamic nucleus (Sauleau et al. 2014), both thought to be related to modulation of limbic mechanisms.

Because of gastrointestinal dysfunction, treatment options that partially or wholly bypass the gastrointestinal tract have been explored. These include liquid levodopa or orally dissolving formulations, although these still require absorption in the proximal small intestine. Parenteral routes include dopamine agonist patch,

apomorphine subcutaneous infusion, intranasal levodopa, and percutaneous endoscopic gastrostomy with a jejunal tube for intrajejunal levodopa infusion. Novel formulations include microspheres for drug delivery, microenemas, and esters or alkyl esters that can be administered by transdermal, intranasal, subcutaneous, and intramuscular routes (Fasano et al. 2015).

Summary

In summary, symptoms of gastrointestinal dysfunction in PD can occur anywhere along the gastrointestinal tract, from mouth to anus. Symptoms are caused by dysfunctional motility due to both central and peripheral nervous system impairment, autonomic dysfunction, and endocrine dysfunction. Symptoms both affect and are affected by treatment. Constipation is the most common complaint and occurs early in the disease process. Although symptomatic treatment is available, there are no PD-specific treatments for gastrointestinal dysfunction. Development of disease-specific treatments requires a better understanding of the underlying disease pathophysiology within the gut.

GASTROINTESTINAL PATHOPHYSIOLOGY IN PARKINSON'S DISEASE

In addition to symptomatic manifestations throughout the gastrointestinal tract, typical neuropathological changes are also seen in the ENS, suggesting that pathophysiological changes in the gut are a critical part of the disease process. We first discuss hypotheses about where the initial pathophysiological changes leading to PD occur and conclude with a discussion on the potential role of ENS neurodegeneration.

Braak Hypothesis

Braak et al. (2003b) postulated that the disease mechanism underlying PD is the neuroinvasion of routes containing vulnerable neurons. This was in part based on the finding that pathology occurs in specific brain regions, notably the dopaminergic neurons of the substantia nigra (Fearnley and Lees 1991; Halliday et al. 1996)

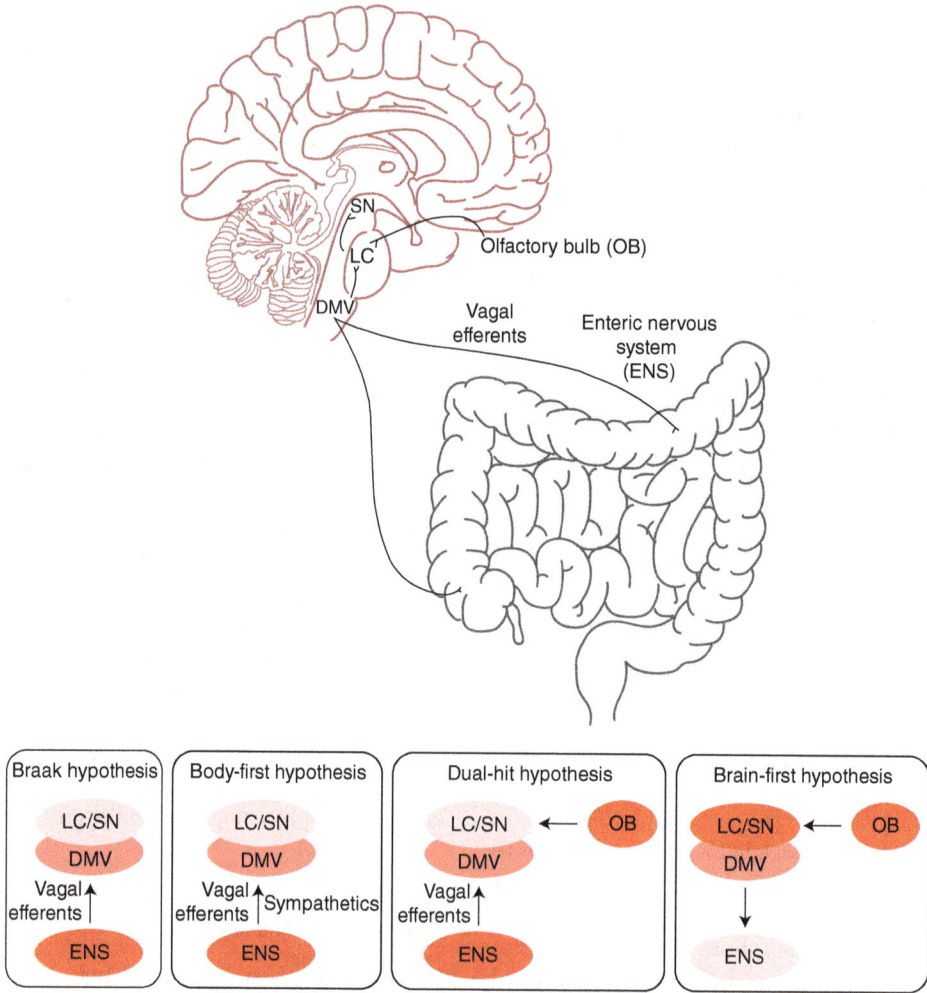

Figure 4. Hypotheses concerning pathological origins and spread of Parkinson's disease (PD). Proposed pathological origins of PD in the olfactory and gastrointestinal system. Sites of hypothesized pathological initiation are in red. (SN) Substantia nigra, (LC) locus coeruleus, (DMV) dorsal motor nucleus of the vagus, (OB) olfactory bulb, (ENS) enteric nervous system.

and the DMV (Fig. 4). The vagus nerve connects the ENS and CNS, with initial pathology occurring in the brain in the DMV and then advancing upward to the medulla, pontine tegmentum, midbrain, basal forebrain, and cerebral cortex. "Incidental cases of idiopathic Parkinson's disease may show involvement of both the enteric nervous system and the dorsal motor nucleus of the vagus nerve," and the hypothesis implies a gastrointestinal origin, prompting "the question whether the disorder might originate outside of the central nervous system … via postganglionic enteric neurons, entering the central nervous system along unmyelinated preganglionic fibers generated from the visceromotor projection cells of the vagus nerve" (Braak et al. 2003b). Indeed, Lewy bodies, the pathological hallmark of PD, composed primarily of α-synuclein (Spillantini et al. 1997), are found in both the brain and the ENS.

Consistent with a gastrointestinal origin of pathology, gastrointestinal symptoms (especially constipation) are a prominent prodromal feature of PD (Fig. 1; Adams-Carr et al. 2016). The

Cite this article as *Cold Spring Harb Perspect Med* doi: 10.1101/cshperspect.a041618

vagal nuclei in the brainstem include the nucleus ambiguus, which innervates the palate, pharynx, larynx, and heart; the nucleus of the solitary tract, which contains afferent innervations from the viscera; and the dorsal motor nucleus, which innervates the postganglionic parasympathetic to the transverse colon. Of these, only the DMV is severely affected in early PD (Del Tredici et al. 2002). Lewy pathology in the ENS has been identified in both presymptomatic and clinically confirmed cases of PD (Braak et al. 2006), from where spread could occur to the preganglionic parasympathetic neurons of the DMV. Supporting the concept that α-synuclein pathology can spread from neuron to neuron, α-synuclein inclusions were seen in a small subset of striatal engrafted neurons in PD patients who had undergone a neuronal transplant a decade earlier (Kordower et al. 2008; Li et al. 2008). The pattern of progression of neuropathology and neurodegeneration in the brain led Braak et al. (2003a) to develop a neuropathological staging system for PD, with later stages corresponding to the clinical progression of the disease.

Dual-Hit Hypothesis

To reconcile the presence of early pathology in both the olfactory bulb and the DMV, a dual-hit hypothesis was subsequently proposed, with pathology proposed to progress retrogradely from both the nasal and gastrointestinal epithelium to the brain (Hawkes et al. 2007). Olfactory dysfunction is an early feature of idiopathic PD (Fig. 1) and can be used to distinguish between PD and other parkinsonism (Fullard et al. 2017). In olfactory areas, Lewy bodies are most numerous in the anterior olfactory nucleus and are also found in mitral cells, the first projection neurons to receive input from olfactory epithelial bipolar neurons (Daniel and Hawkes 1992; Braak et al. 2003a). Further, the extent of neuronal loss in the anterior olfactory nucleus correlated with disease duration (Pearce et al. 1995). From there, additional olfactory areas are affected, including the olfactory tubercle, the pyriform and peri-amygdalar cortex, and the entorhinal cortex (Braak et al. 2003b).

Body-First versus Brain-First Hypothesis

A subsequent detailed pathological study suggested a revision to the dual-hit hypothesis and proposed that the pathological process starts in a single location. Across a number of neuropathological studies, the most frequent single-location-affected areas with incidental findings of Lewy bodies are the olfactory bulb, the DMV, and the sympathetics (Borghammer 2021). Triggered by environmental factors such as infection, inflammation, or toxic exposure to the nasal or enteric epithelium, pathology would most commonly initiate in the olfactory bulb or ENS, but rarely in both simultaneously (Borghammer 2021; Borghammer et al. 2022). Accordingly, a brain-first or body-first clinical phenotype corresponding to this pathology was proposed.

In the brain-first subtype, pathology is postulated to spread from the olfactory bulb to the amygdala or vice versa, and then mono-synaptically from the amygdala to the substantia nigra pars compacta (SNpc) (Poulin et al. 2018), or from the olfactory bulb to the SNpc (Höglinger et al. 2015), with asymmetric dopamine loss at presentation. In these patients, characteristic nonmotor symptoms including constipation and REM sleep behavior disorder (RBD) would appear after the onset of parkinsonism. Another subtype of the brain-first subtype would include original symmetric pathology in the SNpc or locus coeruleus, in which RBD and parkinsonism would appear coincidentally. In the body-first subtype, pathology is proposed to propagate from the ENS to the vagus to the DMV, and via the sympathetics to the celiac or mesenteric ganglia to the sympathetic trunk to the heart. This would be commensurate with the appearance of autonomic symptoms including constipation and orthostatic hypotension before the onset of RBD or parkinsonism and is supported by evidence of prodromal cardiac denervation by ^{123}I-MIBG scintigraphy (Miyamoto et al. 2006; Kashihara et al. 2010; Knudsen et al. 2018; Nishikawa et al. 2022), cardiac noradrenergic deficiency by ^{18}F-dopamine positron emission tomography (Goldstein et al. 2024), and increased colonic transit time (Fedorova et al. 2022). This symmetry of dopamine loss proposed is supported by the lateralization of the nervous system. Ipsilateral connections outnumber contralateral connections in the brain, whereas there is more substantial right-life overlap in the innervation of the ENS (Borghammer 2023). One caveat is that hyposmia is more common in the body-first subtype, but the degree of hyposmia may not correlate with disease burden in the olfactory bulb, but rather in the total CNS (Nag et al. 2019; Tremblay et al. 2022).

Summary

In summary, various hypotheses have proposed pathological origins of PD in the olfactory and/or gastrointestinal system (Fig. 4), with supportive evidence from neuropathology and clinical subtypes. The brain-first clinical phenotype, associated with pathology originating in the olfactory system, is postulated to be characterized by few prodromal symptoms and asymmetric dopamine loss at presentation, whereas a body-first clinical phenotype, associated with pathology originating in the ENS, is postulated to be characterized by prodromal autonomic symptoms, RBD, and more symmetric dopamine loss at presentation.

Neurodegeneration in the ENS

Loss of dopaminergic neurons in the substantia nigra leads to the cardinal motor symptoms of PD in the brain, and data from patients as well as animal models suggest that PD affects distinct subsets of neurons in the ENS, although further work is needed to determine which enteric neurons are most affected. Although VIP-ergic neurons were originally identified as the major cell type harboring LBs in PD (Wakabayashi et al. 1990), this finding was not reproduced in a later study (Annerino et al. 2012). It remains unclear whether PD targets a specific site or neuronal subtype in the ENS the way it targets the substantia nigra dopaminergic neurons in the CNS, and, further, many aspects of enteric neuronal organization remain unknown.

Conflicting data also exist regarding potential enteric neuron degeneration in PD. At least two studies have found enteric ganglion cell de-generation (Qualman et al. 1984; Ohlsson and Englund 2019), mainly in Auerbach's plexus but also in Meissner's plexus (Bloch et al. 2006). Atrophic or pyknotic neurons were found both in the myenteric and submucosal plexuses in the jejunum and colon from some PD patients, and degenerative alterations were accompanied by α-synuclein deposits, which were also present in some preserved neurons (Ohlsson and Englund 2019). However, several others have reported no evidence of neuronal cell loss in the ENS (Wakabayashi et al. 1988, 1990; Annerino et al. 2012; Corbillé et al. 2014; Pfeiffer 2014). Neurodegeneration may also occur during aging, with loss of neurons in both the submucosal and myenteric plexus, with selective vulnerability of cholinergic and intrinsic sensory neurons (Wade and Cowen 2004). Overall, it remains to be determined whether there is neuron loss of enteric neurons and which cell types contain Lewy pathology.

α-SYNUCLEIN IN THE ENTERIC NERVOUS SYSTEM

α-Synuclein is a key protein involved in PD pathology. Therefore, it is not surprising that of the various proteins implicated in PD pathology, α-synuclein has been explored most thoroughly in the ENS, although many open questions about its localization and function remain. Here, we review what is known about the physiological and pathological functions of ENS α-synuclein.

Physiological Functions of ENS α-Synuclein

In the brain, α-synuclein is found in presynaptic terminals and plays critical physiological roles, including SNARE complex assembly and synaptic vesicle clustering (Iwai et al. 1995; Gosavi et al. 2002; Ahn et al. 2006; Burré et al. 2010; Varkey et al. 2010; Diao et al. 2013; Westphal and Chandra 2013). α-Synuclein adopts various conformations, from an α-helical state on synaptic vesicles, to a largely disordered state in the cytosol (Bussell et al. 2005; Burré et al. 2013), to deposits of β-sheet-rich aggregates forming the major components of Lewy bodies (El-Agnaf et al.

1998; Burré et al. 2015). Membrane-binding of α-synuclein defines its physiological and pathological roles. Membrane-bound α-helical α-synuclein is protected from aggregation, whereas cytosolic, natively unstructured α-synuclein has a propensity to misfold and aggregate (Burré et al. 2015). Further, most mutations in α-synuclein associated with disease entities are found in the lipid-binding region of the protein (Jo et al. 2002; Fares et al. 2014; Ghosh et al. 2014; Mohite et al. 2018). Like in the brain, α-synuclein is abundantly expressed in the human ENS in physiological, nonpathological conditions (Böttner et al. 2012). However, the localization and function of α-synuclein in the ENS, and whether it plays a critical synaptic role as in the brain, remains largely unknown.

The expression of α-synuclein throughout the length of the intestinal tract and in the different layers of the intestinal wall has not been systematically examined, although studies suggest that there are regional differences to be elucidated. In rats, expression of α-synuclein in the myenteric plexus showed a rostral–caudal increase from the stomach to the small intestine (Phillips et al. 2008), contrary to the rostral–caudal decrease seen in pathological conditions. There was also a regional difference with more α-synuclein colocalization with NOS more proximally in the stomach, and cholinergic neurons more distally. In the human appendix and nonhuman primate cecum, α-synuclein was expressed through the mucosa, submucosal plexus, and myenteric plexus (Zinnen et al. 2022), and in the human appendix, α-synuclein was particularly enriched in axonal varicosities in the mucosal plexus (Gray et al. 2014). In human colon samples, α-synuclein was abundantly expressed in neuronal cell bodies and processes throughout the intestinal wall, although most abundantly in the longitudinal and circular muscle and in the submucosa (Böttner et al. 2012).

Concerning cell type, a few studies have suggested that α-synuclein is most highly expressed in cholinergic nerve terminals in the small and large intestines. In guinea pig ileum, α-synuclein colocalization was primarily seen with the majority of vesicular acetylcholine transporter (vAChT) varicosities, also colocalized with vari-

cosities containing 5-HT, calretinin, NOS, TH, and VIP, and rarely colocalized with varicosities containing calbindin, CGRP, somatostatin, and substance P (Sharrad et al. 2013b). As was seen in the ileum, α-synuclein colocalization in guinea pig rectum was again seen in the majority of vAChT varicosities, which was confirmed in the human colon as well (Fig. 5, left; Sharrad et al. 2013a). In the mouse colon, α-synuclein was primarily expressed in cholinergic varicosities (Swaminathan et al. 2019).

Several studies have suggested a synaptic localization for α-synuclein in the ENS, similar to the brain. In guinea pig ileum, α-synuclein colocalization was seen in the majority of varicosities containing the synaptic proteins synaptobrevin-2, synaptophysin, synaptotagmin-1, and CSPα (Sharrad et al. 2013b). In rat myenteric plexus culture, α-synuclein was found to colocalize with synaptophysin and accumulate in neuronal varicosities (Böttner et al. 2015). Colocalization with synaptophysin was also seen in the human colon (Fig. 5, left; Böttner et al. 2015).

α-Synuclein may be differentially expressed with age. In middle-aged and aged rats, α-synuclein-positive dystrophic axons were found throughout the gastrointestinal (GI) tract in the myenteric plexus, which colocalized with NOS-, calretinin-, calbindin-, or TH-positive swollen neurites (Phillips et al. 2009). In the colonic submucosal plexus of aging rats, α-synuclein first increased and then decreased over age (Phillips et al. 2013). In contrast, in surgical samples from human colon, α-synuclein was expressed in all intestinal layers and did not change with age, whereas phosphorylated α-synuclein at the residue serine 129 (S129 α-synuclein), a traditional marker for α-synuclein pathology (Fujiwara et al. 2002), was seen predominantly in the submucosa and myenteric plexus and increased with age (Bu et al. 2020). A recent physiological role of phosphorylated S129 α-synuclein has been identified, so its use as a pathological marker may need to be revisited (Parra-Rivas et al. 2023; Ramalingam et al. 2023).

To examine α-synuclein expression in extrinsic inputs to the ENS, vagal terminals were labeled using a retrograde tracer. Although almost all vagal preganglionic efferents express

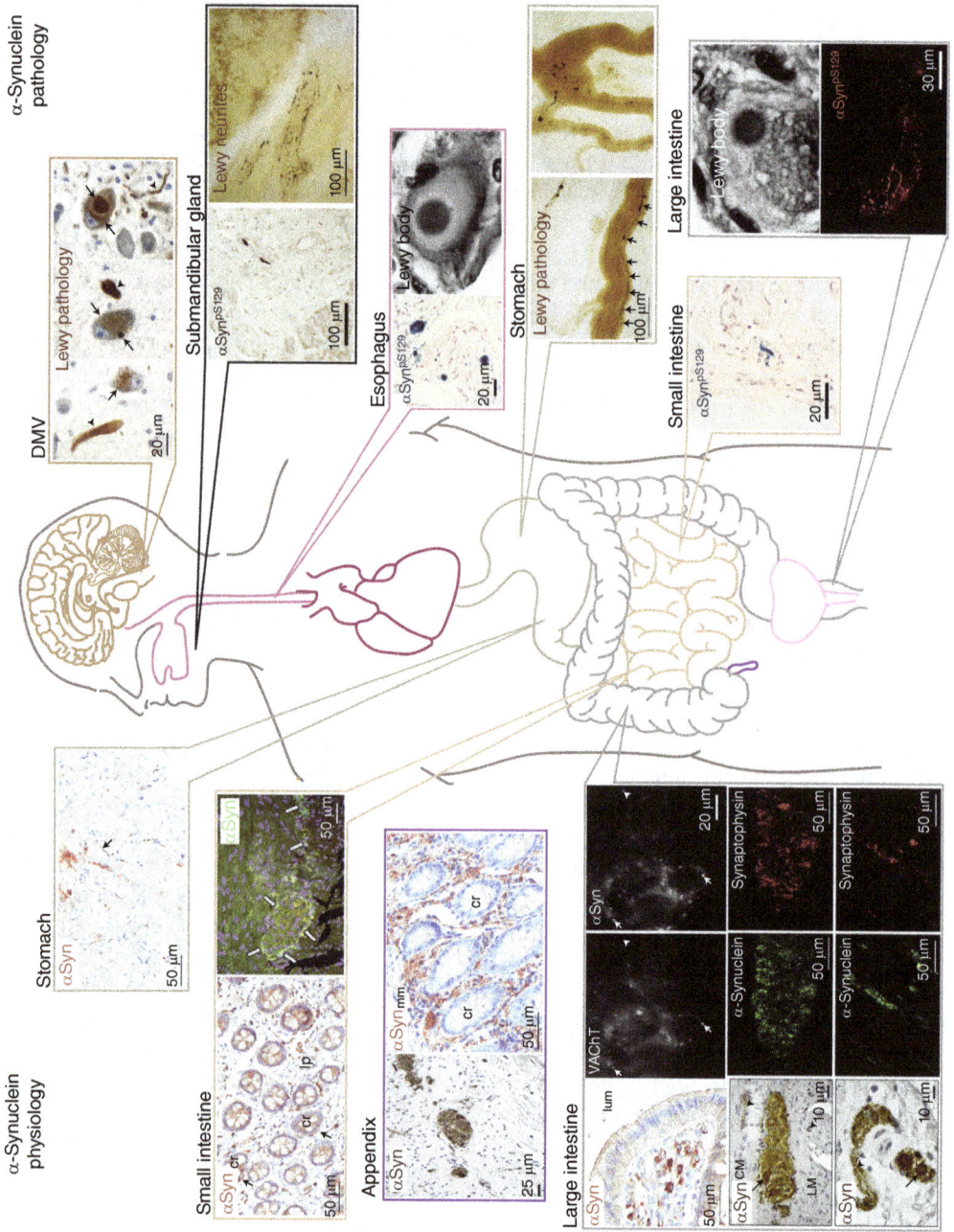

Figure 5. (*See following page for legend.*)

α-synuclein, both in the axons and terminal varicosities, almost no vagal afferents express α-synuclein (Phillips et al. 2008). α-Synuclein is also expressed in enteroendocrine cells in the mouse and human intestine (Chandra et al. 2017), which is presumably a site of vulnerability for initial pathology. In human jejunum samples, α-synuclein was found to be widely expressed and was found to colocalize with enteroendocrine cells expressing 5-HT, but not those expressing choline acetyltransferase (ChAT) (Casini et al. 2021).

All the above studies relied on imaging approaches. Only one study examined α-synuclein biochemically using colonic biopsies taken during routine cancer screening, confirming the expression of α-synuclein and S129-phosphorylated α-synuclein by immunoblotting (Corbillé et al. 2017). Given its likely role in PD pathophysiology, it will be essential to unravel the still unknown functional impact of α-synuclein on enteric neurotransmission and regulation of GI motility.

Pathological Functions of ENS Synuclein

Some evidence supports appearance of peripheral Lewy pathology in the olfactory bulb, spinal cord, peripheral autonomic ganglia, submandibular gland, and cardiac nerves, in addition to the ENS before the appearance of neuron loss and Lewy pathology in the SNpc (Del Tredici et al. 2002, 2010; Braak et al. 2003a, 2006; Ghebremedhin et al. 2009; Del Tredici and Braak 2012). To support an ENS origin of disease, researchers have attempted to identify cases of incidental pathology isolated to the ENS, although there is some controversy about whether a true early-stage "ENS-only" has been convincingly identified (Beach et al. 2010). However, this study was limited to the examination of a single slide for each ENS subdivision, and relevant regions, including the lumbar prevertebral celiac ganglion, the spinal sympathetic preganglionic neurons, the superior mesenteric ganglion, and the appendix, were not included (Warnecke et al. 2022). Specifically, through the examination of

Figure 5. α-Synuclein (α-Syn) physiology and pathology. (*Left*) α-Syn physiology. Stomach: α-Syn immunoreactivity in the mucosa of the gastric corpus in a control human subject. (Reprinted, with permission, from Gray et al. 2014.) Small intestine: α-Syn immunoreactivity in the terminal ileum in a control human subject and α-Syn immunofluorescence (green) in the normal human jejunum myenteric ganglia (white arrows). (*Left*, reprinted, with permission, from Gray et al. 2014; *right*, reprinted, with permission, from Casini et al. 2021, under the terms of the Creative Commons Attribution NonCommercial 4.0 License [CC BYNC 4.0].) Appendix: α-Syn immunoreactivity in the myenteric ganglia and mucosal plexus of the vermiform appendix in a control human subject. (*Left*, reprinted, with permission, from Zinnen et al. 2022, under the terms of the Creative Commons Attribution License; *right*, reprinted, with permission, from Gray et al. 2014.) Large intestine: α-Syn immunoreactivity in the ascending colon (*left top*, reprinted, with permission, from Gray et al. 2014), colonic myenteric plexus (*left middle*), and colonic submucosal plexus (*left bottom*, reprinted, with permission, from Böttner et al. 2012) from control human subjects. Colocalization of vesicular acetylcholine transport (VAChT) and α-Syn in the human colon. (*Top middle* and *right*, reprinted, with permission, from Sharrad et al. 2013a.) Colocalization of α-Syn and synaptophysin in the enteric ganglia and nerve fiber strands. (*Middle middle, middle right, bottom middle*, and *bottom right*, reprinted, with permission, from Böttner et al. 2015.) (*Right*) α-Syn pathology. Dorsal motor nucleus of the vagus (DMV): Lewy neurites (arrowheads), Lewy bodies (large arrows), and α-Syn aggregates (small arrows) in cholinergic neurons in Parkinson's disease (PD). (Reprinted, with permission, from Braak et al. 2004.) Submandibular gland: Phosphorylated α-Syn in PD and Lewy neurites in PD. (*Left*, reprinted, with permission, from Beach et al. 2010; *right*, reprinted, with permission, from Del Tredici and Duda 2011.) Esophagus: Phosphorylated α-Syn in PD and Lewy body in myenteric plexus of the lower esophagus. (*Left*, reprinted, with permission, from Beach et al. 2010; *right*, reprinted, with permission, from Wakabayashi et al. 1989.) Stomach: Lewy neurites (*left*, arrows) and Lewy bodies (*right*) in myenteric plexus of the gastric cardia. (Reprinted, with permission, from Del Tredici and Duda 2011.) Small intestine: Phosphorylated α-Syn in PD. (Reprinted, with permission, from Beach et al. 2010.) Large intestine: Lewy body in submucosal plexus of the rectum and phosphorylated α-Syn in submucosal neurites in PD. (*Top*, reprinted, with permission, from Wakabayashi et al. 1989; *bottom* is reprinted from Lebouvier et al. 2010, under the terms of the Creative Commons Attribution License.)

92 autopsies, S129-phosphorylated α-synuclein was found in the gastrointestinal tract in ~65% of cases, in the vagus nerve in ~73% of cases, and in the sympathetic trunk in ~80% of PD cases (Beach et al. 2010). A large study was conducted in which more than 600 whole-body autopsies were screened, nine biopsies were collected in each gastrointestinal tract, and no gut-only cases were identified (Adler and Beach 2016). However, assuming a small area of involvement and the relative size of the gastrointestinal tract, even with a large number of cases, there is still a high probability of missing a case (Borghammer 2023). For example, If the area affected is commensurate to the other areas where isolated pathology is found (olfactory bulb, DMV), the size of the affected region would be quite small and methodologically challenging given the large surface area of the ENS (Helander and Fändriks 2014). Assuming a small area is affected and given the small number of neurons in a given pathological gut section as compared to a brain section (Borghammer 2023), it is likely that hundreds of slides would have to be examined to identify a gut-only pathology.

Some work supports a rostrocaudal gradient of α-synuclein deposition throughout the nervous system (Visanji et al. 2014). For example, S129-phosphorylated α-synuclein was reported to be highest in the submandibular gland and lower esophagus, lower in the stomach and small intestine, and lowest in the colon and rectum (Beach et al. 2010), following the pattern of innervation from visceromotor projection neurons, which originate in the DMV, with vagal innervation to the upper gastrointestinal tract and predominant sympathetic and pelvic outflow projections in the lower gastrointestinal tract.

Summary

In summary, α-synuclein is neuronally expressed throughout the gastrointestinal tract (Fig. 5), especially in cholinergic neurons, as well as in enteroendocrine cells and vagal efferents. Like in the brain, a synaptic localization of α-synuclein and age-related changes in α-synuclein have been suggested. Pathology containing α-synuclein aggregates is also found throughout the gas-

trointestinal tract (Fig. 5, right), but as in the brain, how α-synuclein physiology and pathophysiology relate remains to be determined.

GUT MICROBIOME

There are bidirectional influences between the gastrointestinal tract and the microbial community within. The majority of microorganisms in the gut are found in the large intestine, and most studies of the human microbiome rely on sequencing of bacterial DNA in stool. Determining whether differences in the microbiome are due to lifestyle, disease, or compensatory responses to disease is challenging to unravel. Aging remains the most significant risk factor for developing PD, and affects the microbiome in and of itself, because of changes in gut physiology with aging. Another caveat in interpreting microbiome data is to note that there is likely strain specificity, which cannot be evaluated with genetic data, and, further, that there is limited knowledge of strain-specific function and microbial interactions, as most microbial species identified have not been functionally characterized in culture (Almeida et al. 2021).

That being stated, there are small, but significant, differences in gut microbiome composition between PD and control individuals, which persist even when taking age, diet, constipation, and medication into account (Nishiwaki et al. 2020b; Tan et al. 2021). Further, genomic sequencing can distinguish PD from control individuals, even for prodromal individuals with RBD (Heintz-Buschart et al. 2018). Results among studies vary, but consistent changes include an increased abundance of *Akkermansia*, *Bifidobacterium*, *Lactobacillus*, and reduced abundance of butyrate producers, including *Roseburia*, *Faecalibacterium*, and *Blautia* in PD (Tan et al. 2022).

Measurements of bacterial metabolites and metabolic pathways using transcriptomics, proteomics, and metabolomics can provide insight into the functional contribution of the microbiome. Consistent with the decreased abundance of butyrate producers in taxonomic descriptions, reduction in short-chain fatty acids (SCFAs), including butyrate, acetate, and propionate, have been widely reported (Aho et al. 2021). Butyrate in particular has been shown to be neuroprotective in animal models of PD (Rane et al. 2012; Sharma et al. 2015; Srivastav et al. 2019). Interestingly, SCFAs are not decreased in prodromal RBD (Nishiwaki et al. 2020a).

Finally, the gut microbiome may affect treatment response because of its role in drug metabolism (Zimmermann et al. 2019). Bacterial tyrosine decarboxylase, predominantly from *Enterococcus faecalis*, converts levodopa to dopamine affecting its bioavailability (Maini Rekdal et al. 2019; van Kessel et al. 2019), which could thus affect individual response to treatment. Further, dopamine itself acts on gut motility. Microbiome-based treatments, including fecal microbiota transplantation (FMT) (DuPont et al. 2023), administration of SCFAs, and targeting of levodopa-metabolizing bacterial enzymes, thus represent promising avenues of treatment, although studies exploring these treatments remain preliminary.

In summary, elucidation of specific mechanisms remains limited by confounding influences on GI function and the microbiome itself, and functional microbiome data remain limited as most studies rely on genetic sequencing. However, the microbiome and related metabolites and pathways are notably altered in PD, including in the prodromal period.

ANIMAL MODELS OF GASTROINTESTINAL DYSFUNCTION IN PARKINSON'S DISEASE

Animal models are useful to dissect some of the open questions raised in the previous sections (Pellegrini and Travagli 2024). Below, we discuss the various animal models of PD that test gastrointestinal involvement, some of which support early gastrointestinal pathophysiology and have investigated the hypothesis of gut-to-brain spread of pathology.

α-Synuclein Models

α-Synuclein overexpression models, including those driven by the ubiquitous Thy1 promotor (Wang et al. 2008, 2012; Hallett et al. 2012), as well as those with pathogenic PD-causing A53T (Kuo et al. 2010; Noorian et al. 2012) and A30P

mutations (Kuo et al. 2010), demonstrate gastrointestinal dysfunction and changes in α-synuclein expression. Age-dependent aggregation of α-synuclein in either enteric neurons or varicosities around enteric neurons is associated with GI motility dysfunction, particularly slowed colonic transit (Wang et al. 2008, 2012; Kuo et al. 2010; Noorian et al. 2012). When earlier ages were examined, colonic changes occurred before loss of striatal dopamine (Wang et al. 2012).

Injection of α-synuclein preformed fibrils or viral overexpression of A53T mutant α-synuclein in the colon of rats and stomach of nonhuman primates led to gastrointestinal dysfunction (Manfredsson et al. 2018). Spread of α-synuclein pathology from gut to brain was demonstrated in rats with intestinal injection of α-synuclein derived from human PD brain lysates and recombinant α-synuclein (Holmqvist et al. 2014) and mice injected with α-synuclein preformed fibrils (Uemura et al. 2018). This spread was abolished with truncal vagotomy and required endogenous expression of α-synuclein (Kim et al. 2019). There is also evidence for the transfer of pathogenic α-synuclein from intestinal epithelial cells to the vagus (Chandra et al. 2023). Further, aging plays a role in facilitating gut-to-brain propagation (Challis et al. 2020; Van Den Berge et al. 2021).

Other Models

Gut dysfunction and/or dysbiosis were found in several neurotoxin models, including 1 methyl-4-phenyl-1,2,3,6-tetrahydropyridine (MPTP) (Lai et al. 2018; Xie et al. 2020), rotenone (Zheng et al. 2011; Yang et al. 2018), and 6-hydroxydopamine (6-OHDA) (Koutzoumis et al. 2020). Neurotoxin models also support early involvement of gut pathophysiology. In MPTP mice (Lai et al. 2018), gut features preceded motor features. In 6-OHDA lesioned animals, delayed gastric emptying was prevented by vagotomy (Zheng et al. 2014). Indeed, neuroanatomical tracing demonstrated a monosynaptic nigro-vagal pathway that modulates gastric tone and is susceptible to both 6-OHDA and paraquat (Anselmi et al. 2017). Furthermore, in rotenone models, there was increased α-synuclein S129-phosphorylation in the ENS and intermediolateral cell column of the spinal cord (Pan-Montojo and Funk 2010). Finally, mice lacking mitochondrial transcription factor A in dopaminergic neurons showed gastrointestinal dysfunction preceding motor dysfunction, as well as changes in microbiome composition (Ghaisas et al. 2019), and a Pink1 knockout rat showed swallowing dysfunction and delayed colonic transit (Krasko et al. 2023).

Summary

In summary, both toxin- and α-synuclein-based models show early gut dysfunction, pathological and microbiome changes, and involvement of the vagus nerve in the spread of pathology and are thus useful for further study of early pathophysiological mechanisms.

DISEASE BIOMARKER AND TREATMENT DEVELOPMENT

Current treatments for PD remain symptomatic and do not target the underlying disease physiology. Diagnosis is based on clinical symptoms; although imaging methods can demonstrate dopamine deficiency, they do not elucidate the underlying pathophysiology of the dopamine deficiency and are not useful for distinguishing between different parkinsonisms or predicting disease progression (Stoessl and Halliday 2014). There is thus a critical need for the development of validated disease biomarkers. Types of useful biomarkers include those that are diagnostic, those that can be used to monitor response to interventions, and those that are predictive and prognostic.

Research studies often use clinical end points to measure disease progression and response to treatment. However, the isolated use of clinical evaluation in PD research is limited by the availability of effective treatments for the motor symptoms, treatment-related symptom fluctuations, disease heterogeneity, and slow progression of disease. Although efforts at clinical subtyping have emerged, reliable biomarkers do not exist to validate these subtypes. Further, biomarkers could be helpful in distinguishing between different par-

kinsonisms, which are currently mostly distinguished by clinical characteristics. Diagnostic biomarkers would also be useful in identifying patients earlier in the disease course, as patients with prodromal or unrecognized disease have been shown to already demonstrate functional impairments in mobility (Miller-Patterson et al. 2023). Finally, although movement disorders specialists can diagnose PD and parkinsonisms with a high degree of accuracy (Hughes et al. 2002), in the United States, >90% of patients with PD are not seen by a movement disorders specialist (Pearson et al. 2023). Biomarkers could thus serve to identify patients throughout the disease course, to track disease progression, and to monitor treatment response.

Because of the convincing involvement of the gastrointestinal system in disease physiology, especially early in the disease course, the gut is an attractive target for the development of disease biomarkers and disease-modifying treatment and warrants further research and development. Previous work exploring the use of colonic mucosal α-synuclein and S129-phosphorylated α-synuclein as a biomarker was not advanced further because of the lack of specificity (Visanji et al. 2015). However, these studies were limited by methodology and extent of tissue sampled. More recently, aggregated α-synuclein was identified in duodenal biopsies of living patients with various stages of disease (Emmi et al. 2023). Seeding of aggregated α-synuclein has also been explored in the human GI tract. In a small study of gastrointestinal biopsies from the stomach (antrum), sigmoid colon, and rectum, biopsies from PD patients were able to seed α-synuclein aggregation, mostly from the sigmoid colon (Fenyi et al. 2019). Another study showed α-synuclein seeding from duodenal biopsies in 22/23 PD patients and 0/6 controls (Vascellari et al. 2023). These studies also have the advantage of being conducted in living patients.

Other peripheral α-synuclein-based biomarkers are currently under development, and advances in peripheral biomarkers have been made with α-synuclein seeding, detection of S129-phosphorylated α-synuclein, and identification of α-synuclein strains (Fricova et al. 2020; Coughlin and Irwin 2023). There is now a commercially available skin biomarker based on S129-phosphorylation of α-synuclein (Gibbons et al. 2024), and significant advances have been made using α-synuclein seeding assays, especially in the cerebrospinal fluid (CSF), and including prodromal patients (Siderowf et al. 2023). The α-synuclein seeding assay is based on the demonstration that fluids and other tissues from PD patients can seed aggregation of recombinant α-synuclein in vitro (Kang et al. 2019), thought to model the prion-like propagation of pathogenic oligomers from native α-synuclein (Jucker and Walker 2013). However, α-synuclein seeding activity has mainly been tested with CSF as this assay currently relies on CSF sampling, and although demonstrating a high diagnostic performance for differentiating synucleinopathies with Lewy bodies from controls, results for the diagnosis of multiple system atrophy have been less robust (Grossauer et al. 2023).

In summary, biomarkers are essential to advance the development of disease-modifying treatments. Peripheral biomarker development remains in early stages for PD, but holds promise for future research and clinical utility. Further work remains to be done to develop an accessible, sensitive, and specific GI biomarker.

CONCLUSIONS

Gastrointestinal dysfunction is an early manifestation of PD, occurs throughout the GI tract, and includes difficulty swallowing, gastroparesis, and constipation. These symptoms contribute to disease morbidity; both affect and are affected by treatments for motor symptoms, and are important to treat. However, there are currently no PD-specific treatments for gastrointestinal dysfunction.

PD-specific mechanisms are important to understand as typical neuropathology is seen throughout the ENS. ENS function is governed by both intrinsic circuits and extrinsic modulation from the CNS. Much progress has been made in understanding neuronal populations and physiology in the ENS, but many open questions remain, including how PD neuropathology contributes to ENS dysfunction, how neuropathology spreads between the ENS and CNS, and

how microbiome changes contribute to disease pathophysiology.

PD is the fastest-growing disease of the brain worldwide, and multiple lines of evidence suggest that the gut is involved in early disease pathophysiology. It is thus an important target when considering both biomarker development and disease-modifying treatment, which remain lacking in PD. Further research on the gut–brain axis and its role in PD pathophysiology is of critical importance.

ACKNOWLEDGMENTS

This work was supported by funding from the Leon Levy Fellowship in Neuroscience and the McGraw Fellowship in Neurology Research (to V.G.) and the NIH (R01NS113960, R01NS121077, RF1NS126342, and UL1TR002384 subaward to J.B.).

REFERENCES

Adams-Carr KL, Bestwick JP, Shribman S, Lees A, Schrag A, Noyce AJ. 2016. Constipation preceding Parkinson's disease: a systematic review and meta-analysis. *J Neurol Neurosurg Psychiatry* **87**: 710–716. doi:10.1136/jnnp-2015-311680

Adler CH, Beach TG. 2016. Neuropathological basis of nonmotor manifestations of Parkinson's disease. *Mov Disord* **31**: 1114–1119. doi:10.1002/mds.26605

Ahn M, Kim S, Kang M, Ryu Y, Kim TD. 2006. Chaperone-like activities of α-synuclein: α-synuclein assists enzyme activities of esterases. *Biochem Biophys Res Commun* **346**: 1142–1149. doi:10.1016/j.bbrc.2006.05.213

Aho VTE, Houser MC, Pereira PAB, Chang J, Rudi K, Paulin L, Hertzberg V, Auvinen P, Tansey MG, Scheperjans F. 2021. Relationships of gut microbiota, short-chain fatty acids, inflammation, and the gut barrier in Parkinson's disease. *Mol Neurodegener* **16**: 6. doi:10.1186/s13024-021-00427-6

Almeida A, Nayfach S, Boland M, Strozzi F, Beracochea M, Shi ZJ, Pollard KS, Sakharova E, Parks DH, Hugenholtz P, et al. 2021. A unified catalog of 204,938 reference genomes from the human gut microbiome. *Nat Biotechnol* **39**: 105–114. doi:10.1038/s41587-020-0603-3

Annerino DM, Arshad S, Taylor GM, Adler CH, Beach TG, Greene JG. 2012. Parkinson's disease is not associated with gastrointestinal myenteric ganglion neuron loss. *Acta Neuropathol* **124**: 665–680. doi:10.1007/s00401-012-1040-2

Anselmi L, Toti L, Bove C, Hampton J, Travagli RA. 2017. A nigro-vagal pathway controls gastric motility and is affected in a rat model of Parkinsonism. *Gastroenterology* **153**: 1581–1593. doi:10.1053/j.gastro.2017.08.069

Arai E, Arai M, Uchiyama T, Higuchi Y, Aoyagi K, Yamanaka Y, Yamamoto T, Nagano O, Shiina A, Maruoka D, et al. 2012. Subthalamic deep brain stimulation can improve gastric emptying in Parkinson's disease. *Brain* **135:** 1478–1485. doi:10.1093/brain/aws086

Arbouw ME, Movig KL, Koopmann M, Poels PJ, Guchelaar HJ, Egberts TC, Neef C, van Vugt JP. 2010. Glycopyrrolate for sialorrhea in Parkinson disease: a randomized, double-blind, crossover trial. *Neurology* **74:** 1203–1207. doi:10.1212/WNL.0b013e3181d8c1b7

Athukorala RP, Jones RD, Sella O, Huckabee ML. 2014. Skill training for swallowing rehabilitation in patients with Parkinson's disease. *Arch Phys Med Rehabil* **95:** 1374–1382. doi:10.1016/j.apmr.2014.03.001

Bayliss WM, Starling EH. 1899. The movements and innervation of the small intestine. *J Physiol* **24:** 99–143. doi:10.1113/jphysiol.1899.sp000752

Beach TG, Adler CH, Sue LI, Vedders L, Lue L, White Iii CL, Akiyama H, Caviness JN, Shill HA, Sabbagh MN, et al. 2010. Multi-organ distribution of phosphorylated α-synuclein histopathology in subjects with Lewy body disorders. *Acta Neuropathol* **119:** 689–702. doi:10.1007/s00401-010-0664-3

Bloch A, Probst A, Bissig H, Adams H, Tolnay M. 2006. α-Synuclein pathology of the spinal and peripheral autonomic nervous system in neurologically unimpaired elderly subjects. *Neuropathol Appl Neurobiol* **32:** 284–295. doi:10.1111/j.1365-2990.2006.00727.x

Borghammer P. 2021. The α-synuclein origin and connectome model (SOC Model) of Parkinson's disease: explaining motor asymmetry, non-motor phenotypes, and cognitive decline. *J Parkinsons Dis* **11:** 455–474. doi:10.3233/JPD-202481

Borghammer P. 2023. The brain-first vs. body-first model of Parkinson's disease with comparison to alternative models. *J Neural Transm* **130:** 737–753. doi:10.1007/s00702-023-02633-6

Borghammer P, Just MK, Horsager J, Skjærbæk C, Raunio A, Kok EH, Savola S, Murayama S, Saito Y, Myllykangas L, et al. 2022. A postmortem study suggests a revision of the dual-hit hypothesis of Parkinson's disease. *NPJ Parkinsons Dis* **8:** 166. doi:10.1038/s41531-022-00436-2

Böttner M, Zorenkov D, Hellwig I, Barrenschee M, Harde J, Fricke T, Deuschl G, Egberts JH, Becker T, Fritscher-Ravens A, et al. 2012. Expression pattern and localization of α-synuclein in the human enteric nervous system. *Neurobiol Dis* **48:** 474–480. doi:10.1016/j.nbd.2012.07.018

Böttner M, Fricke T, Müller M, Barrenschee M, Deuschl G, Schneider SA, Egberts JH, Becker T, Fritscher-Ravens A, Ellrichmann M, et al. 2015. α-Synuclein is associated with the synaptic vesicle apparatus in the human and rat enteric nervous system. *Brain Res* **1614:** 51–59. doi:10.1016/j.brainres.2015.04.015

Braak H, Del Tredici K, Rüb U, de Vos RA, Jansen Steur EN, Braak E. 2003a. Staging of brain pathology related to sporadic Parkinson's disease. *Neurobiol Aging* **24:** 197–211. doi:10.1016/S0197-4580(02)00065-9

Braak H, Rüb U, Gai WP, Del Tredici K. 2003b. Idiopathic Parkinson's disease: possible routes by which vulnerable neuronal types may be subject to neuroinvasion by an unknown pathogen. *J Neural Transm* **110:** 517–536. doi:10.1007/s00702-002-0808-2

Braak H, Ghebremedhin E, Rüb U, Bratzke H, Del Tredici K. 2004. Stages in the development of Parkinson's disease-related pathology. *Cell Tissue Res* **318:** 121–134. doi:10.1007/s00441-004-0956-9

Braak H, de Vos RA, Bohl J, Del Tredici K. 2006. Gastric α-synuclein immunoreactive inclusions in Meissner's and Auerbach's plexuses in cases staged for Parkinson's disease–related brain pathology. *Neurosci Lett* **396:** 67–72. doi:10.1016/j.neulet.2005.11.012

Bu LL, Huang KX, Zheng DZ, Lin DY, Chen Y, Jing XN, Liang YR, Tao EX. 2020. α-Synuclein accumulation and its phosphorylation in the enteric nervous system of patients without neurodegeneration: an explorative study. *Front Aging Neurosci* **12:** 575481. doi:10.3389/fnagi.2020.575481

Burré J, Sharma M, Tsetsenis T, Buchman V, Etherton MR, Südhof TC. 2010. α-Synuclein promotes SNARE-complex assembly in vivo and in vitro. *Science* **329:** 1663–1667. doi:10.1126/science.1195227

Burré J, Vivona S, Diao J, Sharma M, Brunger AT, Südhof TC. 2013. Properties of native brain α-synuclein. *Nature* **498:** E4–E6; discussion E6–7. doi:10.1038/nature12125

Burré J, Sharma M, Südhof TC. 2015. Definition of a molecular pathway mediating α-synuclein neurotoxicity. *J Neurosci* **35:** 5221–5232. doi:10.1523/JNEUROSCI.4650-14.2015

Bussell R Jr, Ramlall TF, Eliezer D. 2005. Helix periodicity, topology, and dynamics of membrane-associated α-synuclein. *Protein Sci* **14:** 862–872. doi:10.1110/ps.041255905

Camilleri M, Shin A. 2013. Novel and validated approaches for gastric emptying scintigraphy in patients with suspected gastroparesis. *Dig Dis Sci* **58:** 1813–1815. doi:10.1007/s10620-013-2715-9

Casini A, Mancinelli R, Mammola CL, Pannarale L, Chirletti P, Onori P, Vaccaro R. 2021. Distribution of α-synuclein in normal human jejunum and its relations with the chemosensory and neuroendocrine system. *Eur J Histochem* **65:** 3310. doi:10.4081/ejh.2021.3310

Challis C, Hori A, Sampson TR, Yoo BB, Challis RC, Hamilton AM, Mazmanian SK, Volpicelli-Daley LA, Gradinaru V. 2020. Gut-seeded α-synuclein fibrils promote gut dysfunction and brain pathology specifically in aged mice. *Nat Neurosci* **23:** 327–336. doi:10.1038/s41593-020-0589-7

Chandra R, Hiniker A, Kuo YM, Nussbaum RL, Liddle RA. 2017. α-Synuclein in gut endocrine cells and its implications for Parkinson's disease. *JCI Insight* **2:** e92295. doi:10.1172/jci.insight.92295

Chandra R, Sokratian A, Chavez KR, King S, Swain SM, Snyder JC, West AB, Liddle RA. 2023. Gut mucosal cells transfer α-synuclein to the vagus nerve. *JCI Insight* **8:** e172192. doi:10.1172/jci.insight.172192

Chang MC, Park JS, Lee BJ, Park D. 2021. Effectiveness of pharmacologic treatment for dysphagia in Parkinson's disease: a narrative review. *Neurol Sci* **42:** 513–519. doi:10.1007/s10072-020-04865-w

Claus I, Muhle P, Czechowski J, Ahring S, Labeit B, Suntrup-Krueger S, Wiendl H, Dziewas R, Warnecke T. 2021. Expiratory muscle strength training for therapy of pharyngeal dysphagia in Parkinson's disease. *Mov Disord* **36:** 1815–1824. doi:10.1002/mds.28552

Cite this article as *Cold Spring Harb Perspect Med* doi: 10.1101/cshperspect.a041618

Cook RD, Burnstock G. 1976. The altrastructure of Auerbach's plexus in the guinea-pig. I: Neuronal elements. *J Neurocytol* 5: 171–194. doi:10.1007/BF01181655

Corbillé AG, Coron E, Neunlist M, Derkinderen P, Lebouvier T. 2014. Appraisal of the dopaminergic and noradrenergic innervation of the submucosal plexus in PD. *J Parkinsons Dis* 4: 571–576. doi:10.3233/JPD-140422

Corbillé AG, Preterre C, Rolli-Derkinderen M, Coron E, Neunlist M, Lebouvier T, Derkinderen P. 2017. Biochemical analysis of α-synuclein extracted from control and Parkinson's disease colonic biopsies. *Neurosci Lett* 641: 81–86. doi:10.1016/j.neulet.2017.01.050

Coughlin DG, Irwin DJ. 2023. Fluid and biopsy based biomarkers in Parkinson's disease. *Neurotherapeutics* 20: 932–954. doi:10.1007/s13311-023-01379-z

Daniel SE, Hawkes CH. 1992. Preliminary diagnosis of Parkinson's disease by olfactory bulb pathology. *Lancet* 340: 186. doi:10.1016/0140-6736(92)93275-R

Del Tredici K, Braak H. 2012. Spinal cord lesions in sporadic Parkinson's disease. *Acta Neuropathol* 124: 643–664. doi:10.1007/s00401-012-1028-y

Del Tredici K, Duda JE. 2011. Peripheral Lewy body pathology in Parkinson's disease and incidental Lewy body disease: four cases. *J Neurol Sci* 310: 100–106. doi:10.1016/j.jns.2011.06.003

Del Tredici K, Rüb U, De Vos RA, Bohl JR, Braak H. 2002. Where does Parkinson disease pathology begin in the brain? *J Neuropathol Exp Neurol* 61: 413–426. doi:10.1093/jnen/61.5.413

Del Tredici K, Hawkes CH, Ghebremedhin E, Braak H. 2010. Lewy pathology in the submandibular gland of individuals with incidental Lewy body disease and sporadic Parkinson's disease. *Acta Neuropathol* 119: 703–713. doi:10.1007/s00401-010-0665-2

Diao J, Burré J, Vivona S, Cipriano DJ, Sharma M, Kyoung M, Südhof TC, Brunger AT. 2013. Native α-synuclein induces clustering of synaptic-vesicle mimics via binding to phospholipids and synaptobrevin-2/VAMP2. *eLife* 2: e00592. doi:10.7554/eLife.00592

Dorsey ER, Sherer T, Okun MS, Bloem BR. 2018. The emerging evidence of the Parkinson pandemic. *J Parkinsons Dis* 8: S3–S8. doi:10.3233/JPD-181474

Drokhlyansky E, Smillie CS, Van Wittenberghe N, Ericsson M, Griffin GK, Eraslan G, Dionne D, Cuoco MS, Goder-Reiser MN, Sharova T, et al. 2020. The human and mouse enteric nervous system at single-cell resolution. *Cell* 182: 1606–1622.e23. doi:10.1016/j.cell.2020.08.003

DuPont HL, Suescun J, Jiang ZD, Brown EL, Essigmann HT, Alexander AS, DuPont AW, Iqbal T, Utay NS, Newmark M, et al. 2023. Fecal microbiota transplantation in Parkinson's disease—a randomized repeat-dose, placebo-controlled clinical pilot study. *Front Neurol* 14: 1104759. doi:10.3389/fneur.2023.1104759

El-Agnaf OM, Jakes R, Curran MD, Middleton D, Ingenito R, Bianchi E, Pessi A, Neill D, Wallace A. 1998. Aggregates from mutant and wild-type α-synuclein proteins and NAC peptide induce apoptotic cell death in human neuroblastoma cells by formation of β-sheet and amyloid-like filaments. *FEBS Lett* 440: 71–75. doi:10.1016/S0014-5793(98)01418-5

Emmi A, Sandre M, Russo FP, Tombesi G, Garrì F, Campagnolo M, Carecchio M, Biundo R, Spolverato G, Macchi V,

et al. 2023. Duodenal α-synuclein pathology and enteric gliosis in advanced Parkinson's disease. *Mov Disord* 38: 885–894. doi:10.1002/mds.29358

Fares MB, Ait-Bouziad N, Dikiy I, Mbefo MK, Jovicic A, Kiely A, Holton JL, Lee SJ, Gitler AD, Eliezer D, et al. 2014. The novel Parkinson's disease linked mutation G51D attenuates in vitro aggregation and membrane binding of α-synuclein, and enhances its secretion and nuclear localization in cells. *Hum Mol Genet* 23: 4491–4509. doi:10.1093/hmg/ddu165

Fasano A, Visanji NP, Liu LW, Lang AE, Pfeiffer RF. 2015. Gastrointestinal dysfunction in Parkinson's disease. *Lancet Neurol* 14: 625–639. doi:10.1016/S1474-4422(15)00007-1

Fearnley JM, Lees AJ. 1991. Ageing and Parkinson's disease: substantia nigra regional selectivity. *Brain* 114: 2283–2301. doi:10.1093/brain/114.5.2283

Fedorova TD, Knudsen K, Andersen KB, Horsager J, Skjærbæk C, Beier CP, Sommerauer M, Svendsen KB, Otto M, Borghammer P. 2022. Imaging progressive peripheral and central dysfunction in isolated REM sleep behaviour disorder after 3 years of follow-up. *Parkinsonism Relat Disord* 101: 99–104. doi:10.1016/j.parkreldis.2022.07.005

Fenyi A, Leclair-Visonneau L, Clairembault T, Coron E, Neunlist M, Melki R, Derkinderen P, Bousset L. 2019. Detection of α-synuclein aggregates in gastrointestinal biopsies by protein misfolding cyclic amplification. *Neurobiol Dis* 129: 38–43. doi:10.1016/j.nbd.2019.05.002

Fricova D, Harsanyiova J, Kralova Trancikova A. 2020. α-Synuclein in the gastrointestinal tract as a potential biomarker for early detection of Parkinson's disease. *Int J Mol Sci* 21: 8666. doi:10.3390/ijms21228666

Fujiwara H, Hasegawa M, Dohmae N, Kawashima A, Masliah E, Goldberg MS, Shen J, Takio K, Iwatsubo T. 2002. α-Synuclein is phosphorylated in synucleinopathy lesions. *Nat Cell Biol* 4: 160–164. doi:10.1038/ncb748

Fullard ME, Morley JF, Duda JE. 2017. Olfactory dysfunction as an early biomarker in Parkinson's disease. *Neurosci Bull* 33: 515–525. doi:10.1007/s12264-017-0170-x

Furness JB. 2007. *The enteric nervous system*. Blackwell, Boston.

Furness JB. 2012. The enteric nervous system and neurogastroenterology. *Nat Rev Gastroenterol Hepatol* 9: 286–294. doi:10.1038/nrgastro.2012.32

Furness JB, Alex G, Clark MJ, Lal VV. 2003. Morphologies and projections of defined classes of neurons in the submucosa of the guinea-pig small intestine. *Anat Rec A Discov Mol Cell Evol Biol* 272: 475–483. doi:10.1002/ar.a.10064

Furness JB, Callaghan BP, Rivera LR, Cho HJ. 2014. The enteric nervous system and gastrointestinal innervation: integrated local and central control. *Adv Exp Med Biol* 817: 39–71. doi:10.1007/978-1-4939-0897-4_3

Ghaisas S, Langley MR, Palanisamy BN, Dutta S, Narayanaswamy K, Plummer PJ, Sarkar S, Ay M, Jin H, Anantharam V, et al. 2019. Mitopark transgenic mouse model recapitulates the gastrointestinal dysfunction and gut-microbiome changes of Parkinson's disease. *Neurotoxicology* 75: 186–199. doi:10.1016/j.neuro.2019.09.004

Ghebremedhin E, Del Tredici K, Langston JW, Braak H. 2009. Diminished tyrosine hydroxylase immunoreactivity in the cardiac conduction system and myocardium in Parkinson's disease: an anatomical study. *Acta Neuropathol* **118**: 777–784. doi:10.1007/s00401-009-0596-y

Ghosh D, Sahay S, Ranjan P, Salot S, Mohite GM, Singh PK, Dwivedi S, Carvalho E, Banerjee R, Kumar A, et al. 2014. The newly discovered Parkinson's disease associated Finnish mutation (A53E) attenuates α-synuclein aggregation and membrane binding. *Biochemistry* **53**: 6419–6421. doi:10.1021/bi5010365

Gibbons CH, Levine T, Adler C, Bellaire B, Wang N, Stohl J, Agarwal P, Aldridge GM, Barboi A, Evidente VGH, et al. Skin biopsy detection of phosphorylated α-synuclein in patients with synucleinopathies. *JAMA* **331**: 1298–1306. doi:10.1001/jama.2024.0792

Gil RA, Hwynn N, Fabian T, Joseph S, Fernandez HH. 2011. Botulinum toxin type A for the treatment of gastroparesis in Parkinson's disease patients. *Parkinsonism Relat Disord* **17**: 285–287. doi:10.1016/j.parkreldis.2011.01.007

Goldman SM. 2014. Environmental toxins and Parkinson's disease. *Annu Rev Pharmacol Toxicol* **54**: 141–164. doi:10.1146/annurev-pharmtox-011613-135937

Goldstein DS, Holmes C, Sullivan P, Lopez G, Gelsomino J, Moore S, Isonaka R, Wu T, Sharabi Y. 2024. Cardiac noradrenergic deficiency revealed by ^{18}F-dopamine positron emission tomography identifies preclinical central Lewy body diseases. *J Clin Invest* **134**: e172460. doi:10.1172/JCI172460

Gosavi N, Lee HJ, Lee JS, Patel S, Lee SJ. 2002. Golgi fragmentation occurs in the cells with prefibrillar α-synuclein aggregates and precedes the formation of fibrillar inclusion. *J Biol Chem* **277**: 48984–48992. doi:10.1074/jbc.M208194200

Gray MT, Munoz DG, Gray DA, Schlossmacher MG, Woulfe JM. 2014. α-Synuclein in the appendiceal mucosa of neurologically intact subjects. *Mov Disord* **29**: 991–998. doi:10.1002/mds.25779

Grossauer A, Hemicker G, Krismer F, Peball M, Djamshidian A, Poewe W, Seppi K, Heim B. 2023. α-Synuclein seed amplification assays in the diagnosis of synucleinopathies using cerebrospinal fluid—a systematic review and meta-analysis. *Mov Disord Clin Pract* **10**: 737–747. doi:10.1002/mdc3.13710

Hallett PJ, McLean JR, Kartunen A, Langston JW, Isacson O. 2012. α-Synuclein overexpressing transgenic mice show internal organ pathology and autonomic deficits. *Neurobiol Dis* **47**: 258–267. doi:10.1016/j.nbd.2012.04.009

Halliday GM, McRitchie DA, Cartwright H, Pamphlett R, Hely MA, Morris JG. 1996. Midbrain neuropathology in idiopathic Parkinson's disease and diffuse Lewy body disease. *J Clin Neurosci* **3**: 52–60. doi:10.1016/S0967-5868(96)90083-1

Hammer MJ, Murphy CA, Abrams TM. 2013. Airway somatosensory deficits and dysphagia in Parkinson's disease. *J Parkinsons Dis* **3**: 39–44. doi:10.3233/JPD-120161

Hardoff R, Sula M, Tamir A, Soil A, Front A, Badarna S, Honigman S, Giladi N. 2001. Gastric emptying time and gastric motility in patients with Parkinson's disease. *Mov Disord* **16**: 1041–1047. doi:10.1002/mds.1203

Hawkes CH, Del Tredici K, Braak H. 2007. Parkinson's disease: a dual-hit hypothesis. *Neuropathol Appl Neurobiol* **33**: 599–614. doi:10.1111/j.1365-2990.2007.00874.x

Heintz-Buschart A, Pandey U, Wicke T, Sixel-Döring F, Janzen A, Sittig-Wiegand E, Trenkwalder C, Oertel WH, Mol-

lenhauer B, Wilmes P. 2018. The nasal and gut microbiome in Parkinson's disease and idiopathic rapid eye movement sleep behavior disorder. *Mov Disord* **33**: 88–98. doi:10.1002/mds.27105

Helander HF, Fändriks L. 2014. Surface area of the digestive tract – revisited. *Scand J Gastroenterol* **49**: 681–689. doi:10.3109/00365521.2014.898326

Höglinger GU, Alvarez-Fischer D, Arias-Carrión O, Djufri M, Windolph A, Keber U, Borta A, Ries V, Schwarting RK, Scheller D, et al. 2015. A new dopaminergic nigro-olfactory projection. *Acta Neuropathol* **130**: 333–348. doi:10.1007/s00401-015-1451-y

Holmqvist S, Chutna O, Bousset L, Aldrin-Kirk P, Li W, Björklund T, Wang ZY, Roybon L, Melki R, Li JY. 2014. Direct evidence of Parkinson pathology spread from the gastrointestinal tract to the brain in rats. *Acta Neuropathol* **128**: 805–820. doi:10.1007/s00401-014-1343-6

Hughes AJ, Daniel SE, Ben-Shlomo Y, Lees AJ. 2002. The accuracy of diagnosis of parkinsonian syndromes in a specialist movement disorder service. *Brain* **125**: 861–870. doi:10.1093/brain/awf080

Hyson HC, Johnson AM, Jog MS. 2002. Sublingual atropine for sialorrhea secondary to parkinsonism: a pilot study. *Mov Disord* **17**: 1318–1320. doi:10.1002/mds.10276

Iwai A, Masliah E, Yoshimoto M, Ge N, Flanagan L, de Silva HA, Kittel A, Saitoh T. 1995. The precursor protein of non-Aβ component of Alzheimer's disease amyloid is a presynaptic protein of the central nervous system. *Neuron* **14**: 467–475. doi:10.1016/0896-6273(95)90302-X

Jo E, Fuller N, Rand RP, St George-Hyslop P, Fraser PE. 2002. Defective membrane interactions of familial Parkinson's disease mutant A30P α-synuclein. *J Mol Biol* **315**: 799–807. doi:10.1006/jmbi.2001.5269

Jost WH, Friedman A, Michel O, Oehlwein C, Slawek J, Bogucki A, Ochudlo S, Banach M, Pagan F, Flatau-Baqué B, et al. 2019. SIAXI: Placebo-controlled, randomized, double-blind study of incobotulinumtoxinA for sialorrhea. *Neurology* **92**: e1982–e1991. doi:10.1212/WNL.0000000000007368

Jucker M, Walker LC. 2013. Self-propagation of pathogenic protein aggregates in neurodegenerative diseases. *Nature* **501**: 45–51. doi:10.1038/nature12481

Kalf JG, de Swart BJ, Bloem BR, Munneke M. 2012. Prevalence of oropharyngeal dysphagia in Parkinson's disease: a meta-analysis. *Parkinsonism Relat Disord* **18**: 311–315. doi:10.1016/j.parkreldis.2011.11.006

Kalia LV, Lang AE. 2015. Parkinson's disease. *Lancet* **386**: 896–912. doi:10.1016/S0140-6736(14)61393-3

Kang UJ, Boehme AK, Fairfoul G, Shahnawaz M, Ma TC, Hutten SJ, Green A, Soto C. 2019. Comparative study of cerebrospinal fluid α-synuclein seeding aggregation assays for diagnosis of Parkinson's disease. *Mov Disord* **34**: 536–544. doi:10.1002/mds.27646

Kashihara K, Imamura T, Shinya T. 2010. Cardiac 123I-MIBG uptake is reduced more markedly in patients with REM sleep behavior disorder than in those with early stage Parkinson's disease. *Parkinsonism Relat Disord* **16**: 252–255. doi:10.1016/j.parkreldis.2009.12.010

Kikuchi A, Baba T, Hasegawa T, Kobayashi M, Sugeno N, Konno M, Miura E, Hosokai Y, Ishioka T, Nishio Y, et al. 2013. Hypometabolism in the supplementary and anterior cingulate cortices is related to dysphagia in Parkinson's disease: a cross-sectional and 3-year longitudinal cohort study. *BMJ Open* 3: e002249. doi:10.1136/bmjopen-2012-002249

Kim S, Kwon SH, Kam TI, Panicker N, Karuppagounder SS, Lee S, Lee JH, Kim WR, Kook M, Foss CA, et al. 2019. Transneuronal propagation of pathologic α-synuclein from the gut to the brain models Parkinson's disease. *Neuron* 103: 627–641.e7. doi:10.1016/j.neuron.2019.05.035

Knudsen K, Krogh K, Østergaard K, Borghammer P. 2017. Constipation in Parkinson's disease: subjective symptoms, objective markers, and new perspectives. *Mov Disord* 32: 94–105. doi:10.1002/mds.26866

Knudsen K, Fedorova TD, Hansen AK, Sommerauer M, Otto M, Svendsen KB, Nahimi A, Stokholm MG, Pavese N, Beier CP, et al. 2018. In-vivo staging of pathology in REM sleep behaviour disorder: a multimodality imaging case-control study. *Lancet Neurol* 17: 618–628. doi:10.1016/S1474-4422(18)30162-5

Kordower JH, Chu Y, Hauser RA, Freeman TB, Olanow CW. 2008. Lewy body-like pathology in long-term embryonic nigral transplants in Parkinson's disease. *Nat Med* 14: 504–506. doi:10.1038/nm1747

Koutzoumis DN, Vergara M, Pino J, Buddendorff J, Khoshbouei H, Mandel RJ, Torres GE. 2020. Alterations of the gut microbiota with antibiotics protects dopamine neuron loss and improve motor deficits in a pharmacological rodent model of Parkinson's disease. *Exp Neurol* 325: 113159. doi:10.1016/j.expneurol.2019.113159

Krasko MN, Szot J, Lungova K, Rowe LM, Leverson G, Kelm-Nelson CA, Ciucci MR. 2023. Pink1$^{-/-}$ rats demonstrate swallowing and gastrointestinal dysfunction in a model of prodromal Parkinson disease. *Dysphagia* 38: 1382–1397. doi:10.1007/s00455-023-10567-0

Kuo YM, Li Z, Jiao Y, Gaborit N, Pani AK, Orrison BM, Bruneau BG, Giasson BI, Smeyne RJ, Gershon MD, et al. 2010. Extensive enteric nervous system abnormalities in mice transgenic for artificial chromosomes containing Parkinson disease–associated α-synuclein gene mutations precede central nervous system changes. *Hum Mol Genet* 19: 1633–1650. doi:10.1093/hmg/ddq038

Lai F, Jiang R, Xie W, Liu X, Tang Y, Xiao H, Gao J, Jia Y, Bai Q. 2018. Intestinal pathology and gut microbiota alterations in a methyl-4-phenyl-1,2,3,6-tetrahydropyridine (MPTP) mouse model of Parkinson's disease. *Neurochem Res* 43: 1986–1999. doi:10.1007/s11064-018-2620-x

Langmore SE. 2003. Evaluation of oropharyngeal dysphagia: which diagnostic tool is superior? *Curr Opin Otolaryngol Head Neck Surg* 11: 485–489. doi:10.1097/00020840-200312000-00014

Lebouvier T, Neunlist M, Bruley des Varannes S, Coron E, Drouard A, N'Guyen JM, Chaumette T, Tasselli M, Paillusson S, Flamand M, et al. 2010. Colonic biopsies to assess the neuropathology of Parkinson's disease and its relationship with symptoms. *PLoS ONE* 5: e12728. doi:10.1371/journal.pone.0012728

Li JY, Englund E, Holton JL, Soulet D, Hagell P, Lees AJ, Lashley T, Quinn NP, Rehncrona S, Björklund A, et al. 2008. Lewy bodies in grafted neurons in subjects with Parkinson's disease suggest host-to-graft disease propagation. *Nat Med* 14: 501–503. doi:10.1038/nm1746

Lin CH, Lin JW, Liu YC, Chang CH, Wu RM. 2014. Risk of Parkinson's disease following severe constipation: a na-

tionwide population-based cohort study. *Parkinsonism Relat Disord* 20: 1371–1375. doi:10.1016/j.parkreldis.2014.09.026

Maini Rekdal V, Bess EN, Bisanz JE, Turnbaugh PJ, Balskus EP. 2019. Discovery and inhibition of an interspecies gut bacterial pathway for Levodopa metabolism. *Science* 364: eaau6323. doi:10.1126/science.aau6323

Manfredsson FP, Luk KC, Benskey MJ, Gezer A, Garcia J, Kuhn NC, Sandoval IM, Patterson JR, O'Mara A, Yonkers R, et al. 2018. Induction of α-synuclein pathology in the enteric nervous system of the rat and non-human primate results in gastrointestinal dysmotility and transient CNS pathology. *Neurobiol Dis* 112: 106–118. doi:10.1016/j.nbd.2018.01.008

Manor Y, Mootanah R, Freud D, Giladi N, Cohen JT. 2013. Video-assisted swallowing therapy for patients with Parkinson's disease. *Parkinsonism Relat Disord* 19: 207–211. doi:10.1016/j.parkreldis.2012.10.004

Marrinan S, Emmanuel AV, Burn DJ. 2014. Delayed gastric emptying in Parkinson's disease. *Mov Disord* 29: 23–32. doi:10.1002/mds.25708

Miller-Patterson C, Hsu JY, Willis AW, Hamedani AG. 2023. Functional impairment in individuals with prodromal or unrecognized Parkinson disease. *JAMA Neurol* 80: 200–204. doi:10.1001/jamaneurol.2022.4621

Miyamoto T, Miyamoto M, Inoue Y, Usui Y, Suzuki K, Hirata K. 2006. Reduced cardiac 123I-MIBG scintigraphy in idiopathic REM sleep behavior disorder. *Neurology* 67: 2236–2238. doi:10.1212/01.wnl.0000249313.25627.2e

Mohite GM, Kumar R, Panigrahi R, Navalkar A, Singh N, Datta D, Mehra S, Ray S, Gadhe LG, Das S, et al. 2018. Comparison of kinetics, toxicity, oligomer formation, and membrane binding capacity of α-synuclein familial mutations at the A53 site, including the newly discovered A53V mutation. *Biochemistry* 57: 5183–5187. doi:10.1021/acs.biochem.8b00314

Mu L, Sobotka S, Chen J, Su H, Sanders I, Adler CH, Shill HA, Caviness JN, Samanta JE, Beach TG, et al. 2012. Altered pharyngeal muscles in Parkinson disease. *J Neuropathol Exp Neurol* 71: 520–530. doi:10.1097/NEN.0b013e318258381b

Mu L, Sobotka S, Chen J, Su H, Sanders I, Nyirenda T, Adler CH, Shill HA, Caviness JN, Samanta JE, et al. 2013. Parkinson disease affects peripheral sensory nerves in the pharynx. *J Neuropathol Exp Neurol* 72: 614–623. doi:10.1097/NEN.0b013e3182965886

Nag S, Yu L, VanderHorst VG, Schneider JA, Bennett DA, Buchman AS, Wilson RS. 2019. Neocortical Lewy bodies are associated with impaired odor identification in community-dwelling elders without clinical PD. *J Neurol* 266: 3108–3118. doi:10.1007/s00415-019-09540-5

Nirenberg MJ, Waters C. 2006. Compulsive eating and weight gain related to dopamine agonist use. *Mov Disord* 21: 524–529. doi:10.1002/mds.20757

Nishikawa N, Murata M, Hatano T, Mukai Y, Saitoh Y, Sakamoto T, Hanakawa T, Kamei Y, Tachimori H, Hatano K, et al. 2022. Idiopathic rapid eye movement sleep behavior disorder in Japan: an observational study. *Parkinsonism Relat Disord* 103: 129–135. doi:10.1016/j.parkreldis.2022.08.011

Nishiwaki H, Hamaguchi T, Ito M, Ishida T, Maeda T, Kashihara K, Tsuboi Y, Ueyama J, Shimamura T, Mori H, et al. 2020a. Short-chain fatty acid-producing gut microbiota is decreased in Parkinson's disease but not in rapid-eye-movement sleep behavior disorder. *mSystems* **5**: e00797-20. doi:10.1128/mSystems.00797-20

Nishiwaki H, Ito M, Ishida T, Hamaguchi T, Maeda T, Kashihara K, Tsuboi Y, Ueyama J, Shimamura T, Mori H, et al. 2020b. Meta-analysis of gut dysbiosis in Parkinson's disease. *Mov Disord* **35**: 1626–1635. doi:10.1002/mds.28119

Nóbrega AC, Rodrigues B, Melo A. 2008a. Is silent aspiration a risk factor for respiratory infection in Parkinson's disease patients? *Parkinsonism Relat Disord* **14**: 646–648. doi:10.1016/j.parkreldis.2007.12.007

Nóbrega AC, Rodrigues B, Melo A. 2008b. Silent aspiration in Parkinson's disease patients with diurnal sialorrhea. *Clin Neurol Neurosurg* **110**: 117–119. doi:10.1016/j.clineuro.2007.09.011

Noorian AR, Rha J, Annerino DM, Bernhard D, Taylor GM, Greene JG. 2012. α-Synuclein transgenic mice display age-related slowing of gastrointestinal motility associated with transgene expression in the vagal system. *Neurobiol Dis* **48**: 9–19. doi:10.1016/j.nbd.2012.06.005

Ohlsson B, Englund E. 2019. Atrophic myenteric and submucosal neurons are observed in Parkinson's disease. *Parkinsons Dis* **2019**: 7935820. doi:10.1155/2019/7935820

Pan-Montojo FJ, Funk RH. 2010. Oral administration of rotenone using a gavage and image analysis of α-synuclein inclusions in the enteric nervous system. *J Vis Exp* **26**: 2123. doi:10.3791/2123

Parkinson J. 2002. An essay on the shaking palsy. 1817. *J Neuropsychiatry Clin Neurosci* **14**: 223–236; discussion 222. doi:10.1176/jnp.14.2.223

Parra-Rivas LA, Madhivanan K, Aulston BD, Wang L, Prakashchand DD, Boyer NP, Saia-Cereda VM, Branes-Guerrero K, Pizzo DP, Bagchi P, et al. 2023. Serine-129 phosphorylation of α-synuclein is an activity-dependent trigger for physiologic protein-protein interactions and synaptic function. *Neuron* **111**: 4006–4023.e10. doi:10.1016/j.neuron.2023.11.020

Pearce RK, Hawkes CH, Daniel SE. 1995. The anterior olfactory nucleus in Parkinson's disease. *Mov Disord* **10**: 283–287. doi:10.1002/mds.870100309

Pearson C, Hartzman A, Munevar D, Feeney M, Dolhun R, Todaro V, Rosenfeld S, Willis A, Beck JC. 2023. Care access and utilization among medicare beneficiaries living with Parkinson's disease. *NPJ Parkinsons Dis* **9**: 108. doi:10.1038/s41531-023-00523-y

Pellegrini C, Travagli RA. 2024. Gastrointestinal dysmotility in rodent models of Parkinson's disease. *Am J Physiol Gastrointest Liver Physiol* doi:10.1152/ajpgi.00225.2023

Pfeiffer RF. 2014. Parkinson's disease and the gut: "the wheel is come full circle." *J Parkinsons Dis* **4**: 577–578. doi:10.3233/JPD-149007

Pfeiffer RF, Isaacson SH, Pahwa R. 2020. Clinical implications of gastric complications on levodopa treatment in Parkinson's disease. *Parkinsonism Relat Disord* **76**: 63–71. doi:10.1016/j.parkreldis.2020.05.001

Pflug C, Bihler M, Emich K, Niessen A, Nienstedt JC, Flügel T, Koseki JC, Plaetke R, Hidding U, Gerloff C, et al. 2018. Critical dysphagia is common in Parkinson disease and occurs even in early stages: a prospective cohort study. *Dysphagia* **33**: 41–50. doi:10.1007/s00455-017-9831-1

Phillips RJ, Walter GC, Wilder SL, Baronowsky EA, Powley TL. 2008. α-Synuclein-immunopositive myenteric neurons and vagal preganglionic terminals: autonomic pathway implicated in Parkinson's disease? *Neuroscience* **153**: 733–750. doi:10.1016/j.neuroscience.2008.02.074

Phillips RJ, Walter GC, Ringer BE, Higgs KM, Powley TL. 2009. α-Synuclein immunopositive aggregates in the myenteric plexus of the aging Fischer 344 rat. *Exp Neurol* **220**: 109–119. doi:10.1016/j.expneurol.2009.07.025

Phillips RJ, Martin FN, Billingsley CN, Powley TL. 2013. α-Synuclein expression patterns in the colonic submucosal plexus of the aging Fischer 344 rat: implications for biopsies in aging and neurodegenerative disorders? *Neurogastroenterol Motil* **25**: e621–e633. doi:10.1111/nmo.12176

Poulin JF, Caronia G, Hofer C, Cui Q, Helm B, Ramakrishnan C, Chan CS, Dombeck DA, Deisseroth K, et al. 2018. Mapping projections of molecularly defined dopamine neuron subtypes using intersectional genetic approaches. *Nat Neurosci* **21**: 1260–1271. doi:10.1038/s41593-018-0203-4

Proulx M, de Courval FP, Wiseman MA, Panisset M. 2005. Salivary production in Parkinson's disease. *Mov Disord* **20**: 204–207. doi:10.1002/mds.20189

Qualman SJ, Haupt HM, Yang P, Hamilton SR. 1984. Esophageal Lewy bodies associated with ganglion cell loss in achalasia. Similarity to Parkinson's disease. *Gastroenterology* **87**: 848–856. doi:10.1016/0016-5085(84)90079-9

Ramalingam N, Jin SX, Moors TE, Fonseca-Ornelas L, Shimanaka K, Lei S, Cam HP, Watson AH, Brontesi L, Ding L, et al. 2023. Dynamic physiological α-synuclein S129 phosphorylation is driven by neuronal activity. *NPJ Parkinsons Dis* **9**: 4. doi:10.1038/s41531-023-00444-w

Rane P, Shields J, Heffernan M, Guo Y, Akbarian S, King JA. 2012. The histone deacetylase inhibitor, sodium butyrate, alleviates cognitive deficits in pre-motor stage PD. *Neuropharmacology* **62**: 2409–2412. doi:10.1016/j.neuropharm.2012.01.026

Sauleau P, Le Jeune F, Drapier S, Houvenaghel JF, Dondaine T, Haegelen C, Lalys F, Robert G, Drapier D, Vérin M. 2014. Weight gain following subthalamic nucleus deep brain stimulation: a PET study. *Mov Disord* **29**: 1781–1787. doi:10.1002/mds.26063

Schneider KM, Kim J, Bahnsen K, Heuckeroth RO, Thaiss CA. 2022. Environmental perception and control of gastrointestinal immunity by the enteric nervous system. *Trends Mol Med* **28**: 989–1005. doi:10.1016/j.molmed.2022.09.005

Schröder JB, Marian T, Claus I, Muhle M, Pawlowski M, Wiendl H, Suntrup-Krueger S, Meuth SG, Dziewas R, Ruck T, Warnecke T. 2019. Substance P saliva reduction predicts pharyngeal dysphagia in Parkinson's disease. *Front Neurol* **10**: 386. doi:10.3389/fneur.2019.00386

Serrano-Duenas M. 2003. Treatment of sialorrhea in Parkinson's disease patients with clonidine. Double-blind, comparative study with placebo. *Neurologia* **18**: 2–6.

Sharma S, Taliyan R, Singh S. 2015. Beneficial effects of sodium butyrate in 6-OHDA induced neurotoxicity and behavioral abnormalities: modulation of histone deacetylase activity. *Behav Brain Res* **291**: 306–314. doi:10.1016/j.bbr.2015.05.052

Sharrad DF, de Vries E, Brookes SJ. 2013a. Selective expression of α-synuclein-immunoreactivity in vesicular acetylcholine transporter-immunoreactive axons in the guinea pig rectum and human colon. *J Comp Neurol* **521:** 657–676. doi:10.1002/cne.23198

Sharrad DF, Gai WP, Brookes SJ. 2013b. Selective coexpression of synaptic proteins, α-synuclein, cysteine string protein-α, synaptophysin, synaptotagmin-1, and synaptobrevin-2 in vesicular acetylcholine transporter-immunoreactive axons in the guinea pig ileum. *J Comp Neurol* **521:** 2523–2537. doi:10.1002/cne.23296

Shin A, Wo JM. 2015. Therapeutic applications of ghrelin agonists in the treatment of gastroparesis. *Curr Gastroenterol Rep* **17:** 430. doi:10.1007/s11894-015-0430-8

Siderowf A, Concha-Marambio L, Lafontant DE, Farris CM, Ma Y, Urenia PA, Nguyen H, Alcalay RN, Chahine LM, Foroud T, et al. 2023. Assessment of heterogeneity among participants in the Parkinson's progression markers initiative cohort using α-synuclein seed amplification: a cross-sectional study. *Lancet Neurol* **22:** 407–417. doi:10.1016/S1474-4422(23)00109-6

Song N, Wang W, Jia F, Du X, Xie A, He Q, Shen X, Zhang J, Rogers JT, Xie J, et al. 2017. Assessments of plasma ghrelin levels in the early stages of Parkinson's disease. *Mov Disord* **32:** 1487–1491. doi:10.1002/mds.27095

Spillantini MG, Schmidt ML, Lee VM, Trojanowski JQ, Jakes R, Goedert M. 1997. α-Synuclein in Lewy bodies. *Nature* **388:** 839–840. doi:10.1038/42166

Srivastav S, Neupane S, Bhurtel S, Katila N, Maharjan S, Choi H, Hong JT, Choi DY. 2019. Probiotics mixture increases butyrate, and subsequently rescues the nigral dopaminergic neurons from MPTP and rotenone-induced neurotoxicity. *J Nutr Biochem* **69:** 73–86. doi:10.1016/j.jnutbio.2019.03.021

Stoessl AJ, Halliday GM. 2014. DAT-SPECT diagnoses dopamine depletion, but not PD. *Mov Disord* **29:** 1705–1706. doi:10.1002/mds.26000

Swaminathan M, Fung C, Finkelstein DI, Bornstein JC, Foong JPP. 2019. α-Synuclein regulates development and function of cholinergic enteric neurons in the mouse colon. *Neuroscience* **423:** 76–85. doi:10.1016/j.neuroscience.2019.10.029

Tan AH, Chong CW, Lim SY, Yap IKS, Teh CSJ, Loke MF, Song SL, Tan JY, Ang BH, Tan YQ, et al. 2021. Gut microbial ecosystem in Parkinson disease: new clinicobiological insights from multi-omics. *Ann Neurol* **89:** 546–559. doi:10.1002/ana.25982

Tan AH, Lim SY, Lang AE. 2022. The microbiome-gut-brain axis in Parkinson disease - from basic research to the clinic. *Nat Rev Neurol* **18:** 476–495. doi:10.1038/s41582-022-00681-2

Thomsen TR, Galpern WR, Asante A, Arenovich T, Fox SH. 2007. Ipratropium bromide spray as treatment for sialorrhea in Parkinson's disease. *Mov Disord* **22:** 2268–2273. doi:10.1002/mds.21730

Travagli RA, Anselmi L. 2016. Vagal neurocircuitry and its influence on gastric motility. *Nat Rev Gastroenterol Hepatol* **13:** 389–401. doi:10.1038/nrgastro.2016.76

Tremblay C, Serrano GE, Intorcia AJ, Sue LI, Wilson JR, Adler CH, Shill HA, Driver-Dunckley E, Mehta SH, et al. 2022. Effect of olfactory bulb pathology on olfactory function in normal aging. *Brain Pathol* **32:** e13075. doi:10.1111/bpa.13075

Troche MS, Okun MS, Rosenbek JC, Musson N, Fernandez HH, Rodriguez R, Romrell J, Pitts T, Wheeler-Hegland KM, Sapienza CM. 2010. Aspiration and swallowing in Parkinson disease and rehabilitation with EMST: a randomized trial. *Neurology* **75:** 1912–1919. doi:10.1212/WNL.0b013e3181fef115

Uemura N, Yagi H, Uemura MT, Hatanaka Y, Yamakado H, Takahashi R. 2018. Inoculation of α-synuclein preformed fibrils into the mouse gastrointestinal tract induces Lewy body-like aggregates in the brainstem via the vagus nerve. *Mol Neurodegener* **13:** 21. doi:10.1186/s13024-018-0257-5

Van Den Berge N, Ferreira N, Mikkelsen TW, Alstrup AKO, Tamgüney G, Karlsson P, Terkelsen AJ, Nyengaard JR, Jensen PH, Borghammer P. 2021. Ageing promotes pathological α-synuclein propagation and autonomic dysfunction in wild-type rats. *Brain* **144:** 1853–1868. doi:10.1093/brain/awab061

van Kessel SP, Frye AK, El-Gendy AO, Castejon M, Keshavarzian A, van Dijk G, El Aidy S. 2019. Gut bacterial tyrosine decarboxylases restrict levels of levodopa in the treatment of Parkinson's disease. *Nat Commun* **10:** 310. doi:10.1038/s41467-019-08294-y

Varkey J, Isas JM, Mizuno N, Jensen MB, Bhatia VK, Jao CC, Petrlova J, Voss JC, Stamou DG, Steven AC, et al. 2010. Membrane curvature induction and tubulation are common features of synucleins and apolipoproteins. *J Biol Chem* **285:** 32486–32493. doi:10.1074/jbc.M110.139576

Vascellari S, Orrù CD, Groveman BR, Parveen S, Fenu G, Pisano G, Piga G, Serra G, Oppo V, Murgia D, et al. 2023. α-Synuclein seeding activity in duodenum biopsies from Parkinson's disease patients. *PLoS Pathog* **19:** e1011456. doi:10.1371/journal.ppat.1011456

Visanji NP, Marras C, Hazrati LN, Liu LW, Lang AE. 2014. Alimentary, my dear Watson? The challenges of enteric α-synuclein as a Parkinson's disease biomarker. *Mov Disord* **29:** 444–450. doi:10.1002/mds.25789

Visanji NP, Marras C, Kern DS, Al Dakheel A, Gao A, Liu LW, Lang AE, Hazrati LN. 2015. Colonic mucosal α-synuclein lacks specificity as a biomarker for Parkinson disease. *Neurology* **84:** 609–616. doi:10.1212/WNL.0000000000001240

Wade PR, Cowen T. 2004. Neurodegeneration: a key factor in the ageing gut. *Neurogastroenterol Motil* **16:** 19–23. doi:10.1111/j.1743-3150.2004.00469.x

Wakabayashi K, Takahashi H, Takeda S, Ohama E, Ikuta F. 1988. Parkinson's disease: the presence of Lewy bodies in Auerbach's and Meissner's plexuses. *Acta Neuropathol* **76:** 217–221. doi:10.1007/BF00687767

Wakabayashi K, Takahashi H, Takeda S, Ohama E, Ikuta F. 1989. Lewy bodies in the enteric nervous system in Parkinson's disease. *Arch Histol Cytol* **52:** 191–194. doi:10.1679/aohc.52.Suppl_191

Wakabayashi K, Takahashi H, Ohama E, Ikuta F. 1990. Parkinson's disease: an immunohistochemical study of Lewy body-containing neurons in the enteric nervous system. *Acta Neuropathol* **79:** 581–583. doi:10.1007/BF00294234

Wang L, Fleming SM, Chesselet MF, Taché Y. 2008. Abnormal colonic motility in mice overexpressing human wild-type α-synuclein. *Neuroreport* **19:** 873–876. doi:10.1097/WNR.0b013e3282ffda5e

Wang L, Magen I, Yuan PQ, Subramaniam SR, Richter F, Chesselet MF, Tache Y. 2012. Mice overexpressing wild-type human α-synuclein display alterations in co-lonic myenteric ganglia and defecation. *Neurogastroenterol Motil* **24:** e425–e436. doi:10.1111/j.1365-2982.2012.01974.x

Warnecke T, Schäfer KH, Claus I, Del Tredici K, Jost WH. 2022. Gastrointestinal involvement in Parkinson's disease: pathophysiology, diagnosis, and management. *NPJ Parkinsons Dis* **8:** 31. doi:10.1038/s41531-022-00295-x

Wedel T, Spiegler J, Soellner S, Roblick UJ, Schiedeck TH, Bruch HP, Krammer HJ. 2002. Enteric nerves and interstitial cells of Cajal are altered in patients with slow-transit constipation and megacolon. *Gastroenterology* **123:** 1459–1467. doi:10.1053/gast.2002.36600

Westphal CH, Chandra SS. 2013. Monomeric synucleins generate membrane curvature. *J Biol Chem* **288:** 1829–1840. doi:10.1074/jbc.M112.418871

Won JH, Byun SJ, Oh BM, Park SJ, Seo HG. 2021. Risk and mortality of aspiration pneumonia in Parkinson's disease: a nationwide database study. *Sci Rep* **11:** 6597. doi:10.1038/s41598-021-86011-w

Xie W, Gao J, Jiang R, Liu X, Lai F, Tang Y, Xiao H, Jia Y, Bai Q. 2020. Twice subacute MPTP administrations induced time-dependent dopaminergic neurodegeneration and inflammation in midbrain and ileum, as well as gut microbiota disorders in PD mice. *Neurotoxicology* **76:** 200–212. doi:10.1016/j.neuro.2019.11.009

Yang X, Qian Y, Xu S, Song Y, Xiao Q. 2018. Longitudinal analysis of fecal microbiome and pathologic processes in a rotenone induced mice model of Parkinson's disease. *Front Aging Neurosci* **9:** 441. doi:10.3389/fnagi.2017.00441

Zheng LF, Wang ZY, Li XF, Song J, Hong F, Lian H, Wang Q, Feng XY, Tang YY, Zhang Y, et al. 2011. Reduced expression of choline acetyltransferase in vagal motoneurons and gastric motor dysfunction in a 6-OHDA rat model of Parkinson's disease. *Brain Res* **1420:** 59–67. doi:10.1016/j.brainres.2011.09.006

Zheng LF, Song J, Fan RF, Chen CL, Ren QZ, Zhang XL, Feng XY, Zhang Y, Li LS, Zhu JX. 2014. The role of the vagal pathway and gastric dopamine in the gastroparesis of rats after a 6-hydroxydopamine microinjection in the substantia nigra. *Acta Physiol* **211:** 434–446. doi:10.1111/apha.12229

Zimmermann M, Zimmermann-Kogadeeva M, Wegmann R, Goodman AL. 2019. Mapping human microbiome drug metabolism by gut bacteria and their genes. *Nature* **570:** 462–467. doi:10.1038/s41586-019-1291-3

Zinnen AD, Vichich J, Metzger JM, Gambardella JC, Bondarenko V, Simmons HA, Emborg ME. 2022. α-Synuclein and tau are abundantly expressed in the ENS of the human appendix and monkey cecum. *PLoS ONE* **17:** e0269190. doi:10.1371/journal.pone.0269190

Adaptive Immunity and Parkinson's Disease

Wassim Elyaman

Division of Translational Neurobiology, Department of Neurology, the Taub Institute for Research on Alzheimer's Disease and the Aging Brain, and the Center for Motor Neuron Biology and Disease, Columbia University Irving Medical Center, New York, New York 10032, USA

Correspondence: we2152@cumc.columbia.edu

Adaptive immunity plays a key role in the pathogenesis of Parkinson's disease (PD) and related conditions. This paper reviews the involvement of CD4$^+$ and CD8$^+$ T cells in PD development as well as the effect of PD genetic susceptibility variants and aging. Specifically, the major histocompatibility complex is associated with PD, influencing antigen presentation and, consequently, the T-cell receptor repertoire, believed to contribute to disease susceptibility and progression. Moreover, aging—a major risk factor for PD—also shapes T-cell dynamics, with immunosenescence impacting the adaptive immune system, and potentially exacerbating neuroinflammatory responses in PD. These T-cell-mediated immune responses hold substantial influence over brain physiopathology, dictating the degenerative processes seen in PD. Understanding these interactions offers insights into early immunotherapy intervention during the prodromal phase using engineered regulatory T cells for antigen-specific immunomodulation against pathogenic proteins such as α-synuclein.

Parkinson's disease (PD) is a prevalent neurodegenerative disease influenced by both inflammatory and genetic factors (Bandres-Ciga et al. 2020; Tansey et al. 2022). Pathologically, PD is characterized by the loss of specific subsets of both dopaminergic and nondopaminergic neurons, the formation of α-synuclein (α-Syn)-containing inclusions called Lewy bodies, and discrete areas of neuroinflammation involving various immune cells.

The immune system comprises two main components: the innate immune system and the adaptive immune system. Although these systems are distinct, they collaborate closely to defend the body against microbial invasion and help in repairing damaged tissues (Dressman and Elyaman 2022). For this reason, while the focus of this work is on the adaptive immune system, references to the innate immune system will be included where relevant, with a more in-depth discussion in an article in this collection on innate immunity (Bradshaw E, in prep.).

In the context of PD, microglia—the innate immune cells—monitor the brain environment for signs of damage and contribute to the inflammatory component of PD. These glial cells may thus regulate the progression of neuronal cell death in PD, albeit through mechanisms that remain uncertain (Kam et al. 2020). In addition to microglia, adaptive immune T-cell lymphocytes have been observed in the substantia nigra pars compacta (SNpc) of PD patients (McGeer et al. 1988a; Brochard et al. 2009), a brain region that often bears the brunt of PD

Copyright © 2025 Cold Spring Harbor Laboratory Press; all rights reserved
Cite this article as *Cold Spring Harb Perspect Med* doi: 10.1101/cshperspect.a041639

neuropathology. The presence of T-cell infiltrates in proximity of activated microglia, which upregulate major histocompatibility complex (MHC)I and II (McGeer et al. 1988b; Imamura et al. 2003), suggests a cross talk between innate and adaptive immune responses in the neuropathogenesis of PD.

Aging, the primary risk factor in PD (Hou et al. 2019), interacts with genetic susceptibility variants and environmental factors to shape an individual's vulnerability to the disease (Pang et al. 2019). This multifaceted interplay of age, genetics, and environment modulates the immune response particularly innate and adaptive T cells, which, in turn, influence PD-related neuropathology.

These points outline the key topics that will be discussed in this article, and readers are encouraged to read it alongside the aforementioned companion article on the innate immune system.

CENTRAL AND PERIPHERAL T CELLS IN HUMAN AND EXPERIMENTAL PD

The adaptive immune response is slower than the innate, but it is specific. This adaptive T-cell response requires programmed activation and differentiation. The two T-cell subtypes, CD4$^+$ and CD8$^+$ T cells, are critical to mediating the humoral and cell-mediated arms of the adaptive immune system, respectively (Sun et al. 2023). CD4$^+$ T cells can differentiate into various subsets, including T helper (Th1), Th2, Th17, and regulatory T (Treg) cells, each with distinct functions. Th1 cells participate in the defense against intracellular pathogens, Th2 cells combat extracellular parasites, and Th17 cells are essential for mucosal immunity and inflammatory responses in autoimmune diseases. Tregs help maintain immune homeostasis by suppressing excessive immune responses and preventing autoimmunity. CD8$^+$ T cells, primarily known as cytotoxic T lymphocytes (CTLs), are critical for targeting and eliminating virus-infected and cancerous cells. The balance between effector T cells and Tregs is vital for maintaining immune homeostasis; an imbalance can lead to various diseases, such as autoimmunity if Treg function is impaired, or cancer and chronic infections if effector functions are insufficiently regulated (Jäger and Kuchroo 2010).

Genome-wide association studies (GWASs) have revealed compelling evidence of the genetic associations between PD and the immune system (Shiina et al. 2009; Hamza et al. 2010; Wissemann et al. 2013; Yu et al. 2021). Specifically, fine mapping of the human leukocyte antigen (HLA) locus in PD patients, a gene-encoding protein involved in antigen presentation, revealed four HLA haplotypes, HLA-DQA1*03:01, HLA-DQB1*03:02, HLA-DRB1*04:01, and HLA-DRB1*04:04, which are significant contributors to PD susceptibility (Yu et al. 2021). Additionally, other genes implicated in the regulation of inflammatory responses on innate immune cells including microglia, including leucine-rich repeat kinase 2 (LRRK2), and possibly bone sialostatin 1 (BST1) (Nalls et al. 2019), have been implicated in PD risk (Table 1). These findings suggest a crucial role of antigen-specific immune responses in the pathogenesis of PD.

The first report on the presence of inflammation in the brain of Parkinson's patients dates back to the nineteenth century. Indeed, in 1817, Dr. James Parkinson, an English surgeon, published a booklet entitled An Essay on the Shaking Palsy, which we now refer to as Parkinson's disease. Dr. Parkinson reported the presence of signs of inflammation in the PD brain stating "I have had occasion to observe the extraordinary energy of the inflammatory process in the brain" (Parkinson 2002). This discovery laid the early foundation for understanding the inflammatory process and its implication in the disease pathogenesis. T cells constitute the central players of the adaptive arm of the immune system maintaining continuous communication with their innate counterparts in the periphery and central compartments in PD (Sun et al. 2023). In contrast to innate immune cells, T cells are highly programmed as they are able to mount specific immune responses to self- and non-self-antigens through the unique expression of T-cell receptors (TCRs) (La Gruta et al. 2018). These reactive T cells differentiate into antigen-experienced cells that exhibit memory, retaining information about encountered antigens for quicker and more robust responses upon

Table 1. List of genes that are genetically associated with Parkinson's disease (PD) and have immune functions

Gene	Association with PD	Function in immune cells
SNCA	Associated with hereditary PD; point mutations and duplications/triplications linked to disease (Chartier-Harlin et al. 2004; Nussbaum 2018)	Involved in T-cell function and maturation; α-synuclein (α-Syn) knockout (KO) mice show alterations in T cells, including changes in thymocyte numbers, T-cell activation, cytokine production, and regulatory T (Treg) levels (Shameli et al. 2016)
		CCR2-KO prevents α-Syn-induced inflammation and neuronal degeneration in a mouse model of α-Syn overexpression in dopaminergic neurons (Harms et al. 2018)
DJ1	Mutations associated with familial PD (Bonifati et al. 2003)	DJ1-KO mice show alterations in Treg populations with higher levels of natural Tregs, impaired iTreg function, and reduced replication (Singh et al. 2015)
PINK1	Mutations linked to familial PD (Kumazawa et al. 2008)	PINK1-KO mice exhibit dysfunctional iTregs with diminished suppressive capacity (Ellis et al. 2013)
GBA	Mutations are high-risk factors for PD (Aharon-Peretz et al. 2004; Sidransky et al. 2009)	Implicated in T-cell maturation and homeostasis; GBA-KO mice exhibit impaired T-cell maturation (Liu et al. 2012)
		Lower levels of Tregs in Gaucher disease, a rare genetic disease caused by GBA mutations (Rodic et al. 2014; Sotiropoulos et al. 2015)
		Higher levels of CD8[+] T cells have also been reported in these patients (Rodic et al. 2014; Sotiropoulos et al. 2015)
LRKK2	Mutations in LRKK2 are the most common cause of familial PD (Paisán-Ruíz et al. 2004; Zimprich et al. 2004)	Proinflammatory cytokine levels are higher in a subset of asymptomatic individuals carrying the G2019S mutation (Dzamko et al. 2016)
	The G2019S mutation is the most common (Di Fonzo et al. 2005)	LRRK2 levels are increased in immune cells of patients with sporadic PD (Cook et al. 2017)
		LRRK2 mutations may sensitize neurons to IFN-γ signaling and trigger neurotoxic responses in microglia (Panagiotakopoulou et al. 2020)
		Associated with Crohn's disease (Hui et al. 2018)
VPS35	Mutations associated with familial PD (Vilariño-Güell et al. 2011)	Involved in IFN type I response and microglial function (Ren et al. 2022)
		Increase in LRRK2-mediated RAB10 phosphorylation in innate immune cells from PD patients carrying the D620N mutation (Chartier-Harlin et al. 2004)
Parkin	79 different PRKN mutations that are the most common cause of autosomal-recessive early-onset cases (Ferreira and Massano 2017)	Reduces STING-induced inflammation (Sliter et al. 2018)
		Repression of mitochondrial antigen presentation (Fahmy et al. 2019)
HLA	HLA-DRB6, HLA-DQA1	Implicated in the regulation of antigen presentation, potentially linking the immune response to genetic susceptibility in PD

Each entry includes the gene name, its primary immune function, and relevant studies or evidence linking the gene to PD. The table aims to provide an overview of the current understanding of the genetic factors involving immune responses in the context of PD.

(PINK1) PTEN-induced kinase 1, (GBA) β-glucocerebrosidase, (LRRK2) leucine-rich repeat kinase 2, (VPS35) vacuolar protein sorting 35, (iTregs) inducible regulatory CD4[+] T cells, (STING) stimulator of interferon genes, (IFN-γ) interferon-γ.

subsequent exposures (Mix and Harty 2022). CD4$^+$ and CD8$^+$ T cells include various regulatory and effector subsets characterized by the expression of master transcription factors, which contribute to immune responses in a specific manner (Dressman and Elyaman 2022).

Early studies reported infiltration of CD4$^+$ and CD8$^+$ T cells in the SNpc of postmortem PD brains as shown using conventional immunohistochemistry techniques (McGeer et al. 1988b; Brochard et al. 2009; Galiano-Landeira et al. 2020). Advancements in molecular technology have enabled deep and unbiased dissection of the phenotype of infiltrating immune cells in the human brain in health and disease. Indeed, single-cell RNA sequencing (scRNA-seq) technology has revolutionized our understanding of the heterogeneity of human tissues and the specificity of the immune responses in human diseases including PD. Moreover, combining scRNA-seq with single-cell TCR sequencing (scTCR-seq) provides invaluable insights into the clonality of infiltrating T cells and their specificity to self- and non-self-antigens that are presented by the MHC (Han et al. 2014). Recently, paired scRNA-seq and scTCR-seq on blood and cerebrospinal fluid (CSF) immune cells from PD patients revealed clonally expanded cytotoxic CD4$^+$ and CD8$^+$ T cells (Wang et al. 2021). Another study reported elevated clonal expansion of numerous TCRs in the CSF in patients with Alzheimer's disease (AD) and PD (Gate et al. 2020). More recently, in attempt to understand the role of T cells in the surveillance of the central nervous system (CNS) in neurodegenerative diseases, my group analyzed the phenotypes and clonotypes of adaptive and innate immune cells in the leptomeninges of PD and other neurodegenerative diseases. Our analysis revealed massive clonal expansion of cytotoxic CD8$^+$ T cells in PD compared to other neurodegenerative disease cases suggesting that T cells play an active role in the (Hobson et al. 2023). These findings suggest that T cells might play a role in initiating or perpetuating neuroinflammation, which is implicated in the progression of PD.

It is noteworthy that neuropathological studies revealed T-cell infiltration into the CNS of PD patients suggests significant involvement of T cells in the pathogenesis of the disease. Indeed, in the earliest stage of the disease, marked by α-Syn aggregates confined to the olfactory bulb, cytotoxic CD8$^+$ T cells were detected in the SNpc, even before any α-Syn aggregation or dopaminergic neuronal death occurred (Galiano-Landeira et al. 2020). In a more advanced stage, where α-Syn aggregates were found in the SNpc and neuronal loss had begun, CD8$^+$ T-cell infiltration was notably milder. These findings suggest that T-cell infiltration in the brain precedes neurodegeneration, indicating a potential functional involvement of CD8$^+$ T cells in triggering neuronal death and synucleinopathy.

In the periphery, several studies have highlighted changes in the balance between effector and regulatory T cells in PD patients. One notable finding is the increase in memory CD4$^+$ CD45RO$^+$ T cells and the reduction in circulating naive CD4$^+$ T cells, as evidenced by multiple studies and meta-analyses (Saunders et al. 2012; Jiang et al. 2017). The decrease in naive CD4$^+$ T cells correlated with a development of a Th1-biased phenotype characterized by the production of interferon γ (IFN-γ) and tumor necrosis factor α (Kustrimovic et al. 2018). However, another study reported an increase in peripheral proinflammatory Th1 and Th17 cells and a decrease in Th2 and Tregs in PD patients (Chen et al. 2015). This finding disagreed with another study where increased Treg suppressive activity in the blood of PD patients was described (Rosenkranz et al. 2007). Interestingly, this observation was linked to dopamine, which is deficient in patients with PD, and lowers Treg suppressive function. Given that dopamine receptors are expressed on T-cell subsets and that correlates with disease severity in patients (Kustrimovic et al. 2016; Elgueta et al. 2019), there has been an interest to study a possible link between dopamine levels in PD patients and T-cell phenotype and function. However, functional studies using blood samples from patients treated with dopamine replacement medication did not show any difference between treated patients and controls (Rosenkranz et al. 2007; Kustrimovic et al. 2018), suggesting that dopamine deficiency does not alter directly T-cell phenotype or function.

Cite this article as *Cold Spring Harb Perspect Med* doi: 10.1101/cshperspect.a041639

While there is no perfect mouse model that fully recapitulates human neurodegenerative diseases, experimental models provide invaluable data for understanding the functional role of immune cells in these conditions. The direct evidence of T-cell involvement in PD pathogenesis was initially demonstrated in the 1-methyl-4-phenyl-1,2,3,6-tetrahydropyridine (MPTP) mouse model of PD where T cells were genetically deleted that resulted in an attenuation of MPTP-induced dopaminergic cell death (Brochard et al. 2009). This study further demonstrated that T-cell-mediated dopaminergic toxicity is almost exclusively $CD4^+$ T-cell-dependent and does not require $CD8^+$ T cells (Brochard et al. 2009). Other studies in PD mouse models reported the infiltration of $CD4^+$ T cells into the brain that contribute to neurodegeneration that require MHC class II expression on microglia (Harms et al. 2013). These subsets exhibit complex interactions, engaging with antigens presented by innate immune cells and CNS-resident T cells, shaping the immune milieu within the PD-afflicted brain (Brochard et al. 2009; Harms et al. 2013; Cebrián et al. 2014; Liu et al. 2017). Additional in vivo studies showed that α-Syn-specific T cells contribute to neurodegeneration in mouse models of PD (Karikari et al. 2022) and in PD dementia (Gate et al. 2021).

ANTIGEN-SPECIFICITY OF T-CELL RESPONSES IN PD

GWAS have identified >90 risk variants that are associated with PD with several of the candidate genes influencing the immune system including the HLA region (Nalls et al. 2019). HLA is one of the most polymorphic regions in the human genome and presents a complex combination of alleles in high linkage disequilibrium (Alter et al. 2017). HLA class I and class II genes encode MHCI and MHCII molecules that present antigens to $CD8^+$ and $CD4^+$ T lymphocytes, respectively, and thereby regulate adaptive immune responses. A combined GWAS of PD with type 1 diabetes, Crohn's disease, ulcerative colitis, rheumatoid arthritis, celiac disease, psoriasis, and multiple sclerosis (MS) identified 17 loci shared between PD and these autoimmune

disorders (Witoelar et al. 2017). This provides compelling evidence of the involvement of antigen presentation in disease pathophysiology and possibly self-reactive T-cell responses in PD. One proposed mechanism is molecular mimicry, where antigens from infectious agents or environmental factors resemble self-antigens, leading to cross-reactive immune responses. In PD, misfolded proteins like α-Syn, which aggregate in neurons, could trigger such autoreactive T-cell responses. Autoantibodies generated against these proteins might also inadvertently target healthy neurons, contributing to neurodegeneration. Indeed, autoreactive T cells have been detected in the blood circulation of PD patients and they are mainly $CD4^+$ T cells and produce both IFN-γ and interleukin 4 (IL-4) (Sulzer et al. 2017; Lindestam Arlehamn et al. 2020). These cells expand before the diagnosis of motor PD and peak shortly after diagnosis and then decline (Lindestam Arlehamn et al. 2020). The temporal dynamics of T-cell expansion in PD highlight the critical role of the immune system and provide valuable information for diagnosis, understanding disease mechanisms, and developing targeted treatments.

The observed early dynamics of T-cell expansion in PD are linked to the concept of prodromal PD, which refers to the stage where early symptoms and biomarkers of the disease are present before the full onset of motor symptoms. In prodromal PD, patients may experience nonmotor symptoms such as sleep disturbances, constipation, and hyposmia (reduced sense of smell) that precede the classical motor manifestations (Berg et al. 2021). Recent research has highlighted the involvement of the gut in the pathogenesis of PD, suggesting that the disease may originate in the gastrointestinal tract before spreading to the brain (Warnecke et al. 2022). This gut–brain axis hypothesis is supported by findings of abnormal α-Syn protein accumulation in the gut, which may trigger an immune response involving α-Syn-reactive T cells (Allen Reish and Standaert 2015). These T cells can cross the blood–brain barrier, contributing to neuroinflammation and subsequent neurodegeneration in the brain, thereby linking peripheral and central mechanisms in the development of PD. It is suggested that gut's connection to the immune system, particularly T cells, forms a critical aspect of the prodromal phase of PD (Campos-Acuña et al. 2019). Understanding how gut alterations impact T-cell responses and their subsequent influence on systemic and CNS immune function offers a compelling perspective on the early phases of PD and potential immunomodulatory interventions targeting T-cell-mediated responses in this phase of the disease. In a recent study, and based on the findings that α-Syn32-46 antigenic epitopes displayed by HLA-DRB1*15:01 and drive helper and cytotoxic T-cell responses in PD patients (Sulzer et al. 2017), the authors developed a mouse model to study the effects of α-Syn32-46 immunization in mice expressing human HLA-DRB1*15:01. The authors found that immunization with α-Syn32-46 induced intestinal inflammation, leading to the loss of enteric neurons, damaged enteric dopaminergic neurons, constipation, and weight loss. This immunization activated immune responses in the gut and altered the CD4$^+$ Th1/Th17 transcriptome, resembling tissue-resident memory cells found in inflamed mucosal barriers. These CD4$^+$ T cells are pathogenic since depleting CD4$^+$, but not CD8$^+$, T cells partially mitigated enteric neurodegeneration (Garretti et al. 2023). The identification of CD4$^+$ T cells as pathogenic highlights potential therapeutic strategies focused on modulating these immune cells to prevent or treat neurodegeneration associated with gut inflammation. Additionally, the TCR repertoire and the transcriptomic changes in CD4$^+$ T cells could serve as biomarkers for diagnosing and monitoring the progression of gut-related Parkinsonism.

AGING AND IMMUNOSENESCENCE IN PD

Aging stands as one of the most significant risk factors for neurodegenerative conditions, notably PD (Hou et al. 2019). The complex processes of aging bring about a myriad of changes in cellular and molecular mechanisms, contributing to increased vulnerability to neurological disorders. Key aspects of immunosenescence include a decrease in the production of new naive T cells due to thymic involution, an accumulation of memory T cells, and a reduced diversity of the TCR repertoire (Thomas et al. 2020). Additionally, the function of other immune cells, such as B cells and natural killer (NK) cells, is impaired. This age-related immune decline is further worsened by chronic inflammation, known as "inflammaging," contributing to the overall decrease in immune competence seen in elderly populations (Liu et al. 2023).

In the context of PD, advancing age correlates with a decline in physiological resilience and an upsurge in molecular dysregulations, fostering an environment conducive to neurodegeneration. Age-related alterations impact various facets crucial in PD pathogenesis, including mitochondrial dysfunction, oxidative stress, protein aggregation, and compromised immune responses (Coleman and Martin 2022). Moreover, aging disrupts the balance of cellular repair and degradation mechanisms, amplifying the susceptibility to aberrant protein accumulation, such as α-Syn, hallmarking PD pathology (Ho et al. 2020). This amalgamation of age-associated changes primes the brain for neurodegenerative processes, underscoring the pivotal role of aging as a predominant risk factor in the onset and progression of PD.

While aging is a key factor in the development of PD, it also profoundly impacts the immune system. Immunosenescence, the gradual deterioration of immune function, impacts the adaptive arm of immunity, particularly T-cell-mediated responses (Lee et al. 2022). In this process, alterations in T-cell subsets, including decreased diversity, functional impairment, and dysregulation of Tregs, have been observed. These changes in the adaptive immune system with age not only influence general immune response (Cabezudo et al. 2023) but also potentially contribute to the neuroinflammatory milieu associated with PD pathogenesis, underscoring the relationship between aging, the immune system, and the development of PD. This process reduces the capacity to respond to infections and develops long-term immune memory, increasing susceptibility to infections, cancer, and autoimmune diseases in older individuals. Indeed, not only has the role of infection become plausible due to Braak's hypothesis, but cases of PD and parkinsonism have been described following virus infections, including the more recent severe acute respiratory syndrome coronavirus 2 (SARS-CoV-2) (Iacono et al. 2023). One compelling hypothesis to explain this link is molecular mimicry, where non-self-antigens from microbial proteins, such as those from viruses and bacteria, closely resemble brain antigens. This resemblance can trigger an autoimmune response, as the immune system may mistakenly target brain-resident cells while attacking microbial antigens. In PD, this autoimmune response could lead to the destruction of dopaminergic neurons in the substantia nigra, contributing to disease progression. Understanding the mechanisms behind molecular mimicry and its role in PD may open new avenues for therapeutic interventions aimed at modulating the immune response to prevent or halt neurodegeneration.

The impact of aging on PD intertwines deeply with the effects of human genetics, orchestrating a complex interplay that shapes both innate and adaptive immune responses in PD pathogenesis (Pang et al. 2019). Human genetic variants linked to PD influence immune regulation, exacerbating the age-driven changes in immune function. These genetic factors modulate the activation and functionality of innate immune cells, like microglia, affecting their ability to clear aberrant protein aggregates and regulate neuroinflammation. Simultaneously, aging exerts its influence on adaptive immunity, leading to alterations in T-cell responses, including dysregulation of CD4$^+$ and CD8$^+$ T-cell functions. The amalgamation of genetic predisposition and age-related immune alterations converges in PD to create a milieu that compromises immune surveillance, perpetuating neuroinflammation and contributing to neuronal damage. This complex interplay between aging and genetic variants in shaping immune responses elucidates the multifaceted nature of immune dysregulation in PD. In aging individuals with specific PD-associated genetic variants, there is an increased likelihood of dysfunctional T-cell responses. For example, the presence of the LRRK2 mutation, commonly found in PD patients, can exacerbate age-related T-cell dysfunction. This mutation can lead to hyperactive

T cells, further driving neuroinflammatory processes (Wallings et al. 2020).

EMERGING IMMUNOTHERAPY APPROACHES

Mechanisms of dopaminergic cell loss in PD are not fully understood; however, α-Syn plays a toxic role in the neurodegenerative process in PD (Negi et al. 2024). There are no approved disease-modifying therapies for PD. Immunotherapy has emerged as a promising approach for treating neurodegenerative diseases, aiming to harness the body's immune system to combat the pathological processes underlying these conditions. In the context of neurodegenerative diseases, such as Alzheimer's, Parkinson's, and amyotrophic lateral sclerosis (ALS), immunotherapy seeks to target and modulate the immune response to reduce inflammation, clear toxic proteins, and protect neuronal function. Specifically for PD, immunotherapy holds the potential to address the accumulation of α-Syn by promoting its clearance and mitigating the resulting neuroinflammation, ultimately slowing disease progression and improving patient outcomes. An intriguing question arises: Could PD have an autoimmune component? The response of T cells to α-Syn and the presence of inflammation in PD suggest that autoimmune mechanisms may play a role in the disease's progression. This possibility opens new avenues for research and treatment, highlighting the importance of immune modulation in managing PD. It is noteworthy that recent studies have explored the repurposing of immune-based drugs originally developed for MS to potentially protect against PD. For instance, dimethyl fumarate (DMF), known for its nuclear factor erythroid 2–related factor 2 (NRF2)-activating properties in MS treatment, has shown promise in preclinical models of PD. Daily oral administration of DMF in animal models of α-synucleinopathy significantly protected nigral dopaminergic neurons from toxicity induced by α-Syn. This protection was accompanied by reductions in astrocytosis and microgliosis, key markers of neuroinflammation. Importantly, these effects were absent in Nrf2-knockout mice, underscoring the role of NRF2 in mediating DMF's neuroprotective mechanisms (Lastres-Becker et al. 2016). These findings highlight the potential of immune modulators as a disease-modifying therapy for PD, leveraging its ability to enhance endogenous brain defense mechanisms and mitigate α-Syn-induced neurotoxicity and inflammation.

Studies investigating the TCR repertoire in PD patients have revealed alterations characterized by clonal expansion, indicating the proliferation of specific T-cell populations (Gate et al. 2021; Hobson et al. 2023). This clonal expansion suggests a selective and focused T-cell response toward certain antigens, potentially including those related to PD pathology, such as α-Syn. Understanding the nuances of TCR repertoire changes and clonal expansion in PD contributes

to unraveling the specificity and nature of T-cell responses implicated in the disease. These findings hold promise in identifying specific antigen targets and shaping potential immunotherapeutic strategies aimed at modulating these T-cell responses for therapeutic benefit in PD. Indeed, emerging immune-based therapies for PD are showing promise, particularly those that target α-Syn. Traditional approaches have focused on passive and active immunotherapies, such as monoclonal antibodies and vaccines, which have shown potential in preclinical models and early clinical trials. However, recent phase 2 clinical trials with monoclonal antibodies against α-Syn did not demonstrate efficacy (Lang et al. 2022; Pagano et al. 2022), emphasizing the need for refined trial designs and robust biomarkers to measure target engagement and clinical out-

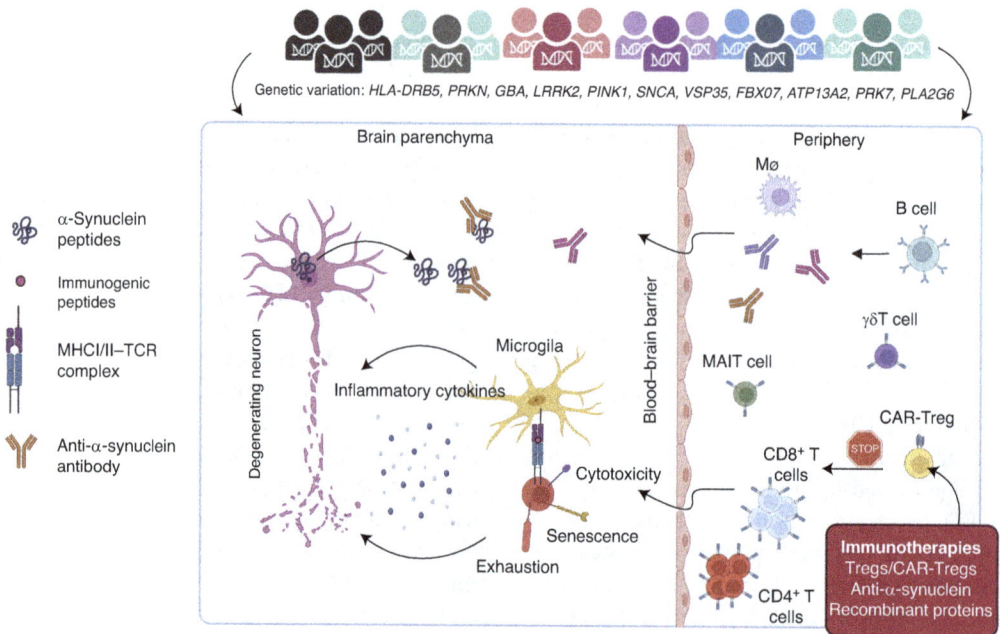

Figure 1. Schematic representation of adaptive T-cell responses in Parkinson's disease (PD) pathogenesis. CD4[+] and CD8[+] T cells are depicted interacting with antigens presented by innate immune cells and central nervous system (CNS)-resident T cells, specifically α-synuclein. This figure shows the influence of several factors, such as population genetics, aging, or senescence, on the regulation of this immune response. The consequences of these regulatory mechanisms on neuronal degeneration are highlighted, emphasizing the interplay between adaptive T-cell responses and the pathogenesis of PD. Immunotherapy strategies are advancing tools to enhance regulatory T-cell (Treg) expansion or use chimeric antigen receptor (CAR) Tregs, aiming to restore immune balance and mitigate the harmful immune responses in PD. (TCR) T-cell receptor. (Figure generated with BioRender; https://biorender.com.)

Cite this article as *Cold Spring Harb Perspect Med* doi: 10.1101/cshperspect.a041639

comes. Nonetheless, α-Syn remains a critical target due to its role in PD pathology and its genetic associated with PD. T cells have been shown to respond to α-Syn (Sulzer et al. 2017; Lindestam Arleham et al. 2020), indicating its potential for immune-based therapeutic strategies.

One novel approach in this field is the use of T-cell-based therapies. These therapies can be designed to be antigen-specific, targeting α-Syn and potentially other self- and non-self-antigens that contribute to PD. For example, TCR-engineered regulatory T cells (TCR-Tregs) and chimeric antigen receptor regulatory T cells (CAR-Tregs) can be used to modulate the immune response selectively. TCR-Tregs are engineered to express a TCR specific to an antigen (Aharon-Peretz et al. 2004), such as α-Syn, allowing them to precisely target and regulate the immune response against this protein. This specificity helps in reducing inflammation and potentially halting the progression of PD. CAR-Tregs, on the other hand, are modified to express chimeric antigen receptors that can recognize specific antigens such as β-amyloid in AD (Saetzler et al. 2023) and downregulate unwanted inflammation. In the context of PD, CAR-Tregs can be engineered to target α-Syn aggregates in the brain, reducing neuroinflammation and possibly slowing disease progression (Fig. 1). These T-cell-based therapies hold the potential to not only target α-Syn but also address other antigens involved in PD, providing a multifaceted approach to treatment.

Early intervention, particularly in the prodromal phase of PD, holds significant promise, with the success of clinical trials often hinging on the timing of the intervention. Intervening during this phase could prevent or delay the progression of PD. Immune-based therapies targeting α-Syn in the prodromal phase could mitigate early α-Syn aggregation and its associated neuroinflammation, preserving neuronal function and delaying disease progression.

In summary, immune-based therapies for PD, particularly those targeting α-Syn, are evolving with the development of antigen-specific T-cell therapies like TCR-Tregs and CAR-Tregs. These therapies, which are discussed in greater detail in the aforementioned article on innate immunity, offer the potential to modulate the immune response precisely, reduce neuroinflammation, and address the underlying pathology of PD. The potential for early intervention in the prodromal phase further enhances the promise of these innovative treatments. As the field progresses, the refinement of biomarkers and clinical trial designs will be crucial in realizing the full potential of these therapies.

CONCLUDING REMARKS

This paper describes aspects of the altered adaptive immune response observed in PD, underscoring the role of T cells in the pathogenesis of this neurodegenerative disorder. Dysregulated immune responses contribute to neuronal damage and disease progression, emphasizing the need for a deeper understanding of these mechanisms. Immune therapy emerges as a promising avenue, with potential to modulate the immune system, reduce neuroinflammation, and ultimately slow or halt disease progression. Continued human and experimental research leveraging genetics and genomics data are essential to identify specific targets of immune therapies and integrate them into early and comprehensive treatment strategies for PD.

ACKNOWLEDGMENTS

The work in the author's laboratory is supported by the Parkinson's Foundation and by U.S. National Institutes of Health grants (R01AG067581 and 5R01AG076018).

REFERENCES

Aharon-Peretz J, Rosenbaum H, Gershoni-Baruch R. 2004. Mutations in the glucocerebrosidase gene and Parkinson's disease in Ashkenazi Jews. *N Engl J Med* **351:** 1972–1977. doi:10.1056/NEJMoa033277

Allen Reish HE, Standaert DG. 2015. Role of alpha-synuclein in inducing innate and adaptive immunity in Parkinson disease. *J Parkinsons Dis* **5:** 1–19. doi:10.3233/JPD-140491

Alter I, Gragert L, Fingerson S, Maiers M, Louzoun Y. 2017. HLA class I haplotype diversity is consistent with selection for frequent existing haplotypes. *PLoS Comput Biol* **13:** e1005693. doi:10.1371/journal.pcbi.1005693

Bandres-Ciga S, Diez-Fairen M, Kim JJ, Singleton AB. 2020. Genetics of Parkinson's disease: an introspection of its journey towards precision medicine. *Neurobiol Dis* **137**: 104782. doi:10.1016/j.nbd.2020.104782

Berg D, Borghammer P, Fereshtehnejad SM, Heinzel S, Horsager J, Schaeffer E, Postuma RB. 2021. Prodromal Parkinson disease subtypes—key to understanding heterogeneity. *Nat Rev Neurol* **17**: 349–361. doi:10.1038/s41582-021-00486-9

Bonifati V, Rizzu P, van Baren MJ, Schaap O, Breedveld GJ, Krieger E, Dekker MC, Squitieri F, Ibanez P, Joosse M, et al. 2003. Mutations in the *DJ-1* gene associated with autosomal recessive early-onset parkinsonism. *Science* **299**: 256–259. doi:10.1126/science.1077209

Brochard V, Combadiere B, Prigent A, Laouar Y, Perrin A, Beray-Berthat V, Bonduelle O, Alvarez-Fischer D, Callebert J, Launay JM, et al. 2009. Infiltration of CD4⁺ lymphocytes into the brain contributes to neurodegeneration in a mouse model of Parkinson disease. *J Clin Invest* **119**: 182–192. doi:10.1172/JCI36470

Cabezudo D, Tsafaras G, Van Acker E, Van den Haute C, Baekelandt V. 2023. Mutant LRRK2 exacerbates immune response and neurodegeneration in a chronic model of experimental colitis. *Acta Neuropathol* **146**: 245–261. doi:10.1007/s00401-023-02595-9

Campos-Acuña J, Elgueta D, Pacheco R. 2019. T-cell-driven inflammation as a mediator of the gut–brain axis involved in Parkinson's disease. *Front Immunol* **10**: 239. doi:10.3389/fimmu.2019.00239

Cebrián C, Zucca FA, Mauri P, Steinbeck JA, Studer L, Scherzer CR, Kanter E, Budhu S, Mandelbaum J, Vonsattel JP, et al. 2014. MHC-I expression renders catecholaminergic neurons susceptible to T-cell-mediated degeneration. *Nat Commun* **5**: 3633. doi:10.1038/ncomms4633

Chartier-Harlin MC, Kachergus J, Roumier C, Mouroux V, Douay X, Lincoln S, Levecque C, Larvor L, Andrieux J, Hulihan M, et al. 2004. Alpha-synuclein locus duplication as a cause of familial Parkinson's disease. *Lancet* **364**: 1167–1169. doi:10.1016/S0140-6736(04)17103-1

Chen Y, Qi B, Xu W, Ma B, Li L, Chen Q, Qian W, Liu X, Qu H. 2015. Clinical correlation of peripheral CD4⁺-cell subsets, their imbalance and Parkinson's disease. *Mol Med Rep* **12**: 6105–6111. doi:10.3892/mmr.2015.4136

Coleman C, Martin I. 2022. Unraveling Parkinson's disease neurodegeneration: does aging hold the clues? *J Parkinsons Dis* **12**: 2321–2338. doi:10.3233/JPD-223363

Cook DA, Kannarkat GT, Cintron AF, Butkovich LM, Fraser KB, Chang J, Grigoryan N, Factor SA, West AB, Boss JM, et al. 2017. LRRK2 levels in immune cells are increased in Parkinson's disease. *NPJ Parkinsons Dis* **3**: 11. doi:10.1038/s41531-017-0010-8

Di Fonzo A, Rohé CF, Ferreira J, Chien HF, Vacca L, Stocchi F, Guedes L, Fabrizio E, Manfredi M, Vanacore N, et al. 2005. A frequent LRRK2 gene mutation associated with autosomal dominant Parkinson's disease. *Lancet* **365**: 412–415. doi:10.1016/S0140-6736(05)17829-5

Dressman D, Elyaman W. 2022. T cells: a growing universe of roles in neurodegenerative diseases. *Neuroscientist* **28**: 335–348. doi:10.1177/10738584211024907

Dzamko N, Rowe DB, Halliday GM. 2016. Increased peripheral inflammation in asymptomatic leucine-rich repeat kinase 2 mutation carriers. *Mov Disord* **31**: 889–897. doi:10.1002/mds.26529

Elgueta D, Contreras F, Prado C, Montoya A, Ugalde V, Chovar O, Villagra R, Henríquez C, Abellanas MA, Aymerich MS, et al. 2019. Dopamine receptor D3 expression is altered in CD4⁺ T-cells from Parkinson's disease patients and its pharmacologic inhibition attenuates the motor impairment in a mouse model. *Front Immunol* **10**: 981. doi:10.3389/fimmu.2019.00981

Ellis GI, Zhi L, Akundi R, Büeler H, Marti F. 2013. Mitochondrial and cytosolic roles of PINK1 shape induced regulatory T-cell development and function. *Eur J Immunol* **43**: 3355–3360. doi:10.1002/eji.201343571

Fahmy AM, Boulais J, Desjardins M, Matheoud D. 2019. Mitochondrial antigen presentation: a mechanism linking Parkinson's disease to autoimmunity. *Curr Opin Immunol* **58**: 31–37. doi:10.1016/j.coi.2019.02.004

Ferreira M, Massano J. 2017. An updated review of Parkinson's disease genetics and clinicopathological correlations. *Acta Neurol Scand* **135**: 273–284. doi:10.1111/ane.12616

Galiano-Landeira J, Torra A, Vila M, Bové J. 2020. CD8 T cell nigral infiltration precedes synucleinopathy in early stages of Parkinson's disease. *Brain* **143**: 3717–3733. doi:10.1093/brain/awaa269

Garretti F, Monahan C, Sloan N, Bergen J, Shahriar S, Kim SW, Sette A, Cutforth T, Kanter E, Agalliu D, et al. 2023. Interaction of an alpha-synuclein epitope with HLA-DRB1*15:01 triggers enteric features in mice reminiscent of prodromal Parkinson's disease. *Neuron* **111**: 3397–3413.e5. doi:10.1016/j.neuron.2023.07.015

Gate D, Saligrama N, Leventhal O, Yang AC, Unger MS, Middeldorp J, Chen K, Lehallier B, Channappa D, De Los Santos MB, et al. 2020. Clonally expanded CD8 T cells patrol the cerebrospinal fluid in Alzheimer's disease. *Nature* **577**: 399–404. doi:10.1038/s41586-019-1895-7

Gate D, Tapp E, Leventhal O, Shahid M, Nonninger TJ, Yang AC, Strempfl K, Unger MS, Fehlmann T, Oh H, et al. 2021. CD4⁺ T cells contribute to neurodegeneration in Lewy body dementia. *Science* **374**: 868–874. doi:10.1126/science.abf7266

Hamza TH, Zabetian CP, Tenesa A, Laederach A, Montimurro J, Yearout D, Kay DM, Doheny KF, Paschall J, Pugh E, et al. 2010. Common genetic variation in the HLA region is associated with late-onset sporadic Parkinson's disease. *Nat Genet* **42**: 781–785. doi:10.1038/ng.642

Han A, Glanville J, Hansmann L, Davis MM. 2014. Linking T-cell receptor sequence to functional phenotype at the single-cell level. *Nat Biotechnol* **32**: 684–692. doi:10.1038/nbt.2938

Harms AS, Cao S, Rowse AL, Thome AD, Li X, Mangieri LR, Cron RQ, Shacka JJ, Raman C, Standaert DG. 2013. MHCII is required for alpha-synuclein-induced activation of microglia, CD4 T cell proliferation, and dopaminergic neurodegeneration. *J Neurosci* **33**: 9592–9600. doi:10.1523/JNEUROSCI.5610-12.2013

Harms AS, Thome AD, Yan Z, Schonhoff AM, Williams GP, Li X, Liu Y, Qin H, Benveniste EN, Standaert DG. 2018. Peripheral monocyte entry is required for alpha-synuclein induced inflammation and neurodegeneration in a model of Parkinson disease. *Exp Neurol* **300**: 179–187. doi:10.1016/j.expneurol.2017.11.010

Ho PW, Leung CT, Liu H, Pang SY, Lam CS, Xian J, Li L, Kung MH, Ramsden DB, Ho SL. 2020. Age-dependent accumulation of oligomeric SNCA/alpha-synuclein from impaired degradation in mutant LRRK2 knockin mouse model of Parkinson disease: role for therapeutic activation of chaperone-mediated autophagy (CMA). *Autophagy* **16:** 347–370. doi:10.1080/15548627.2019.1603545

Hobson R, Levy SHS, Flaherty D, Xiao H, Ciener B, Reddy H, Singal C, Teich AF, Shneider NA, Bradshaw EM, et al. 2023. Clonal CD8 T cells in the leptomeninges are locally controlled and influence microglia in human neurodegeneration. bioRxiv doi:10.1101/2023.07.13.548931

Hou Y, Dan X, Babbar M, Wei Y, Hasselbalch SG, Croteau DL, Bohr VA. 2019. Ageing as a risk factor for neurodegenerative disease. *Nat Rev Neurol* **15:** 565–581. doi:10.1038/s41582-019-0244-7

Hui KY, Fernandez-Hernandez H, Hu J, Schaffner A, Pankratz N, Hsu NY, Chuang LS, Carmi S, Villaverde N, Li X, et al. 2018. Functional variants in the *LRRK2* gene confer shared effects on risk for Crohn's disease and Parkinson's disease. *Sci Transl Med* **10:** eaai7795. doi:10.1126/scitranslmed.aai7795

Iacono S, Schirò G, Davì C, Mastrilli S, Abbott M, Guajana F, Arnao V, Aridon P, Ragonese P, Gagliardo C, et al. 2023. COVID-19 and neurological disorders: what might connect Parkinson's disease to SARS-CoV-2 infection. *Front Neurol* **14:** 1172416. doi:10.3389/fneur.2023.1172416

Imamura K, Hishikawa N, Sawada M, Nagatsu T, Yoshida M, Hashizume Y. 2003. Distribution of major histocompatibility complex class II–positive microglia and cytokine profile of Parkinson's disease brains. *Acta Neuropathol* **106:** 518–526. doi:10.1007/s00401-003-0766-2

Jäger A, Kuchroo VK. 2010. Effector and regulatory T-cell subsets in autoimmunity and tissue inflammation. *Scand J Immunol* **72:** 173–184. doi:10.1111/j.1365-3083.2010.02432.x

Jiang S, Gao H, Luo Q, Wang P, Yang X. 2017. The correlation of lymphocyte subsets, natural killer cell, and Parkinson's disease: a meta-analysis. *Neurol Sci* **38:** 1373–1380. doi:10.1007/s10072-017-2988-4

Kam TI, Hinkle JT, Dawson TM, Dawson VL. 2020. Microglia and astrocyte dysfunction in Parkinson's disease. *Neurobiol Dis* **144:** 105028. doi:10.1016/j.nbd.2020.105028

Karikari AA, McFleder RL, Ribechini E, Blum R, Bruttel V, Knorr S, Gehmeyr M, Volkmann J, Brotchie JM, Ahsan F, et al. 2022. Neurodegeneration by alpha-synuclein-specific T cells in AAV-A53T-alpha-synuclein Parkinson's disease mice. *Brain Behav Immun* **101:** 194–210. doi:10.1016/j.bbi.2022.01.007

Kumazawa R, Tomiyama H, Li Y, Imamichi Y, Funayama M, Yoshino H, Yokochi F, Fukusako T, Takehisa Y, Kashihara K, et al. 2008. Mutation analysis of the PINK1 gene in 391 patients with Parkinson disease. *Arch Neurol* **65:** 802–808. doi:10.1001/archneur.65.6.802

Kustrimovic N, Rasini E, Legnaro M, Bombelli R, Aleksic I, Blandini F, Comi C, Mauri M, Minafra B, Riboldazzi G, et al. 2016. Dopaminergic receptors on CD4$^+$ T naive and memory lymphocytes correlate with motor impairment in patients with Parkinson's disease. *Sci Rep* **6:** 33738. doi:10.1038/srep33738

Kustrimovic N, Comi C, Magistrelli L, Rasini E, Legnaro M, Bombelli R, Aleksic I, Blandini F, Minafra B, Riboldazzi G, et al. 2018. Parkinson's disease patients have a complex phenotypic and functional Th1 bias: cross-sectional studies of CD4$^+$ Th1/Th2/T17 and Treg in drug-naive and drug-treated patients. *J Neuroinflammation* **15:** 205. doi:10.1186/s12974-018-1248-8

La Gruta NL, Gras S, Daley SR, Thomas PG, Rossjohn J. 2018. Understanding the drivers of MHC restriction of T cell receptors. *Nat Rev Immunol* **18:** 467–478. doi:10.1038/s41577-018-0007-5

Lang AE, Siderowf AD, Macklin EA, Poewe W, Brooks DJ, Fernandez HH, Rascol O, Giladi N, Stocchi F, Tanner CM, et al. 2022. Trial of cinpanemab in early Parkinson's disease. *N Engl J Med* **387:** 408–420. doi:10.1056/NEJMoa2203395

Lastres-Becker I, García-Yagüe AJ, Scannevin RH, Casarejos MJ, Kügler S, Rábano A, Cuadrado A. 2016. Repurposing the NRF2 activator dimethyl fumarate as therapy against synucleinopathy in Parkinson's disease. *Antioxid Redox Signal* **25:** 61–77. doi:10.1089/ars.2015.6549

Lee KA, Flores RR, Jang IH, Saathoff A, Robbins PD. 2022. Immune senescence, immunosenescence and aging. *Front Aging* **3:** 900028. doi:10.3389/fragi.2022.900028

Lindestam Arlehamn CS, Dhanwani R, Pham J, Kuan R, Frazier A, Rezende Dutra J, Phillips E, Mallal S, Roederer M, Marder KS, et al. 2020. Alpha-synuclein-specific T cell reactivity is associated with preclinical and early Parkinson's disease. *Nat Commun* **11:** 1875. doi:10.1038/s41467-020-15626-w

Liu J, Halene S, Yang M, Iqbal J, Yang R, Mehal WZ, Chuang WL, Jain D, Yuen T, Sun L, et al. 2012. Gaucher disease gene *GBA* functions in immune regulation. *Proc Natl Acad Sci* **109:** 10018–10023. doi:10.1073/pnas.1200941109

Liu Z, Huang Y, Cao BB, Qiu YH, Peng YP. 2017. Th17 cells induce dopaminergic neuronal death via LFA-1/ICAM-1 interaction in a mouse model of Parkinson's disease. *Mol Neurobiol* **54:** 7762–7776. doi:10.1007/s12035-016-0249-9

Liu Z, Liang Q, Ren Y, Guo C, Ge X, Wang L, Cheng Q, Luo P, Zhang Y, Han X. 2023. Immunosenescence: molecular mechanisms and diseases. *Signal Transduct Target Ther* **8:** 200. doi:10.1038/s41392-023-01451-2

McGeer PL, Itagaki S, Akiyama H, McGeer EG. 1988a. Rate of cell death in parkinsonism indicates active neuropathological process. *Ann Neurol* **24:** 574–576. doi:10.1002/ana.410240415

McGeer PL, Itagaki S, Boyes BE, McGeer EG. 1988b. Reactive microglia are positive for HLA-DR in the substantia nigra of Parkinson's and Alzheimer's disease brains. *Neurology* **38:** 1285–1291. doi:10.1212/WNL.38.8.1285

Mix MR, Harty JT. 2022. Keeping T cell memories in mind. *Trends Immunol* **43:** 1018–1031. doi:10.1016/j.it.2022.10.001

Nalls MA, Blauwendraat C, Vallerga CL, Heilbron K, Bandres-Ciga S, Chang D, Tan M, Kia DA, Noyce AJ, Xue A, et al. 2019. Identification of novel risk loci, causal insights, and heritable risk for Parkinson's disease: a meta-analysis of genome-wide association studies. *Lancet Neurol* **18:** 1091–1102. doi:10.1016/S1474-4422(19)30320-5

Negi S, Khurana N, Duggal N. 2024. The misfolding mystery: alpha-synuclein and the pathogenesis of Parkinson's disease. *Neurochem Int* 177: 105760. doi:10.1016/j.neuint .2024.105760

Nussbaum RL. 2018. Genetics of synucleinopathies. *Cold Spring Harb Perspect Med* 8: a024109. doi:10.1101/cshperspect.a024109

Pagano G, Taylor KI, Anzures-Cabrera J, Marchesi M, Simuni T, Marek K, Postuma RB, Pavese N, Stocchi F, Azulay JP, et al. 2022. Trial of prasinezumab in early-stage Parkinson's disease. *N Engl J Med* 387: 421–432. doi:10 .1056/NEJMoa2202867

Paisán-Ruíz C, Jain S, Evans EW, Gilks WP, Simón J, van der Brug M, de Munain L ópez A, Aparicio S, Gil AM, Khan N, et al. 2004. Cloning of the gene containing mutations that cause PARK8-linked Parkinson's disease. *Neuron* 44: 595–600. doi:10.1016/j.neuron.2004.10.023

Panagiotakopoulou V, Ivanyuk D, De Cicco S, Haq W, Arsić A, Yu C, Messelodi D, Oldrati M, Schöndorf DC, Perez MJ, et al. 2020. Interferon-gamma signaling synergizes with LRRK2 in neurons and microglia derived from human induced pluripotent stem cells. *Nat Commun* 11: 5163. doi:10.1038/s41467-020-18755-4

Pang SY, Ho PW, Liu HF, Leung CT, Li L, Chang EES, Ramsden DB, Ho SL. 2019. The interplay of aging, genetics and environmental factors in the pathogenesis of Parkinson's disease. *Transl Neurodegener* 8: 23. doi:10.1186/ s40035-019-0165-9

Parkinson J. 2002. An essay on the shaking palsy. 1817. *J Neuropsychiatry Clin Neurosci* 14: 223–236; discussion 222. doi:10.1176/jnp.14.2.223

Ren X, Yao L, Wang Y, Mei L, Xiong WC. 2022. Microglial VPS35 deficiency impairs Aβ phagocytosis and Aβ-induced disease-associated microglia, and enhances Aβ associated pathology. *J Neuroinflammation* 19: 61. doi:10 .1186/s12974-022-02422-0

Rodic P, Kraguljac Kurtovic N, Suvajdzic Vukovic N, Djordjevic M, Mitrovic M, Sumarac Z, Janic D. 2014. Flow cytometric assessment of lymphocyte subsets in Gaucher type 1 patients. *Blood Cells Mol Dis* 53: 169–170. doi:10 .1016/j.bcmd.2014.07.020

Rosenkranz D, Weyer S, Tolosa E, Gaenslen A, Berg D, Leyhe T, Gasser T, Stoltze L. 2007. Higher frequency of regulatory T cells in the elderly and increased suppressive activity in neurodegeneration. *J Neuroimmunol* 188: 117–127. doi:10.1016/j.jneuroim.2007.05.011

Saetzler V, Riet T, Schienke A, Henschel P, Freitag K, Haake A, Heppner FL, Buitrago-Molina LE, Noyan F, Jaeckel E, et al. 2023. Development of beta-amyloid-specific CAR-Tregs for the treatment of Alzheimer's disease. *Cells* 12: 2115. doi:10.3390/cells12162115

Saunders JA, Estes KA, Kosloski LM, Allen HE, Dempsey KM, Torres-Russotto DR, Meza JL, Santamaria PM, Bertoni JM, Murman DL, et al. 2012. CD4⁺ regulatory and effector/memory T cell subsets profile motor dysfunction in Parkinson's disease. *J Neuroimmune Pharmacol* 7: 927–938. doi:10.1007/s11481-012-9402-z

Shameli A, Xiao W, Zheng Y, Shyu S, Sumodi J, Meyerson HJ, Harding CV, Maitta RW. 2016. A critical role for alpha-synuclein in development and function of T lymphocytes. *Immunobiology* 221: 333–340. doi:10.1016/j .imbio.2015.10.002

Shiina T, Hosomichi K, Inoko H, Kulski JK. 2009. The HLA genomic loci map: expression, interaction, diversity and disease. *J Hum Genet* 54: 15–39. doi:10.1038/jhg .2008.5

Sidransky E, Nalls MA, Aasly JO, Aharon-Peretz J, Annesi G, Barbosa ER, Bar-Shira A, Berg D, Bras J, Brice A, et al. 2009. Multicenter analysis of glucocerebrosidase mutations in Parkinson's disease. *N Engl J Med* 361: 1651–1661. doi:10.1056/NEJMoa0901281

Singh Y, Chen H, Zhou Y, Föller M, Mak TW, Salker MS, Lang F. 2015. Differential effect of DJ-1/PARK7 on development of natural and induced regulatory T cells. *Sci Rep* 5: 17723. doi:10.1038/srep17723

Sliter DA, Martinez J, Hao L, Chen X, Sun N, Fischer TD, Burman JL, Li Y, Zhang Z, Narendra DP, et al. 2018. Parkin and PINK1 mitigate STING-induced inflammation. *Nature* 561: 258–262. doi:10.1038/s41586-018-0448-9

Sotiropoulos C, Theodorou G, Repa C, Marinakis T, Verigou E, Solomou E, Karakantza M, Symeonidis A. 2015. Severe impairment of regulatory T-cells and Th1-lymphocyte polarization in patients with Gaucher disease. *JIMD Rep* 18: 107–115. doi:10.1007/8904_2014_357

Sulzer D, Alcalay RN, Garretti F, Cote L, Kanter E, Agin-Liebes J, Liong C, McMurtrey C, Hildebrand WH, Mao X, et al. 2017. T cells from patients with Parkinson's disease recognize α-synuclein peptides. *Nature* 546: 656–661. doi:10.1038/nature22815

Sun L, Su Y, Jiao A, Wang X, Zhang B. 2023. T cells in health and disease. *Signal Transduct Target Ther* 8: 235. doi:10 .1038/s41392-023-01471-y

Tansey MG, Wallings RL, Houser MC, Herrick MK, Keating CE, Joers V. 2022. Inflammation and immune dysfunction in Parkinson disease. *Nat Rev Immunol* 22: 657–673. doi:10.1038/s41577-022-00684-6

Thomas R, Wang W, Su DM. 2020. Contributions of age-related thymic involution to immunosenescence and inflammaging. *Immun Ageing* 17: 2. doi:10.1186/s12979-020-0173-8

Vilariño-Güell C, Wider C, Ross OA, Dachsel JC, Kachergus JM, Lincoln SJ, Soto-Ortolaza AI, Cobb SA, Wilhoite GJ, Bacon JA, et al. 2011. VPS35 mutations in Parkinson disease. *Am J Hum Genet* 89: 162–167. doi:10.1016/j .ajhg.2011.06.001

Wallings RL, Herrick MK, Tansey MG. 2020. LRRK2 at the interface between peripheral and central immune function in Parkinson's. *Front Neurosci* 14: 443. doi:10.3389/ fnins.2020.00443

Wang P, Yao L, Luo M, Zhou W, Jin X, Xu Z, Yan S, Li Y, Xu C, Cheng R, et al. 2021. Single-cell transcriptome and TCR profiling reveal activated and expanded T cell populations in Parkinson's disease. *Cell Discov* 7: 52. doi:10 .1038/s41421-021-00280-3

Warnecke T, Schäfer KH, Claus I, Del Tredici K, Jost WH. 2022. Gastrointestinal involvement in Parkinson's disease: pathophysiology, diagnosis, and management. *NPJ Parkinsons Dis* 8: 31. doi:10.1038/s41531-022-00295-x

Wissemann WT, Hill-Burns EM, Zabetian CP, Factor SA, Patsopoulos N, Hoglund B, Holcomb C, Donahue RJ, Thomson G, Erlich H, et al. 2013. Association of Parkinson disease with structural and regulatory variants in the

HLA region. *Am J Hum Genet* **93:** 984–993. doi:10.1016/j .ajhg.2013.10.009

Witoelar A, Jansen IE, Wang Y, Desikan RS, Gibbs JR, Blauwendraat C, Thompson WK, Hernandez DG, Djurovic S, Schork AJ, et al. 2017. Genome-wide pleiotropy between Parkinson disease and autoimmune diseases. *JAMA Neurol* **74:** 780–792. doi:10.1001/jamaneurol.2017.0469

Yu E, Ambati A, Andersen MS, Krohn L, Estiar MA, Saini P, Senkevich K, Sosero YL, Sreelatha AAK, Ruskey JA, et al.

2021. Fine mapping of the HLA locus in Parkinson's disease in Europeans. *NPJ Parkinsons Dis* **7:** 84. doi:10 .1038/s41531-021-00231-5

Zimprich A, Biskup S, Leitner P, Lichtner P, Farrer M, Lincoln S, Kachergus J, Hulihan M, Uitti RJ, Calne DB, et al. 2004. Mutations in LRRK2 cause autosomal-dominant parkinsonism with pleomorphic pathology. *Neuron* **44:** 601–607. doi:10.1016/j.neuron .2004.11.005

Innate Immunity and Parkinson's Disease

Davina B. Oludipe, Xiaoqing Du, Samia Akter, Chen Zhang, R. Lee Mosley, Howard E. Gendelman, and Susmita Sil

Department of Pharmacology and Experimental Neuroscience, College of Medicine, University of Nebraska Medical Center, Omaha, Nebraska 68198, USA

Correspondence: hegendel@unmc.edu

Parkinson's disease (PD) is a progressive neurodegenerative disorder characterized by the loss of both dopaminergic and non-dopaminergic neurons associated with the accumulation of α-synuclein aggregates and signs of neuroinflammation. This inflammatory aspect of PD neuropathology has led to the hypothesis that the immune system, both adaptive and innate, contributes to the neurodegenerative process. While the adaptive immune system is discussed in detail in another article in this collection, this review focuses on the innate immune system, which includes monocytes, macrophages, microglia, and dendritic cells. We will also discuss the increasingly recognized link between genetic and immune response and the cross talk between peripheral and central immune cells, and its contribution to the overall immune response in PD. Finally, we will propose therapeutic strategies aimed at modulating immunity for neuroprotective and disease-modifying benefits in PD and related disorders.

Innate immunity is the primary and immediate defense mechanism against invading pathogens and harmful foreign substances. Unlike adaptive immunity, which is reviewed in detail elsewhere and requires days to mount a specific response, innate immunity can be activated within hours. This rapid response is critical for providing initial protection, allowing time for the adaptive immune system to develop a more targeted and sustained response (Medzhitov and Janeway 2000; Janeway et al. 2001). Innate immunity relies on a limited but highly conserved set of pattern recognition receptors (PRRs), such as Toll-like receptors (TLRs) expressed on innate immune cells that can detect pathogen-associated molecular patterns (PAMPs) and damage-associated molecular patterns (DAMPs) (West et al. 2006; Beutler 2009). Despite their limited specificity, these receptors enable the innate immune system to recognize a broad range of pathogens, including bacteria, viruses, and fungi. Thus, a generalized and rapid response is essential, as this first response often may be the decisive barrier to systemic infection, especially in the context in which adaptive immunity is compromised or delayed (Turvey and Broide 2010).

Various investigations have highlighted the potential role of innate immune processes not only in the aforementioned initial immune responses but also in the pathogenesis of neurodegenerative diseases (Heneka et al. 2014; Labzin et al. 2018). In neurodegenerative disorders such as Parkinson's disease (PD), Alzheimer's

Copyright © 2025 Cold Spring Harbor Laboratory Press; all rights reserved
Cite this article as *Cold Spring Harb Perspect Med* doi: 10.1101/cshperspect.a041640

disease (AD), and amyotrophic lateral sclerosis, the innate immune response in the central nervous system (CNS) was reported to play a pivotal role in the accumulation of misfolded proteins and cellular debris (Perry et al. 2010; Zhang et al. 2021; Schwartz and Cahalon 2022). Although adaptive immunity is traditionally associated with high specificity, the fixed specificity of the innate immune system allows for rapid responses to changes in homeostasis and pathological insults to the brain, which are potentially critical in the early stages of neurodegeneration (Choi et al. 2022; Rajesh and Kanneganti 2022). A well-developed network of microglia is positioned throughout the CNS, and under homeostatic conditions, they exhibit a morphology with a small soma and long ramified processes, which provide a large surrounding territory for dynamic surveillance and direct interaction with neurons, astrocytes, and blood vessels (Fig 1; Petry et al. 2023). Under these conditions, microglia express low levels of activation markers such as major histocompatibility complex II (MHCII), cluster of differentiation (CD)45 (CD45), and CD68. The movement of microglia and processes are primarily purinergic-directed responses toward sites of injury and ATP/ADP gradients that are detected by the microglial purine receptor P2Y G-protein 12 (P2RY12) receptors. With prolonged exposure to proinflammatory stimuli such as bacteria, lipopolysaccharide (LPS), and interferon γ (IFN-γ)-like IFNs, microglia become more ameboid with retracted processes indicating a heightened activated state with increased expression of activation markers. In the presence of efficient stimuli, microglia can rapidly transition from a homeostatic state to an activated state. However, under continuous exposure to stimulating elements, such as posttranslationally modified proteins (e.g., nitrated, phosphorylated, or aggregated α-synuclein), microglia can transition to a heightened state of activation, secreting cytotoxic factors like tumor necrosis factor α (TNF-α). This leads to a chronic neuroinflammation environment, which is speculated to contribute to the initiation and even progression of neurodegenerative diseases, such as PD and related conditions.

Recent research has increasingly focused on the interactions between the peripheral immune system and CNS to better understand how these systems interact to affect the progression of neurodegenerative diseases (Fakhoury 2015; Stephenson et al. 2018; Berriat et al. 2023). Peripheral innate immune cells such as monocytes and macrophages infiltrate the CNS in response to signals from damaged neurons and activated glial cells (Weiss et al. 2022). This infiltration further amplifies CNS inflammation, promoting hyperactivation of microglia, which in turn leads to the production of proinflammatory cytokines and neurotoxic factors, potentially exacerbating neuronal damage and attracting more peripheral immune cells to migrate across the blood–brain barrier (BBB). This cycle of peripheral and CNS innate immune interactions has been proposed as a mechanism that leads to the characteristic loss of vulnerable neurons in PD such as dopaminergic neurons within the substantia nigra and their respective striatal termini (Roodveldt et al. 2024). In addition, other peripheral innate immune cells, such as dendritic cells (DCs), natural killer (NK) cells, and mast cells, also infiltrate the CNS compartment in PD models and are discussed in further detail. Thus, understanding the cross talk between peripheral and CNS innate immunity may prove crucial for developing therapeutic strategies that target both compartments to slow or halt PD progression. Research indicates a significant link between the immune system and PD genetics, with several genes associated with PD risk directly affecting the immune function. This suggests that genetic variations in immune-related genes can contribute to the development of disease through mechanisms such as neuroinflammation and altered microglial activity.

In this work, we will discuss the connection between genetic factors and immune responses in PD, emphasizing how genetic variations may influence susceptibility to neuroinflammation and disease pathogenesis. We will review the composition of the innate immune system, including an overview of its various cellular constituents. Additionally, we will examine the complex cross talk between peripheral and central immune cells and its role in shaping the overall immune response in PD. Finally, we will propose therapeutic strategies aimed at modulating immune responses to achieve neuroprotective and disease-modifying benefits in PD and related disorders.

GENETIC AND INNATE IMMUNE SYSTEM

Since the discovery of the α-synuclein mutations causing a familial form of PD, the genetics of this specific disease has made enormous strides, and this exciting topic is discussed in detail in Westenberger et al. (2024). However, here we will revisit specific aspects of PD genetics, focusing on how certain genetic traits may modulate immune responses and contribute to the predisposition of carriers of specific genetic variants to neurodegenerative disorders such as PD, with particular emphasis on their connection to the innate immune system.

Genome-wide association studies have identified mutations in genes like human leukocyte antigen (HLA) and bone marrow stromal cell antigen 1 (BST1), which are crucial components of the immune system, as potential risk factors for PD (Schulte and Gasser 2011). Genetic variations influence microglial function, leading to abnormal activation and neuroinflammation. Approximately 25% of the PD risk is attributed to genetic variations (Goldman et al. 2019; Nalls et al. 2019) that are implicated in immune system dysfunction and chronic neuroinflammation. For instance, pathogenic leucine-rich repeat kinase 2 (LRRK2) mutations (e.g., G2019S, R1441C/G/H, and Y1699C) in LRRK2 p.G2019S transgenic mice are associated with increased kinase activity and disruption of microglial motility and phagocytosis, which, in turn, affects intercellular communication and downstream cellular responses (Bailey and Cookson 2024). Microglia phagocytose misfolded proteins, cellular debris, and dying cells to maintain homeostasis in the CNS (Gao et al. 2023). In PD, α-synuclein aggregation triggers microglial proliferation and activation, involving immune receptors such as TLR2 in rat and mice models, Fcγ receptor IIB (FcγRIIB or CD32B), and CD36, as well as intracellular pathways such as nuclear factor-κB (NF-κB) and nucleotide-binding oligomerization domain-like receptor family pyrin domain containing 3 (NLRP3) inflammasome (Kam et al. 2020).

Extracellular α-synuclein oligomers, acting as DAMPs, bind to TLR2 on microglia in mice, initiating NF-κB signaling and promoting the release of proinflammatory cytokines, including interleukin 6 (IL-6), IL-1β, and TNF-α (Cui et al. 2024).

Additionally, *SNCA*-encoded α-synuclein interacts with FcγRIIB in A53T α-synuclein transgenic mice, triggering the activation of the enzyme Src homology 2 domain–containing phosphatase-1 (SHP-1), which inhibits microglial phagocytosis and reduces the clearance of α-synuclein aggregates (Choi et al. 2015; Zhang et al. 2023). The uptake of α-synuclein fibrils by microglia in A53T and viral α-synuclein mouse models is regulated by Fyn kinase and CD36, which leads to NLRP3 inflammasome activation and subsequent IL-1β release (Panicker et al. 2019; Kam et al. 2020). These processes have been shown to contribute to the relationship between α-synuclein pathology and chronic inflammation (Calabresi et al. 2023).

Vacuolar protein sorting 35 (VPS35), whose mutations have also been linked to a familial form of PD, is a critical component of the retromer complex that plays a role in endosomal sorting (Sargent and Moore 2021). Dysfunctional VPS35 in animal models disrupts autophagy-lysosomal pathways in dopaminergic neurons, cortical neurons, and microglia, leading to α-synuclein accumulation and impaired mitochondrial dynamics through interactions with dynamin-like protein 1 and mitochondrial ubiquitin ligase 1 (Sargent and Moore 2021). Although the understanding of the actual role of DJ-1 (gene *PARK7*) remains limited, its deficiency results in diminished microglial activation and attenuated IFN pathway signaling in DJ-1 knockout mice compared to wild-type, suggesting a role for DJ-1 in chronic neuroinflammation (Lind-Holm Mogensen et al. 2024).

RESPONSIBLE CELLS

As previously mentioned, the innate immune system comprises a variety of cells in both the CNS and the periphery. It is important to detail these cells to provide a better understanding of the innate immune landscape and through

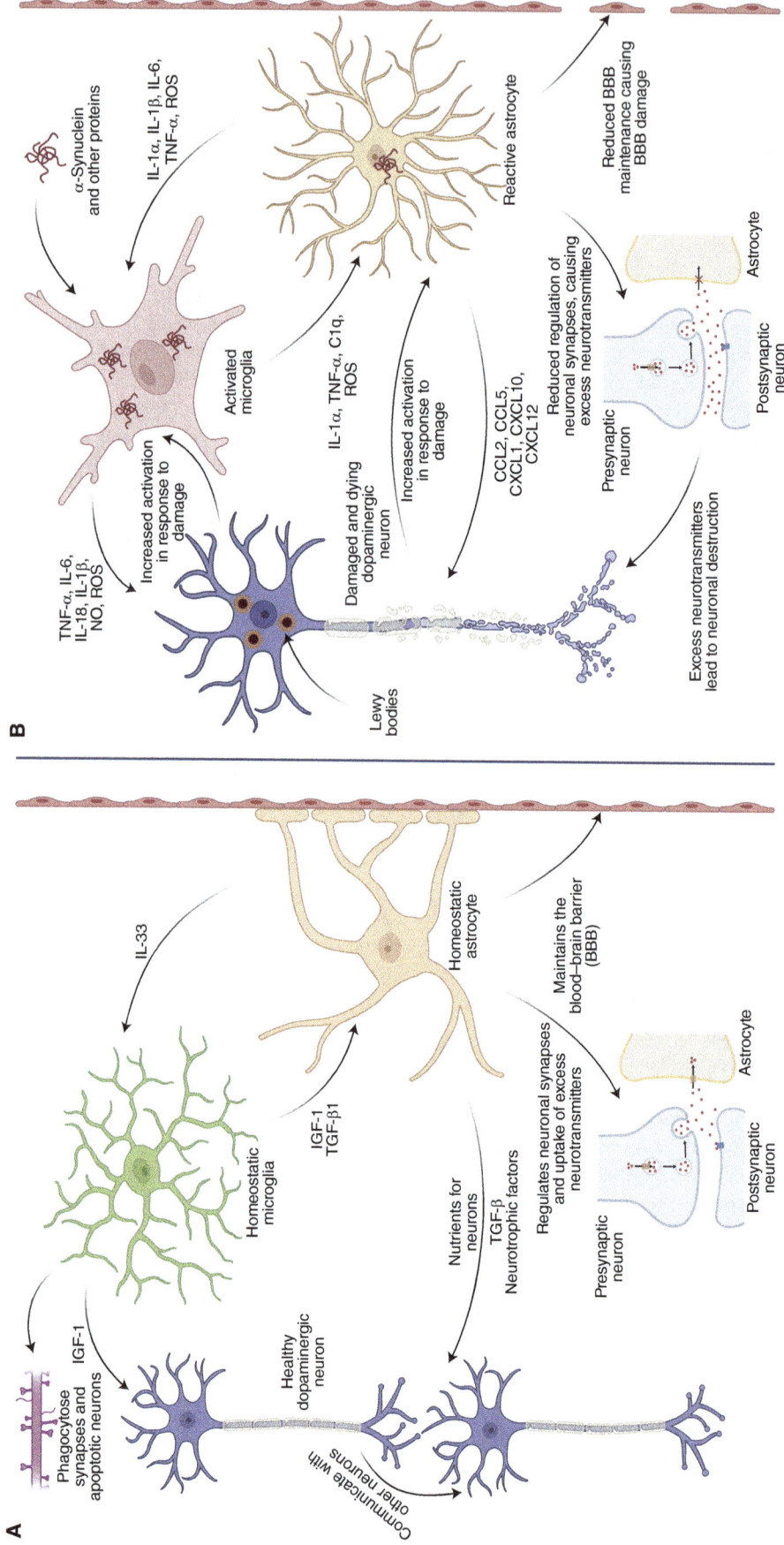

Figure 1. (*See following page for legend.*)

which cells and mechanisms it may contribute to neurodegeneration.

Microglia

Microglia represent the resident innate immune cells of the brain, comprising ~5%–20% of brain glial cells, and play a crucial role in the first-line defense of the brain (Chabot and Yong 2002; Arcuri et al. 2017; Dermitzakis et al. 2023). Early in development, erythromyeloid progenitors from the embryonic yolk sac migrate to the CNS and take up residence throughout the brain (Pessac et al. 2001; Ginhoux et al. 2013). Microglia mature from these progenitors largely under the influence of the colony-stimulating factor 1 receptor (CSF1R) and provide a broadly dispersed network throughout the CNS. As progenitor migration to the brain is limited to early development, mature microglia after early development prove to be long-lived, with an average life span of 4.2 years, but undergo slow self-renewal under normal homeostatic conditions, with 28% being replaced every year (Réu et al. 2017). Microglia play a similar role in the CNS to macrophages in the peripheral tissues. Not only do they provide a first line of defense for the CNS, they are also crucial for maintaining brain homeostasis, immune surveillance, and neuro-development by processes that include synaptic pruning and neural circuit sculpting (Arcuri et al. 2017). Their cellular processes are equipped with surface receptors such as PRRs that sense changes in the proximal environment and include PAMPs and DAMPs, which allow microglia to sample and survey the surrounding environment and directly respond to foreign or harmful substances by increasing their function to a more active state (Fig. 2; Kigerl et al. 2014).

Microglial activation is characterized by morphological changes from ramified morphology with long, mobile processes to phagocytic amoeboid shape with retracted processes, both regulated by cytoskeletal components, ion channels, and small Rho guanosine triphosphatases (GTPases) (Eder 2005; Neubrand et al. 2014; Melo et al. 2021). Several complex signaling pathways that include NF-κB, mitogen-activated protein kinase (MAPK), Janus kinase/signal transducer and activator of transcription (JAK/STAT), Notch homolog 1 (NOTCH1), Ras-related C3 botulinum toxin substrate, and phosphoinositide 3-kinase/protein kinase B (PI3/PKB) pathways regulate microglial activation (Imai and Kohsaka 2002; Popiolek-Barczyk and Mika 2016; Yao et al. 2020; Chu et al. 2021). Microglia can transition between proinflammatory and anti-inflammatory/repair phenotypes,

---

Figure 1. Central nervous system (CNS) under homeostatic and disease conditions. Microglia are the primary innate immune cells of the CNS and share many of the properties associated with peripheral macrophages including defense of foreign or infectious agents. (*A*) Under homeostatic conditions, microglia take on a ramified morphology and are responsible for synaptic pruning of neuronal dendritic processes necessary for synaptic plasticity, which allow neurons to maintain structures necessary for proper interneuronal communication. Growth factors such as insulin-like growth factor 1 (IGF-1) produced by microglia are crucial for neuronal survival and myelination. IGF-1 and transforming growth factor β (TGF-β) also interact with receptors on astrocyte to induce neurotrophic factors. In turn, astrocytes produce interleukin (IL)-33, which primarily signals microglia to promote synaptic depletion. In the neurovascular unit, astrocytic end feet maintain blood–brain barrier (BBB) tight junction integrity and structure and regulate nutrient transport to the neuron. Astrocytes also regulate dopamine levels in synaptic cleft of dopaminergic neurons. (*B*) Under disease conditions, extracellular, misfolded, and aggregated α-synuclein is phagocytosed by microglia, which elevates its activation level. Activated microglia take on an amoeboid morphology and produce proinflammatory cytokines, nitric oxide (NO), and reactive oxygen species (ROS). These interact with dopaminergic neurons and modify intracellular α-synuclein to aggregate and form Lewy bodies and α-synuclein released by injured or killed neurons and taken up by microglia increases microglial activation. Interactions with proinflammatory cytokines and α-synuclein activate astrocytes and lead to dysregulated astrocytic functions including disruption of the BBB and loss of dopamine removal from the synaptic cleft. Increased dopamine levels cause dysregulated dopaminergic neurons leading to neurodegeneration. Additionally, proinflammatory cytokines and chemokines produced by astrocytes sensitize dopaminergic neurons to neurodegenerative processes and hyperactivate microglia, both leading to increased Lewy body formation and more modified and aggregated α-synuclein into the CNS environment. (TNF) Tumor necrosis factor.

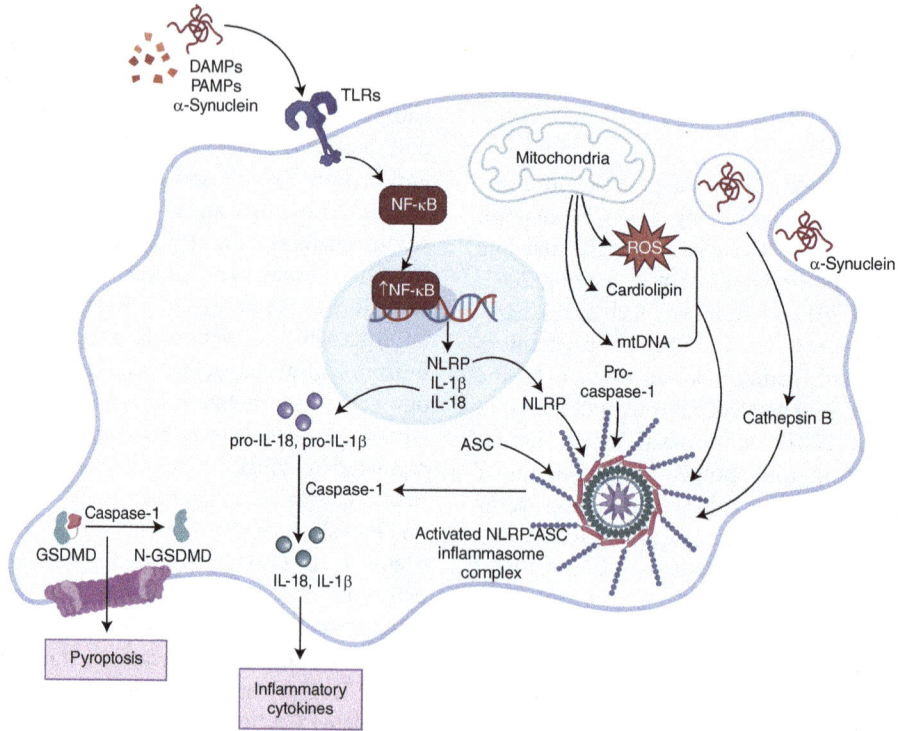

Figure 2. Microglia activation processes in Parkinson's disease (PD). Microglia recognize damage-associated molecular patterns (DAMPs), pathogen-associated molecular patterns (PAMPs), and aggregated α-synuclein via pattern recognition receptors (PRRs) primarily Toll-like receptors (TLRS), purinergic receptors such as P2X and P2Y, and scavenger receptors such as cluster of differentiation 38 (CD38). Additionally, aggregated α-synuclein can be phagocytosed and bind mitochondrial receptors that translocate with α-synuclein across the mitochondrial (mt) membrane, which disrupts calcium exchange and ATP production. Fibril forms of α-synuclein can also bind the membrane and disrupt mitochondrial function and autophagy. This increases reactive oxygen species (ROS) production and releases inner membrane cardiolipin and mtDNA. Receptor signaling through adaptor proteins and recruitment of secondary mediators such as TIRAP-inducing IFN-β (TRIF), myeloid differentiation primary response protein 88 (MyD88), and a family of TNF receptor-associated factors (TRAFs), activate mitogen-activated protein kinase (MAPK) and extracellular signal-regulated kinase (ERK) pathways that in turn activate downstream signaling pathways such as nuclear factor (NF)-κB and nuclear factor of activated T cells (NFAT). NF-κB and NFAT then translocate to the nucleus, dimerize, and bind to DNA elements upstream of proinflammatory cytokine genes such as interleukin (IL)-1β, IL-6, IL-8, TNF-α, and the nucleotide-binding oligomerization domain, leucine-rich repeat and pyrin domain-containing (NLRP) family of inflammasomes. This primes the cell for assembly of the inflammasome complex, and release of ROS, cardiolipin, and mtDNA activate the NLRPs and induce assembly of the inflammasome. With activation of caspase, pro-IL-8 and pro-IL-18 are cleaved to yield IL-18 and IL-8 leading to more proinflammatory cytokine production. Caspase-mediated pyroptosis is triggered by cleavage of gasdermin proteins that serve to form pores in the cell membrane. (GSDMD) Gasdermin D.

influenced by various modulators and signaling pathways (Luo and Chen 2012; Li et al. 2021; Guo et al. 2022; Biswas 2023). These phenotypes cover a spectrum of activation and metabolic states, which were initially combined and over-simplified as either M1 or M2 phenotypes based on earlier defined macrophage phenotypes. M1 microglia release proinflammatory mediators, whereas M2 microglia release anti-inflammatory or reparative factors and neurotrophins (Orihuela et al. 2016). LPS, IFN-γ, amyloid-β, and aggregated α-synuclein promote M1 polarization, whereas IL-4 and IL-13 drive M2 polarization (Xie et al. 2020; Biswas 2023). With age, microglia skew toward M1 phenotypes, potentially contributing to age as an associated risk factor for several neurodegenerative conditions, such as PD and AD (Varnum and Ikezu 2012). The classical M1 microglia activation produces proinflammatory cytokines, chemokines, and reactive oxygen species (ROS) (Kim and Joh 2006; Schlachetzki and Winkler 2015). Although crucial for the elimination of foreign invaders and protection of the CNS, chronic activation and expression of these protective proinflammatory mediators can lead to sustained neuroinflammation, and may thus contribute to neurotoxicity and neurodegeneration. Conversely, microglia release neurotrophic factors that support neuronal health and survival (Singh et al. 2011; Smith et al. 2012). The dual nature of microglial function in PD and other neurodegenerative diseases highlights the complex role of both the protector and perpetrator of neuronal homeostasis. With the increased resolution of microglial activation status by transcriptomics and proteomics, M1/M2 polarization has been expanded to include M0, M1, M2a, M2b, M2c, and M2d phenotypes that are dependent on the activation state and microenvironment (Franco and Fernández-Suárez 2015). More recently, bioinformatic stratification has refined microglial phenotypes to reflect a spectrum of transcriptional and proteomic states of activation, which may reflect the different homeostatic or disease states of the patient (Boche and Gordon 2022; Paolicelli et al. 2022; Sun et al. 2023).

The presence of activated human leukocyte antigen–antigen D-related (HLA-DR) microglia in the brains of patients with PD provided an impetus for studying the role of neuroinflammation in PD (McGeer et al. 1988). Since then, PD-related studies have confirmed these findings and have shown that microglia can respond directly to α-synuclein aggregates and neuronal injury by acquiring an activated state, even prior to neuronal death (Tansey and Romero-Ramos 2019). This suggests that microglia sense neuronal insults well before the neuron dies, possibly by using DAMP signaling mechanisms. The peripheral immune system has also been impli-

cated in changes observed in monocytes and T cells (Mosley et al. 2012; Harms et al. 2021). Activated microglia recruit and induce cells of both innate and adaptive immune systems by releasing proinflammatory mediators, such as C-C motif chemokine ligand 2 (CCL2), C-C motif chemokine ligand 8 (CCL8), C-X-C motif chemokine ligand 9 (CXCL9), and C-X-C motif chemokine ligand 10 (CXCL10) that upregulate MHC class I and II molecules, as well as costimulatory molecules that are necessary for T-cell responses (Lehnardt 2010; Unger et al. 2018). Moreover, adaptive immune responses have been demonstrated to facilitate the misfolding of nitrated proteins, which perpetuates chronic microglial activation, neuroinflammation, and neurodegeneration in experimental models of PD (Benner et al. 2008; Reynolds et al. 2009; Mosley et al. 2012).

Astrocytes

Astrocytes are a critical glial cell type in the CNS. They comprise most glial cells and play essential roles in maintaining CNS homeostasis, supporting neuronal function, and regulating the BBB (Rahman et al. 2022). These star-shaped cells arise from neural progenitor cells during embryogenesis and are indispensable for neurovascular coupling, neurotransmitter recycling, and synaptic support (Reemst et al. 2016). Astrocytes are characterized by the expression of glial fibrillary acidic protein (GFAP) and exhibit a highly dynamic morphology, allowing close interactions with neurons, microglia, and endothelial cells comprising the neurovascular unit (Sofroniew and Vinters 2010; Lawrence et al. 2023).

Under physiological conditions, astrocytes maintain the CNS environment by regulating ion balance, recycling neurotransmitters such as glutamate, and providing metabolic support to neurons via lactate shuttle mechanisms (Mahmoud et al. 2019; Lee et al. 2022; Edison 2024). They also secrete neurotrophic factors such as brain-derived neurotrophic factor, which promotes neuronal survival and plasticity (Albini et al. 2023). Additionally, astrocytes contribute to the formation and maintenance of the BBB by releasing factors that modulate tight junction integrity in endothelial cells (Cabezas et al. 2014).

Reactive astrogliosis, a hallmark of neuroinflammation, is characterized by the hypertrophy of astrocytic processes and upregulation of GFAP expression (Pekny and Pekna 2016). Astrocytes in PD are implicated in several key pathological processes. The disruption of glutamate metabolism, a vital function of astrocytes, has been linked to disease progression (Wang et al. 2023). Under normal conditions, astrocytes regulate extracellular glutamate levels to prevent excitotoxicity (Shen et al. 2022), a process compromised in PD. Mutations in PD-related genes, such as LRRK2 and DJ-1, have been shown to affect the expression and function of glutamate transporters like excitatory amino acid transporter 2 (EAAT2 or GLT1) (Luo et al. 1989; Iovino et al. 2022). Additionally, altered lipid raft formation, due to DJ-1 mutations, further impairs glutamate transport function, exacerbating the neurotoxic environment in the striatum (Butchbach et al. 2004; Kim et al. 2016). In addition to altered glutamate metabolism, reactive astrocytes are also involved in the secretion of inflammatory cytokines such as TNF-α, IL-1β, and IL-6. Elevated levels of these cytokines promote neuroinflammation contributing to the degeneration of dopaminergic neurons (Wang et al. 2023).

Monocyte–Macrophages

Like microglia, monocytes and macrophages play a critical role in mediating the immune response through the activation of their receptors by DAMPs or PAMPs. In PD, neuromelanin and modified and aggregated α-synuclein are recognized by PRRs (e.g., TLRs), which are also expressed by these cells and serve as potent activators (Beutler 2009). The engagement of TLRs initiates a cascade of signaling elements, including the recruitment of myeloid differentiation primary response protein 88 (MyD88), TIRAP-inducing IFN-β (TRIF), and a family of TNF receptor-associated factors (TRAFs), which lead to kinase-activated factors that bind DNA elements to upregulate proinflammatory cytokine expression and exacerbate neuroinflammatory processes (Kustrimovic et al. 2019).

Several studies have demonstrated that patients with PD exhibit a reduced number of peripheral blood monocytes, which may be attributed to increased differentiation and migration of these cells into the CNS (Schröder et al. 2018). In experimental rat models of PD, monocytes have been shown to infiltrate the brain parenchyma and contribute to the neuroinflammatory environment (Parillaud et al. 2017; Konstantin Nissen et al. 2022; Weiss et al. 2022). However, distinguishing between microglia and activated monocytes derived from the peripheral blood is challenging. Specific markers, such as the scavenger receptor CD163 (Polfliet et al. 2006) and chemokine receptor CCR2 (Mizutani et al. 2012), which are highly expressed on monocytes, but not on resident microglia, have been used to identify infiltrating monocytes in the CNS (Parillaud et al. 2017; Weiss et al. 2022). Alternatively, transmembrane protein 119 (TMEM119) and P2RY12 have been reported as differential markers of microglia (Zrzavy et al. 2017; Jurga et al. 2020).

In various animal models of PD, using monocyte-specific markers, CCR2[+] monocytes have been identified following the administration of neurotoxic agents such as N-methyl-4-phenyl-1,2,3,6-tetrahydropyridine (MPTP), a known dopaminergic neurotoxin (Jackson-Lewis and Przedborski 2007; Langston 2017; Parillaud et al. 2017; Harms et al. 2018). Similarly, after the administration of 6-hydroxydopamine (6-OHDA) or misfolded α-synuclein injection, the number of CD163[+] cells increased in the brains of rats and remained elevated for up to 6 months, further implicating these monocytes in the pathophysiology of PD (Harms et al. 2018; Weiss et al. 2022). Interestingly, flow cytometric and immunohistochemical analyses have also revealed that infiltration of peripheral monocytes and macrophages occurs shortly after α-synuclein fibril exposure in the brain (Harms et al. 2017).

Although the exact role of infiltrating monocytes in PD remains to be determined, emerging evidence suggests that their activation may have detrimental effects on neuronal viability. Studies have indicated that inhibition of monocytic infiltration into the substantia nigra reduces dopaminergic neuronal death (Cai et al. 2021; Weiss et al. 2022) and may serve as potential targets for the mitigation of neuroinflammation and neurodegeneration in PD.

Dendritic Cells

DCs play a central role in bridging the innate and adaptive immune systems. As primary antigen-presenting cells (APCs), DCs secrete cytokines that activate different subtypes of T cells. DCs may affect different pathological processes in PD through the presentation of modified α-synuclein to T cells (Mula et al. 2024). Peripheral decreases in DC populations from patients with PD compared to healthy individuals have been reported (Konstantin Nissen et al. 2022) and correlated with motor symptom severity (Ciaramella et al. 2013). The ability of tolerogenic DCs (tDCs) to induce regulatory T cells (Tregs) and the neuroprotective capacity of both tDCs and tDC-induced Tregs in MPTP-treated mice demonstrated a pivotal role for DCs and suggests DCs as putative therapeutic targets (Schutt et al. 2018). Another recent study reported that DC-mediated α-synuclein presentation to activated Tregs induced Tregs that migrated to inflammatory foci in MPTP-intoxicated mice, modulated microglia polarization, and improved motor deficits compared to controls (Park et al. 2023). In contrast, under neuroinflammatory conditions, parenchymal DCs may also contribute to neuroinflammation through cross talk between T cells and glial cells (Almolda et al. 2015; Negi and Das 2018; Schutt et al. 2018; Kustrimovic et al. 2019). Thus, while DCs contribute to CNS immune responses, it is unclear whether they serve to potentiate immune tolerance and homeostasis or exacerbate neurodegenerative processes.

Natural Killer and Mast Cells

NK cells are another critical rapid-acting component of the innate immune system that can target tumors, infectious agents, or infected cells. NK cells target cells and infectious agents via direct or indirect target recognition. Direct targeting is mediated by killer cell lectin-like receptors and killer cell immunoglobulin-like receptors. Indirect target recognition is mediated via Fc receptor (FcR) recognition of an antibody-bound target. This mechanism of killing is referred to as antibody-dependent cellular cytotoxicity (ADCC). Both target recognition events activate NK cells to release cytoplasmic granules containing perforins and granzymes that form pores in the target, leading to caspase activation and apoptosis. NK cells also release cytokines such as IFN-γ and TNF-α, which coordinate adaptive immune responses (Ochoa et al. 2017). Recently, NK cells in the CNS of patients with PD have gained increased attention. Infiltration of CD244 (2B4) expressing NK cells in the brain was found in eight of 11 PD and dementia with Lewy bodies (DLB) patients (Earls et al. 2019) and was colocalized with α-synuclein aggregates (Earls et al. 2020). Consistent with these data, NK cells were found in the brain parenchyma of mice 5 months after intrastriatal injection of α-synuclein fibrils (Earls and Lee 2020). Depletion of NK cells led to increased phosphorylated α-synuclein and motor deficits in a PD mouse model using preformed fibrils (PFFs) of α-synuclein-induced pathology, suggesting that NK cells may play a role in mitigating α-synuclein accumulation and neurodegeneration by increasing the internalization and degradation of α-synuclein aggregates through the endosomal/lysosomal pathway (Earls et al. 2020). These findings suggest that NK cells may play a beneficial role in PD by attenuating chronic inflammation and promoting the clearance of α-synuclein.

Mast cells are tissue-resident hematopoietic cells with primary defense mechanisms against extracellular pathogens such as parasites and some bacteria. However, they are also involved in allergic reactions and can lead to chronic inflammation in many diseases (Kolkhir et al. 2022). Although most mast cells are distributed throughout the peripheral compartment of the body, CNS involvement has been reported under certain conditions and detected in the meninges, area postrema, choroid plexus, and parenchyma of the thalamic hypothalamic region (Kempuraj et al. 2019; Traina 2019). Notably, >90% of mast cells are present on the abluminal side of blood vessels and are positioned in close proximity to neurons, astrocytes, microglia, and the extracellular matrix within the neurovascular unit (Medeiros et al. 2019). Mast cells respond to foreign stimuli, such as allergens, antigens, and complement factors, and trauma, with the capacity to produce proinflammatory cytokines and chemokines, such as TNF-α, IL-6, IL-1β, and monocyte chemotactic protein (Silver and Curley 2013). Most notably, after activation, mast cells rapidly release histamine-containing granules, inducing the characteristic symptoms associated with an allergic response. Moreover, histamine functions as a neurotransmitter, affecting nearby neurons by binding to histamine receptors and influencing histaminergic neuronal pathways in the thalamus and cortex. Thus, mast cells not only profoundly influence the proximal microenvironment but also influence neighboring neurons and glial cells and contribute to neuroinflammation, neurogenesis, neurodegeneration, and disruption of the BBB (Jones et al. 2019). Postmortem brain sections of patients with PD showed increased numbers of mast cells compared to healthy controls (Kempuraj et al. 2019; Zhang et al. 2021). In mouse mast cell and glia-neuron mixed cultures, activation of mast cells by mouse mast cell protease-6 (MMCP-6) and MMCP-7 triggers the release of inflammatory mediators such as IL-33, which is implicated in neuroinflammation and dopaminergic neurodegeneration via activation of multiple signaling pathways such as p38, extracellular signal-regulated kinase (ERK) 1/2, MAPK, and NF-κB (Kempuraj et al. 2019). Other studies have reported that the recruitment of mast cells into the substantia nigra, mediated by CCL2 and activated transglutaminase 2, leads to the release of inflammatory mediators such as histamine, lymphotoxins, and cytokines, which contribute to neuroinflammation and dopaminergic neurodegeneration (Rocha et al. 2016; Hong et al. 2018). Additionally, several in vitro studies have indicated that mast cells affect BBB permeability and promote diapedesis by releasing MMPs and degrading microvascular basement structures using tryptase and chymase, rendering the CNS more vulnerable to inflammatory cell and mediator infiltration (Huang et al. 2024). Although evidence of mast cell infiltration

in human postmortem PD studies is limited, in vitro and in vivo studies suggest that mast cells may have potential as therapeutic targets (Kempuraj et al. 2018).

Granulocytes

Granulocytes, comprising neutrophils, basophils, and eosinophils, are essential effector cells of the innate immune system and are characterized by their granule-rich cytoplasm and rapid response to pathogens and injuries (Vorobjeva et al. 2023). Each subtype plays a distinct role in immune defenses and inflammation. Neutrophils are the most abundant granulocytes and are known for their phagocytic activity and formation of neutrophil extracellular traps (NETs) to capture and neutralize pathogens (Hahn et al. 2016; Schoen et al. 2022). Although they are less abundant, basophils and eosinophils are pivotal in allergic responses and parasitic infections, releasing histamine, cytokines, and other mediators upon activation (Costa et al. 1997). Emerging evidence has highlighted the involvement of granulocytes in PD. Neutrophils produce nitric oxide, which contributes to oxidative stress and inflammation in PD by forming peroxynitrite and triggering the release of extracellular traps (Chakraborty et al. 2023). Elevated neutrophil-to-lymphocyte ratios in the peripheral blood have also been associated with disease severity and progression in patients with PD, suggesting their potential role as biomarkers (Grillo et al. 2023; Li et al. 2024; Zhang et al. 2024). Although less studied in PD, basophils and eosinophils may contribute to neuroinflammation through the release of histamine and other proinflammatory mediators, potentially affecting the integrity of the BBB and neuronal health (Bañuelos-Cabrera et al. 2014; Borriello et al. 2017; Kempuraj et al. 2017).

COMPLEMENT SYSTEM

The complement system is a critical component of innate immunity and consists of a series of proteins that are activated in a cascade-like manner to eliminate pathogens and damaged cells (Ricklin et al. 2010). Activation occurs through three primary pathways—the classical, alternative, and lectin pathways—all converging on the generation of C3 convertase and culminating in the formation of the membrane attack complex (Ricklin et al. 2010). In the CNS, complement proteins are primarily synthesized by microglia, astrocytes, and neurons and contribute to synaptic pruning and immune surveillance (Tenner et al. 2018).

Dysregulation of the complement system has been implicated in the pathogenesis of PD. Elevated levels of complement proteins such as C3 have been detected in postmortem PD brains (Liddelow et al. 2017). The presence of Lewy bodies and neuromelanized neurons in the substantia nigra has been shown to activate the classical complement pathway, and immunohistochemical (IHC) staining has revealed complement components, such as C3b, C4d, C7, C9, C1q, and C5a, deposited on CNS cells (Loeffler et al. 2006; Baidya et al. 2021). Activation of the complement cascade by misfolded α-synuclein aggregates in human neuroblastoma cell line SH-SY5Y and by intrastriatal injection of α-synuclein PFFs and overexpression of human mutant A53T-α-synuclein from adeno-associated virus constructs in mice can trigger microglial phagocytosis and release of proinflammatory cytokines, contributing to dopaminergic neurodegeneration (Gregersen et al. 2021; Ma et al. 2021).

Furthermore, C5a, a complement protein fragment and potent proinflammatory mediator, recruits immune cells and amplifies the local inflammatory responses in the CNS (Fonseca et al. 2013). Additionally, membrane attack complex deposition on neuronal membranes may exacerbate neurodegenerative processes (McGeer and McGeer 2004). The complement receptor 3 (CR3 or CD11/CD18 or Mac-1) on microglia also stimulates NADPH oxidase, suggesting a direct role of complement activation in neuroinflammation and dopaminergic neuronal loss (Hou et al. 2018).

PROTECTIVE BARRIERS

Aside from the different cells and complement system discussed above, the innate immune system relies on various protective barriers, which are briefly discussed here to provide a comprehensive view of this complex, multifaceted system.

Lymphatic System

The lymphatic system, which is traditionally associated with peripheral immune surveillance (Liao and Padera 2013), plays a critical role in CNS homeostasis through glial-associated lymphatic (glymphatic) system and meningeal lymphatic vessels (Li et al. 2022). The glymphatic system facilitates the clearance of interstitial solutes, including misfolded proteins such as α-synuclein, by perivascular spaces and astrocytic aquaporin-4 (AQP4) channels (Iliff et al. 2012; Jessen et al. 2015; Lian et al. 2025). Meningeal lymphatic vessels further drain cerebrospinal fluid (CSF) and interstitial fluid into the peripheral lymph nodes, linking CNS immunity with systemic immune responses (Louveau et al. 2015; Li et al. 2022).

Dysfunction of these drainage pathways has been suggested to occur in PD. Indeed, impaired glymphatic clearance has been observed in PD models, correlating with the accumulation of α-synuclein and the exacerbation of neuroinflammation (Zhang et al. 2023). Reduced CSF flow and lymphatic drainage have been observed in preclinical models of PD, correlating with increased α-synuclein aggregation and neuroinflammation (Zou et al. 2019). Impairment of lymphatic drainage pathways may also influence the integrity of the BBB, facilitating the infiltration of peripheral immune cells into the CNS (Da Mesquita et al. 2018). The interplay between disrupted lymphatic clearance and BBB dysfunction creates a feedback loop that may amplify neuroinflammation and neuronal degeneration (Da Mesquita et al. 2018).

Epithelium

Epithelial barriers, including the skin and mucosal surfaces of the gastrointestinal, respiratory, and genitourinary tracts, represent the first line of defense of the innate immune system (Moens and Veldhoen 2012; Park and Lee 2018). These barriers act as physical and biochemical shields that prevent the entry of pathogens and non-activated host immune cells and produce anti-microbial peptides (Moens and Veldhoen 2012; Artis and Spits 2015; Johnstone and Herzberg 2022; Berni Canani et al. 2024). The intestinal epithelium is integral to maintaining gut homeostasis, with tight junction proteins and mucosal immune responses playing critical roles in the regulation of permeability and microbial composition (Turner 2009).

Emerging evidence supports that epithelial barrier dysfunction is involved in the pathogenesis of PD. Along this line, the "gut–brain axis" has garnered significant attention, as gastrointestinal symptoms often precede motor manifestations in PD by years (Benvenuti et al. 2020), and is discussed in detail in Gao et al. (2025).

Blood–Brain Barrier

The BBB is a semipermeable capillary membrane composed of endothelial cells, pericytes, basement membrane, and astrocytic end feet. In 1967, the fine structure of the BBB was delineated by electron microscopy as primarily endothelial cells fused together by tight junctions, absence of intracellular fenestrations, and low levels of pinocytosis (Reese and Karnovsky 1967). Under normal physiological conditions, the BBB serves as an obstacle to harmful substances, pathogens, and peripheral immune cells, yet it allows necessary nutrients to cross from the blood to supply the brain (Knox et al. 2022). In addition to impeding harmful substances from entering the brain, it represents a formidable barrier to drug delivery.

A major function of the BBB in CNS and peripheral immune system interactions requires diapedesis of immune cells when needed for invasion of pathogens or neoplasia. Extensive evidence has shown that peripheral immune cells, including macrophages, T cells, and B cells, enter the brain through the BBB and enter the CNS under certain conditions (Wilson et al. 2010). The rate of diapedesis varies with the condition, as determined by the specific disease and level of associated inflammation (Marchetti and Engelhardt 2020). In the context of neuro-inflammation, BBB endothelial cells undergo alterations characterized by the upregulation of adhesion molecules, proinflammatory cytokines, and chemokines, accompanied by the downregulation of tight junction proteins. These changes facilitate enhanced migration of circulating leukocytes across the BBB (Liebner et al. 2018). In PD, evidence from neuroimaging and postmortem samples suggests that the loss of BBB integrity is widely observed in patients and correlates with disease progression (Sweeney et al. 2018). Although this correlation may simply reflect a parallel, unrelated progressive defect in both the BBB and neurodegeneration in PD, preclinical studies have suggested possible mechanisms by which compromised BBB integrity could exacerbate PD phenotype. The role of BBB integrity in PD development has recently focused on immune cell infiltration. Increased leukocyte infiltration into the substantia nigra has been suggested by the presence of activated leukocytes, microglia, and astrocytes in proximity to the α-synuclein aggregates (Cardinale et al. 2021; Roodveldt et al. 2024). α-Synuclein aggregates induce a reactive state in microglia and astrocytes, leading to the release of proinflammatory cytokines and chemokines that upregulate adhesion molecules, such as vascular cell adhesion molecule 1 (VCAM-1) and intercellular adhesion molecule 1 (ICAM-1), by vascular endothelial cells (Sweeney et al. 2018). This change in the endothelial environment allows immune cells to adhere more easily and translocate to the BBB into the brain parenchyma. Following endothelial activation, peripheral immune cells, including T cells, NK cells, mast cells, and monocytes, infiltrate the brain more easily. Once in the brain, activated T cells interact through the activation of the T-cell receptor by the presentation of antigens via the MHC expressed by APCs, which include monocytes, macrophages, microglia, and DCs. Moreover, these mononuclear phagocytes are responsible for phagocytosing α-synuclein aggregates and neuronal debris, leading to further activation (Lv et al. 2023). This cycle of perpetual activation sustains a chronic inflammatory state marked by the release of cytokines, such as TNF-α, IL-1β, IL-6, IFN-γ, and ROS, which amplify neuronal sensitivity and damage that drive the degeneration of dopaminergic neurons (Sweeney et al. 2018; Weiss et al. 2022). This persistent inflammatory state with increased BBB permeability is characterized by interactions between infiltrating immune cells, resident glial cells, modified α-synuclein, and cellular debris, which foster continuous neuroinflammation and neuronal degeneration.

Genetic predisposition to BBB-associated proteins plays a crucial role in the susceptibility to PD. Several studies have emphasized that hereditary genetic variants of *ABCB1*, *HMOX*, and *MDR1* influence the expression of key proteins essential for BBB transporter function, facilitating immune cell infiltration (Ayuso et al. 2014; Narayan et al. 2015; Tansey et al. 2022; Cossu et al. 2023). In addition, a series of BBB alterations have been observed in patients with PD, including altered tight junction protein expression, endothelial cell activation, and upregulation of matrix metalloproteases, which are crucial for barrier integrity and cohesion. These changes may increase vulnerability to neuroinflammatory processes in PD (Bendig et al. 2024; Chen et al. 2024; Lau et al. 2024).

PROTECTIVE IMMUNITY: CELL AND TISSUE COMMUNICATIONS

Cell–Cell Interactions

The interactions between cells of the CNS and innate immunity serve as the main driver that potentially amplifies the neuroinflammatory processes in PD. To understand this neuroimmune cross talk, the loop of activation of glial cells, particularly microglia and astrocytes, along with peripheral immune responses mediated by cytokines and other neurotoxic factors together contribute to the pathological features of PD.

Released by activated microglia, proinflammatory cytokines not only perpetuate inflammation, but also contribute to neuronal damage (González et al. 2014). Sustained upregulated levels of proinflammatory mediators have been documented in the brains of patients with PD and in animal models, supporting the idea that

neuroinflammation can contribute to the pathogenesis of PD (Wang et al. 2015; Blaylock 2017). In addition to cytokines, alterations in neurotoxic processes can aggravate neurodegeneration. For example, α-synuclein has been well-documented for causing neuroinflammation through its effects on glial cells and neuronal health (Yi et al. 2022). The neurotoxic effects of α-synuclein occur not only through its aggregation but also through its ability to induce microglial activation, thus perpetuating a cycle of inflammation and neurodegeneration (Roodveldt et al. 2024).

Emerging evidence underlines the role of glia in neuroimmune communication modeling. Microglia not only serve as immune sentinels but also engage in bidirectional communication with astrocytes, which play crucial roles in maintaining homeostasis and responding to impaired neurons (Bhusal et al. 2023). In PD, glial activation leads to astrogliosis, and activated astrocytes further release proinflammatory cytokines, which can exacerbate neuronal damage through the production of neurotoxic factors (Schwartz and Deczkowska 2016). In addition, glutamate uptake is inhibited in astrocytes, leading to glutamate toxicity and excitotoxic neuronal damage (Miyazaki and Asanuma 2020).

The cross talk between CNS cells and innate immune cells is not confined to the central nervous system; rather, systemic immune responses significantly affect neuroinflammatory status in PD. Peripheral immune cells, including T lymphocytes, monocytes, and mast cells, contribute to the inflammatory environment within the brain, triggering responses that aggravate neurodegeneration (Passaro et al. 2021). The phenomenon of "peripheral immune cell infiltration" across the BBB clarifies how these peripheral immune cells can infiltrate the CNS and induce local inflammatory responses, thus connecting systemic inflammation with neurodegenerative processes (Troncoso-Escudero et al. 2018). The role of neuroimmune cross talk between the CNS and innate immune cells in the regulation of inflammation and progression of PD cannot be overstated. This highlights the need for a holistic understanding of neuroimmune interactions

(Fig. 1) in which local and peripheral immune responses converge to exacerbate the degeneration of dopaminergic neurons (Kempuraj et al. 2018; Yang et al. 2020; Balestri et al. 2024). This intricate interaction between innate immune responses and CNS cell activation underlies a cascade of neuroinflammatory mechanisms that may contribute to neurodegeneration in PD and related conditions.

BRIDGE BETWEEN INNATE AND ADAPTIVE IMMUNITY

Compared with innate immunity, the adaptive immune system has significant specificity and rapidly reacts to re-exposure to the initiating pathogen, a phenomenon referred to as "immunological memory" (Vitetta et al. 1991; Netea et al. 2019). In the past, the CNS was thought to be an immune-privileged organ that lacked a lymphatic system, thus physically isolating the CNS from the peripheral immune system by the BBB (Proulx and Engelhardt 2022). However, these concepts are no longer consistent with current findings that describe a complex glymphatic system and abundant T and B cells associated with the lymphatic elements of the brain (Engelhardt et al. 2017; Mezey et al. 2021; Mogensen et al. 2021). As microglia represent an abundant portion of the CNS glial population with a wide distribution and excellent mobility, interactions with infiltrating T and B cells are quite likely (Chhatbar and Prinz 2021). Moreover, microglia can generate significant immune responses in the CNS, like those initiated by peripheral macrophages and DCs in the periphery. Typically, CNS adaptive immune responses involve reactivation by microglia or by infiltrating macrophages or DCs presenting antigens from a variety of CNS pathogens and pathologies (Cardinale et al. 2021). Evidence from patients with PD and preclinical models has recently provided a plausible connection between the adaptive immune response and disease severity (Benner et al. 2008; Saunders et al. 2012; Gendelman et al. 2017; Sulzer et al. 2017; Saleh et al. 2022). Therefore, a clear understanding of adaptive immune responses and interactions with innate CNS immunity in PD is urgently needed (Fig. 3), a topic that is discussed in Elyaman (2025).

CONCLUSION

PD is increasingly recognized as a disorder with significant immune system involvement, particularly through innate immune mechanisms. This review highlights the crucial role of innate immune cells, including microglia, monocytes, macrophages, DCs, NK cells, and mast cells, in shaping the neuroinflammatory landscape of PD. The activation of those immune cells, in response to genetic predisposition and environmental triggers, contributes to chronic inflammation and progressive neuronal loss (Heneka et al. 2014; Labzin et al. 2018; Choi et al. 2022; Rajesh and Kanneganti 2022).

The intricate interplay between peripheral and central immune responses emphasizes how infiltrating immune cells aggravate neuroinflammation and neurodegeneration. Peripheral monocytes and macrophages infiltrate the CNS in response to neuronal injury, amplifying inflammatory responses (Weiss et al. 2022). Even before neuronal loss, activated microglia expressing HLA-DR in PD brains further support a key role in innate immune activation in disease progression (McGeer et al. 1988; Tansey and Romero-Ramos 2019). In addition, NK and mast cells, which are often overlooked in PD pathology, have been implicated in α-synuclein clearance and immune regulation, although their roles require further investigation (Kempuraj et al. 2019; Earls et al. 2020).

Moreover, the breakdown of protective barriers such as the BBB and dysfunction in the glymphatic and lymphatic systems further contribute to immune dysregulation and α-synuclein pathology (Da Mesquita et al. 2018; Zou et al. 2019; Zhang et al. 2023). Genetic factors, including mutations in LRRK2, HLA, and VPS35, have also been implicated in altering immune functions, reinforcing the link between immune dysregulation and PD pathogenesis (Kam et al. 2020; Bailey and Cookson 2024; Cui et al. 2024). These genetic alterations influence microglial activation, antigen presentation, and the inflammatory response, ultimately con-

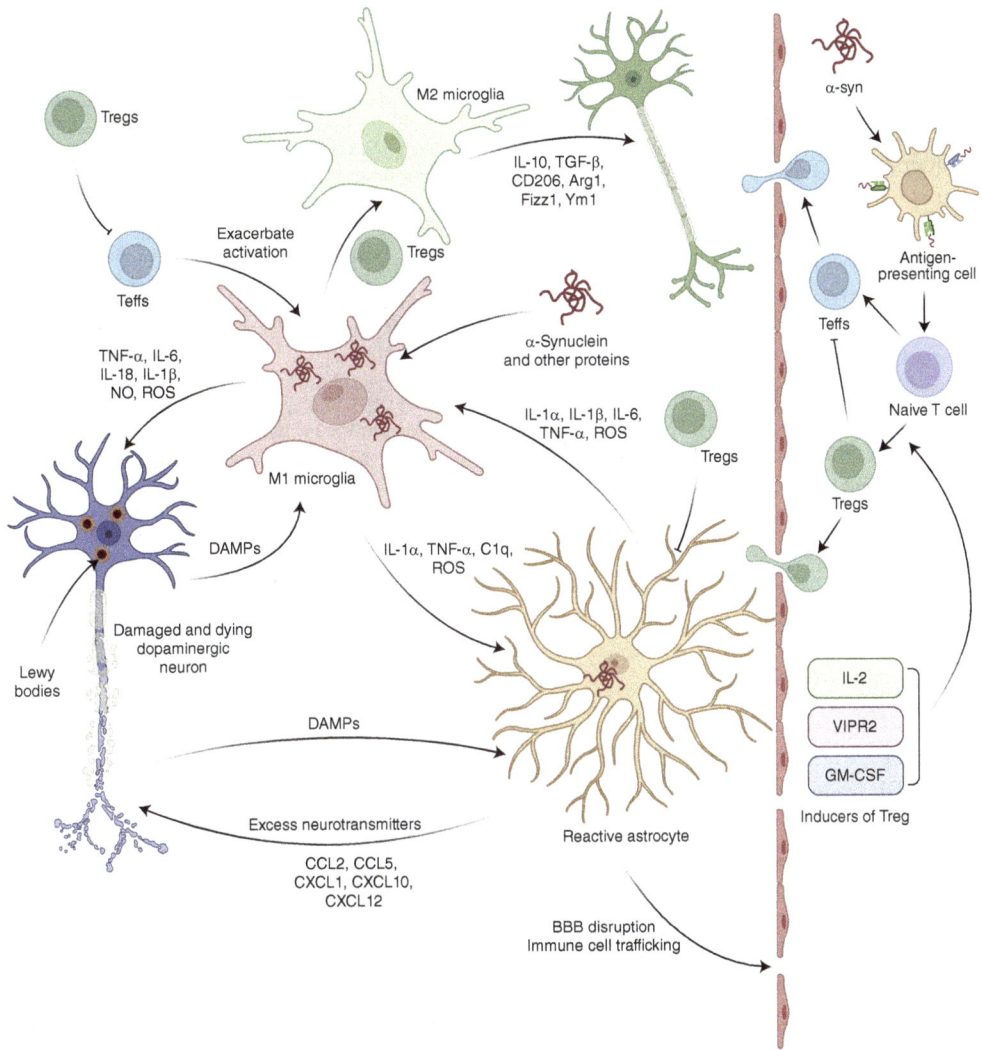

Figure 3. Innate and adaptive immune cross talk in Parkinson's disease (PD). The inflammatory and oxidatively stressful environment found in the PD central nervous system (CNS) leads to modification and misfolding of α-synuclein within dopaminergic neurons. Failure to remove misfolded α-synuclein leads to oligomerization and aggregation in the form of Lewy bodies. Injured or dead neurons release α-synuclein and damage-associated molecular patterns (DAMPs) to the extraneuronal environment and activate microglia and astrocytes. Activated microglia produce neurotoxic proinflammatory factors that spread to surrounding neurons and induce modification and misfolding of α-synuclein in those neurons. Additionally, microglial proinflammatory factors increase activation levels of astrocytes, which feedback to sensitize neurons to further injury, hyperactivate microglia, and dysregulate blood–brain barrier (BBB) integrity. Eventually, the inflammatory milieu and modified self-proteins such as nitrated α-synuclein drains to the peripheral lymphoid tissues to activate antigen-presenting cells (APCs) and induce effector T cells (Teffs) that recognize modified proteins such as nitrated-α-synuclein as foreign. Those T cells are recruited to inflammatory foci in the brain, where representation to the nitrated antigen initiates the Teff effector program to secrete Teff specific cytokines such as interferon γ (IFN-γ), interleukin 1 (IL-1), and tumor necrosis factor α (TNF-α) to further exacerbate neuroinflammation and neurotoxicity. Over time, a perpetual cascade of protein modification, glial activation, T-cell exacerbation, and neuronal death is formed. One therapeutic strategy to counter cascade is intervention with anti-inflammatory T cells, and a most potent anti-inflammatory T cell is the regulatory T cell (Treg). Tregs suppress neuroinflammation using anti-inflammatory cytokines such as IL-10 and transforming growth factor β (TGF-β), which drive M1 microglia to M2 microglia, and suppress Teffs and Teff induction or can transform Teffs into Tregs. Treg inducing agents that have been shown to have neuroprotective capabilities via Treg induction include low-dose IL-2, vasoactive intestinal peptide receptor 2 (VIPR2) agonists, and granulocyte–macrophage colony-stimulating factor (GM-CSF). (ROS) Reactive oxygen species.

tributing to neurodegeneration (Benner et al. 2008; Sulzer et al. 2017).

Understanding the mechanisms by which the innate immune system contributes to PD pathophysiology offers valuable insights into potential therapeutic interventions. Modulating immune responses through anti-inflammatory strategies, immune checkpoint modulation, and targeting neuroimmune interactions holds promise for disease modification (Schutt et al. 2018; Alfaidi et al. 2024). Future research strategies should refine our understanding of the balance between protective and pathological immune responses in PD, paving the way for precision medicine approaches to halt or slow disease progression.

ACKNOWLEDGMENTS

We acknowledge the support of the National Institutes of Health under multiple grants, including R01 AI145342, R01 AI158160, T32 NS105594, P01 DA028555, and R01 NS36126, awarded to H.E.G. Additional funding was provided by the University of Nebraska Foundation through generous donations from the Carol Swarts, M.D. Emerging Neuroscience Research Laboratory; the Margaret R. Larson Professorship; and the Frances and Louie Blumkin, and Harriet Singer Endowments.

REFERENCES

*Reference is also in this subject collection.

Albini M, Krawczun-Rygmaczewska A, Cesca F. 2023. Astrocytes and brain-derived neurotrophic factor (BDNF). Neurosci Res 197: 42–51. doi:10.1016/j.neures.2023.02.001

Alfaidi M, Barker RA, Kuan WL. 2024. An update on immune-based alpha-synuclein trials in Parkinson's disease. J Neurol 272: 21. doi:10.1007/s00415-024-12770-x

Almolda B, González B, Castellano B. 2015. Are microglial cells the regulators of lymphocyte responses in the CNS? Front Cell Neurosci 9: 440. doi:10.3389/fncel.2015.00440

Arcuri C, Mecca C, Bianchi R, Giambanco I, Donato R. 2017. The pathophysiological role of microglia in dynamic surveillance, phagocytosis and structural remodeling of the developing CNS. Front Mol Neurosci 10: 191. doi:10.3389/fnmol.2017.00191

Artis D, Spits H. 2015. The biology of innate lymphoid cells. Nature 517: 293–301. doi:10.1038/nature14189

Ayuso P, Martínez C, Pastor P, Lorenzo-Betancor O, Luengo A, Jiménez-Jiménez FJ, Alonso-Navarro H, Agúndez JA, García-Martín E. 2014. An association study between Heme oxygenase-1 genetic variants and Parkinson's disease. Front Cell Neurosci 8: 298. doi:10.3389/fncel.2014.00298

Baidya F, Bohra M, Datta A, Sarmah D, Shah B, Jagtap P, Raut S, Sarkar A, Singh U, Kalia K, et al. 2021. Neuroimmune crosstalk and evolving pharmacotherapies in neurodegenerative diseases. Immunology 162: 160–178. doi:10.1111/imm.13264

Bailey HM, Cookson MR. 2024. How Parkinson's disease-linked LRRK2 mutations affect different CNS cell types. J Parkinsons Dis 14: 1331–1352. doi:10.3233/JPD-230432

Balestri W, Sharma R, da Silva VA, Bobotis BC, Curle AJ, Kothakota V, Kalantarnia F, Hangad MV, Hoorfar M, Jones JL, et al. 2024. Modeling the neuroimmune system in Alzheimer's and Parkinson's diseases. J Neuroinflammation 21: 32. doi:10.1186/s12974-024-03024-8

Bañuelos-Cabrera I, Valle-Dorado MG, Aldana BI, Orozco-Suárez SA, Rocha L. 2014. Role of histaminergic system in blood–brain barrier dysfunction associated with neurological disorders. Arch Med Res 45: 677–686. doi:10.1016/j.arcmed.2014.11.010

Bendig J, Frank A, Reichmann H. 2024. Aging and Parkinson's disease: a complex interplay of vulnerable neurons, the immune system and the blood–brain barrier. Ageing Neur Dis 4: 5. doi:10.20517/and.2023.36

Benner EJ, Banerjee R, Reynolds AD, Sherman S, Pisarev VM, Tsiperson V, Nemachek C, Ciborowski P, Przedborski S, Mosley RL, et al. 2008. Nitrated alpha-synuclein immunity accelerates degeneration of nigral dopaminergic neurons. PLoS One 3: e1376. doi:10.1371/journal.pone.0001376

Benvenuti L, D'Antongiovanni V, Pellegrini C, Antonioli L, Bernardini N, Blandizzi C, Fornai M. 2020. Enteric glia at the crossroads between intestinal immune system and epithelial barrier: implications for Parkinson disease. Int J Mol Sci 21: 9199. doi:10.3390/ijms21239199

Berni Canani R, Caminati M, Carucci L, Eguiluz-Gracia I. 2024. Skin, gut, and lung barrier: physiological interface and target of intervention for preventing and treating allergic diseases. Allergy 79: 1485–1500. doi:10.1111/all.16092

Berriat F, Lobsiger CS, Boillée S. 2023. The contribution of the peripheral immune system to neurodegeneration. Nat Neurosci 26: 942–954. doi:10.1038/s41593-023-01323-6

Beutler BA. 2009. TLRs and innate immunity. Blood 113: 1399–1407. doi:10.1182/blood-2008-07-019307

Bhusal A, Afridi R, Lee WH, Suk K. 2023. Bidirectional communication between microglia and astrocytes in neuroinflammation. Curr Neuropharmacol 21: 2020–2029. doi:10.2174/1570159X21666221129121715

Biswas K. 2023. Microglia mediated neuroinflammation in neurodegenerative diseases: a review on the cell signaling pathways involved in microglial activation. J Neuroimmunol 383: 578180. doi:10.1016/j.jneuroim.2023.578180

Blaylock RL. 2017. Parkinson's disease: microglial/macrophage-induced immunoexcitotoxicity as a central mechanism of neurodegeneration. Surg Neurol Int 8: 65. doi:10.4103/sni.sni_441_16

Boche D, Gordon MN. 2022. Diversity of transcriptomic microglial phenotypes in aging and Alzheimer's disease. *Alzheimers Dement* **18:** 360–376. doi:10.1002/alz.12389

Borriello F, Iannone R, Marone G. 2017. Histamine release from mast cells and basophils. *Handb Exp Pharmacol* **241:** 121–139. doi:10.1007/164_2017_18

Butchbach ME, Tian G, Guo H, Lin CL. 2004. Association of excitatory amino acid transporters, especially EAAT2, with cholesterol-rich lipid raft microdomains: importance for excitatory amino acid transporter localization and function. *J Biol Chem* **279:** 34388–34396. doi:10.1074/jbc.M403938200

Cabezas R, Avila M, Gonzalez J, El-Bachá RS, Báez E, García-Segura LM, Jurado Coronel JC, Capani F, Cardona-Gomez GP, Barreto GE. 2014. Astrocytic modulation of blood brain barrier: perspectives on Parkinson's disease. *Front Cell Neurosci* **8:** 211. doi:10.3389/fncel.2014.00211

Cai HY, Fu XX, Jiang H, Han S. 2021. Adjusting vascular permeability, leukocyte infiltration, and microglial cell activation to rescue dopaminergic neurons in rodent models of Parkinson's disease. *NPJ Parkinsons Dis* **7:** 91. doi:10.1038/s41531-021-00233-3

Calabresi P, Mechelli A, Natale G, Volpicelli-Daley L, Di Lazzaro G, Ghiglieri V. 2023. Alpha-synuclein in Parkinson's disease and other synucleinopathies: from overt neurodegeneration back to early synaptic dysfunction. *Cell Death Dis* **14:** 176. doi:10.1038/s41419-023-05672-9

Cardinale A, Calabrese V, de Iure A, Picconi B. 2021. Alpha-synuclein as a prominent actor in the inflammatory synaptopathy of Parkinson's disease. *Int J Mol Sci* **22:** 6517. doi:10.3390/ijms22126517

Chabot S, Yong VW. 2002. Microglia in the CNS. In *The neuronal environment: brain homeostasis in health and disease* (ed. Walz W), pp. 379–400. Humana, Totowa, NJ.

Chakraborty S, Tabrizi Z, Bhatt NN, Franciosa SA, Bracko O. 2023. A brief overview of neutrophils in neurological diseases. *Biomolecules* **13:** 743. doi:10.3390/biom13050743

Chen T, Dai Y, Hu C, Lin Z, Wang S, Yang J, Zeng L, Li S, Li W. 2024. Cellular and molecular mechanisms of the blood-brain barrier dysfunction in neurodegenerative diseases. *Fluids Barriers CNS* **21:** 60. doi:10.1186/s12987-024-00557-1

Chhatbar C, Prinz M. 2021. The roles of microglia in viral encephalitis: from sensome to therapeutic targeting. *Cell Mol Immunol* **18:** 250–258. doi:10.1038/s41423-020-00620-5

Choi YR, Kang SJ, Kim JM, Lee SJ, Jou I, Joe EH, Park SM. 2015. FcgammaRIIB mediates the inhibitory effect of aggregated alpha-synuclein on microglial phagocytosis. *Neurobiol Dis* **83:** 90–99. doi:10.1016/j.nbd.2015.08.025

Choi I, Heaton GR, Lee YK, Yue Z. 2022. Regulation of alpha-synuclein homeostasis and inflammasome activation by microglial autophagy. *Sci Adv* **8:** eabn1298. doi:10.1126/sciadv.abn1298

Chu E, Mychasiuk R, Hibbs ML, Semple BD. 2021. Dysregulated phosphoinositide 3-kinase signaling in microglia: shaping chronic neuroinflammation. *J Neuroinflammation* **18:** 276. doi:10.1186/s12974-021-02325-6

Ciaramella A, Salani F, Bizzoni F, Pontieri FE, Stefani A, Pierantozzi M, Assogna F, Caltagirone C, Spalletta G, Bossù P. 2013. Blood dendritic cell frequency declines in idiopathic Parkinson's disease and is associated with motor symptom severity. *PLoS One* **8:** e65352. doi:10.1371/journal.pone.0065352

Cossu D, Hatano T, Hattori N. 2023. The role of immune dysfunction in Parkinson's disease development. *Int J Mol Sci* **24:** 16766. doi:10.3390/ijms242316766

Costa JJ, Weller PF, Galli SJ. 1997. The cells of the allergic response: mast cells, basophils, and eosinophils. *JAMA* **278:** 1815–1822. doi:10.1001/jama.1997.03550220021005

Cui H, Sun F, Yu N, Cao Y, Wang X, Zhang D, Chen Z, Wang N, Yuan B, Liu P, et al. 2024. TLR2/NF-kappaB signaling in macrophage/microglia mediated COVID-pain induced by SARS-CoV-2 envelope protein. *iScience* **27:** 111027. doi:10.1016/j.isci.2024.111027

Da Mesquita S, Fu Z, Kipnis J. 2018. The meningeal lymphatic system: a new player in neurophysiology. *Neuron* **100:** 375–388. doi:10.1016/j.neuron.2018.09.022

Dermitzakis I, Manthou ME, Meditskou S, Tremblay ME, Petratos S, Zoupi L, Boziki M, Kesidou E, Simeonidou C, Theotokis P. 2023. Origin and emergence of microglia in the CNS—an interesting (hi)story of an eccentric cell. *Curr Issues Mol Biol* **45:** 2609–2628. doi:10.3390/cimb45030171

Earls RH, Lee JK. 2020. The role of natural killer cells in Parkinson's disease. *Exp Mol Med* **52:** 1517–1525. doi:10.1038/s12276-020-00505-7

Earls RH, Menees KB, Chung J, Barber J, Gutekunst CA, Hazim MG, Lee JK. 2019. Intrastriatal injection of preformed alpha-synuclein fibrils alters central and peripheral immune cell profiles in non-transgenic mice. *J Neuroinflammation* **16:** 250. doi:10.1186/s12974-019-1636-8

Earls RH, Menees KB, Chung J, Gutekunst CA, Lee HJ, Hazim MG, Rada B, Wood LB, Lee JK. 2020. NK cells clear alpha-synuclein and the depletion of NK cells exacerbates synuclein pathology in a mouse model of alpha-synucleinopathy. *Proc Natl Acad Sci* **117:** 1762–1771. doi:10.1073/pnas.1909110117

Eder C. 2005. Regulation of microglial behavior by ion channel activity. *J Neurosci Res* **81:** 314–321. doi:10.1002/jnr.20476

Edison P. 2024. Astroglial activation: current concepts and future directions. *Alzheimers Dement* **20:** 3034–3053. doi:10.1002/alz.13678

* Elyaman W. 2025. Adaptive immunity and Parkinson's disease. *Cold Spring Harb Perspect Med* doi:10.1101/cshperspect.a041639

Engelhardt B, Vajkoczy P, Weller RO. 2017. The movers and shapers in immune privilege of the CNS. *Nat Immunol* **18:** 123–131. doi:10.1038/ni.3666

Fakhoury M. 2015. Role of immunity and inflammation in the pathophysiology of neurodegenerative diseases. *Neurodegener Dis* **15:** 63–69. doi:10.1159/000369933

Fonseca MI, McGuire SO, Counts SE, Tenner AJ. 2013. Complement activation fragment C5a receptors, CD88 and C5L2, are associated with neurofibrillary pathology. *J Neuroinflammation* **10:** 25. doi:10.1186/1742-2094-10-25

Franco R, Fernández-Suárez D. 2015. Alternatively activated microglia and macrophages in the central nervous system. *Prog Neurobiol* **131:** 65–86. doi:10.1016/j.pneurobio.2015.05.003

Gao C, Jiang J, Tan Y, Chen S. 2023. Microglia in neurodegenerative diseases: mechanism and potential therapeutic targets. *Signal Transduct Target Ther* **8:** 359. doi:10.1038/s41392-023-01588-0

* Gao V, Crawford CV, Burré J. 2025. The gut–brain axis in Parkinson's disease. *Cold Spring Harb Perspect Med* **15:** a041618. doi:10.1101/cshperspect.a041618

Gendelman HE, Zhang Y, Santamaria P, Olson KE, Schutt CR, Bhatti D, Shetty BLD, Lu Y, Estes KA, Standaert DG, et al. 2017. Evaluation of the safety and immunomodulatory effects of sargramostim in a randomized, double-blind phase 1 clinical Parkinson's disease trial. *NPJ Parkinsons Dis* **3:** 10. doi:10.1038/s41531-017-0013-5

Ginhoux F, Lim S, Hoeffel G, Low D, Huber T. 2013. Origin and differentiation of microglia. *Front Cell Neurosci* **7:** 45. doi:10.3389/fncel.2013.00045

Goldman SM, Marek K, Ottman R, Meng C, Comyns K, Chan P, Ma J, Marras C, Langston JW, Ross GW, et al. 2019. Concordance for Parkinson's disease in twins: a 20-year update. *Ann Neurol* **85:** 600–605. doi:10.1002/ana.25441

González H, Elgueta D, Montoya A, Pacheco R. 2014. Neuroimmune regulation of microglial activity involved in neuroinflammation and neurodegenerative diseases. *J Neuroimmunol* **274:** 1–13. doi:10.1016/j.jneuroim.2014.07.012

Gregersen E, Betzer C, Kim WS, Kovacs G, Reimer L, Halliday GM, Thiel S, Jensen PH. 2021. Alpha-synuclein activates the classical complement pathway and mediates complement-dependent cell toxicity. *J Neuroinflammation* **18:** 177. doi:10.1186/s12974-021-02225-9

Grillo P, Sancesario GM, Bovenzi R, Zenuni H, Bissacco J, Mascioli D, Simonetta C, Forti P, Degoli GR, Pieri M, et al. 2023. Neutrophil-to-lymphocyte ratio and lymphocyte count reflect alterations in central neurodegeneration-associated proteins and clinical severity in Parkinson disease patients. *Parkinsonism Relat Disord* **112:** 105480. doi:10.1016/j.parkreldis.2023.105480

Guo S, Wang H, Yin Y. 2022. Microglia polarization from M1 to M2 in neurodegenerative diseases. *Front Aging Neurosci* **14:** 815347. doi:10.3389/fnagi.2022.815347

Hahn J, Knopf J, Maueroder C, Kienhofer D, Leppkes M, Herrmann M. 2016. Neutrophils and neutrophil extracellular traps orchestrate initiation and resolution of inflammation. *Clin Exp Rheumatol* **34:** 6–8.

Harms AS, Delic V, Thome AD, Bryant N, Liu Z, Chandra S, Jurkuvenaite A, West AB. 2017. α-Synuclein fibrils recruit peripheral immune cells in the rat brain prior to neurodegeneration. *Acta Neuropathol Commun* **5:** 85. doi:10.1186/s40478-017-0494-9

Harms AS, Thome AD, Yan Z, Schonhoff AM, Williams GP, Li X, Liu Y, Qin H, Benveniste EN, Standaert DG. 2018. Peripheral monocyte entry is required for alpha-synuclein induced inflammation and Neurodegeneration in a model of Parkinson disease. *Exp Neurol* **300:** 179–187. doi:10.1016/j.expneurol.2017.11.010

Harms AS, Ferreira SA, Romero-Ramos M. 2021. Periphery and brain, innate and adaptive immunity in Parkinson's disease. *Acta Neuropathol* **141:** 527–545. doi:10.1007/s00401-021-02268-5

Heneka MT, Kummer MP, Latz E. 2014. Innate immune activation in neurodegenerative disease. *Nat Rev Immunol* **14:** 463–477. doi:10.1038/nri3705

Hong GU, Cho JW, Kim SY, Shin JH, Ro JY. 2018. Inflammatory mediators resulting from transglutaminase 2 expressed in mast cells contribute to the development of Parkinson's disease in a mouse model. *Toxicol Appl Pharmacol* **358:** 10–22. doi:10.1016/j.taap.2018.09.003

Hou L, Wang K, Zhang C, Sun F, Che Y, Zhao X, Zhang D, Li H, Wang Q. 2018. Complement receptor 3 mediates NADPH oxidase activation and dopaminergic neurodegeneration through a Src-Erk-dependent pathway. *Redox Biol* **14:** 250–260. doi:10.1016/j.redox.2017.09.017

Huang X, Lan Z, Hu Z. 2024. Role and mechanisms of mast cells in brain disorders. *Front Immunol* **15:** 1445867. doi:10.3389/fimmu.2024.1445867

Iliff JJ, Wang M, Liao Y, Plogg BA, Peng W, Gundersen GA, Benveniste H, Vates GE, Deane R, Goldman SA, et al. 2012. A paravascular pathway facilitates CSF flow through the brain parenchyma and the clearance of interstitial solutes, including amyloid beta. *Sci Transl Med* **4:** 147ra111. doi:10.1126/scitranslmed.3003748

Imai Y, Kohsaka S. 2002. Intracellular signaling in M-CSF-induced microglia activation: role of Iba1. *Glia* **40:** 164–174. doi:10.1002/glia.10149

Iovino L, Giusti V, Pischedda F, Giusto E, Plotegher N, Marte A, Battisti I, Di Iacovo A, Marku A, Piccoli G, et al. 2022. Trafficking of the glutamate transporter is impaired in LRRK2-related Parkinson's disease. *Acta Neuropathol* **144:** 81–106. doi:10.1007/s00401-022-02437-0

Jackson-Lewis V, Przedborski S. 2007. Protocol for the MPTP mouse model of Parkinson's disease. *Nat Protoc* **2:** 141–151. doi:10.1038/nprot.2006.342

Janeway CA Jr, Travers P, Walport M, Shlomchik MJ. 2001. Principles of innate and adaptive immunity. In *Immunobiology: the immune system in health and disease*, 5th ed. Garland Science, New York.

Jessen NA, Munk AS, Lundgaard I, Nedergaard M. 2015. The glymphatic system: a beginner's guide. *Neurochem Res* **40:** 2583–2599. doi:10.1007/s11064-015-1581-6

Johnstone KF, Herzberg MC. 2022. Antimicrobial peptides: defending the mucosal epithelial barrier. *Front Oral Health* **3:** 958480. doi:10.3389/froh.2022.958480

Jones MK, Nair A, Gupta M. 2019. Mast cells in neurodegenerative disease. *Front Cell Neurosci* **13:** 171. doi:10.3389/fncel.2019.00171

Jurga AM, Paleczna M, Kuter KZ. 2020. Overview of general and discriminating markers of differential microglia phenotypes. *Front Cell Neurosci* **14:** 198. doi:10.3389/fncel.2020.00198

Kam TI, Hinkle JT, Dawson TM, Dawson VL. 2020. Microglia and astrocyte dysfunction in Parkinson's disease. *Neurobiol Dis* **144:** 105028. doi:10.1016/j.nbd.2020.105028

Kempuraj D, Thangavel R, Selvakumar GP, Zaheer S, Ahmed ME, Raikwar SP, Zahoor H, Saeed D, Natteru PA, Iyer S, et al. 2017. Brain and peripheral atypical inflammatory mediators potentiate neuroinflammation and neurodegeneration. *Front Cell Neurosci* **11:** 216. doi:10.3389/fncel.2017.00216

Kempuraj D, Selvakumar GP, Zaheer S, Thangavel R, Ahmed ME, Raikwar S, Govindarajan R, Iyer S, Zaheer A. 2018. Cross-talk between glia, neurons and mast cells in neuroinflammation associated with Parkinson's disease. *J Neuroimmune Pharmacol* **13:** 100–112. doi:10.1007/s11481-017-9766-1

Kempuraj D, Thangavel R, Selvakumar GP, Ahmed ME, Zaheer S, Raikwar SP, Zahoor H, Saeed D, Dubova I, Giler G, et al. 2019. Mast cell proteases activate astrocytes and glia-neurons and release interleukin-33 by activating p38 and ERK1/2 MAPKs and NF-κB. *Mol Neurobiol* **56:** 1681–1693. doi:10.1007/s12035-018-1177-7

Kigerl KA, de Rivero Vaccari JP, Dietrich WD, Popovich PG, Keane RW. 2014. Pattern recognition receptors and central nervous system repair. *Exp Neurol* **258:** 5–16. doi:10.1016/j.expneurol.2014.01.001

Kim YS, Joh TH. 2006. Microglia, major player in the brain inflammation: their roles in the pathogenesis of Parkinson's disease. *Exp Mol Med* **38:** 333–347. doi:10.1038/emm.2006.40

Kim JM, Cha SH, Choi YR, Jou I, Joe EH, Park SM. 2016. DJ-1 deficiency impairs glutamate uptake into astrocytes via the regulation of flotillin-1 and caveolin-1 expression. *Sci Rep* **6:** 28823. doi:10.1038/srep28823

Knox EG, Aburto MR, Clarke G, Cryan JF, O'Driscoll CM. 2022. The blood–brain barrier in aging and neurodegeneration. *Mol Psychiatry* **27:** 2659–2673. doi:10.1038/s41380-022-01511-z

Kolkhir P, Elieh-Ali-Komi D, Metz M, Siebenhaar F, Maurer M. 2022. Understanding human mast cells: lesson from therapies for allergic and non-allergic diseases. *Nat Rev Immunol* **22:** 294–308. doi:10.1038/s41577-021-00622-y

Konstantin Nissen S, Farmen K, Carstensen M, Schulte C, Goldeck D, Brockmann K, Romero-Ramos M. 2022. Changes in CD163[+], CD11b[+], and CCR2[+] peripheral monocytes relate to Parkinson's disease and cognition. *Brain Behav Immun* **101:** 182–193. doi:10.1016/j.bbi.2022.01.005

Kustrimovic N, Marino F, Cosentino M. 2019. Peripheral immunity, immunoaging and neuroinflammation in Parkinson's disease. *Curr Med Chem* **26:** 3719–3753. doi:10.2174/0929867325666181009161048

Labzin LI, Heneka MT, Latz E. 2018. Innate immunity and neurodegeneration. *Annu Rev Med* **69:** 437–449. doi:10.1146/annurev-med-050715-104343

Langston JW. 2017. The MPTP story. *J Parkinsons Dis* **7:** S11–S19. doi:10.3233/JPD-179006

Lau K, Kotzur R, Richter F. 2024. Blood-brain barrier alterations and their impact on Parkinson's disease pathogenesis and therapy. *Transl Neurodegener* **13:** 37. doi:10.1186/s40035-024-00430-z

Lawrence JM, Schardien K, Wigdahl B, Nonnemacher MR. 2023. Roles of neuropathology-associated reactive astrocytes: a systematic review. *Acta Neuropathol Commun* **11:** 42. doi:10.1186/s40478-023-01526-9

Lee HG, Wheeler MA, Quintana FJ. 2022. Function and therapeutic value of astrocytes in neurological diseases. *Nat Rev Drug Discov* **21:** 339–358. doi:10.1038/s41573-022-00390-x

Lehnardt S. 2010. Innate immunity and neuroinflammation in the CNS: the role of microglia in Toll-like receptor-mediated neuronal injury. *Glia* **58:** 253–263. doi:10.1002/glia.20928

Li J, Shui X, Sun R, Wan L, Zhang B, Xiao B, Luo Z. 2021. Microglial phenotypic transition: signaling pathways and influencing modulators involved in regulation in central nervous system diseases. *Front Cell Neurosci* **15:** 736310. doi:10.3389/fncel.2021.736310

Li G, Cao Y, Tang X, Huang J, Cai L, Zhou L. 2022. The meningeal lymphatic vessels and the glymphatic system: potential therapeutic targets in neurological disorders. *J Cereb Blood Flow Metab* **42:** 1364–1382. doi:10.1177/0271678X221098145

Li F, Weng G, Zhou H, Zhang W, Deng B, Luo Y, Tao X, Deng M, Guo H, Zhu S, et al. 2024. The neutrophil-to-lymphocyte ratio, lymphocyte-to-monocyte ratio, and neutrophil-to-high-density-lipoprotein ratio are correlated with the severity of Parkinson's disease. *Front Neurol* **15:** 1322228. doi:10.3389/fneur.2024.1322228

Lian X, Liu Z, Gan Z, Yan Q, Tong L, Qiu L, Liu Y, Chen JF, Li Z. 2025. Targeting the glymphatic system to promote alpha-synuclein clearance: a novel therapeutic strategy for Parkinson's disease. *Neural Regen Res* **21:** 233–247. doi:10.4103/NRR.NRR-D-24-00764

Liao S, Padera TP. 2013. Lymphatic function and immune regulation in health and disease. *Lymphat Res Biol* **11:** 136–143. doi:10.1089/lrb.2013.0012

Liddelow SA, Guttenplan KA, Clarke LE, Bennett FC, Bohlen CJ, Schirmer L, Bennett ML, Münch AE, Chung WS, Peterson TC, et al. 2017. Neurotoxic reactive astrocytes are induced by activated microglia. *Nature* **541:** 481–487. doi:10.1038/nature21029

Liebner S, Dijkhuizen RM, Reiss Y, Plate KH, Agalliu D, Constantin G. 2018. Functional morphology of the blood–brain barrier in health and disease. *Acta Neuropathol* **135:** 311–336. doi:10.1007/s00401-018-1815-1

Lind-Holm Mogensen F, Sousa C, Ameli C, Badanjak K, Pereira SL, Muller A, Kaoma T, Coowar D, Scafidi A, Poovathingal SK, et al. 2024. PARK7/DJ-1 deficiency impairs microglial activation in response to LPS-induced inflammation. *J Neuroinflammation* **21:** 174. doi:10.1186/s12974-024-03164-x

Loeffler DA, Camp DM, Conant SB. 2006. Complement activation in the Parkinson's disease substantia nigra: an immunocytochemical study. *J Neuroinflammation* **3:** 29. doi:10.1186/1742-2094-3-29

Louveau A, Smirnov I, Keyes TJ, Eccles JD, Rouhani SJ, Peske JD, Derecki NC, Castle D, Mandell JW, Lee KS, et al. 2015. Structural and functional features of central nervous system lymphatic vessels. *Nature* **523:** 337–341. doi:10.1038/nature14432

Luo XG, Chen SD. 2012. The changing phenotype of microglia from homeostasis to disease. *Transl Neurodegener* **1:** 9. doi:10.1186/2047-9158-1-9

Luo SQ, Li DZ, Zhang MZ, Wang ZC. 1989. Occipital transtentorial approach for removal of pineal region tumors: report of 64 consecutive cases. *Surg Neurol* **32:** 36–39. doi:10.1016/0090-3019(89)90032-3

Lv QK, Tao KX, Wang XB, Yao XY, Pang MZ, Liu JY, Wang F, Liu CF. 2023. Role of α-synuclein in microglia: autophagy and phagocytosis balance neuroinflammation in Parkinson's disease. *Inflamm Res* **72:** 443–462. doi:10.1007/s00011-022-01676-x

Ma SX, Seo BA, Kim D, Xiong Y, Kwon SH, Brahmachari S, Kim S, Kam TI, Nirujogi RS, Kwon SH, et al. 2021. Complement and coagulation cascades are potentially involved in dopaminergic neurodegeneration in alpha-synuclein-based mouse models of Parkinson's disease. *J Proteome Res* **20:** 3428–3443. doi:10.1021/acs.jproteome.0c01002

Mahmoud S, Gharagozloo M, Simard C, Gris D. 2019. Astrocytes maintain glutamate homeostasis in the CNS by controlling the balance between glutamate uptake and release. *Cells* **8:** 184. doi:10.3390/cells8020184

Marchetti L, Engelhardt B. 2020. Immune cell trafficking across the blood–brain barrier in the absence and presence of neuroinflammation. *Vasc Biol* **2:** H1–H18. doi:10.1530/VB-19-0033

McGeer PL, McGeer EG. 2004. Inflammation and neurodegeneration in Parkinson's disease. *Parkinsonism Relat Disord* **10:** S3–S7. doi:10.1016/j.parkreldis.2004.01.005

McGeer PL, Itagaki S, Boyes BE, McGeer EG. 1988. Reactive microglia are positive for HLA-DR in the substantia nigra of Parkinson's and Alzheimer's disease brains. *Neurology* **38:** 1285–1291. doi:10.1212/WNL.38.8.1285

Medeiros WLGJ, Bandeira IP, Franzoi AEA, Brandão WN, Santos Durão A, Gonçalves MVM. 2019. Mast cells: a key component in the pathogenesis of neuromyelitis optica spectrum disorder? *Immunobiology* **224:** 706–709. doi:10.1016/j.imbio.2019.05.010

Medzhitov R, Janeway C Jr. 2000. Innate immunity. *N Engl J Med* **343:** 338–344. doi:10.1056/NEJM200008033430506

Melo PN, Souza da Silveira M, Mendes Pinto I, Relvas JB. 2021. Morphofunctional programming of microglia requires distinct roles of type II myosins. *Glia* **69:** 2717–2738. doi:10.1002/glia.24067

Mezey É, Szalayova I, Hogden CT, Brady A, Dósa Á, Sótonyi P, Palkovits M. 2021. An immunohistochemical study of lymphatic elements in the human brain. *Proc Natl Acad Sci* **118:** e2002574118. doi:10.1073/pnas.2002574118

Miyazaki I, Asanuma M. 2020. Neuron–astrocyte interactions in Parkinson's disease. *Cells* **9:** 2623. doi:10.3390/cells9122623

Mizutani M, Pino PA, Saederup N, Charo IF, Ransohoff RM, Cardona AE. 2012. The fractalkine receptor but not CCR2 is present on microglia from embryonic development throughout adulthood. *J Immunol* **188:** 29–36. doi:10.4049/jimmunol.1100421

Moens E, Veldhoen M. 2012. Epithelial barrier biology: good fences make good neighbours. *Immunology* **135:** 1–8. doi:10.1111/j.1365-2567.2011.03506.x

Mogensen FL, Delle C, Nedergaard M. 2021. The glymphatic system (en)during inflammation. *Int J Mol Sci* **22:** 7491. doi:10.3390/ijms22147491

Mosley RL, Hutter-Saunders JA, Stone DK, Gendelman HE. 2012. Inflammation and adaptive immunity in Parkinson's disease. *Cold Spring Harb Perspect Med* **2:** a009381. doi:10.1101/cshperspect.a009381

Mula A, Yuan X, Lu J. 2024. Dendritic cells in Parkinson's disease: regulatory role and therapeutic potential. *Eur J Pharmacol* **976:** 176690. doi:10.1016/j.ejphar.2024.176690

Nalls MA, Blauwendraat C, Vallerga CL, Heilbron K, Bandres-Ciga S, Chang D, Tan M, Kia DA, Noyce AJ, Xue A, et al. 2019. Identification of novel risk loci, causal insights, and heritable risk for Parkinson's disease: a meta-analysis of genome-wide association studies. *Lancet Neurol* **18:** 1091–1102. doi:10.1016/S1474-4422(19)30320-5

Narayan S, Sinsheimer JS, Paul KC, Liew Z, Cockburn M, Bronstein JM, Ritz B. 2015. Genetic variability in ABCB1, occupational pesticide exposure, and Parkinson's disease. *Environ Res* **143:** 98–106. doi:10.1016/j.envres.2015.08.022

Negi N, Das BK. 2018. CNS: not an immunoprivilaged site anymore but a virtual secondary lymphoid organ. *Int Rev Immunol* **37:** 57–68. doi:10.1080/08830185.2017.1357719

Netea MG, Schlitzer A, Placek K, Joosten LAB, Schultze JL. 2019. Innate and adaptive immune memory: an evolutionary continuum in the host's response to pathogens. *Cell Host Microbe* **25:** 13–26. doi:10.1016/j.chom.2018.12.006

Neubrand VE, Pedreño M, Caro M, Forte-Lago I, Delgado M, Gonzalez-Rey E. 2014. Mesenchymal stem cells induce the ramification of microglia via the small RhoGTPases Cdc42 and Rac1. *Glia* **62:** 1932–1942. doi:10.1002/glia.22714

Ochoa MC, Minute L, Rodriguez I, Garasa S, Perez-Ruiz E, Inogés S, Melero I, Berraondo P. 2017. Antibody-dependent cell cytotoxicity: immunotherapy strategies enhancing effector NK cells. *Immunol Cell Biol* **95:** 347–355. doi:10.1038/icb.2017.6

Orihuela R, McPherson CA, Harry GJ. 2016. Microglial M1/M2 polarization and metabolic states. *Br J Pharmacol* **173:** 649–665. doi:10.1111/bph.13139

Panicker N, Sarkar S, Harischandra DS, Neal M, Kam TI, Jin H, Saminathan H, Langley M, Charli A, Samidurai M, et al. 2019. Fyn kinase regulates misfolded alpha-synuclein uptake and NLRP3 inflammasome activation in microglia. *J Exp Med* **216:** 1411–1430. doi:10.1084/jem.20182191

Paolicelli RC, Sierra A, Stevens B, Tremblay ME, Aguzzi A, Ajami B, Amit I, Audinat E, Bechmann I, Bennett M, et al. 2022. Microglia states and nomenclature: a field at its crossroads. *Neuron* **110:** 3458–3483. doi:10.1016/j.neuron.2022.10.020

Parillaud VR, Lornet G, Monnet Y, Privat AL, Haddad AT, Brochard V, Bekaert A, de Chanville CB, Hirsch EC, Combadière C, et al. 2017. Analysis of monocyte infiltration in MPTP mice reveals that microglial CX3CR1 protects against neurotoxic over-induction of monocyte-attracting CCL2 by astrocytes. *J Neuroinflammation* **14:** 60. doi:10.1186/s12974-017-0830-9

Park YJ, Lee HK. 2018. The role of skin and orogenital microbiota in protective immunity and chronic immune-mediated inflammatory disease. *Front Immunol* **8:** 1955. doi:10.3389/fimmu.2017.01955

Park SY, Yang H, Kim S, Yang J, Go H, Bae H. 2023. Alpha-synuclein-specific regulatory T cells ameliorate Parkinson's disease progression in mice. *Int J Mol Sci* **24:** 15237. doi:10.3390/ijms242015237

Passaro AP, Lebos AL, Yao Y, Stice SL. 2021. Immune response in neurological pathology: emerging role of central and peripheral immune crosstalk. *Front Immunol* **12:** 676621. doi:10.3389/fimmu.2021.676621

Pekny M, Pekna M. 2016. Reactive gliosis in the pathogenesis of CNS diseases. *Biochim Biophys Acta* **1862**: 483–491. doi:10.1016/j.bbadis.2015.11.014

Perry VH, Nicoll JA, Holmes C. 2010. Microglia in neurodegenerative disease. *Nat Rev Neurol* **6**: 193–201. doi:10.1038/nrneurol.2010.17

Pessac B, Godin I, Alliot F. 2001. Microglia: origin and development. *Bull Acad Natl Med* **185**: 337–346.

Petry P, Oschwald A, Kierdorf K. 2023. Microglial tissue surveillance: the never-resting gardener in the developing and adult CNS. *Eur J Immunol* **53**: e2250232. doi:10.1002/eji.202250232

Polfliet MM, Fabriek BO, Daniëls WP, Dijkstra CD, van den Berg TK. 2006. The rat macrophage scavenger receptor CD163: expression, regulation and role in inflammatory mediator production. *Immunobiology* **211**: 419–425. doi:10.1016/j.imbio.2006.05.015

Popiolek-Barczyk K, Mika J. 2016. Targeting the microglial signaling pathways: new insights in the modulation of neuropathic pain. *Curr Med Chem* **23**: 2908–2928. doi:10.2174/0929867323666160607120124

Proulx ST, Engelhardt B. 2022. Central nervous system zoning: how brain barriers establish subdivisions for CNS immune privilege and immune surveillance. *J Intern Med* **292**: 47–67. doi:10.1111/joim.13469

Rahman MM, Islam MR, Yamin M, Islam MM, Sarker MT, Meem AFK, Akter A, Emran TB, Cavalu S, Sharma R. 2022. Emerging role of neuron-glia in neurological disorders: at a glance. *Oxid Med Cell Longev* **2022**: 3201644. doi:10.1155/2022/3201644

Rajesh Y, Kanneganti TD. 2022. Innate immune cell death in neuroinflammation and Alzheimer's disease. *Cells* **11**: 1885. doi:10.3390/cells11121885

Reemst K, Noctor SC, Lucassen PJ, Hol EM. 2016. The indispensable roles of microglia and astrocytes during brain development. *Front Hum Neurosci* **10**: 566. doi:10.3389/fnhum.2016.00566

Reese TS, Karnovsky MJ. 1967. Fine structural localization of a blood–brain barrier to exogenous peroxidase. *J Cell Biol* **34**: 207–217. doi:10.1083/jcb.34.1.207

Réu P, Khosravi A, Bernard S, Mold JE, Salehpour M, Alkass K, Perl S, Tisdale J, Possnert G, Druid H, et al. 2017. The lifespan and turnover of microglia in the human brain. *Cell Rep* **20**: 779–784. doi:10.1016/j.celrep.2017.07.004

Reynolds AD, Stone DK, Mosley RL, Gendelman HE. 2009. Nitrated alpha-synuclein-induced alterations in microglial immunity are regulated by CD4+ T cell subsets. *J Immunol* **182**: 4137–4149. doi:10.4049/jimmunol.0803982

Ricklin D, Hajishengallis G, Yang K, Lambris JD. 2010. Complement: a key system for immune surveillance and homeostasis. *Nat Immunol* **11**: 785–797. doi:10.1038/ni.1923

Rocha SM, Saraiva T, Cristóvão AC, Ferreira R, Santos T, Esteves M, Saraiva C, Je G, Cortes L, Valero J, et al. 2016. Histamine induces microglia activation and dopaminergic neuronal toxicity via H1 receptor activation. *J Neuroinflammation* **13**: 137. doi:10.1186/s12974-016-0600-0

Roodveldt C, Bernardino L, Oztop-Cakmak O, Dragic M, Fladmark KE, Ertan S, Aktas B, Pita C, Ciglar L, Garraux

G, et al. 2024. The immune system in Parkinson's disease: what we know so far. *Brain* **147**: 3306–3324. doi:10.1093/brain/awae177

Saleh M, Markovic M, Olson KE, Gendelman HE, Mosley RL. 2022. Therapeutic strategies for immune transformation in Parkinson's disease. *J Parkinsons Dis* **12**: S201–s222. doi:10.3233/JPD-223278

Sargent D, Moore DJ. 2021. Mechanisms of VPS35-mediated neurodegeneration in Parkinson's disease. *Int Rev Mov Disord* **2**: 221–244. doi:10.1016/bs.irmvd.2021.08.005

Saunders JA, Estes KA, Kosloski LM, Allen HE, Dempsey KM, Torres-Russotto DR, Meza JL, Santamaria PM, Bertoni JM, Murman DL, et al. 2012. CD4+ regulatory and effector/memory T cell subsets profile motor dysfunction in Parkinson's disease. *J Neuroimmune Pharmacol* **7**: 927–938. doi:10.1007/s11481-012-9402-z

Schlachetzki JC, Winkler J. 2015. The innate immune system in Parkinson's disease: a novel target promoting endogenous neuroregeneration. *Neural Regen Res* **10**: 704–706. doi:10.4103/1673-5374.156958

Schoen J, Euler M, Schauer C, Schett G, Herrmann M, Knopf J, Yaykasli KO. 2022. Neutrophils' extracellular trap mechanisms: from physiology to pathology. *Int J Mol Sci* **23**: 12855. doi:10.3390/ijms232112855

Schröder JB, Pawlowski M, Meyer Zu Hörste G, Gross CC, Wiendl H, Meuth SG, Ruck T, Warnecke T. 2018. Immune cell activation in the cerebrospinal fluid of patients with Parkinson's disease. *Front Neurol* **9**: 1081. doi:10.3389/fneur.2018.01081

Schulte C, Gasser T. 2011. Genetic basis of Parkinson's disease: inheritance, penetrance, and expression. *Appl Clin Genet* **4**: 67–80. doi:10.2147/TACG.S11639

Schutt CR, Gendelman HE, Mosley RL. 2018. Tolerogenic bone marrow-derived dendritic cells induce neuroprotective regulatory T cells in a model of Parkinson's disease. *Mol Neurodegener* **13**: 26. doi:10.1186/s13024-018-0255-7

Schwartz M, Cahalon L. 2022. The vicious cycle governing the brain–immune system relationship in neurodegenerative diseases. *Curr Opin Immunol* **76**: 102182. doi:10.1016/j.coi.2022.102182

Schwartz M, Deczkowska A. 2016. Neurological disease as a failure of brain–immune crosstalk: the multiple faces of neuroinflammation. *Trends Immunol* **37**: 668–679. doi:10.1016/j.it.2016.08.001

Shen Z, Xiang M, Chen C, Ding F, Wang Y, Shang C, Xin L, Zhang Y, Cui X. 2022. Glutamate excitotoxicity: potential therapeutic target for ischemic stroke. *Biomed Pharmacother* **151**: 113125. doi:10.1016/j.biopha.2022.113125

Silver R, Curley JP. 2013. Mast cells on the mind: new insights and opportunities. *Trends Neurosci* **36**: 513–521. doi:10.1016/j.tins.2013.06.001

Singh S, Swarnkar S, Goswami P, Nath C. 2011. Astrocytes and microglia: responses to neuropathological conditions. *Int J Neurosci* **121**: 589–597. doi:10.3109/00207454.2011.598981

Smith JA, Das A, Ray SK, Banik NL. 2012. Role of proinflammatory cytokines released from microglia in neurodegenerative diseases. *Brain Res Bull* **87**: 10–20. doi:10.1016/j.brainresbull.2011.10.004

Sofroniew MV, Vinters HV. 2010. Astrocytes: biology and pathology. *Acta Neuropathol* **119**: 7–35. doi:10.1007/s00401-009-0619-8

Stephenson J, Nutma E, van der Valk P, Amor S. 2018. Inflammation in CNS neurodegenerative diseases. *Immunology* **154**: 204–219. doi:10.1111/imm.12922

Sulzer D, Alcalay RN, Garretti F, Cote L, Kanter E, Agin-Liebes J, Liong C, McMurtrey C, Hildebrand WH, Mao X, et al. 2017. T cells from patients with Parkinson's disease recognize α-synuclein peptides. *Nature* **546**: 656–661. doi:10.1038/nature22815

Sun N, Victor MB, Park YP, Xiong X, Scannail AN, Leary N, Prosper S, Viswanathan S, Luna X, Boix CA, et al. 2023. Human microglial state dynamics in Alzheimer's disease progression. *Cell* **186**: 4386–4403.e29. doi:10.1016/j.cell.2023.08.037

Sweeney MD, Sagare AP, Zlokovic BV. 2018. Blood–brain barrier breakdown in Alzheimer disease and other neurodegenerative disorders. *Nat Rev Neurol* **14**: 133–150. doi:10.1038/nrneurol.2017.188

Tansey MG, Romero-Ramos M. 2019. Immune system responses in Parkinson's disease: early and dynamic. *Eur J Neurosci* **49**: 364–383. doi:10.1111/ejn.14290

Tansey MG, Wallings RL, Houser MC, Herrick MK, Keating CE, Joers V. 2022. Inflammation and immune dysfunction in Parkinson disease. *Nat Rev Immunol* **22**: 657–673. doi:10.1038/s41577-022-00684-6

Tenner AJ, Stevens B, Woodruff TM. 2018. New tricks for an ancient system: physiological and pathological roles of complement in the CNS. *Mol Immunol* **102**: 3–13. doi:10.1016/j.molimm.2018.06.264

Traina G. 2019. Mast cells in gut and brain and their potential role as an emerging therapeutic target for neural diseases. *Front Cell Neurosci* **13**: 345. doi:10.3389/fncel.2019.00345

Troncoso-Escudero P, Parra A, Nassif M, Vidal RL. 2018. Outside in: unraveling the role of neuroinflammation in the progression of Parkinson's disease. *Front Neurol* **9**: 860. doi:10.3389/fneur.2018.00860

Turner JR. 2009. Intestinal mucosal barrier function in health and disease. *Nat Rev Immunol* **9**: 799–809. doi:10.1038/nri2653

Turvey SE, Broide DH. 2010. Innate immunity. *J Allergy Clin Immunol* **125**: S24–S32. doi:10.1016/j.jaci.2009.07.016

Unger MS, Schernthaner P, Marschallinger J, Mrowetz H, Aigner L. 2018. Microglia prevent peripheral immune cell invasion and promote an anti-inflammatory environment in the brain of APP-PS1 transgenic mice. *J Neuroinflammation* **15**: 274. doi:10.1186/s12974-018-1304-4

Varnum MM, Ikezu T. 2012. The classification of microglial activation phenotypes on neurodegeneration and regeneration in Alzheimer's disease brain. *Arch Immunol Ther Exp (Warsz)* **60**: 251–266. doi:10.1007/s00005-012-0181-2

Vitetta ES, Berton MT, Burger C, Kepron M, Lee WT, Yin XM. 1991. Memory B and T cells. *Annu Rev Immunol* **9**: 193–217. doi:10.1146/annurev.iy.09.040191.001205

Vorobjeva NV, Chelombitko MA, Sud'ina GF, Zinovkin RA, Chernyak BV. 2023. Role of mitochondria in the regulation of effector functions of granulocytes. *Cells* **12**: 2210. doi:10.3390/cells12182210

Wang Q, Liu Y, Zhou J. 2015. Neuroinflammation in Parkinson's disease and its potential as therapeutic target. *Transl Neurodegener* **4**: 19. doi:10.1186/s40035-015-0042-0

Wang T, Sun Y, Dettmer U. 2023. Astrocytes in Parkinson's disease: from role to possible intervention. *Cells* **12**: 2336. doi:10.3390/cells12192336

Weiss F, Labrador-Garrido A, Dzamko N, Halliday G. 2022. Immune responses in the Parkinson's disease brain. *Neurobiol Dis* **168**: 105700. doi:10.1016/j.nbd.2022.105700

West AP, Koblansky AA, Ghosh S. 2006. Recognition and signaling by toll-like receptors. *Annu Rev Cell Dev Biol* **22**: 409–437. doi:10.1146/annurev.cellbio.21.122303.115827

* Westenberger A, Brüggemann N, Klein C. 2024. Genetics of Parkinson's disease: from causes to treatment. *Cold Spring Harb Perspect Med* doi:10.1101/cshperspect.a041774

Wilson EH, Weninger W, Hunter CA. 2010. Trafficking of immune cells in the central nervous system. *J Clin Invest* **120**: 1368–1379. doi:10.1172/JCI41911

Xie L, Zhang N, Zhang Q, Li C, Sandhu AF, Iii GW, Lin S, Lv P, Liu Y, Wu Q, et al. 2020. Inflammatory factors and amyloid beta-induced microglial polarization promote inflammatory crosstalk with astrocytes. *Aging* **12**: 22538–22549. doi:10.18632/aging.103663

Yang L, Mao K, Yu H, Chen J. 2020. Neuroinflammatory responses and Parkinson' disease: pathogenic mechanisms and therapeutic targets. *J Neuroimmune Pharmacol* **15**: 830–837. doi:10.1007/s11481-020-09926-7

Yao YY, Ling EA, Lu D. 2020. Microglia mediated neuroinflammation—signaling regulation and therapeutic considerations with special reference to some natural compounds. *Histol Histopathol* **35**: 1229–1250. doi:10.14670/HH-18-239

Yi S, Wang L, Wang H, Ho MS, Zhang S. 2022. Pathogenesis of alpha-synuclein in Parkinson's disease: from a neuron-glia crosstalk perspective. *Int J Mol Sci* **23**: 14753. doi:10.3390/ijms232314753

Zhang X, Shao Z, Xu S, Liu Q, Liu C, Luo Y, Jin L, Li S. 2021. Immune profiling of Parkinson's disease revealed its association with a subset of infiltrating cells and signature genes. *Front Aging Neurosci* **13**: 605970. doi:10.3389/fnagi.2021.605970

Zhang Y, Zhang C, He XZ, Li ZH, Meng JC, Mao RT, Li X, Xue R, Gui Q, Zhang GX, et al. 2023. Interaction between the glymphatic system and alpha-synuclein in Parkinson's disease. *Mol Neurobiol* **60**: 2209–2222. doi:10.1007/s12035-023-03212-2

Zhang F, Chen B, Ren W, Yan Y, Zheng X, Jin S, Chang Y. 2024. Association analysis of dopaminergic degeneration and the neutrophil-to-lymphocyte ratio in Parkinson's disease. *Front Aging Neurosci* **16**: 1377994. doi:10.3389/fnagi.2024.1377994

Zou W, Pu T, Feng W, Lu M, Zheng Y, Du R, Xiao M, Hu G. 2019. Blocking meningeal lymphatic drainage aggravates Parkinson's disease-like pathology in mice overexpressing mutated alpha-synuclein. *Transl Neurodegener* **8**: 7. doi:10.1186/s40035-019-0147-y

Zrzavy T, Hametner S, Wimmer I, Butovsky O, Weiner HL, Lassmann H. 2017. Loss of "homeostatic" microglia and patterns of their activation in active multiple sclerosis. *Brain* **140**: 1900–1913. doi:10.1093/brain/awx113

Index

www.ingramcontent.com/pod-product-compliance
Lightning Source LLC
Chambersburg PA
CBHW061929190326
41458CB00009B/2699

* 9 7 8 1 6 2 1 8 2 5 0 3 6 *